CW01142645

The Early Earth: Physical, Chemical and Biological Development

Geological Society Special Publications

Society Book Editors

A. J. FLEET (CHIEF EDITOR)

P. DOYLE

F. J. GREGORY

J. S. GRIFFITHS

A. J. HARTLEY

R. E. HOLDSWORTH

A. C. MORTON

N. S. ROBINS

M. S. STOKER

J. P. TURNER

Special Publication reviewing procedures

The Society makes every effort to ensure that the scientific and production quality of its books matches that of its journals. Since 1997, all book proposals have been refereed by specialist reviewers as well as by the Society's Books Editorial Committee. If the referees identify weaknesses in the proposal, these must be addressed before the proposal is accepted.

Once the book is accepted, the Society has a team of Book Editors (listed above) who ensure that the volume editors follow strict guidelines on refereeing and quality control. We insist that individual papers can only be accepted after satisfactory review by two independent referees. The questions on the review forms are similar to those for *Journal of the Geological Society*. The referees' forms and comments must be available to the Society's Book Editors on request.

Although many of the books result from meetings, the editors are expected to commission papers that were not presented at the meeting to ensure that the book provides a balanced coverage of the subject. Being accepted for presentation at the meeting does not guarantee inclusion in the book.

Geological Society Special Publications are included in the ISI Science Citation Index, but they do not have an impact factor, the latter being applicable only to journals.

More information about submitting a proposal and producing a Special Publication can be found on the Society's web site: www.geolsoc.org.uk.

GEOLOGICAL SOCIETY SPECIAL PUBLICATION NO. 199

The Early Earth: Physical, Chemical and Biological Development

EDITED BY

C. M. R. FOWLER
University of London, UK

C. J. EBINGER
University of London, UK

C. J. HAWKESWORTH
University of Bristol, UK

2002

Published by

The Geological Society

London

THE GEOLOGICAL SOCIETY

The Geological Society of London (GSL) was founded in 1807. It is the oldest national geological society in the world and the largest in Europe. It was incorporated under Royal Charter in 1825 and is Registered Charity 210161.

The Society is the UK national learned and professional society for geology with a worldwide Fellowship (FGS) of 9000. The Society has the power to confer Chartered status on suitably qualified Fellows, and about 2000 of the Fellowship carry the title (CGeol). Chartered Geologists may also obtain the equivalent European title, European Geologist (EurGeol). One fifth of the Society's fellowship resides outside the UK. To find out more about the Society, log on to www.geolsoc.org.uk.

The Geological Society Publishing House (Bath, UK) produces the Society's international journals and books, and acts as European distributor for selected publications of the American Association of Petroleum Geologists (AAPG), the American Geological Institute (AGI), the Indonesian Petroleum Association (IPA), the Geological Society of America (GSA), the Society for Sedimentary Geology (SEPM) and the Geologists' Association (GA). Joint marketing agreements ensure that GSL Fellows may purchase these societies' publications at a discount. The Society's online bookshop (accessible from www.geolsoc.org.uk) offers secure book purchasing with your credit or debit card.

To find out about joining the Society and benefiting from substantial discounts on publications of GSL and other societies world-wide, consult www.geolsoc.org.uk, or contact the Fellowship Department at: The Geological Society, Burlington House, Piccadilly, London W1J 0BG: Tel. +44 (0)20 7434 9944; Fax +44 (0)20 7439 8975; Email: enquiries@geolsoc.org.uk.

Published by The Geological Society from:
The Geological Society Publishing House
Unit 7, Brassmill Enterprise Centre
Brassmill Lane
Bath BA1 3JN, UK

(*Orders*: Tel. +44 (0)1225 445046
        Fax +44 (0)1225 442836)
Online bookshop: http://bookshop.geolsoc.org.uk

The publishers make no representation, express or implied, with regard to the accuracy of the information contained in this book and cannot accept any legal responsibility for any errors or omissions that may be made.

© The Geological Society of London 2002. All rights reserved. No reproduction, copy or transmission of this publication may be made without written permission. No paragraph of this publication may be reproduced, copied or transmitted save with the provisions of the Copyright Licensing Agency, 90 Tottenham Court Road, London W1P 9HE. Users registered with the Copyright Clearance Center, 27 Congress Street, Salem, MA 01970, USA: the item-fee code for this publication is 0305-8719/02/$15.00.

British Library Cataloguing in Publication Data
A catalogue record for this book is available from the British Library.

ISBN 1-86239-109-2
ISSN 0305-8719

Distributors

USA
AAPG Bookstore
PO Box 979
Tulsa
OK 74101-0979
USA
Orders: Tel. +1 918 584-2555
        Fax +1 918 560-2652
        E-mail: *bookstore@aapg.org*

India
Affiliated East-West Press PVT Ltd
G-1/16 Ansari Road, Daryaganj,
New Delhi 110 002
India
Orders: Tel. +91 11 327-9113
        Fax +91 11 326-0538
        E-mail: *affiliat@nda.vsnl.net.in*

Japan
Kanda Book Trading Co.
Cityhouse Tama 204
Tsurumaki 1-3-10
Tama-shi
Tokyo 206-0034
Japan
Orders: Tel. +81 (0)423 57-7650
        Fax +81 (0)423 57-7651

Typeset by Aarontype Ltd, Bristol, UK

Printed by Cromwells, Trowbridge, UK

Contents

Preface   vii

Geophysical and petrological constraints on Archaean lithosphere

JAMES, D. E. & FOUCH, M. J. Formation and evolution of Archaean cratons: insights from southern Africa   1

KENDALL, J.-M., SOL, S., THOMSON, C. J., WHITE, D. J., ASUDEH, I., SNELL, C. S. & SUTHERLAND, F. H. Seismic heterogeneity and anisotropy in the Western Superior Province, Canada: insights into the evolution of an Archaean craton   27

PRIESTLY, K. & MCKENZIE, D. The structure of the upper mantle beneath southern Africa   45

PEARSON, D. G., IRVINE, G. J., CARLSON, R. W., KOPYLOVA, M. G. & IONOV, D. A. The development of lithospheric keels beneath the earliest continents: time constraints using PGE and Re–Os isotope systematics   65

ARNDT, N. T., LEWIN, É. & ALBARÈDE, F. Strange partners: formation and survival of continental crust and lithospheric mantle   91

LUAIS, B. & HAWKESWORTH, C. J. Pb isotope variations in Archaean time and possible links to the sources of certain Mesozoic–Recent basalts   105

MUSACCHIO, G. & MOONEY, W. D. Seismic evidence for a mantle source for mid-Proterozoic anorthosites and implications for models of crustal growth   125

Models of cratonic evolution and modification

SLEEP, N. H., EBINGER, C. J. & KENDALL, J.-M. Deflection of mantle plume material by cratonic keels   135

BLEEKER, W. Archaean tectonics: a review, with illustrations from the Slave Craton   151

JELSMA, H. A. & DIRKS, P. H. G. M. Tectono-magmatic evolution of the Zimbabwe Craton   183

Constraints on the Archaean environment

MARTY, B. & DAUPHAS, N. Formation and early evolution of the atmosphere   213

ZAHNLE, K. & SLEEP, N. H. Carbon dioxide cycling through the mantle and implications for the climate of ancient Earth   231

KRAMERS, J. D. Global modelling of continent formation and destruction through geological time and implications for CO_2 drawdown in the Archaean Eon   259

NISBET, E. G. Fermor Lecture: the influence of life on the face of the Earth: garnets and moving continents   275

GRASSINEAU, N. V., NISBET, E. G., FOWLER, C. M. R., BICKLE, M. J., LOWRY, D., CHAPMAN, H. J., MATTEY, D. P., ABELL, P., YONG, J. & MARTIN, A. Stable isotopes in the Archaean Belingwe belt, Zimbabwe: evidence for a diverse microbial mat ecology   309

ROLLINSON, H. The metamorphic history of the Isua Greenstone Belt, West Greenland   329

Index   351

It is recommended that reference to all or part of this book should be made in one of the following ways:

FOWLER, C. M. R., EBINGER, C. J. & HAWKESWORTH, C. J. (eds) *The Early Earth: Physical, Chemical and Biological Development*. Geological Society, London, Special Publications, **199**.

PRIESTLY, K. & MCKENZIE, D. The structure of the upper mantle beneath southern Africa *In*: FOWLER, C. M. R., EBINGER, C. J. & HAWKESWORTH, C. J. (eds) *The Early Earth: Physical, Chemical and Biological Development*. Geological Society, London, Special Publications, **199**, 45–64.

Preface

The early Earth, to borrow from the term used by Queen Elizabeth I to describe the Archaean of Labrador, is Terra Meta Incognita. This is still largely correct as much of its character is still unrecognized. Some of the most interesting problems in geology and biology lie in the study of the planet's early history. This book arose from a Discussion Meeting in February 2000, sponsored by the Royal Astronomical Society and the Geological Society. The meeting was designed to cover a wide range of different aspects of the early Earth. Much was discussed, arguments were detailed, and perhaps some insight was gained.

Though much is published about the Archaean, there are few recent books that offer, in a single volume, a wide-ranging access to the subject. Conference volumes can be little more than random assortments of papers by people who happened to be there, while single-author textbooks have thematic unity but can be very out of date by the time they are published. This book, we hope, bridges that gap to some extent. It stems from the meeting but incorporates more recent work by the speakers, and also some work by authors not presented at the meeting. The intent is to cover a wide range of Archaean topics at research level (and so be useful to graduate students and those seeking an introduction to the subject), while retaining some of the work-in-progress vigour of a discussion meeting.

The papers are divided into three sections:

(1) Geophysical and petrological constraints on archaean lithosphere.
(2) Models of cratonic evolution and modification.
(3) Constraints on the Archaean environment.

The first section contains a series of papers concerned with the differences in overall structure and composition between Archaean and post-Archaean lithosphere, and has contributions from seismologists, petrologists, and geochemists. **James & Fouch** present seismological evidence from southern African, particularly the Kaapvaal craton, for the formation and evolution of cratons: Archaean cratons have relatively thin crust and high velocity mantle roots that extend to depths of at least 200 km, in contrast to younger terrains which have a thicker complex crust and relatively low seismic velocities in the mantle. Then **Kendall *et al.*** turn to North America and present seismic evidence for the evolution of the Archaean Superior province. **Priestley & McKenzie** present seismic and geochemical constraints on the formation of Archaean lithosphere and its current structure. The next three papers cover the geochemical and petrological constraints on the origin and development of Archaean lithosphere. **Pearson *et al.*** consider the time constraints imposed by Re and Os; **Arndt *et al.*** present evidence on the origin of continental lithosphere from studies of high Mg olivine and the residue left beneath continents after eruption of flood basalts; and **Luais & Hawkesworth** discuss the isotopic evolution of Pb in the Archaean. In the final paper in this section, **Musacchio & Mooney** use P-wave and S-wave velocities. to suggest mid-Proterozoic anothosites had a mantle source and that the mid-Proterozoic lower crust is mafic.

The second section contains papers modelling cratonic evolution and accretion. **Sleep *et al.*** present numerical models of cratonic roots in normal mantle flow and in the presence of plumes. Cratons may suffer lateral erosion rather thinning. **Bleeker** reviews the tectonic evolution of Archaean granite–greenstone terrains. Then, taking a rather different approach, **Jelsma & Dirks** provide a geochemical and age database and propose a tectono-magmatic evolution model for the Zimbabwe craton.

The third section contains papers concerned with details of the formation, maintenance and development of the Archaean surface environment, and linkage between the evolution of the surface and the planetary interior. This is a very wide field and the papers deal with matters such as the timing of hydrogen loss from the Earth, the shift to an oxygen-rich atmosphere, carbon dioxide build-up in the atmosphere, the appearance of life, early biogenic controls on the carbon cycle and the origins of photosynthesis. **Marty & Dauphas** discuss rare gases and major volatiles in the Hadaean, followed by **Zahnle & Sleep**, who investigate the climate of the ancient Earth and the role of carbon dioxide cycling. Then **Kramers** considers the rates of formation and destruction of continental crust in the Archaean and the drawdown of a massive CO_2 atmosphere. **Nisbet** discusses the physical and biological controls on the Earth in the late Hadean and Archaean. This paper was the Geological Society's 2000 Fermor Lecture, commemorating Sir Lewis Fermor, of the Geological Survey of India. In keeping with Fermor's wide-ranging interests, the brief for this paper was very broad, and looks back to a meeting in Harare, Zimbabwe 50 years ago when Fermor and A. M. Macgregor explored physical and biological controls on the Earth's evolution. The topic of Archaean life is continued in the following paper by **Grassineau *et al.***, which presents detailed C and S isotopic data from the Belingwe greenstone belt. Finally, **Rollinson** sets out the metamorphic history of the 3.8 Ga Isua Greenstone Belt in Greenland, for which carbon isotope values suggest that some of the sediments may be of biogenic origin.

We much are indebted to all the reviewers who patiently worked through versions of the manuscripts: M. J Bickle, W. Bleeker, R. Buick, D. Demaitte, R. Griffiths, S. Hanmer, G. Helffrich, B. Kamber, J. Kasting, J. Kramers, C. A. Langston, D. P. McKenzie, M. A. Menzies, J. S. Myers, A. Nyblade, J. Ritsema, H. Rollinson, C. Tiberi, P. J. Treloar, D. Waters, Y. L. Yung, K. Zahnle, G. Zandt, W. McDonough, R. Arculus, P. Rey, R. Buick, E. G. Nisbet, K. Zahnle, B. Marty, G. Pearson, D. James and others. Finally we gratefully acknowledge the financial and logistical assistance from the Royal Astronomical Society and the Geological Society, which enabled us to hold the Discussion Meeting that has given rise to this book.

Mary Fowler, Cindy Ebinger & Chris Hawkesworth
June 2002

Formation and evolution of Archaean cratons: insights from southern Africa

D. E. JAMES[1] & M. J. FOUCH[2]

[1] *Carnegie Institution of Washington, Department of Terrestrial Magnetism, 5241 Broad Branch Road, NW, Washington, DC 20015, USA (e-mail: james@dtm.ciw.edu)*
[2] *Present address: Arizona State University, Department of Geological Sciences, PO Box 871404, Tempe, AZ 85287, USA*

Abstract: Archaean cratons are the stable remnants of Earth's early continental lithosphere, and their structure, composition and survival over geological time make them unique features of the Earth's surface. The Kaapvaal Project of southern Africa was organized around a broadly diverse scientific collaboration to investigate fundamental questions of craton formation and mantle differentiation in the early Earth. The principal aim of the project was to characterize the physical and chemical nature of the crust and mantle of the cratons of southern Africa in geological detail, and to use the 3D seismic and geochemical images of crustal and mantle heterogeneity to reconstruct the assembly history of the cratons. Seismic results confirm that the structure of crust and tectospheric mantle of the cratons differs significantly from that of post-Archaean terranes. Three-dimensional body-wave tomographic images reveal that high-velocity mantle roots extend to depths of at least 200 km, and locally to depths of 250–300 km beneath cratonic terranes. No low-velocity channel has been identified beneath the cratonic root. The Kaapvaal Craton was modified approximately 2.05 Ga by the Bushveld magmatic event, and the mantle beneath the Bushveld Province is characterized by relatively low seismic velocities. The crust beneath undisturbed Archaean craton is relatively thin (*c.* 35–40 km), unlayered and characterized by a strong velocity contrast across a sharp Moho, whereas post-Archaean terranes and Archaean regions disrupted by large-scale Proterozoic magmatic or tectonic events are characterized by thicker crust, complex Moho structure and higher seismic velocities in the lower crust. A review of Re–Os depletion model age determinations confirms that the mantle root beneath the cratons is Archaean in age. The data show also that there is no apparent age progression with depth in the mantle keel, indicating that its thickness has not increased over geological time. Both laboratory experiments and geochemical results from eclogite xenoliths suggest that subduction processes played a central role in the formation of Archaean crust, the melt depletion of Archaean mantle and the assembly of early continental lithosphere. Co-ordinated geochronological studies of crustal and mantle xenoliths have revealed that both crust and mantle have experienced a multi-stage history. The lower crust in particular retains a comprehensive record of the tectonothermal evolution of the lithosphere. Analysis of lower-crustal xenoliths has shown that much of the deep craton experienced a dynamic and protracted history of tectonothermal activity that is temporally associated with events seen in the surface record. Cratonization thus occurred not as a discrete event, but in stages, with final stabilization postdating crustal formation.

Archaean cratons have long been prime targets for a broad array of scientific studies, partly because they form the oldest cores of the continents, but also because they have economic significance as a major source of the world's mineral wealth. (The term craton is confined here to Archaean crust, although the word is commonly applied to Proterozoic shields as well. The Archaean–Proterozoic boundary is, as always, imprecisely defined, but in southern Africa is marked by a sufficient hiatus in time that there is no difficulty in distinguishing Archaean terranes from all others (e.g. Windley 1995).) Cratons are underlain by a thick mantle root that is both chemically and physically distinct from the rest of the mantle, suggesting formation by processes or under conditions unique to Archaean time (e.g. Jordan 1975, 1978; Richardson *et al.* 1984; Pearson *et al.* 1995; McDonough & Rudnick 1998; Rudnick *et al.* 1998; Shirey & Walker 1998; Carlson *et al.* 1999, 2000). Jordan (1975) adopted the term 'tectosphere' to describe the deep conductive (non-convecting) layer of the Archaean mantle that remains attached to the craton through geological time. The tectosphere is composed dominantly of highly depleted peridotites with low normative density and high seismic velocities (Boyd & McCallister 1976; Jordan 1979; Boyd 1987). These refractory roots extend into the mantle to depths of at least 200 km and perhaps more (e.g. Jordan 1975; Lerner-Lam & Jordan 1987; Van der Lee & Nolet 1997; Jaupart & Mareschal 1999; Rudnick & Nyblade 1999; Shapiro et al. 1999; Zhao et al. 1999; Ritsema & van Heijst 2000; James et al. 2001b). The tectosphere is characterized by low heat flow (Jones 1988) and a very low geothermal gradient relative to Proterozoic mantle (McDonough & Rudnick 1998; Jaupart & Mareschal 1999; Nyblade 1999; Rudnick & Nyblade 1999).

Although attention over the years has tended to focus on the distinctive character of cratonic mantle, the nature of the crust and the crust–mantle boundary further distinguishes Archaean from post-Archaean terranes. First, Archaean crust is typically thinner than that of adjacent Proterozoic terranes (Durrheim & Mooney 1991, 1994; Clitheroe et al. 2000; Nguuri et al. 2001; Assumpção et al. 2001), although not all agree with this assertion (e.g. Rudnick & Fountain 1995). The lower crust beneath cratons may be less mafic than that beneath Proterozoic terranes (Griffin & O'Reilly 1987; Durrheim & Mooney 1994). Moreover, the Archaean crust–mantle discontinuity is characteristically sharper than that beneath post-Archaean regions (Nguuri et al. 2001; Assumpção et al. 2001). These observations, together with the unique character of the cratonic root, suggest that the processes of craton formation, or the physical conditions that controlled those processes, differed in important ways from continent-forming processes in post-Archaean times. In this paper we examine these issues in light of results from the Kaapvaal Project of southern Africa.

The Kaapvaal Project

The Kaapvaal Project was undertaken to study the formation, stabilization and evolution of cratons, and to image the deep structure of the tectosphere (Carlson et al. 1996) (see also http://www.ciw.edu/kaapvaal for participants and a description of the project). A cornerstone of the Kaapvaal Project was a large-scale broadband seismic experiment designed specifically for geological-scale imaging of the crust and upper mantle beneath the cratons and adjacent Proterozoic provinces of southern Africa (Fig. 1). The seismic studies were undertaken in concert with a host of complementary investigations in geology, geochemistry and petrology of the cratons of southern Africa and their relationship to adjacent Proterozoic belts (see project summary on http://www.ciw.edu/kaapvaal).

The Kaapvaal and Zimbabwe Cratons of southern Africa form one of the pre-eminent natural laboratories of the world for the study of early continental formation. Indeed, southern Africa is something of a 'type locality' for probing fundamental questions of cratonic formation and evolution. Preserved within the Kaapvaal Craton is a nearly continuous Archaean geological record, from 3.6 to 2.6 Ga (e.g. De Wit et al. 1992). From an experimental point of view, the terrain, coupled with the logistical and academic resources of the region, made it ideal for a large-scale seismic deployment. The seismic array covered a wide range of age provinces spanning more than 3 Ga (Fig. 1). A significant area of the region covered by the seismic array, both on and off craton, has been perforated by thousands of kimberlite pipes from which a wealth of crustal and mantle xenoliths were erupted. These xenoliths, arguably the most extensively studied of any in the world, have been derived from depths up to 200 km or more. The chemical and mineralogical compositions of these nodules provide powerful constraints on the interpretation of results from the seismic analyses.

This paper is, first, a compilation and summary of initial results drawn from the Kaapvaal Project studies published as a special section in *Geophysical Research Letters* (**28**(13), 2001) and, second, a preliminary attempt to integrate those results into a comprehensive set of constraints on craton formation and evolution. The multi-disciplinary studies discussed below represent the contributions of an army of students and scientists, cited in the text and in the acknowledgements section at the end of this paper. We first review recent geochemical and petrological results and then focus upon two central imaging aspects of the seismic investigations: (1) delay time (tomographic) analysis of mantle structure beneath southern Africa (James et al. 2001b); (2) receiver function analysis of the depth and topography of the Moho beneath the seismic array (Nguuri et al. 2001).

Tomographic results show that high-velocity mantle roots extend locally to depths of at least 250 km beneath undisturbed Archaean craton, with no comparable root structures beneath post-Archaean terranes. Neither body- nor surface-wave analyses have yet produced evidence of a low-velocity zone beneath the cratonic keel (Ritsema & van Heijst 2000; Freybourger et al. 2001;

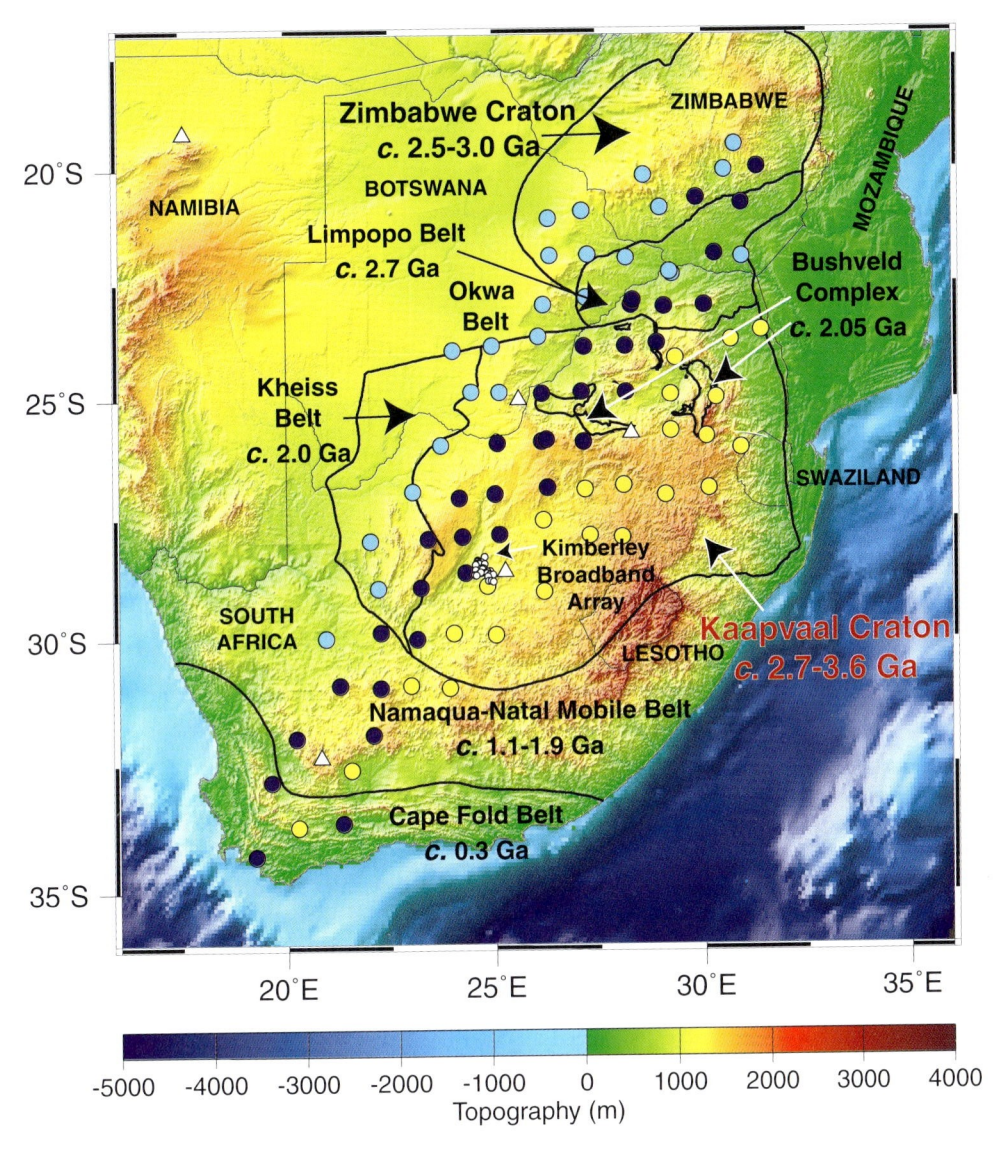

Fig. 1. Map showing station locations, topography and principal geological provinces in the region of study within southern Africa. Fifty-five broadband (REFTEK/STS-2) stations were installed in April 1997 in South Africa, Botswana and Zimbabwe. Stations in light blue were redeployed in April 1998 to sites indicated in yellow. A total of 82 sites were occupied over the two year deployment. In addition, three global seismic network broadband stations (white triangles) are located in the region and their data incorporated in the analysis. The array extends from the Cape Fold Belt in the south, through the Proterozoic Namaqua–Natal Mobile Belt, across the Kaapvaal Craton and Bushveld Province, through the Archaean Limpopo Mobile Belt and into the Zimbabwe Craton. On the west, the array covers part of the Kheis and Okwa Proterozoic Fold and Thrust Belts of Botswana and western South Africa. To the east, the array extends into the Early Archaean Barberton terrane, near the NW border with Swaziland. Published in James *et al.*, GRL28, 2001, fig. 1.

James *et al.* 2001*b*). Strong variations in crustal thickness based on receiver functions reveal significant differences in the nature of the crust and the crust–mantle boundary between Archaean and post-Archaean geological terranes. Both seismic and geochemical results show that the Kaapvaal keel was modified *c.* 2.05 Ga by a massive Bushveld event that affected both crust and mantle across a broad east–west swath of southern Africa.

Geological outline of southern Africa

Southern Africa is a complex collage of geological provinces. In the summary below we provide a thumbnail geological sketch of those provinces that bear directly on this study. In categorizing and describing the surface geology we have made a number of simplifications in the interest of characterizing features on a scale appropriate to our studies.

Cratons

The Archaean Kaapvaal and Zimbabwe Cratons form the nucleus of southern Africa. The Kaapvaal Craton, which is the better studied of the two, is composed of a mosaic of distinct geological terranes covering more than 10^6 km^2, with the oldest units generally in the eastern part of the craton and the youngest in the western part (de Wit *et al.* 1992). These terranes of disparate geological histories were assembled over a 1 Ga period from early Archaean (*c.* 3.6 Ga) to late Archaean time (*c.* 2.6 Ga) (de Wit *et al.* 1992; de Wit & Hart 1993; Carlson *et al.* 2000).

The nearly 0.5×10^6 km^2 continental mass that stabilized in the early Archaean time was termed the 'Kaapvaal shield' by de Wit *et al.*, and forms the eastern part of the craton. The term shield was used by de Wit *et al.* to distinguish the geological character and mechanism of formation of the early Archaean regions from late Archaean stabilization. The oldest geological sub-domains, the Barberton and the Ancient Gneiss terranes, crop out in the northeastern part of the craton, close to the most easterly extent of the seismic array near the NW border with Swaziland (Fig. 1). The mafic and ultramafic volcanic rocks of the Barberton region have been interpreted by de Wit and coworkers (de Wit *et al.* 1992; de Wit & Hart 1993) as the products of mid-ocean ridge (MOR) magmatism, preserved by subsequent obduction onto an arc-like terrane. Recent field studies combined with laboratory experiments suggest, however, that the Barberton komatiites and associated basalts were the products of wet melting in an Archaean subduction zone, a process that can also lead to high degrees of depletion in the upper mantle (Parman *et al.* 1997, 2001; Grove *et al.* 2000).

The second period of cratonic development (onset *c.* 3.1 Ga) was viewed by de Wit and co-workers (de Wit *et al.* 1992; de Wit & Hart 1993) as a period of continental growth through a combination of tectonic accretion of smaller crustal terranes and subduction-related igneous and tectonic processes. Craton development culminated in the final mantle stabilization *c.* 2.6 Ga. The growth of the craton during this period was dominantly in the northern (including the Limpopo Belt) and in the western regions, which consist largely of agglomerated granite–greenstone fragments. The onset of this second period of craton evolution witnessed widespread extensional volcanism, followed by the development of enormous depositional basins. Notable among these was the Witwatersrand basin, supplied by sediments from a predominantly northern provenance, and the Pongola basin, at least the southern sector of which was part of a passive continental margin facing open ocean on the south (de Wit *et al.* 1992). A craton-wide extensional cycle of Ventersdorp volcanism and sedimentation between 2.7 and 2.6 Ga marked the end of craton formation in Archaean time.

Limpopo Belt

The Limpopo Belt is a high-grade metamorphic terrane formed by collision in late Archaean time between the Kaapvaal and Zimbabwe Cratons. The Limpopo Belt is divided into a Northern Marginal Zone, a Central Zone and a Southern Marginal Zone (Van Reenen *et al.* 1992). Seismic evidence summarized below confirms that the two marginal zones are overthrusts atop cratonic crust. The Central Zone, on the other hand, has a long and complex tectonic history that culminated in considerable crustal thickening during the collisional phase, followed by the uplift and exhumation of at least 20 or 30 km of crust (Van Reenen *et al.* 1987, 1992; Treloar *et al.* 1992; Windley 1995).

Proterozoic mobile belts

Mobile belts, accreted to southern Africa in mid- to late-Proterozoic time, surround the cratons and act to buffer them from later plate-tectonic events. The Kaapvaal is bounded on the south and east by the subduction-related Namaqua–Natal Belt of Proterozoic age and on the west by the Kheis overthrust belt (de Wit *et al.* 1992). The complex region flanking the Zimbabwe Craton west and NW of the Limpopo Belt is here termed the Okwa–Magondi terrane, although the regional geology remains enigmatic as a result of very complex geological relationships made all the more obscure by extensive Kalahari sand cover (Carney *et al.* 1994). The *c.* 2 Ga events of the Okwa and Magondi Belts were recorded in the western Limpopo Belt during the same time period and imposed a strong overprint on the western sector of the Zimbabwe Craton where it extends into Botswana. The region we have termed Okwa–Magondi Belt is the site of the richest diamond mines in the world (Fig. 1), and from Re–Os studies of xenoliths and diamond inclusions the mantle beneath that region is known to be of Archaean age (Shirey et al. 2001). The presence of abundant diamonds with Archaean-age inclusions suggests that despite the Proterozoic overprint, the Archaean geotherm was not significantly disturbed. As we will show below, the Proterozoic-age events in the western sector of the Zimbabwe Craton are clearly recorded in the seismic crustal structure, and to a lesser extent in the mantle velocity structure.

Bushveld Province

The Bushveld Complex is the largest layered mafic intrusion in the world, with a total volume of magma intruded into the crust of the order of 0.6×10^6 km^3 (Von Gruenewaldt et al. 1985). The 2.05 Ga events associated with the Bushveld intrusion had a profound impact on the seismic and geological character of crust and mantle over a region far more extensive than that implied by outcrops of the Complex itself. Thus, we adopt here the term Bushveld Province in place of Bushveld Complex to connote the fact that Bushveld age correlations (2.05 Ga) are found along a broad east–west swath extending from the easternmost outcrops of the Bushveld Complex westward well into southern Botswana (Hatton & Von Gruenewaldt 1990). For the purposes of discussing the seismic results presented here, the Bushveld Province encompasses the greater region of Bushveld age exposures.

The Vredefort structure: a cross-section of Archaean crust

The Vredefort dome is the deeply eroded remains of a large meteor impact crater formed c. 2.0 Ga in the central part of the Kaapvaal Craton. The crater is roughly 40 km in diameter and reveals a crustal section turned on edge, a virtually complete and well-preserved crustal cross-section exposed on the floor of the crater (e.g. Hart et al. 1990; Moser et al. 2001). The mid-crust (below about 14 km of Witwatersrand sedimentary rocks) is composed of about 10 km of granodiorite composition rocks (the outer granite–gneiss domain), which shows a transition into an incomplete section of lower-crustal granulite facies gneiss of mixed felsic and mafic components. Peridotites believed to be mantle samples and with Re-depletion model ages of 3.3–3.5 Ga have been obtained from a drill hole near the centre of the impact structure (Tredoux et al. 1999).

Primary results from geochemistry and petrology

Comprehensive Re–Os isotope studies show that peridotite xenoliths brought to the surface in kimberlite pipes both on and off craton in southern Africa have ages similar to the overlying crust (Carlson et al. 2000; Irvine et al. 2001) (see Fig. 2). The mantle root of the craton is of Archaean age at least to the maximum depth ($\geq$200 km) from which nodules are derived. There is no correlation between the age of the xenolith and its depth of origin, an observation that rules out keel formation by progressive underplating of material to the continent over long time scales. Most peridotite nodules associated with undisturbed craton are now known to be ancient, including the sheared nodules once thought to have been part of the asthenosphere (Boyd 1987), but which are now considered to be metasomatized samples of highly depleted Archaean mantle (Carlson et al. 2000, with reference to previous work).

Recent studies have also shown that eclogitic materials, occurring both as inclusions in diamond and as whole-rock samples, are as old as or older than the peridotites (Richardson et al. 2001; Shirey et al. 2001). This finding suggests that the eclogites and, by inference, subduction zones, were an integral part of the formation process of the craton. In addition, stable isotope analyses suggest a MOR basalt protolith for the eclogites, providing further evidence for a subduction-zone source for the eclogites. This evidence that subduction processes were involved in earliest craton formation is buttressed by laboratory petrological experiments (Parman et al. 1997, 2001; Grove et al. 2000). Those studies show that the high-MgO komatiite magmas of the Kaapvaal probably formed within an Archaean subduction zone via a single melt generation process similar to that by which high Mg-number boninites are formed in very young modern subduction zones. Petrological examination of Archaean ultramafic magmas (komatiites) from South Africa indicates that some komatiitic magmas contained substantial quantities of water (>4 wt%). This finding strengthens the possibility that the cratonic lithosphere formed initially in subduction-zone settings whose demise led to accretion of the arc crust and thickening of refractory mantle to create a stable, thick, continental

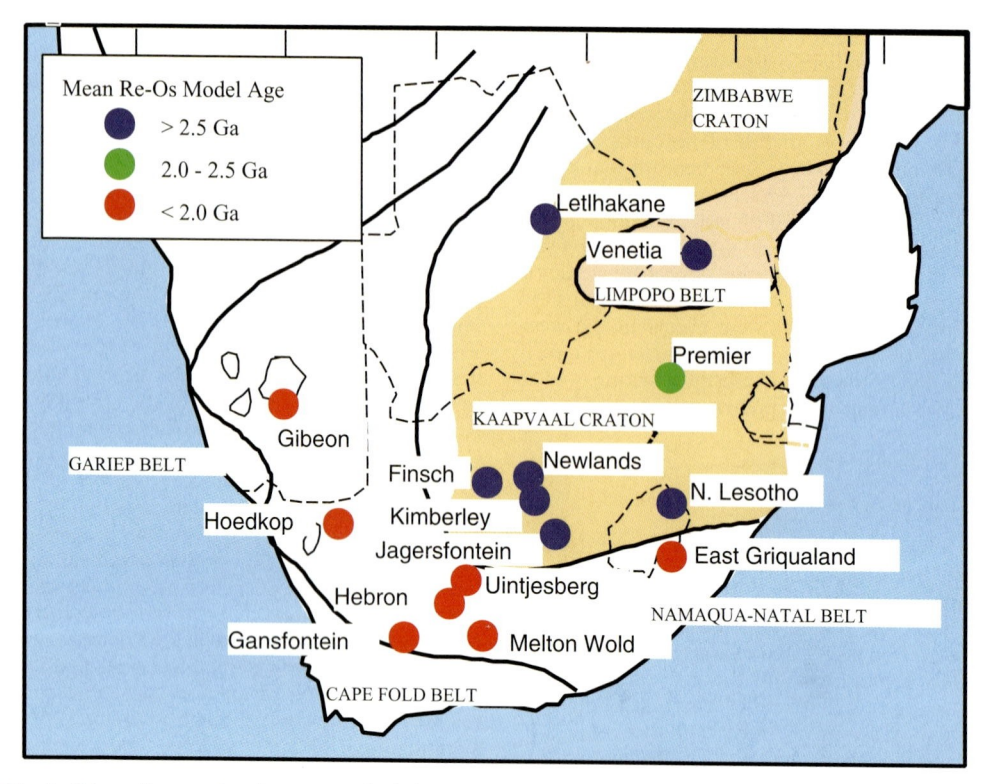

Fig. 2. Schematic map showing mean redepletion model ages determined for peridotite nodules from kimberlite pipes erupted in southern Africa. Cratonic regions are outlined in yellow, the Limpopo Belt in brown. Solid coloured circles indicate individual kimberlite pipes from which several xenoliths have been analysed. Published in Carlson *et al.*, GSA Today, 2000, February issue, fig. 4.

lithosphere (Carlson *et al.* 2000; Grove *et al.* 2000). The paucity of eclogite xenoliths (<1% (Schulze 1989) indicates that essentially all of the descending oceanic plates during Archaean time were subducted into the deeper mantle, with little eclogitic material incorporated into the cratonic root.

The co-ordinated Kaapvaal Project geochronological studies of crustal and mantle xenoliths reveal that both crust and mantle have experienced a multi-stage history, and that a simple view of cratonization as a discrete event is not a viable model for craton formation (Schmitz *et al.* 1998; Schmitz & Bowring 2000; Moser *et al.* 2001). The lower crust in particular retains a comprehensive record of the tectonothermal evolution of the lithosphere. The study of lower-crustal samples has shown that much of the deep craton experienced a dynamic and protracted history of tectonothermal activity that is temporally associated with events seen in the surface record, including late Archaean magmatism (Ventersdorp) and even Proterozoic deformation (Namaqua–Natal) (Schmitz *et al.* 1998). Thermal events are reflected in ages of 2 Ga or even less for some eclogitic diamond inclusions from tectospheric mantle originally of Archaean age (Carlson *et al.* 2000; Shirey *et al.* 2001).

Moser *et al.* (2001) suggested that stabilization of the mantle keel in the region of the Vredefort impact structure may have taken place roughly 100 Ma after formation of the overlying crust. They based this assertion on the notion that late Archaean thermotectonic events recorded in the lower crust argue against the presence of a thick mantle root that would have served to buffer the crust from such events. A separate evolutionary history for crust and underlying mantle is not inconsistent with the peridotite results that show the majority of peridotite xenoliths have Re–Os model ages younger than 3.0 Ga, compared with overlying crustal formation ages of 3.2–3.4 Ga. However, if this result is correct, and if it is craton-wide, it implies that cratonic crust and its underlying mantle root need not have a synegenetic relationship.

Lithospheric mantle xenoliths of Proterozoic age are derived from depths up to about

150 km beneath the surrounding Proterozoic mobile belts, indicating crust–mantle coupling and long-term stability of the upper mantle beneath the Proterozoic belts (Carlson et al. 2000). The Proterozoic mantle root is not only thinner than that present beneath the cratons, but the peridotite components are more fertile in composition. Interestingly, however, the geothermal gradients beneath the southernmost Kaapvaal Craton and the adjacent Namaqua–Natal Belt are similar based on mantle xenolith geothermometry and geobarometry (P. Janney, pers. comm. 2000).

Southern Africa seismic experiment

The southern Africa seismic experiment is one of the largest broadband portable array seismic investigations ever undertaken. The experiment has produced images of crustal and upper-mantle structure in unprecedented detail beneath the cratons and adjacent Proterozoic provinces. Fifty-five portable broadband REFTEK/STS-2 seismic stations were deployed from April 1997 to July 1999 in a roughly 100 km grid along a NNE–SSW transect about 1800 km long by 600 km wide in southern Africa (Fig. 1). Approximately half the instruments were redeployed to new sites in April–May 1998 for a total of 82 station locations. The matched three-component sensors were installed in semi-permanent low-noise vaults on bedrock and signals were recorded with 24-bit dynamic range and a continuous sampling rate of 20 samples s^{-1}. The array data were supplemented by data from three global digital stations in the region of the array (Fig. 1). The experiment was augmented by a 6 month deployment of 32 broadband telemetered stations installed in a tight ($c.\,65 \times 50$ km) array in the region around and to the NW of Kimberley.

The seismic results discussed here are organized into two major topics: (1) tomographic images of the upper mantle beneath southern Africa; (2) receiver function characterization of crustal thickness and the crust–mantle interface beneath the stations of the array. Included also from work in progress is S-wave velocity structure from two-station surface-wave phase and group velocity inversion for a propagation path across undisturbed Kaapvaal Craton (Gore 2002; Nguuri 2002).

Upper-mantle structure

Tomographic results. Upper-mantle velocity structure (James et al. 2001b) was determined by tomographic techniques based on the analysis of delay times from teleseismic broadband waveform data. Relative arrival times of phases P, PKPdf, S and SKS were retrieved via a multichannel cross-correlation procedure using all possible pairs of waveforms (VanDecar & Crosson 1990). This procedure produces highly accurate delay times, with typical standard errors for the

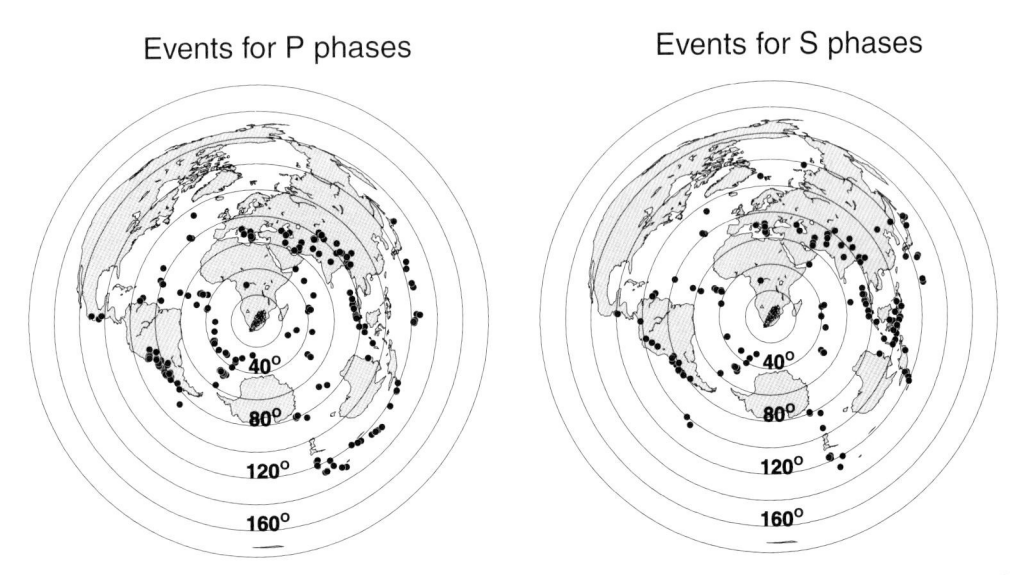

Fig. 3. Location map of events used for tomographic inversions. The map is centred on the approximate centre of the southern Africa array. Locations are from the National Earthquake Information Centre bulletin as archived by the Incorporated Research Institutions for Seismology (IRIS) Data Management Center (James et al. 2001b). Published in James et al., GRL28, 2001, fig. 2.

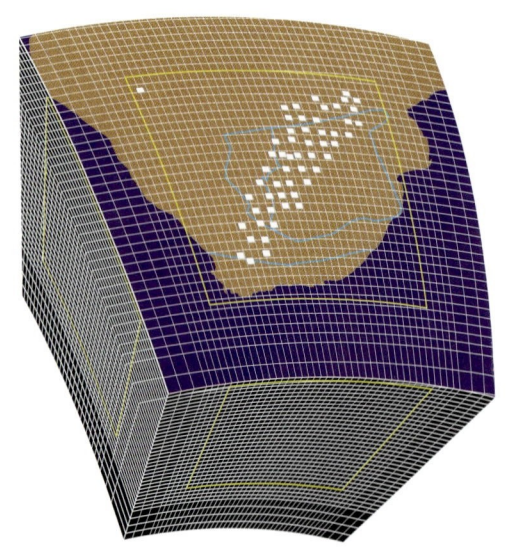

Fig. 4. A 3D perspective view of grid knots for splines under tension that constrain the velocity perturbation model. Knot positions occur at line crossing. The yellow lines indicate regions to which the velocity maps and sections are confined.

southern Africa data of c. 0.03 s for P waves and 0.06 s for S waves. Results for P waves are of higher resolution than those results for S with 8693 rays from 234 P-wave events compared with 4834 rays from 148 S-wave events. Epicentres of earthquakes used in the analysis are shown in Figure 3.

The inversion method for obtaining velocity structure has been fully described by (VanDecar 1991). By this method, P- and S-wave delay times are inverted independently for structure beneath the array. The model is parameterized identically for the P- and the S-wave inversion with splines under tension constrained by a series of regular knots (Fig. 4). Within the interior portion of the model, the knots are spaced 50 km apart in depth and $\frac{1}{2}$ degree apart in latitude and longitude. Corrections for station elevation and crustal thickness from Nguuri (2002) are applied to the data before inversion. The data are inverted simultaneously for the slowness perturbation field, earthquake relocations and station corrections. The inclusion of earthquake relocations and station corrections ensures that the resulting velocity model will be constrained to contain the least amount of structure required to satisfy the observations within their estimated standard errors.

The tomographic images presented in Figures 5 and 6 were determined using linear inversion and are therefore preliminary (James et al. 2001b). Results of the non-linear inversion and a detailed analysis of resolution tests will be provided in a future work. Although the non-linear inversion will undoubtedly affect results in the deeper parts of the model, the shallow (<300 km), relatively mild, velocity perturbations obtained from the linear inversion are unlikely to change significantly. We have designed simple resolution tests that approximately mimic the observed structures. The output from the tests and a description of our procedures can be found in the electronic supplement to James et al. (2001b), or as a downloadable file at http://www.ciw.edu/kaapvaal/pubs/tomography/grl_supplement.pdf. The resolution tests indicate that both laterally and vertically the cratonic roots are well recovered. As expected, recovered velocity perturbations are lower by varying degrees than those of the input model; this effect increases with depth in the keel. The maximum and minimum values of the recovered velocities are, however, approximately equal to those of the input model, suggesting that the maximum and minimum velocity perturbations obtained in the data inversions are not greatly different from those in the real Earth. Downward smearing of structure does occur, but the effect is relatively small and does not preclude reasonably accurate estimates of keel thickness. Because station spacing of the southern Africa array is about 100 km, shallow structures above about 50 km are sampled by relatively few crossing paths from teleseismic events and are thus included largely in the station terms. A further qualitative assessment of resolution is provided by the ray density as seen in Figure 7, which shows the ray 'hit' count in both P- and S-velocity horizontal and vertical sections shown in Figures 5 and 6. Of particular relevance is the high and rather uniform ray density along the central profile B–B'.

Fig. 5. Cross-sections through the P-wave velocity perturbation models obtained by inversion of delay times corrected for elevation and crustal thickness. (**a**), (**b**), and (**c**) show plan views of velocity perturbations at depths of 150, 200 and 300 km, respectively. (**d**), (**e**) and (**f**) show vertical cross-sections along profiles A–A', B–B' and C–C', respectively, as shown in the horizontal sections. Surface topography is plotted at 20 times actual scale. Uppermost 50 km (shaded area) in vertical sections denotes regions where station delay time residuals are incorporated in model calculations. Colour scale shows the velocity perturbation in percent. Colours fade to black for ray hit counts less than 10 (see Fig. 7).

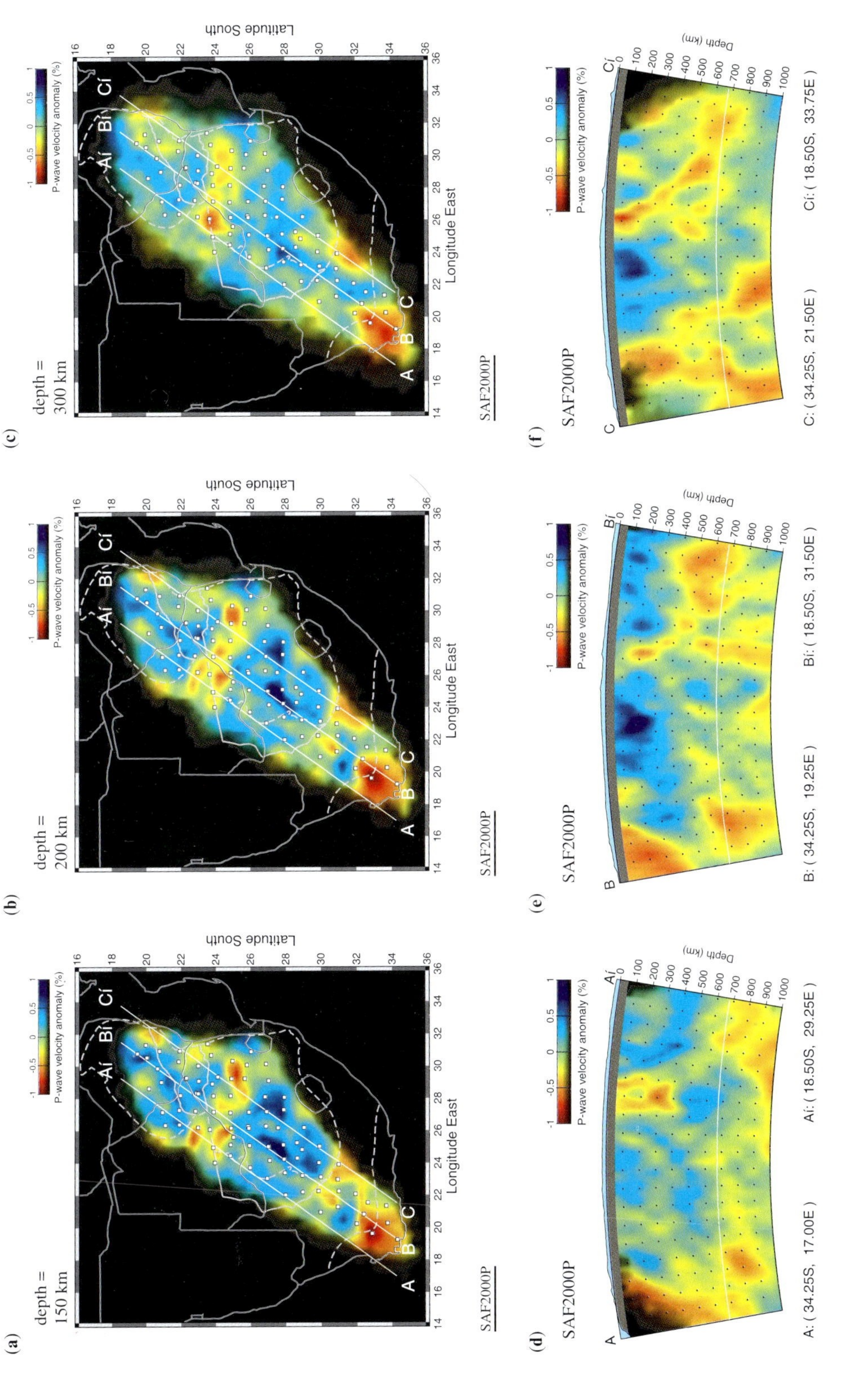
(a)
depth = 150 km
(b)
depth = 200 km
(c)
depth = 300 km
SAF2000P
Longitude East
Latitude South
P-wave velocity anomaly (%)
A
B
C
Aí
Bí
Cí
(d)
(e)
(f)
Depth (km)
A: (34.25S, 17.00E)
Aí: (18.50S, 29.25E)
B: (34.25S, 19.25E)
Bí: (18.50S, 31.50E)
C: (34.25S, 21.50E)
Cí: (18.50S, 33.75E)

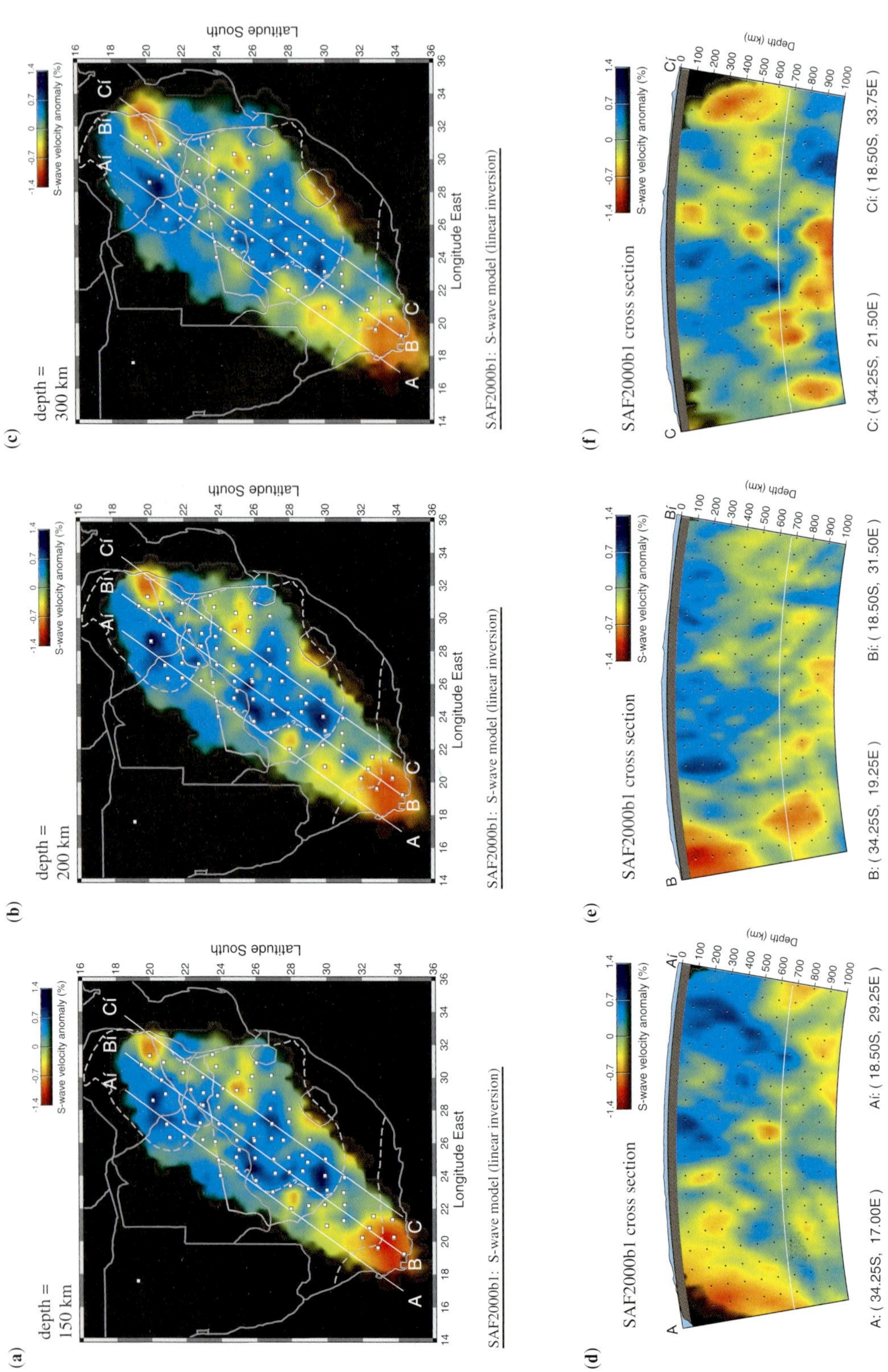
(a)
depth = 150 km
SAF2000b1: S-wave model (linear inversion)
(b)
depth = 200 km
SAF2000b1: S-wave model (linear inversion)
(c)
depth = 300 km
SAF2000b1: S-wave model (linear inversion)
Longitude East
Latitude South
S-wave velocity anomaly (%)
A
B
C
Aí
Bí
Cí
(d)
SAF2000b1 cross section
A: (34.25S, 17.00E)
Aí: (18.50S, 29.25E)
(e)
SAF2000b1 cross section
B: (34.25S, 19.25E)
Bí: (18.50S, 31.50E)
(f)
SAF2000b1 cross section
C: (34.25S, 21.50E)
Cí: (18.50S, 33.75E)
Depth (km)

CRATON FORMATION IN SOUTHERN AFRICA

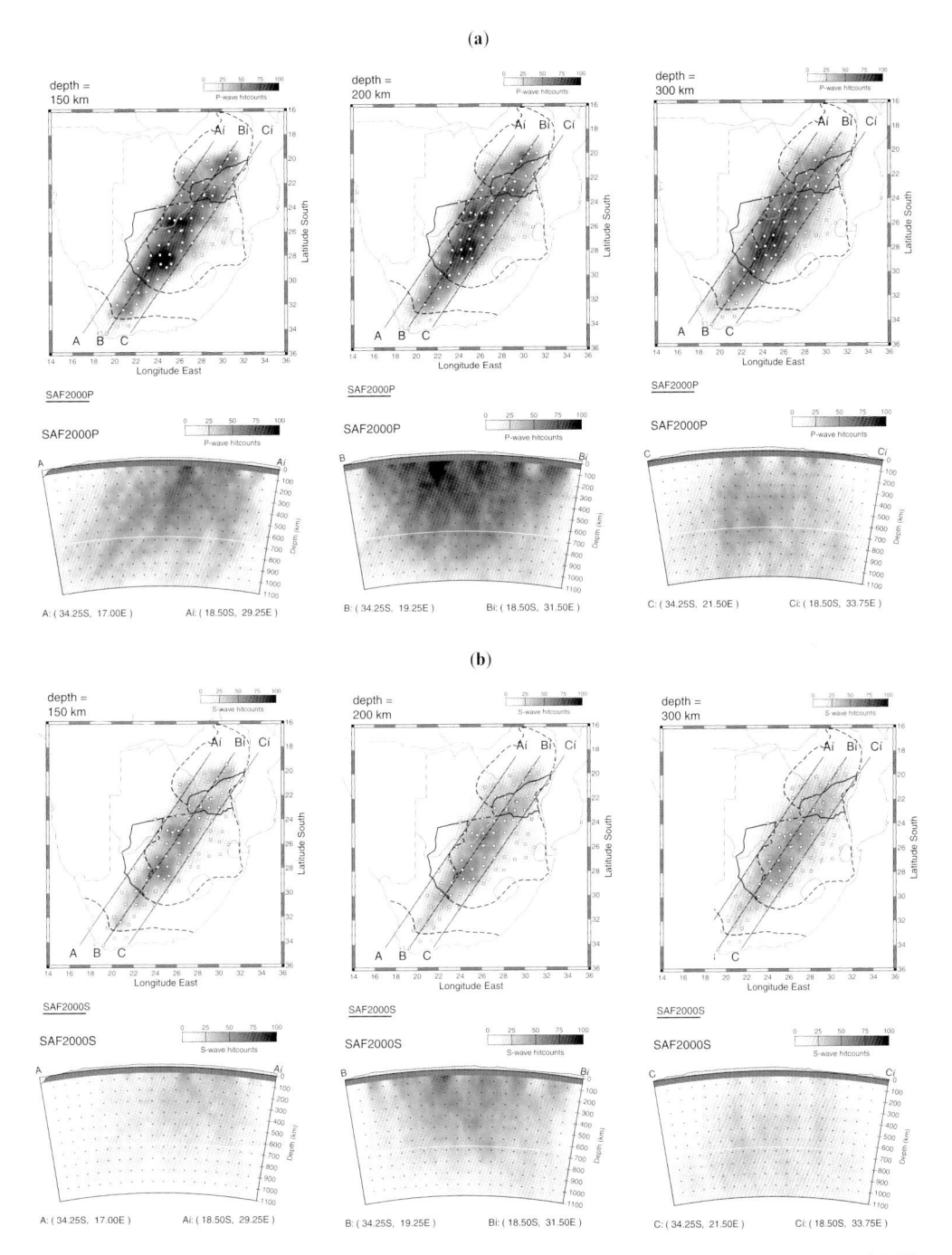

Fig. 7. Grey-scaled ray density maps of horizontal and vertical velocity sections shown in Figures 5 and 6. The scale grades from 0 (white) to 100 (black), with the scale reflecting the number of ray 'hits' in each node of the velocity perturbation model. Ray densities along the axis of the array are typically >50 hits per node.

Fig. 6. Cross-sections through the S-wave velocity perturbation models. (See caption for Figure 5 for a description of the cross-sections.)

Initial results for the linear, smoothed travel time inversion for P waves are summarized in Figure 5 and for S waves in Figure 6 (see also James *et al.* 2001*b*). The independently computed P and S models are remarkably similar within the limits of resolution. High mantle velocities in both P and S models coincide with the interiors of the Kaapvaal and Zimbabwe Cratons. An extensive region of maximum positive velocity perturbations (blue regions, Figs 5 and 6) is concentrated in the present-day heart of the Kaapvaal Craton, which extends from the southern edge of the Bushveld Province SSW to the contact with the Namaqua–Natal Mobile Belt. In this region, the undisturbed cratonic keel may attain depths of 250–300 km and perhaps more (Fig. 5). The evidence for root structures deeper than approximately 300 km beneath the array is sparse, although excepting regions of disrupted craton (as in the Bushveld), the tectospheric root attains a thickness of about 200–250 km almost everywhere beneath the cratons, including the Archaean Limpopo Mobile Belt.

The most remarkable 'disrupted' feature of the Kaapvaal Craton is associated with the Bushveld Province. Distinctly lower mantle velocities, strikingly evident for P but also seen in S waves, are associated with the larger Bushveld Province, extending at least into the southern part of the Okwa Belt of Botswana and probably into the Magondi–Limpopo zone in NE Botswana as well. Although these low mantle velocities are well resolved overall, the localized 'patchiness' of the low-velocity perturbations seen in Figures 5 and 6 is not. The tomographic results are consistent with the fact that whereas surface geology of the Okwa–Magondi region registers a strong Proterozoic overprint, the Re–Os signatures of the mantle xenoliths and diamond inclusions there are clearly of Archaean age (Richardson *et al.* 2001; Shirey *et al.* 2001).

Within the resolution of the data, the mantle structure of the Archaean Limpopo Belt does not differ significantly from that of the adjacent cratons. The similarity with cratonic mantle structure contrasts sharply with the results of crustal structure determinations (Nguuri *et al.* 2001), which show the Central Zone of the Limpopo Belt to be characterized by thick crust and poorly developed Moho relative to the adjacent cratons. Interestingly, the SKS splitting results for the southern Africa array show that the Limpopo Belt exhibits a consistent east–west mantle fabric, presumably acquired at the time of craton collision (Silver *et al.* 2001).

The Proterozoic Namaqua–Natal Mobile Belt, thought to be the remnants of a major north–south convergent margin that extended as far north as the Zimbabwe Craton (De Wit *et al.* 1992), is characterized by velocity perturbations uniformly lower than those observed beneath the craton. The lower velocities are in keeping with the observation that the off-craton Proterozoic mantle tends to be somewhat more fertile (higher Fe) than that of the adjacent craton (Carlson *et al.* 2000; Pearson *et al.* 2002). Patches of higher-velocity material are seen in the 200–400 km depth range beneath the belt, however, and these higher velocities typically exhibit continuity with the high-velocity material beneath the adjacent Kaapvaal Craton.

Surface-wave results. Comprehensive studies of inter-station surface-wave dispersion across the southern Africa array are included in studies currently in progress (Gore 2002; Nguuri 2002). We cite here some preliminary results based on both phase and group velocity dispersion curves from the high-quality southern Africa data because they provide a relatively unambiguous test for low-velocity zones in the upper mantle. The question of a low-velocity zone at depths <200 km beneath the craton is an issue because it has purportedly been detected beneath southern Africa (Qiu *et al.* 1996; Priestley 1999). Initial results from the Kaapvaal experiment, including velocity estimates from mantle xenoliths to depths of 200 km, provide no support for a low-velocity zone beneath the craton. Surface-wave studies using much larger global and regional datasets (Zhao *et al.* 1999; Ritsema & van Heijst 2000) and the Kaapvaal array data using different methods (Freybourger *et al.* 2001) also failed to reveal a low-velocity zone beneath the craton (also see Discussion section below).

A typical example of inter-station Rayleigh-wave phase and group velocity dispersion from an Andean event for a pure cratonic path (between stations 16 and 32 (see Fig. 9)) across the southern Kaapvaal Craton near Kimberley is shown in Figure 8. Inversion of the surface-wave dispersion curves yields a model with high shear-wave velocities (>4.6 km s^{-1}) in the mantle to a depth of at least 200 km. No models incorporating a significant low-velocity zone fit the data. The results shown in Figure 8 are typical of many dozens of similar inversions of inter-station phase and group velocities of both Rayleigh and Love waves across the southern Africa array. The surface-wave data are entirely consistent with S-wave velocity estimates from mantle xenoliths (James *et al.* 2001*a*), which show velocities reaching a maximum of nearly 4.75 km s^{-1} in the uppermost mantle, decreasing slightly to about 4.6–4.65 km s^{-1} at 200 km depth. No samples, including several high-T lherzolites from

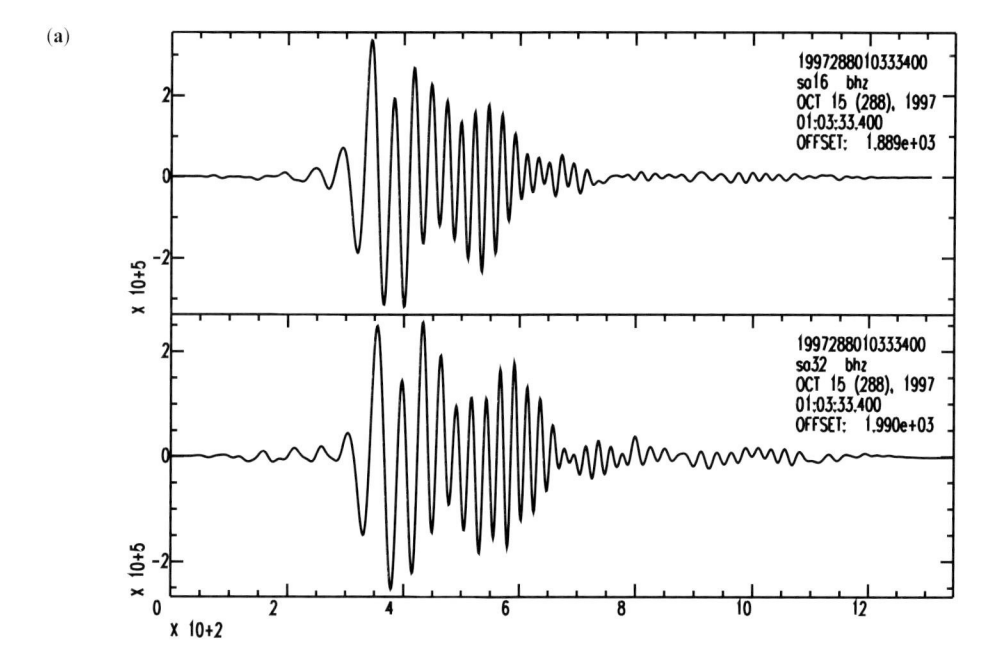

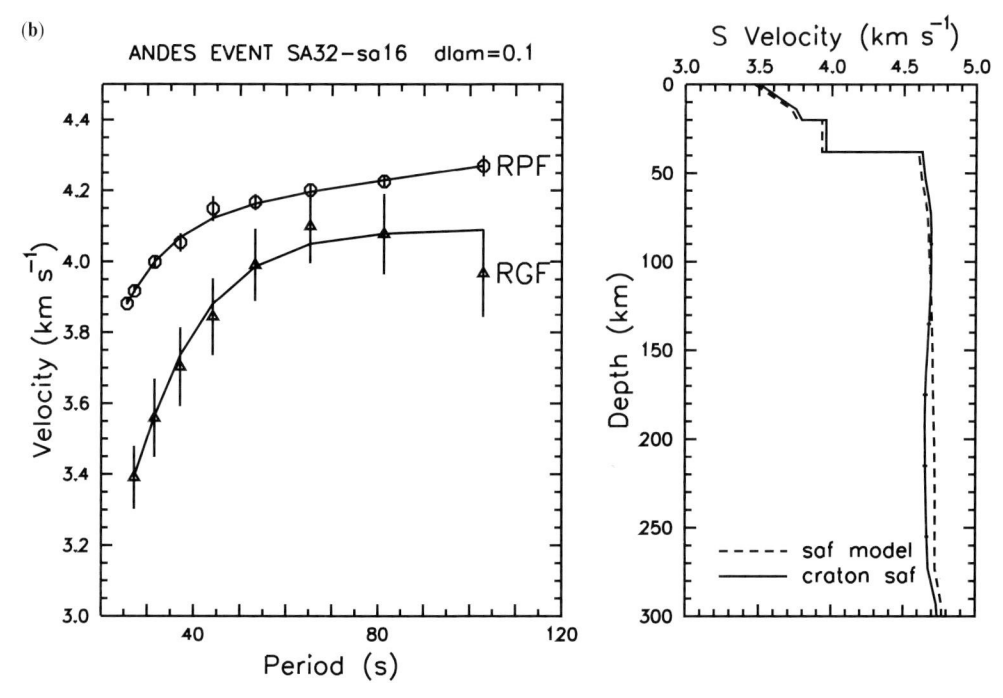

Fig. 8. (**a**) Rayleigh waves recorded on the vertical component at stations sa16 and sa32 for an on-azimuth magnitude 6.8 event on 15 October 1977 in central Chile. Total time 1350 s on the horizontal axis. Records have been decimated and instrument corrected for displacement. Seismograms have been low-pass filtered at 0.05 Hz. (**b**) Two-station Rayleigh wave phase and group velocities v. period for pure craton path across the southern Kaapvaal Craton (left panel). Left-hand panel shows measured phase velocities (RPF) and group velocities (RGF) for the 464 km interstation path between sa16 and sa32. Data, with phase velocities more heavily weighted than group velocities, have been inverted to obtain the best-fit S velocity–depth model shown as a continuous line in the right-hand panel. The dashed line was the starting model. The line curves in the right-hand panel are based on the best-fit S velocity model (adapted from Nguuri (2002)).

about 200 km depth, have calculated S-waves velocities <4.6 km s^{-1}.

Crustal structure

Initial results from the analysis of P- to S-wave converted phases (Ps) from the M-discontinuity were reported by Nguuri *et al.* (2001) and are summarized below. The results reported by Nguuri *et al.* were based upon 35 teleseisms processed at 75 stations (Fig. 9) to yield a comprehensive set of high-quality receiver functions (see Ammon (1991) for review and other references to receiver function techniques). Individual receiver functions were corrected for travel time moveout, binned by station and stacked by phasing depth at depth intervals of 0.5 km between 1 and 101 km. The resulting stacked receiver function images provide a display of the discontinuity structure beneath each station based on Ps conversions (e.g. Gurrola *et al.* 1994; Dueker & Sheehan 1997, 1998). The crustal model for the moveout correction is based on seismic refraction models for the Kaapvaal Craton, with an average P-wave velocity for the crust of 6.5 km s^{-1}, Poisson's ratio of 0.25, and Moho depth of 38 km (Durrheim & Green 1992). Crustal multiples are suppressed in this procedure by virtue of the fact that the moveout correction

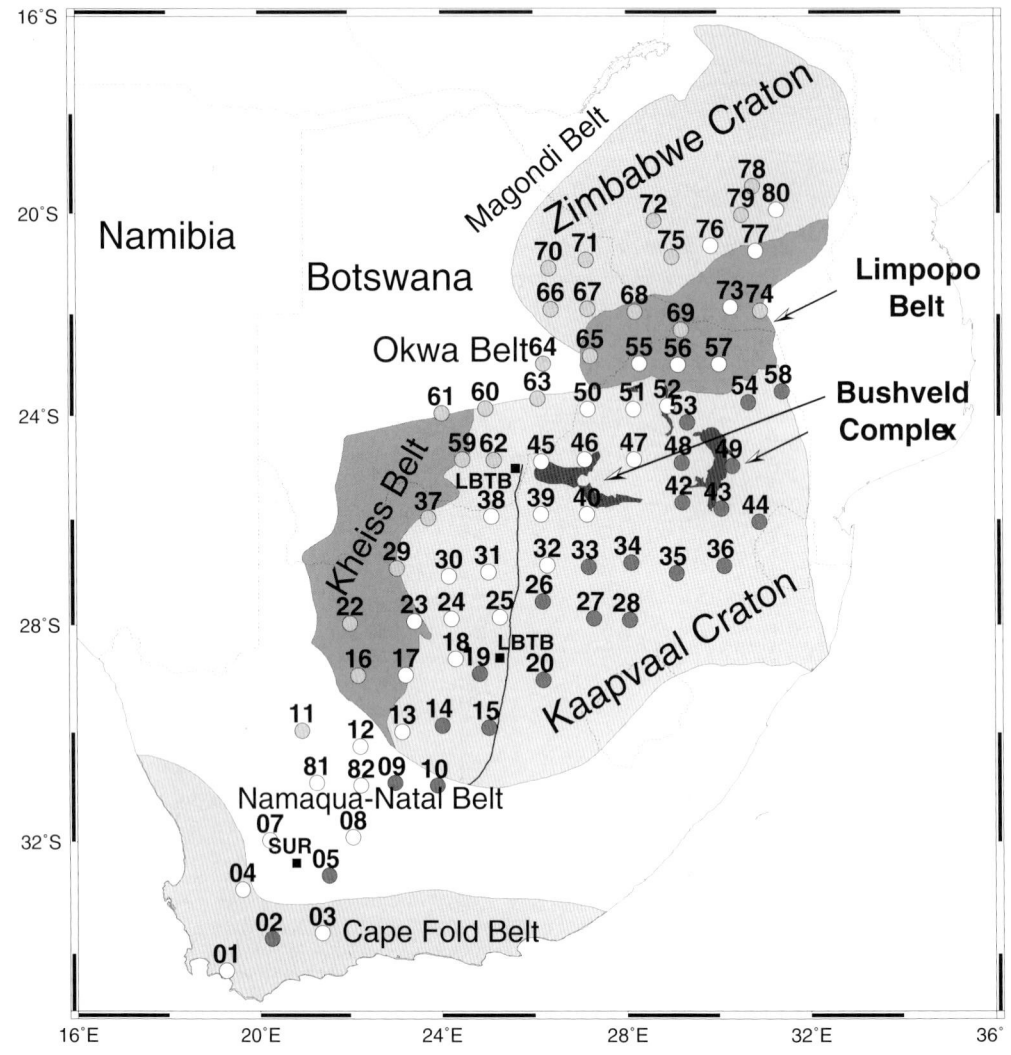

Fig. 9. Location map for the southern Africa seismic array, plotted on an outline map of southern Africa geology (from Nguuri *et al.* 2001). Published in Nguuri *et al.*, GRL28, 2001, fig. 1 (corresponding author D. E. James).

is appropriate only for the direct P- to S-wave conversion. Moreover, the amplitudes of the Ps signals in the stacked traces (phasing depth images) plotted in Figure 10 are not strictly interpretable, as they average events with differing ray parameters. As seen from Figure 10, only one consistent Ps signal occurs, and it is readily associated with the M-discontinuity. These results are summarized in the gridded map of crustal thickness shown in Figure 11 (from Nguuri et al. 2001). As the velocity–depth model was constructed by averaging seismic refraction results for the Kaapvaal Craton, the model may underestimate mean crustal velocity and crustal thickness for off-craton or modified cratonic regions (Nguuri et al. 2001).

Stations located within undisturbed Kaapvaal or Zimbabwe Craton typically have sharp, large-amplitude depth images for the Moho. Among the more distinctive results are those from stations located in the Zimbabwe Craton, where Moho depths with one exception cluster tightly

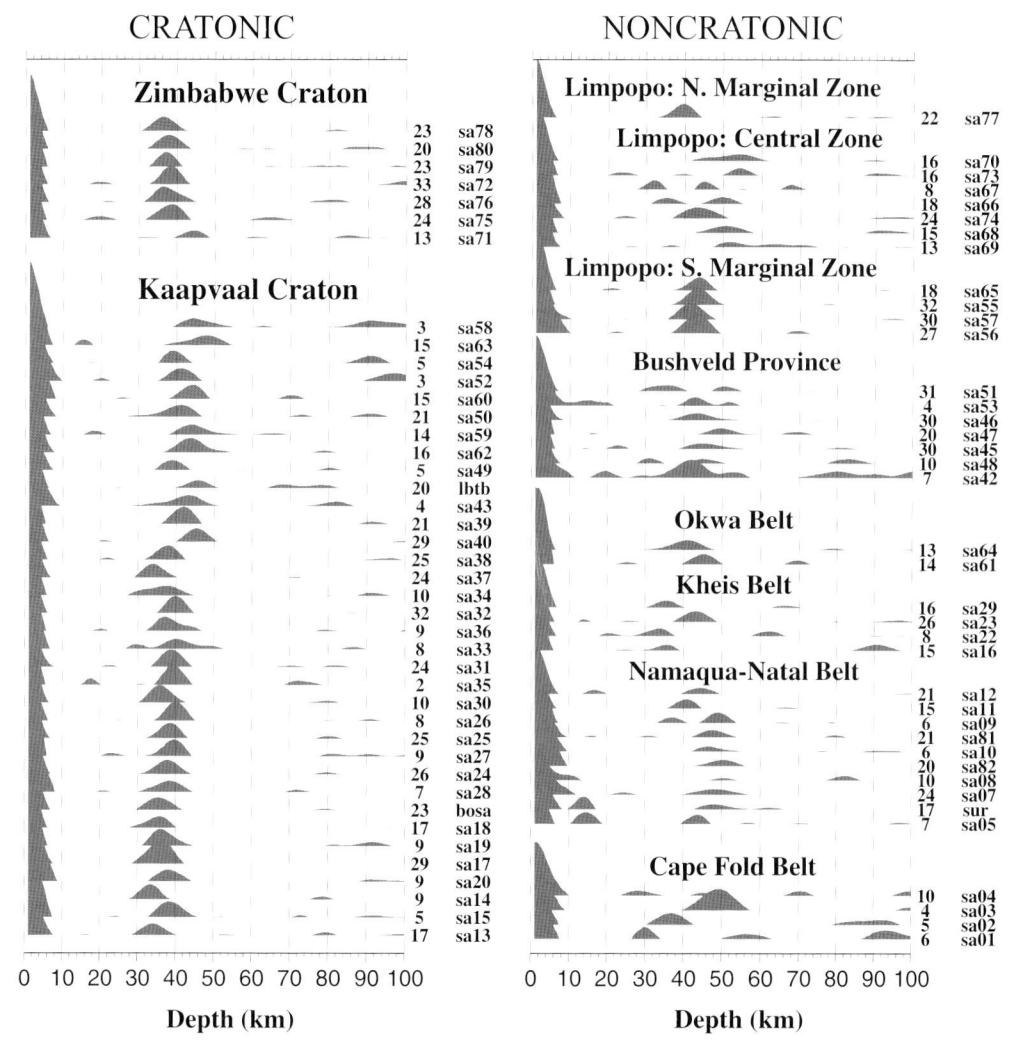

Fig. 10. One-dimensional phasing depth images for southern Africa, organized by geological province. The number of events included in the stack and the station name are shown to the right of each trace. The dominant signal on most of the depth images shown is the Ps conversions from the Moho. Relatively thin (35–40 km) crust and sharp, well-defined Ps Moho conversions are associated with undisturbed craton. Ps arrivals associated with modified regions of the craton and post-Archaean terranes tend to be more diffuse and of smaller amplitude (from Nguuri et al. 2001, published in GRL28, 2001, fig,. 2).

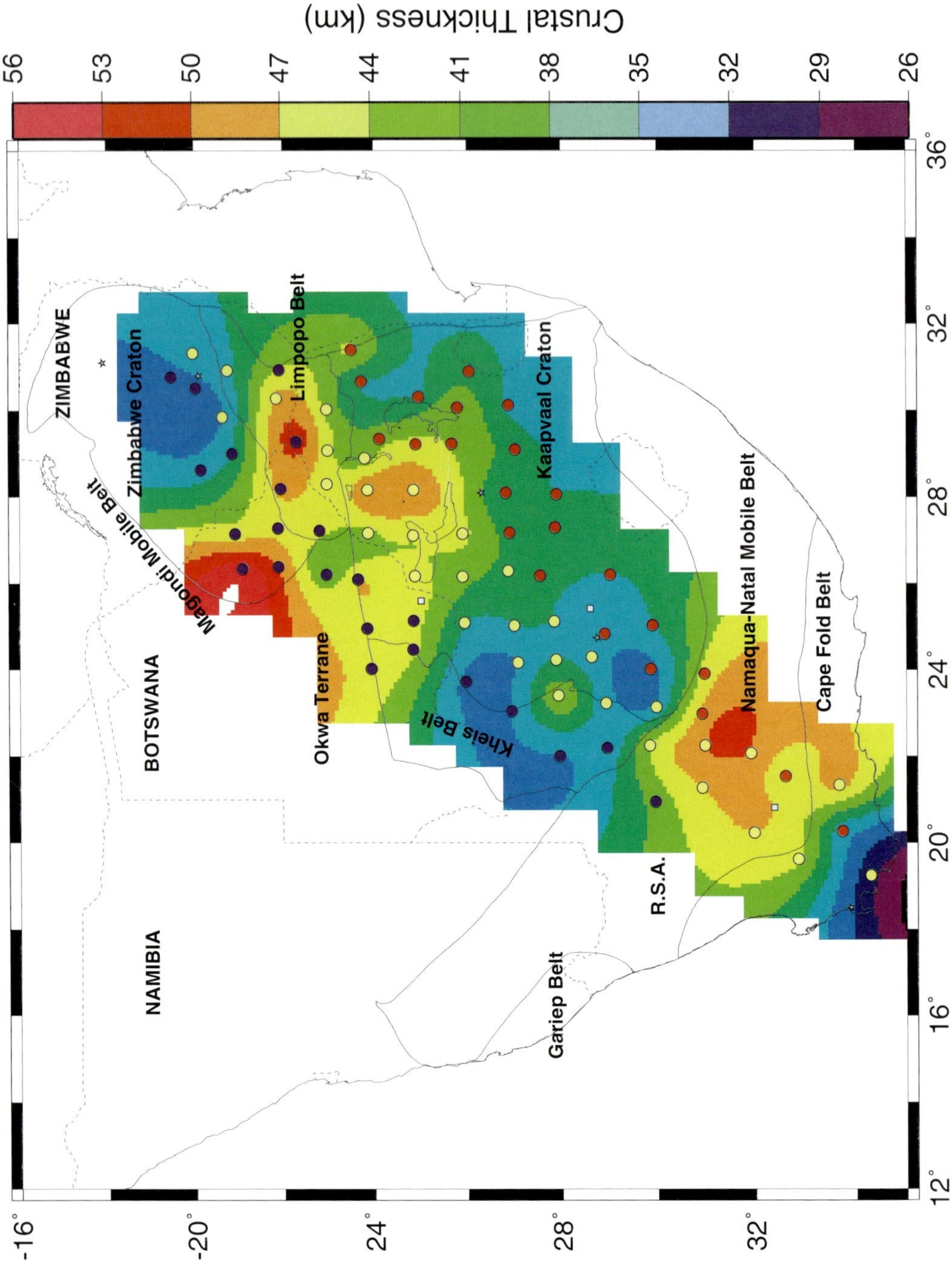
Crustal Thickness (km)
56
53
50
47
44
41
38
35
32
29
26
ZIMBABWE
Zimbabwe Craton
Limpopo Belt
Kaapvaal Craton
Magondi Mobile Belt
BOTSWANA
Okwa Terrane
Kheis Belt
Namaqua-Natal Mobile Belt
Cape Fold Belt
R.S.A.
NAMIBIA
Gariep Belt
12°
16°
20°
24°
28°
32°
36°
-16°
-20°
-24°
-28°
-32°

between 34 and 37 km. The one exception, sa71, is in the western craton near the zone of Proterozoic (c. 1–2 Ga) overprinting. Results for crustal thickness beneath the Kaapvaal Craton exhibit more variability, with thickness varying between about 33 and 45 km, and averaging about 38 km.

By comparison with stations within regions of undisturbed craton, those within post-Archaean and modified cratonic regions typically exhibit much more poorly defined Ps converted phases. Moreover, the crust beneath post-Archaean and modified cratonic regions is in almost all cases thicker than that beneath the cratons. Of particular interest is the broad region of the Bushveld and Okwa–Magondi terranes that extends from the Bushveld Complex in the northern Kaapvaal Craton westward into eastern and northeastern Botswana. The Archaean craton through this entire zone has been overprinted by tectonomagmatic events of Proterozoic age (e.g. Carney *et al.* 1994; Shirey *et al.* 2001) and the effects on crustal structure have been profound. The crust thickens systematically and the Moho signature degrades, for example, across the boundary between undisturbed craton and the Bushveld Province and between undisturbed Zimbabwe Craton and the Magondi Belt. The change in crustal structure from undisturbed craton into the Bushveld and Okwa–Magondi terranes is consistent with the results from upper-mantle tomographic studies described above, which show a significant low-velocity perturbation in the uppermost mantle extending beneath the entire region (James *et al.* 2001*b*).

The intercratonic Limpopo Belt is of particular interest. Depth images for the marginal zones are typical of cratonic structure, both in character of the Ps conversion and in crustal thickness. The northern marginal zone is underlain by crust about 37 km thick, typical of the adjacent Zimbabwe Craton, and the southern marginal zone has a crust around 40–42 km thick, consistent with that of the adjacent Kaapvaal Craton (see Fig. 10). The central Limpopo Belt, the site of pervasive deformation during the collision of the Kaapvaal and Zimbabwe Cratons in Archaean time, displays particularly complex structure. In some instances (e.g. stations sa66 and sa67) the identification of the Moho is ambiguous and could be resolved only with constraints from two-station surface-wave phase velocity inversions, which show unambiguously that thicker crust is required to satisfy the dispersion data (Gore 2002; Nguuri 2002). The broad, poorly defined Ps images have depth maxima occurring between 40 and 53 km and are indicative of a structurally complex Moho. These values for crustal thickness beneath the central zone of the Limpopo Belt obtained by Nguuri *et al.* (2001) differ significantly from those of some previous seismic and gravity studies, where a Moho depth of about 30 km was postulated (Stuart *et al.* 1986; De Beer & Stettler 1992; Gwavava *et al.* 1992). Although our results show that there may be a deep crustal discontinuity at about 30–35 km depth beneath some stations, inter-station surface-wave phase velocity inversion results show unambiguously that the crust must be at least 40–45 km thick beneath the Central Zone (Gore 2002; Nguuri 2002). As with the Bushveld Complex, the relatively thick crust beneath the Limpopo Belt does not translate to higher elevations relative to the adjacent cratons.

The Kheis Belt is similar to the Northern and Southern Marginal Zones of the Limpopo Belt in that it is an overthrust sheet atop Archaean crust of the Kaapvaal Craton. Not surprisingly, therefore, the crustal thickness of the Kheis Belt is similar to that of the adjacent Kaapvaal Craton. Indeed, the only evidence for crustal thickening beneath the Kheis Belt is in its northern extent near its contact with the Okwa Belt of southern Botswana. The large amplitudes of the Ps Moho conversions observed in the marginal zones of the Limpopo Belt, however, are not so apparent in the Kheis Belt. The lack of a purely 'cratonic' Moho signature is in some sense consistent with tomographic images of the upper mantle beneath the Kheis Belt, where large positive velocity perturbations observed in the cratonic mantle of the Kimberley region tend to decrease westward (James *et al.* 2001*b*).

Discussion

Tectospheric roots

The depth extent of cratonic roots has long been an issue of some controversy, dating back to Jordan's seminal work in the mid-1970s (Jordan 1975). Although the tectosphere hypothesis of deep, cold, and chemically distinct keels beneath

Fig. 11. Colour-coded contour map of depth to Moho beneath the southern Africa array based on phasing depth images of Figure 2 (from Nguuri *et al.* 2001). Crustal thickness colour scale is shown on right. Thin crust is associated with undisturbed areas of craton, particularly in the southern and eastern parts of the Kaapvaal Craton and in the Zimbabwe Craton north of the Limpopo Belt. Greater crustal thickness is associated with the Bushveld region and its westward extension into the Okwa and Magondi Belts and with the Central Zone of the Limpopo Belt and the Proterozoic Namaqua–Natal Mobile Belt.

continents was widely challenged when first proposed (e.g. Okal & Anderson 1975; Anderson 1979; Sclater et al. 1980), the model has gained widespread acceptance, buttressed by petrological and geochemical studies of mantle xenoliths, particularly those from southern Africa. Notable among these studies are Re–Os age determinations that show that mantle nodules erupted from even the greatest depths beneath the craton (about 200 km) are of Archaean age (Pearson et al. 1995; Carlson et al. 2000). Moreover, recent analyses of xenolith P–T data suggest that the intersection between the craton geotherm and the mantle adiabat occurs between depths of 160 and 300 km, with a best estimate in the range of 220–250 km beneath the cratons (Rudnick et al. 1998; Rudnick & Nyblade 1999). These estimates are entirely consistent with our results, which show that positive velocity perturbations beneath the cratons extend to depths of at least 250–300 km, and with results from other detailed regional seismic studies (VanDecar et al. 1995; Van der Lee & Nolet 1997; Ritsema et al. 1998; Ritsema & van Heijst 2000).

In general, low geotherms and refractory mantle compositions contribute to the high velocities associated with the tectospheric keels of cratons (Jordan 1975, 1981). The rocks that make up the bulk of the tectospheric mantle are highly depleted peridotites. The sheared nodules formerly believed to be enriched are now known to be depleted peridotites that were heavily metasomatized shortly before eruption (Pearson et al. 1995; McDonough & Rudnick 1998; Carlson et al. 2000). Although it is possible that metasomatized rocks are widespread in the deep cratonic mantle, pervasive metasomatism would significantly reduce seismic velocities, which is not observed in the seismic data. Jordan showed, for example, that fertile cratonic samples contain significant weight percentages of both clinopyroxene and garnet, resulting in seismic velocities up to 1% lower and densities 1–2% higher than the depleted nodular peridotites (Jordan 1979). Similarly, eclogitic mantle, if present (Shirey et al. 2001), will have both lower velocity and higher density than depleted peridotitic mantle at the same temperature, although an inventory of mantle xenoliths of the Kaapvaal Craton suggests that the cratonic mantle is <1% eclogite by volume (Schulze 1989).

Results to date from the southern Africa seismic experiment, including the surface-wave results cited above, fail to reveal a low-velocity asthenospheric layer beneath the Archaean keel. This result is particularly relevant in light of studies published by Qiu et al. (1996) and modified by Priestley (1999), which fit surface waveform data from regional earthquakes recorded at global digital stations in southern Africa. Their interpretation of the data indicated a high-velocity lid in the uppermost mantle to a depth of 160 km, underlain by a significant low-velocity zone with S-wave velocity about 4.32–4.45 km s^{-1} (Qiu et al. 1996; Priestley 1999). As indicated in the discussion of tomographic results, however, these results are inconsistent with other surface-wave results from the Kaapvaal experiment as well as studies based on much more comprehensive global and regional datasets (Zhao et al. 1999; Ritsema & van Heijst 2000; Freybourger et al. 2001). Perhaps more significantly, velocity estimates based on modal abundances in mantle xenoliths and P–T conditions determined from geothermometry and geobarometry reveal not a single sample out of the 50 or so studied that produces a predicted S-wave velocity less than 4.6 km s^{-1} at ambient mantle conditions to depths of 200 km (James et al. 2001a).

The low P-velocity anomaly (c. 0.5%) in the mantle associated with the extended Bushveld and Okwa–Magondi terranes suggests considerable modification of cratonic mantle by Proterozoic tectonomagmatic events. That the mantle has been modified chemically is evidenced in the Re–Os data, where mantle nodules from the Bushveld Province have been reset to Proterozoic (about 2.05 Ga) ages (Carlson et al. 2000; Pearson et al. 2001). The isotopic resetting of a large volume of Archaean Kaapvaal mantle apparently required material addition (Carlson et al. 2000). The observed seismic velocity reduction is small, perhaps c. 0.5% in P and c. 0.8% in S. Although a thermal anomaly of c. 100°C could produce the velocity effect observed (Christensen 1982), there is no evidence for higher geotherms in the region of the Bushveld Province either from the observed heat-flow measurements or from thermobarometric determinations in mantle nodules (Danchin 1979; Jones 1988). For standard P–T conditions, Jordan showed that observed chemical variations in rocks of the cratonic mantle, including 'refertilized' samples, can account for up to 1% total velocity variation, with typical heterogeneity of the order of 0.5% (Jordan 1979). The seismic velocity perturbations that are observed between undisturbed Archaean cratonic mantle and cratonic mantle that has been modified by Proterozoic events are comfortably within this velocity range. The lower velocities observed in the adjacent Proterozoic belts are probably the result of more fertile compositions and higher geothermal gradients combined.

The regions of modified cratonic mantle such as that associated with the Bushveld Province are cautionary in terms of a central assumption of the geochronological studies discussed in a previous section (see Results from geochemistry and petrology). There it was postulated that a thick mantle root, had it been present during the whole of late Archaean time, would have shielded the crust from tectonomagmatic activity (Moser et al. 2001). Although plausible, this is not proven and may not be generally applicable, as we have seen from the Bushveld example. The abundance of diamonds brought to the surface in kimberlite pipes of post-Bushveld age within the Bushveld Province indicates that massive volumes of mafic magma can rise through cratonic mantle and intrude the crust without raising the geotherm of the cratonic root sufficiently to move it out of the diamond stability field. The inference from this observation, therefore, is that widespread magmatic events recorded in the lower crust do not necessarily preclude the presence of a tectospheric root. On the other hand, in the case of the Bushveld Province, the through-flow of magma seems apparently to have refertilized the mantle involved. If late Archaean events similar to the Bushveld event had occurred within the craton, their geochemical signature should remain visible in the seismic image, as it does beneath the Bushveld Province. This is not observed elsewhere in the craton.

Crustal structure

Among the most significant findings presented by Nguuri et al. (2001) is evidence for pervasive Proterozoic (c. 2 Ga) modification and thickening of Archaean crust across a broad east–west zone bounded on the east by the Bushveld and on the west by the Okwa–Magondi terranes. The area of thickened crust corresponds closely to the zone of reduced upper-mantle velocities found from body-wave tomography (James et al. 2001b). Moho Ps conversions for stations in this region of disturbed craton tend to be low in amplitude and in some cases ambiguous, suggesting that the Moho is a weak and/or transitional (e.g. >5 km) boundary. One possible interpretation of the poor Ps signals is that they reflect Proterozoic age magmatic underplating or reworking of Archaean crust (e.g. Griffin & O'Reilly 1987). Although magmatic addition to the crust is plausible beneath the Bushveld Complex, the cause of increased crustal thickness elsewhere in the region of Proterozoic overprinting is not so apparent. Both crustal thickness and the Moho signature observed in the region of modified Archaean crust are similar to those observed at stations in the Proterozoic Namaqua–Natal Belt, although a comparative analysis of structure between disturbed Archaean and Proterozoic terranes has yet to be carried out.

In a review of Precambrian crustal structure worldwide, Durrheim & Mooney (1994) found that Archaean crust is typically 27–40 km thick, whereas Proterozoic crust is about 40–55 km thick, with higher-velocity material (P waves >7 km s^{-1}) at its base. For southern Africa, the thinnest crust (35–40 km) is found beneath those regions of the Kaapvaal and Zimbabwe Cratons that have been undisturbed since Archaean time. A prominent exception is the crustal thickness beneath the Limpopo Belt. The Central Zone of the Limpopo Belt is not only characterized by thick crust (up to 50 km or more) and complex Moho, but there also is strong geological evidence that a total of 20–30 km of crustal uplift and exhumation has taken place since Archaean time (Treloar et al. 1992). If correct, this implies that the crustal section beneath the Limpopo Belt in Archaean time was comparable with the thickest crust observed today in the Himalayan and Andean convergent margins. It also may indicate that those Archaean terranes where the crust is thin and the Moho simple were not shaped to any significant degree by continent–continent collisional processes.

The unexpected relationship between thin cratonic crust and high elevations and large negative Bouguer gravity anomalies has been the subject of a number of recent studies (Webb et al. 1999; Webb 2002). Although thin crust and high elevation have been observed elsewhere in the world (Braile et al. 1989; Durrheim & Mooney 1991, 1994; Assumpção et al. 2002), the question remains of how the higher elevations are gravitationally compensated at depth. Compensation produced by the low-density upper mantle beneath the craton is only part of the story, particularly if the lower crust beneath cratons is less mafic on average (and hence less dense and lower velocity) than is post-Archaean lower crust (Griffin & O'Reilly 1987; Durrheim & Mooney 1994; Rudnick & Fountain 1995). Poisson's ratios (a measure of the elastic parameters of a solid) for the crust determined both from receiver function analysis and from the relationship between P-wave and S-wave travel times (Wadati diagrams) tend to be low for undisturbed regions of the Kaapvaal, around 0.25, whereas values for off-craton areas and disturbed craton are in the range 0.27–0.29 (Nguuri 2002). The average Poisson's ratio of 0.25 for the cratonic crust suggest that it may have an intermediate or even felsic bulk composition. Similarly, the large-amplitude Ps conversions reported by Nguuri et al. (2001) are consistent with a less mafic lower crust in the Kaapvaal Craton. Whereas high-amplitude

Ps conversions are not alone sufficient to establish an intermediate composition for the lower crust beneath the cratons, that feature coupled with low Poisson's ratio favours a less mafic lower crust relative to that of the adjacent Proterozoic belts. The Ps signatures for the Proterozoic terranes or overprinted Archaean terranes, on the other hand, are characteristic of a complex and/or gradational Moho discontinuity, a significantly smaller velocity contrast across the Moho, or both. It has been suggested that such regions may reflect a history of underplating through successive or episodic intrusion of basaltic melts into the lowermost crust (e.g. Griffin & O'Reilly 1987), so that the M-discontinuity itself becomes a complex interlayering of mafic and ultramafic rocks. Such a model is consistent with the results from southern Africa. A mafic or mafic-to-ultramafic lower crust would also mean a smaller density contrast across the Moho, perhaps as little as 300 kg m^{-3} (Webb et al. 1999), helping to account for the apparently anomalous negative correlation between elevation and crustal thickness in southern Africa.

The evidence reported by Nguuri et al. (2001) for a thick crust and a relatively diffuse Moho in the Bushveld Province supports the argument for a broadly continuous mafic body at depth beneath the Bushveld Complex (Cawthorn et al. 1998; Webb et al. 2000; Cawthorn & Webb 2001). The substantial load of material added to the crust during the Bushveld event was interpreted by Webb and coworkers to have produced a downward crustal flexure of up to 6 km with a resultant crustal thickness of 45–50 km beneath stations located in the central Bushveld Province (Webb et al. 2000).

Conclusions and implications for formation of cratons

The results described above place a number of important constraints on the structure and composition of cratons and provide insights into the processes by which they formed. Several conclusions may be enumerated with some confidence from the results cited above.

(1) Re–Os studies of mantle nodules demonstrate that the craton keel is Archaean in age and that there is no correlation with depth in the keel. The keel, therefore, did not thicken over time. Many eclogite inclusions in diamond are the same age as depleted peridotite or even slightly older. The age data on eclogite inclusions, coupled with evidence from stable isotope measurements that the eclogites formed from seawater-altered basalts, indicate that subduction processes were probably important in Archaean time. Subduction processes have also been invoked from both laboratory and field studies for the formation of komatiites by wet melting, suggesting that subduction was active at least as far back as early Archaean time, c. 3.6 Ga. The paucity of eclogite in the xenolith samples, however, suggests that the descending plate in Archaean time was, as it is today, almost always subducted into the deeper mantle.

(2) Tomographic images of the upper mantle exhibit a clear correspondence between seismic structures and geological boundaries. The boundaries of the higher-velocity root structures that define the cratons are typically well correlated with surface geological contacts. Adjacent Proterozoic mobile belts are in almost all cases characterized by slightly lower mantle velocities. Deep keel structures within the cratons are irregular, with evidence for maximum keel depths of at least 250–300 km in the southern and apparently most undisturbed part of the Kaapvaal Craton and in regions of the Zimbabwe Craton. The contrast in velocity perturbation within cratonic regions is <1% and thus could be accounted for entirely by variations in mantle composition.

(3) The northern part of the Kaapvaal Craton was profoundly affected by the massive magmatic events associated with the emplacement of the 2.054 Ga Bushveld mafic igneous complex. The mantle beneath the larger Bushveld Province exhibits lower seismic velocities, particularly in the P-wave results. The seismic data, which suggest widespread compositional modification (metasomatic enrichment) of the mantle beneath the Bushveld Complex, are consistent with evidence of younger (reset) Re–Os model ages of mantle nodules brought up in the Bushveld region.

(4) The thickness of the crust in undisturbed Archaean terranes is typically 35–40 km, some 5–15 km less than that found for adjacent Proterozoic terranes and in disturbed craton. Perhaps the most remarkable result of the receiver function studies to date in southern Africa is the pervasive evidence for crustal thickening and Moho disruption of Archaean terranes associated with tectonomagmatic disturbances of Proterozoic age. The most significant example of crustal modification is in the broad region of the Bushveld and Okwa–Magondi terranes. Although the evidence is not conclusive, the relatively large amplitude of Ps conversions at the Moho beneath undisturbed craton may suggest lower seismic velocities in the lowermost crust. This observation, coupled with low Poisson's ratios, may in turn indicate less mafic compositions for the lower crust in Archaean terranes relative to those of Proterozoic age. Systematic

variations in lower-crustal composition may be one reason why crustal thickness across southern Africa does not correlate with elevation (Webb *et al.* 1999, 2000).

(5) The Limpopo Mobile Belt appears to be underlain by a mantle root of cratonic seismic character, consistent with its Archaean age. The Northern and Southern Marginal Zones, which are thrust belts overlying cratonic crust, exhibit typical cratonic Moho signatures. These marginal zones of the Limpopo Belt contrast markedly in terms of both crustal thickness and Moho structure with the Central Zone of the Limpopo Belt, which is a major collisional terrane deeply exhumed through uplift and erosion (Treloar *et al.* 1992). Here, despite at least 10–20 km of erosional unroofing, the crust remains anomalously thick, suggestive of a Himalayan-scale crustal thickening at the time of the Kaapvaal–Zimbabwe Craton collision in the Archaean eon. The Ps signal from the Moho is complex, as would be expected for a highly foreshortened crust, and could be due to mantle–crust interlayering, dipping layers, or a host of other structural complications in the lower crust.

(6) Geochronological studies demonstrate that a simple view of cratonic root formation (depleted peridotite) as a discrete event is not a viable model for craton formation. Crust and mantle have both experienced a multi-stage tectono-magmatic history. Measurements on lower-crustal xenoliths and exposed rocks of the Vredefort crustal section reveal an intricate record of lithospheric evolution clearly associated with events seen in the surface geology. Of considerable relevance are the late Archaean magmatic events (*c.* 3.1–2.6 Ga), which overprinted earlier ages in much of the lower crust over vast regions of the western part of the Kaapvaal Craton. Widespread magmatism and lower-crustal heating events seem to preclude the presence of an underlying cold tectospheric root that would act to shield the crust from thermal events (Schmitz *et al.* 1998; Schmitz & Bowring 2000; Moser *et al.* 2001). A possible implication is that the crust and mantle may share an allochthonous relationship and that stabilization of a tectospheric keel (cratonization) and crust–root fusion occurred rapidly in late Archaean time (Moser *et al.* 2001). Although many other lines of evidence support the independent formation of crust and tectospheric mantle of the present-day craton, there are caveats, including the possibility that significant volumes of magma can rise through cratonic mantle and intrude the crust without altering the original geotherm sufficiently to move it out of the diamond stability field, as observed in the Bushveld Province.

(7) Although the results reported here represent only an initial examination of a vast seismic dataset, there is little evidence to date in body-wave tomography, surface-wave inversion, or receiver function analysis of deep discontinuity structure for a low-velocity zone beneath the cratonic root. High-resolution seismic studies of the upper mantle are continuing, however, and more sophisticated analysis may yet reveal a velocity reversal at the base of the tectosphere.

Remarks on craton formation

The constraints summarized above have numerous and important ramifications for the formation and stabilization of cratons. It is now clear that at least three things must occur to form tectospheric crust and mantle in Archaean time: (1) continental crust of intermediate composition must be generated in substantial volume, probably through arc-like processes; (2) significant volumes of upper mantle must be thoroughly depleted of both their basaltic fraction and volatile content; (3) fragments of continental crust and blocks of buoyant depleted mantle must be accreted together in late Archaean time, with the refractory mantle agglomerated into a coherent tectosphere at least 250 km thick.

Two key features of the cratonic mantle are critical to understanding its formation: its very low geotherm relative to that in other parts of the upper mantle, and the absence of any radiogenic age progression in mantle xenoliths with depth. The Archaean age of many diamonds suggests that the cratonic geotherm has always been low, as has been postulated previously (Burke & Kidd 1978; Jordan 1988). The fact that the keel in any given region is the same age within error over the depth range from which nodules were derived confirms that the chemical boundary layer (CBL) that forms the tectospheric keel did not thicken by progressive cooling. By inference, then, any process advanced to explain the formation of tectospheric mantle must include a mechanism by which the resultant keel possessed a low geotherm from the time of its formation.

Jordan (1988) investigated much this same question in considerable detail. The conclusions he drew more than a decade ago from available seismic, geochemical, petrological and geological data remain viable today. Jordan concluded that the most plausible mechanism of keel formation involved subduction-zone depletion of mantle rocks and subsequent advective thickening through large-scale compressional (collisional) processes where the refractory residues are swept together to form a thickened CBL beneath a consolidated Archaean crust. The model he proposed is a multi-stage one, in which basalt (or komatiite) (Boyd 1987) melt depletion and upper-mantle cooling precede cratonization. From the results of the Kaapvaal Project, it appears most plausible that the bulk of basaltic and komatiitic melt extraction and a considerable degree of mantle cooling must have occurred in supra-subduction zones, although mantle differentiation along mid-ocean ridges may also have been a factor. A corollary to the model of very wet subduction-zone melting for formation of komatiites and the depletion of the mantle is that mantle temperatures need have been only slightly higher in Archaean time than they are today. The relatively low temperature required to produce komatiite in wet subduction-zone melting (Parman *et al.* 2001) means both that komatiites are not necessarily evidence for a hot mantle in Archaean time (as suggested, for example, by Nisbet & Fowler (1983)) nor that they had an origin in Archaean hotspots (e.g. Abbott *et al.* 1994). (Our conclusion here is that mantle temperatures were not necessarily much higher in Archaean time than at the present day, perhaps $c. 150°C$. There are suggestions, however, based on thermal modelling that even in a hot Archaean mantle thermal blanketing could maintain a cool tectospheric root within the central zone of a cratonic volume substantially more massive than that remaining today (e.g. Ballard & Pollack 1988). The principal question remains in that case of how large cool cratonic masses stabilized in Archaean time in the presence of very hot mantle.) In light of our current understanding of cratonic structure and composition, Jordan's advective thickening model provides a satisfactory explanation for the geothermal gradient and the observed seismic mantle velocities.

Abbott (1991) faulted Jordan's advective thickening model, citing lack of geological evidence for an abundance of large-scale collisional terranes in the Archaean geological record. Thus, whereas the high-grade metamorphic gneiss terranes are consistent with crustal thickening and subsequent exhumation during major tectonic events, the more abundant low-grade greenstone belts provide no comparable evidence of either the crustal thickening or large-scale foreshortening characteristic of continental collisions. Although this remains a serious objection to the hypothesis of advective thickening via plate-scale compressional events, it may be that successions of smaller accretionary events, such as those found in the westward-stepping subduction zones of the northern Andes or in the allochthonous collage along the margin of western North America, played a larger role in the process of root thickening than is generally recognized.

Two principal alternatives to advective thickening for formation of the CBL have garnered varying degrees of support in the past (Abbott 1991). The first, the so-called 'plum pudding' model (Davies 1979), involves the gradual accretion of buoyant refractory material from the deep mantle to a thickening cratonic root. The model depends on conductive cooling, however, and can reasonably be excluded, mostly on the grounds that the keel exhibits no evidence of temporal growth vertically. The second model, which retains a strong following, is based on the notion of progressive imbrication or 'trench jumping' of subduction zones (Helmstaedt & Schulze 1989; Abbott 1991; de Wit *et al.* 1992) and the successive stacking of layers of oceanic lithosphere beneath the cratonic crust. Several variants of this mechanism have been put forward, two of which we discuss here. As envisioned by Helmstaedt & Schulze, continental lithosphere proto-plates progressively override continental margin subduction zones, shearing off the descending oceanic plate and leaving a trail of stacked and shallowly dipping remnants beneath the continent to form a root structure composed of imbricated slabs. Although little is certain about how subduction processes operated in Archaean time, the Helmstaedt and Schulze slab imbrication model finds no support in modern subduction analogues. Continents simply do not override subduction zones in the absence of external factors, such as collision of a continental mass with the trench. The trenchward motion of a continental plate is readily compensated by trench rollback, a common, dynamically predicted and unremarkable phenomenon (e.g. Forsyth & Uyeda 1975; Meijer 1995). Abbott (1991) proposed another model involving imbricated slabs, but suggested instead that the Archaean mantle was sufficiently hot to melt the basaltic crust of the descending plate before it reached the eclogite stability field. Under those conditions, the buoyant refractory (harzburgite) mantle of the oceanic lithosphere would survive to underplate the continent rather than sink under the gravitational effect of the basalt-to-eclogite conversion. Thus, the model postulates that successive layers of depleted, relatively cool, oceanic lithospheric slabs are stacked atop one another and at low angles to form a thickened root beneath the cratonic crust. A host of difficulties can be cited for this model, among them the virtual improbability that the cold oceanic crust would melt before its water had infiltrated and fluxed the hot overlying mantle (as in modern subduction zones). Full melting of the basaltic crust creates an even more fundamental problem: it eliminates gravitational sinking induced by the basalt-to-eclogite transition as a driving force in subduction (Forsyth & Uyeda 1975). Without gravitational sinking the principal force that drives subduction goes to zero and subduction ceases. The issue is further muddied by the fact that the MOR basaltic crust may have been thicker in Archaean time (Bickle 1986), making wholesale crustal melting an even less viable hypothesis. Finally, it should be noted that for southern Africa it is estimated that <1% of the volume of the cratonic mantle is made up of eclogite (Schulze 1989).

We conclude by reiterating and re-emphasizing the fact that the above discussion leaves many fundamental questions unanswered. Among the outstanding issues touched on above but that remain yet to be fully resolved, we include the following.

(1) Why and how was the Archaean different, and what does it mean in terms of lithosphere formation, plate processes and differentiation in the early Earth?
(2) Is cratonization (i.e. mantle stabilization) separate from crust formation? How and over what time period did craton stabilization occur?
(3) How do images of mantle heterogeneity and crustal composition and thickness relate to the assembly history of the cratons?
(4) What is the influence of post-Archaean events on the composition, seismic structure, and geothermal gradient of the craton?

The Kaapvaal Project has been an adventure. It has involved the efforts of more than 100 scientists, students and technicians whose names and affiliations along with a project summary can be found on the website http://www.ciw.edu/kaapvaal. Contributions too numerous to list have been made in the way of blood, sweat and thought to the Project. We owe a particular debt of thanks to those whose work constitutes the scientific underpinning for this paper: J. Gore, T. Nguuri, S. Webb, S. Parman, T. Grove, M. Schmitz, S. Bowring, R. Carlson, S. Shirey, S. Richardson, S. van der Lee, J. VanDecar, T. Jordan, M. Freybourger, R. Saltzer and A. Snoke. All have provided ideas and stimulation. The southern Africa seismic experiment owes much of its success first and foremost to the efforts of R. Green of Green's Geophysics, who laid out so much of the logistical groundwork for the experiment. As always, the project could not have proceeded at all without the extraordinary talents of seismology supertech R. Kuehnel. Others who made crucial contributions to the field operations besides those already mentioned above include J. Robey J. Harvey and L. Kennedy, all of de Beers Geology Division; F. Reichhardt and M. Jutz of Rio Tinto Zimbabwe; M. de Wit, T. Zengeni, C. Wright, T. Kwadiba, P. Burkholder and M. Nkwaane. Finally, special thanks go to J. Fowler, C. Ebeling, S. Hellman, P. Friberg and the rest of the very able crew at the PASSCAL instrument center. We also thank R. Benson and the IRIS DMC for invaluable help in archiving the data. The paper benefited greatly from comments by A. Nyblade and J. Boyd, as well as an anonymous reviewer. This work was supported by the National Science Foundation Continental Dynamics Program with grants to Carnegie Institution and MIT; the National Research Foundation of South Africa; and by universities, geological surveys and exploration companies in South Africa, Zimbabwe and Botswana. Map figures were produced with GMT (Wessel & Smith 1991) and plotting codes are from J. VanDecar.

References

ABBOTT, D. H. 1991. The case for accretion of the tectosphere by buoyant subduction. *Geophysical Research Letters*, **18**(4), 585–588.

ABBOTT, D. H., BURGESS, L. & LONGHI, J. 1994. An empirical thermal history of the Earth's upper mantle. *Journal of Geophysical Research*, **99**, 13835–13850.

AMMON, C. J. 1991. The isolation of receiver effects from teleseismic P waveforms. *Bulletin of the Seismological Society of America*, **81**, 2504–2510.

ANDERSON, D. L. 1979. The deep structure of continents. *Journal of Geophysical Research*, **84**(B13), 7555–7560.

ASSUMPÇÃO, M., JAMES, D. E. & SNOKE, J. A. 2002. Crustal thickness in SE Brazilian shield by receiver function analysis: Implications for isostatic compensation. *Journal of Geophysical Research*, **107**(1).

BALLARD, S., III & POLLACK, H. N. 1988. Diversion of heat by Archean cratons; a model for Southern Africa. *Earth and Planetary Science Letters*, **85**(1–3), 253–264.

BICKLE, M. J. 1986. Implications of melting for stabilization of the lithosphere and heat loss in the Archean. *Earth and Planetary Science Letters*, **80**, 314–324.

BOYD, F. B. & MCCALLISTER, R. H. 1976. Densities of fertile and sterile garnet peridotites. *Geophysical Research Letters*, **3**(9), 509–512.

BOYD, F. R. 1987. High- and low-temperature garnet peridotite xenoliths and their possible relation to the lithosphere–asthenosphere boundary beneath southern Africa. *In*: NIXON, P. H. (ed.) *Mantle Xenoliths*, Wiley–Interscience, Chichester, 403–412.

BRAILE, L. W., HINZE, W. J., VON FRESE, R. R. B. & KELLER, G. R. 1989. Seismic properties of the crust and uppermost mantle of the conterminous United States and adjacent Canada. *In*: PAKISER, L. C. & MOONEY, W. D. (eds) *Geophysical Framework of the Continental United States*. Geological Society of America, Boulder, CO, 655–680.

BURKE, K. & KIDD, W. S. F. 1978. Were Archean continental geothermal gradients much steeper than today? *Nature*, **272**, 240–241.

CARLSON, R. W., BOYD, F. R., SHIREY, S. B. & 14 OTHERS 2000. Continental growth, preservation and modification in southern Africa. *GSA Today*, **10**, 1–7.

CARLSON, R. W., GROVE, T. L., DE WIT, M. J. & GURNEY, J. J. 1996. Anatomy of an Archean craton: a program for interdisciplinary studies of the Kaapvaal craton, southern Africa. *EOS Transactions, American Geophysical Union*, **77**, 273–277.

CARLSON, R. W., PEARSON, D. G., BOYD, F. R., SHIREY, S. B., IRVINE, G., MENZIES, A. H. & GURNEY, J. J. 1999. Re–Os systematics of lithospheric peridotites: implications for lithosphere formation and preservation. *In*: GURNEY, J. J., GURNEY, J. L., PASCOE, M. D. & RICHARDSON, S. H. (eds) *7th International Kimberlite Conference*. Red Roof Design, Cape Town, 99–108.

CARNEY, J. N., ALDISS, D. T. & LOCK, N. P. 1994. *The Geology of Botswana*. Geological Survey Department, Gaborone.

CAWTHORN, R. G. & WEBB, S. J. 2001. Connectivity between the western and eastern limbs of the Bushveld Complex. *Tectonophysics*, **330**, 195–209.

CAWTHORN, R. G., COOPER, G. R. J. & WEBB, S. J. 1998. Connectivity between the western and eastern limbs of the Bushveld Complex. *South African Journal of Geology*, **101**(4), 291–298.

CHRISTENSEN, N. I. 1982. Seismic velocities. *In*: CARMICHAEL, R. S. (ed.) *Handbook of Physical Properties of Rocks*. CRC Press, Boca Raton, FL, **2**, 228.

CLITHEROE, G., GUDMUNDSSON, O. & KENNETT, B. L. N. 2000. The crustal thickness of Australia. *Journal of Geophysical Research*, **105**, 13 697–13 713.

DANCHIN, R. V. 1979. Mineral and bulk chemistry of garnet lherzolite and garnet harzburgite xenoliths from the Premier Mine, South Africa. *In*: BOYD, F. R. & MEYER, H. O. A. (eds) *The Mantle Sample: Inclusions in Kimberlites and Other Volcanics, Proceedings of the Second International Kimberlite Conference*. American Geophysical Union, Washington, DC, 104–126.

DAVIES, G. F. 1979. Thickness and thermal history of continental crust and root zones. *Earth and Planetary Science Letters*, **44**, 231–238.

DE BEER, J. H. & STETTLER, E. H. 1992. The deep structure of the Limpopo Belt from geophysical studies. *In*: VAN REENEN, D. D., ROERING, C., ASHWAL, L. D. & DE WIT, M. J. (eds) *The Archaean Limpopo Granulite Belt; Tectonics and Deep Crustal Processes*. Precambrian Research, Special Issue, Elsevier, Amsterdam, 173–186.

DE WIT, M. J. & HART, R. A. 1993. Earth's earliest continental lithosphere, hydrothermal flux and crustal recycling. *Lithos*, **30**, 309–335.

DE WIT, M. J., ROERING, C., HART, R. J. & 6 OTHERS 1992. Formation of an Archaean continent. *Nature*, **357**(6379), 553–562.

DUEKER, K. G & SHEEHAN, A. F. 1997. Mantle discontinuity structure from midpoint stacks of converted P to S waves across the Yellowstone hotspot track. *Journal of Geophysical Research*, **102**(4), 8313–8327.

DUEKER, K. G. & SHEEHAN, A. F. 1998. Mantle discontinuity structure beneath the Colorado Rocky Mountains and High Plains. *Journal of Geophysical Research*, **103**, 7153–7169.

DURRHEIM, R. J. & GREEN, R. W. E., 1992. A seismic refraction investigation of the Archaean Kaapvaal Craton, South Africa, using mine tremors as the energy source. *Geophysical Journal International*, **108**, 812–832.

DURRHEIM, R. J. & MOONEY, W. D. 1991. Archean and Proterozoic crustal evolution; evidence from crustal seismology. *Geology (Boulder)*, **19**(6), 606–609.

DURRHEIM, R. J. & MOONEY, W. D. 1994. Evolution of the Precambrian lithosphere; seismological and geochemical constraints. *Journal of Geophysical Research*, **99**, 15 359–15 374.

FORSYTH, D. W. & UYEDA, S. 1975. On the relative importance of the driving forces of plate motion. *Geophysical Journal of the Royal Astronomical Society*, **43**, 163–200.

FREYBOURGER, M., GAHERTY, J. B., JORDAN, T. H. & KAAPVAAL SEISMIC GROUP 2001. Structure of the Kaapvaal craton from surface waves. *Geophysical Research Letters*, **28**(13), 2489–2492.

GORE, J. 2002. *Seismological structure of the crust and upper mantle of the Zimbabwe craton and Limpopo belt, southern Africa*. PhD thesis, University of Zimbabwe, Harare.

GRIFFIN, W. L. & O'REILLY, S. Y. 1987. The composition of the lower crust and the nature of the continental Moho–xenolith evidence. *In*: NIXON, P. H. (ed.) *Mantle Xenoliths*. Wiley, Chichester, 413–432,

GROVE, T. L., PARMAN, S. W. & DANN, J. C. 2000. Conditions of magma generation for Archean koma-tiites from the Barberton Mountainland, South Africa. *In*: FEI, Y., BERTKA, C. M. & MYSEN, B. O. (eds) *Mantle Petrology: Field Observations and High Pressure Experimentation*. Geochemical Society, Houston, TX, 155–170.

GURROLA, H., MINSTER, J. B. & OWENS, T. J. 1994. The use of velocity spectrum for stacking receiver functions and imaging upper mantle discontinuities. *Geophysical Journal International*, **117**, 427–440.

GWAVAVA, O., SWAIN, C. J., PODMORE, F. & FAIRHEAD, J. D. 1992. Evidence of crustal thinning beneath the Limpopo Belt and Lebombo monocline of southern Africa based on regional gravity studies and implications for the reconstruction of Gondwana. *In*: VON FRESE, R. B. & TAYLOR, P. (eds) *Lithospheric Analysis of Magnetic and Related Geophysical Anomalies*. Elsevier, Amsterdam, 1–20.

HART, R., DE WIT, M., MARCO, A. & TREDOUX, M. 1990. Formation of the Archaean Kaapvaal Craton; II: The late Archaean (2.5–3.0 Ga); evidence from a 'crust on edge' section at Vredefort. *In*: ROCCI, G. & DESCHAMPS, J. (eds) *Études Récentes sur la Géologie de l'Afrique; 15e Colloque de Géologie africaine; Résumés détaillés*. Centre International pour la Formation et les Échanges Géologiques (CIFEG), Paris, 7–9.

HATTON, C. J. & VON GRUENEWALDT, G. 1990. Early Precambrian layered intrusions. *In*: HALL, R. P. & HUGHES, D. J. (eds) *Early Precambrian Basic Magmatism*. Chapman and Hall, New York, 56–82.

HELMSTAEDT, H. & SCHULZE, D. J. 1989. Southern African kimberlites and their mantle sample: implications for Archean tectonics and lithosphere evolution. *In: Kimberlites and Related Rocks. Their Composition, Occurrence, Origin and Emplacement. Proceedings of the 4th International Kimberlite Conference, Perth, Australia 1986.* Geological Society of Australia, Victoria, 358–360.

IRVINE, G. J., PEARSON, D. G. & CARLSON, R. W. 2001. Evolution of the Kaapvaal lithospheric mantle: a Re–Os isotope study of peridotite xenoliths from Lesotho kimberlites, *Geophysical Research Letters,* **28**(13), 2505–2508.

JAMES, D. E., CARLSON, R. W., BOYD, F. B. & JANNEY, P. E. 2001a. Petrologic constraints on seismic velocity variations in the upper mantle beneath southern Africa. *EOS Transactions, American Geophysical Union,* **82**(20, Abstract Suppl.), S247.

JAMES, D. E., FOUCH, M. J., VANDECAR, J. C., VAN DER LEE, S. & KAAPVAAL SEISMIC GROUP 2001b. Tectospheric structure beneath southern Africa. *Geophysical Research Letters,* **28**(13), 2485–2488.

JAUPART, C. & MARESCHAL, J. C. 1999. The thermal structure and thickness of continental roots. *In:* VAN DER HILST, R. D. & MCDONOUGH, W. F. (eds) *Composition, Deep Structure and Evolution of Continents.* Elsevier, Amsterdam, 93–114.

JONES, M. Q. W. 1988. Heat flow in the Witwatersrand Basin and environs and its significance for the South African shield geotherm and lithosphere thickness. *Journal of Geophysical Research,* **93**, 3243–3260.

JORDAN, T. H. 1975. The continental tectosphere. *Reviews of Geophysics,* **13**, 1–12.

JORDAN, T. H. 1978. Composition and structure of the continental tectosphere. *Nature,* **274**, 544–548.

JORDAN, T. H. 1979. Mineralogies, densities and seismic velocities of garnet lherzolites and their geophysical implications. *In:* BOYD, F. R. & MEYER, H. O. A. *The Mantle Sample: Inclusions in Kimberlites and Other Volcanics, Proceedings of the Second International Kimberlite Conference.* American Geophysical Union, Washington, DC, 1–14.

JORDAN, T. H. 1981. Continents as a chemical boundary layer. *In: The Origin and Evolution of the Earth's Continental Crust.* Transactions of the Royal Society of London, London, **301**, 359–373.

JORDAN, T. H. 1988. Structure and formation of the continental tectosphere. *Journal of Petrology, Special Lithosphere Issue,* 11–37.

LERNER-LAM, A. L. & JORDAN, T. H. 1987. How thick are the continents? *Journal of Geophysical Research,* **92**(13), 14007–14026.

MCDONOUGH, W. F. & RUDNICK, R. L. 1998. Mineralogy and composition of the upper mantle. *In:* HEMLEY, R. (ed.) *Ultrahigh-Pressure Mineralogy; Physics and Chemistry of the Earth's Deep Interior.* Mineralogical Society of America, Washington, DC, 139–164.

MEIJER, P. T. 1995. *Dynamics of active continental margins: the Andes and the Aegean region.* PhD thesis, University of Utrecht.

MOSER, D. E., FLOWERS, R. M. & HART, R. J. 2001. Birth of the Kaapvaal tectosphere 3.08 billion years ago. *Science,* **291**, 465–468.

NGUURI, T. 2002. *Crustal structure of the Kaapvaal craton and surrounding mobile belts: analysis of teleseismic P-waveforms and surface wave inversions.* PhD thesis, University of the Witwatersrand, Johannesburg.

NGUURI, T., GORE, J., JAMES, D. E. & 6 OTHERS 2001. Crustal structure beneath southern Africa and its implications for the formation and evolution of the Kaapvaal and Zimbabwe cratons. *Geophysical Research Letters,* **28**(13), 2501–2504.

NISBET, E. G. & FOWLER, C. M. R. 1983. Model for Archean plate tectonics. *Geology (Boulder),* **11**, 376–379.

NYBLADE, A. A. 1999. Heat flow and the structure of Precambrian lithosphere. *Lithos,* **48**, 81–91.

OKAL, E. & ANDERSON, D. L. 1975. A study of lateral inhomogeneities in the upper mantle by multiple ScS travel-time residuals. *Geophysical Research Letters,* **2**(8), 313–316.

PARMAN, S. W., DANN, J. C., GROVE, T. L. & DE WIT, M. J. 1997. Emplacement conditions of komatiite magmas from the 3.49 Ga Komati formation, Barberton Greenstone Belt, South Africa. *Earth and Planetary Science Letters,* **150**, 303–323.

PARMAN, S. W., GROVE, T. L. & DANN, J. C. 2001. The production of Barberton komatiites in an Archean subduction zone. *Geophysical Research Letters,* **28**(13), 2513–2516.

PEARSON, D. G., CARLSON, R. W., SHIREY, S. B., BOYD, F. R. & NIXON, P. H. 1995. Stabilisation of Archaean lithospheric mantle; a Re–Os isotope study of peridotite xenoliths from the Kaapvaal Craton. *Earth and Planetary Science Letters,* **134**(3–4), 341–357.

PEARSON, D. G., IRVINE, G. J., CARLSON, R. W., KOPYLOVA, M. G. & IONOV, D. A. 2002. The development of lithospheric keels beneath the earliest continents: time constraints using PGE and Re–Os isotope systematics. *In:* FOWLER, C. M. R., EBINGER, C. J. & HAWKESWORTH, C. J. (ed.) *The Early Earth: Physical, Chemical and Biological Development.* Geological Society, London, Special Publications, **199**, 65–90.

PRIESTLEY, K. 1999. Velocity structure of the continental upper mantle: evidence from southern Africa. *Lithos,* **48**, 45–56.

QIU, X., PRIESTLEY, K. & MCKENZIE, D. 1996. Average lithospheric structure of southern Africa. *Geophysical Journal International,* **127**, 563–587.

RICHARDSON, S. H., GURNEY, J. J., ERLANK, A. J. & HARRIS, J. W. 1984. Origin of diamonds in old enriched mantle. *Nature,* **310**, 198–202.

RICHARDSON, S. H., SHIREY, S. B., HARRIS, J. W. & CARLSON, R. W. 2001. Archaean subduction recorded by Re–Os isotopes in sulphide inclusions in diamonds. *Earth and Planetary Science Letters,* **191**, 239–248.

RITSEMA, J. & VAN HEIJST, H. 2000. New seismic model of the upper mantle beneath Africa. *Geology (Boulder),* **28**(1), 63–66.

RITSEMA, J., NYBLADE, A. A., OWENS, T. J., LANGSTON, C. A. & VANDECAR, J. C. 1998. Upper mantle seismic velocity structure beneath Tanzania, East Africa: implications for the stability of cratonic lithosphere. *Journal of Geophysical Research*, **103**, 21 201–21 213.

RUDNICK, R. L. & FOUNTAIN, D. M. 1995. Nature and composition of the continental crust: a lower crustal perspective. *Reviews of Geophysics*, **33**(3), 267–309.

RUDNICK, R. L. & NYBLADE, A. A. 1999. The thickness and heat production of Archean lithosphere: constraints from xenolith thermobarometry and surface heat flow. *In*: FEI, Y., BERTKA, C. & MYSEN, B. O. (eds) *Mantle Petrology: Field Observations and High Pressure Experimentation: a Tribute to Francis R. (Joe) Boyd*. Geochemical Society, Houston, TX, 3–12.

RUDNICK, R. L., MCDONOUGH, W. F. & O'CONNELL, R. J. 1998. Thermal structure, thickness and composition of continental lithosphere. *In*: ALBARÈDE, F., BLICHERT, T., STAUDIGEL, H. & WHITE, W. (eds) *Geochemical Earth Reference Model (GERM)*. Elsevier, Amsterdam, 395–411.

SCHMITZ, M. D. & BOWRING, S. A. 2000. The significance of U–Pb zircon ages from lower crustal xenoliths of the southwestern margin, Kaapvaal Craton, southern Africa. *Chemical Geology*, **172**, 59–76.

SCHMITZ, M. D., BOWRING, S. A. & ROBEY, J. V. A. 1998. Constraining the thermal history of an Archean craton: U–Pb thermochronology of lower crustal xenoliths from the Kaapvaal craton, southhern Africa. *In: 7th International Kimberlite Conference, Extended Abstracts*. University of Cape Town, Cape Town, 766–768.

SCHULZE, D. J. 1989. Constraints on the abundance of eclogite in the upper mantle. *Journal of Geophysical Research*, **94**, 4205–4212.

SCLATER, J. G., JAUPART, C. & GALSON, D. 1980. The heat flow through oceanic and continental crust and the heat loss of the Earth. *Reviews of Geophysical Space Physics*, **18**(1), 269–311.

SHAPIRO, S. S., HAGER, B. H. & JORDAN, T. H. 1999. Stability and dynamics of the continental tectosphere. *In*: VAN DER HILST, R. D. & MCDONOUGH, W. F. (eds) *Composition, Deep Structure and Evolution of Continents*. Elsevier, Amsterdam, 115–133.

SHIREY, S. B. & WALKER, R. J. 1998. The Re–Os isotopic system in cosmochemistry and igneous geochemistry. *Annual Reviews of Earth and Planetary Sciences*, **26**, 425–500.

SHIREY, S. B., CARLSON, R. W., RICHARDSON, S. H. & 5 OTHERS 2001. Emplacement of eclogite components into the lithospheric mantle during craton formation. *Geophysical Research Letters*, **28**(13), 2509–2512.

SILVER, P. G., GAO, S. S., LIU, K. H. & KAAPVAAL SEISMIC GROUP 2001. Mantle deformation beneath southern Africa. *Geophysical Research Letters*, **28**(13), 2493–2496.

STUART, G. W., ZENGENI, T. & CLARK, R. A. 1986. Crustal structure of the Limpopo mobile belt, Zimbabwe. *In: Abstracts, the Tenth UK Geophysical Assembly*. Royal Astronomical Society, London, 261.

TREDOUX, M., HART, R. J., CARLSON, R. W. & SHIREY, S. B. 1999. Ultramafic rocks at the center of the Vredefort structure: further evidence for the crust on edge model. *Geology (Boulder)*, **27**, 923–926.

TRELOAR, P. J., COWARD, M. P. & HARRIS. N. B. W. 1992. Himalayan–Tibetan analogies for the evolution of the Zimbabwean Craton and Limpopo belt. *Precambrian Research*, **55**, 571–587.

VANDECAR, J. C. 1991. *Upper-mantle structure of the Cascadia subduction zone from non-linear teleseismic travel-time inversion*. PhD thesis, University of Washington, Seattle.

VANDECAR, J. C. & CROSSON, R. S. 1990. Determination of teleseismic relative phase arrival times using multi-channel cross-correlation and least squares. *Bulletin of the Seismological Society of America*, **80**(1), 150–159.

VANDECAR, J. C., JAMES, D. E. & ASSUMPÇAO, M. 1995. Seismic evidence for a fossil mantle plume beneath South America and implications for plate driving forces. *Nature*, **378**(6552), 25–31.

VAN DER LEE, S. & NOLET, G. 1997. Upper mantle S-velocity structure of North America. *Journal of Geophysical Research*, **102**, 22 815–22 838.

VAN REENEN, D. D., BARTON, J. M., ROERING, C. A., SMITH, C. A. & VAN SCHALKWYK, J. F. 1987. Deep crustal response to continental collision: the Limpopo belt of southern Africa. *Geology (Boulder)*, **15**, 11–14.

VAN REENEN, D. D., ROERING, C., ASHWAL, L. D. & DE WIT, M. J. 1992. Regional geological setting of the Limpopo belt. *Precambrian Research*, **55**, 1–5.

VON GRUENEWALDT, G., SHARPE, M. R. & HATTON, C. J. 1985. The Bushveld Complex; introduction and review. *Economic Geology*, **80**, 803–812.

WEBB, S. J. 2002. *The use of potential field and seismological data to analyse the structure of the lithosphere beneath southern Africa*. PhD thesis, University of the Witwatersrand, Johannesburg.

WEBB, S. J., JAMES, D. E. & NGUURI, T. 1999. The integration of gravity data with seismological results from the Kaapvaal Craton Seismic Experiment: an assessment of isostatic balance. *In: 6th Biennial Conference of the South African Geological Association*. South African Council on Geosciences, Cape Town, 1–6.

WEBB, S. J., NGUURI, T., JAMES, D. E. & JORDAN, T. H. 2000. Crustal thickness supports Bushveld continuity. *EOS Transactions, American Geophysical Union, Supplement*, **81**(19), S175.

WESSEL, P. & SMITH, W. H. F. 1991. Free software helps map and display data. *EOS Transactions, American Geophysical Union*, **72**, 445–446.

WINDLEY, B. F. 1995. *The Evolving Continents*. Wiley, Chichester.

ZHAO, M., LANGSTON, C. A., NYBLADE, A. A. & OWENS, T. J. 1999. Upper mantle velocity structure beneath southern Africa from modeling regional seismic data. *Journal of Geophysical Research*, **104**, 4783–4794.

Seismic heterogeneity and anisotropy in the Western Superior Province, Canada: insights into the evolution of an Archaean craton

J.-M. KENDALL[1], S. SOL[2], C. J. THOMSON[2], D. J. WHITE[3], I. ASUDEH[3], C. S. SNELL[1] & F. H. SUTHERLAND[1]

[1] *School of Earth Sciences, University of Leeds, Leeds LS2 9JT, UK (e-mail: kendall@earth.leeds.ac.uk)*
[2] *Department of Geological Sciences, Queen's University, Kingston, Ontario K7L 3N6, Canada*
[3] *Geological Survey of Canada, Booth Street, Ottawa, Ontario K1A 0E9, Canada*

Abstract: The Superior Province, which forms the nucleus of North America, is the largest preserved Archaean crustal block in the world and may have originated as the result of a widespread crustal accretion event (*c.* 2.7 Ga) manifested in Archaean cratons worldwide. An understanding of the accretionary evolution of this craton is the objective of the continuing Lithoprobe transect in the Western Superior Province, Canada. The geophysical components of the transect include seismic reflection and refraction, magnetotellurics and teleseismic experiments. The Teleseismic Western Superior Transect (TWiST) was designed to explore the structural and physical properties of the subcrustal lithosphere and their implications for proposed accretionary models. A north–south-trending array of 17 broadband three-component seismometers was deployed between May and November 1997. Surface-wave analyses, SKS-splitting studies and travel-time tomography show variations in the velocity structure and anisotropy between the southern end of the transect, a region affected by Keweenawan rifting, and the northern part, which lies in the Proterozoic Trans-Hudson shear zone. Surface waves reveal evidence for a thin high-velocity layer, 5–20 km thick, beneath a 37–43 km thick crust and above *c.* 250 km of high-velocity continental root. This thin layer is also visible in wide-angle refraction data from the southern end of the line and may be evidence of underplating during terrane accretion. Discrepancies in the Love and Rayleigh waves and surface-wave particle motions show evidence for an anisotropic mantle. SKS analysis shows large amounts (up to 2 s) of shear-wave splitting with a roughly east–west trend in the fast-shear-wave polarization direction for most stations. This conforms with crustal deformation trends. Stations in the younger Trans-Hudson orogen show much less splitting. Detailed analysis at a permanent station in the Western Superior shows evidence for two layers of anisotropy. A thinner upper layer is aligned with the surface geology, indicating crust–mantle coupling during craton formation, whereas a thicker lower layer is aligned with the direction of absolute plate motion. Tomographic results show a featureless mantle beneath the Sachigo proto-craton and more heterogeneity towards the south end of the line. A steeply dipping slab-like feature in the lithosphere correlates with wide-angle refraction and deep-reflection seismic profiles. A similar high-velocity feature continues well into the transition zone, but its origin remains to be understood. Towards the southern end of the line there is a deep-seated low-velocity anomaly, which may be associated with Keweenawan plume activity. As a whole, the seismic results show many features that support ideas of subduction-related accretion of a thick stable Archaean tectosphere. There are, however, interesting details that are to date unique to the Western Superior Province. These include thicker than normal Archaean crust, a slab-like velocity anomaly in the mantle transition zone, and large SKS splitting in the Archaean Superior Province but little splitting in the surrounding Trans-Hudson Proterozoic shear zone.

An enduring question in the study of the Earth concerns the origin and stability of Archaean cratons (Jordan 1978; Bickle 1986; Helmstaedt & Schulze 1989; Hoffman 1990; Abbott 1991). They drift around the globe with remarkable resistance to tectonic deformation and thermal erosion. A key to understanding the nature of these cratons lies in the continental mantle and its degree of coupling with the crust (Silver 1996). Central to this is Jordan's concept of a tectosphere, a thick rigid continental root of depleted upper mantle (Jordan 1978, 1981). Our primary means of 'viewing' such mantle is by using seismic methods (Bostock 1999). Such studies combined with xenolith studies (e.g. Helmstaedt & Schulze 1989), geochemical or isotope analysis (e.g. Bell *et al.* 1982) and other geophysical techniques (e.g. electromagnetic techniques, Jones & Ferguson 2001) are providing new insights into continental evolution. Here we investigate the seismic structure of the Archaean Western Superior Province in Canada.

Continental interiors are thought to be constructed from welded pieces of island arcs, metamorphosed sediments, oceanic plateaux and continental fragments (e.g. Hoffman 1989). Modern-day analogues for this include the convergent zones of the Western Pacific (Hamilton 1979), and the South American–Caribbean plate boundary, where the Leeward Antilles arc is colliding with South America (Avé Lallemant 1997). To what extent these processes were occurring in the Archaean, when the mantle was much warmer, is a subject of debate.

The Western Superior Province is composed of a series of roughly east–west-trending granite–greenstone and metasedimentary belts (Fig. 1), which represent the largest exposure of Archaean craton in the world. At the heart of the craton is the Sachigo block, composed primarily of 3.0 Ga tonalitic gneisses (Thurston & Chivers 1990). The structural and geochemical patterns of the belts (Langford & Morin 1976) and the younging towards the south (2.7–3.0 Ga; Corfu & Davis 1992) have led to the suggestion of subduction-related accretion against the Sachigo proto-craton (Card 1990; Calvert & Ludden 1999). This contrasts with old ideas of Archaean crustal formation dominated by vertical tectonics (see discussion by Card 1990). Although

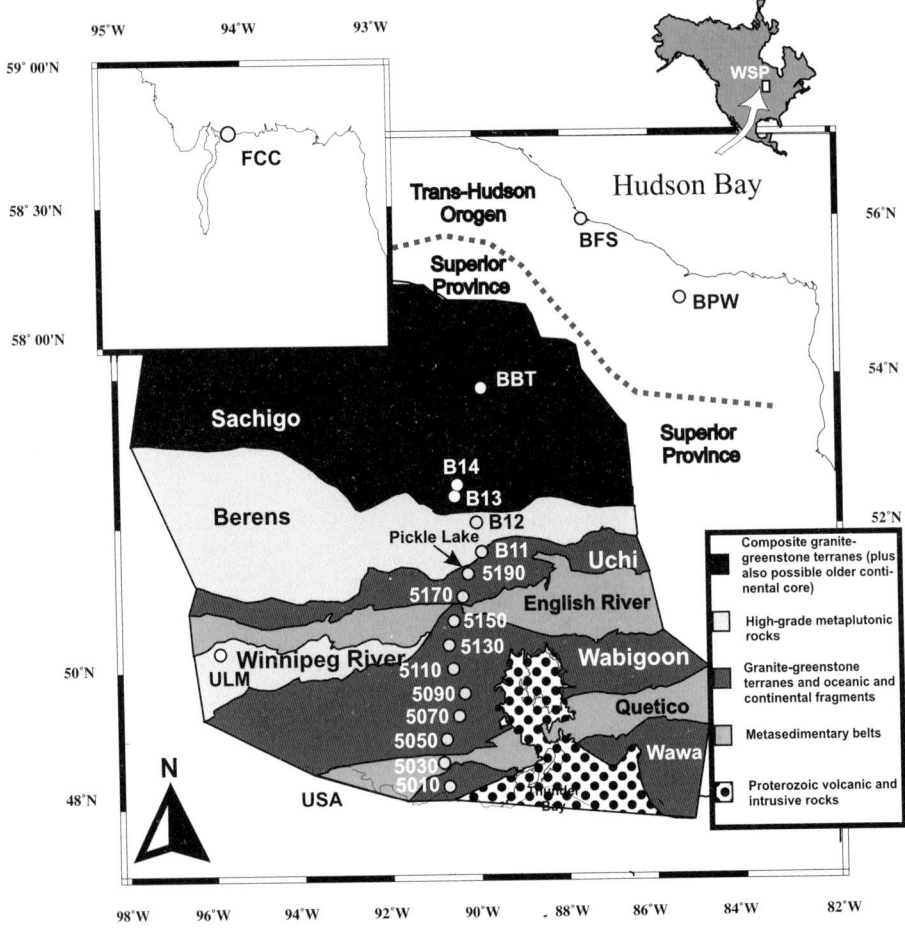

Fig. 1. Map of the Western Superior Province (WSP). ○, locations of the broadband seismic stations of the TWiST experiment, and of the two permanent CNSN stations, ULM and FCC. The dashed line marks the border between the Archaean Superior Province and the Proterozoic Trans-Hudson Orogen.

subduction-driven accretion is now generally accepted there is still some question as to how this occurs. In comparison with today, a warmer Earth during the Archaean would have led to thicker oceanic crust (Bickle 1978), thin buoyant plates, and more prolific subduction and plume activity. The enhanced subduction-driven accretion of volcanic arcs, sedimentary wedges and continental fragments would tend to choke subduction and lead to oceanward jumps in trench location. The result is a lithospheric root constructed from imbricated shallow-dipping slabs (e.g. Helmstaedt & Schulze 1989; Hoffman 1990). Slab-like structures extending into the mantle beneath the Canadian Shield have been imaged in seismic reflection profiles (e.g. in the Abitibi belt, Calvert et al. 1995; Slave Province, Cook et al. 1997; Western Superior, White et al. 2000), but the depth extent of these palaeo-slabs is unclear.

Here we present highlights from a seismological investigation of the upper mantle beneath the Western Superior Province. The Teleseismic Western Superior Transect (TWiST) is a component of the Canadian Lithoprobe Western Superior Transect (Lithoprobe phase V proposal; Clowes 1997). Between May and November 1997, nearly 1100 teleseismic events were recorded by an array of 14 broadband (BB) three-component instruments, deployed along a 600 km line from west of Thunder Bay to north of Pickle Lake, Ontario (Fig. 1). Three more BB sites operated at northern communities, extending the profile into the Proterozoic Trans-Hudson shear zone. Permanent stations of the Canadian National Seismograph Network (CNSN) were additionally incorporated into our network. Concurrent geophysical experiments were also used to study the crust and uppermost mantle beneath the southern part of the Western Superior (Kay et al. 1999a; Musacchio et al. 1999; Craven et al. 2000; White et al. 2000).

Regional tomography studies have shown that the Canadian Shield, which is composed of the Superior and other Archaean cratons and Proterozoic orogens of North America, is underlain by a high-velocity root (Grand et al. 1997) (Fig. 2). Our experiment lies in the centre of this high-velocity region and we are therefore looking at both the lateral and vertical variations in heterogeneity and anisotropy in the heart of a craton. A number of seismic techniques are potentially useful for mapping variations in the velocity structure in this region; here we review the results from three. The first is an investigation using surface waves, the second is a study of the upper-mantle anisotropy using observations of SKS splitting and the final one is an investigation of the P- and S-wave structure using travel-time tomography.

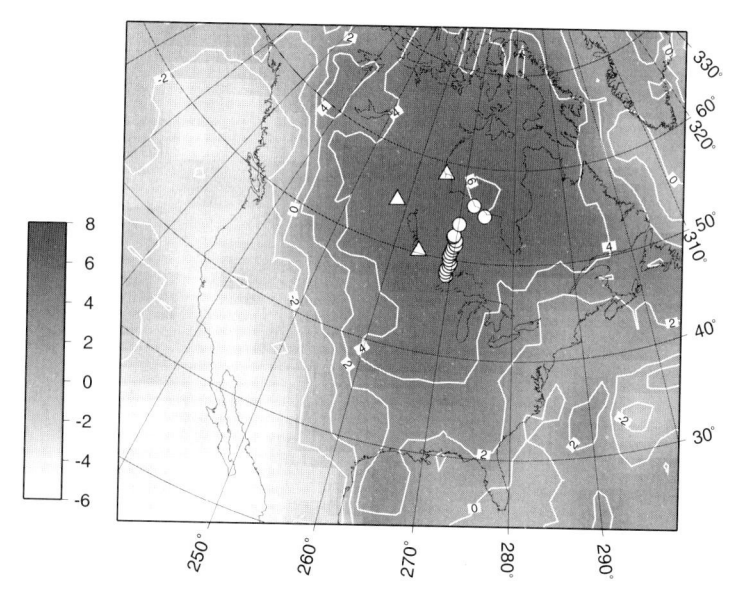

Fig. 2. Large-scale velocity structure beneath the Canadian Shield as determined by Grand et al. (1997). The scale indicates the shear-wave velocity anomaly as a percentage relative to IASP91 (Kennett & Engdahl 1991). ○, locations of the TWiST seismic stations; △, from south to north, the permanent seismic stations ULM (CNSN), FFC (IRIS) and FCC (CNSN).

Surface-wave analyses

Surface-wave analysis is useful for obtaining a 1D profile (i.e. variation with depth) of upper-mantle velocity structure. The Canadian Shield was the subject of the first broad-spectrum (3-90 s period) surface-wave study of the crust and upper mantle (Brune & Dorman 1963). This and subsequent studies using Love-, Rayleigh- and body-wave travel times (e.g. Grand & Helmberger 1984) suggested tht the Canadian Shield is underlain by a high-velocity root that extends to depths between 120 and nearly 150 km. However, more recently, Van der Lee & Nolet (1997) have shown evidence for high-velocity structures extending to nearly 350 km in places beneath the Shield, by using waveform tomography.

Surface waves also provide a means for estimating upper-mantle anisotropy. There are three canonical ways of detecting anisotropy using surface waves. One comes from discrepancies in isotropic inversions of Love and Rayleigh phase velocities (e.g. Anderson 1961) and leads to transversely isotropic models or polar anisotropy. Another comes from azimuthal variations in phase velocities (e.g. Forsyth 1975) and leads to models of azimuthal or radial anisotropy. Finally, Kirkwood & Crampin (1981) showed that particle-motion anomalies are diagnostic of anisotropy and that the variations with period can be used to estimate the approximate depth of the anisotropic layer. This is best done using fundamental quasi-Love waves visible on the radial and vertical components, as they will not be contaminated by the later arriving Rayleigh waves (Yu & Park 1994).

Shear-wave velocity structure

Surface-wave group velocities and phase velocities have been calculated for three Caribbean events that lie within 5° of the great-circle path that runs through the TWiST (Snell 1999). Phase velocities are calculated along three segments of the line, roughly equal in length, which are defined by stations with the same seismic sensors and hence common instrument response. Traces are then windowed around group-velocity arrival times and filtered, and cross-correlation is used to estimate the delay time of a given phase across a segment of the profile (each roughly 250 km long). Interstation phase velocities are higher than those measured for the Canadian Shield in

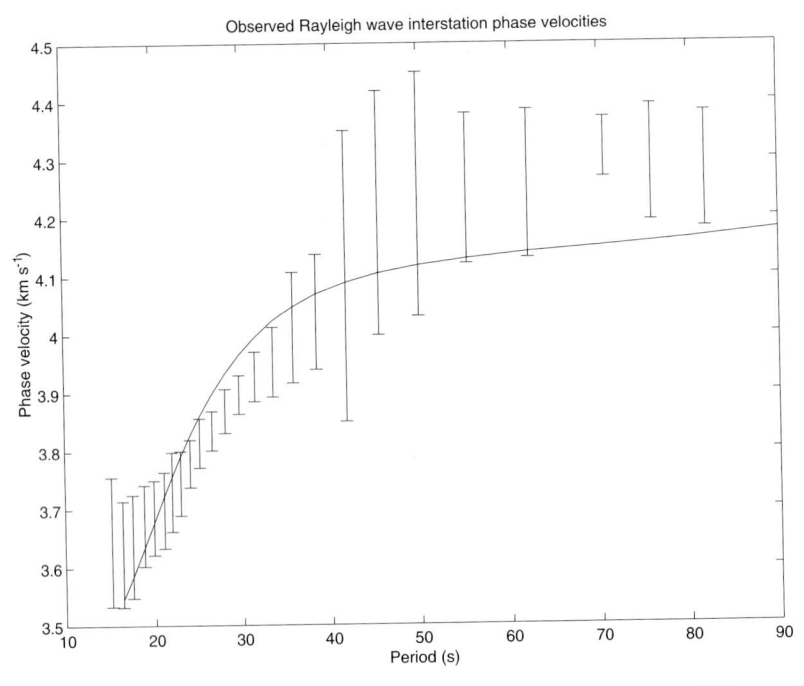

Fig. 3. Interstation phase velocities calculated from surface waves that traverse the TWiST array. The Rayleigh-wave phase velocities (plotted as error bars 2 SD either side of the mean) are clearly higher than those measured in an older study of the Canadian Shield (continuous lines; Brune & Dorman 1963), which was not confined to the Superior Craton.

the Brune & Dorman (1963) study (Fig. 3). This is perhaps not surprising as the Brune & Dorman (1963) study used long propagation paths that sample a number of Archaean and Proterozoic provinces.

Inversions for shear-wave velocity models that best fit the TWiST data are compared with previous models (Fig. 4). Although these models are still somewhat preliminary, there is clear evidence for a subcrustal layer, 5–20 km thick, and a high-velocity uppermost mantle that extends to the resolution depth of this array ($c.$ 250 km). These results support the notion of a high-velocity cratonic root that extends to depth in excess of 200 km. In contrast, the models of Brune & Dorman (1963) and Grand & Helmberger (1984) show a lower-velocity asthenosphere below a $c.$ 110—150 km thick lithosphere. Neither study is confined to the Western Superior, they average mantle structure from a wider area. Our results are in agreement with those of Van der Lee & Nolet (1997), who showed a well-defined continental root beneath the Superior Province to depths in excess of 200 km and sometimes reaching depths >300 km. Interestingly, their results do not show evidence for a continental root beneath the nearby Wyoming Craton.

Snell (1999) has also shown that variations in the interstation phase velocities in the 15–30 s period range can be explained by variations in crustal thickness that range from 37 to 44 km moving southward across the array. Refraction data showed similar variations in Moho depth, ranging between 39.5 and 42 km along a $c.$ 400 km line aligned with the southern end of our transect (Kay *et al.* 1999*a*). The refraction experiment also shows evidence for the thin sub-Moho layer (Musacchio *et al.* 1999).

Evidence for anisotropy in surface waves

Figure 5 shows the Rayleigh- and Love-wave interstation phase velocities predicted from the model derived from Rayleigh waves. The Love-wave velocities sensitive to upper-mantle structure (i.e. >40 s periods) are clearly higher than those predicted, suggesting anisotropy. However, because we are dealing with a single propagation path we cannot distinguish between polar anisotropy (transverse isotropy with a vertical symmetry axis) and azimuthal anisotropy where east–west velocities in the uppermost mantle are faster than north–south velocities.

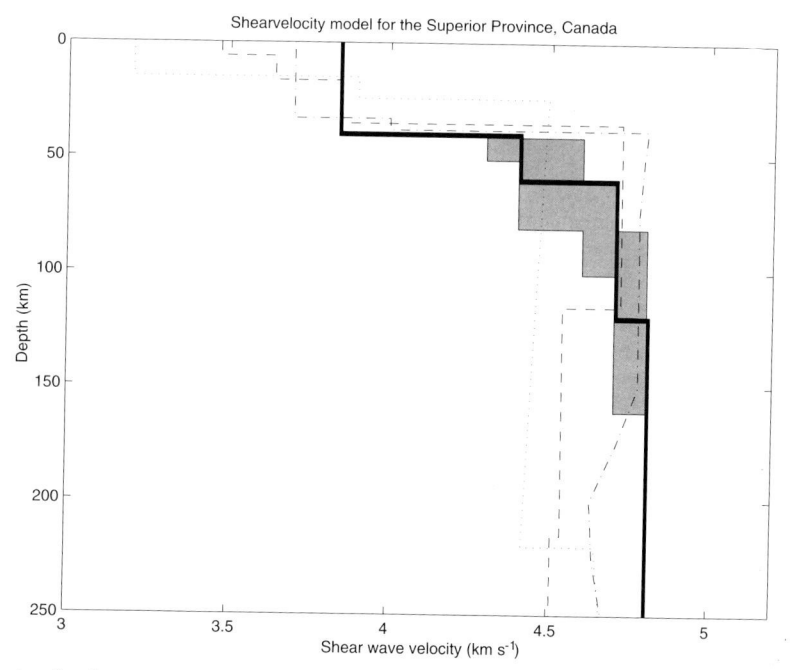

Fig. 4. Inversion for shear-wave velocity model (continuous line) which best fits the Rayleigh-wave interstation phase velocities. The grey area shows range of acceptable solutions. Three other velocity models are shown for comparison: dashed line, Brune & Dorman (1963); dot–dashed line, Grand & Helmberger (1984); dotted line, PREM (Dziewonski & Anderson 1981). It should be noted that the TWiST model shows no indication of a low-velocity zone below a continental root.

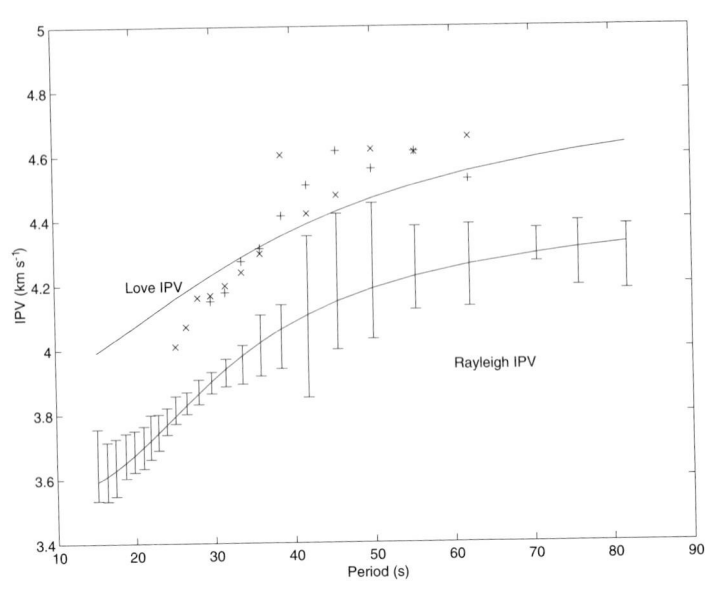

Fig. 5. Love- and Rayleigh-wave interstation phase velocities (IPV) (continuous lines) predicted from the best-fit model derived from the Rayleigh-wave observation. The observed Love- and Rayleigh-wave IPVs are shown for comparison. It should be noted that the Love waves have higher velocities than those predicted in the period range 38–62 s (i.e. in the upper mantle).

Results from the Western Superior wide-angle refraction experiment support an interpretation of azimuthal anisotropy. Musacchio et al. (1999) found evidence for a thin high-velocity anisotropic subcrustal layer, where the P-wave velocities in the east–west direction are faster than those in the north–south direction. The results from SKS splitting (see below) also indicate azimuthal anisotropy with a fast east–west direction. As a whole, the results are consistent with a preferred olivine *a*-axis alignment in a roughly east–west direction.

Evidence for anisotropy in the uppermost mantle also comes from observations of elliptical particle-motion anomalies in Love waves. Such anomalies are indicative of anisotropic structure within a few wavelengths of the receiver (Kirkwood & Crampin 1981). The advantage of using the fundamental Love wave is that it is the first surface-wave arrival and should not be contaminated by the Rayleigh wave. The other advantage of this approach is that interstation calculations are not required. Events from the Western Pacific were examined as the large epicentral distances give a clear separation between the Rayleigh and Love modes.

Our analysis has shown that not all frequency bands show elliptical motion (Fig. 6), suggesting vertical variations in anisotropy. The clearest ellipticity is visible in the 33–50 s period. As the highest frequencies (<33 s) are more sensitive to the 40 km thick crust and show linear particle motion, the crust as a whole is inferred to be isotropic. The particle-motion anomalies for the 33–50 s pass band suggest an anisotropic region below 50 km, but we cannot further constrain the depth extent of the anisotropy at this time.

SKS splitting

Anisotropy in the upper mantle is primarily attributed to the preferred alignment of olivine and, to a lesser extent, the alignment of other upper-mantle minerals. Evidence comes from measurements on mantle xenoliths (Mainprice & Silver 1993; Ben Ismaïl & Mainprice 1998) and ophiolites (Nicolas & Christensen 1987) and laboratory deformation studies (Zhang & Karato 1995). These measurements help guide the interpretation of seismic observations. It is conventionally assumed that the anisotropy is c. 4–5% in magnitude and has hexagonal symmetry with a horizontal symmetry axis (azimuthal anisotropy) (Mainprice & Silver 1993). Numerical simulations of the lattice preferred orientation

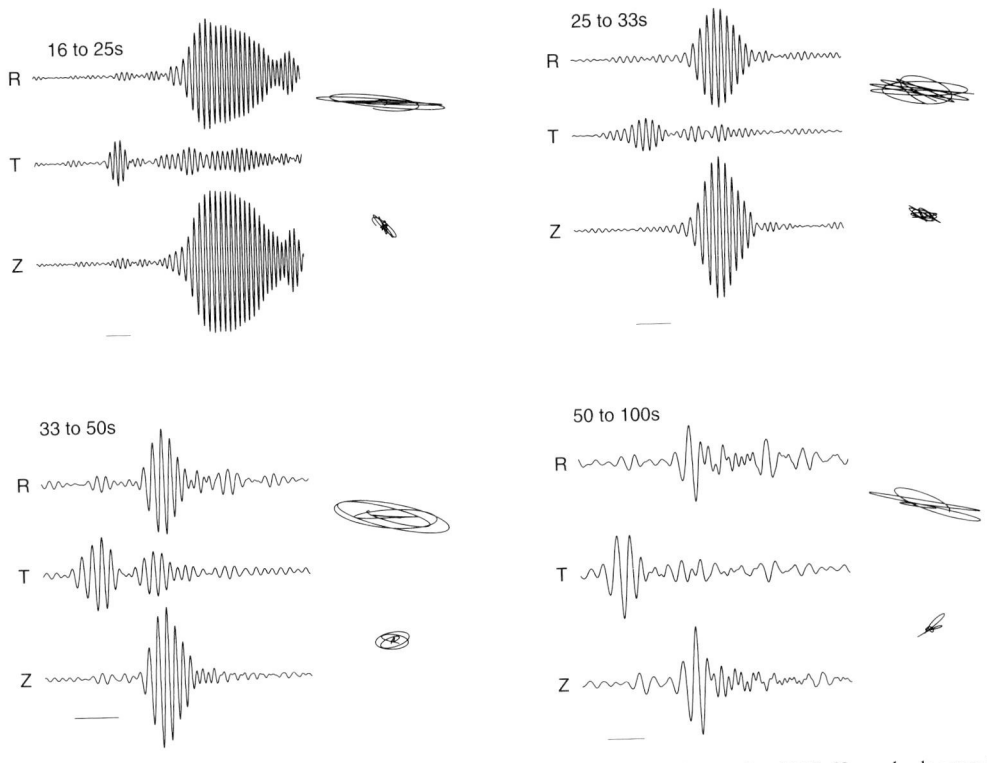

Fig. 6. Particle motion analysis of an event from the Western Pacific recorded at station 5110. Upper hodogram is in the horizontal plane; lower one is in the sagittal plane. Ellipticity is greatest in the 33–50 s period. R, T and Z, radial, transverse and vertical components, respectively.

(LPO) of olivine suggest that upper-mantle anisotropy may be locally much stronger in magnitude and more complex in symmetry (Blackman et al. 1996; Tommasi 1998). There have also been suggestions for other mechanisms for anisotropy in the mantle, such as preferentially aligned inclusions (e.g. melt) (Schlue & Knopoff 1977; Kendall 1994) and grain shape fabric or foliation (Wendt et al. 1999).

An important interpretation of observed anisotropy centres on whether or not the crust is coupled to the mantle during orogenic episodes (Silver 1996). Silver and coworkers (Silver & Chan 1988; Savage & Silver 1993) referred to the coupled case as vertically coherent deformation (VCD) and the decoupled case as simple asthenospheric flow (SAF). In the VCD case it is argued that anisotropy will be frozen into the lithosphere at the time of deformation and the orientation of the anisotropy will mimic trends in the surface geology. In contrast, Vinnik and coworkers have argued that in the SAF case the orientation of the anisotropy will reflect absolute plate motions (Vinnik et al. 1992). Analysis of the TWiST dataset, in conjunction with detailed analysis of permanent CNSN stations, enables us to test these two hypotheses.

The perhaps most unambiguous indicator of anisotropy is the observation of two roughly orthogonal shear waves with different arrival times (commonly known as shear-wave splitting). Although splitting can be studied in any shear phase, core phases such as SKS are attractive to study as the anisotropy can be confined to the mantle path on the receiver side. The P–S conversion at the core–mantle boundary generates a radially polarized shear wave. Therefore, evidence of transverse-component energy indicates that the receiver-side shear wave may have passed either through an anisotropic region or through a heterogeneous region with strong velocity gradients or dipping interfaces. Furthermore, lateral variations in the splitting over relatively short distances at the surface further confine the anisotropy to the upper mantle where there is little overlap in Fresnel zones between adjacent stations (Alsina & Snieder 1995). The disadvantage of this technique is that there is

Upper-mantle anisotropy: splitting analysis

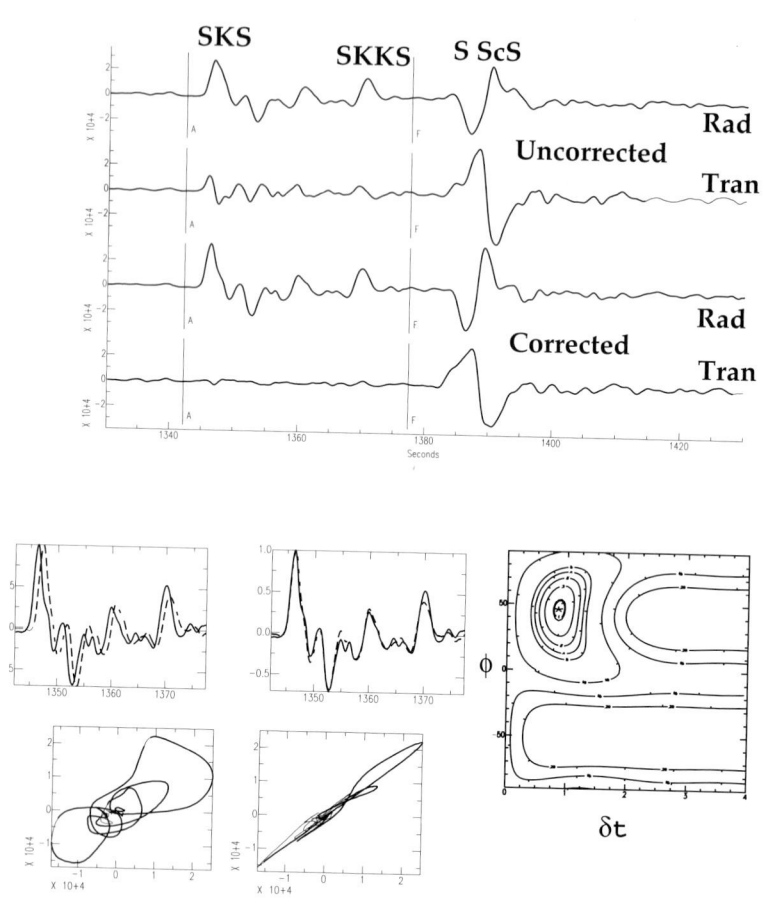

Fig. 7. An example of upper-mantle splitting analysis for a Solomon Islands' event recorded at the Canadian station ULM. The top traces show the radial and transverse component before and after the correction for SKS and SKKS splitting. The vertical bars marked A and F indicate the analysis window. It should be noted that after the correction the SKS and SKKS energy is minimized on the transverse component. The four lower-left panels show the isolated slow (continous line) and fast (dashed line) shear waves and the particle motion within the analysis window before and after the splitting correction. It should be noted also that the particle motion becomes linear after the correction. The lower right panel shows the confidence intervals for the splitting parameters. The 95% confidence interval is the innermost circle around the asterisk at 50° and 0.96 s.

poor vertical resolution. Comprehensive reviews of splitting measurements through continental upper mantle have been given by Silver (1996) and Savage (1999). An example of SKS–SKKS splitting analysis at the CNSN station ULM is shown in Figure 7.

Figure 8 shows the magnitude of SKS splitting, δt, and the orientation of the fast shear wave, ϕ, at the TWiST stations (Kay *et al.* 1999b), surrounding permanent CNSN stations (Sutherland 2000), stations of the APT89 experiment (Silver & Kaneshima 1993) and at the IRIS station FFC (Bostock & Cassidy 1995). Also shown is the absolute plate motion (APM) for the Superior Province as estimated by Gripp & Gordon (1990) (computer program written by K. Fischer).

There is a remarkable correspondence between trends in the surface geology and ϕ (compare with Fig. 1), thereby supporting the VCD model. This agreement and the large δt values on the Archaean part of the craton have been used to

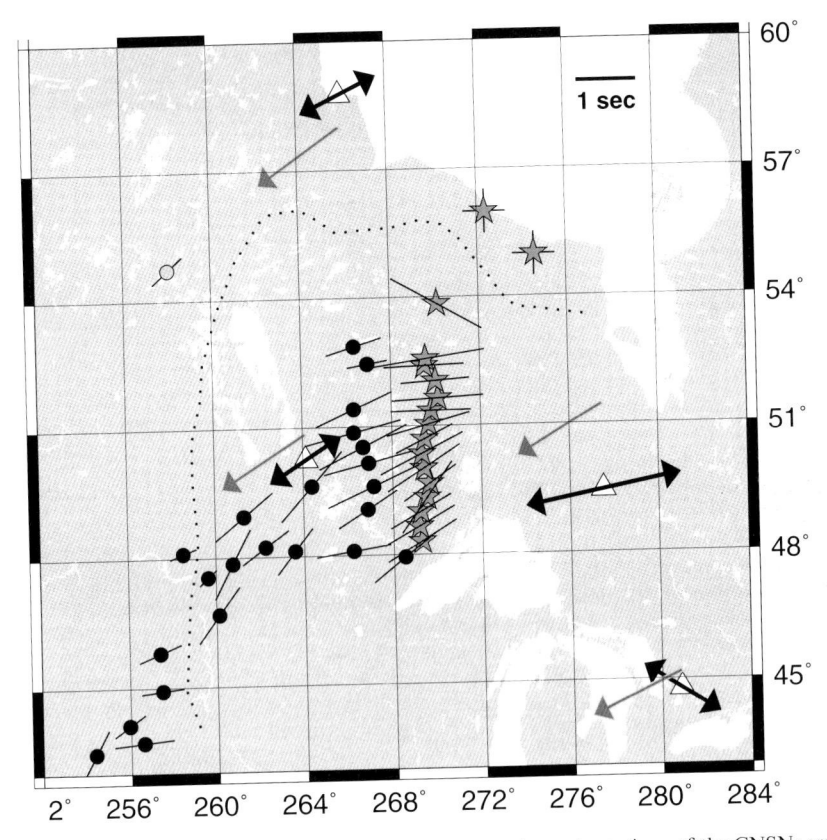

Fig. 8. Compilation of SKS-splitting results. Grey stars, TWiST stations, △, stations of the CNSN; grey circle, IRIS station FFC; ●, stations of the APT89 experiment (Silver & Kaneshima 1993). Vectors at each station show the polarization of the fast shear wave and their length is proportional to the degree of splitting. A null measurement is indicated by a '+' sign. Large grey arrows show the direction of absolute plate motion as predicted by Gripp & Gordon (1990) and the dotted outline shows the Superior–Trans-Hudson boundary.

argue for the existence of a thick continental root where the anisotropy was frozen in at the time of accretion (Silver & Chan 1988). However, most stations also show good agreement between ϕ and the APM direction, thereby supporting the SAF model. Upon closer inspection there are, however, a number of stations where the APM does not align with ϕ. For example, at the station BBT (53.8°N, 89.9°W) in the heart of the Sachigo sub-Province there is nearly a 40° discrepancy between ϕ and the APM direction. At the station SADO (44.8°N, 79.1°W) the APM and ϕ are nearly orthogonal. Finally, the null measurements at the stations on Hudson Bay are difficult to explain in terms of either the SAF or VCD models.

The other striking feature is that δt is largest in the centre of the Western Superior and seems to diminish in magnitude towards the Trans-Hudson shear zone. In fact, there are a number of null measurements (no splitting) in the Trans-Hudson shear zone (Silver & Kaneshima 1993; Kay et al. 1999b). One would perhaps expect more uniform δt values if the anisotropy is solely a result of SAF. Given the fact that the Trans-Hudson Orogeny was an event of Himalayan proportion, it is surprising that so little evidence for associated anisotropy exists. An alternate explanation is that the mantle is anisotropic but oriented in such a way that there is no apparent splitting (e.g. vertical orientation of the olivine a-axes).

Most of these results are based on 1–5 SKS measurements at each station, the exception being the measurements made at the CNSN stations. At ULM, for example, the result is based on 45 measurements. Figure 9 shows that these measurements have a clear dependence on the back-azimuth to the source. Two anisotropic layers with different orientations in symmetry axes lead to a 90° periodicity in the delay time, δt,

ULM 2-Layer Model

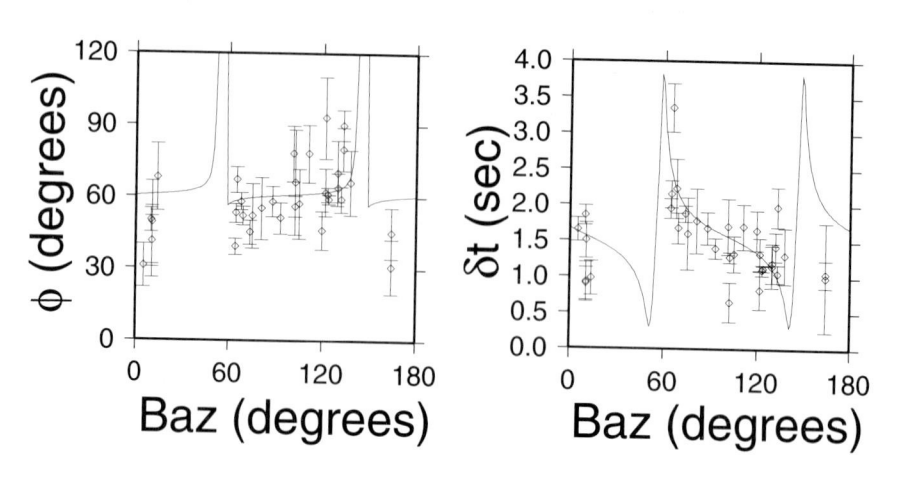

Fig. 9. Best fit two-layer model for ULM. Left shows the fast shear-wave polarization, ϕ, as a function of the back-azimuth (Baz) to individual events. Right shows the delay time, δt, between the fast and slow shear waves as a function of back-azimuth. The continuous line shows the predicted variation for a two-layer model where $\phi = 85.0°$ and $\delta t = 0.4$ s in the upper layer and $\phi = 50.0°$ and $\delta t = 1.3$ s in the lower layer and for a frequency of 0.15 Hz.

as a function of back-azimuth (Silver & Savage 1994) whereas a dipping anisotropic symmetry axis leads to a 180° periodicity (Babuska & Plomerova 1989; Levin & Park 1997). Figure 9 shows that there is an apparent 90° variation in the data, suggesting at least two layers of anisotropy.

Following the method of Silver and Savage (1994), we determine a two-layer model that explains the observations. Figure 9 shows the 0.15 Hz predictions for an upper layer with $\phi = 85.0°$ and $\delta t = 0.40$ s, underlain by another anisotropic layer with $\phi = 50.0°$ and $\delta t = 1.30$ s. It should be noted that the underlying layer has an anisotropy oriented roughly in the APM direction, whereas the upper layer is oriented with the surface geology. Assuming 5% anisotropy, the upper layer need only be 40 km thick (i.e. crustal thickness). The lower layer, roughly 130 km thick, may be frozen into the continental root or may represent a layer of shear flow that is forced to deviate around the root. In the latter case the flow-induced strain may be higher than normal. This would lead to a well-developed LPO and higher degrees of anisotropy in a relatively thinner layer. The surface-wave particle motion anomalies suggest anisotropy in regions deeper than the crust. Thus the upper layer must also be in the mantle, suggesting crust–mantle coupling. It therefore appears that both the VCD

and aspects of SAF models could be invoked to explain the observations.

Tomography

Seismic tomography has been used to provide a more detailed picture of the velocity structure beneath the Western Superior. Here we review the results and interpretations presented by Sol et al. (2002).

The 5 month seismometer deployment yielded 1423 P-wave and 651 S-wave travel-time residuals with good coverage in both epicentral distance (slowness) and azimuth (Fig. 10). The model encompasses a region between 48° and 56°N in latitude and 85° and 92°W in longitude, extending to a depth of 900 km, and is parameterized on a grid 35 nodes in depth, 43 in latitude and 29 in longitude. The inversion uses the technique and software developed by Vandecar (1991) and details of this application have been given by Sol et al. (2002). It is a non-linear inversion, where 3D ray tracing is used to update the velocity model with each inversion iteration, although Sol et al. (2002) showed that a linear inversion yields similar results. The iterations are carried out until stability in the solution is reached. The method accounts for station corrections, event

P-waves

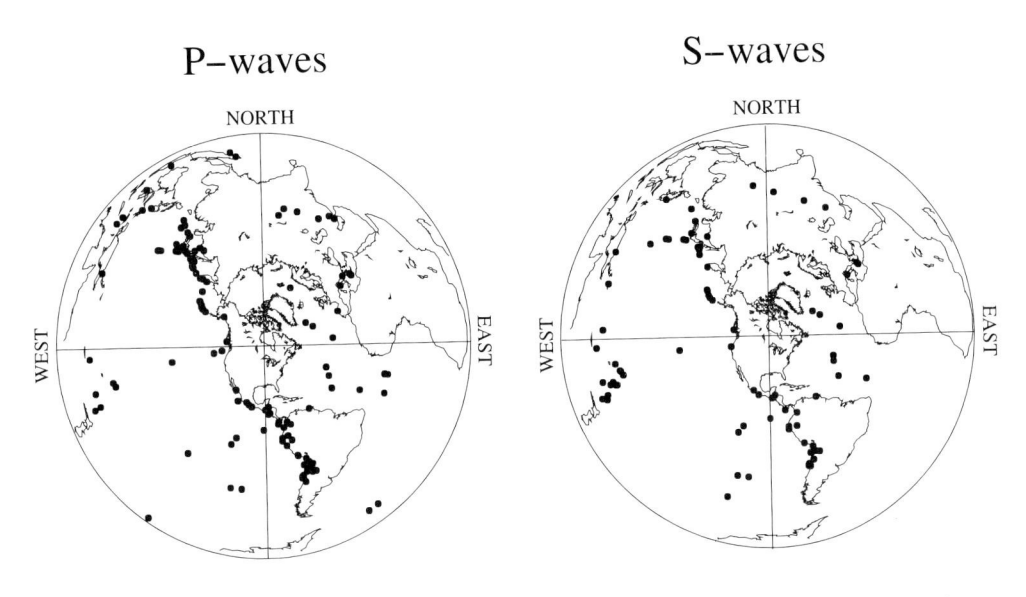

S-waves

Fig. 10. Azimuthal projection, centred on the TWiST array, of 160 P-wave and 82 S-wave events used in tomography study.

mislocation and difference in station elevations (statics). The inversion finds the smoothest model, which fits the data to a required criterion. Both the P- and S-wave inversions explain >92% of the r.m.s. residual in the data.

Sol *et al.* (2002) have conducted a series of resolution tests to assess the reliability of the resulting images. Chequerboard tests reveal P-wave resolution of the order of 50 km between depths of 70 and 500 km in the region to the south of the Sachigo Province. S waves show less resolution, especially below 400 km, because of the significantly lower number of S-wave raypaths through the model. The real-data inversions also prompted a series of resolution tests for isolated anomalies and a slab-like feature. Although there is some vertical smearing, especially towards the edges of the model, a slab-like anomaly with a thickness of the order of 50 km and extending to depths *c.* 500 km is resolvable.

The resolution tests are based on ray tracing that implicitly assumes infinitely high frequencies. Resolution is also controlled by the size of the Fresnel zone (the width of a finite-frequency ray). At a given depth, this can be calculated by estimating the lengths of raypaths that connect points on a 2D surface at this depth to the source and to the receiver. The Fresnel zone is defined by those rays whose lengths are within a quarter wavelength of the length of the geometrical raypath. The Fresnel zone at the midpoint of a 1 Hz P wave, 10 000 km in length is roughly

240 km in diameter, but this will narrow towards the source and receiver. At a depth of 300 km the Fresnel zone is *c.* 60 km, which is roughly the resolution expected based on our P-wave ray coverage. The Fresnel zone for S waves is nearly the same, because although S waves are generally lower in dominant frequency, they are also lower in velocity.

Figure 11 shows horizontal slices through the P- and S-wave models at a depth of 150 km. The results show most variation in velocity beneath the southern part of the transect and that there is a reasonable correlation between the P-wave and S-wave images, despite their independent treatment. The P-wave image shows a series of high- and low-velocity anomalies in the southern part, whereas the S-wave image shows a similar but less detailed picture. In these and the other images presented by Sol *et al.* (2002), the Sachigo sub-Province in the northern part of the line is underlain by comparatively uniform material. There is insufficient coverage to say much about the transition from the Superior Province to the younger Trans-Hudson Orogen to the north.

Vertical cross-sections oriented in a SW–NE direction (Fig. 12) reveal a number of interesting features. Striking features in the P-wave image are the high-velocity anomalies labelled X and Y (Fig. 12). The tabular anomaly, Y, extends to the resolution depth limits (*c.* 500 km). It is not clear how connected these anomalies are, but resolution tests suggest it would be difficult to resolve

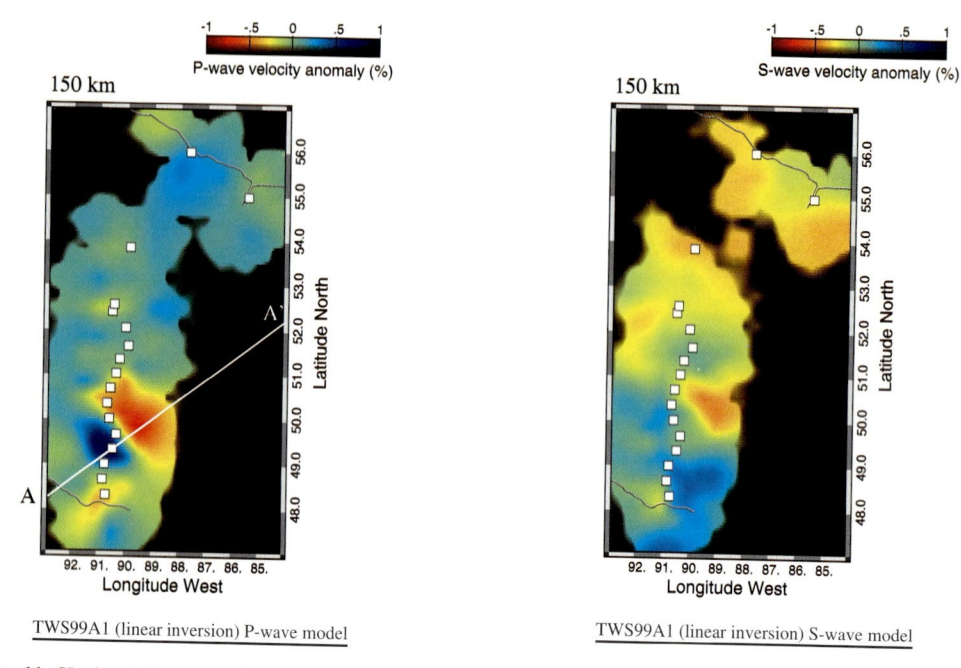

Fig. 11. Horizontal cross-sections of the P- and S-wave tomography models at a depth of 150 km. □, stations; velocity anomaly scale (with respect to the reference model IASP91, Kennett & Engdahl 1991) is indicated above. The NE–SW-trending white line marks the location of the depth profile shown in Figure 12.

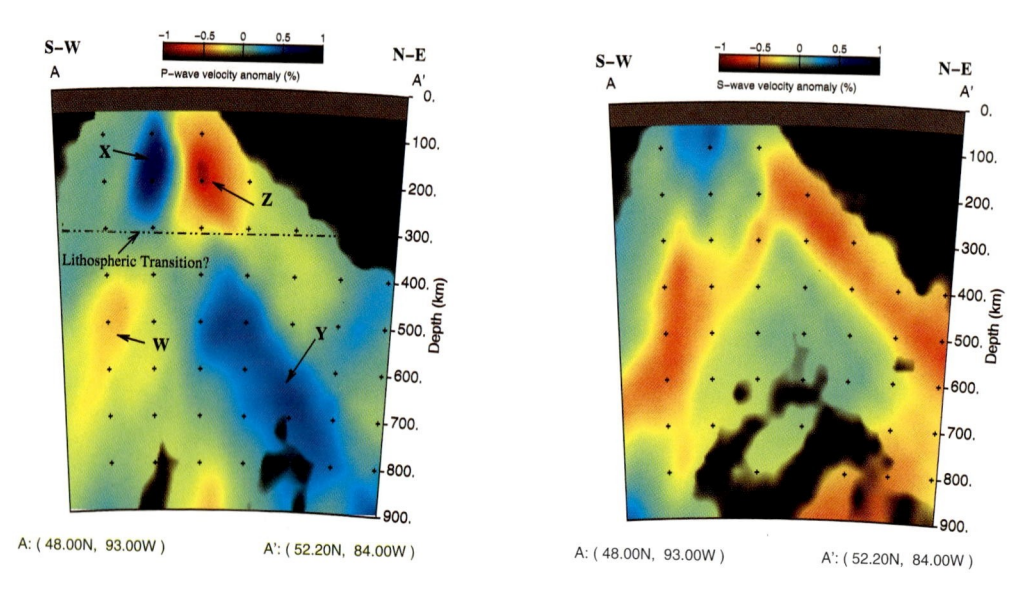

Fig 12. Vertical cross-section of the P- and S-wave tomography models oriented along the traverse marked in Figure 11. The latitude and longitude of the end points of the traverse, A and A′, are indicated below.

this one way or another. A well-resolved low-velocity anomaly sits above the slab feature in the uppermost 250 km of the model (labelled Z). There is a second low-velocity anomaly c. 500 km beneath the southern end of the line, but its shape is poorly resolved. Finally, there is a distinct interruption in the coherency of these images at a depth of roughly 300 km. The shear-wave images show similar but more diffuse features, reflecting poorer ray coverage.

The anomaly labelled X (Fig. 12) can be interpreted in terms of a remnant slab of subducted oceanic lithosphere that is crosscut by a low-velocity feature near 300 km. The results suggest subduction in a SE–NW sense, oblique to the surface trends in the geology. Given the age of the craton, it is highly unlikely that this feature is thermal in origin. A candidate mechanism for a chemical anomaly is former oceanic crust that is now eclogitic (Cloos 1993) and high in velocity (Gubbins et al. 1994). The wide range of seismic velocities estimated and measured in eclogites is controlled by garnet v. clinopyroxene content. Our interpretation requires an oceanic crust somewhat higher than average in garnet content, as is the suggested explanation for a high-velocity channel in the Tonga slab (Gubbins et al. 1994). The velocity contrast is perhaps accentuated by a more depleted, hence more dunitic Archaean oceanic subcrustal lithosphere, which would have higher seismic velocities than modern harzburgitic mantle (Ringwood 1991; Sol et al. 2002). The slab must be at least 50–60 km in thickness to be visible in our inversions. This is much thicker than modern-day oceanic crust, but Bickle (1978) has argued that oceanic crust may have been this thick in warmer Archaean times. It is tempting to interpret the low-velocity anomaly adjacent to the slab in terms of processes associated with slab volatiles frozen into the upper-mantle wedge.

It is unclear to what extent the structure below 300 km is related to Archaean processes. The resolution tests of Sol et al. (2002) show that a slab structure that is continuous to >500 km would produce an image very similar to that in Figure 12 and may suggest that the transition-zone high-velocity anomaly (Y) is connected or related to the shallower anomaly (X). There are at least four possible explanations for the deeper high-velocity anomaly, as follows. (1) Slab material extends continuously from the surface to the base of the transition zone. This palaeo-slab has survived for well over 1 Ga and therefore the anomaly cannot be thermal in origin and for some reason it has not been mechanically eroded from the base of the continental root. (2) Material has delaminated from the base of the craton in more recent time, in which case the seismic anomaly may be compositional or thermal in origin. (3) The Western Superior Province has overridden a relic piece of material from a more recent subduction episode elsewhere. It is difficult to explain this with a known subduction event, but similar relic features have been observed elsewhere (e.g. Vandecar 1991). (4) The transition-zone anomalies are a signature of small-scale convection beneath the craton. The slab-like feature is a cold downwelling and the flanking low-velocity anomalies mark the upwellings. Convection modelling shows that this is possible, especially if there are lateral variations in thermal conductivity in the overlying insulat-ing lid (Mimouni & Rabinowicz 1988). There is also the further complication of the effect of the Keweenawan plume at roughly 1 Ga. The low-velocity anomaly below the southern end of the profile may be related to this plume, which affected the Lake Superior region. Testing which of these ideas is most likely is the subject of future studies.

Discussion and conclusions

A variety of seismic techniques have been used to study the upper mantle beneath the Western Superior Province. Each technique has its own advantages and disadvantages, but often one technique compensates for another's deficiency. For example, SKS-splitting analysis has poor vertical resolution but good lateral resolution, whereas surface-wave studies have poor lateral resolution but better vertical resolution.

Surface waves show evidence for a thick continental root or tectosphere that extends to depths in excess of 200 km. Although this conflicts with earlier studies of the Canadian Shield (Brune & Dorman 1963; Grand & Helmberger 1984), those investigations were not confined to the Western Superior. There is support for thick Western Superior root from other studies. Using surface-wave tomography, Van der Lee & Nolet (1997) showed that the Superior Craton may reach depths of nearly 350 km in places. Heat-flow measurements suggest a thermally stable root to roughly 240 km (Jaupart et al. 1998). Although not conclusive in themselves, our tomographic images also show an interruption in the velocity structure at a depth of nearly 300 km.

We see evidence for velocity structure and anisotropy within the tectosphere. Surface waves reveal evidence for a thin high-velocity layer, 5–20? km thick, which lies beneath a 37–43 km thick crust. Wide-angle refractions also show these features (Kay et al. 1999a; Musacchio et al.

1999). The refraction experiment and surface-wave analyses show evidence for anisotropy in a sub-Moho layer, which suggest faster velocities in a direction that roughly follows the geological terrane of the Western Superior Province. SKS-splitting results are also compatible with this interpretation. This layer may be a result of eclogitic underplating associated with subduction and may not be peridotitic in composition. Recent studies of P-wave anisotropy in eclogites suggest that compositional layering contributes more than LPO to an observed 6% anisotropy (Mauler *et al.* 2000). Bostock (1997) has shown evidence for anisotropic layers in the uppermost 250 km of mantle beneath the Slave Craton and has suggested that compositional layering may be a contributing factor to the anisotropy. Such ideas are compatible with ideas of tectosphere accretion via low-angle subduction.

SKS-splitting measurements from the Western Superior Province and the surrounding Trans-Hudson Proterozoic shear zone provide further insights into upper-mantle anisotropy. It seems that a portion of the upper-mantle continental root must be aligned with the surface geology, suggesting coupling between the crust and mantle at the time of cratonic accretion some 2.7–3.0 Ga. It also seems that there could be a component of anisotropy oriented with current mantle flow patterns. A suggestion for the variation in magnitude and orientation of sub-craton anisotropy is that the asthenospheric flow is forced to divert around the rigid keel. Such flow around continental roots has been suggested by Fouch *et al.* (2000). There may in fact be regions where the flow is locally downwards or upwards, thereby explaining low amounts of splitting in the Trans-Hudson. These results suggest that Silver & Chan's (1988) interpretation of the large splitting values as evidence of a deep root beneath the Western Superior may be somewhat oversimplified.

Tomographic images of the P- and S-wave structure beneath the Western Superior Province reveal a more uniform mantle in the north beneath the proto-cratonic Sachigo sub-Province. Towards the south well-resolved high- and low-velocity anomalies are apparent. These may be a product of Archaean subduction and/or the later effects of Keweenawan rifting. A steeply dipping high-velocity anomaly extends well into the mantle, with low-velocity anomaly on the updip side of it. Reflection experiments in the Superior Province have shown evidence for shallow subduction into the upper mantle (Calvert *et al.* 1995; Calvert & Ludden 1999; White *et al.* 2000). Through comparison with models of mantle stratigraphy obtained from Ps conversion in teleseismic phases, Bostock (1999) has also demonstrated that reflection profile images of slabs (Cook *et al.* 1997) may extend well into the upper mantle beneath the Slave Province. It is not clear at this stage how the interpreted remnant oceanic lithosphere would remain as a coherent structure for $c.\,2.7$ Ga. The anomaly would not have survived if thermal in origin. A thicker oceanic crust in the Archaean may delaminate from the subducting lithosphere, but it should be buoyant (Hoffman & Ranalli 1988). However, a phase transition to eclogite phase could not only explain the high velocities but also increase the density (Aherns & Schubert 1975; Gubbins *et al.* 1994; Bostock 1999) and perhaps lead to steeper dips. Beneath the continental root, the tomographic images show a well-resolved tabular high-velocity anomaly that aligns with the shallower slab structure, but its origin is still unclear.

Figure 13 shows a schematic compilation of the seismic results. A more detailed picture of the crust has been given by White *et al.* (2000). Support for the idea that the Sachigo Block may serve as the proto-craton comes from the fact that the seismic images show little lateral variation beneath this area. Electromagnetic studies also suggest this, and, like the seismic studies, show most of the heterogeneity towards the southern end of the transect (Craven *et al.* 2000). The heterogeneity in the south may be the deep signature of island arcs, metasediments and oceanic plateaux, which have been accreted against the Sachigo Block. As pointed out by Bostock (1999), it is still not clear how a succession of imbricated slabs results in an Archaean craton with such long-term stability. The Superior Province has been subjected to the repeated opening and closing of the Atlantic and plume activity. It is not known, for example, to what extent the Keweenawan rifting episode has affected the southern part of the transect; the deep low-velocity anomaly in the south may be related to this. There is also the question of the slab-like high-velocity anomaly in the transition zone. Is this related in some way to the formation of the Western Superior Province, has the Western Superior migrated over a remnant slab or is this an image of small-scale convection beneath the continental root?

In some ways the Superior Province appears different from other Archaean cratons worldwide. For example, it exhibits thicker than average crust and long uninterrupted geological belts, not commonly seen on other cratons. The Superior Province is also unusual in the sense that it exhibits very large shear-wave splitting values in the heart of the craton and very little in the surrounding Proterozoic Trans-Hudson shear zone. It is more normal for the surrounding mobile

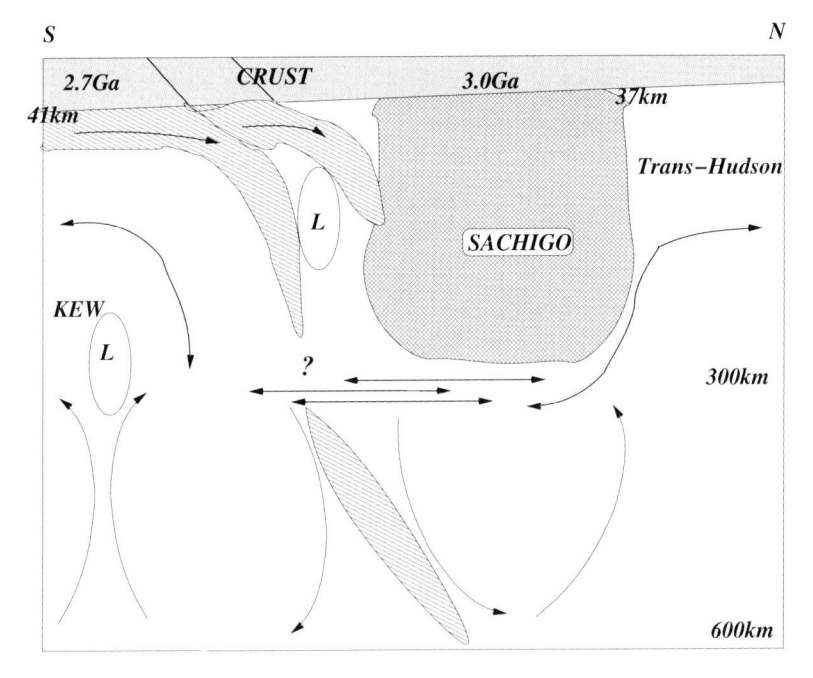

Fig. 13. Schematic representation of the upper mantle beneath the Western Superior Province. Lined or hatched regions mark high velocities; L indicates low-velocity region; KEW refers to region potentially affected by Keweenawan rifting. Arrows show potential regions of localized flow around and below the tectosphere.

belts to exhibit higher levels of anisotropy. For example, the São Francisco Craton of South America shows very little anisotropy in comparison with the surrounding belts (James & Assumpção 1996). Only small amounts of SKS splitting are in general observed beneath both Archaean and Proterozoic provinces in southern Africa (Silver *et al.* 2001). Beneath the Kaapvaal Craton of South Africa, the splitting is smallest beneath the oldest region and increases into the adjacent Limpopo belt. Cratonic mantle beneath Australia also shows very little SKS anisotropy (Clitheroe & Van der Hilst 1998), but interestingly significant surface-wave anisotropy (Simons *et al.* 2001). There is also a wide variability in the thickness of continental roots. For example, whereas the Superior Province is underlain by a nearly 300 km thick root, the nearby Archaean Wyoming Craton does not show evidence for a root (Van der Lee & Nolet 1997). All this suggests that not all Archaean cratons were created in the same way. There is, however, mounting support for Archaean continental evolution via plate-tectonic processes. No doubt plate-boundary geometry played an important role in shaping the Archaean Earth, just as it does with the present-day Earth.

Success in the TWiST experiment would not have been possible without the help of the First Nations people of Peawanuck, Big Trout Lake and Fort Severn. We also thank the staff of Thunder Airlines in Pickle Lake for logistic support. Funding was generously provided by Lithoprobe (NSERC-Canada), the Royal Society of London, NERC grant GR9/03469, Monopros Ltd, and the Geological Survey of Canada. J. Vandecar provided considerable assistance with his tomographic inversion code. The IRIS-PASSCAL programme is thanked for the loan of six seismic stations. We are indebted to H. Helmstaedt and R. Harrap for helpful advice and coordinating the Western Superior Transect. Constructive reviews from G. Helffrich and D. James are gratefully appreciated.

References

ABBOTT, D. 1991. The case for accretion of the tectosphere by buoyant subduction. *Geophysical Research Letters*, **18**, 585–588.

AHERNS, T. J. & SCHUBERT, G. 1975. Gabbro–eclogite reaction rate and its geophysical significance. *Reviews of Geophysics and Space Physics*, **13**, 383–400.

ALSINA, D. & SNIEDER, R. K. 1995. Small-scale sub-lithospheric continental deformation: constraints from SKS splitting observations. *Geophysical Journal International*, **123**, 431–448.

AVÉ LALLEMANT, H. G. 1977. Transpression, displacement partitioning and exhumation in the eastern Caribbean/South American plate boundary zone. *Tectonics*, **16**, 272–289.

BABUSKA, V. & PLOMEROVA, J. 1989. Seismic anisotropy of the subcrustal lithosphere in Europe: another clue to recognition of accreted terranes? *In*: HILLHOUSE, J. W. (ed.) *Deep Structure and Past Kinematics of Accreted Terranes*. Geophysical Monograph, American Geophysical Union, **50**, 209–217.

BELL, K., BLENKINSOP, T., COLE, J. S. & MENAGH, D. P. 1982. Evidence from Sr isotopes for long-lived heterogeneities in the upper mantle. *Nature*, **298**, 251–225.

BEN ISMAÏL, W. & MAINPRICE, D. 1998. An olivine fabric database: an overview of upper mantle fabrics and seismic anisotropy. *Tectonophysics*, **296**, 145–157.

BICKLE, M. J. 1978. Heat loss from the Earth: a constraint on Archaean tectonics from the relation between geothermal gradients and the rate of plate production. *Earth and Planetary Science Letters*, **40**, 301–315.

BICKLE, M. J. 1986. Implications of melting for stabilization of the lithosphere and heat loss in the Archaean, *Earth and Planetary Science Letters*, **80**, 314–324.

BLACKMAN, D. K., KENDALL, J.-M., DAWSON, P., WENK, H.-R., BOYCE, D. & PHIPPS MORGAN, J. 1996. Teleseismic imaging of subaxial flow at mid-ocean ridges: travel-time effects of anisotropic mineral texture in the mantle. *Geophysical Journal International*, **127**, 415–426.

BOSTOCK, M. G. 1997. Anisotropic upper-mantle stratigraphy and architecture of the Slave craton. *Nature*, **390**, 392–395.

BOSTOCK, M. G. 1999. Seismic imaging of lithospheric discontinuities and continental evolution. *Lithos*, **48**, 1–16.

BOSTOCK, M. G. & CASSIDY, J. F. 1995. Variations in SKS splitting across western Canada. *Geophysical Research Letters*, **22**, 5–8.

BRUNE, J. & DORMAN, J. 1963. Seismic waves and earth structure in the Canadian shield. *Bulletin of the Seismological Society of America*, **53**, 167–210.

CALVERT, A. J. & LUDDEN, J. N. 1999. Archean continental assembly in the southeastern Superior Province of Canada. *Tectonics*, **18**, 412–429.

CALVERT, A. J., SAWYER, E. W., DAVIS, W. J. & LUDDEN, J. N. 1995. Archean subduction inferred from seismic images of a mantle suture in the Superior Province. *Nature*, **375**, 670–674.

CARD, K. D. 1990. A review of the Superior Province of the Canadian Shield, a product of Archean accretion. *Precambrian Research*, **48**, 99–156.

CLOOS, M. 1993. Lithospheric buoyancy and collisional orogenesis: subduction of oceanic plateaus, continental margins, island arcs, spreading ridges, and seamounts. *Geological Society of American Bulletin*, **105**, 715–737.

CLITHEROE, G. & VAN DER HILST, R. D. 1998. Complex anisotropy in the Australian lithosphere from shear-wave splitting in broad-band SKS-records. *In*: BRAUN, J., DOOLEY, J. C., GOLEBY, B., VAN DER HILST, R. D. & KLOOTWIJK, C. (eds) *Structure and Evolution of the Australian Continent*. Geodynamics Series, American Geophysical Union, **26**, 73–78.

CLOWES, R. M. (ed.) 1977. *Lithoprobe Phase V Proposal – Evolution of a Continent Revealed*. Lithoprobe Secretariat, University of British Columbia, Vancouver, B.C.

COOK, F. A., VAN DER VELDEN, A. J. & HALL, K. W. 1997. Upper mantle reflectors beneath the SNORCLE transect – images of the base of the lithosphere? *In*: COOK, F. & ERDMER, P. (eds) *Slave–Northern Cordillera Lithospheric Evolution (SNORCLE) Transect and Cordilleran Tectonics Workshop Meeting, Lithoprobe Report 56*. Lithoprobe Secretariat, University of British Columbia, Vancouver, B.C., 58–62.

CORFU, F. & DAVIS, D. W. 1992. A U–Pb geochronological framework for the western Superior Province, Ontario. *In*: THURSTON, P. C. *ET AL*. (eds) *Geology of Ontario, Special Volume 4, Part 2*. Ontario Geological Survey, Toronto, 1334–1348.

CRAVEN, J. A., KURTZ, R. D., BOERNER, D. E., FERGUSON, I. J. & BAILEY, R. C. 2000. Overview of 1999 activities – Electromagnetic component. *In*: HARRAP, R. M. & HELMSTAEDT, H. H. (eds) *Western Superior Transect Sixth Annual Workshop, Lithoprobe Report 77*. Lithoprobe Secretariat, University of British Columbia, Vancouver, B.C., 137.

DZIEWONSKI, A. M. & ANDERSON, D. L. 1981. Preliminary reference earth model, *Physics of Earth and Planetary Interiors*, **25**, 297–356.

FORSYTH, D. W. 1975. The early structural evolution and anisotropy of the oceanic upper mantle. *Geophysical Journal of the Royal Astrological Society*, **43**, 103–162.

FOUCH, M. J., FISCHER, K. M., PARMENTIER, E. M., WYSESSION, M. E. & CLARKE, T. J. 2000. Shear wave splitting, continental keels and patterns of mantle flow. *Journal of Geophysical Research*, **105**, 6255–6276.

GRAND, S. P. & HELMBERGER, D. V. 1984. Upper mantle shear structure of North America. *Journal of Geophysical Research*, **76**, 399–438.

GRAND, S. P., VAN DER HILST R. D. & WIDIYANTORO, S. 1977. Global seismic tomography: a snapshot of convection in the Earth. *GSA*, **7**, 1–7.

GRIPP, A. E. & GORDON, R. G. 1990. Current plate velocities relative to the hotspots incorporating the NUVEL-1 global plate motion model. *Geophysical Research Letters*, **17**, 1109–1112.

GUBBINS, D., BARNICOAT, A. & CANN, J. 1994. Seismological constraints on the gabbro–eclogite transition in subducted oceanic-crust. *Earth and Planetary Science Letters*, **122**, 89–101.

HAMILTON, W. B. 1979. *Tectonics of the Indonesian Region*. US Geological Survey Professional Paper, **P1078**.

HELMSTAEDT, H. H. & SCHULZE, D. J. 1989. Southern African kimberlites and the mantle sample: implications for Archean tectonics and lithospheric evolution. *In*: ROSS, J. *ET AL*. (eds) *Kimberlites and*

Related Rocks. Geological Society of Australia, Sydney, Special Publication, **14**(2), 358–368.

HOFFMAN, P. F. 1989. Precambrian geology and tectonic history of North America. *In*: BALLY, E. W. & PALMER, A. R. (eds) *The Geology of North America – An Overview, A.* Geological Society of America, Boulder, CO, 447–512.

HOFFMAN, P. F. 1990. Geological constraints on the origin of the mantle root beneath the Canadian Shield. *Philosophical Transactions of the Royal Society of London, Series A*, **331**, 523–532.

HOFFMAN, P. F. & RANALLI, G. 1988. Archean oceanic flake tectonics. *Geophysical Research Letters*, **15**, 1077–1080.

JAMES, D. E. & ASSUMPÇÃO, M. 1966. Tectonic implications of S-wave anisotropy beneath SE Brazil. *Geophysical Journal International*, **126**, 1–10.

JAUPART, C., MARESCHAL, J. C., GUILLOU-FROTTIER, L. & DAVAILLE, A. 1998. Heat flow and thickness of the lithosphere in the Canadian Shield. *Journal of Geophysical Research*, **103**, 15 269–15 286.

JONES, A. G. & FERGUSON, I. J. 2001. The electric Moho. *Nature*, **409**, 331–333.

JORDAN, T. H. 1978. Composition and development of the continental tectosphere. *Nature*, **274**, 544–548.

JORDAN, T. H. 1981. Continents as a chemical boundary layer. *Philosoppchical Transactions of the Royal Society of London, Series A*, **301**, 359–373.

KAY, I., MUSACCHIO, G., WHITE, D. & 6 OTHERS 1999a. Imaging the Moho and V_p/V_s ratio in the Western Superior Archean craton with wide angle reflections. *Geophysical Research Letters*, **26**, 2585–2588.

KAY, I., SOL, S., KENDALL, J.-M. & 5 OTHERS 1999b. Shear wave splitting observations in the Archean craton of Western Superior. *Geophysical Research Letters*, **26**, 2669–2672.

KENDALL, J.-M. 1994. Teleseismic arrivals at a midocean ridge: effects of mantle melt and anisotropy, *Geophysical Research Letters*, **21**, 301–304.

KENNETT, B. L. N. & ENGDAHL, E. R. 1991. Travel times for global earthquake location and phase identification. *Geophysical Journal International*, **105**, 429–465.

KIRKWOOD, S. C. & CRAMPIN, S. 1981. Surface-wave propagation in an ocean basin with an anisotropic upper mantle: observations of polarization anomalies. *Geophysical Journal of the Royal Astronomical Society*, **64**, 487–497.

LANGFORD, F. F. & MORIN, J. A. 1976. The development of the Superior Province of Northwestern Ontario by merging island arcs. *American Journal of Science*, **276**, 1023–1034.

LEVIN, V. & PARK, J. 1977. P–SH conversions in a flat-layered medium with anisotropy of arbitrary orientation. *Geophysical Journal International*, **131**, 253–266.

MAINPRICE, D. & SILVER, P. G. 1993. Interpretation of SKS-waves using samples from the sub-continental lithosphere, *Physics of the Earth and Planetary Interiors*, **78**, 257–280.

MAULER, A. M., BURLINI, L., KUNZE, K., PHILIPPOT, P. & BURG, J. P. 2000. P-wave anisotropy in eclogites and relationship to the omphacite crystallographic fabric. *Physics and Chemistry of the Earth, A*, **25**, 119–126.

MIMOUNI, A. & RABINOWICZ, M. 1988. The old continental shields' stability related to mantle convection, *Geophysical Research Letters*, **15**, 68–71.

MUSACCHIO, G., THOMSON, C. J., WHITE, D. J. & ASUDEH, I. 1999. the upper 100 km of the lithosphere within an Archean terrane: hints from the western Superior Province. *Journal of Conference Abstracts*, **4**(1), 150.

NICOLAS, A. & CHRISTENSEN, N. I. 1987. Formation of anisotropy in upper mantle peridotites – a review. *In*: FUCHS, K. & FROIDEVAUX, C. (eds) *Composition, Structure and Dynamics of the Lithosphere–Asthenosphere System.* Geodynamics Series, American Geophysical Union, **16**, 111–123.

RINGWOOD, A. E. 1991. Phase transitions and their bearing on the constitution and dynamics of the mantle. *Geochimica et Cosmochimica Acta*, **55**, 2083–2110.

SAVAGE, M. 1999. Seismic anisotropy and mantle deformation: What have we learned from shear wave splitting? *Reviews of Geophysics*, **37**, 65–106.

SAVAGE, M. & SILVER, P. G. 1993. Mantle deformation and tectonics: constraints from seismic anisotropy in western United States. *Physics of the Earth and Planetary Interiors*, **78**, 207–227.

SCHLUE, J. W. & KNOPOFF, L. 1977. Shear wave polarization anisotropy in the Pacific Basin. *Geophysical Journal of the Royal Astronomical Society*, **49**, 145–165.

SILVER, P. G. 1996. Seismic anisotropy beneath the continents: probing the depths of geology. *Annual Review of Earth and Space Sciences*, **24**, 385.

SILVER, P. G. & CHAN, W. W. 1988. Implications for continental structure and evolution from seismic anisotropy. *Nature*, **335**, 34–39.

SILVER, P. G. & KANESHIMA, S. 1993. Constraints on mantle anisotropy beneath Precambrian North America from a transportable experiment. *Geophysical Research Letters*, **20**, 1127–1130.

SILVER, P. G. & SAVAGE, M. K. 1994. The interpretation of shear wave splitting parameters in the presence of two anisotropic layers, *Geophysical Journal International*, **199**, 949–963.

SILVER, P. G., GAO, S. S., LIU, K. H. ET AL. 2001. Mantle deformation beneath southern Africa, *Geophysical Research Letters*, **28**, 2493–2496.

SIMONS, F. J., VAN DER HILST, R. D., MONTAGNER J.-P. & ZIELHUIS, A. 2001. Multimode Rayleigh wave inversion for shear wave speed heterogeneity and azimuthal anisotropy of the Australian upper mantle. *Geophysical Journal International*, **203**, 1–17.

SNELL, C. S. 1999. *Surface wave analysis of upper mantle velocity structure, Western Superior Province, Canada.* MGeophy thesis, University of Leeds.

SOL, S., THOMSON, C. J., KENDALL, J.-M., WHITE, D., VANDECAR, J. C. & ASUDEH, I. 2002. Seismic tomographic images of the cratonic upper mantle beneath the Western Superior Province of the Canadian Shield – a remnant Archean slab? *Physics of the Earth and Planetary Interiors*, (in press).

SUTHERLAND, F. H. 2000. *Characterising upper-mantle seismic anisotropy in Archaean and Proterozoic Provinces beneath eastern Canada.* MGeophy thesis, University of Leeds.

THURSTON, P. C. & CHIVERS, K. M. 1990. Secular variation in greenstone development. *Precambrian Research*, **46**, 21–58.

TOMMASI, A. 1998. Forward modelling of the development of seismic anisotropy in the upper mantle, *Earth and Planetary Science Letters*, **160**, 1–13.

VANDECAR, J. C. 1991. *Upper mantle structure of the Cascadia subduction zone from non-linear teleseismic travel time inversion.* PhD thesis, University of Washington, Seattle.

VAN DER LEE, S. & NOLET G. 1977. Upper mantle S velocity structure of North America, *Journal of Geophysical Research*, **102**, 22 815–22 838.

VINNIK, L. P., MAKEYEVA, L. I., MILEV, A. & USENKO, Y. 1992. Global patterns of azimuthal anisotropy and deformation in the continental mantle, *Geophysical Journal International*, **111**, 433–447.

WENDT, A. S., COVEY-CRUMP, S. J., CHESNOKOV, E. M. & MAINPRICE, D. 1999. Constraining the seismic anisotropy of mantle rocks: a comparison between experiments and numerical modelling. *Journal of Conference Abstracts*, **4**, 844.

WHITE, D., HELMSTAEDT, H. H., HARRAP, R. M. & THURSTON, P. 2000. Crustal sutures preserved in the Western Superior Province: evidence from deep seismic profiling. *In*: HARRAP, R. M. & HELMSTAEDT, H. H. (eds) *Western Superior Transect Sixth Annual Workshop, Lithoprobe Report 77.* Lithoprobe Secretariat, University of British Columbia, Vancouver, B.C., 155.

YU, Y. & PARK, J. 1994. Hunting for azimuthal anisotropy beneath the Pacific Ocean region, *Journal of Geophysical Research*, **99**, 15 399–15 421.

ZHANG, S. & KARATO, S. 1995. Lattice preferred orientation of olivine aggregates in simple shear. *Nature*, **375**, 774–777.

The structure of the upper mantle beneath southern Africa

KEITH PRIESTLEY & DAN McKENZIE

Department of Earth Sciences, Bullard Laboratories, University of Cambridge, Cambridge CB3 0EZ, UK (e-mail: keith@madingley.org; keith@esc.cam.ac.uk)

Abstract: A large number of velocity models derived from a variety of seismic data and using different seismic techniques have been published for the Archaean and Proterozoic shields. Here, we focus on the structure beneath southern Africa, where velocity models derived from most regional seismic data find the thickness of the seismic lithosphere to be less than 200 km. In contrast, velocity models derived from teleseismic body-wave and long-period surface-wave data determine the seismic lithosphere to be as much as 400 km thick. We believe that this disagreement is due to the ways in which the various datasets average the velocity structure. Our analysis of regional seismograms from propagation paths largely confined to the stable region shows that the average thickness of the seismic lithosphere beneath southern Africa does not exceed 160 km. We compare the vertical S-wave travel time of our velocity model derived from regional seismic data and those models derived from teleseismic data and find no significant difference. We determine the *in situ* velocities and densities from nodules from beneath southern Africa using a recently derived geobarometer and geothermometer; these are in excellent agreement with the velocity found in the high-velocity lid from the analysis of the regional seismic data. The lithospheric model that best fits the nodule data has a mechanical boundary layer thickness of 156 km and a lithosphere thickness of 176 km. However, the shear-wave velocity decrease at the base of the lid does not correspond to a change in mineralogy. Recent experimental studies of the shear-wave velocity in olivine as a function of temperature and period of oscillation demonstrate that this decrease can result from grain boundary relaxation at high temperatures at the period of seismic waves. This decrease in velocity occurs where the mantle temperature is closest to the melting temperature.

In the past 40 years, a large number of velocity models derived from a variety of seismic data and using different seismic techniques have been published for the Archaean and Proterozoic shields. Some are from the analysis of horizontally (Brune & Dorman 1963; Cara 1979) or vertically (Ritsema *et al.* 1999; James *et al.* 2001) travelling waves with propagation paths confined to structure beneath the shields, whereas others have been extracted from larger global velocity models (Ritsema & van Heijst 2000). These shield models come from: (1) high-frequency ($>c.\,0.1$ Hz) body-wave data (Zhao *et al.* 1999); (2) intermediate frequency ($c.\,0.1$–0.01 Hz) body- and surface-wave data (Grand & Helmberger 1984; Debayle & Kennett 2000a); (3) low-frequency (<0.01 Hz) surface-wave and normal mode data (Masters *et al.* 1996). Although almost all studies agree that the upper mantle beneath the shields is higher velocity than that beneath the ocean basins and tectonically active areas, they disagree on important details. Some models have high shear-wave velocities ($c.\,4.75\,\text{km s}^{-1}$) in the shallow uppermost mantle ($<c.\,200$ km) and lower shear-wave velocity ($c.\,4.45\,\text{km s}^{-1}$) at greater depths (Cara 1979; Priestley 1999), which do not differ significantly from the global average. Others have lower velocities ($c.\,4.50\,\text{km s}^{-1}$) in the shallow mantle, which are close to the global average, but have high-velocity ($c.\,4.75\,\text{km s}^{-1}$) roots extending to depths as great as 400 km (Zhao *et al.* 1999). These very different velocity models have led various groups to propose completely different geological models for the deep structure of shields.

To a geologist who has not worked with seismic data, this situation is confusing. Even among seismologists, there is disagreement as to what features of the different models are genuine characteristics of the velocity structures beneath shields. For this reason, we thought it would be helpful to discuss what features can and cannot be well resolved by the various seismic methods as well as the seismic data that have been used. However, it was unexpectedly difficult in some cases to make accurate comparisons of the various models because they use different radial

From: FOWLER, C. M. R., EBINGER, C. J. & HAWKESWORTH, C. J. (eds) *The Early Earth: Physical, Chemical and Biological Development*. Geological Society, London, Special Publications, **199**, 45–64. 0305-8719/02/$15.00
© The Geological Society of London 2002.

velocity models, some of which are not tabulated. We found that most of the discrepancies between the published velocity models can result from differences in resolution alone.

For example, global models of the velocity structure (Su et al. 1994) based on long period surface waves and free oscillations cannot resolve short-wavelength features in either the vertical or horizontal velocity structure. If the Earth does in fact contain such short-wavelength variations, their seismic images are smeared over a large depth range by the method used for the inversion of long-period surface-wave and free oscillation data. Regional velocity models have better vertical resolution than do the global models because they use higher-frequency seismic waves. But until recently (Debayle & Kennett 2000a), they often resolved lateral variations poorly because they use few earthquakes with propagation paths that are largely horizontal. In contrast, teleseismic travel time tomography (James et al. 2001) can resolve horizontal velocity variations, but is almost entirely insensitive to the vertical velocity structure. We believe that these differences in resolution largely account for the various seismological models.

Archaean and Proterozoic shields form the oldest parts of the crust and lithosphere, and are therefore of particular concern to geologists. They are also of special interest to seismologists as they are the only regions where we have samples of mantle material from depths of at least 200 km whose wave speeds can be measured directly and can be compared with velocities estimated from seismic studies of shields. These samples come from kimberlite eruptions that bring mantle nodules to the surface from a variety of depths within the lithosphere. Because the lithosphere beneath the shields is thick and cold and the magma that transports the nodules is generated at its base, kimberlites bring up nodules from greater depths than do alkali basalts, which constitute the typical magma-type of eruptions found throughout the younger parts of continents.

A particularly important feature of the shields that is still in dispute is the depth to which the high seismic velocities beneath shields extend. Because the shields must carry their high-velocity roots with them as the continents drift, this question is also of major geodynamic significance. In the discussion below, we are principally concerned with the shields of southern Africa, partly because their seismic structure has been recently intensively studied by us and others (Qiu et al. 1996; Priestley 1999; Zhao et al. 1999; Ritsema & van Heijst 2000; James & Fouch 2002) and also because a great deal of petrological and geochemical work has been carried out on the mantle nodules from the kimberlite pipes.

The principal conclusions of Qiu et al. (1996) are that a high-velocity lid shown by seismic and petrological data exists in the upper mantle beneath southern Africa with lower shear-wave velocities at greater depths in the upper mantle. Although these findings have not changed, more recent seismic and laboratory studies have led to a better understanding of the physical state of the upper mantle beneath southern Africa. Priestley (1999) found from a re-examination of the data studied by Qiu et al. (1996) and the analysis of additional data that the seismic lithosphere beneath southern Africa could extend to an average depth of c. 160 km and still fit the data well. In this paper we use Brey & Kohler's (1990) expressions to estimate the pressure and temperature at which kimberlite nodules equilibrated, in contrast to Qiu et al. (1996), who used Finnerty & Boyd (1987) expressions to estimate pressure, and Bertrand & Mercier (1985) to estimate temperature. Brey & Kohler's expressions are based on laboratory experiments alone, and are consistent with the phase boundary between graphite and diamond measured by Kennedy & Kennedy (1976). Using Brey & Kohler's expressions, rather than those used earlier, slightly reduces the thickness of the mechanical boundary layer to 156 km (165 km) and that of the lithosphere to 176 km (185 km), where the values that Qiu et al. (1996) obtained using petrological arguments are given in parentheses.

Perhaps the most important change that has occurred since Qiu et al.'s study is in our understanding of seismic velocities at high temperature and seismic periods, (studied by Gribb & Cooper (1998, 2000) and Jackson (2000). Before those studies, laboratory measurements of seismic velocities were made at ultrasonic frequencies and the ultrasonic results extrapolated to seismic frequencies. New laboratory experiments made at seismic frequencies strongly suggest that the decrease in shear-wave velocity beneath the seismic lithosphere of the shields results from elevated temperature and is not an indication of the presence of melt (Gribb & Cooper 2000). This question is discussed in more detail below.

Seismic constraints on the upper mantle beneath southern Africa from regional seismic data

The Kalahari Craton of southern Africa, consisting of the Kaapvaal and Zimbabwe Cratons and the Limpopo Mobile Belt, has formed a stable unit for the past 2.3 Ga (McElhinny &

McWilliams 1977). Qiu et al. (1996) used earthquakes in southern Africa recorded at stations in Zimbabwe and South Africa (Fig. 1) to obtain an average velocity model for southern Africa. The main features of their model are a high shear-wave velocity lid in the upper mantle shown by both seismic and petrological data, below which there is a decrease in the shear-wave velocity shown by the seismic data. Priestley (1999) re-examined the seismograms studied by Qiu et al. (1996) and included additional data to determine how well constrained was the thickness of the high-velocity lid. He showed that the lid thickness could be increased to no more than 160 km (Fig. 2) and still fit the observations. The crustal structure of the model is constrained using published seismic refraction results and receiver function analyses, and the upper-mantle structure of the model by S_n velocity measurements, surface-wave dispersion and waveform modelling of seismograms of regional earthquakes. The average velocity model for southern Africa

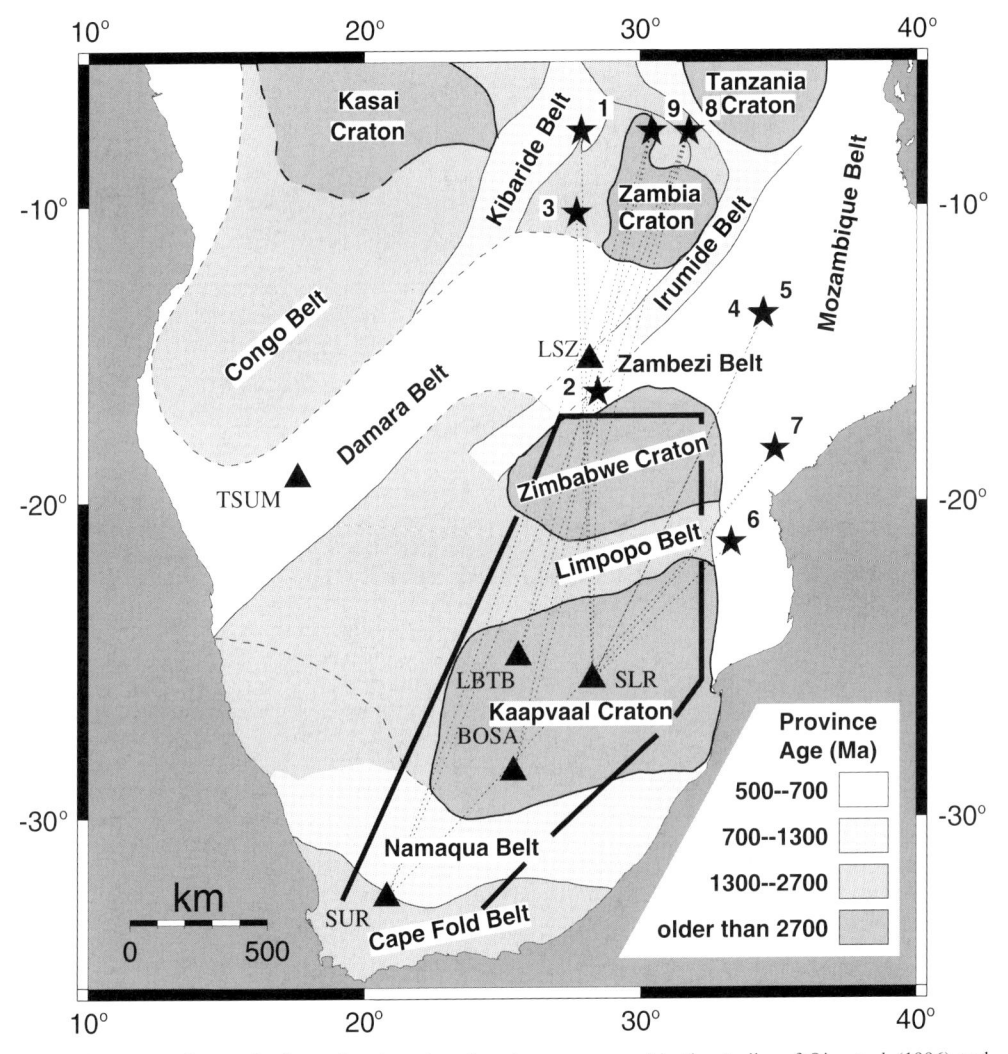

Fig. 1. Source–receiver paths for regional earthquake seismograms used in the studies of Qiu et al. (1996) and Priestley (1999) superimposed on the major crustal subdivisions of southern Africa. ▲, seismograph stations; ★, earthquake locations; dotted lines denote paths for the events used in the regional waveform modelling. Events 1–8 were used by Qiu et al. (1996); event 9 was used by Priestley (1999). Fundamental mode Rayleigh wave phase velocity dispersion from Priestley (1999) was measured for the SUR–BOSA and BOSA–LBTB paths. The bold lines denote the extent of the South African array.

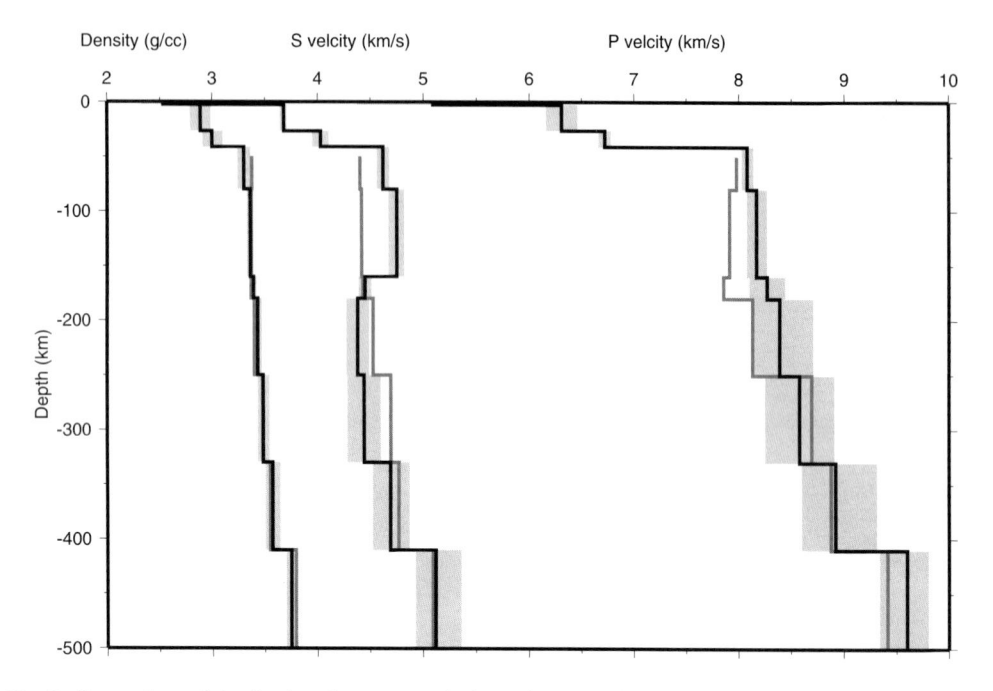

Fig. 2. Comparison of the density, shear-wave velocity and compressional-wave velocity profiles beneath southern Africa from Priestley (1999) (bold continuous lines), and the density and velocity profiles for PREM (fine continuous lines). The shaded area denote estimates of the uncertainties in the density and velocity model of Priestley (1999) derived from the waveform fitting tests, the earthquake location errors as described by Qiu et al. (1996), and c. 2% anisotropy as proposed by Vinnik et al. (1995).

(Fig. 2) from this analysis has a 42 km thick crust and a 120 km thick high-velocity upper-mantle lid, giving a total thickness of 162 km for the seismic lithosphere. The P and S velocities beneath the Moho are 8.09 km s^{-1} and 4.62 km s^{-1}, respectively and the compressional and shear velocity gradients in the lid are 0.0008 s^{-1} and 0.0013 s^{-1}, respectively. Below the lid, the S-wave velocity drops to at least 4.45 km s^{-1} at 250 km depth, but no decrease in the P-wave velocity is required by the data. Below 160 km depth, the P-wave gradient increases to 0.0015 s^{-1} and increases again to 0.0035 s^{-1} between 250 km depth and the 410 km discontinuity.

The upper-mantle velocity structure shown in Figure 2 is rougher in the vertical direction than are those of the global tomographic models (Su et al. 1994; Masters et al. 1996). The high velocity mantle lid is required to match both the high S$_n$ velocity observed in southern Africa and the propagation of intermediate period fundamental mode Love and Rayleigh waves from both teleseismic and regional earthquakes. However, lower velocity must exist beneath the lid structure to account for the waveform and arrival time of the higher-mode Love and Rayleigh waves, which are sensitive to deeper structure.

This sensitivity is illustrated in Figure 3, which shows a comparison of synthetic seismograms for four lid models with the seismograms observed at three distance ranges. When the base of the lid is at 160 km, there is a good match of the synthetic and observed higher-mode waveform at 1038 km distance (Fig. 3a). However, there is only a small additional increase in amplitude and no significant change in the travel time at this distance if the base of the lid is increased to 180 or 220 km depth. There is a correspondence between mode and ray representations of the seismic wavefield. In terms of the ray representation, the lack of sensitivity of the synthetic waveform to the velocity structure below 180 km occurs because the S waves that arrive in this distance range consist of rays turning at depths of less than 180 km. A similar comparison at 1480 km distance (Fig. 3b) shows that extending the base of the lid to 180 or 200 km depth advances the travel time and increases the amplitude of the synthetic waveform with respect to the observed waveform, resulting in a poor fit. The higher modes observed in this distance range are equivalent to an S wave turning near the base of the lid. Increasing the lid thickness increases both the arrival time and the amplitude of the S wave.

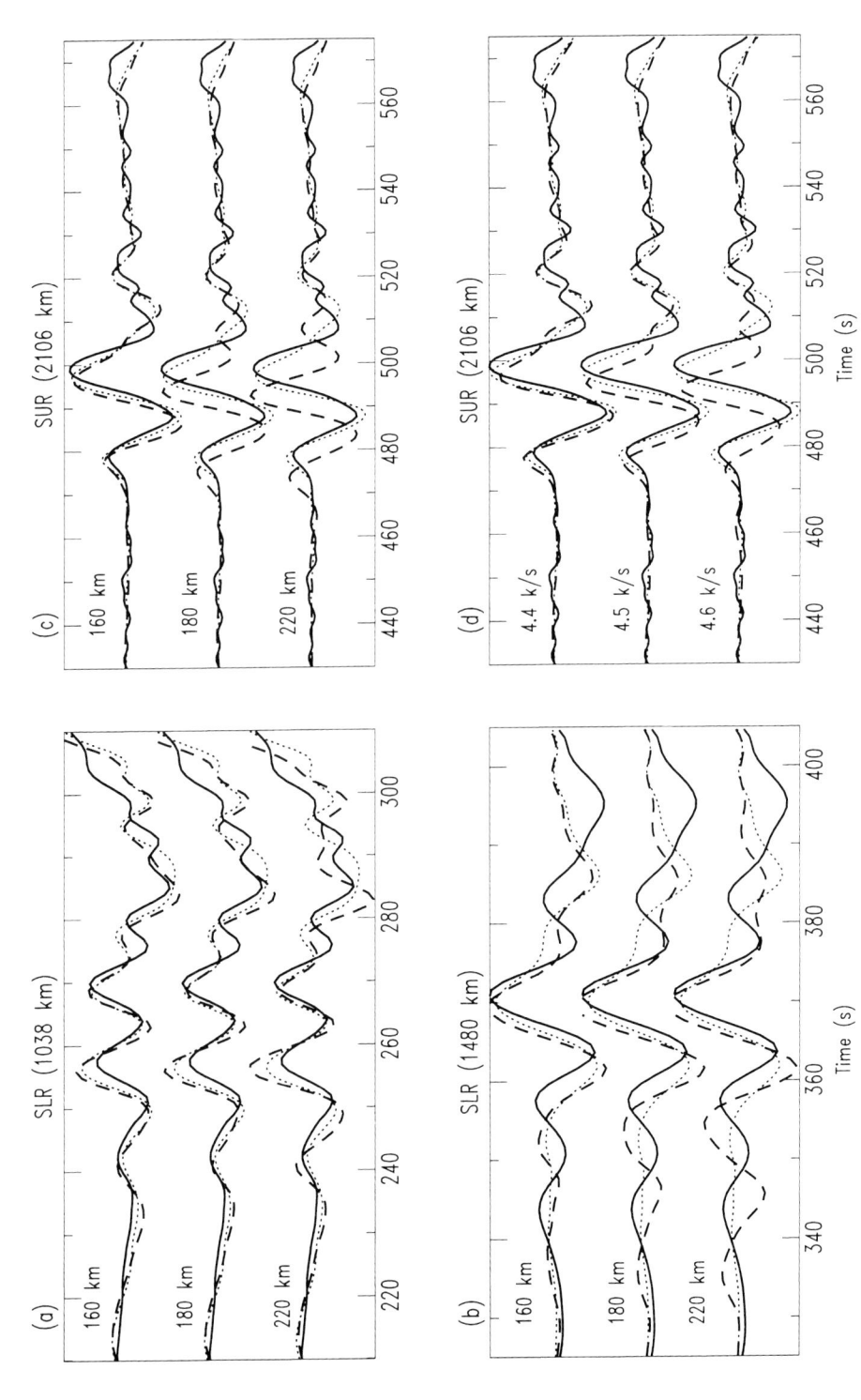

Fig. 3. (**a**) Sensitivity test of the higher-mode waveforms to the depth to the base of the upper-mantle lid for the SLR seismogram of the 18 July 1986 earthquake (Fig. 1, event 2). The continuous line is the observed waveform, the dotted line is the synthetic for the southern Africa velocity model of Qiu *et al.* (1996), and the dashed line is the synthetic for the same velocity model but with the lid base increased to the depth indicated at the left of each seismogram. (**b**) Same as (**a**) but for the SLR seismogram of the 10 March 1989 earthquake (Fig. 1, event 5). (**c**) Same as (**a**) but for the SUR seismogram of the 24 July 1991 earthquake to the minimum S-wave velocity of the low-velocity zone (LVZ).

Rayleigh modes displacement amplitues T=15s

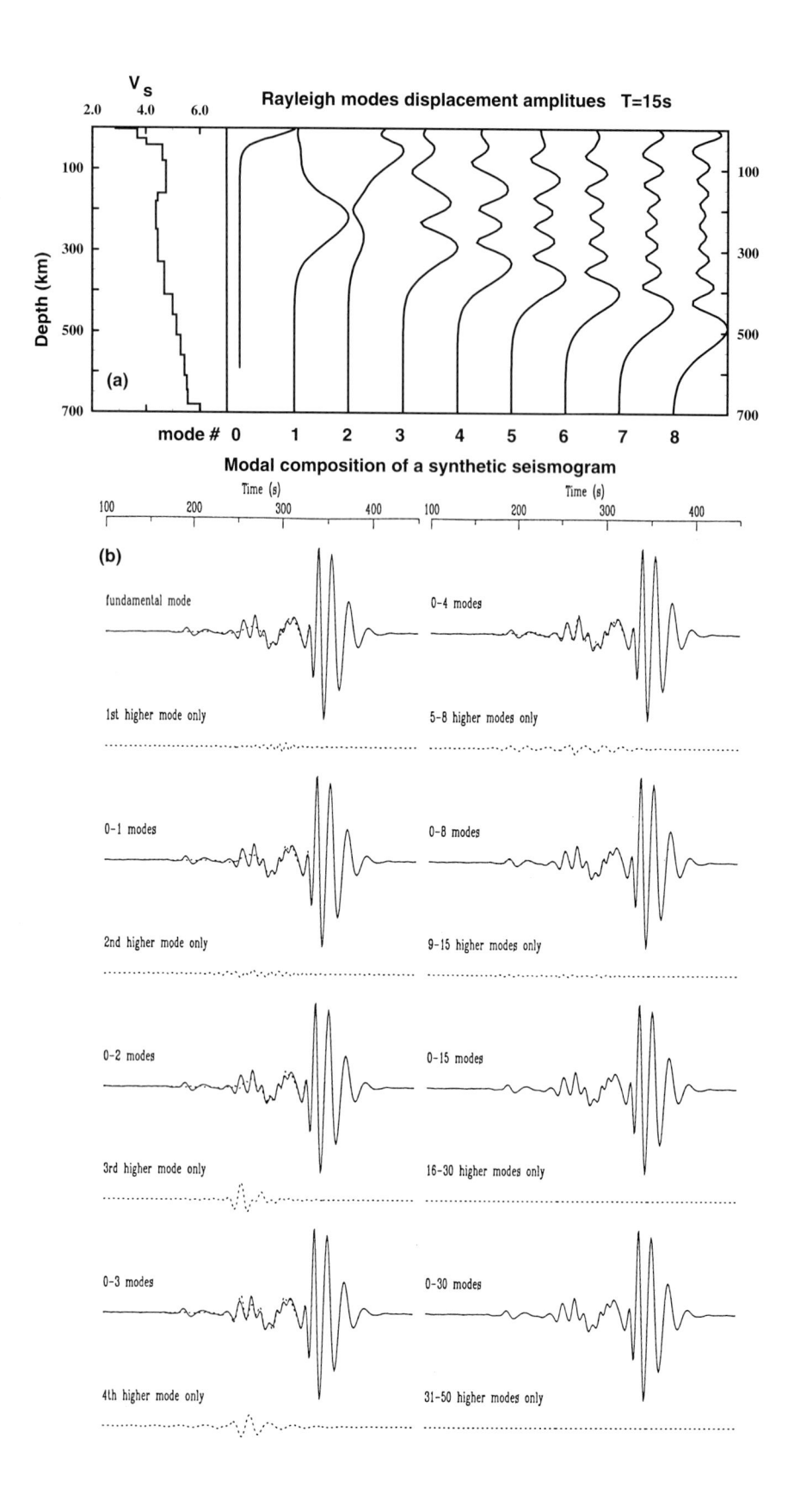

(a)

V_s

mode # 0 1 2 3 4 5 6 7 8

Depth (km)

Modal composition of a synthetic seismogram

(b)

fundamental mode

1st higher mode only

0-1 modes

2nd higher mode only

0-2 modes

3rd higher mode only

0-3 modes

4th higher mode only

0-4 modes

5-8 higher modes only

0-8 modes

9-15 higher modes only

0-15 modes

16-30 higher modes only

0-30 modes

31-50 higher modes only

The comparison at 2106 km distance (Fig. 3c) shows an acceptable match of the synthetic and observed waveforms with the base of the lid at 160 km depth. Thicker lids advance the arrival time of the wavepacket, but the waveform shape is not altered. At this distance range, the higher modes are equivalent to an S wave turning in the mantle transition zone. Increasing the lid thickness reduces the delay time caused by the lower shear-wave velocities beneath the lid, resulting in an advance in the arrival time but no change in the amplitude at this epicentral distance.

However, the details of the velocity structure beneath the lid are not well constrained. Figure 3d shows that the minimum velocity at these depths can range from 4.35 to 4.45 km s^{-1} and has only a small effect on the synthetic fits to the observed waveforms at 2106 km distance, but increasing the minimum velocity to 4.5 km s^{-1} or greater results in an unacceptably early higher-mode arrival for the synthetic seismograms compared with the observed seismograms at this distance.

These tests show that the average thickness of the seismic lithosphere (crust plus upper-mantle lid) can be as much as c. 160 km and the minimum S-wave velocity beneath the lid can be as high as c. 4.45 km s^{-1} and still produce synthetic waveforms that match the observed waveforms. A thicker lid or higher S-wave velocities at depth below the lid are not consistent with the regional seismic waveforms. It is the high S-wave velocity lid that is unique to the upper mantle of the shield; below that, the S-wave velocity is not significantly different from PREM (Preliminary Reference Earth Model) (Fig. 2).

These tests demonstrate the sensitivity of the higher-mode data to the upper-mantle structure. Figure 4a shows the displacement amplitude v. depth for the fundamental and first eight higher Rayleigh modes at 15 s period for the southern African model shown at the left in the figure. The fundamental mode at this period is sensitive to the velocity structure in the top c. 50 km of the model, whereas the higher modes are sensitive to the velocity structure as deep as the 410 km discontinuity. Figure 4b shows the modal composition of the synthetic seismogram that matches the observed seismogram for the 860718 earthquake (event 2, Fig. 1) recorded at SLR at 1038 km distance. At least the first eight higher modes are required to synthesize the observed higher-mode waveform, and this, together with the displacement amplitude v. depth in Figure 4a, indicates that the waveform is sensitive to structures as deep as the upper-mantle transition zone or the turning point of the equivalent S wave. The ray paths that Priestley (1999) used are shown in Figure 1 and have little horizontal resolution over this region. Hence, the velocity model for the Kalahari Craton (Fig. 2) will be satisfactory only if the lateral velocity variations are small. In contrast, the resolution in depth is good because of the higher-frequency content of the multimode regional seismograms.

Zhao *et al.* (1999) derived a velocity model for the upper mantle and transition zone beneath southern Africa from regional seismic data. Their primary data were travel times and waveforms from the 30 October 1994 mine tremor in South Africa recorded at broadband stations of the Tanzania Broadband Seismic Experiment. These data were supplemented with arrival time data from other events in the region taken from the Bulletin of the International Seismological Centre. Zhao *et al.* (1999) determined the P-wave velocity model from these travel time data using Herglotz–Wiechert inversion and compared synthetic waveforms for this model with the observed waveforms. An S-wave model was formed from the P-wave model assuming a constant Poisson's ratio with depth of 0.27 and travel time curve for the S-wave model was compared with the observed travel time data. The S-wave model has a velocity 4.55 km s^{-1} just below the Moho and high velocity in the upper mantle. Figure 5 compares mode summation synthetic seismograms for the Zhao *et al.* (1999) and Priestley (1999) southern Africa models with

---

Fig. 4. (**a**) Normalized eigenfunction amplitudes for the fundamental and first eight higher Rayleigh modes at 15 s period (0.067 Hz) for the southern African crust and upper-mantle velocity model shown at the left in the figure. The first and second higher modes are channel waves with their displacements largely confined to the upper-mantle LVZ. (**b**) Model composition of the vertical component synthetic seismograms for the South African model which match the observed seismogram at SLR for the 860718 earthquake (Event 2, Fig. 1). The waveform denoted by the continuous line is a sum of the fundamental and first 50 higher modes. The waveform denoted by the dotted line superimposed on the continuous trace is a sum of the number of modes indicated at the upper right of the synthetic siemogram. The waveform denoted by the single dotted line consists just of the indicated modes and shows the amount of vertical displacement by which two adjacent seismograms differ. The channel waves (higher modes 1 and 2) contribute very little to the surface seismogram. There is no difference between the synthetic seismogram with eight and 50 higher modes. The 860718 seismogram at SLR can be modelled with a synthetic seismogram consisting primarily of the fundamental and first eight higher modes.

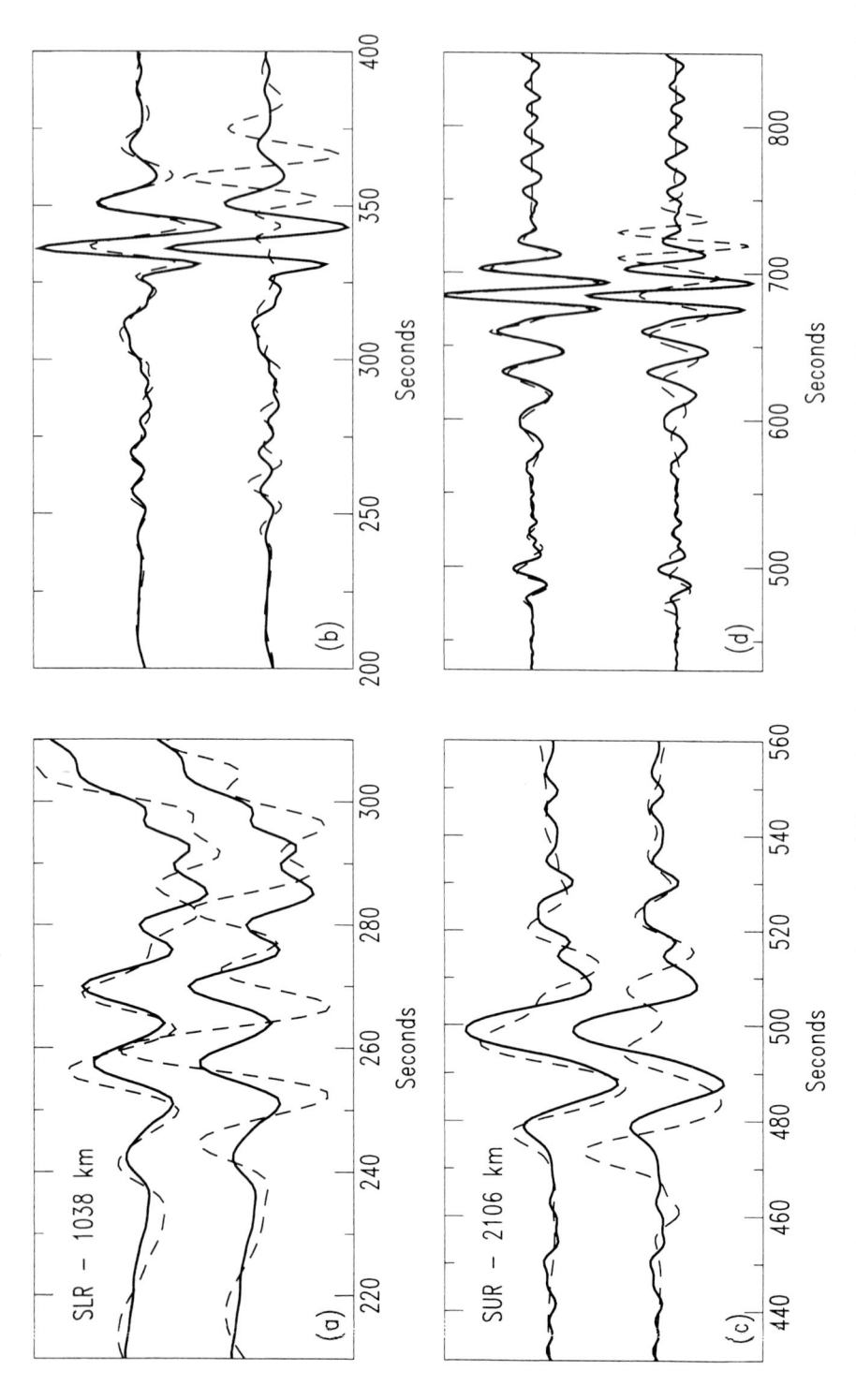

Fig. 5. Comparison of vertical component seismograms for southern Africa earthquakes recorded at two distances: (**a**), (**b**) 1038 km (event 2 recorded at SLR, Fig. 1) and (**c**), (**d**) 2106 km (event 7 recorded at SUR, Fig. 1), and synthetic seismograms computed for the southern Africa models of Priestley (1999) and Zhao *et al.* (1999). (**b**) and (**d**) show the match of the observed waveform (continuous line) to the synthetic waveforms (dashed lines); (**a**) and (**c**) show an enlarged plot of the higher-mode wave train. In each panel the upper waveforms are the comparison for the Priestley (1999) model and the lower waveforms are the comparison for the Zhao *et al.* (1999) model.

observed seismograms at c. 1000 and c. 2000 km distance. At both distances the synthetic higher modes computed for the Zhao et al. (1999) model arrive early and have larger amplitude, and the fundamental mode model arrives late with respect to the observed seismogram, indicating that velocities in the shallow upper-mantle part of their model are too slow and velocities in the deeper upper mantle are too fast.

James & Fouch (2002) showed intermediate period fundamental mode Rayleigh wave phase and group velocity data measured across the Kalahari Craton and a shear-wave velocity model from inversion of these data. They find evidence for a weak low-velocity zone below 120–130 km depth. The phase velocity data of James & Fouch (2002) are not significantly different from the Rayleigh wave phase velocity measured by Priestley (1999). However, such intermediate period fundamental mode dispersion data do not provide stringent constraints on mantle velocities below c. 200 km depth.

The effect of anisotropy on the seismic velocity structure

Significant anisotropy in the lithosphere or upper mantle beneath southern Africa could bias the velocity model in Figure 2. There are two forms of anisotropy: azimuthal anisotropy in which waves of the same type have a wave speed that is a function of azimuth, and polarization anisotropy or transverse isotropy, in which waves of different polarization (e.g. Love and Rayleigh waves) propagate with different wave speeds along the same great circle path. Low-frequency surface-wave observations (Nataf et al. 1984; Montagner 1996) show weak polarization anisotropy in the upper mantle beneath southern Africa. Vinnik et al. (1995) measured shear-wave splitting at seven sites on the Kaapvaal Craton and found an average delay of 0.9 ± 0.5 s. From this they concluded that the anisotropy is c. 2%, and is localized between 150 and 400 km depth, with the fast direction approximately parallel to the direction of the current plate motion. This result implies that the material in this depth range has been deformed by plate motion in geologically recent time. A similar result has been observed for Australia (Debayle 1999; Debayle & Kennett 2000a, b) from surface-wave analysis where the depth constraint of the anisotropy is better than that of the shear-wave splitting observation.

In many studies, isotropic Earth models cannot fit both the Love and Rayleigh wave data, and Anderson & Dziewonski (1983) pointed out that inverting Rayleigh wave data separately in such cases can produce a low-velocity zone in the resulting velocity model. The simultaneous match of the Love (SH) and Rayleigh (P–SV) waveforms (Fig. 6) with the same isotropic velocity model shows that, if polarization anisotropy is present beneath southern Africa, its effects are not sufficiently strong to create a pronounced discrepancy between the Love and Rayleigh waves. Thus, our data suggest that polarization anisotropy is low or not present, and that the velocity model shown in Figure 2 is not strongly biased by our assumption of isotropic velocity structure. This result is consistent with that of Bloch et al. (1969), who also found no discrepancy between the Love and Rayleigh waves dispersion in southern Africa, and with that of Brune & Dorman (1963), who fitted Love and Rayleigh wave dispersion for the Canadian Shield with a single upper-mantle velocity model. However, the earthquakes we examined all lie north to NE of the seismographs, and the paths studied sample only a small range of azimuths. Consequently, we have no constraint on azimuthal anisotropy. If c. 2% anisotropy exists in the 150–400 km depth range as suggested by Vinnik et al. (1995), its presence could modify somewhat the velocity structure shown in Figure 2. Their observation of anisotropy in this depth range with the fast direction approximately aligned with the current plate motion direction implies that the material involved is being deformed by mantle flow and is therefore part of the asthenosphere.

Comparison with global mantle models for southern Africa

Figure 4 shows that the sensitivity of the theoretical seismogram to the detailed vertical S-wave velocity structure arises only because the modelling has included higher modes whose wavelength in the vertical direction is short enough (c. 100 km) to be sensitive to the velocity structure in Figure 2. If we had used only longer-period fundamental mode surface waves, we would have been unable to resolve the lid structure and its underlying low-velocity zone, and would have smeared the high-velocity lid over a vertical distance controlled by the vertical wavelength of the higher modes at the shortest period used. The global tomographic models (Su et al. 1994; Masters et al. 1996) use near-vertically travelling body waves and long-period surface waves. Both produce more smearing of the velocity structure in the vertical direction than do to the higher-frequency regional waves we

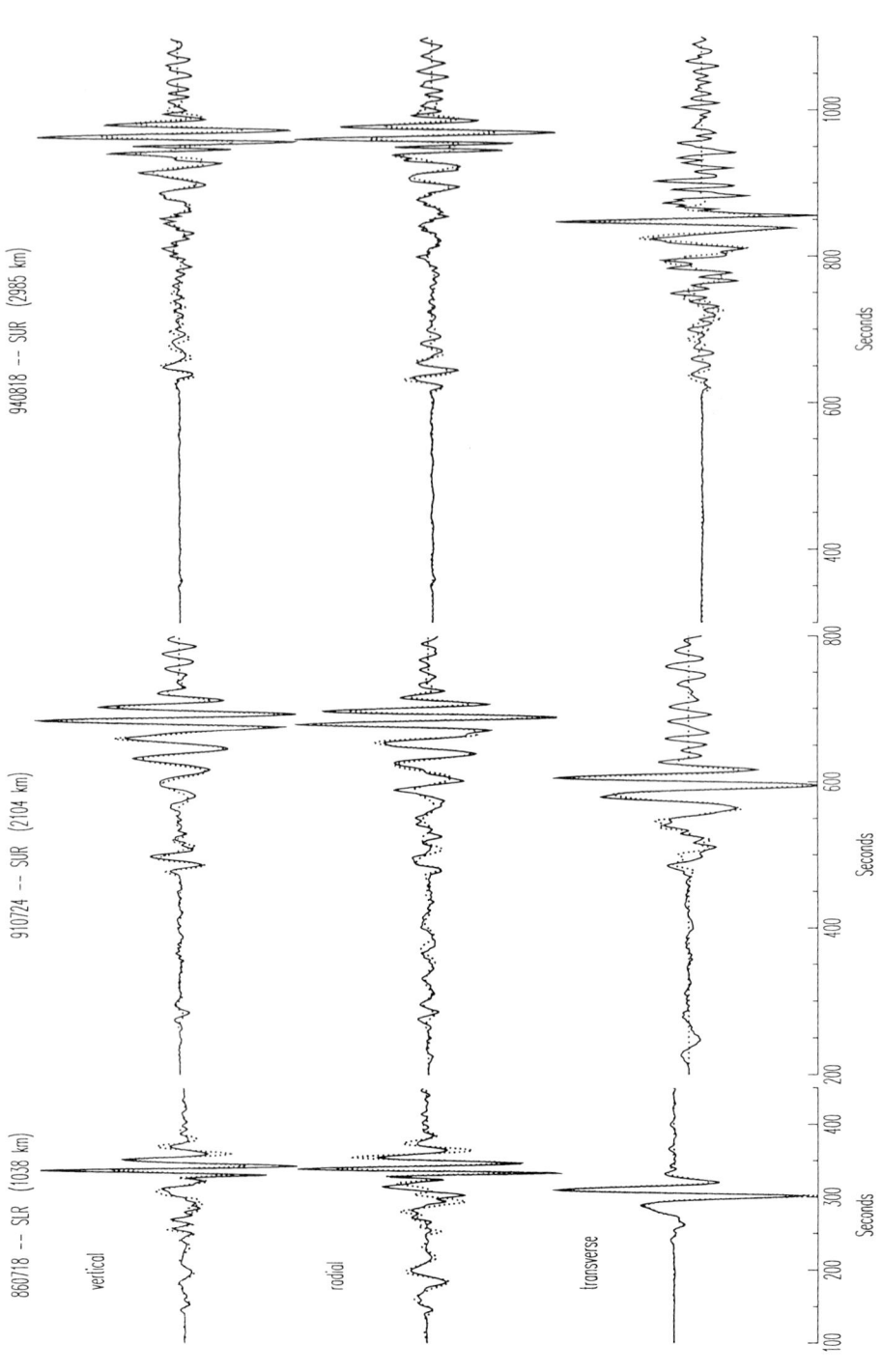

Fig. 6. Three-component waveform fits at three distance ranges for synthetic seismograms (dotted line) computed from the southern African velocity model of Priestley (1999) and the observed seismograms (continuous lines). The Love and Rayleigh wave seismograms are fitted with the same velocity model, implying that at least the upper-mantle lid is isotropic. Event 860718 is event 2, Figure 1; 910724 is event 7, Figure 1; and 940818 is event 8, Figure 1.

have used. Figure 7 shows that features in the upper mantle are smeared about 200 km vertically in the global tomographic models. This is the reason global tomographic models have higher-than-average shear-wave velocities that extend to 400–600 km depth beneath the cratons. Such structures have been cited as evidence for deep, high-velocity cratonic roots and as support of the deep tectosphere model (Jordan 1975). However, models for the cratons based on regional seismic data (i.e. Brune & Dorman 1963; Cara 1979; Qiu et al. 1996; Priestley 1999) with better vertical resolution show that the high S-wave velocity upper-mantle lid overlies a low S-wave velocity layer whose velocity is similar to PREM, the world-wide average.

If the variation among the velocity structures results from differences in resolution, then at low resolution these structures should be similar. A convenient measure of such an averaged structure is the difference between the travel time of an S-wave travelling vertically through the velocity model and through PREM. Although this method of averaging is not exactly equivalent to that achieved by using long-period modes, the equivalence between the mode and ray descriptions suggests it is a suitable approximation, and the vertical travel time is easily calculated. It also provides a useful method of comparing models obtained from regional studies with velocity models derived from global inversions of P- and S-wave travel times.

Model S12WM13 (Su et al. 1994) was obtained from the inversion of body and mantle waveforms and differential and absolute travel time residuals. The model is expanded to spherical harmonic degree 12 to describe the horizontal variation and to Chebyshev polynomial order 13 to describe the radial variation. Model S12WM13 is given in terms of a relative deviation from PREM. It shows higher shear-wave velocities than in PREM, extending to about 500 km depth beneath a broad region around southern Africa. We constructed an average southern African upper-mantle velocity model for S12WM13 by averaging six velocity profiles computed from the model for an area beneath the cratonic region of southern Africa. The resulting vertical S-wave travel time between depths of 42 and 400 km is 76.33 s, and the mean S-wave vertical travel time through a number of shields is 77.26($\pm$0.82) s. Making the same crustal correction as that made for S12WM13 (Woodhouse & Dziewonski 1984), we find that the vertical S-wave travel time through the model in Figure 2 is 77.65 s. There is, therefore, no significant difference between these global velocity models and that in Figure 2.

Figure 8 shows synthetic seismograms computed for a modified S12WM13 model compared with seismograms of the 14 August 1994 earthquake recorded at three distance ranges between 2000 and 3000 km across southern Africa. These synthetics are computed for the velocity model in Figure 2 for depths less than 160 km and the southern African upper-mantle model from S12WM13 below 160 km depth. The synthetic seismogram for this model matches the observed fundamental mode surface waves as they are primarily controlled by the crust and shallow mantle structure. At 2000–3000 km distance, the higher-mode surface waves correspond to an S-wave turning in the mantle transition zone, and hence their arrival times are controlled by the S-wave delay time across the upper mantle. Figure 8 shows that the synthetic S wave arrives

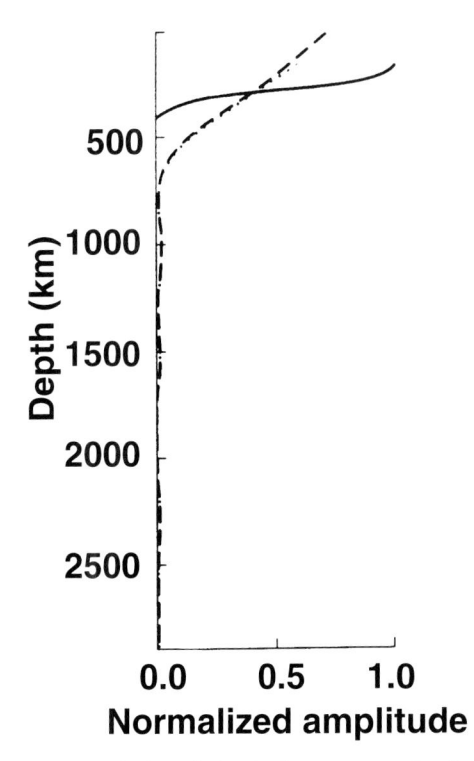

Fig. 7. Vertical resolution in the upper mantle of model S16B30 of Masters et al. (1996). The input velocity anomaly is a linear combination of three natural B-splines in radius and Y_7^4 spherical harmonics laterally. The continuous curve shows the amplitude of the input anomaly as a function of depth, normalized to unit amplitude. The dashed curve is the peak recovered value. The data inversion tends to smear the anomaly c. 200 km radially. Adapted from Masters et al. (1996).

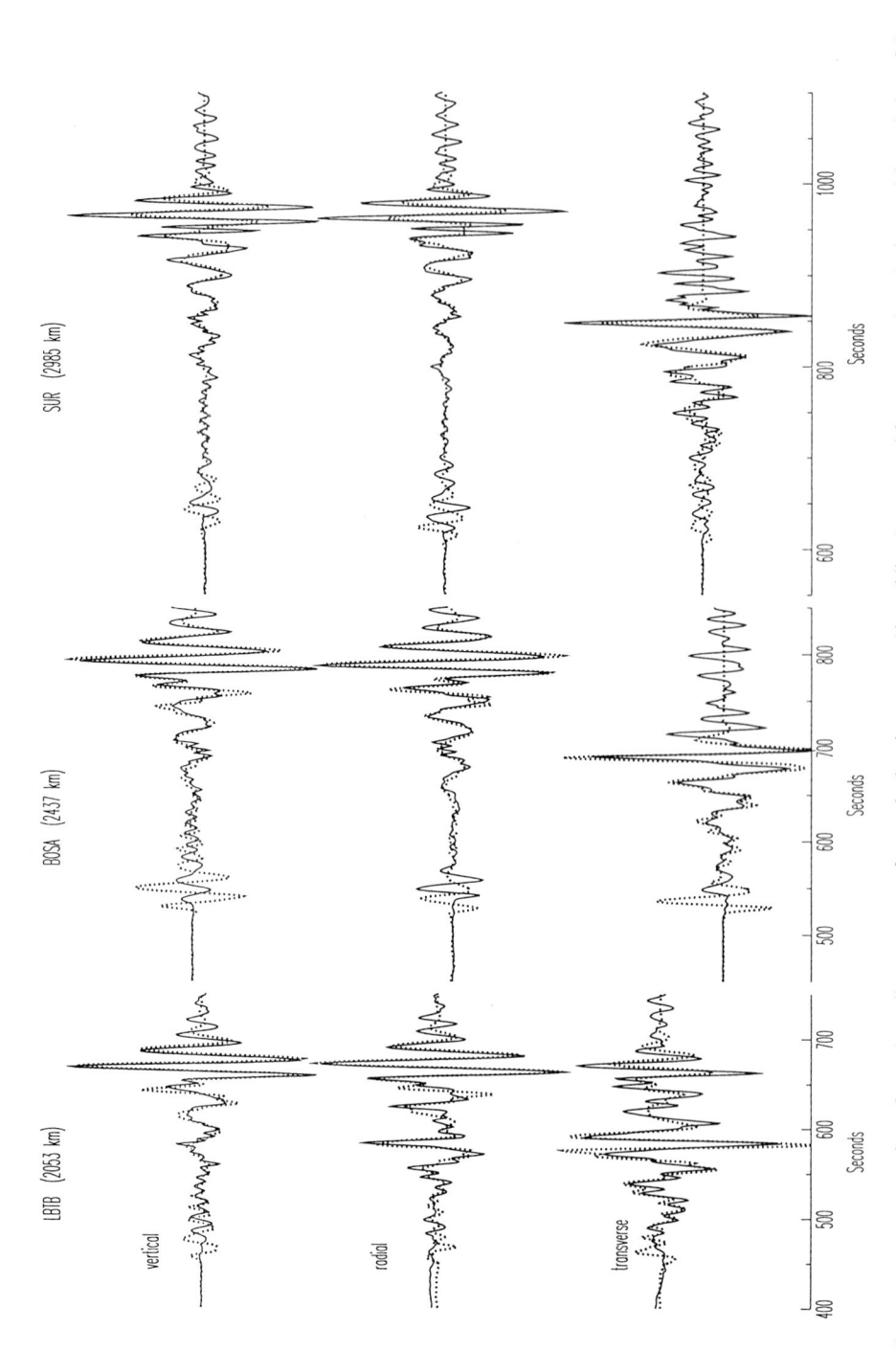

Fig. 8. Three-component waveform fits at three distance ranges for synthetic seismograms (dotted lines) computed from composite velocity model consisting of the seismic lithosphere of the southern African model of Priestley (1999) above 160 km depth and the southern Africa upper-mantle model derived from the global tomographic model SI2WM13 below 160 km depth compared with the observed seismograms (continuous lines) of the 14 August 1994 earthquake (Fig. 1, event 8) recorded at three distance ranges.

early with respect to the observed S wave. Lowering the shear-wave speed of the upper-mantle lid to $c.\ 4.55\ \mathrm{km\ s^{-1}}$ as in S12WM13 would improve the match of the synthetic and observed mantle S wave but results in a poor fit of the fundamental mode surface waves and also disagrees with the high-frequency S_n travel time observations. Therefore, the smooth upper-mantle velocity model of S12WM13 does not produce synthetic seismograms that match the observed ones.

Model S16B30 (Masters *et al.* 1996) was derived from the inversion of absolute and differential travel time residuals, surface-wave dispersion and polarization measurements, and mode structure coefficients. The model is parameterized laterally by spherical harmonics expanded up to degree 16, and by 30 natural cubic B-splines in radius. Model S16B30 is specified in terms of the percent deviation from an average spherical velocity model, which is not retained in the inversion. Therefore, a direct comparison of the integrated travel times cannot be made for S16B30. Only indirect comparisons are possible, two of which are made below. The velocity profile for southern Africa from S16B30 was produced in the same way as S12WM13 except Earth model 1066a (Gilbert & Dziewonski 1975) was used as the reference Earth model. Model 1066a was used rather than PREM because it is more similar to the average spherical reference model of S16B30 (Laske, pers. comm., 1998). For this case, the vertical S-wave travel time between 42 and 400 km depth is 77.95 s. Masters *et al.* (1996) corrected their data using the global crustal model of Mooney *et al.* (1998). Assuming this crustal correction is the same as that of Woodhouse & Dziewonski (1984), and using Mooney *et al.*'s (1998) value for the thickness of the South African crust, rather than that of PREM-C (Dziewonski *et al.* 1975), the vertical S-wave travel time for the velocity model in Figure 2 is 78.12 s. The mean vertical S-wave travel time for the shields in model S16B30, assuming the average spherical model 1066a is the reference model, is 77.98($\pm$0.89) s. Again, the difference between the two is not significant.

Masters *et al.*'s (1996) model and that in Figure 2 can also be compared by computing the difference in the vertical S-wave travel time between southern Africa and an old ocean basin for each model. Using Nishimura & Forsyth's (1989) model for an ocean basin whose age is greater than 110 Ma, the difference between the travel times between 42 and 400 km is -1.00 s for model S16B30 and -2.01 s for the velocity model shown in Figure 2.

These comparisons show that the vertical S-wave travel times through the southern Africa shear-wave models derived from the regional and global data are not significantly different. What variations exist may be a result of the way in which the crustal corrections are made. However, the model derived from regional seismic data is rougher in the vertical direction, having a high-velocity upper-mantle lid overlying lower shear-wave velocities at depth, whereas the global models smooth this variation over the whole of the upper mantle. Neither the global nor the regional models have good horizontal resolution in southern Africa.

The similarity of the vertical S-wave travel times in the regional and global models for southern Africa suggests that the deep, broad, high-velocity root beneath this region in the global tomographic models results from vertical averaging. Such averaging of the vertical velocity structure is evident in other parts of the global models. For example, both S12WM13 and S16B30 show low S-wave velocity structures associated with the mid-ocean ridges extending to depths of more than 300 km. Mid-ocean ridges are passive features caused by lithospheric stretching. Hot asthenospheric material rises to fill the void resulting from the stretching, and the magmatism observed along the ridge crest is caused by decompression melting at depths shallower than the base of the plate ($c.\ 100$ km, McKenzie & Bickle 1988). Hence, the low S-wave velocities associated with the mid-ocean ridges should not extend to depths of $c.\ 350$ km.

Recent global tomographic models (Ritsema & van Heijst 2000) include intermediate-frequency surface-wave data. These models show the high shear-wave velocities extending to no more than 200 km depth beneath the Kalahari Craton with PREM-like velocities below 200 km depth, consistent with the model shown in Figure 2.

Teleseismic travel time tomography analysis of the South African Craton

Recent work in southern Africa has focused on developing a teleseismic travel time tomography image for the upper mantle (James *et al.* 2001; James & Fouch 2002). The objective of teleseismic travel time tomography is to determine a 3D seismic image for a volume of Earth beneath the seismic array based on the arrival times from a large number of source–receiver combinations. The 3D model is normally expressed in terms of perturbations to a reference model in which the velocity is a function only of radius, where the 3D perturbations account for that part of the travel time not explained by the reference model.

The depth to which a tomographic image extends is dependent upon the dimension of the seismic array and the resolution of the tomographic image depends on the density of crossing rays. The tomographic images of southern Africa (James et al. 2001; James & Fouch 2002) show a high-velocity contrast to 400–600 km depth.

Teleseismic travel time tomographic images are determined from the inversion of relative travel time residuals so as to minimize bias resulting from errors in the earthquake location and origin time and the effects of 3D structures outside the Earth volume of interest. A second reason for using relative arrival times is that they can be determined more precisely than absolute arrival times in the presence of noise. The relative travel time is given by

$$\Delta t_{ij} = t_{ij}^{\text{obs}} - t_{ij}^{\text{ref}} - t_{ij}^{\text{elev}} - \langle t_j \rangle$$

where t_{ij}^{obs} is the observed arrival time for the ith station for the the jth event, t_{ij}^{ref} is the predicted arrival time for some reference Earth model, t_{ij}^{elev} is the elevation correction for the ith station, and $\langle t_j \rangle$ is chosen separately for the event such that the sum of the travel time residual associated with each event is zero. The earthquake location and origin time are also redetermined in travel time tomography inversion to absorb any additional effects of location errors and structures outside the Earth volume under consideration.

The use of relative travel time residuals and of the relocated hypocentres of earthquakes causes the travel time contribution of any horizontal velocity structure common to the whole array to be absorbed in the origin time correction $\langle t_j \rangle$. Therefore, only lateral differences in the velocity structure can be resolved. For this reason, travel time tomography at best recovers the effects of lateral velocity variations beneath an array of seismographs. The South African array shown in Figure 1 is on the Kalahari Craton, and thus any horizontally averaged feature of the velocity model will not affect the tomographic image. Hence, the velocity model in Figure 2 which is the horizontally averaged velocity structure of the whole craton, has no effect on the travel time anomalies between different stations in the array and is therefore completely invisible to this method of studying velocity variations within the Earth.

Positive regions of the tomographic image are normally interpreted as high-velocity regions in the Earth, and negative regions are seen as low-velocity regions. But, as both Aki et al. (1977) and Leveque & Masson (1999) have pointed out, teleseismic travel time tomography does not retrieve the 3D structure of a volume beneath the array. It can only recover the velocity contrast relative to the horizontal average of the velocity. Tomographic images can therefore be misleading because the velocity contrast shown in the image is relative to an often unknown horizontally averaged velocity that varies with depth. Thus, a uniform contrast in a tomographic image does not imply a uniform velocity, and a positive velocity contrast in a tomographic image at one depth may signify a lower absolute velocity than a negative velocity contrast at a different depth.

Another consideration in interpreting tomographic images is that of estimating how close the retrieved tomographic image is to the actual Earth structure. The effects of smearing of structure along ray paths when there is insufficient crossing ray coverage are well known. Formal resolution parameters for tomographic models are not often determined because of the size of the computational problem. The usual approach is to estimate resolution by inverting synthetic data for a simple velocity model; the similarity of the synthetic input model and inversion output image is then taken as an estimate of the data resolution. It is generally accepted that if the inversion of the synthetic data can retrieve small-scale structure in, for example, a chequerboard test, then large-scale structure will be well resolved. However, Leveque et al. (1993) have shown that this is not the case, and that such synthetic experiments can be misleading. They used simple models to demonstrate that large-scale structures are not always more easily retrieved than are small-scale structures.

An important advantage of body-wave tomography using travel times is that, under some conditions, it can provide better horizontal resolution than the global and regional studies. The regional paths in Figure 1 were chosen to sample cratons, but there is no *a priori* reason why the lithospheric structure should correspond to the surface geology; however, the match between the synthetic and observed waveforms in Figure 6 and the fits shown by Qiu et al. (1996) and Priestley (1999) implies that there is, in general, such a correspondence. Travel time tomography can, in principal, provide a more severe test than can seismograms from regional earthquakes. In practice, however, this is not the case in southern Africa because the rays used for such tomography come largely from distant earthquakes and therefore have steeply dipping ray paths beneath the southern African stations. Such datasets have few rays that intersect at shallow depths, and therefore have poor vertical resolution. In these circumstances, it is difficult to distinguish travel time anomalies that arise from horizontal velocity variations on the vertical part of the ray path in the top 200 km from anomalies produced by deeper velocity variations. Doing so in southern Africa is especially difficult because almost all the seismic stations are on the Kalahari Craton.

Petrological constraints on the structure of the Kalahari Craton

Upper-mantle seismic velocities depend on the composition, pressure and temperature of the rocks through which the seismic waves pass. Many kimberlites erupted through the Kalahari Craton have brought up mantle nodules. The mineralogy of such nodules can be used to estimate the pressure and temperature at which they equilibrated, and their density and seismic velocities can then be calculated. Qiu *et al.* (1996) used Finnerty & Boyd's (1987) expression for the solubility of aluminium in enstatite in the presence of garnet to estimate the pressure and Bertrand & Mercier's (1985) expression for the intersolubility of enstatite and diopside to obtain the temperature. Brey & Kohler (1990) used the observed mineral compositions from a large number of laboratory experiments carried out at known pressure and temperature to obtain improved estimates of the reaction constants. They showed that Finnerty & Boyd's (1987) expression overestimates the pressure by about 0.5 GPa between pressures of 3 and 5 GPa, corresponding to a depth error of about 15 km. Brey & Kohler's (1990) expressions were used to obtain the geotherm beneath the Kaapvaal Craton shown in Figure 9a. The resulting thickness of the mechanical boundary layer is 156 km, and the lithospheric thickness is 176 km. These are both 9 km less than the values obtained by Qiu *et al.* (1996) using Finnerty & Boyd's expression. The uncertainty in the mean thickness of the mechanical boundary layer and lithosphere beneath the Kaapvaal Craton are probably both about 10 km.

A useful check on the accuracy of pressure and temperature estimates from nodules is to obtain such estimates from graphite- and diamond-bearing nodules. Pearson *et al.* (1994) carried out such a test, which is shown in Figure 9c. The phase boundary between graphite and diamond shown in the plot is that of Kennedy & Kennedy (1976). The plot shows that two nodules containing graphite plot in the diamond stability field, and one containing diamond in the graphite field. However, in all three cases the error is comparable with the standard deviation of 7 km

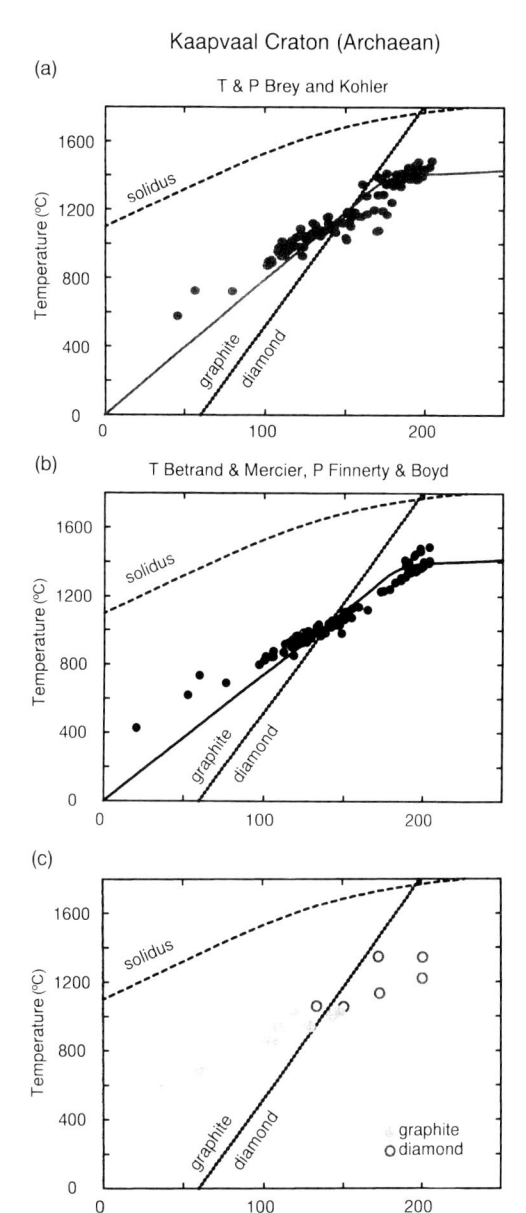

Fig. 9. Pressure and temperature estimates for 114 nodules (see McKenzie 1989) from the Kaapvaal Craton calculated using (**a**) Brey & Kohler's (1990) geothermometer and geobarometer, and (**b**) Bertrand & Mercier's (1985) geothermometer and Finnerty & Boyd's (1984) geobarometer. The solidus is from McKenzie & Bickle (1988), and the diamond–graphite phase boundary from Kennedy & Kennedy (1976). (**c**) shows the pressure and temperature estimates obtained from the silicate phases using Brey & Kohler's (1990) expressions for nodules containing graphite (Pearson *et al.* 1994) and diamond (see McKenzie 1989). The continuous lines in (**a**) and (**b**) show the best fitting geotherms.

estimated by Brey & Kohler (1990), whose expressions are therefore consistent with all the available data.

Velocities and density from seismology and nodules

The seismic velocities and densities can be calculated if the mineral proportions (called the modal mineralogy) have been measured in a nodule whose mineral compositions are known. Surprisingly few such measurements have all been made on the same nodule. Qiu *et al.* (1996) therefore used Tainton & McKenzie's (1994) melting models to estimate the modal mineralogy and hence ρ, V_p and V_s. They used four nodule suites: phlogopite K-richterite peridotite (PKP), garnet peridotite (GP) and garnet phlogopite peridotite (GPP) from Erlank *et al.* (1987), and the depleted and fertile nodules from Nixon *et al.* (1981). They obtained the modal mineralogy by minimizing the misfit between the bulk composition and that calculated from the mineral proportions and compositions, and then used the mineral compositions and modal mineralogy to estimate ρ, V_p and V_s at the depths and temperatures from which the nodules came. The nodule compositions from southern Africa vary with depth. Brey & Kohler's (1990) expressions give depths of 100–130 km for the PKP nodules, 100–150 km for the GP and GPP nodules, and 160–200 km for the fertile nodules. The velocities and the densities calculated from the nodules are compared with those estimated from seismology in Figure 10, using the same approach as Qiu *et al.* (1996, appendix). The agreement between the seismological and petrological models in Figure 10 is better than that in the corresponding figure 11 of Qiu *et al.* (1996), partly because the thickness of the high-velocity lid has been increased, and partly because the pressure estimates from petrology have been

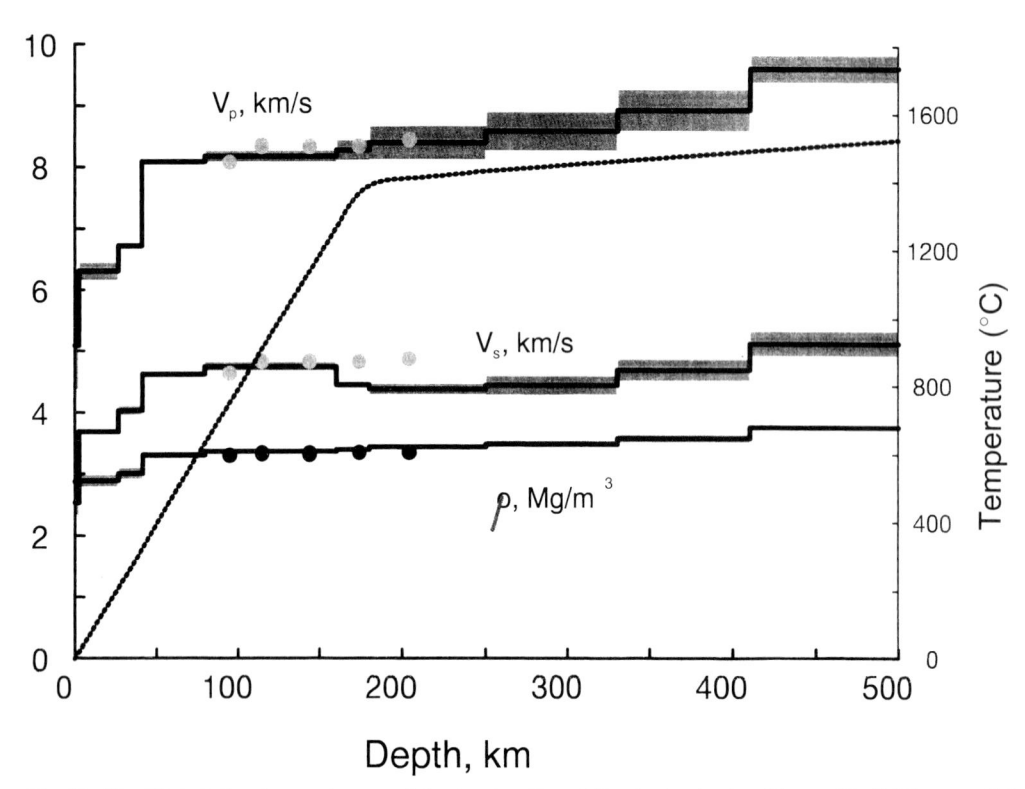

Fig. 10. The filled circles show estimates of the density, V_s and V_p obtained using Qiu *et al.*'s (1996, appendix) expressions and models of nodule compositions and modal mineralogy from Tainton & McKenzie (1994) for PKP nodules (92 km), depleted and GPP nodules (112 and 142 km), and fertile nodules (172 and 202 km). The continuous lines show the seismological estimates of the density, V_s and V_p from Appendix C, with the shaded region corresponding to one standard deviation. The scale for these quantities is on the left. the dotted line shows the geotherm from Fig. 7a, obtained using Brey & Kohler's (1990) expressions, with the scale on the right.

slightly reduced. However, one major and important difference between the seismological and petrological estimates remains; the depth estimates from the highest-pressure nodules require them to come from within the low-velocity zone, yet the estimates of V_p and V_s for nodules increase monotonically with depth. Qiu et al. (1996) found the same discrepancy, which they suggested might be due to the presence of a small melt fraction that prevented the transmission of shear stress across grain boundaries.

Important progress has occurred in our understanding of the processes that control the dependence of the shear modules on temperaure and on the frequency of the seismic waves. At temperatures close to the melting point, the shear modulus depends strongly on frequency (Gribb & Cooper 1998; Jackson 2000). Gribb & Cooper's measurements show a decrease of more than a factor of two between periods of 3 s and 100 s at a temperature of 1250°C for the fine-grained polycrystalline sample of olivine that they used (Fig. 11). More recently, the same workers (Gribb & Cooper 2000) have shown experimentally that this behaviour does not require the presence of melt. Instead, they argue that it results from transient diffusive effects that relax the stresses on subgrain boundaries (Gribb & Cooper 1998), and that the time constant involved depends on d^3, where d is the size of the subgrains. The observed attenuation in the upper mantle requires a subgrain size of 35 μm, (Gribb & Cooper 1998) compared with a grain size of 3 μm used in their experiments.

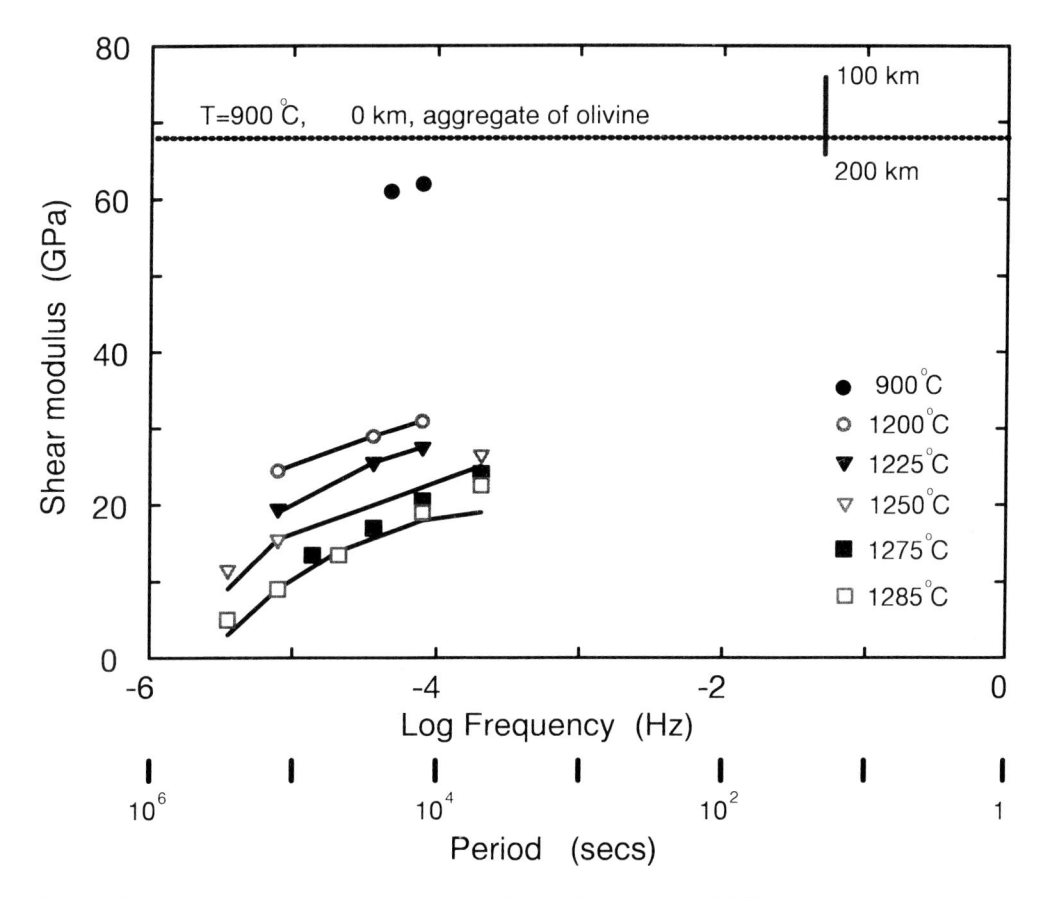

Fig. 11. The experimental points and lines are taken from Gribb & Cooper (1998), but are plotted at frequencies appropriate for a mantle grain size by dividing the experimental frequencies by $(35/3)^3$. The horizontal line shows the shear modulus of an olivine aggregate with Fo$_{92}$ calculated at $P = 0$ GPa, $T = 900$°C using the method of Qiu et al. (1996, appendix). The vertical line shows the shear moduli at depths of 100 and 200 km estimated from seismological observations.

Their laboratory results can therefore be applied to the mantle if the periods are increased by $(35/3)^3$ (Fig. 11).

Also shown in Figure 11 is the shear modulus for an olivine aggregate that consists of 92% forsterite and 8% fayalite (Fo$_{92}$), and the seismological estimates at depths of 100 and 200 km, plotted at a period of 20 s. Even though none of the aggregates used by Gribb & Cooper (1998) contained melt, they show a large decrease in the shear modulus at high temperature and low frequencies. As Figure 11 shows, the experimental results are at effective periods that are longer than those used to study the structure of the crust and mantle beneath the Kalahari Craton, and so it is not yet possible to use the geotherms in Figure 9 to calculate the seismic velocities when grain boundary relaxation occurs. However, Figure 11 clearly shows that the changes in the shear modulus caused by such relaxation are much larger than that required to account for the decrease in V_s between 100 and 200 km. Therefore, this decrease does not require the presence of melt. The geotherm from the nodules is also plotted in Figure 9 and shows that the low shear-wave velocities beneath the high-velocity lid occur where the temperature gradient decreases sharply at the base of the lithosphere.

Conclusions

A variety of seismic methods have been used to map the velocity structure beneath the shields of southern Africa. Propagation of waves from regional earthquakes to stations on the same shields provides the best vertical resolution, but only of the horizontally averaged velocity. The analysis of the regional seismic waveform data shows the existence of a high-velocity lid that can extend to an average depth of no more than $c.$ 160 km. Velocity structures obtained from analysis of near-vertical travelling body waves and longer-period fundamental mode data show high velocities extending to depths as great as 400 km beneath the Kalahari Shield. Because of the limited vertical resolution of such waves, such models are consistent with those from regional wave propagation, and result from the high lid velocities being smeared over a greater depth range. Body-wave tomography using teleseismic events can resolve lateral velocity variations, but is completely insensitive to variations in the mean vertical velocity. We therefore believe that all seismic measurements are consistent with the horizontally averaged velocity and density structure shown in Figure 2.

The best resolved average vertical V_p and V_s structure within the high-velocity lid beneath southern Africa agrees well with velocities estimated from the mineralogy of mantle nodules. However, the shear-wave velocity decrease at the base of the lid does not correspond to a change in mineralogy. Recent experimental studies of the shear-wave velocity in olivine as a function of temperature and period of oscillation show that this decrease can result from grain boundary relaxation at elevated temperatures at the period of seismic waves. Because the decrease in velocity occurs where the mantle temperature is closest to the melting temperature, such an explanation is consistent with thermal models of the temperature structure beneath the Kaapvaal Craton. The high-velocity depleted upper-mantle lid beneath southern Africa extends to an average depth of no more than $c.$ 160 km. It is underlain by material that becomes steadily less depleted with increasing depth, where the shear modulus also decreases because of the increasing temperature. Modelling of regional waveform data for the Australian Shield suggests that the high-velocity continental roots there extend to 200–225 km depth (Kennett *et al.* 1994; Debayle & Kennett 2000a, b), similar to or perhaps a little greater than that observed in southern Africa.

This paper has benefited from thoughtful reviews by C. A. Langston and J. Ritsema, and from discussions with J. Ritsema. D.M. acknowledges support from NERC and the Royal Society. This is Cambridge University Department of Earth Sciences contribution 6517.

References

AKI, K., CHRISTOFFERSON, A. & HUSEBYE, E. S. 1977. Determination of the three-dimensional seismic structure of the lithosphere. *Journal of Geophysical Research*, **82**, 277–296.

ANDERSON, D. L. & DZIEWONSKI, A. M. 1983. Upper mantle anisotropy: evidence from free oscillations. *Geophysical Journal of the Royal Astronomical Society*, **69**, 383–404.

BERTRAND, P. & MERCIER, J.-C. C. 1985. The mutual solubility of coexisting ortho- and clinopyroxene: toward an absolute geothermometer for the natural system? *Earth and Planetary Science Letters*, **76**, 109–122.

BLOCH, S., HALES, A. L. & Landisman, M. 1969. Velocities in the crust and upper mantle of southern Africa from multi-mode surface wave dispersion. *Bulletin of the Seismological Society of America*, **59**, 1599–1629.

BREY, G. P. & KOHLER, T. 1990. Geothermobarometry in four-phase lherzolites II. New thermobarometers, and practical assessment of existing thermobarometers. *Journal of Petrology*, **31**, 1353–1378.

BRUNE, J. N. & DORMAN, J. 1963. Seismic waves and earth structure in the Canadian shield. *Bulletin of the Seismological Society of America*, **53**, 167–210.

CARA, M. 1979. Lateral variation of S velocity in the upper mantle from higher Rayleigh modes. *Geophysical Journal of the Royal Astronomical Society*, **57**, 646–670.

DEBAYLE, E. 1999. SV-wave azimuthal anisotropy in the Australian upper mantle: preliminary results from automated Rayleigh waveform inversions. *Geophysical Journal International*, **137**, 747–754.

DEBAYLE, E. & KENNETT, B. L. N. 2000a. The Australian continental upper mantle: structure and deformation inferred from surface waves. *Journal of Geophysical Research*, **105**, 25 423–24 450.

DEBAYLE, E. & KENNETT, B. J. N. 2000b. Anisotropy in the Australian upper mantle from Love and Rayleigh waveform inversion. *Earth and Planetary Science Letters*, **184**, 339–351.

DZIEWONSKI, A. M., HALES, A. L. & LAPWOOD, E. R. 1975. Parametrically simple Earth models consistent with geophysical data. *Physics of the Earth and Planetary Interiors*, **10**, 12–48.

ERLANK, A. J., WATERS, F. G., HAWKESWORTH, C. J., HAGGERTY, S. E., ALLSOP, H. L., RICKARD, R. S. & MENZIES, M. A. 1987. Evidence for mantle metasomatism in peridotite nodules from the Kimberly Pipes, South Africa. *In*: MENZIES, M. A. & HAWKESWORTH, C. J. (eds) *Mantle Metasomatism*. Academic Press, London, 221–311.

FINNERTY, A. A. & BOYD, F. R. 1987. Thermabarometry for garnet peridotites: basis for the determination of thermal and compositional structure of the upper mantle. *In*: NIXON, P. H. (ed.) *Mantle Xenoliths*. Wiley, New York, 381–402.

GILBERT, F. & DZIEWONSKI, A. M. 1975. An application of normal mode theory to the retrieval of structural parameters and source mechanism from seismic spectra. *Philosophical Transactions of the Royal Society of London, Series A*, **278**, 187–269.

GRAND, S. P. & HELMBERGER, D. V. 1984. Upper mantle shear structure of North America. *Geophysical Journal of the Royal Astronomical Society*, **76**, 399–438.

GRIBB, T. T. & COOPER, R. F. 1998. Low-frequency shear attenuation in polycrystalline olivine: grain boundary diffusion and the physical significance of the Andrade model for viscoelastic rheology. *Journal of Geophysical Research*, **103**, 27 267–27 279.

GRIBB, T. T. & COOPER, R. F. 2000. The effect of an equilibrated melt phase on the shear creep and attenuation behaviour of polycrystalline olivine. *Geophysical Research Letters*, **27**, 2341–2344.

JACKSON, I. 2000. Laboratory measurement of seismic wave dispersion and attenuation: recent progress. *In*: *The Earth's Deep Interior: Mantle Physics and Tomography from the Atomic to the Global Scale*. Geophysical Monogaph, American Geophysical Union, **117**, 265–289.

JAMES, D. E. & FOUCH, M. J. 2002. Formation and evolution of Archaean cratons: insights from southern Africa. *In*: FOWLER, C. M. R., EBINGER, C. J. & HAWKESWORTH, C. J. (eds) *The Early Earth: Physical, Chemical and Biological Development*. Geological Society, London, Special Publication, **199**, 1–26.

JAMES, D. E., FOUCH, M. J., VANDECAR, J. C., VAN DER LEE, S. & KAAPVAAL SEISMIC GROUP 2001. Tectospheric structure beneath southern Africa. *Geophysical Research Letters*, **28**, 2485–2488.

JORDAN, T. H. 1975. The continental tectosphere. *Reviews of Geophysics*, **13**, 1–12.

KENNEDY, C. S. & KENNEDY, G. C. 1976. The equilibrium boundary between graphite and diamond. *Journal of Geophysical Research*, **81**, 2467–2470.

KENNETT, B. L. N., GUDMUNDSSON, O. & TONG, C. 1994. The upper-mantle S and P velocity structure beneath northern Australia from broad-band observations. *Physics of the Earth and Planetary Interiors*, **86**, 85–98.

LEVEQUE, J. J. & MASSON, F. 1999. From ACH tomographic models to absolute velocity models. *Geophysical Journal*, **137**, 621–629.

LEVEQUE, J. J., RIVERA, L. & WITTLINGER, G. 1993. On the use of the checker-board test to assess the resolution of tomographic inversions. *Geophysical Journal*, **115**, 313–318.

MCELHINNY, M. W. & MCWILLIAMS, M. O. 1977. Precambrian geodynamics – a paleomagnetic view. *Tectonophysics*, **40**, 137–159.

MCKENZIE, D. 1989. Some remarks on the movement of small melt fractions in the mantle. *Earth and Planetary Science Letters*, **95**, 53–72.

MCKENZIE, D. & BICKLE, M. J. 1988. The volume and composition of melt generated by extension of the lithosphere. *Journal of Petrology*, **29**, 625–679.

MASTERS, T. G., JOHNSON, S., LASKE, G. & BOLTON, H. 1996. A shear-velocity model of the mantle. *Philosophical transactions of the Royal Society of London, Series A*, **354**, 1385–1411.

MONTAGNER, J.-P. 1996. Can seismology tell us anything about convection in the mantle? *Reviews of Geophysics*, **32**, 115–137.

MOONEY, W. D., LASKE, G. & MASTERS, T. G. 1998. Crust 5.1: a global crustal model at $5° \times 5°$. *Journal of Geophysical Research*, **103**, 727–747.

NATAF, H. C., NAKANISHI, I. & ANDERSON, D. L. 1984. Anisotropy and shear-velocity heterogeneities in the upper mantle. *Geophysical Research Letters*, **11**, 109–112.

NISHIMURA, C. & FORSYTH, D. 1989. The anisotropic structure of the upper mantle in the Pacific. *Geophysical Journal*, **94**, 193–218.

NIXON, P. H., ROGERS, N. W., GIBSON, I. L. & GREY, A. 1981. Depleted and fertile mantle xenoliths from southern Africa kimberlites. *Annual Review of Earth and Planetary Sciences*, **9**, 285–309.

PEARSON, D. G., BOYD, F. R., HAGGERTY, S. E., PASTERIS, J. D., FIELD, S. W., NIXON, P. H. & POKHILENKO, N. P. 1994. The characterisation and origin of graphite in cratonic lithospheric mantle: a petrological carbon isotope and Raman spectroscopic study. *Contributions to Mineralogy and Petrology*, **115**, 449–466.

PRIESTLY, K. 1999. Velocity structure of the continental upper mantle: evidence from southern Africa. *Lithos*, **48**, 45–56.

QIU, X., PRIESTLY, K. & MCKENZIE, D. 1996. Average lithospheric structure of southern Africa. *Geophysical Journal International*, **127**, 563–587.

RITSEMA, J. & VAN HEIJST, H. 2000. New seismic model of the upper mantle beneath Africa. *Geology*, **28**, 63–66.

RITSEMA, J., NYBLADE, A. A., OWENS, T. J., LANGSTON, C. A. & VANDECAR, J. C. 1999. Upper mantle seismic velocity structure beneath Tanzania, east Africa: Implications for the stability of cratonic lithosphere. *Journal of Geophysical Research*. **103**, 21 201–21 213.

SU, W., WOODWARD, R. L. & DZIEWONSKI, A. M. 1994. Degree 12 model of shear velocity heterogeneity in the mantle. *Journal of Geophysical Research*, **99**, 6945–6980.

TAINTON, K. M. & MCKENZIE, D. 1994. The generation of kimberlites, lamproites, and their source rock. *Journal of Petrology*, **35**, 787–817.

VINNIK, L. P., GREEN, R. W. E. & NICOLAYSEN, L. O. 1995. Recent deformation of the deep continental root beneath southern Africa. *Nature*, **375**, 50–52.

WOODHOUSE, J. H. & DZIEWONSKI, A. M. 1984. Mapping the upper mantle: three-dimensional modeling of Earth structure by inversion of seismic waveforms. *Journal of Geophysical Research*, **89**, 5953–5986.

ZHAO, M., LANGSTON, C. A., NYBLADE, A. A. & OWENS, T. J. 1999. Upper mantle velocity structure beneath southern Africa from modeling regional seismic data. *Journal of Geophysical Research*, **104**, 4783–4794.

The development of lithospheric keels beneath the earliest continents: time constraints using PGE and Re–Os isotope systematics

D. G. PEARSON[1], G. J. IRVINE[1], R. W. CARLSON[2], M. G. KOPYLOVA[3] & D. A. IONOV[4]

[1] *Department of Geological Sciences, Durham University, South Road, Durham DH1 3LE, UK (e-mail: d.g.pearson@durham.ac.uk)*
[2] *Department of Terrestrial Magnetism, Carnegie Institution of Washington, 5241 Broad Branch Road, NW, Washington DC 20015, USA*
[3] *Geological Sciences Division, Earth and Ocean Sciences, The University of British Columbia, 6339 Stores Road, Vancouver, B.C. V6T 1R9, Canada*
[4] *Department of Earth and Environmental Sciences, Université Libre de Bruxelles, 50 Roosevelt av., B-1050 Brussels, Belgium*

Abstract: Continued studies of xenolith suites found in kimberlites on and around the Kaapvaal Craton, together with those from newly discovered localities on other cratons, are providing new insights into the generation and evolution of the Earth's oldest continents. Comparison of modal abundance data with melt depletion models, together with trace element and isotope systematics in Kaapvaal low-temperature peridotites, suggest that much or all of the diopside and garnet in these rocks may have formed significantly after initial melt depletion. The Re–Os isotope system has been instrumental in providing an improved understanding of the timing of the formation of cratonic lithospheric keels. New studies that focus on carefully selected whole-rock peridotites and use combined platinum group element (PGE) and Re–Os isotope analysis provide better constraints on the significance of Re–Os model ages. The large database of Re–Os isotope analyses for peridotites for the Kaapvaal Craton indicate formation of significant amounts of lithospheric mantle in Neoarchaean time, associated with voluminous mafic magmatism. Formation of lithospheric mantle in Neoarchean time (3.0–2.5 Ga) follows the cessation of major crustal differentiation events at *c.* 3.1 Ga and marks the onset of craton stabilization. Some lithospheric mantle was produced in Palaeo- to Mesoarchaean time (3.8–3.0 Ga) in southern Africa, which preserved ancient crustal fragments. Large-scale preservation of Archaean continental masses was effective only after the formation of substantial, buoyant, rigid, deep lithospheric keels and their stabilization in Neoarchean time. Formation of lithospheric mantle beneath the surrounding Proterozoic crustal regions occurred in Mesoproterozoic time, with lower degrees of mantle melting than associated with the cratonic peridotites. This circum-cratonic mantle is of similar age to the oldest overlying crust and has been coupled to the margins of the craton since its formation. Major magmatic events, some coincident with the formation of circum-cratonic mantle, added new lithosphere to the Kaapvaal mantle root but failed to destroy it. The mechanically strong, buoyant lithospheric keels beneath cratons protect their crust from subduction and recycling over 3 Ga time periods.

The evolution of the earliest continents remains an enigmatic problem in Earth sciences. Particular problems revolve around issues such as the timing and stabilization of the first large continental masses and the subsequent roles of crust generation v. crustal recycling in the evolution of the continents (e.g. Armstrong 1981; Abbott *et al.* 2000). A key factor in the preservation of continental crust is its protection by thick continental keels. How extensive these keels were, and whether they stabilized simultaneously, beneath the whole of a cratonic area, remains an unknown yet important factor in understanding the evolution of continents (de Wit *et al.* 1992). To understand these problems further requires a more complete understanding of the genetic relationship between the deep lithospheric keels present beneath cratons and their overlying crust, and more comprehensive and definitive knowledge of the timing of formation of the lithospheric keels underlying the crust.

Some of the first isotopic studies of lithospheric mantle recognized its antiquity (Holmes & Paneth 1936; Kramers 1979; Menzies & Murthy

From: FOWLER, C. M. R., EBINGER, C. J. & HAWKESWORTH, C. J. (eds) *The Early Earth: Physical, Chemical and Biological Development.* Geological Society, London, Special Publications, **199**, 65–90. 0305-8719/02/$15.00 © The Geological Society of London 2002.

1980; Hawkesworth et al. 1983; Richardson et al. 1984) but the lithophile character of the isotope systems employed did not allow the time of initial differentiation from the convecting mantle to be well constrained. Application of the Re–Os isotope system to dating lithospheric peridotites has given us the opportunity to better constrain the timing of the melt extraction creating these peridotites, the key step in establishing their chemically induced buoyancy, and hence their role in lithosphere formation and stabilization (Walker et al. 1989; Pearson 1999a,b). The aim of this paper is to explore the time scales of craton generation and the relationship between the crust and the mantle part of the continental lithosphere by examining the timing of formation of the peridotites that form the lithospheric keels.

Samples of the mantle beneath and around cratons are mostly available as xenoliths erupted by deep-seated kimberlitic volcanism, and Fermor (1913) was one of the first geologists to recognize the value in studying this material to understand the evolution of the continents. Such samples have played a leading role in understanding the formation and evolution of continental keels (Nixon 1963, 1987; Nixon & Boyd 1973; Boyd 1989; Griffin et al. 1999; Pearson 1999a, b). Until recently, samples of sub-cratonic mantle were dominated by the extensive xenolith collections available from the Kaapvaal Craton in southern Africa, where diamond mining has provided abundant material for study. In particular, the style of early mining activity in this region was conducive to the separation and preservation of numerous large peridotite xenoliths. Recent prospecting and mining activity on other cratons has produced a much more representative suite of cratonic peridotites from around the world that now allows a broader view of lithospheric keel formation. Below we present a detailed summary of available age criteria for the lithosphere beneath southern Africa and also discuss age information available for selected other regions. We concentrate on age information obtained using the Re–Os isotope system, and try to evaluate various models for craton or circum-craton lithospheric stabilization. In addition, we try to provide an improved basis for the understanding of Re–Os model ages using combined Re–Os isotope and platinum group element (PGE) concentration measurements.

Sample suites

Melt depletion histories

Kaapvaal peridotites that have low equilibration temperatures, coarse grain size and relatively undeformed textures (the low-T suite) have high olivine mg-numbers (mean = 92.6) and widely varying modal proportions of olivine to orthopyroxene (Boyd & Mertzman 1987; Boyd 1989). Modal orthopyroxene contents vary between 20 and 50% producing lower olivine contents than 'fertile mantle' in rocks that are highly depleted in Al and Ca. Many of the Kaapvaal low-T peridotites are anomalously enriched in SiO_2, manifest as orthopyroxene (Maaloe & Aoki 1977; Boyd 1989), and, as such, up to 80% of these specimens cannot be generated directly via simple partial melting from normal mantle compositions (Herzberg 1999; Walter 1999). SiO_2 may have been introduced during melt–rock reaction (Keleman et al. 1992, 1998; Rudnick et al. 1994), or cumulate mixing (Herzberg 1993). The reader is referred to Herzberg (1999) and Walter (1999) for more detailed discussion of these issues. If the FeO, MgO and Al_2O_3 contents of the average low-T Kaapvaal peridotite are primary, their compositions are consistent with a considerable amount of melt extraction ($c.$ 40–45%) at high pressures (4 to >7 GPa depending on the nature of the melt models used (Walter 1999).

Some low-T peridotites from the Siberian Craton (Boyd et al. 1997) also show SiO_2 enrichment, to the extent that 30% of the population studied cannot be generated by simple melt extraction models; however, their average composition can be produced by high degrees ($c.$ 40%) melt extraction at high pressures (Walter 1999). Peridotites from the recently discovered Slave Craton kimberlites appear to be, in general, lower in SiO_2 than Kaapvaal or Siberian peridotites and on average have lower mg-numbers (Pearson et al. 1999; Kopylova & Russell 2000). Some peridotite suites from Slave Craton kimberlites can have higher SiO_2 contents (McKenzie & Canil 1999) and further sampling of other localities on this craton will reveal a more complete picture.

Peridotites from the Tanzanian and Greenland (North Atlantic) Cratons do not show SiO_2 enrichments (Rudnick et al. 1994; Bernstein et al. 1998) and their compositions can be readily explained by moderate to large degrees of partial melting at pressures of 3–4 GPa in the case of Tanzania (Walter 1999) and $c.$ 40% melt extraction at between 7 and 2–3 GPa for the East Greenland peridotites (Bernstein et al. 1998). Kopylova & Russell (2000) pointed out that compositional distinction between other cratonic sample suites and the Tanzanian and East Greenland peridotite suites might be expected. They reasoned that Siberian and Kaapvaal peridotites are clearly cratonic in their setting whereas for Tanzania and East Greenland, alkali basalt magmatism has sampled mainly shallow mantle following major

Cenozoic rifting events at craton margins and this significantly elevated the mantle geotherms beneath these regions. To this extent their evolution might be expected to be different from the areas of the Kaapvaal, Siberian and Slave cratons sampled by kimberlites (Kopylova & Russell 2000). None the less, both Tanzanian and East Greenland peridotites show clear evidence of Archaean melt depletion (see below). Although the details of the chemical evolutionary histories of these contrasting areas remain a subject for further research, it is clear that the mantle beneath cratonic regions varies in composition and history of melt extraction or interaction and so all cratonic areas cannot be expected to be alike.

Despite differences in their recent tectonic evolution, a common aspect of the chemistry of the various peridotite suites available for study is their high level of depletion of 'basaltic components', those elements that are removed with the melt during mantle melting. This indicates a likely common origin for lithospheric peridotites as residues of melting at varying pressure. The subtly differing bulk compositions of peridotites from different cratons is an indication of the variation in their age of stabilization (Griffin *et al.* 1998) and differences in their subsequent evolution (Kopylova & Russell 2000).

Metasomatic histories and post-melting mineral growth

The effects of post melt-depletion interaction with fluid or melt components in the lithospheric mantle has been extensively documented (e.g. Menzies & Hawkesworth 1987, and references therein; Harte *et al.* 1993; Pearson 1999*b*) and it is widely accepted that these phenomena dominate the minor element geochemistry of cratonic peridotites. Extensive studies of the effect of metasomatism on the major element chemistry of lithospheric peridotites have also been made (Boyd & Mertzman 1987; Keleman *et al.* 1992, 1998; Walter 1999). To date, most of the discussion has centred around the apparent excess of orthopyroxene, especially in Kaapvaal peridotites. However, major and trace element studies show that it is likely that the abundances of garnet and clinopyroxene are also grossly affected (Burgess & Harte 1999; Shimizu 1999).

To illustrate the effects of metasomatism on the silicate mineralogy of peridotite xenoliths it is instructive to compare the modal abundances of garnet and clinopyroxene, expected to be the first minerals to be exhausted during partial melting, with indices of melt extraction such as the mg-number of olivine (Fig. 1). The observed abundances of garnet plus clinopyroxene with olivine mg-number in low-T peridotites from the Kaapvaal Craton compared with the variation expected from partial melt extraction as simulated experimentally by Kinzler & Grove (1999; Fig. 1) reveals that: (1) approximately one-third of Kaapvaal rocks have total modal abundances of garnet plus clinopyroxene in excess of that predicted for simple melt removal, especially at high olivine mg-number; (2) if only garnet is considered, all of the garnet-bearing low-T Kaapvaal peridotites have an excess of garnet over that expected from simple melt removal, with some samples having much more garnet than predicted.

Although factors such as the starting composition of the experiments used to simulate melt removal will affect the details of predicted garnet and clinopyroxene-'out' points, similar rates of consumption of these phases and thus essentially the same result is obtained using the mineral reaction mass balance equations of Walter (1999). The simplest conclusion from these observations is that the 'excess' garnet and clinopyroxene grew from infiltrating melt or hydrous fluid phases during later metasomatic events as suggested by Burgess & Harte (1999) and Shimizu (1999). Boyd & Mertzman (1987) inferred a metasomatic origin via a hydrous fluid for some diopside in Jagersfontein peridotites on the basis of the correlation between modal diopside and mica. In contrast, Burgess & Harte (1999) and Shimizu (1999) suggested that silicate magmas were responsible for the deposition of much of the diopside and garnet in Kaapvaal low-T lherzolites and harzburgites. What proportion of the garnets and clinopyroxenes formed from hydrous (supercritical?) fluids or magmas requires further investigation but we suggest that the majority of these phases in Kaapvaal low-T peridotites are of metasomatic origin. This is consistent with the younger, generally Proterozoic Nd model ages of many of the garnets and clinopyroxenes (Menzies & Murthy 1980; Richardson *et al.* 1985; Pearson 1999*b*) and their commonly incompatible element rich nature (e.g. Shimizu 1999). As such, the anhydrous Kaapvaal garnet harzburgites and lherzolites are just as much 'metasomites' as those rocks containing phases of more obvious metasomatic origin such as amphibole and mica. The important point is that metasomatism is probably ubiquitous in cratonic lithospheric mantle and has had dramatic and observable effects on modal mineralogy as well as trace element and isotope geochemistry. Because modal mineralogy, particularly the abundance of garnet, affects density, it is likely that the bulk density of large portions of the lithospheric mantle has varied substantially since initial lithosphere stabilization. In terms of determining formation ages,

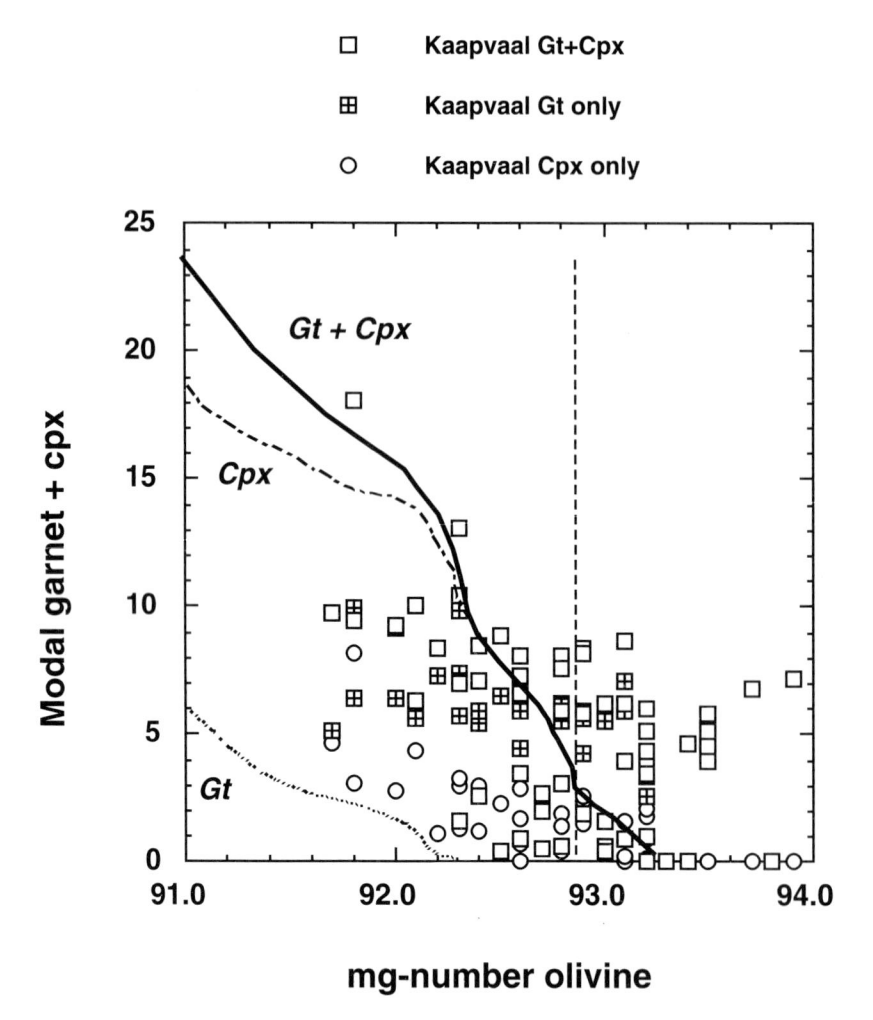

Fig. 1. Modal abundance of garnet and diopside in low-*T* Kaapvaal garnet peridotite xenoliths v. olivine mg-number. Curves display the experimentally determined consumption of garnet and diopside during progressive melting of garnet peridotite from Kinzler & Grove (1999) and the combined total of garnet and diopside. Some of the 'Gt + Cpx' points plot over samples containing garnet only. The starting composition using by Kinzler & Grove (1999) was very rich in diopside in particular and so probably represents a situation most favourable to preservation of diopside to high degrees of melting. Similar rates of mineral consumption are obtained using the melting reactions of Walter (1999). The very early removal of garnet should be noted; this implies that most Kaapvaal low-*T* garnet peridotites have excess garnet for the amount of melt removal that they have experienced. Data supplied by F. R. Boyd.

seeing past the veil of metasomatism to the age of the melt depletion event can be particularly difficult, and this should be kept in mind in any discussion of the chemical evolution of these rocks.

Dating lithospheric peridotites: Re–Os isotope systematics

Extensive studies of peridotite xenoliths using the Rb–Sr, Sm–Nd and U–Pb isotope systems have provided much information on the processes that modify lithospheric mantle keels. These systems, however, have been of more limited value in estimating the overall age of the lithospheric mantle because they primarily track metasomatic enrichment events rather than the melt-depletion event that established the major element composition of the peridotites (see reviews by Menzies & Hawkesworth 1987; Menzies 1990; Pearson 1999*a, b*). Developments in our technical ability to exploit the Re–Os isotope system together with the recognition that the compatible nature of Os makes the Os isotope composition a powerful time-integrated monitor of melt-depletion events (Allègre & Luck 1980) have led to its application in dating cratonic mantle peridotites (Walker et al. 1989; Carlson & Irving 1994; Pearson et al. 1995a, b); numerous further studies have been made and a growing body of knowledge is emerging about the timing of differentiation of lithospheric mantle on different cratons.

The Re–Os isotope system differs from isotope systems based on lithophile elements (Sr–Nd–Hf–Pb) because Os is a compatible element during mantle melting and thus remains in the residue following melt extraction. Re is moderately incompatible and is preferentially partitioned into the melt phase with increasing degree of melting. For the large degrees of melting experienced by cratonic peridotites (Boyd 1989; Walter 1999), most Re should be removed from a residual peridotite, such that the $^{187}Os/^{188}Os$ ratio of the mantle at the time of melting is locked in and hence may be used to estimate the age of the peridotite if we know the mantle Os isotope evolution curve (Walker et al. 1989; Fig. 2). The resulting model age is a minimum age estimate based on the assumption that all Re is removed from the peridotite during mantle melting and that none re-enters as a result of later melt or fluid infiltration events. Such an age is known as a Re

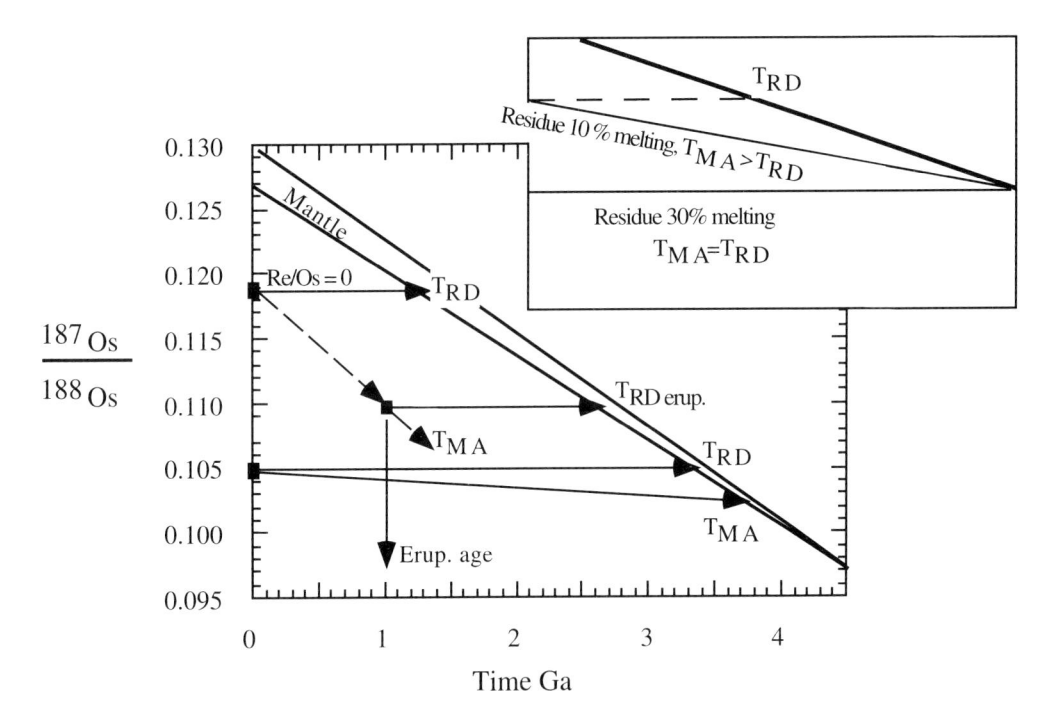

Fig. 2. Os isotope evolution diagram showing details of Re–Os model age systematics. The T_{RD} age assumes that all Re was removed during a single stage of melting whereas the T_{MA} age uses the measured Re concentration to calculate a model age with respect to the mantle evolution line. The equations for each age calculation are:

$$T_{RD} = \frac{1}{\lambda} \ln \left\{ \left[\left(\frac{^{187}Os}{^{188}Os_{BE(t)}} - \frac{^{187}Os}{^{188}Os_{sample(t)}} \right) \Big/ \frac{^{187}Re}{^{188}Os_{BE(t)}} \right] + 1 \right\}$$

$$T_{MA} = \frac{1}{\lambda} \ln \left\{ \left[\left(\frac{^{187}Os}{^{188}Os_{BE(t)}} - \frac{^{187}Os}{^{188}Os_{sample(t)}} \right) \Big/ \left(\frac{^{187}Re}{^{188}Os_{BE(t)}} - \frac{^{187}Re}{^{188}Os_{sample(t)}} \right) \right] + 1 \right\}$$

where BE is the Bulk Earth reservoir, usually taken as chondrite, and t is the time of eruption. Following Shirey & Walker (1998), parameters are: $\lambda^{187}Re = 1.666 \times 10^{-11}$ a^{-1}; $^{187}Os/^{188}Os_{BE}$ at a given time is calculated from a present-day $^{187}Os/^{188}Os$ value of 0.1287 and a bulk-mantle $^{187}Re/^{188}Os$ of 0.4243. $T_{RD\,erup}$ shows the effect of subtracting post-emplacement radiogenic Os in-growth of ^{187}Os from xenoliths erupted by ancient kimberlite pipes (e.g. Premier) on the normal T_{RD} model age. For young kimberlites the difference between T_{RD} and T_{RDerup} is insignificant unless the xenolith has very high Re/Os. Inset shows hypothetical effects of various degrees of melt extraction on the Os isotope evolution of a residual peridotite.

depletion age or T_{RD} age (Walker et al. 1989; Fig. 1). This age should closely approximate the traditional model age of the peridotite (T_{MA} age), calculated using the measured Re/Os ratio of the sample (Fig. 2), if most Re is removed by melting and none re-enters. Given the complex history experienced by lithospheric peridotites, as revealed by their Rb–Sr and Sm–Nd isotope systematics, the assumption of a closed system since melting seems unlikely for most samples. If Re is introduced to the peridotite in significant quantities after the initial melting event then the calculated T_{MA} age will be artificially old and the true age will lie somewhere between the two age estimates (Fig. 2). Because of the possibility of disturbance, suites of approximately 10 or more samples are commonly analysed to examine the range in Re–Os behaviour in a peridotite suite and the possible effect of metasomatism on the age information. The tendency has been to select the oldest T_{RD} age as an indication of the age of the suite and to assume that other samples giving younger ages in some way reflect open-system behaviour (e.g. Walker et al. 1989; Carlson & Irving 1994; Pearson et al. 1995a,b). In the following discussion our aim will be to evaluate the validity of this assumption when made on a craton-wide scale (e.g. Pearson et al. 1995a,b) by examining more extensive data from the Kaapvaal Craton in particular.

Using the above approach, numerous studies have produced Archaean age estimates for cratonic peridotite sample suites (Walker et al. 1989; Carlson & Irving 1994; Pearson et al. 1995a,b; Carlson et al. 1999; Chesley et al. 1999). Recent detailed studies of mineral separates from peridotites, including sulphide separates (Burton et al. 1999) together with in situ laser-ablation measurements of sulphide Re–Os isotope systematics (Alard et al. 2000; Pearson et al. 2000) indicate the presence of multiple generations of sulphides in some mantle peridotites. Because sulphides contain the majority of the Re and Os in a fertile peridotite, the introduced intergranular sulphides probably account for some of the spread in Os isotope ratios observed in cratonic peridotite suites. This second-generation sulphide introduction may have occurred during metasomatic events that are thought to alter the silicate modal mineralogy of cratonic rocks. Re–Os model ages (T_{MA} ages) of primary sulphides included in silicate minerals, protected from subsequent interaction with percolating fluids, are significantly older than those for interstitial sulphides. In this case, T_{MA} ages are used in the belief that the sulphides have experienced a single-stage evolution and reflect the melt-depletion event responsible for their incorporation into the lithospheric mantle. A whole-rock analysis of a sample containing primary and later-generation sulphides would thus be an integration of the various generations of sulphides in the rock. Many cratonic whole-rock peridotites, however, have exceedingly low Re concentrations, an indication that they have not experienced sulphide introduction. For example, T_{MA} ages of 2.6–3.6 Ga for sulphides included in olivine macrocrysts from the Udachnaya kimberlite, Siberia (Pearson et al. 2000), are within error of the oldest T_{RD} ages (3.1 Ga) obtained from a suite of whole-rock peridotite xenoliths from the same kimberlite (Pearson et al. 1995b; Fig. 3). T_{MA} ages of 3.1–3.5 Ga have been obtained for sulphides included in Siberian diamonds (Pearson et al. 1999; Fig. 3), making the oldest xenolith T_{RD} age a reasonable estimate of the age of lithospheric stabilization. In contrast, the majority of the Siberian peridotites have much younger T_{RD} ages, possibly reflecting multiple phases of metasomatism (Pearson et al. 1995b) that disturbed their Re–Os systematics. Multiple generations of sulphides have been documented in numerous non-cratonic peridotites that have significantly disturbed whole-rock Re–Os systematics (Burton et al. 1999; Alard et al. 2000).

These studies highlight the importance of finding other ways to more systematically evaluate whole-rock peridotite Re–Os isotope model ages. Ideally, Re–Os dating studies on peridotites should be carried out on primary sulphides included in minerals such as olivine. However, as already indicated, most cratonic peridotites are residues of large degrees of melting. The removal

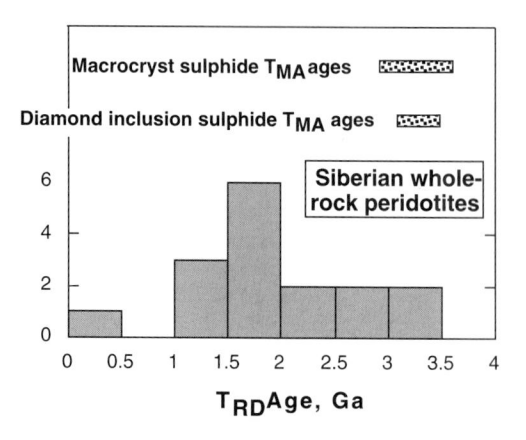

Fig. 3. T_{RD} ages of Siberian whole-rock peridotites (Pearson et al. 1995b), mostly from the Udachnaya kimberlite, compared with T_{MA} ages of sulphides from Udachnaya olivine macrocrysts (Pearson et al. 2000) and T_{MA} ages from Udachnaya diamond inclusions (Pearson et al. 1999).

of >30% partial melt from a peridotite acts to remove the primary monosulphide solid-solution phase (*mss*). Hence, primary sulphides are very scarce in such rocks. Brenan & Andrews (2001) have suggested that *mss* in a peridotite undergoing melt extraction may break down to very small sulphides such as laurite, or to an Ru–Ir–Os alloy phase, with much higher solidi than *mss*. In this way Os and associated Ir group PGEs (I-PGEs; Os, Ir, Ru) remain in the peridotite, but as much more finely dispersed grains, and PGE patterns become highly fractionated. Preliminary work aimed at searching for such phases in cratonic peridotites has not yet revealed their presence (Brenan & Pearson unpub. data).

Combined Re–Os and PGE systematics

Because of the scarcity of primary sulphide inclusions in cratonic peridotites, as evidenced by their generally very low bulk S contents, we have investigated the use of PGE data to evaluate possible open-system behaviour in the Re–Os systematics of bulk-rock cratonic peridotites. We use a chemistry technique that allows determination of the Re–Os isotopic composition and PGE abundances (except Rh) from a single dissolution (Pearson & Woodland 2000). The rationale of this approach is that chondrite-normalized PGE patterns can be used to track melt depletion v. enrichment events in peridotites, in the same way that rare earth element patterns reflect these processes (e.g. Handler & Bennett 1999; Lorand *et al.* 1999; Rehkamper *et al.* 1999). Obtaining the PGE abundances and Os isotope composition on a single dissolution avoids potential problems from 'nugget effects'. Once sulphde has been removed from a residue, the difference in solid silicate–liquid silicate partition coefficients ($D^{Sil/melt}$) between the I-PGEs such as Ir, and the Pt group PGEs (P-PGEs; Rh, Pt, Pd) leads to fractionated PGE patterns, with progressive depletion of P-PGEs whereas I-PGEs remain in the residue. This leads to a decrease in ratios such as Pd/Ir with progressive melting (Handler & Bennett 1999; Lorand *et al.* 1999; Rehkamper *et al.* 1999; Fig. 4). The progressive interelement PGE fractionation during partial melting can be correlated with variation in major element composition expected during partial melting, as monitored by Al_2O_3 for instance (Fig. 4). If cratonic

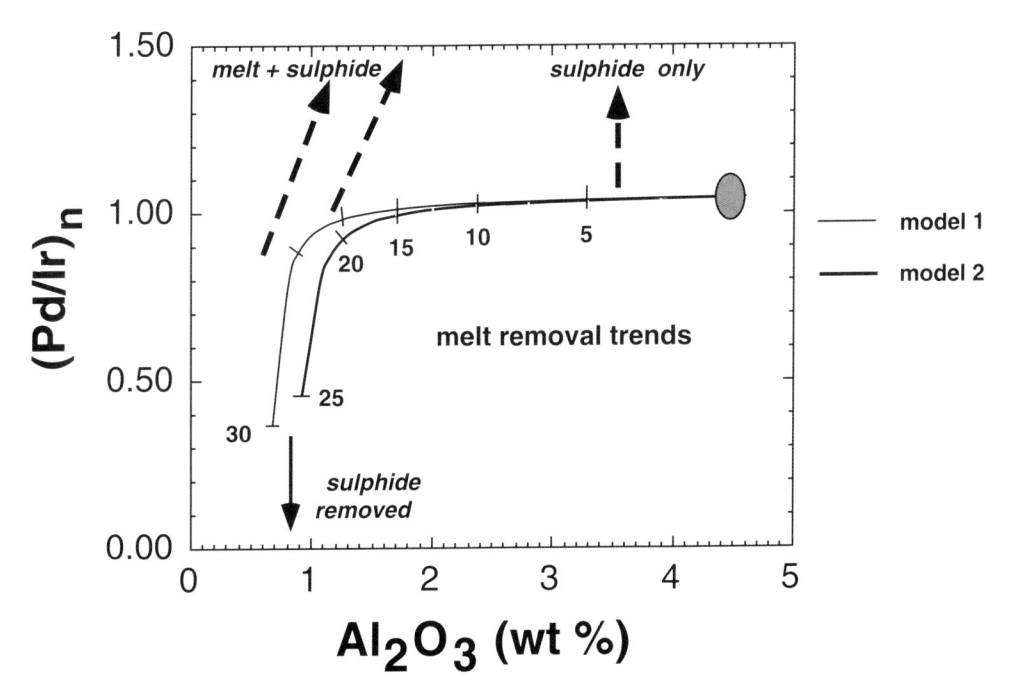

Fig. 4. Fractionation of PGE pattern $(Pd/Ir)_n$ v. Al_2O_3 for progressive melt removal and the effects of sulphide addition as a result of melt–rock interaction. Calculation of the melting trends follows the method of Lorand *et al.* (1999). Tick marks denote 5% melting intervals. The two curves shown depict starting compositions with 250 ppm (model 1) and 300 ppm (model 2) total sulphur. The trends terminate when residual sulphide disappears at 25 and 30% melting, respectively. The sulphide addition trends are general trends taken from Rehkamper *et al.* (1999).

Table 1. *PGE and Os isotope data for selected whole-rock peridotites from the Slave Craton, Namibia and Vitim*

Location:	Namibia Louwrencia	Namibia Louwrencia	Namibia Louwrencia	Namibia Louwrencia	Vitim	Vitim	Vitim	Slave Jericho	Slave Jericho	Slave Jericho	Kaapvaal Kimberley	Kaapvaal Premier
Sample:	1181	5304	5315	2514	313-105	314-5	314-71	8-7	10-12A	40-21	FRB1431	FRB1382
Os (ppb)	5.11	5.01	4.75	3.23	0.94	2.56	2.24	2.47	3.10	2.43	1.761	3.7911
Ir (ppb)	4.96	4.56	3.60	2.81	2.17	4.53	2.39	2.03	2.53	2.52	2.04	2.89
Ru (ppb)	10.02	7.69	9.08	4.37	4.08	8.08	3.91	3.47	4.16	5.47	2.93	4.64
Pt (ppb)	5.87	2.94	4.85	2.78	3.06	3.40	1.34	0.253	0.338	2.26	0.315	5.94
Pd (ppb)	1.56	1.21	3.43	0.64	1.75	0.63	1.30	0.049	0.073	0.856	0.299	4.65
Re (ppb)	0.058	0.319	0.143	0.136	0.016	0.019	0.042	0.006	0.036	0.0337	0.011	0.1288
(Pd/Ir)n	0.275	0.233	0.832	0.198	0.704	0.121	0.478	0.021	0.025	0.297	0.119	1.38
$^{187}Os/^{188}Os$	0.1136	0.1130	0.1237	0.1179	0.1253	0.1150	0.1174	0.1068	0.1074	0.1184	0.1100	0.1167
T_{RD} (Ga)	2.1	2.2	0.7	1.5	0.48	1.9	1.6	3	3.0	1.4	2.6	2.1
T_{MA} (Ga)	2.4	7.5	1.1	2.9	0.6	2.1	2.0	3.1	3.4	1.7	2.8	2.7
Al_2O_3 (wt%)	0.89	1.39	2.05	0.27	3.41	1.43	2.05	0.64	1.24	0.91	1.13	1.89

Fertile mantle values of $^{187}Re/^{188}Os = 0.4243$ and $^{187}Os/^{188}Os = 0.1287$ used for model age calculations. PGE normalization values after McDonald *et al.* (1995).

peridotites represent highly depleted melt residues then they should have very low Pd/Ir ratios to complement the depleted nature of their major element compositions. Drastic departures from these theoretically expected trends should then indicate that processes other than partial melting have affected the PGE and hence Re–Os isotope evolution of a rock, and hence may be used to evaluate the significance of the different Os isotope model ages.

If melt or fluid re-enrichment or some other form of disturbance affects a peridotite subsequent to melting then this should be reflected in PGE patterns and PGE–major element systematics. Modelling of melt re-enrichment involving the addition of new sulphides to variably depleted melt residues during magma–solid interaction produces elevation of the Pd/Ir ratio to supra-chondritic levels, and elevated P-PGE contents (Rehkamper et al. 1999; Fig. 4). Combined major element, PGE and Re–Os isotope systematics therefore provide a potential means to evaluate the validity of Re–Os isotope model ages (T_{RD} v. T_{MA}).

To illustrate the application of these methods we have analysed selected samples from different suites of cratonic peridotites for PGE abundances and Re–Os isotopic composition that had been previously analysed for major elements (Table 1). The sample set comprises three peridotites from the Jericho kimberlite, Northern Slave Craton (Irvine et al. 1999; Kopylova & Russell 2000); two peridotites from the Kaapvaal Craton (Pearson et al. 1995a; Carlson et al. 1999); three peridotites from the Farm Louwrencia kimberlite, Southern Namibia, on the periphery of the Kaapvaal Craton (Pearson et al. 1994, 1998a; Hoal et al. 1995; Frantz et al. 1996); three peridotites from the Vitim alkali basalt field, on the southern margin of the Siberian Craton (Ionov et al. 1993; Pearson et al. 1998b).

Examples: the Slave Craton

T_{RD} ages for Slave peridotites show a considerable spread, from c. 500 Ma to 3.1 Ga (Irvine et al. 1999; Fig. 5). A comparable spread, but a

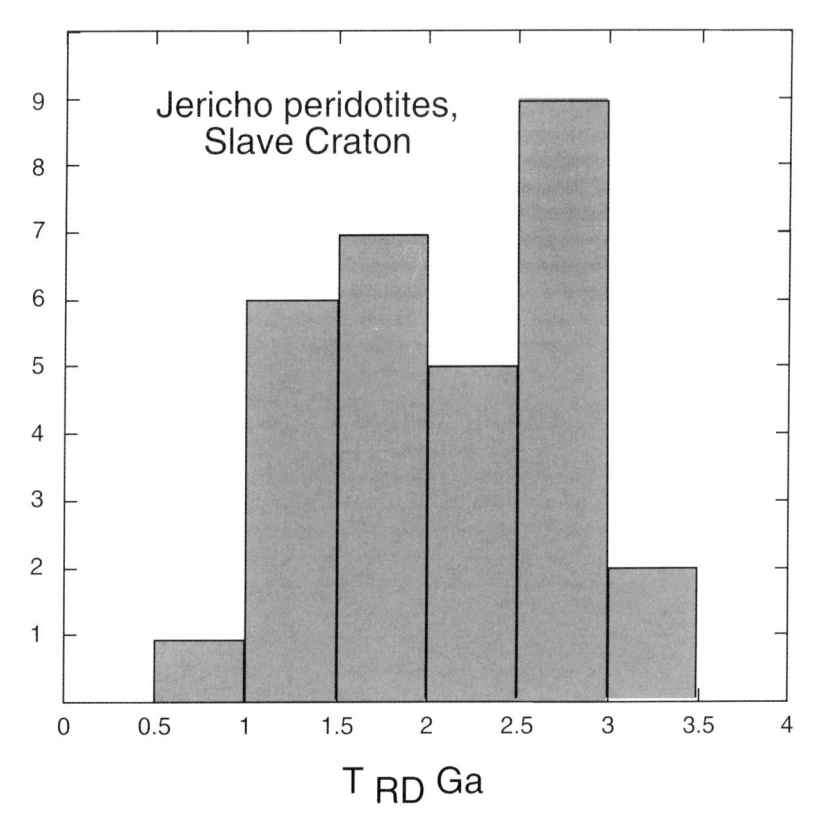

Fig. 5. Histogram of T_{RD} ages for peridotites from the Jericho kimberlite, Nothern Slave Craton. Data from Irvine et al. (2001).

distribution much more peaked in Archaean time, is shown by peridotites from the Kaapvaal Craton (Carlson et al. 1999; Irvine et al. 2001). The meaning of the wide age range in these datasets is not certain and it has been suggested (e.g. Pearson et al. 1995a) that the maximum T_{RD} ages represent the likely formation age of much of the lithosphere whereas the younger ages represent disturbance of primary Re–Os isotope systematics that may, or may not, be of geological significance.

Three peridotites selected from the Jericho suite were analysed for PGEs (Fig. 6a). Re was included with the PGE patterns, to the right of Pd, on the basis of the similar compatibility of Re and Pd. Sample 8-7 shows a very fractionated extended-PGE pattern, with marked depletion of Pt, Pd and Re. This type of PGE pattern is that expected for a peridotite that has experienced extensive melt extraction beyond the level of sulphide stability, as is also reflected in the depleted major element composition of this sample (Fig. 7). The very low S content of this peridotite (c. 10 ppm) and its depleted major element composition indicate that primary sulphides are unlikely to be present and so cannot be analysed. The very low level of Re in sample 8-7 means that the T_{RD} age agrees closely with the T_{MA} age (3.0 v. 3.1 Ga), as the T_{RD} age is calculated on the assumption that Re/Os = 0 whereas T_{MA} uses the measured, in this case very low, Re/Os ratio of the sample. The depleted PGE pattern of this sample, with low Pd concentration and low $(Pd/Ir)_n$ (0.021), together with the low bulk rock Al_2O_3 content, all indicate that this sample is a residue from partial melting and that it has probably

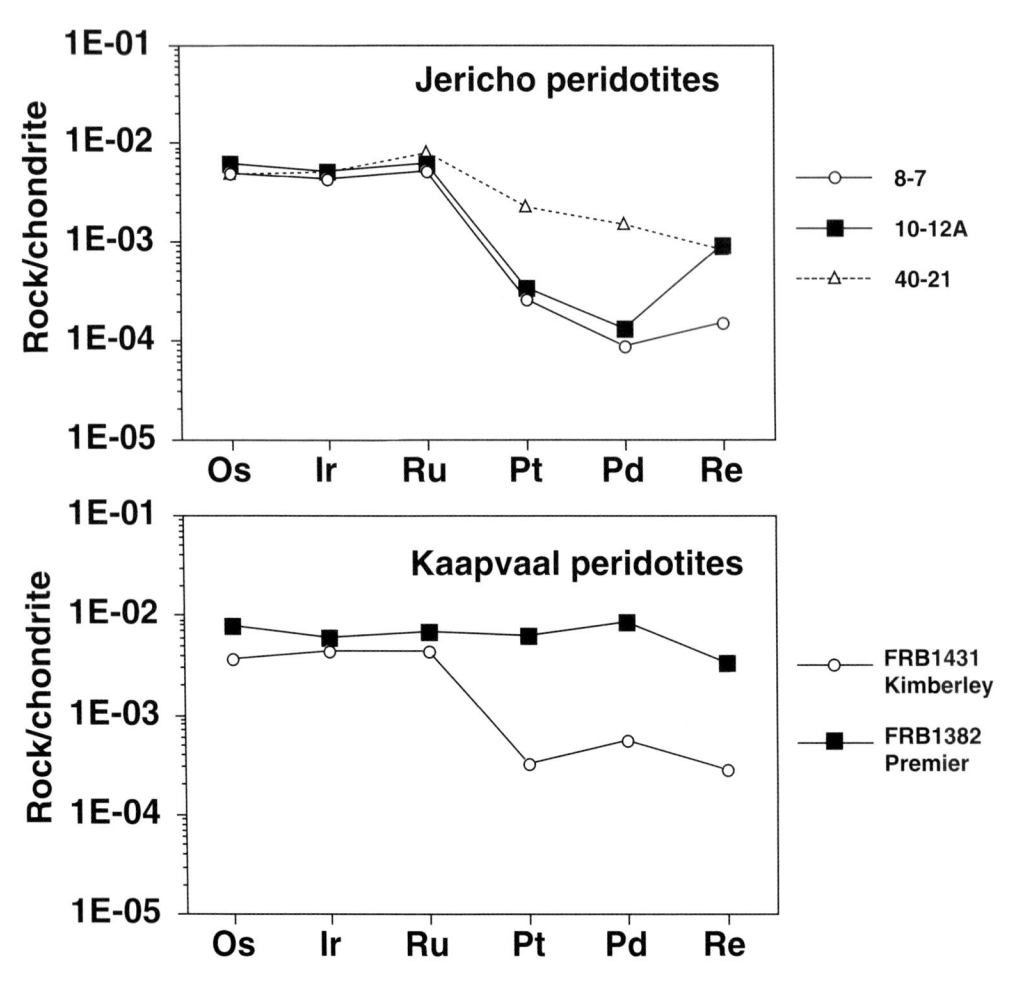

Fig. 6. Chondrite-normalized PGE patterns of whole-rock peridotites from the Jericho kimberlite, Northern Slave Craton and for Kaapvaal peridotites. Data from Table 1. Normalizing values taken from McDonald et al. (1995).

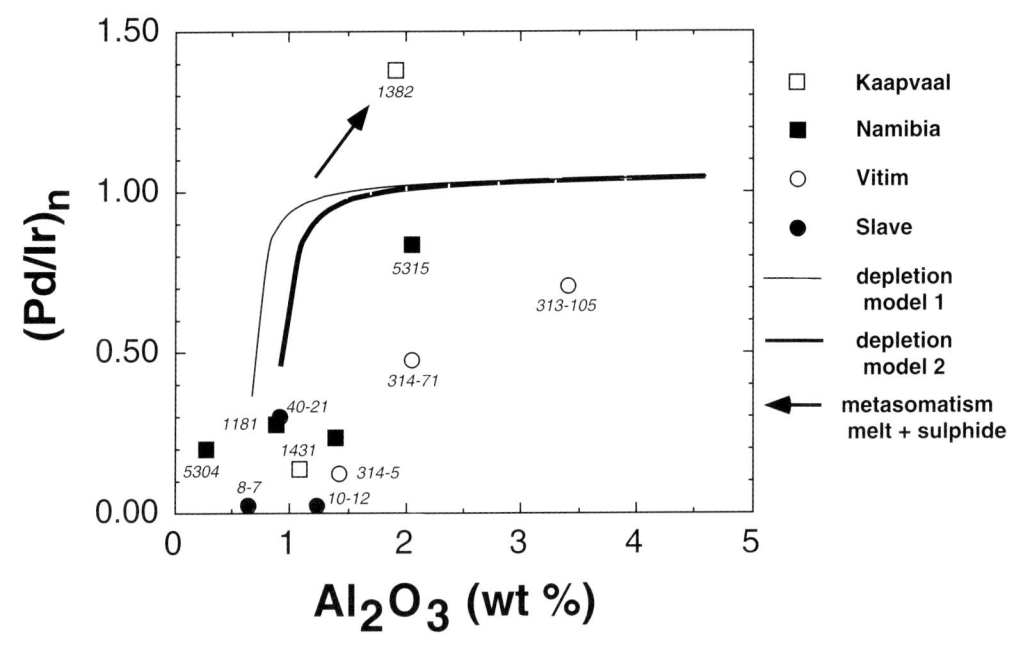

Fig. 7. Platinum group element fractions $(Pd/Ir)_n$ v. bulk-rock Al_2O_3 for selected cratonic and circum-cratonic peridotites. Also plotted are the two melt-depletion curves from Fig. 3. It should be noted that some peridotites from Table 1 plot close to the theoretical melting trends (FRB1181; 40–21) or on an extension of the trends, at very low Pd/Ir, following complete removal of sulphide. These samples have T_{RD} and T_{MA} ages in reasonably close agreement. Other samples are dispersed away from the trends (see text for details).

remained a closed system since the time of melting. The low Re content is thus likely to be a primary feature of the sample and thence the T_{MA} age is a likely approximation of the age of melting.

In contrast to 8-7, sample 10-12A, also from Jericho, has distinctly higher Re content despite having a similar $(Pd/Ir)_n$ ratio (0.025), resulting in a marked upward deflection in the extended PGE pattern at Re (Fig. 6a). As a consequence, there is a substantial difference between the T_{RD} age (2.9 Ga) and the T_{MA} age (3.4 Ga), probably because Re has been introduced recently into the whole-rock peridotite, perhaps by infiltration by the host magma during kimberlite eruption. Significantly, the T_{RD} age is within error of that for sample 8-7, and consistent with a melting event in the Northern Slave lithosphere at $c.$ 3.0 Ga (Irvine et al. 1999). Sample 10-12A has a factor of two higher Al_2O_3 content than 8-7 and plots to the right of the theoretical $(Pd/Ir)_n$ v. Al_2O_3 trend, suggesting that Re addition was also accompanied by Al addition, highlighting the possible complications of using correlations between Al_2O_3 and $^{187}Os/^{188}Os$ to estimate peridotite ages (Reisberg & Lorand 1995). We will show below that in most cases it is better to use the Re–Os systematics of the most depleted samples in suites showing Al_2O_3 v. $^{187}Os/^{188}Os$ correlations, rather than in 'initial' $^{187}Os/^{188}Os$, extrapolated to $Al_2O_3 = 0$.

Excesses of Re and Al of the magnitude shown by sample 10-12A are expected for bulk addition of a metasomite of broadly basaltic composition to the peridotite (Carlson et al. 1999). In contrast, most Kaapvaal peridotites that show Re addition do not have excess Al, suggesting that the Re addition in this case was caused by an Al-poor melt such as carbonatite, the host kimberlite, or similar composition magmas or fluids (Carlson et al. 1999; Pearson 1999b).

Finally, sample 40-21 from Jericho has a much less fractionated PGE pattern than either 8-7 or 10-12A (Fig. 6a), with more moderate depletion of Pt, Pd and Re, relative to Ir and Os; $(Pd/Ir)_n = 0.3$. These characteristics are consistent with the less depleted major element composition of the sample compared with sample 8-7, and the observation that it plots on or close to an extension to the theoretical trend for $(Pd/Ir)_n$ v. Al_2O_3 variation during melting (Fig. 7). There is no obvious sign of open-system behaviour for PGEs in this system. Re is approximately an order of magnitude higher than in sample 8-7, consistent with the possibility that melting was close to the point of sulphide breakdown (Fig. 7), rather than well beyond this point as for sample 8-7. Hence, it is likely that the T_{MA} age (1.7 Ga) is a more accurate reflection of the age of melting experienced by this sample than the T_{RD} age (1.4 Ga), although there is fairly close agreement between the two. For this sample, the T_{MA} age may have geological significance and, as pointed out by Irvine et al. (1999), may correspond to one of the numerous crustal differentiation events that has affected this part of the Slave Craton.

The Kaapvaal Craton

Two central cratonic localities were considered for initial PGE–Os isotope studies, Kimberley and the Premier Mine. The results of a recent large-scale broadband seismic experiment across southern Africa (James & Fouch 2002) identify Kimberley as having very thick (at least 200 km), cool cratonic mantle beneath the area that may have remained relatively undisturbed since Archaean time. In contrast, a prominent low-velocity anomaly is located directly below the Premier area and extending beneath an area covering the areal extent of the large 2050 Ma Bushveld igneous intrusion. These two cratonic mantle 'end-members' are thus expected to show markedly different geochemistry and this has already been suggested on the basis of Re–Os isotope systematics (Pearson et al. 1995a; Carlson et al. 1999).

Peridotites from the 'Old Boshof Road' dump at Kimberley have consistently very unradiogenic Os isotopes and generally low Re contents, indicative of Re (= melt?) depletion and relatively little subsequent disturbance (Carlson et al. 1999). We selected a spinel-facies lherzolite, FRB1431, from this locality to analyse for PGEs and Re–Os isotopes. This sample is rather depleted in Al_2O_3 (1.1 wt%; Table 1) and has a highly fractionated chondrite-normalized PGE pattern (Fig. 6b) similar to the very depleted Slave sample, 8-7 (Fig. 6a). The overall levels of I-PGEs in FRB1431 are slightly lower than in the Slave sample, but there is no significant inter-element fractionation between Os–Ir–Ru. In contrast, the P-PGEs, Pt and Pd, are strongly fractionated from I-PGEs, resulting in low $(Pd/Ir)_n$ (0.12). This sample plots close to the extension of the $(Pd/Ir)_n$–Al_2O_3 melting trends in Figure 7 and this, together with the relatively low Re content, suggests that the bulk-rock system has remained relatively undisturbed since melting. The levels of Re and Al_2O_3 in FRB1431 suggest that it has not experienced the same level of melt depletion as the two most depleted Slave samples. FRB1431 has a moderately unradiogenic $^{187}Os/^{188}Os$ giving a T_{RD} age of 2.6 Ga. This is similar to the mean T_{RD} age for previously published Kaapvaal samples (Carlson et al. 1999; Table 2). The T_{MA} age for 1431 is 2.8 Ga, within error of the mean T_{MA} age for other Kimberley samples (Table 2). Using the criteria outlined above (i.e. the slightly less depleted nature of this rock combined with its relatively undisturbed PGE pattern) suggests that the T_{MA} age provides

Table 2. *Summary of Re–Os model age information for peridotite xenolith suites on and around the Kaapvaal Craton*

Location	Setting	n	Mean T_{RD} (Ga)	Median T_{RD}	Mean T_{MA} (Ga)	Data source
Northern Lesotho	SE Kaapvaal	36	2.5 ± 0.5	2.8	4.7 ± 3.2	1, 2
Northern Lesotho*	SE Kaapvaal	25	2.8 ± 0.1	2.8		1
Kimberley	SW Kaapvaal	9	2.6 ± 0.2	2.7	2.9† ± 0.1	3; Table 1
Letlhakane	Kaapvaal–Zimbabwe	10	2.6 ± 0.2	2.7	3.2 ± 0.5	3
Newlands	SW Kaapvaal	13	2.6 ± 0.4	2.6	3.0 ± 0.3	4
Finsch	SW Kaapvaal	3	2.4 ± 0.4	2.7	3.2 ± 0.9	1
Venetia	Limpopo–Kaapvaal	7	2.5 ± 0.8	2.7	2.5 ± 0.9	3
Premier	Central Kaapvaal	9	2.2 ± 0.4	2.3	2.5 ± 0.6	1, 3
Eastern Griqualand	Off SE margin, Kaapvaal	14	1.7 ± 0.4	1.6	‡	5
Southern Namibia	Off Western margin Kaapvaal	16	1.4 ± 0.6	1.4	4.0 ± 2.7	6; Table 1

The ± values are 1 standard deviation of the mean. Data sources: 1, Irvine et al. (2001); 2, Pearson et al. (1995a); 3, Carlson et al. (1999), 4, Menzies et al. (1999); 5, Pearson et al. (1994); 6, Pearson et al. (1998a)
* Lesotho samples selected for depleted character and with T_{RD} in range 2.5 to 3.0 Ga.
† Not including one sample >9 Ga and including FRB1431 in Table 1.
‡ Continuing study; Re concentrations not all available yet.

the most accurate indication of the age of melting and this is in line with other Kimberley peridotites, implying a major melt removal event at 2.8–2.9 Ga, similar to that advocated for the mantle beneath Northern Lesotho, in the east of the Kaapvaal Craton.

In contrast to the relative uniformity of Re–Os model ages observed for Kimberley peridotites, those from Premier appear essentially bimodal (Pearson et al. 1995a; Carlson et al. 1999). Some samples have T_{RD} ages that are clearly Archaean and in the range of other ancient T_{RD} ages from Kaapvaal peridotites (Pearson et al. 1995a; Carlson et al. 1999). In contrast, a number of Premier peridotites have T_{RD} ages that cluster close to 2 Ga. These ages are significant in that the Premier kimberlite is the only locality studied on the Kaapvaal Craton that penetrates the Bushveld layered intrusion and it is likely that substantial proportions of the lithospheric mantle beneath this area have been considerably modified by magmatic activity associated with this intrusion.

Previous discussions of the Re–Os isotope systematics of Premier peridotites have suggested that the Proterozoic (c. 2 Ga) T_{RD} ages result from either substantial metasomatism of pre-existing lithosphere during Bushveld magmatic activity (Pearson et al. 1995a) and/or new cratonic mantle created by melt removal at this time (Carlson et al. 1999). Boyd & Mertzman (1987) pointed out that the Mg/Si ratios (or modal orthopyroxene contents) of Premier peridotites span a similar range to those of 'typical' Kaapvaal peridotites and hence may have formed by the same process, in Archaean time. This observation would imply that the younger peridotite model ages represent major disturbance of pre-existing lithospheric material. In addition to modification of the Re–Os isotope system, Shimizu (1999) has documented clear evidence of trace element variability within Premier peridotites that can only reflect disturbance or silicate mineral growth just before kimberlite eruption at c. 1200 Ma. Clearly, Premier peridotites have experienced a multi-stage geochemical evolution, making interpretation of Re–Os isotope systematics and model ages difficult without other criteria to help.

The Premier sample selected for PGE study (FRB1382) is typical of the dilemma faced in interpreting the Re–Os isotope results in isolation. This sample has a T_{RD} age of 2.2 Ga, close to the Bushveld age, but a T_{MA} age of 2.9 Ga, typical of the bulk of Kaapvaal peridotites. Thus, one explanation is that the sample formed as a melt residue during the formation of the Bushveld Complex at c. 2 Ga and has been subsequently enriched in Re, giving an anomalously old T_{MA} age. Alternatively, this sample may have experienced only moderate depletion of Re during a melting event at 2.9 Ga, allowing significant ^{187}Os growth since then and producing a T_{RD} that is too young. In this case, if the Re content was a simple function of melting then the T_{MA} age would be the real age. Yet another explanation for the Re–Os isotope systematics of FRB1382 is that both Re and radiogenic Os have been introduced to this rock during metasomatism related to Bushveld or host kimberlite magmatism, as found for Tanzanian peridotites (Chesley et al. 1999) and hence neither T_{RD} nor T_{MA} age is accurate.

The PGE data for FRB1382 are illuminating in terms of differentiating between these alternatives. FRB1382 has a very different chondrite-normalized PGE pattern from the Kimberley sample discussed above (Fig. 6b). The sample does not show substantial depletion of P-PGEs from I-PGEs and in fact has a supra-chondritic $(Pd/Ir)_n$ (1.4) such that it plots significantly above the $(Pd/Ir)_n$–Al_2O_3 melting trends of Fig. 7 in an area likely to be occupied by residues that have interacted with sulphide-bearing melt (e.g. Rehkamper et al. 1999). The Re abundance is also distinctly elevated compared with Kimberley samples (Carlson et al. 1999). This sample thus appears to have experienced a multistage evolution of its PGE geochemistry and is strongly influenced by some type of melt addition process. This means that model age approaches employing measured elemental ratios will lead to inaccurate results and that the T_{MA} age is unlikely to be accurate. In addition to Re and P-PGE addition, it is possible that radiogenic Os has been added to FRB1382, as suggested for peridotites from Tanzania (Chesley et al. 1999). Hence the T_{RD} age of FRB1382 would be biased towards the age of the event introducing younger Os, in this case resulting in a T_{RD} age close to the age of the Bushveld event. Hence, the relatively young T_{RD} ages of some of the peridotite xenoliths derived from Premier may reflect substantial disturbance of Archaean peridotites during Bushveld magmatism. Keleman et al. (1998) predicted that such extensive melt–rock reaction, and the growth of new silicate phases, would generate a correlation between the Ni content of olivine and the modal olivine content. Of various peridotite xenolith suites studied from the Kaapvaal Craton, the suite from Premier is the only one to show such a correlation as yet (Boyd 1997), thus supporting the notion of extensive magma–peridotite interaction at this locality, superimposed on older protoliths.

Circum-craton examples:
Vitim and Namibia

Compared with cratonic peridotites, xenoliths from circum-cratonic peridotite suites, erupted through Proterozoic crust, have elemental compositions characteristic of lower levels of melt depletion (Hawkesworth *et al.* 1990; Boyd 1996; Griffin *et al.* 1999). This is clearly the case for the two examples selected here, and our aim is to establish whether the more fertile compositions correspond to younger ages, as is commonly presumed. Compositions of the Vitim peridotite suite (Ionov *et al.* 1993) are substantially more fertile than those of peridotites derived from the adjacent Siberian Craton (Boyd *et al.* 1997). This is also the situation with peridotite xenoliths derived from the south–central Namibian kimberlites, which have significantly lower average olivine mg-numbers than peridotites from within the Kaapvaal Craton (Boyd *et al.* 1994).

The less melt-depleted character of circum-cratonic peridotites is reflected in their less fractionated PGE patterns, $(Pd/Ir)_n$ values of the Namibian suite varying from 0.2 to 0.83 (Figs 7 and 8). Within both suites of off-craton peridotites there are samples with PGE patterns that are not consistent with single-stage melt extraction histories (Fig. 8). Some samples have either $(Re/Pd)_n > 1$ or $(Pd/Pt)_n > 1$, indicating complex, multistage histories perhaps involving the type of melt–rock reaction process identified by Rehkamper *et al.* (1999) in abyssal peridotites. One sample from each of the Vitim and Namibian suites has a PGE pattern that appears undisturbed for P-PGEs and Re (Fig. 8). Both of these samples (314-5 from Vitim and FRB 1181 from Namibia) have T_{RD} and T_{MA} ages that are in close

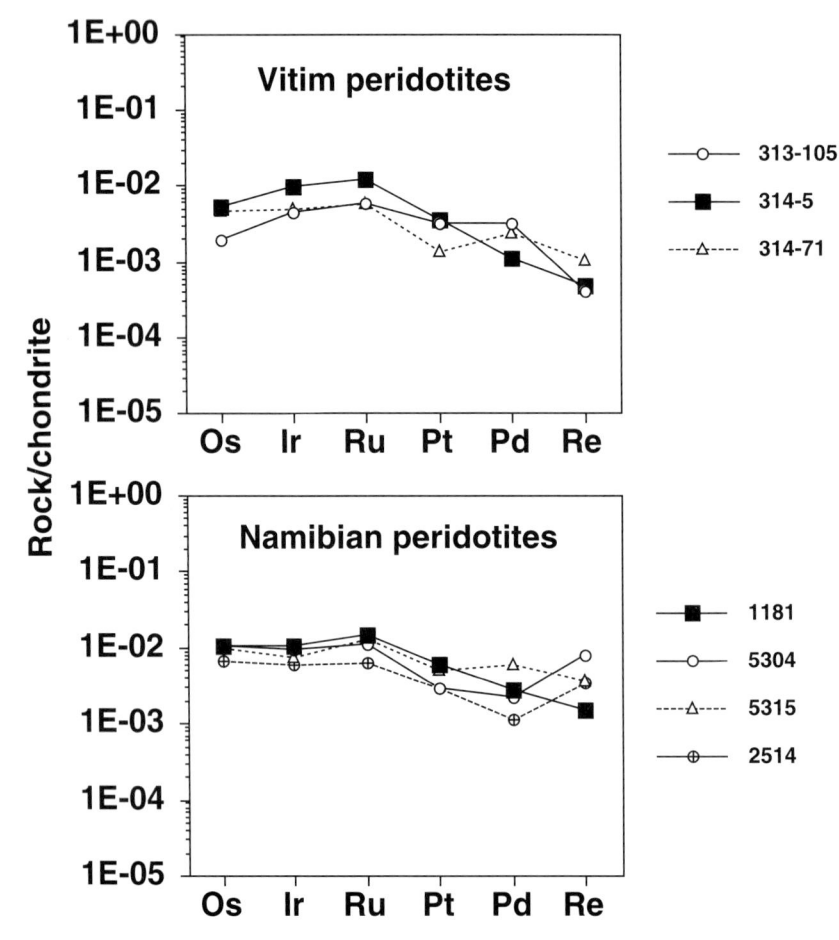

Fig. 8. Chondrite-normalized PGE patterns of whole-rock peridotites from the Vitim region and Namibia (see data in Table 1). Normalizing values taken from McDonald *et al.* (1995).

agreement (Table 1). The T_{RD} age of sample FRB 1181 is amongst the oldest of this suite and approximates very well the likely formation age defined by the 2.2 Ga T_{MA} age (Table 1). Such an age for the lithospheric mantle beneath this region of South Africa overlaps the age of the basement of the Rehoboth Subprovince, from which this peridotite was derived (Hoal et al. 1995), which is thought to be between 1.7 and 2.3 Ga on the basis of U–Pb zircon dating and Nd model age constraints (Ziegler & Stoessel 1991).

The Vitim peridotite with the least disturbed extended-PGE pattern in Fig. 8 sample 314-5, also has similar T_{RD} and T_{MA} ages (1.8 and 2.0 Ga, respectively). The T_{RD} age of 1.8 Ga is the oldest of this suite of peridotites and is similar to the Sm–Nd clinopyroxene model ages in these rocks (Ionov & Jagoutz 1989). In contrast to cratonic rocks, the Sm–Nd isotope systematics of most Vitim peridotites are characteristic of melt depletion, i.e. clinopyroxenes have Sm/Nd > chondrite and mineral REE patterns do not show the complexity of enriched cratonic rocks (Ionov & Jagoutz 1989; Ionov et al. 1993). Hence, in contrast to cratonic rocks (Pearson 1999b), the Sm–Nd model ages of Vitim rocks may reflect the partial melting event that produced them. This possibility is supported by the coincidence of Sm–Nd and Re–Os model ages for sample 314-5 and indicates a Palaeoproterozoic age (between 1.8 and 2.0 Ga) as a likely differentiation age for these rocks.

The Vitim peridotite suite shows a very good correlation between Os isotope composition and Al_2O_3 content (Pearson et al. 1998b), especially compared with other circum-cratonic peridotite xenolith suites such as those from Namibia (Fig. 9). Estimation of the differentiation age using the technique of Reisberg & Lorand (1995), by extrapolation of the Al–Os isotope trend to the intercept, produces an 'age' of c. 3 Ga (Fig. 9), clearly much older than the estimates from Re–Os and Sm–Nd model ages. Given the dominantly fairly fertile composition of these peridotites (Ionov et al. 1993) and the trace element chemistry of their minerals, it appears unlikely that they could be of Archean age (Ionov et al. 1993; Griffin et al. 1999) and the Archaean age for this suite is probably an artefact of the problems associated with the Al–Os isotope method. Pearson (1999a) has discussed the limitations of extrapolating Al–Os isotope correlations to obtain age information. The most significant problem is that it is possible to remove Re from a system completely as a result of *mss* breakdown while still retaining aluminous phases. Thus, Re can approximate to zero well before the unlikely event of Al being completely removed from the

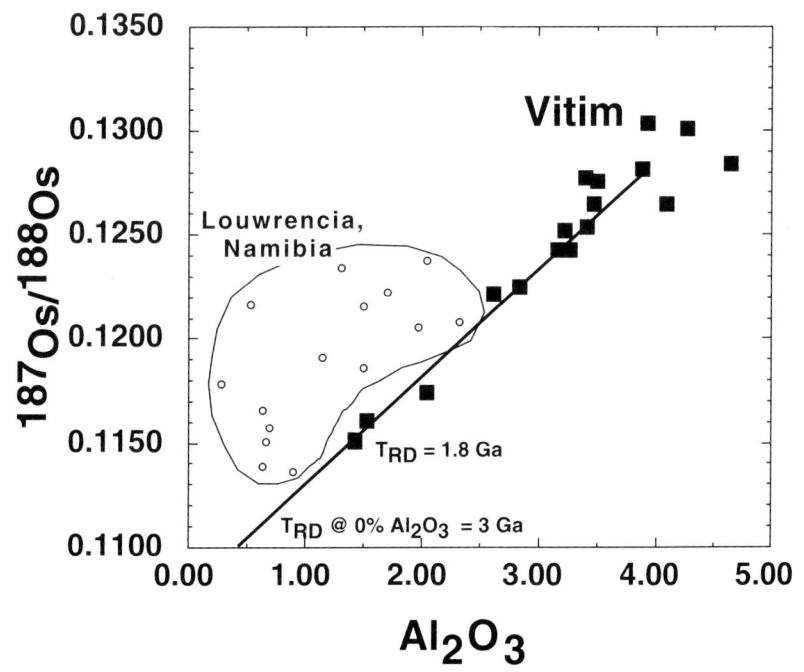

Fig. 9. Al_2O_3–$^{187}Os/^{188}Os$ correlation for the Vitim peridotites suite (Pearson et al. 1998b). The y-intercept of this correlation gives a $^{187}Os/^{188}Os$ value of 0.107, $r^2 = 0.974$, for samples with <4 wt% Al_2O_3. Louwrencia data from Pearson et al. 1998a.

system; hence, such extrapolations are likely to overestimate the melting age of most peridotite suites. In the case of the Vitim xenoliths, the T_{RD}–T_{MA} model age approach for one of the most melt depleted samples, with undisturbed PGE–Re systematics, probably gives the more reliable estimation of the differentiation event, in the absence of single-sulphide studies. This age is 1.8–2.0 Ga and is 1 Ga younger than the age produced using the Al_2O_3–Os isotope correlation. The scattered nature of the $(Pd/Ir)_n$ v. Al_2O_3 relationships for Vitim samples (Fig. 6) indicates the possibility of multiple events affecting Al–PGE–Re systematics and this must be borne in mind when interpreting the significance of the Al–Os isotope relationships shown by this suite.

The above examples illustrate the potential for combined PGE and Re–Os isotope studies to evaluate the significance of Re–Os isotope model ages in peridotite xenolith suites. Coherence of samples with theoretical major-element–PGE fractionation trends provide a firmer basis for evaluating the significance of the oldest T_{RD} model ages of a given peridotite suite. It also allows more confident assessment of the significance of younger T_{RD}–T_{MA} model ages within a peridotite suite, which may signify later additions of new lithosphere where samples appear to exhibit the PGE–major element chemistry expected for melt residues. Further understanding of PGE behaviour during melting will improve this approach and when combined with analysis of sulphides included in silicates wherever possible or available, should do much to improve our understanding of the timing of the initial differentiation events that have formed continental lithospheric mantle.

Kaapvaal Craton case study

On-craton peridotites

Despite the peridotites forming the Kaapvaal Craton keel being of anomalous composition, particularly with respect to SiO_2, far more Re–Os isotope data are available for xenoliths forming this cratonic keel than any other. This is largely because of the excellent availability of xenoliths as a result of diamond mining and also because of intensive study of all aspects of this craton as part of a major, multi-disciplinary, multi-institutional international research project (Carlson et al. 1999). In particular, the high-resolution seismic tomography study (James & Fouch 2002) allows direct comparison of xenolith observations and craton structure evident today. In addition, because of the abundance of data, including Re–Os isotope data, for several circumcratonic areas in this region (Pearson et al. 1994, 1998; Janney et al. 1999; Pearson 1999b), the Kaapvaal Craton allows an un-rivalled opportunity to study the formation of lithospheric mantle beneath and around a craton (Nixon 1987).

Significant numbers of peridotites from six kimberlites distributed across the Kaapvaal Craton have been analysed (Table 2). Except for the Premier kimberlite (discussed below), mean T_{RD} and T_{MA} ages from each suite are exclusively Archaean (Table 2) and a marked peak in the T_{RD} histogram occurs between 2.5 and 3 Ga (median 2.6 Ga), with >60 samples giving T_{RD} ages in this range (Fig. 10). As pointed out by Irvine et al. (2001), initial Re–Os isotope work on Kaapvaal peridotites (Walker et al. 1989; Pearson et al. 1995a) selected from a group of samples that had previously been studied as part of thermobarometry investigations (e.g. Nixon & Boyd 1973) because of the desirability to study well-characterized samples. This selection process imparted a bias towards cpx–garnet-rich samples that lend themselves to mineral thermobarometry and against highly depleted samples. Recent elemental and isotopic studies combined with modal analyses (Fig. 1) suggest that some of the garnet and diopside in cratonic low-T lherzolites is likely to have grown recently, as a result of melt–rock interaction (Keleman et al. 1992; Gunther & Jagoutz 1994; Pearson et al. 1995b). Hence, selection of lherzolites in preference to harzburgites in earlier Re–Os studies has probably oversampled peridotites that have experienced melt–rock interaction and undersampled those that have a more simple history, dominated by melt depletion. The analysis of potentially disturbed or metasomatized lherzolites probably accounts for much of the skewness to younger ages seen in Figure 10.

Irvine et al. (2002) analysed a suite of peridotites from the Letseng, Thaba Putsua and Matsoku kimberlites, Northern Lesotho, on the SE margin of the Kaapvaal Craton. These samples were mostly selected from cpx-free or cpx-poor peridotites that have compositions dominated by melt depletion, with some samples being obviously more fertile. The mean T_{RD} age for the whole suite, unfiltered for very garnet–cpx-rich samples, is 2.5 ± 0.5 Ga ($n = 36$), median 2.7 Ga (Table 2). If only the most melt-depleted (those with low bulk-rock Al_2O_3), nondeformed samples are included in the average, a much more tightly clustered mean T_{RD} age of 2.8 Ga ± 0.1 ($n = 36$; Table 2) is calculated. The very tight clustering of these T_{RD} ages suggests that a large part of the lithospheric mantle beneath this part of the Kaapvaal Craton stabilized during a differentiation event c. 2.8 Ga ago. Other, younger

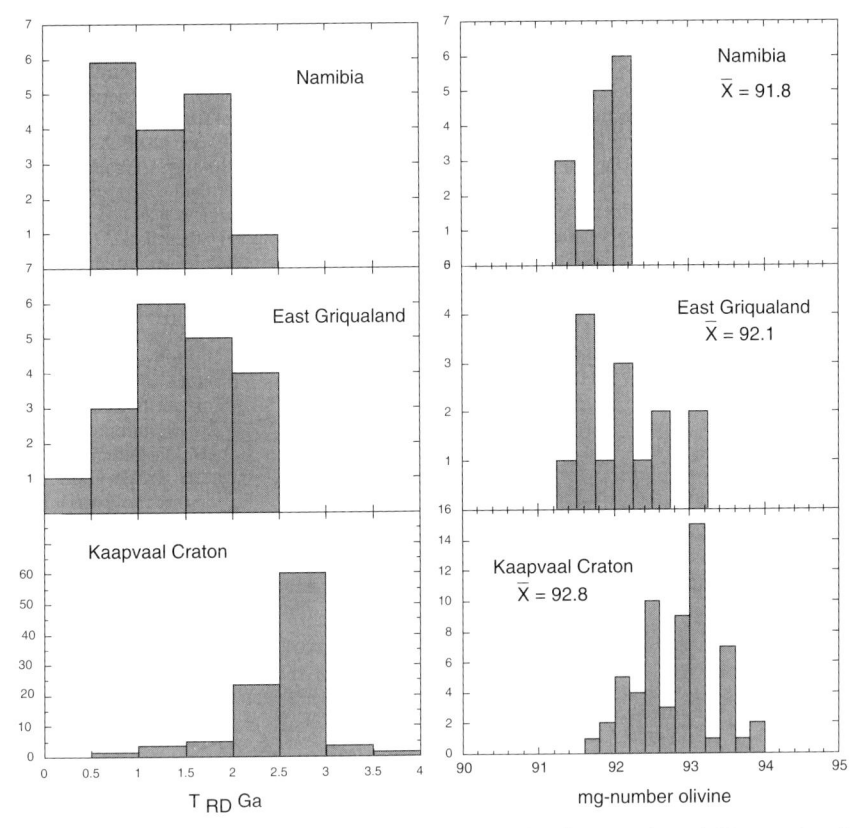

Fig. 10. Histograms of T_{RD} ages for on- and off-craton peridotites from southern Africa co-plotted with histograms of olivine mg-numbers. Kaapvaal low-T peridotites xenoliths, $n = 96$. Data from Nixon *et al.* (1983), Walker *et al.* (1989), Pearson *et al.* (1995*a*), Carlson *et al.* (1999), Menzies *et al.* (1999) and Irvine *et al.* (2002). T_{RD} ages for peridotite xenoliths from kimberlites from East Griqualand, 170 km from the Lesotho on-craton kimberlites (Pearson *et al.* 1998) and for peridotites from the Farm Louwrencia kimberlite, Southern Namibia (Pearson *et al.* 1994; Pearson 1999*a*). It should be noted that no Archaean T_{RD} ages are observed for off-craton peridotites. Olivine compositional data from Boyd & Nixon (1979), Nixon (1987) and Boyd (unpubl. data).

T_{RD} ages from the Northern Lesotho peridotite suite coincide with the various U–Pb zircon ages of lower-crustal xenoliths from this region (Schmitz & Bowring 2000; Irvine *et al.* 2001) indicating that major underplating or metamorphic events affecting the lower Kaapvaal crust were accompanied by additions of new material to the lithospheric mantle.

Peridotites from the Newlands kimberlite, in the southern part of the Kaapvaal Craton, have a similar, but less tightly clustered mean T_{RD} age of 2.6 ± 0.4 Ga, median 2.6 Ga (Table 2) compared with the Northern Lesotho peridotites. Those from the Kimberley area, in the SW of the craton, show a tightly clustered distribution of T_{RD} and T_{MA} ages with a mean T_{RD} age of 2.6 ± 0.2 Ga, median 2.7 Ga, and a mean T_{MA} age of 2.9 ± 0.1 Ga (Table 2). Peridotites forming lithosphere in the north of the craton (Letlhakane, Botswana) are also clearly of Archaean age (Carlson *et al.* 1999). Peridotites from the Venetia mine in South Africa have predominantly Archaean T_{RD} ages ranging up to 3.68 Ga with a mean of 2.96 Ga. Two Venetia samples give Mesoproterozoic T_{RD} and T_{MA} ages (1.3 and 1.5 Ga) indicating the presence of younger material in this mantle section. The Venetia kimberlite erupts through the Archaean Limpopo Metamorphic Belt, which is often interpreted as the suture between the Kaapvaal and Zimbabwean Cratons. The Re–Os systematics of the Venetia peridotites clearly indicates the presence of Archaean mantle beneath this area, but also the presence of a Mesoproterozoic component (Carlson *et al.* 1999).

Of the on-craton peridotites studied so far, the suite from the Premier mine, in the centre of the Kaapvaal Craton, is the only one to have an average T_{RD} age that is not Archaean (Table 2).

As discussed previously, numerous Premier peridotites give T_{RD} ages close to 2.0 Ga (Pearson et al. 1995a; Carlson et al. 1999). The Premier kimberlite erupted near the 2.05 Ga Bushveld intrusion and the Palaeoproterozoic model ages probably reflect substantial modification of the lithosphere beneath this region at c. 2 Ga plus possibly some new lithospheric material (Carlson et al. 1999).

Although many of the T_{MA} ages for the above peridotite suites are also Archaean, there is considerably more scatter in these ages than in the T_{RD} ages (Table 2), probably as a result of later Re disturbance. Detailed PGE–Os isotope studies have not been carried out on all these peridotite suites as yet. Several samples have clearly unrealistic T_{MA} ages (>4.5 Ga) and numerous samples have T_{MA} ages >4 Ga. No samples have been found with >4 Ga T_{RD} ages. It is our view that the >4 Ga T_{MA} ages are probably the result of varying levels of Re and PGE (including Os isotope) disturbance and should be treated with caution. Combined use of PGE data plus Re–Os isotope systematics together with an increasingly large peridotites database indicate that previous estimates of the age of cratonic lithosphere beneath the Kaapvaal Craton may not apply to the craton as a whole. Selection of samples that are depleted in major elements and that show highly fractionated PGE patterns indicative of a single-stage, large-degree melt removal episode are the most reliable samples to obtain age information pertaining to melt depletion. Sulphides, when available, can also be used, providing that their petrological context is well understood and that they are primary. Few of the cratonic peridotites that we have analysed contain primary sulphides. Where a sample does contain abundant sulphide it is often of a secondary nature and can be used to understand the metasomatic history of the lithosphere (Alard et al. 2000; Pearson et al. 2000).

Crust–mantle evolution on the Kaapvaal Craton.
A large number of T_{RD} model ages for whole-rock peridotites erupted across the Kaapvaal Craton cluster around a median value of 2.7 Ga, with small numbers of samples having T_{RD} ages >3 Ga (Fig. 10). Previous studies (e.g. Pearson et al. 1995a) noting the fact that T_{RD} ages are minimum estimates of the time of melt depletion have suggested that the oldest T_{RD} ages, i.e. those c. 3.5 Ga, may represent the stabilization age of the bulk of the cratonic lithosphere. In contrast, the distribution of T_{RD} ages shown in Figure 10 and Table 2 together with the close agreement of T_{RD} and T_{MA} ages at Kimberley, suggests that much of the lithospheric mantle beneath the Kaapvaal Craton formed or was stabilized in the period 2.7–3.0 Ga, i.e. in Neoarchaean time. The Kimberley results are particularly intriguing in this regard in that these peridotites show a very restricted range in both T_{RD} and T_{MA} ages, with a mean T_{MA} of 2.9 ± 0.1 Ga (Table 2). Kimberley lies at the centre of the best-developed seismic lithosphere of the Kaapvaal Craton (James & Fouch 2002), i.e. the seismically fast cratonic lithosphere is thickest and least disrupted beneath the Kimberley area. Such lithospheric mantle might have been expected to be the oldest and least disturbed of the entire craton. The mean peridotite T_{MA} age from Kimberley is nearly identical to the 2.89 ± 0.06 Ga Re–Os isochron age recently obtained for eclogite-suite sulphide inclusions in Kimberley diamonds (Richardson et al. 2001). Mid-crustal gneisses from the Kimberley area have zoned zircons with cores of 3.25 Ga age and rims of 2.94 Ga age (Drennen et al. 1990). Such age systematics for Kimberley suggests that initial crust formation (≥3.25 Ga) substantially predates lithospheric mantle formation at c. 2.9 Ga. The c. 2.9 Ga age is recorded throughout the lithosphere at Kimberley, from the crust through the peridotitic lithospheric mantle at least to depths within the diamond stability field where eclogitic materials, perhaps introduced through subduction at 2.89 Ga (Richardson et al. 2001), provide the source of sulphide-bearing diamonds. In this context it is interesting to note that Re–Os isotope model age estimates of lithospheric mantle formation beneath the Tanzanian Craton are also 2.8–2.9 Ga (Chesley et al. 1999), within error of our estimate of the bulk of the Kaapvaal lithospheric mantle 1000 km to the south. A <3 Ga age for much of the cratonic lithosphere beneath southern Africa introduces the interesting possibility that cratonic lithosphere formation, i.e. cratonization, significantly postdates initial, Mesoarchaean crust formation and reflects a separate tectonothermal event possibly associated with convergent tectonics, magmatism and the amalgamation of discrete crustal blocks into a rigid, thick, cratonic keel.

Recent U–Pb studies of lower-crustal xenoliths, from kimberlites where Re–Os isotope studies have been undertaken on mantle peridotites, and on lower-crustal samples from the Vredefort impact structure, further illuminate the timing of lithosphere stabilization and support the notion of widespread lithospheric mantle stabilization in Neoarchaean time. These studies reveal major episodes of Meso- to Neoarchaean thermal activity in the lower crust (Schmitz & Bowring 2000; Moser et al. 2002). Some of the lower-crustal xenoliths from the craton margins record later, Proterozoic thermal or magmatic events (Schmitz & Bowring 2000) that are also observed in the peridotites (Irvine *et al.* 2001). Many of the new data indicate a very widespread, craton-wide differentiation event that affected the entire crustal thickness at *c.* 3.1 Ga (Moser *et al.* 2002). Such a major, widespread crustal differentiation event is inconsistent with the presence of a protective, thick (>150 km) lithospheric mantle keel beneath the entire craton at this time, and Moser *et al.* (2002) have suggested that this event represents a maximum age for the establishment of a thick, protective lithospheric root beneath the Kaapvaal Craton. Waning of major crustal differentiation and the onset of more stable continental conditions may have coincided with the stabilization of a thick (200 km; Boyd & Gurney 1986), protective lithospheric mantle keel beneath much of the craton at this time. Our new data from Lesotho, combined with scrutiny of the existing dataset, particularly from Kimberley, make a strong case for the largest fraction of lithospheric mantle being stabilized in Neoarchaean time (Fig. 10). Neoarchaean rather than Mesoarchaean lithospheric formation has also been suggested for the Tanzanian Craton (Chesley *et al.* 1999). This model is consistent with the likely absence of a protective keel beneath large parts of the cratonic crust during major intra-crustal differentiation at *c.* 3.1 Ga (Moser *et al.* 2002). At this point it is important to bear in mind that the Os isotope evolution of the Earth's mantle is not well constrained and that different model age input parameters can alter ages by up to 200 Ma. Hence, it is possible that a number of the *c.* 2.9 Ga Kaapvaal model ages are actually closer to 3.1 Ga. None the less, these ages remain later than the main crust-forming events and do not precede the last major central cratonic intracrustal differentiation proposed by Moser *et al.* (2002).

A Neoarchaean or end Mesoarchaean age for much of the lithospheric mantle beneath the Kaapvaal Craton does not preclude the prior existence of significantly older, shallow or deep lithospheric mantle beneath certain areas, as suggested by data for silicate diamond inclusions (e.g. Richardson *et al.* 1984), Mesoarchaean T_{RD} ages from some xenoliths (Pearson *et al.* 1995a; Carlson *et al.* 1999) and very ancient (3.8 Ga) peridotites from Zimbabwe (Nagler *et al.* 1997), but requires amalgamation of these terranes by Neoarchaean time. It might also be that some diamonds were formed in deep lithospheric mantle (assuming our knowledge of diamond formation is correct) that was then tectonically emplaced or mixed with younger mantle during final amalgamation of the craton in Neoarchean time. The mantle beneath the Kimberley area might be a good example of this.

Off-craton peridotites

Although the majority of the peridotite xenoliths erupted through Archaean crust indicate formation in Archaean time, T_{RD} ages from xenolith suites surrounding the Kaapvaal Craton are invariably of Proterozoic age (Pearson *et al.* 1994, 1998a; Janney *et al.* 1999; Pearson 1999a; Figs 10 and 11; Table 2). For the East Griqualand and Namibian suites, no samples give Archaean T_{RD} ages (Fig. 9) and this is also the case for other off-craton localities (Fig. 11; Janney *et al.* 1999). The consistency of Archaean T_{RD} ages for on-craton peridotites and Proterozoic T_{RD} ages for off-craton peridotites is another observation that lends some confidence that T_{RD} ages, when cautiously interpreted, provide a good indication of the age of lithosphere formation.

The lack of Archaean T_{RD} ages in circum-cratonic regions of the Kaapvaal Craton indicates that, at least in the areas studied, the cratonic lithospheric mantle does not extend laterally very far beneath younger Proterozoic crust and hence boundaries between cratons and their surrounds are relatively steep in most instances. The symmetry of Archaean ages for mantle immediately beneath the cratons and Proterozoic mantle beneath circum-cratonic regions indicates long-term crust–mantle coupling over billions of years during plate movement. As noted above, the correspondence of the older peridotite T_{RD} ages from Farm Louwrencia (Namibia), within an extension of the Namaqua–Natal Crustal Belt, corresponds to the age of the oldest crustal rocks known in that region. This age relationship is also the case at localities to the SW of the craton margin, within the Namaqua–Natal Belt (Janney *et al.* 1999) and suggests that the crust and lithospheric mantle in these regions, and within the Namaqua–Natal Belt as a whole, could have form simultaneously over large distances. Janney *et al.* (1999) noted younger T_{RD} ages in xenolith suites approaching the edge of the Cape Fold Belt.

The earliest continental lithospheric mantle

The results of intensive study of Kaapvaal Craton peridotites have implications for different craton generation models and hence for the stabilization of the earliest continents. It has been suggested that the opx-rich Kaapvaal lithospheric mantle could have formed from primitive, Si-enriched material in Hadean time (Herzberg 1993). There are no clear indications of Hadean

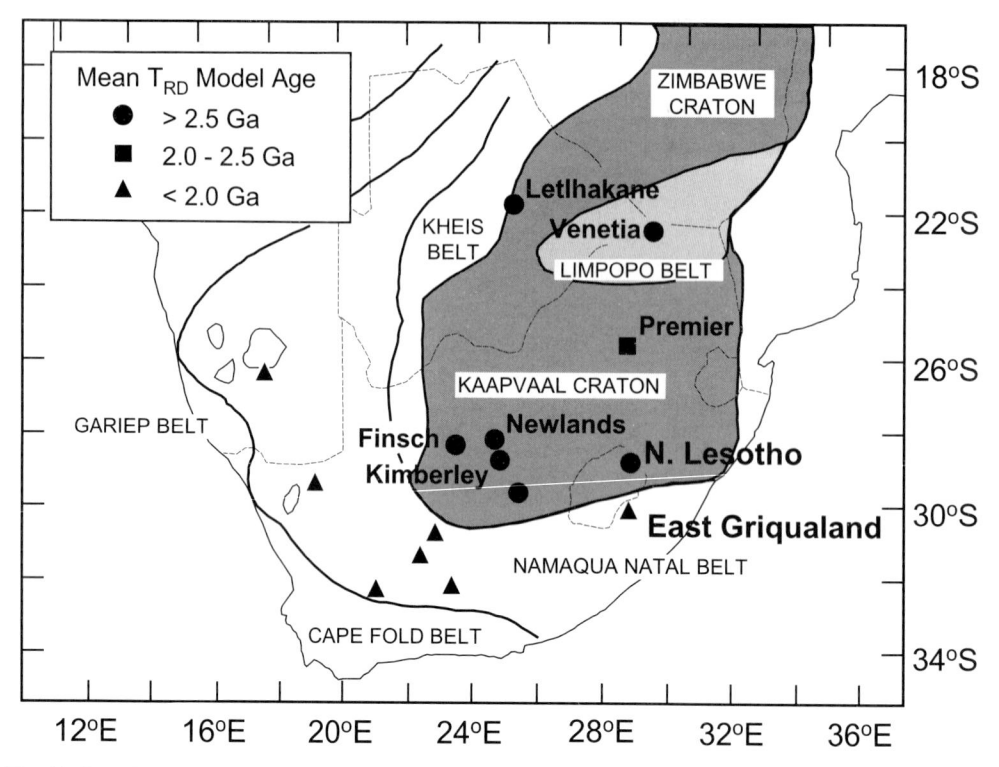

Fig. 11. Location map of kimberlites on and around the Kaapvaal Craton from which peridotite xenolith suites have been dated by Re–Os isotopes (after Carlson et al. 2000; Irvine et al. 2001). Data sources: Pearson et al. (1994, 1995a, 1998); Pearson (1999); Carlson et al. (1999); Janney et al. (1999); Irvine et al. (2001). Kaapvaal Craton is outlined in dark grey with the Archean Limpopo Metamorphic Belt shaded in light grey.

ages in the large number (>100) of lithospheric peridotites analysed for Re–Os isotopic composition so far and thus no support is provided for such a model. Another explanation of the opx-rich nature of some Kaapvaal peridotites is that some may be cumulates or even trapped melts (Herzberg 1999). Continuing PGE characterization of Kaapvaal samples, including several samples with high opx contents, has not yet revealed any samples with PGE systematics that might be consistent with a cumulate origin.

It is now accepted that the bulk compositions of Archaean lithospheric peridotites, including many Kaapvaal peridotites (SiO_2 problems aside), are consistent with an origin as residues of varying degrees of partial melting, over a range of pressures. (Herzberg 1999; Walter 1999). The abundances of PGEs in cratonic peridotites, in particular I-PGEs, and the fractionation of these elements from P-PGEs and Re, is consistent with a partial melting residue interpretation for most samples. Bulk composition data indicate that melting was initiated at between 6 and 7 GPa (Herzberg 1999; Walter 1999). Such pressures are in excess of those associated with the melting regime beneath modern mid-ocean ridges. Whether these depths of melting could have pertained beneath Archaean mid-ocean ridges is important in constraining whether melting and formation of large volumes of lithospheric mantle took place at ridges, followed by subduction, or as a result of mantle upwelling associated with plume activity. Alternatively, melting might have taken place within the mantle wedge at Archaean subduction zones if cratonic mantle represents the amalgamation of Archaean arcs; however, 6–7 GPa may be excessively deep for such an environment, even in Archaean time, when slab dip angles may not have been as steep (de Wit et al. 1992). Other workers have argued that some Kaapvaal peridotites, notably those containing sub-calcic garnets, are probably the residua of shallow melting processes, at <2.5 GPa (Bulatov et al. 1991; Canil & Wei 1992). However, such compositions are rare (Boyd et al. 1993) and could represent minor components of the lithosphere introduced by subduction processes.

The overabundance of garnet and clinopyroxene in most Kaapvaal peridotites (Fig. 1) indicates that these phases probably crystallized some time after melt depletion, during melt–rock (Keleman et al. 1998) or hydrous fluid–rock reaction (as proposed for some Kaapvaal clinopyroxenes (Boyd & Mertzman 1987)). The post-depletion addition of substantial garnet and clinopyroxene would explain the discrepancies in melt models for Kaapvaal peridotites based on comparison of Fe–Mg systematics v. Al–Mg systematics (Walter 1999).

The recently enlarged Re–Os database for Kaapvaal samples indicates that significant volumes of Kaapvaal lithosphere may have been generated in Neoarchaean time (Fig. 10), with major intracrustal differentiation events ceasing after this time (Moser et al. 2002). This is consistent with thermal modelling of cratonic keel evolution (De Smet et al. 2000) and with models of crust generation that suggest progressive growth of the continents through Archaean and into Proterozoic time (Abbott et al. 2000). One interpretation of the cluster of peridotite T_{RD} ages between 2.7 and 3.0 Ga for Kaapvaal Craton peridotites is that this may represent the generation of buoyant lithospheric mantle during a major, widespread and voluminous magmatic event. The most likely products of this event seen in the southern African crust are the Ventersdorp lavas. It should also be noted that volumetrically significant basaltic and komatiitic magmatism was also occurring north of the Kaapvaal Craton, to form the Belingwe Greenstone Belt (Nisbit et al. 1977) at c. 2.7 Ga, and hence this time could have represented a major differentiation event just before final stabilization of the Kaapvaal Craton as it appears today. The mismatch between Kaapvaal peridotite compositions and the likely coeval melt products required in such a scenario (Boyd 1989) could be produced by subsequent melt addition events to the lithosphere (Keleman et al. 1992; Burgess & Harte 1999).

The 3.8 Ga T_{RD} ages of tectonically emplaced chromite-bearing peridotites from the Kalahari Craton (Nagler et al. 1997), together with the few Mesoarchaean xenolith T_{RD} ages (Pearson et al. 1995a) and diamond inclusion data (Richardson et al. 1984), are evidence that melting and formation of lithospheric mantle was occurring beneath southern Africa significantly earlier than Neoarchaean time. However, the volume of material forming at this time appears to be small and may have been restricted to either shallow, or laterally inextensive but deep lithospheric keels that later amalgamated and stabilized in late Archaean time. This scenario for craton evolution is similar to that suggested for the stabilization of the Kaapvaal Craton by de Wit et al. (1992) on the basis of crustal geology, although those workers emphasized the role of subduction processes over plumes, a possibility supported by recent evidence that most of the eclogite xenoliths from the Kaapvaal are of Archaean age (Shirey et al. 2001). Continuing studies of other cratons should soon indicate how typical the sequence of differentiation events recorded by the Kaapvaal Craton is for continent formation generally. Recent Re–Os data for East Greenland peridotites indicates that the formation of some shallow lithospheric mantle on the North Atlantic craton was initiated in Palaeoarchaean time (Hanghoj et al. 2001).

Griffin et al. (1999) have suggested that only the buoyant products of lithospheric mantle formation are preserved beneath cratons. If earlier events formed lithosphere that was denser, these products will not founder beneath the crust. Such nonbuoyant lithosphere would not be preserved at relatively shallow levels for sampling and would play no role in the preservation of large volumes of very ancient crust. Hence, only large-scale crust-forming differentiation events that produced buoyant melt residua appear capable of preserving large areas of ancient crust. These nuclei then amalgamate to form the cratons that we see today. Density calculations for average cratonic peridotite compositions show that they are 1.5% less dense than Phanerozoic lithospheric mantle (Griffin et al. 1999) and thus despite their cooler geothermal gradients, Archaean lithospheric compositions are capable of supporting buoyant lithospheric mantle thicknesses up to, or in excess of 200 km (Jordan 1988). De Smet et al. (2000) have pointed out that the increased rheological strength as a result of the temperature dependence of mantle viscosity arising from cool Archaean lithospheric mantle geotherms, when combined with their compositional buoyancy, makes Archaean cratonic roots very difficult to destroy once formed. This makes them very effective protectors of crustal material over billion year time scales.

Recent work on the effects of mantle metasomatism on water solubility in nominally anhydrous minerals (Peslier et al. 2000) has shown that the (OH) contents of these minerals decrease during the increase in oxidation state accompanying metasomatism. This has the possible implication that the rheological strength of cratons actually increases as a result of metasomatism. The ubiquitous nature of metasomatism in cratonic peridotites (Menzies & Hawkesworth 1987; Pearson 1999b) may have the result of providing additional strength to an already robust lithospheric keel. There are specific instances when such lithosphere is destroyed or removed from beneath cratons (Eggler *et al.* 1988; Menzies *et al.* 1998), and numerous instances where rifting occurs around craton margins (e.g. Thompson & Gibson 1991; Ebinger & Sleep 1998), but many cratons appear to preserve much of their original Archaean lithospheric keels, probably because of the ability of cratonic roots to deflect hot upwelling plume material to its margins (Sleep *et al.* 2002).

Extensive studies of many xenolith suites over the past 20 years now reveal clear secular trends in the composition of lithospheric mantle through time (Boyd 1989; Menzies 1990; Griffin *et al.* 1998, 1999). The compositions of Proterozoic peridotite suites that occur around cratons such as southern Africa (Figs 10 and 11) are also buoyant relative to present-day asthenosphere. This, and their relatively cool geotherms (Boyd *et al.* 1994; Frantz *et al.* 1996) make them robust features of continental plates. The symmetry in apparent ages on and around the Kaapvaal Craton, with Archaean crust underlain by Archaean mantle, and Proterozoic crust being underlain by Proterozoic mantle, shows the long-term coupling of crust plus lithospheric mantle in continental plates over billion year periods. Only around their margins do cratons appear capable of substantial disruption or reworking. If post-Archaean effects are seen inside cratons these may be manifest at suture lines where discrete blocks had joined together to initially form the craton. Delamination of the basal lithosphere may occur where it has been refertilized by melt introduction, making it more dense, or in younger, more fertile post-Archaean mantle (Hawkesworth *et al.* 1990; Griffin *et al.* 1999).

To date, Re–Os isotope studies of peridotite xenolith suites, taken to be representative of the lithospheric mantle beneath continents, reveal that although some lithospheric mantle began to form beneath continents in Palaeo- to Mesoarchaean time, major crustal differentiation events continued until Neoarchaean time when, at least in southern Africa, significant volumes of lithospheric mantle formed and stabilized the newly accreted continent. On the Kaapvaal Craton, major magmatic events such as that which formed the Bushveld intrusion, substantially modified or introduced new material into the lithospheric mantle (Carlson *et al.* 1999) and the lower crust (Schmitz & Bowring 2000), but the craton retained its integrity for another 2 Ga.

Much of this research resulted from the interdisciplinary Kaapvaal Craton Project funded by NSF (EAR-9526840). G. J. I.'s contribution arose during tenure of NERC PhD studentship GT04/97/73/ES. J. Boyd generously shared his bulk analysis database and ideas. We thank F. R. Boyd, M. Menzies, C. Hawkesworth and an anonymous reviewer for helpful comments on the manuscript. M. Fowler, C. Ebinger and C. Hawkesworth are thanked for their invitation to participate in the Fermor 'Early Earth' meeting.

References

ABBOTT, D., SPARKS, D., HERZBERG, C., MOONEY, W., NIKISHIN, A. & ZHANG, Y. S. 2000. Quantifying Precambrian crustal extraction: the root is the answer. *Tectonophysics*, **322**, 163–190.

ALARD, O., GRIFFIN, W. L., LORAND, J. P., JACKSON, S. E. & O'REILLY, S. Y. 2000. Non-chondritic distribution of the highly siderophile elements in mantle sulphides. *Nature*, **407**, 891–894.

ALLÈGRE, C. J. & LUCK, J. M. 1980. Os isotopes as petrogenetic and geologic tracers. *Earth and Planetary Science Letters*, **48**, 148–154.

ARMSTRONG, R. L. 1981. Radiogenic isotopes: the case for crustal recycling on a near-steady-state no-continental-growth Earth. *Philosophical Transactions of the Royal Society of London Series A*, **301**, 443–472.

BERNSTEIN, S., KELEMAN, P. B. & BROOKS, C. K. 1998. Highly depleted spinel harzburgite xenoliths in Tertiary dikes from east Greenland. *Earth and Planetary Science Letters*, **154**, 221–235.

BOYD, F. R. 1989. Compositional distinction between oceanic and cratonic lithosphere. *Earth and Planetary Science Letters*, **96**, 15–26.

BOYD, F. R. 1996. Origins of peridotite xenoliths: major element considerations: high pressure and high temperature research on lithosphere and mantle materials. *Proceedings of the International School of Earth and Planetary Sciences, Siena*, 89–106.

BOYD, F. R. 1997. Correlation of orthopyroxene abundance with the Ni-content of coexisting olivine in cratonic peridotites. *EOS Transactions, American Geophysical Union*, **78**, 746.

BOYD, F. R. & GURNEY, J. J. 1986. Diamonds and the African lithosphere. *Science*, **232**, 472–477.

BOYD, F. R., & MERTZMAN, S. A. 1987. Composition and structure of the Kaapvaal lithosphere, Southern Africa. *In*: MYSEN, B. O. (ed.) *Magmatic Processes: Physicochemical Principles*. Geochemical Society, Houston, TX, 3–12.

BOYD, F. R. & NIXON, P. H. 1979. Garnet lherzolite xenoliths from the kimberlites of East Griqualand, southern Africa. *Carnegie Institute of Washington Yearbook*, **80**, 328–336.

BOYD, F. R., PEARSON, D. G., NIXON, P. H. & MERTZMAN, S. A. 1993. Low Ca garnet harzburgites from southern Africa: their relation to craton structure and diamond crystallisation. *Contributions to Mineralogy and Petrology*, **113**, 352–366.

BOYD, F. R., PEARSON, D. G., OLSON, H. K. E. & HOAL, B. G. 1994. Composition and age of Namibian peridotite xenoliths; a comparison of cratonic and non-cratonic lithosphere. *EOS Transactions, American Geophysical Union*, **75**, 192.

BOYD, F. R., POKHILENKO, N. P., PEARSON, D. G., MERTZMAN, S. A., SOBOLEV, N. V. & FINGER, L. W.

1997. Composition of the Siberian cratonic mantle: evidence from Udachnaya peridotite xenoliths. *Contributions to Mineralogy and Petrology*, **128**, 228–246.

BRENAN, J. M. & ANDREWS, D. 2001. High temperature stability of laurite and Ru–Os–Ir alloy and their role in PGE fractionation in mafic magmas. *Canadian Mineralogist*, **39**, 573–592.

BULATOV, V., BREY, G. P. & FOLEY, S. F. 1991. Origin of low-Ca high-Cr garnets by recrystallisation of low-pressure harzburgites. *Extended Abstracts, Fifth International Kimberlite Conference, Araxa, Brazil*. CPRM Special Publication **2/91**, 29–31.

BURGESS, S. R. & HARTE, B. 1999. Tracing lithospheric evolution through the analysis of heterogeneous G9/G10 garnets in peridotite xenoliths, I: Major element chemistry. *In*: GURNEY, J. J., GURNEY, J. L., PASCOE, M. D. & RICHARDSON, S. H. (eds) *Proceedings of the 7th International Kimberlite Conference, Cape Town*. Red Roof Design, Cape Town, 66–80.

BURTON, K. W., SCHIANO, P., BIRK, J.-L. & ALLÈGRE, C. J. 1999. Osmium isotope disequilibrium between mantle minerals in a spinel-lherzolite. *Earth and Planetary Science Letters*, **172**, 311–322.

CANIL, D. & WEI, K. 1992. Constraints on the origin of mantle-derived low Ca garnets. **109**, 421–430.

CARLSON, R. W. & IRVING, A. J. 1994. Depletion and enrichment history of sub-continental lithospheric mantle: Os, Sr, Nd and Pb evidence for xenoliths from the Wyoming Craton. *Earth and Planetary Science Letters*, **126**, 457–472.

CARLSON, R. W., BOYD, F. R., SHIREY, S. B. & 14 OTHERS 2000. Continental growth, preservation and modification in southern Africa. *GSA Today*, **10**, 1–7.

CARLSON, R. W., PEARSON, D. G., BOYD, F. R., SHIREY, S. B., IRVINE, G., MENZIES, A. H. & GURNEY, J. J. 1999. Regional age variation of the southern African mantle: significance for models of lithospheric mantle formation. *In*: GURNEY, J. J., GURNEY, J. L., PASCOE, M. D. & RICHARDSON, S. H. (eds) *Proceedings of the 7th International Kimberlite Conference, Cape Town*. Red Roof Design, Cape Town, 99–108.

CHESLEY, J. T., RUDNICK, R. L. & LEE, C. T. 1999. Re–Os systematics of mantle xenoliths from the East African Rift: age, structure and history of the Tanzanian craton. *Geochimica et Cosmochimica Acta*, **63**, 1203–1217.

DE SMET, J., VAN DE BERG, A. P. & VLAAR, N. J. 2000. Early formation and long-term stability of continents resulting from decompression melting in a convecting mantle. *Tectonophysics*, **322**, 19–33.

DE WIT, M. J., ROERING, C., HART, R. J. & 6 OTHERS 1992. Formation of an Archaean continent. *Nature*, **357**, 553–562.

DRENNEN, G. R., ROBB, L. J., MEYER, F. M., ARMSTRONG, R. A. & DE BRUIYN, H. 1990. The nature of the Archaean basement in the hinterland of the Witwatersrand Basin: II. A crustal profile west of the Welkom Goldfield and comparisons with the Vredefort crustal profile. *South African Journal of Geology*, **93**, 41–53.

EBINGER, C. J. & SLEEP, N. H. 1998. Cenozoic magmatism throughout east Africa resulting from the impact of a single plume. *Nature*, **395**, 788–791.

EGGLER, D. H., MEEN, J. K., WELT, R., DUDAS, F. O., FURLONG, K. P., MCCALLUM, M. E. & CARLSON, R. W. 1988. Tectonomagmatism of the Wyoming Province. *Colorado School of Mines Quarterly*, **83**, 25–40.

FERMOR, L. L. 1913. Preliminary note on garnet as a geological barometer and on an infra-plutonic zone in the Earth's crust. *Reconnaissances of the Geological Survey of India*, **43**, 41–47.

FRANTZ, L., BREY, G. P. & OKRUSCH, M. 1996. Steady state geotherm, thermal disturbances, and the tectonic development of the lower lithosphere underneath the Gibeon kimberlite Province, Namibia. *Contributions to Mineralogy and Petrology*, **126**, 181–198.

GRIFFIN, W. L., O'REILLY, S. Y., RYAN, C. G., GAUL, O. & IONOV, D. 1998. Secular variations in the composition of subcontinental lithospheric mantle: geophysical and geodynamic implications, *In*: BRAUN, J., DOOLEY, J., GOLEBY, B., VAN DER HILST, R. & KLOOTWIJK, C. (eds) *Structure and Evolution of the Australian Continent*. Geophysical Monograph, American Geophysical Union, 1–26.

GRIFFIN, W. L., O'REILLY, S. Y. & RYAN, C. G. 1999. The composition and origin of sub-continental lithospheric mantle. *In*: FEI, Y., BERTHKA, C. M. & MYSEN, C. M. (eds) *Mantle Petrology: Field Observations and High Pressure Experimentation: a Tribute to Francis R. (Joe) Boyd*. Geochemical Society, Special Publications, **6**, 13–45.

GUNTHER, M. & JAGOUTZ, E. 1994. Isotopic disequilibria (Sm/Nd, Rb/Sr) between minerals of coarse grained, low temperature peridotites from Kimberley floors, southern Africa. *In*: MEYER, H. O. A. & LEONARDOS, O. H. (eds) *Kimberlites, Related Rocks and Mantle Xenoliths. Proceedings of the 5th International Kimberlite Conference, Araxa, Brazil*. CPRM Special Publications, **1/a**, 354–365.

HANDLER, M. R. & BENNETT, V. C. 1999. Behaviour of platinum-group elements in the subcontinental mantle of eastern Australia during variable metasomatism and melt depletion. *Geochimica et Cosmochimica Acta*, **63**, 3597–3618.

HANGHOJ, K., KELEMAN, P. B., BERNSTEIN, S., BLUSTZTAJN, J. & FREI, R. 2001. Osmium isotopes in the Weidemann Fjord mantle xenoliths, a unique record of cratonic mantle formation by melt depletion in the Archean. *Geochemistry, Geophysics and Geosystems*, **2**, 85–92.

HARTE, B., HUNTER, R. H. & KINNY, P. D. 1993. Melt geometry, movement and crystallisation in relation to mantle dykes, veins and metasomatism. *Philosophical Transactions of the Royal Society of London, Series A*, **A342**, 1–21.

HAWKESWORTH, C. J., ERLANK, A. J., MARSH, J. S., MENZIES, M. A. & VAN CALSTEREN, P. 1983. Evolution of the continental lithosphere: evidence from volcanics and xenoliths from southern Africa. *In*: NORRY, M. J. & HAWKESWORTH, C. J. (eds) *Continental Basalts and Mantle Xenoliths*. Shiva, Nantwich, 111–138.

HAWKESWORTH, C. J., KEMPTON, P. D., ROGERS, N. W., ELLAM, R. M. & VAN CALSTEREN, P. W. 1990. Continental mantle lithosphere, and shallow level enrichment processes in the Earth's mantle. *Earth and Planetary Science Letters*, **96**, 256–268.

HERZBERG, C. T. 1993. Lithosphere peridotites of the Kaapvaal craton. *Earth and Planetary Science Letters*, **120**, 13–29.

HERZBERG, C. T. 1999. Phase equilibrium constraints on the formation of cratonic mantle. *In*: FEI, Y., BERTHKA, C. M. & MYSEN, B. O. (eds) *Mantle Petrology: Field Observations and High Pressure Experimentation: a Tribute to Francis R. (Joe) Boyd*. Geochemical Society, Special Publication, **6**, 241–257.

HOAL, B. G., HOAL, K. E. O., BOYD, F. R. & PEARSON, D. G. 1995. Age constraints on crustal and mantle lithosphere beneath the Gibeon kimberlite field, Namibia. *South African Journal of Geology*, **98**, 112–118.

HOLMES, A. & PANETH, F. A. 1936. Helium ratios of rocks and minerals from the diamond pipes of South Africa. *Proceedings of the Royal Society of London, Series A*, **154**, 385–413.

IONOV, D. A. & JAGOUTZ, E. 1989. Sr and Nd isotopic composition in minerals of garnet and spinel peridotite xenoliths from the Vitim Highland: first data for mantle inclusions in the USSR. *Transactions (Doklady) of the USSR Academy Sciences, Earth Science Section*, **301**, 232–236.

IONOV, D. A., ASHCHEPKOV, I. V., STOSCH, H.-G., WITT-EICKSCHEN, G. & SECK, H. A. 1993. Garnet peridotite xenoliths from the Vitim volcanic field, Baikal Region: the nature of the garnet–spinel peridotite transition zone in the continental mantle. *Journal of Petrology*, **34**, 1141–1175.

IRVINE, G. I., CARLSON, R. W., KOPYLOVA, M. G., PEARSON, D. G., SHIREY, S. B. & KJARSGAARD, B. A. 1999, Age of the lithospheric mantle beneath and around the Slave craton: a rhenium-osmium isotopic study of peridotite xenoliths from the Jericho and Somerset Island kimberlites. *Abstracts Ninth Annual V. M. Goldschmidt Conference*. Lunar and Planetary Institute Contribution, **971**, 134–135.

IRVINE, G. I., PEARSON, D. G. & CARLSON, R. W. 2001. Lithospheric mantle evolution in the Kaapvaal craton. A Re–Os isotope study of peridotite xenoliths from Lesotho kimberlites. *Geophysical Research Letters*, **28**, 2505–2508.

JAMES, D. E. & FOUCH, M. J. 2002. Formation and evolution of Archaean cratons: insights from southern Africa. *In*: FOWLER, C. M. R., EBINGER, C. J. & HAWKESWORTH, C. J. (eds) *The Early Earth: Physical, Chemical and Biological Development*. Geological Society, London, Special Publications, **199**, 1–26.

JANNEY, P. A., CARLSON, R. W., SHIREY, S. B., BELL, D. R. & LE ROUX, A. P. 1999. Temperature, pressure and rhenium–osmium age systematics of off-craton peridotite xenoliths from the Namaqua-Natal belt, western South Africa. *Abstracts, Ninth Annual V. M. Goldschmidt Conference*. Lunar and Planetary Institute Contribution, **971**, 139.

JORDAN, T. H. 1988. Structure and formation of the continental tectosphere. *Journal of Petrology, special issue*, 11–37.

KELEMAN, P. B., DICK, H. J. B. & QUICK, J. E. 1992. Formation of harzburgite by pervasive melt/rock reaction in the upper mantle. *Nature*, **358**, 635–641.

KELEMAN, P. B, HART, S. R. & BERNSTEIN, S. 1998. Silica enrichment in the continental upper mantle via melt/rock reaction. *Earth and Planetary Science Letters*, **164**, 387–406.

KINZLER, R. J. & GROVE, T. L. 1999. Origin of depleted cratonic harzburgite by deep fractional melt extraction and shallow olivine cumulate infusion. *In*: GURNEY, J. J., GURNEY, J. L., PASCOE, M. D. & RICHARDSON, S. H. (eds) *Proceedings of the 7th International Kimberlite Conference, Cape Town*. Red Roof Design, Cape Town, 437–443.

KOPYLOVA, M. G. & RUSSELL, J. K. 2000. Chemical stratification of cratonic lithosphere: constraints from the Northern Slave craton, Canada. *Earth and Planetary Science Letters*, **181**, 71–87.

KRAMERS, J. D. 1979. Lead, uranium, strontium, potassium and rubidium in inclusion-bearing diamonds and mantle derived xenoliths from southern Africa. *Earth and Planetary Science Letters*, **42**, 58–70.

LORAND, J.-P., PATTOU, L. & GROS, M. 1999. Fractionation of platinum-group elements and gold in the upper mantle. A detailed study in Pyrenean orogenic lherzolites. *Journal of Petrology*, **40**, 957–981.

MAALOE, S. & AOKI, K. 1977. The major element composition of the upper mantle estimated from the composition of lherzolites. *Contribution to Mineralogy and Petrology*, **63**, 161–173.

MCDONALD, I., DE WIT, M. J., SMITH, C. B., BIZZI, L. A. & VILJOEN, K. S. 1995. The geochemistry of the platinum-group elements in Brazilian and southern African kimberlites. *Geochimica et Cosmochimica Acta*, **59**, 2883–2903.

MCKENZIE, J. M. & CANIL, D. 1999. Composition and thermal evolution of cratonic mantle beneath the central Archean Slave Province, NWT, Canada. *Contributions to Mineralogy and Petrology*, **134**, 313–324.

MENZIES, A. H., SHIRLEY, S. B., CARLSON, R. W. & GURNEY, J. J. 1999. Re–Os systematics of Newlands peridotite xenoliths: implications for diamond and lithosphere formation. *In*: GURNEY, J. J., GURNEY, J. L., PASCOE, M. D. & RICHARDSON, S. H. (eds) *Proceedings of the 7th International Kimberlite Conference*. Red Roof Design, Cape Town, 566–583.

MENZIES, M. & MURTHY, R. V. 1980. Enriched mantle: Nd and Sr isotopes in diopsides from kimberlite nodules. *Nature*, **283**, 634–636.

MENZIES, M. A. 1990. Archaean, Proterozoic and Phanerozoic lithospheres. *In*: MENZIES, M. A. (ed.) *Continental Mantle*. Clarendon Press, Oxford, 67–86.

MENZIES, M. A. & HAWKESWORTH, C. J. 1987. *Mantle Metasomatism*. Academic Press, London.

MENZIES, M. A., XU, Y., FLOWER, M. F. J., CHUNG, S. L., LO, C. H. & LEE, T. Y. 1998. Geodynamics of the North China Craton. *In*: CHUNG, S. L., LO, C. H. & LEE, T. Y. (eds) *Mantle Dynamics and*

Plate Interactions in East Asia. American Geophysical Union, Washington, DC, 155-165.

MOSER, D. E., FLOWERS, R. M. & HART, R. J. 2002. Birth of the Kaapvaal tectosphere 3.08 billion years ago. *Science*, **291**, 465-468.

NAGLER, T. F., KRAMERS, J. D., KAMBER, B. S., FREI, R. & PRENDERGAST, M. D. A. 1997. Growth of subcontinental lithospheric mantle beneath Zimbabwe started at, or before 3.8 Ga: Re-Os study on chromites. *Geology*, **25**, 983-986.

NISBET, E. G., BICKLE, M. J. & MARTIN, A. 1977. The mafic and ultramafic lavas of the Belingwe Greenstone Belt, Rhodesia. *Journal of Petrology*, **18**, 521-566.

NIXON, P. H. 1987. *Mantle Xenoliths*, Wiley, Chichester.

NIXON, P. H. & BOYD, F. R. 1973. Petrogenesis of the granular and sheared ultrabasic nodule suite in kimberlite. *In*: NIXON, P. H. (ed) *Lesotho Kimberlites*. Cape and Transvaal, Cape Town, 48-56.

NIXON, P. H., VON KNORRING, O. & ROOKE, J. M. 1963. Kimberlites and associated inclusions of Basutoland. A mineraological and geochemical study. *American Mineralogist*, **48**, 1090-1132.

NIXON, P. H., BOYD, F. R. & BOCTOR, N. Z. 1983. East Griqualand kimberlites. *Transactions of the Geological Society of South Africa*, **86**, 221-236.

PEARSON, D. G. 1999a. The age of continental roots. *Lithos*, **48**, 171-194.

PEARSON, D. G. 1999b. Evolution of cratonic lithospheric mantle: an isotopic perspective. *In*: FEI, Y., BERTKA, C. M. & MYSEN, B. O. (eds) *Mantle Petrology: Field Observations and High Pressure Experimentation: a Tribute to Francis R. (Joe) Boyd.* Geochemical Society, Special Publications, **6**, 57-78.

PEARSON, D. G. & WOODLAND, S. J., 2000. Carius tube digestion and solvent extraction/ion exchange separation for the analysis of PGEs (Os, Ir, Pt, Pd, Ru) and Re-Os isotopes in geological samples by isotope dilution ICP-mass spectrometry. *Chemical Geology*, **165**, 87-107.

PEARSON, D. G., BOYD, F. R., HOAL, K. E. O., HOAL, B. G., NIXON, P. H. & ROGERS, N. W. 1994. A Re-Os isotopic and petrological study of Namibian peridotites: contrasting petrogenesis and composition of on- and off-craton lithospheric mantle. *Mineralogical Magazine*, **58A**, 703-704.

PEARSON, D. G., CARLSON, R. W., BOYD, F. R., SHIREY, S. B. & NIXON, P. H. 1998a. Lithospheric mantle growth around cratons: a Re-Os isotope study of peridotite xenoliths from East Griqualand. *Extended Abstracts, 7th International Kimberlite Conference, Cape Town.* University of Cape Town, Cape Town, 658-660.

PEARSON, D. G., CARLSON, R. W., SHIREY, S. B., BOYD, F. R. & NIXON, P. H. 1995a. Stabilisation of Archaean lithospheric mantle; a Re-Os isotope study of peridotite xenoliths from the Kaapvaal Craton. *Earth and Planetary Science Letters*, **134**, 341-357.

PEARSON, D. G., IONOV, D., CARLSON, R. W. & SHIREY, S. B. 1998b. Lithospheric evolution in circum-cratonic settings: a Re-Os isotope study of peridotite xenoliths from the Vitim region, Siberia. *Mineralogical Magazine*, **62A**, 1147-1148.

PEARSON, D. G., SHIREY, S. B., BULANOVA, G. P., CARLSON, R. W. & MILLEDGE, H. J. 1999. Single crystal Re-Os isotope study of sulphide inclusions from a zoned Siberian diamond. *Geochimica et Cosmochimica Acta*, **63**, 703-712.

PEARSON, D. G., SHIREY, S. B., CARLSON, R. W., BOYD, F. R., POKHILENKO, N. P. & SHIMIZU, N. 1995b. Re-Os, Sm-Nd, and Rb-Sr isotope evidence for thick Archaean lithospheric mantle beneath the Silurian craton modified by multistage metasomatism. *Geochimica et Cosmochimica Acta*, **59**, 959-977.

PEARSON, N. J., ALLARD, O., GRIFFIN, W. L., GRAHAM, S. & JACKSON, S. E. 2000. LAM-MC-ICPMS analysis of mantle-derived sulphides: the key to Re-Os systematics of mantle peridotites. *Journal of Conference Abstracts*, **5(2)**, 777.

PESLIER, A. H., LUHR, J. & POST, J. 2000, Water in mantle xenolith pyroxenes from Mexico and Simcoe (WA USA): the role of water in nominally anhydrous minerals from the mantle wedge. *Geological Society of America Abstracts with Programs*, **32**, A387.

REHKAMPER, M., HALLIDAY, A. N., ALT, J., FITTON, J. G., ZIPFEL, J. & TAKAZAWA, E. 1999. Non-chondritic platinum-group element ratios in oceanic mantle lithosphere: petrogenetic signature or melt percolation? *Earth and Planetary Science Letters*, **172**, 65-81.

REISBERG, L. & LORAND, J. P. 1995. Longevity of subcontinental mantle lithosphere from osmium isotope systematics in orogenic peridotite massifs. *Nature*, **376**, 159-162.

RICHARDSON, S. H., ERLANK, A. J. & HART, S. R. 1985. Kimberlite-borne garnet peridotite xenoliths from old enriched subcontinental lithosphere. *Earth and Planetary Science Letters*, **75**, 116-128.

RICHARDSON, S. H., GURNEY, J. J., ERLANK, A. J. & HARRIS, J. W. 1984. Origin of diamonds in old enriched mantle. *Nature*, **310**, 198-202.

RICHARDSON, S. H., SHIREY, S. B., HARRIS, J. W. & CARLSON, R. W. 2001. Archean subduction recorded by Re-Os isotopes in eclogitic sulphide inclusions in Kimberley diamonds. *Earth and Planetary Science Letters*, **191**, 257-266.

RUDNICK, R. L., MCDONOUGH, W. F. & ORPIN, A. 1994. Northern Tanzanian peridotite xenoliths: a comparison with Kaapvaal peridotites and inferences on metasomatic interactions. *In*: MEYER, H. O. A. & LEONARDOS, O. H. (eds) *Kimberlites, Related Rocks and Mantle Xenoliths. Proceedings of the 5th International Kimberlite Conference, Araxa, Brazil.* CPRM Special Publications, **1/a**, 336-353.

SCHMITZ, M. D. & BOWRING, S. A. 2000. The significance of U-Pb zircon ages from lower crustal xenoliths of the southwestern margin, Kaapvaal craton, southern Africa. *Chemical Geology*, **172**, 59-76.

SHIMIZU, N. 1999. Young geochemical features in cratonic peridotites from southern Africa and Siberia. *In*: FEI, Y., BERTHKA, C. M., MYSEN, B. O. (eds) *Mantle Petrology: Field Observations and High Pressure Experimentation: a Tribute to Francis R. (Joe) Boyd.* Geochemical Society, Special Publications, **6**, 47-55.

SHIREY, S. B. & WALKER, R. J. 1998. The Re–Os isotope system in cosmochemistry and high-temperature geochemistry. *Annual Review of Earth and Planetary Sciences*, **26**, 423–500.

SHIREY, S. B., CARLSON, R. W., RICHARDSON, S. H., MENZIES, A., GURNEY, J. J., PEARSON, D. G., HARRIS, J. W. & WIECHERT, U. 2001. Archaean emplacement of eclogitic components into the lithospheric mantle during formation of the Kaapvaal craton. *Geophysical Research Letters*, **28**, 2509–2512.

SLEEP, N. H., EBINGER, C. J. & KENDALL, J.-M. 2002. Deflection of mantle plume material by cratonic keels. *In*: FOWLER, C. M. R., EBINGER, C. J. & HAWKESWORTH, C. J. (eds) *The Early Earth: Physical, Chemical and Biological Development*. Geological Society, London, Special Publications, **199**, 135–150.

THOMPSON, R. N. & GIBSON, S. A. 1991. Subcontinental mantle plumes, hotspots and pre-existing thinspots. *Journal of the Geological Society, London*, **148**, 973–979.

WALKER, R. J., CARLSON, R. W., SHIREY, S. B. & BOYD, F. R. 1989. Os, Sr, Nd, and Pb isotope systematics of southern African peridotite xenoliths: Implications for the chemical evolution of sub-continental mantle. *Geochimica Cosmochimica Acta*, **53**, 1583–1595.

WALTER, M. J. 1999. Melting residue of fertile peridotite and the origin of cratonic lithosphere. *In*: FEI, Y. BERTHKA, C. M. & MYSEN, B. O. (eds) *Mantle Petrology: Field Observations and High Pressure Experimentation: a Tribute to Francis R. (Joe) Boyd*. Geochemical Society, Special Publication, **6**, 225–239.

ZIEGLER, U. R. F. & STOESSEL, G. F. U. 1991. New constraints on the age of the Weener Intrusive Suite, the Gamsberg Granite and the crustal evolution of the Rehoboth Basement Inlier, Namibia. *Communications of the Geological Survey of Namibia*, **7**, 75–78.

Strange partners: formation and survival of continental crust and lithospheric mantle

NICHOLAS T. ARNDT[1], ÉRIC LEWIN[1] & FRANCIS ALBARÈDE[2]

[1] *LCGA, UMR 5025 CNRS, 1381 rue de la Piscine, 38400 Grenoble, France*
(e-mail: arndt@ujf-grenoble.fr)
[2] *ENS-Lyon, place d'Italie, Lyon, France*

Abstract: Continental lithosphere is a sandwich of two layers, each composed of materials that are rare in the upper parts of the Earth. Continental crust consists of low-temperature distillates produced during a succession of melting events; the underlying lithosphere is a remarkably pure concentrate of high-temperature, highly refractory minerals. Material with intermediate compositions, which should have been far more abundant, is missing. The two dominant components of Archaean lithospheric mantle, olivine with 92–94% forsterite and similarly magnesian orthopyroxene, form only a small proportion of the residue of large-scale mantle melting. Their accumulation to form the lithosphere requires efficient sorting, to separate them from the products of lower-degree melting. This sorting, which is driven by the buoyancy of these low-density phases and their high viscosity, takes place during: (1) plume ascent through the segregation of residues of high- and low-degree melting; (2) recycling of the residues through the mantle; (3) crystallization in the crust of a Hadean magma ocean. The higher-than-normal orthopyroxene content in the Kaapvaal and other cratons is due in part to density sorting and in part to exsolution of majorite, a residual phase during komatiite melting. Secular variation in lithosphere composition (a decrease in the proportion of magnesian olivine and orthopyroxene and an increase in clinopyroxene and spinel) reflects a progressive decline in high-degree melting, a consequence of falling mantle temperatures.

Subcontinental lithosphere is the cool and relatively rigid layer of mantle immediately beneath the continents. We have a reasonable understanding of its nature and composition from geophysical surveys and from the study of xenoliths brought to the surface in kimberlites or alkaline lavas. We know that it is several hundred degrees cooler than surrounding asthenosphere and consequently has higher viscosity. We know too that it is composed primarily of peridotite, with minor, irregularly distributed eclogite and pyroxenite.

The mineralogical and chemical composition of peridotite from subcontinental lithosphere differs from that of peridotite from other parts of the mantle (Boyd 1989; Berstein *et al.* 1997). Peridotites from subcontinental lithosphere is 'depleted', which means it contains only a small amount of clinopyroxene and an aluminous phase, which together make up the so-called basaltic component. The lithosphere beneath the oldest Archaean cratons has a composition markedly different from that of younger subcontinental lithosphere (Boyd & Mertzman 1987; Griffin *et al.* 1999). Old unmetasomatized lithosphere is harzburgitic, a mixture of olivine and orthopyroxene (the clinopyroxene in minor quantities is thought to have exsolved from orthopyroxene solid solution (O'Hara *et al.* 1975; Cox 1987). Both the olivine and the orthopyroxene are unusually magnesian, with MgO/(MgO + FeO) molecular ratios in the range 92–94. The well-sampled lithosphere beneath the *c.* 3.5 Ga Kaapvaal Craton of South Africa is further characterized by an unusually high ratio of orthopyroxene to olivine (Maaloe & Aoki 1977; Herzberg 1983; Boyd & Mertzman 1987; Hawkesworth *et al.* 1990). These mineralogical features are reflected in major-element compositions: in samples of lithosphere from beneath several old cratons (Kaapvaal, Siberia, Tanzania, Greenland), MgO contents are relatively high, and FeO, CaO and Al_2O_3 are low. The peculiar mineralogy of the Kaapvaal lithosphere results in elevated Ca/Al and particularly high Si/Mg ratios. The depletion of basalt-forming components of the lithosphere beneath younger parts of the continents is less extreme, but still exceeds that of oceanic lithosphere and the asthenosphere.

Radiometric age dating indicates that in many regions, the lithospheric mantle formed at about the same time as the overlying continent (Pearson 1999, and references therein). A combination of Sm–Nd ages from inclusions in diamonds and Re–Os ages from minerals and whole rocks shows, for example, that lithosphere beneath the Kaapvaal Craton is c. 3 Ga old, like the crust itself. For the lithosphere to have survived for so long requires that it had inherently low density. Were this not so, the lithosphere would have become gravitationally unstable at its present temperature, which is several hundred degrees lower than that of underlying asthenosphere. The stability of old lithosphere is probably related to its unusual mineralogy. The low proportion of Fe in ferromagnesian minerals and the low modal abundance of garnet both reduce its density; and the absence of water increases its viscosity (Hirth & Kohlstedt 1996; Wallace 1998). Old lithosphere is therefore relatively buoyant and more rigid than the asthenosphere that underlies it.

Current models for the formation and evolution of subcontinental lithosphere

The origin of subcontinental lithosphere is not well understood. Downward freezing of asthenosphere, to form a mechanical boundary layer like the oceanic lithosphere, is not an acceptable explanation because this process would produce lithosphere with about the same composition as normal asthenosphere. Such lithosphere would lack inherent buoyancy. Many workers (O'Hara et al. 1975; Maaloe & Aoki 1977; Boyd & Mertzman 1987; Herzberg 1999; Walter 1999) have noted that the dominant minerals in old lithosphere (Fo-rich olivine and magnesian orthopyroxene) are the same as those in the solid residue left during high-degree melting of normal mantle peridotite. Abouchami et al. (1990), Hill (1990), Albarède (1998) and Herzberg (1999) have suggested that formation of the continental lithosphere might be related to the emplacement of oceanic plateaux. They have proposed that such plateaux, which are buoyant and resist subduction, became the nuclei of continents, and that the relict mantle from which plateau basalts were derived becomes trapped beneath the continents. On cooling the relict mantle transforms into subcontinental lithosphere.

Lithospheric mantle that formed in this manner would bear the imprint of two processes: (1) partial melting and magma extraction, which creates the oceanic plateau and a depleted residue; (2) metasomatism associated with subduction, the process primarily responsible for the growth of continental crust (Taylor & McLennan 1985; Arculus 1999).

Following its formation, the base of the lithosphere is subject to a persistent flux of melt and fluids from deeper sources (the equivalent of the sources of the oceanic islands and seamounts that pepper the ocean basins). This flux introduces material rich in Fe, Al and incompatible trace elements, and adds another metasomatic component to the base of the lithosphere.

Problems with current models

Normal mantle melting does not produce a residue with the mineralogical composition of old subcontinental lithosphere

Normal mantle melting, such as that which leads to the formation of oceanic crust and oceanic islands, produces magma of basaltic to picritic composition and leaves a residue consisting of olivine, orthopyroxene, clinopyroxene and an aluminous phase. The compositions of minerals in this residue, and their relative proportions, are unlike those in old subcontinental lithosphere; Mg–Fe ratios of the ferromagnesian minerals are too low, and the amount of clinopyroxene and spinel or garnet is too high.

The particular composition of old lithosphere is more like that of the residue of high-degree melting. Such melting produces magma of ultramafic or komatiitic compositions and highly magnesian olivine and orthopyroxene as residual minerals. If the melting takes place at great depth, a large amount of garnet may also be present in the residue. Boyd (1989), Herzberg (1999) and Walter (1999) have stated specifically that Archaean lithosphere consists of the residue of partial melting that formed komatiitic lava.

A problem with this hypothesis becomes apparent when we consider the mechanics of mantle melting. There are two reasons to believe that, in normal circumstances, the residue of high-degree melting makes up only a small fraction of the total residue (Fig. 1). First, the proportion of komatiite in Archaean volcanic successions is never very high, generally less than 5% and never much more than about 20% (Viljoen & Viljoen 1969; Arndt & Nisbet 1982; de Wit & Ashwal 1997). Basalts, which form through lower degrees of melting, are far more abundant. We do recognize that these figures are minima, however, because a large proportion of primary komatiite magma would be trapped within the crust or fractionally crystallize to less magnesian magma before reaching the surface.

The second, more compelling reason comes from consideration of the shape of melting zones

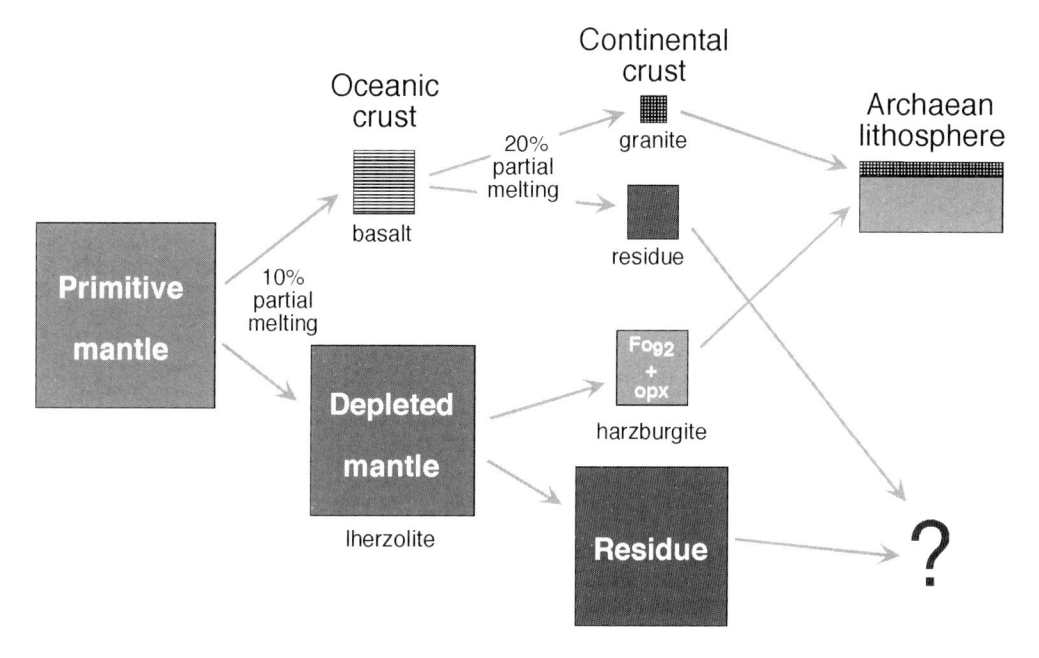

Fig. 1. Diagram illustrating the series of processes that produce continental crust and lithospheric mantle from primitive mantle. The following features should be noted: (1) the crust and lithospheric mantle form through independent pathways; (2) the high proportion of residue that accumulates somewhere in the mantle.

in the mantle, all of which contain a central region where temperatures are highest, surrounded by a larger region where temperatures are lower. We estimated the relative proportion of refractory residue in the hot central region using the procedure given in the Appendix. The results are summarized in Figure 2 and Table 1. We considered two situations. The first is a cylinder with a hot core surrounded by a relatively narrow sheath of cooler material; a situation that corresponds to the upwelling tail or conduit of a mantle plume. The second, a large sphere of cooler material surrounding a small hot core, corresponds to the head of a starting mantle plume. As can be seen from the results listed in Table 1, the volume of refractory residues varies from 60% to as low as 20% of the total volume of the residue.

To extract the refractory constituents to form a thick, uniform layer of refractory components like the subcontinental lithospheric mantle requires an efficient sorting process. To emphasize the problem we need only to consider that if high-temperature residue constitutes 20% of the total residue of melting, to accumulate a 200 km thick layer of Kaapvaal-type lithosphere requires the elimination of five times the volume (the equivalent of a 1000 km thick layer) of less depleted material.

Elimination of the residues of crust formation

There are other waste-disposal problems. As discussed by Arndt & Goldstein (1989), Kay & Kay (1991) and Arculus (1999) and illustrated in Figure 1 formation of continental crust generates large volumes of residue. Although the process is complicated, crust formation boils down to the extraction of material of granitic composition from a source of basaltic composition. Much of this differentiation takes place in the uppermost mantle or within the crust itself. Whether by crystallization or partial melting, the derivation of granitoid magma from basalt produces a residue that is far more voluminous than the product. If we accept a figure of 80% fractional crystallization or 20% partial melting, the formation of 40 km thick crust generates a 200 km thick layer of mafic cumulate or restite. This material, a mixture of Fe-rich pyroxene or amphibole, plus feldspar that converts to garnet at subcrustal depths, has a density significantly higher than that of normal mantle. It cannot remain at subcrustal levels to form a major portion of subcontinental lithosphere, and it too must be eliminated.

Since it formed over 3 Ga ago, the Kaapvaal Craton has been affected by at least three major

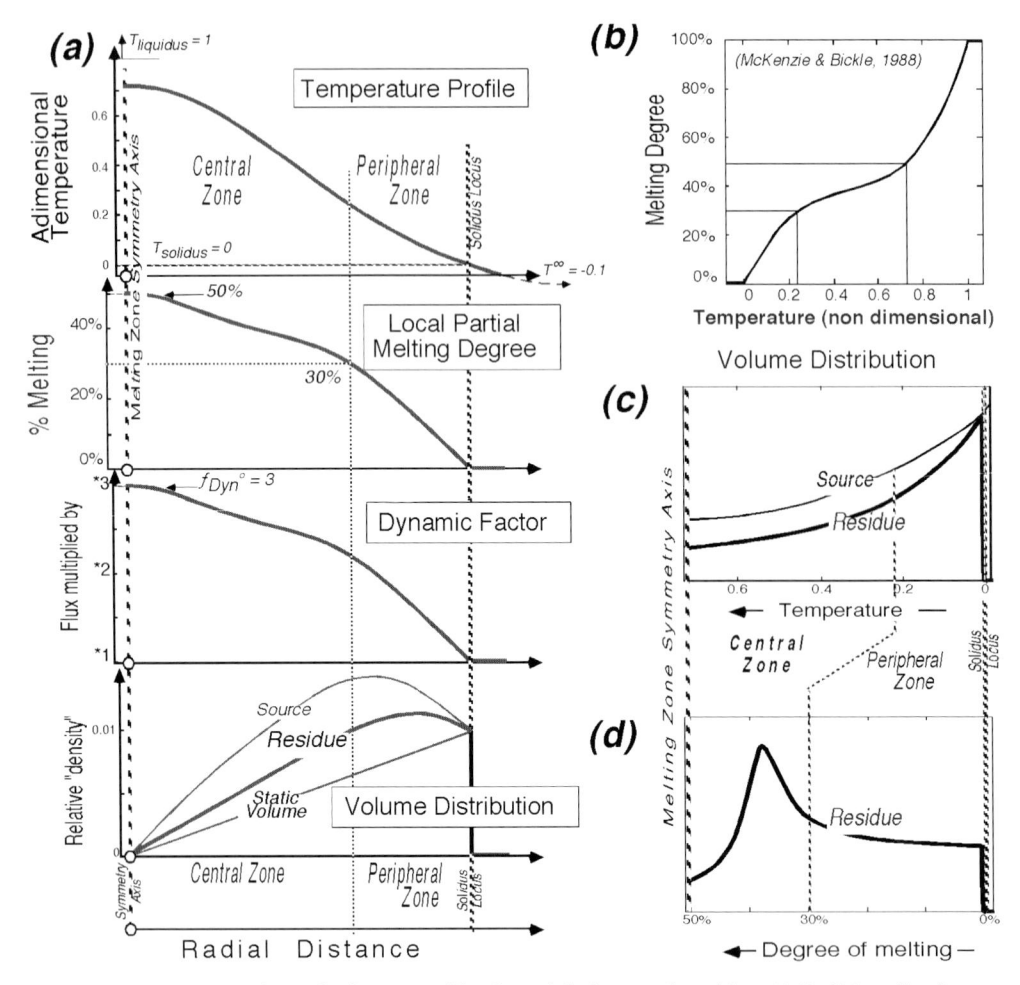

Fig. 2. Calculated proportions of refractory residue formed during mantle melting. (a) Radial profiles for cylindrical symmetry and the radial distribution of residue, of source and of geometric volume; (b) the McKenzie & Bickle (1988) relationship between temperature and local degree of melting; (c) and (d) the distribution as function of temperature and degree of melting. It should be noted that because the two horizontal scales are not linearly related, the distribution maximum around 35–40% melting corresponds to the flattest part of the McKenzie & Bickle curve.

magmatic events: Ventersdorp flood volcanism at 2.7 Ga, emplacement of the Bushveld Complex at 2.1 Ga, and Karoo volcanism at 184 Ma. During each event, enormous volumes of hot mantle arrive at the base of the lithosphere, but only a small fraction of the magma that forms through partial melting erupts at the surface. Large volumes are trapped near the Moho, where they differentiated into low-density gabbro, which underplates or is added to the crust, and olivine ± clinopyroxene ± plagioclase cumulates (Herzberg et al. 1983; Mahlburg Kay & Kay 1985; Arndt & Goldstein 1989). The mafic minerals in these cumulates are relatively Fe rich, and if plagioclase converts to garnet as it cools, the density of the resultant rock (mafic granulite or eclogite) is far greater than that of normal subcontinental lithosphere. A gravitationally unstable layer of high-density material thus forms at the top of the lithosphere. Xenoliths with compositions corresponding to this material are very rare and it appears that this layer does not remain in place. This material as well must be eliminated.

The problem of high orthopyroxene–olivine ratios

The unusually high proportion of orthopyroxene in the Kaapvaal subcontinental lithosphere

Table 1. *Results of model quantification of the average melting residue proportions*

Symmetry	Dynamic case	f°_{Dyn}	Thermal profile	T^∞	Melting model	Global $\langle f_{Dyn}\rangle$	Mean $\langle F\rangle$ (%)	(Static) vol.% CZ (%)	$\langle f_{Dyn}\rangle$ CZ	$\langle f_{Dyn}\rangle$ PZ	$\langle F\rangle$ (%) CZ	$\langle F\rangle$ (%) PZ	Rsd% CZ (%)	Ψ_{vol}	Ψ_{rsd}
Cylindrical	Static		Gaussian	−0.1	McK&B '88	/	24.6	42.7	/	/	38.0	14.7	35.1	/	0.727
	No. 1	1				1.99	28.4	42.7	2.52	1.59	38.3	16.7	46.6	1.59	0.740
	No. 2	3				1.67	28.9	42.7	2.23	1.26	38.8	16.0	48.9	1.76	0.729
	No. 1	3				5.43	30.8	42.7	7.83	3.65	38.5	18.6	54.7	2.15	0.756
	No. 2	10				4.03	32.7	42.7	6.52	2.18	39.2	18.1	62.3	2.99	0.742
Cylindrical	No. 1	10	Gaussian	−0.3	McK&B '88	1.99	28.4	42.7	2.52	1.59	38.3	16.7	46.6	1.59	0.740
		3		−0.2		2.07	30.2	49.2	2.53	1.63	38.5	17.8	52.8	1.55	0.747
				−0.1		2.12	31.1	52.8	2.53	1.65	38.7	18.2	56.3	1.53	0.750
				−0.05		1.89	26.5	36.5	2.51	1.54	38.2	15.5	40.7	1.64	0.731
				−0.03		1.82	25.1	32.4	2.51	1.50	38.1	14.5	36.8	1.68	0.724
Cylindrical	No. 1	3	Gaussian	−0.1	McK&B '88	1.99	28.4	42.7	2.52	1.59	38.3	16.7	46.6	1.59	0.740
			Gaussian	−0.1	Linear	1.72	22.5	22.6	2.57	1.47	39.8	13.6	26.4	1.76	0.696
			Linear		McK&B '88	2.03	28.5	45.5	2.46	1.67	36.6	18.6	48.8	1.47	0.778
			Linear		Linear	1.67	20.0	16.0	2.47	1.51	37.0	14.7	18.6	1.63	0.738
Spherical	Static		Gaussian	−0.1	McK&B '88	/	20.0	27.9	/	/	36.3	13.7	22.8	/	0.738
	No. 1	1				1.80	23.6	27.9	2.45	1.55	36.6	15.6	31.5	1.59	0.752
	No. 2	3				1.48	23.6	27.9	2.09	1.24	36.9	14.9	32.6	1.69	0.741
	No. 1	3				4.59	26.4	27.9	7.53	3.46	36.7	17.7	39.3	2.18	0.769
	No. 2	10				3.15	27.7	27.9	5.91	2.08	37.3	17.1	45.3	2.84	0.756
Spherical	No. 1	10	Gaussian	−0.1	McK&B '88	1.80	23.6	27.9	2.45	1.55	36.6	15.6	31.5	1.59	0.752
		3	Gaussian	−0.1	Linear	1.52	16.4	10.8	2.49	1.40	37.7	11.8	13.2	1.78	0.706
			Linear		McK&B '88	1.87	24.6	30.7	2.40	1.63	35.3	17.7	33.9	1.47	0.786
			Linear		Linear	1.50	15.0	6.4	2.40	1.44	35.2	12.7	7.8	1.67	0.742

The first six columns give the model parameters, as described in the Appendix. The last 10 give the results. Global $\langle f_{Dyn}\rangle$ is the average dynamic factor over the whole melting zone, and Mean $\langle F\rangle$ is the average melting degree. CZ relates to the Central Zone ($F \geq 30\%$) and PZ to the Peripheral Zone, therefore the respective subregional mean f_{Dyn} and mean melting degrees. (Static) vol.% CZ is the volume of the central zone relative to that of the whole melting zone. Rsd% CZ is the amount of residue that comes from the central zone, relative to the global amount of residue. The last two columns are the two Ψ indices defined in the Appendix. $T_{\text{solidus}} = 0$, $T_{\text{liquidus}} = 1$; $F(\text{axis}) = 50\%$, $F(\text{boundary}) = 30\%$. NB: no length scale necessary. McK&B '88 is McKenzie & Bickle (1988).

poses a separate problem. Kelemen et al. (1998), Herzberg (1999) and Walter (1999) have shown that at the high degrees of melting required to form highly magnesian olivine (Fo$_{92-94}$) as a residual phase, the amount of orthopyroxene left in the residue is relatively small, far less than that recorded in xenoliths from the Kaapvaal lithosphere. Two models have been proposed to account for the excess orthopyroxene. Herzberg (1999) suggested that komatiite magma, which had been extracted from a deeper mantle source, partially crystallized to orthopyroxene-rich cumulates that were then added to primitive, olivine-rich lithosphere. Kelemen et al. (1998) suggested that the excess orthopyroxene results from reaction between lithosphere and Si-rich melts emanating from subduction zones.

Neither explanation is entirely satisfactory. It is not clear to us why komatiite should crystallize such a high proportion of orthopyroxene. The position of the phase boundary between olivine and orthopyroxene is strongly pressure dependent, which means that high pressures stabilize orthopyroxene relative to olivine. If komatiite magma forms by deep melting, then ascends to shallower levels before crystallizing, the cumulate will be poorer, not richer, in orthopyroxene than the residue of initial melting. Normally the mineral assemblage that crystallizes from magma from a deeper source will be richer, not poorer, in olivine than the residue of initial melting. Herzberg's explanation requires the opposite effect.

Kelemen et al.'s explanation requires that the subduction-derived melts add only Si to the lithosphere, but produce none of the other changes normally associated with the movement of fluids into the mantle wedge above a subduction zone. It requires that fluids traversing the lithosphere through fractures or by percolation efficiently remove the low-melting phases to leave a refractory harzburgite devoid of 'subduction component'. Xenoliths from old subcontinental lithosphere that have not been metasomatized by post-subduction processes show no systematic enrichment in fluid-soluble elements such as Rb, Sr and K, and lack the negative high field strength element (HFSE) anomalies that characterize the mantle wedge and subduction-related magmas (McDonough 1990, 1992). It is difficult to imagine that Si-rich fluids changed the orthopyroxene–olivine ratio but had no effect on other major and trace-element characteristics. Indeed, the analyses by Zhang et al. (2001) of enstatites that replace diopside in xenoliths from the Kaapvaal Craton demonstrate that the composition of secondary orthopyroxene is very different from that of normal orthopyroxene in Archaean harzburgites. The secondary orthopyroxenes are relatively rich in Fe (mg-number 88–91) and have hump-shaped rare earth element (REE) patterns that contrast with the depleted light REE (LREE) patterns of orthopyroxene in normal cratonic harzburgites.

Models for the formation of subcontinental lithosphere

Accumulation of magnesian olivine and orthopyroxene

Old subcontinental lithosphere consists of two components, highly magnesian olivine and highly magnesian orthopyroxene, that are absent in fertile mantle and rare in the residues of partial melting. To construct this type of lithosphere requires an efficient sorting process that concentrates these components and eliminates all others. The likely driving force is the buoyancy and high viscosity of the dominant constituents. The density of olivine and orthopyroxene depends on their Mg–Fe ratios. Olivine with the composition Fo$_{92-94}$ has a density of 3320–3300 kg m^{-3}. Orthopyroxene with the same Mg/(Mg + Fe) is very slightly less dense (En$_{94}$ = 3250 kg m^{-3}). These values are lower that of olivine from normal asthenospheric mantle, which has the composition Fo$_{89-90}$ and a density of 3360–3340 kg m^{-3}. The density of clinopyroxene is like that of olivine with similar Mg-Fe ratio; garnet is far denser.

The viscosity of mantle peridotite depends strongly on its water content. Hirth & Kohlstedt (1996) have argued that water acts as an incompatible element during peridotite melting and that partial melting efficiently extracts water from the source. The viscosity of the water-poor residue of partial melting is far greater than that of unmelted peridotite, and viscosity of the residue of refractory, anhydrous high-degree melting is greater than that of the residue of low-degree melting.

For the subcontinental lithosphere to form, the least dense and most viscous mantle phases must segregate and accumulate in a layer at or near the surface. Exactly how this occurs is a matter of speculation. We imagine three processes. In the first (Fig. 3a), upwelling residue in the hot central core of a mantle plume separates from the cooler exterior portions and gradually accumulates as the plume nears the surface. The residue of low-degree melting is swept away by mantle flow. In the second (Fig. 3b), blobs of residue segregate slowly from material cycling through the mantle. The dominant lithosphere

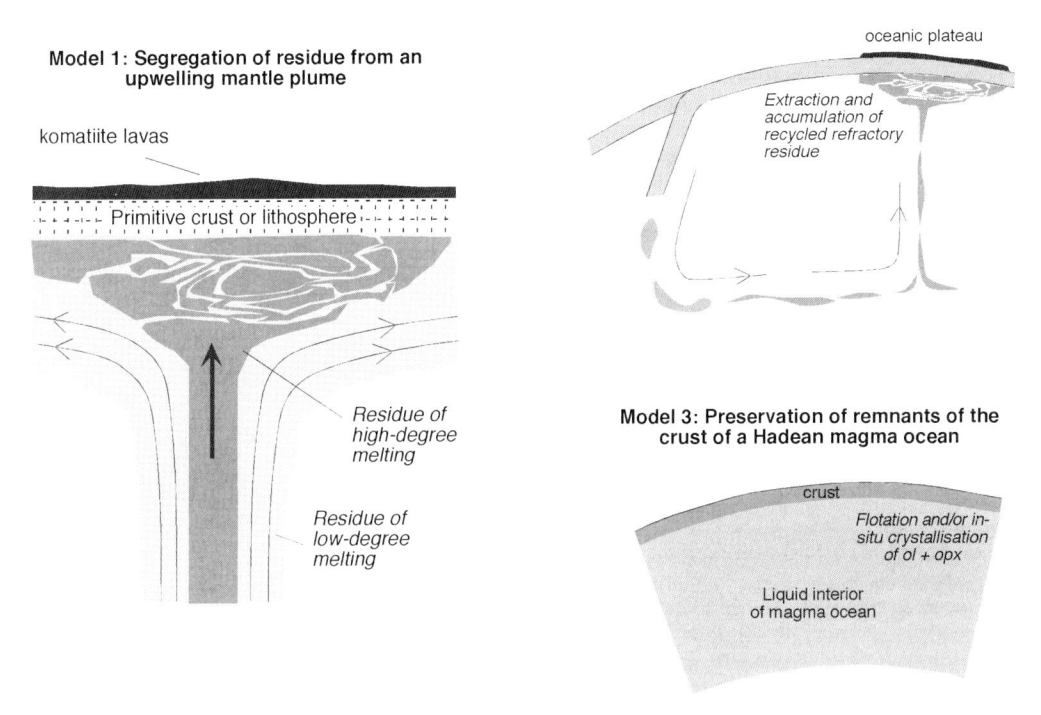

Fig. 3. Three possible mechanisms that could allow the segregation and accumulation near the surface of the Earth of highly magnesian olivine and orthopyroxene.

phases are refractory and would survive unscathed their transportation down subduction zones and onward through the mantle.

In our third, more radical, suggestion, we suggest that the subcontinental lithosphere could be the remnants of the initial crust of a Hadean magma ocean (Fig. 3c). In this model, we assume that a magma ocean formed during the final stages of Earth accretion. The cause of melting was heat released by impacts, core segregation and radioactivity; the consequence was the formation of a magma ocean of ultramafic composition that occupied at least the upper part of the mantle. We know from experimental studies of the solidification of komatiite that the liquidus minerals at low pressures are highly magnesian olivine and orthopyroxene, joined at lower temperature by clinopyroxene and spinel (Green et al. 1974; Arndt 1976; Bickle et al. 1977; Herzberg & Ohtani 1988). We suggest that olivine and orthopyroxene, because they crystallized first and because of their low densities, would be the dominant constituents of the crust that formed at the surface during progressive solidification of the magma ocean (Nisbet & Walker 1982; Miller et al. 1991). Once formed, this crust would have a high survival potential, being refractory, viscous and less dense than other parts of the mantle. Although buffeted by impacts and convection in the underlying mantle, the crust would survive in place at the surface. Some time after it first solidified, around 4 Ga ago or even earlier, remnants of this crust may have provided the focus of subduction zones that then became factories for the production of continental crust.

An accumulation of olivine and orthopyroxene contains a moderate amount of Os, a compatible element, but very little Re, the parent isotope of the Re–Os system (Pearson 1999). The Os isotope composition of the original crust of a magma ocean (and that of olivine and orthopyroxene cumulates formed by the other mechanisms) would have changed little during the initial period after its formation. Only when Re was added, perhaps during fluid transport associated with subduction, would the Re–Os isotopic clock be started. It is for this reason that Re–Os ages measured in mantle xenoliths broadly coincide with the ages of overlying continental crust (Pearson 1999, and references therein).

The excess orthopyroxene problem

The slight difference in density between magnesian olivine and orthopyroxene may contribute to the high abundance of orthopyroxene in old lithosphere, but it cannot be the entire explanation. In particular, it cannot explain the high ratio of orthopyroxene to olivine in peridotites from the Kaapvaal and some other old cratons, which contrast with more normal ratios in samples from other younger cratons. A peculiarity of the greenstone belts of the Kaapvaal Craton is the presence of a high proportion of komatiite of unusual composition (Viljoen & Viljoen 1969; Nesbitt et al. 1979; Arndt & Nisbet 1982). The Barberton greenstone belt is the type locality of Al-depleted komatiite, magma that is believed to form by moderate degrees of partial melting at great depths under conditions that left majorite garnet as a residual phase (Green 1974; Ohtani et al. 1988; Herzberg 1992). We propose here that the orthopyroxene now found in Kaapvaal subcontinental lithosphere may have been the product of exsolution from residual majorite.

According to Ohtani et al. (1988), Herzberg (1992) and Nisbet et al. (1993), Barberton komatiite forms through a moderate degree of partial melting at very high pressure, around 9–10 GPa. This melting leaves a residue of olivine and majorite garnet. The partial melting probably takes place in an extremely hot mantle plume. Because of the high temperature, the plume has a density far less than that of cooler surrounding mantle. The komatiite magma, which initially forms at high pressures (c. 10 GPa), has a density similar to that of unmelted olivine but less than that of majorite garnet. Because of the similar densities of magma and olivine, segregation of melt from residue may initially be sluggish, but as the plume rises to shallower levels, the density contrast between magma and olivine increases. Once the density contrast becomes large enough, at around 8 GPa, the magma segregates from the plume leaving behind an olivine–majorite matrix. For a Barberton-type komatiite, the temperature of melting at 8 GPa is about 1950°C, some 600°C greater than that of surrounding mantle (c. 1350°C) (Herzberg 1992, 1995; Nisbet 1993). This difference is sufficiently large that the effect of thermal expansion compensates for the high inherent density of majorite. The matrix remains buoyant relative to cooler surrounding mantle, and the plume continues to ascend, transporting majorite to regions of ever lower pressure. Because the solution of orthopyroxene and garnet in majorite is pressure dependent, the drop of pressure leads to exsolution of the two phases; low-density magnesian orthopyroxene thereby separates from high-density pyrope garnet. The assemblage of magnesian olivine and orthopyroxene segregates to form the lithosphere; the garnet remains behind and cycles back into the deeper mantle.

How did crust and lithosphere come together, and what happened to the waste?

Figure 1 illustrates schematically the complex series of events required to produce the two contrasting layers that make up the continental lithosphere. The continental crust consists of granitoid rocks that formed through a complex series of events, which includes partial melting of peridotite to form basalt and reprocessing of basalt in a subduction environment. At each step, the low-temperature liquid is extracted, leaving a larger volume of solid residue. The harzburgitic lithospheric mantle, as discussed above, represents a concentrate of the most refractory mantle phases. Again, the volume of residue is greater than that of the material that accumulated in the continental lithosphere.

This view of lithosphere formation has several implications.

(1) The two layers in continental lithosphere represent extreme end-members in the compositional spectrum of magmatic rocks and each forms through independent processes. Archaean continental crust has an allochthonous relationship to underlying lithospheric mantle.

(2) The two layers accumulate together because they share important physical properties: both have relatively low densities, and, at the conditions in which they accumulate at the surface of the Earth, relatively high viscosities. Given their common tendency to congregate at the surface, it makes little sense to speak of one layer providing a refuge that allowed the accumulation of the other layer (Griffin et al. 1999). Continental crust overlies continental lithosphere simply because it is made up of the lighter of the two type of 'surface-seeking' materials.

(3) Although continental crust and lithosphere are stable at the surface of the Earth, neither layer is invulnerable. Material is lost and gained by both layers. Part of the continental crust cycles back into the mantle after transformation to sediment, and new crust is added through melting in subduction zones. The subcontinental lithospheric mantle is subject to a continuous flux of magmas or fluids from deeper mantle sources. These fluids metasomatize the base of the lithosphere, contributing low-temperature and volatile components, most importantly increasing its iron content. The

addition of Fe increases the density, rendering the metasomatized layer unstable (Griffin *et al.* 1999; Poudjom-Djomani *et al.* 2001). This basal layer is also subject to erosion from underlying hotter convecting asthenosphere and must be progressively lost. Some new Fo-rich olivine may be added through the accumulation of the residues of high-degree melting. However, because the amount of high-degree melting has declined throughout Earth history, and because the metasomatizing flux is constant, the total volume of primitive harzburgitic mantle must be decreasing.

(4) As shown in Figure 1, an enormous volume of material with intermediate composition is missing from the continental lithosphere. If we accept the values used in the figure, the volume of residue from crust formation is about five times the volume of the continental crust; and the volume of residue left after segregation of harzburgitic lithosphere is maybe twice the volume of lithospheric mantle. The outstanding characteristic of both types of residue is their richness in Fe. They have higher densities than normal mantle and would tend to sink, either to be mixed into deeper mantle or to congregate at deeper levels. If material with a volume three times that of the continental lithosphere accumulated at the base of the mantle, it would form a layer about 400 km thick. We speculate that this may be the origin of the layer or blobs of material with distinct chemical or seismic characteristics that may exist in the lower part of the mantle (Kellogg *et al.* 1999; van der Hilst & Karason 1999).

The residue of harzburgite segregation, a mixture of Fe-rich olivine, pyroxene and garnet, contains low concentrations of incompatible elements, including heat-producing elements and the parent isotopes of geochemically important systems. Its presence in the source of magmas would leave little obvious trace, except perhaps to increase the Fe contents to levels greater than that of normal mantle. Such material might explain the high Fe contents inferred for the sources of picrites and other high-degree mantle melts (Francis 1995).

The residue of crust formation, on the other hand, has a distinct trace element and isotopic composition. This material, a mixture of Fe-rich pyroxene and garnet under upper-mantle conditions, would contain higher concentrations of incompatible elements, particularly if it had become contaminated with crustal rocks. As proposed by Arndt & Goldstein (1989), the presence of this material in the source of basalts would impart a distinct isotopic signature to resultant melts.

What does the secular variation of lithosphere composition tell us about mantle temperatures?

Boyd (1989), Menzies (1990), O'Reilly & Griffin (1996) and Pearson (1999) have demonstrated that the composition of subcontinental lithosphere varies as a function of its age. Compared with old ($c. 3.5$ Ga) lithosphere, young lithosphere is less depleted in incompatible trace elements, the mafic minerals are less magnesian, and the amount of clinopyroxene and garnet is higher. There are two ways of interpreting these differences. They can be attributed either to a decrease the efficiency and/or duration of the sorting process postulated in the previous section, or to a decrease in the overall amount of high-degree mantle melt.

The composition of liquids in equilibrium with the minerals in the subcontinental lithosphere can be calculated if a series of assumptions are made about Mg–Fe partitioning, the oxygen fugacity and the effect of pressure. The results depend strongly on these assumptions and the absolute numbers cannot be considered reliable, but the relative values are certainly significant. We calculate that the liquid in equilibrium with olivine from Archaean lithosphere (Fo_{92-94}) contained between 20 and 28% MgO. For Proterozoic lithosphere (Fo_{90-92}), we calculate 15–20% MgO, and for Phanerozoic lithosphere (Fo_{88-90}) we calculate 10–15% MgO. Although the constituents of the lithosphere represent only a small fraction of the total melting region (as argued above), it appears inescapable that a large amount of highly magnesian melt must have formed to produce the residue that became the Archaean lithosphere.

If the 200 km thick Kaapvaal lithosphere is the accumulated residue of say 33% partial melting, a 100 km thickness of high-degree melt must have formed. Why do we see no sign of such large volumes of komatiite in Archean greenstone belts? If the lithosphere formed directly by accumulation from the upwelling source (model 1, Fig. 3a), we must explain the fate of these komatiites (Boyd & Mertzman 1987; Herzberg 1999). A large proportion of these magmas may not have reached the surface and would have been trapped within the crust. Part of the magma that erupted onto the continents would have been lost through erosion, and part may have flowed out into or erupted within ocean basins, as was the case during more recent flood volcanism of the Deccan and North Atlantic provinces. Whatever the process, the formation of such large amounts of komatiite, which represent only a fraction of the total magma volume, implies rates of magmatism far greater than those of today. The alternative is that the accumulation of low-density, refractory olivine and orthopyroxene was decoupled from the segregation of magmas (model 2, Fig. 3b), or that the lithosphere is the remnants of the crust of a Hadean magma ocean (model 3, Fig. 3c).

If we accept that the Archaean lithosphere was the direct result of high-degree melting, does this necessarily imply higher temperatures in the Archaean mantle, or could it have been due to higher water contents? Could the formation of abundant high-degree melts be a consequence of a wetter Archaean mantle, as suggested by Parman et al. (1996) and Grove et al. (1999)? Wet melting at low pressures produces high-Si magmas, leaving an orthopyroxene-depleted residue that does not have the composition of old subcontinental lithosphere; wet melting at greater depths, however, produces magmas that are Mg rich and Si poor, and residues that consist of olivine and orthopyroxene ± garnet (Inoue 1994; Kawamoto & Holloway 2002). These magmas have kimberlitic compositions. We see no evidence for the formation of large volumes of erupted kimberlite in Archaean successions, but the eruption of such volatile-rich magmas would have been explosive and the resultant fragmental deposits have a very low preservation potential. On the other hand, very few intrusive rocks with kimberlitic compositions appear to have been emplaced during greenstone belt volcanism. The rarity of such rocks suggests that deep-seated melting of wet mantle was not common in Archaean time. The relative abundance of komatiite, which normally forms through melting of an essentially anhydrous source (Arndt et al. 1998), and the rarity of kimberlite suggest strongly that the secular change in lithosphere composition reflects declining mantle temperatures.

Conclusions

(1) The highly magnesian olivine and orthopyroxene in old subcontinental lithosphere are not the normal products of mantle melting. They formed in regions of high-degree melting and segregated from other mantle phases because of their low density and high viscosity. Continental crust forms through another independent process; this crust has an allochthonous relationship to underlying lithospheric mantle.

(2) The unusually high proportion of orthopyroxene in the lithosphere beneath the Kaapvaal and some other old cratons may result from exsolution from majoritic garnet, as residual phase during komatiitic melting.

(3) Secular variation in the compositions of subcontinental lithosphere probably results from declining mantle temperatures.

We thank W. F. McDonough for his peridotite database and for his very constructive review of the submitted manuscript. We also acknowledge very useful comments from C. Herzberg, M. Cheadle, W. L. Griffin and an anonymous reviewer.

Appendix: calculation of the relative proportions of residue left by high- and low-degree melting

Procedure

The aim of the following calculations was to estimate the relative amount of refractory residue produced by high-degree melting, compared with that produced by lower degrees of melting. Temperature varies across the melting zone, as does the degree of melting. Our comparison focuses on two complementary subregions, the first corresponding to high degrees of melting, which leave a relatively small amount of refractory residue, and the second to the complementary subregion where the melting degree is lower and the proportion of residue higher. The first region has a composition like that of old continental lithosphere, the second more like that of the asthenosphere. Two different geometries will be considered for the melting zone: the first with cylindrical symmetry, which corresponds to a 'plume tail'; the second with spherical symmetry, which corresponds to a 'plume head'. The system is dynamic, which means that varying amounts of material pass through each melting subregion. When comparing the relative amounts of residual material produced in each subregion during a given time period, we have to take into account the relative fluxes of incoming material to weigh the relative production of residual material. The following results show that this dynamic factor controls the relative proportions of residual material.

A more formal description follows. We assume an initial volume δV of mantle material in the melting, which melts to a degree F. In the static model, the amount of residue is $(1 - F)\delta V$. In the dynamic model, the amount of mantle that passes through the melting region is enhanced by what we call the factor f_{Dyn}. For the 'static' case, $f_{Dyn} = 1$. The relative amount of residue is $f_{Dyn}(1 - F)\delta V$. Technically, the calculation consists in taking into account the geometrical symmetry of the model to integrate the amount of material that passes through each subregion.

We illustrate the calculation using the plume-tail model (Figs 2 and 3). The symmetry is cylindrical about a vertical axis. The height of the cylinder does not matter, as only relative quantities are important, and we have assumed the height to be equal to the length unit. Thus, the mathematical calculation 'degenerates' into a 1D problem, with the radial distance to the cylinder axis being the sole descriptive parameter. The elementary volume is expressed by $\delta V = 2\pi r\, dr$. (For spherical symmetry, the expression is $\delta V = 4\pi r^2\, dr$.) The melting region corresponds to the region where temperature exceeds the solidus. The temperature field inherits the cylindrical symmetry of the model. Because the problem of the thermal profile is not simple, we use two hypotheses: in the first we assume that the degree of melting varies linearly with temperature; in the second we use the McKenzie & Bickle (1988) relationship. In view of the objectives of the calculations, it is unnecessary to define the absolute temperatures. Instead, we use non-dimensional temperature such that $T = 0$ at the solidus and $T = 1$ at the liquidus. The temperature profile is Gaussian, decreasing from a maximum value at the symmetry axis corresponding to $F = 50\%$. The temperature at infinite distance is an input parameter, mostly taken as $T_\infty = -0.1$. Other values were tested but were found to have little effect on the results when compared with other degrees of freedom of the calculation. The melting zone is partitioned into two subregions: the central one where the melting degree is higher than 30%, and the peripheral one where the melting degree decreases from 30 to 0%. Finally, apart the static case, two alternatives are used for the radial profile of the dynamic factor, which is taken to be proportional either to the local degree of melting (Case 1), or to the temperature (Case 2). In both cases, the dynamic factor increases from a value of unity (no effect) at the boundary of the melting zone, to a given maximum value $f°_{Dyn}$ along the symmetry axis. The maximum value of $f°_{Dyn} = 3$ is normally used, with the extreme alternative of $f°_{Dyn} = 10$ (Watson & McKenzie 1991).

Results (Table 1)

In the case with cylindrical symmetry, the central subregion occupies 43% of the entire melting region. Using the standard hypothesis leads to an average melting degree of 38%, which is almost independent of the dynamic regime. The average melting degree in the peripheral subregion ranges from 15 to 18%. The effective volume of material that passes through the central subregion per time unit is equal to the volume of this region in the static case. It is increased by a factor of about two with dynamic enhancement (2.5 for Case 1, 2.2 for Case 2). Increasing the maximum dynamic factor from $f°_{Dyn} = 3$ to $f°_{Dyn} = 10$ changes the average volume factor from about two to seven (7.8 for Case 1, 6.5 for Case 2). In the peripheral subregion, the average dynamic factor is 1.5 for $f°_{Dyn} = 3$, and about three for $f°_{Dyn} = 10$, with minor differences between Case 1 and Case 2. Finally, the residue that passes through the central region corresponds to 35% of the total residue in the static case, and to 47% and 49%, respectively, for the two dynamic cases and $f°_{Dyn} = 3$, increasing to 55% and 62%, respectively, with $f°_{Dyn} = 10$.

We summarize the calculations with reference to two indices. The first, Ψ_{vol}, is the ratio of the two average dynamic factors, those of the central and peripheral subregions. This index, which depends on the relative volumes of source mantle that passes through each subregion, characterizes the dynamic enhancement. The second index, Ψ_{rsd}, is the ratio of the amounts of residue produced in each subregion, related to the dynamic volumes of source mantle passed through each subregion. It is equal to the ratio of the respective percentages of residue for each subregion and it characterizes the relative production ratio of residual matter. The results are as follows: when $f°_{Dyn} = 3$, the dynamic enhancement factor increases from $\Psi_{vol} = 1$ for the static case to $\Psi_{vol} = 1.6$ and $\Psi_{vol} = 1.8$ for Cases 1 and 2 respectively; when $f°_{Dyn} = 10$, the enhancement factor increases to $\Psi_{vol} = 2.2$ and $\Psi_{vol} = 3.0$ for the two cases. This indicates that the dynamic regime has a crucial influence on the volumes of residual materials. On the other hand, the relative production ratio of residual matter (Ψ_{red}) is almost constant, changing only from 0.727 to 0.756. It should be recalled that this parameter is related to the effective volume of mantle material that passes through the melting zone, and not to the size of the melting zone; its relative constancy in all these calculations confirms that residue productivity is controlled by the relative fluxes of material with different melting behaviours, rather than by the variability of the melting phenomenon itself.

This result is confirmed when we consider the sensitivity of these results to the various hypotheses, using the alternatives previously described. Changing the surrounding mantle temperature T_∞ slightly modifies the temperature profile (because of the scaling to the distance from the symmetry axis to the solidus), but has only a minute effect on Ψ_{vol} and Ψ_{rsd}. It does change the relative volumes passing through each subzone, and thus it affects the relative amounts of residue coming from each zone: the colder the surrounding mantle, the greater the proportion of refractory residue from the central region. Similarly, changing the thermal model, and more sensitively, changing the melting curve (the relationship between temperature and degree of melting), leads to major changes in the relative amounts of residue. On the other hand, their average petrological characteristics, which are controlled by the mean melting degree in each subregion, do not change much.

Finally, the calculation for the plume-head model is strongly influenced by its spherical symmetry. This dramatically decreases the importance of the central zone, and increases the proportion of refractory residue. However, the general results found with the cylindrical symmetry continue to apply.

References

ABOUCHAMI, W., BOHER, M., MICHARD, A. & ALBARÈDE, F. 1990. A major 2.1 Ga old event of mafic magmatism in West Africa. *Journal of Geophysical Research*, **95**, 17 605–17 629.

ALBARÈDE, F. 1998. The growth of continental crust. *Tectonophysics*, **296**, 1–14.

ARCULUS, R. J. 1999. Origins of the continental crust. *Journal of Proceedings of the Royal Society of New South Wales*, **132**, 83–110.

ARNDT, N. T. 1976. Melting relations of ultramafic lavas (komatiites) at 1 atm and high pressure. *Carnegie Institution of Washington Yearbook*, **75**, 555–562.

ARNDT, N. T. & GOLDSTEIN, S. L. 1989. An open boundary between lower continental crust and mantle: its role in crust formation and crustal recycling. *Tectonophysics*, **161**, 201–212.

ARNDT, N. T. & NISBET, E. G. 1982. *Komatiites*. George Allen & Unwin, London.

BERSTEIN, S., KELEMEN, P. B. & BROOK, C. K. 1997. Highly depleted spinel harzburgite xenoliths in Tertiary dykes from East Greenland. *Earth and Planetary Science Letters*, **154**, 221–235.

BICKLE, M. J., FORD, C. E. & NISBET, E. G. 1977. The petrogenesis of peridotitic komatiites; evidence from high-pressure melting experiments. *Earth and Planetary Science Letters*, **37**, 97–106.

BOYD, F. R. 1989. Compositional distinction between oceanic and cratonic lithosphere. *Earth and Planetary Science Letters*, **96**, 15–26.

BOYD, F. R. & MERTZMAN, S. A. 1987. Composition and structure of the Kaapvaal lithosphere, southern Africa. *In*: MYSEN, B. O. (ed.) *Magmatic Processes: Physicochemical Principles*. Geochemical Society, Special Publication, **1**, 13–24.

COX, K. G. 1987. Textural studies of garnet lherzolites: evidence of exsolution origin of high-temperature harzburgites. *In*: NIXON, P. H. (ed.) *Mantle Xenoliths*. Wiley, New York, 537–550.

DE WIT, M. J. & ASHWAL, L. D. 1997. *Greenstone Belts*. Oxford Scientific, Oxford.

FRANCIS, D. 1995. The implications of picritic lavas for the mantle sources of terrestrial magmatism. *Lithos*, **34**, 89–106.

GREEN, D. H. 1974. Genesis of Archaean peridotitic magmas and constraints on Archaean geothermal gradients and tectonics. *Geology*, **3**, 15–18.

GREEN, J. C., NICHOLLS, I. A., VILJOEN, M. J. & VILJOEN, R. P. 1974. Experimental demonstration of the existence of peridotitic liquids in earliest Archaean magmatism. *Geology*, **3**, 11–14.

GRIFFIN, W. L, O'REILLY, S. Y. & RYAN, C. G. 1999. The composition and origin of sub-continental lithospheric mantle. *In*: FEI, Y., BERTKA, C. M. & MYSEN, B. O. (eds) *Mantle Petrology: Field Observations and High-Pressure Experimentation: a Tribute to Francis R. (Joe) Boyd*. Geochemical Society, Special Publications, **6**, 13–46.

GROVE, T. L., PARMAN, S. W. & DANN, J. C. 1999. Conditions of magma generation for Archean komatiites from the Barberton Mountainland, South Africa. *In*: FEI, Y., BERTKA, C. M. & MYSEN, B. O. (eds) *Mantle Petrology: Field Observations and High-Pressure Experimentation: a Tribute to Francis R. (Joe) Boyd*. Geochemical Society, Special Publications, **6**, 155–167.

HAWKESWORTH, C. J., KEMPTON, P. D., ROGERS, N. W., ELLAM, R. M. & VAN CALSTEREN, P. W. 1990. Continental mantle lithosphere, and shallow level enrichment processes in the Earth's mantle. *Earth and Planetary Science Letters*, **96**, 256–268.

HERZBERG, C. T. 1983. Lithosphere peridotites of the Kaapvaal craton. *Earth and Planetary Science Letters*, **120**, 13–29.

HERZBERG, C. 1992. Depth and degree of melting of komatiite. *Journal of Geophysical Research*, **97**, 4521–4540.

HERZBERG, C. 1995. Generation of plume magmas through time: an experimental perspective. *Chemical Geology*, **126**, 1–17.

HERZBERG, C. 1999. Phase equilibrium constraints on the formation of cratonic mantle. *In*: FEI, Y., BERTKA, C. M. & MYSEN, B. O. (eds) *Mantle Petrology: Field Observations and High-Pressure Experimentation: a Tribute to Francis R. (Joe) Boyd*. Geochemical Society, Special Publication, **6**, 13–46.

HERZBERG, C. & OHTANI, E. 1988. Origin of komatiite at high pressures. *Earth and Planetary Science Letters*, **88**, 321–329.

HERZBERG, C. T., FYFE, W. S. & CARR, M. J. 1983. Density constraints on the formation of continental Moho and crust. *Contributions to Mineralogy and Petrology*, **84**, 1–5.

HILL, R. I. 1990. Starting plumes and continental breakup. *Earth and Planetary Science Letters*, **104**, 398–416.

HIRTH, G. & KOHLSTEDT, D. L. 1996. Water in the oceanic upper mantle: implications for rheology, melt extraction and the evolution of the lithosphere. *Earth and Planetary Science Letters*, **144**, 93–108.

INOUE, T. 1994. Effect of water on melting phase relations and melt composition in the system Mg_2SiO_4–$MgSiO_3$–H_2O up to 15 GPa. *Physics of the Earth and Planetary Interiors*, **85**, 237–263.

KAWAMOTO, T. & HOLLOWAY, J. R. 2002. Melting temperature and partial melt chemistry of H_2O-saturated mantle peridotite to 11 GPa: implications for the stability of H_2O fluid in the Earth's mantle and kimberlite generation. *Science*, in press.

KAY, R. W. & KAY, S. M. 1991. Creation and destruction of lower continental crust. *Geologische Rundschau*, **80**, 259–278.

KELEMEN, P. B., HART, S. R. & BERSTEIN, S. 1998. Silica enrichment in the continental upper mantle via melt/rock reaction. *Earth and Planetary Science Letters*, **164**, 387–406.

KELLOGG, L. H., HAGER, B. H. & VAN DER HILST, R. D. 1999. Compositional stratification in the deep mantle. *Science*, **283**, 1881–1884.

MAALOE, S. & AOKI, K.-I. 1977. The major element composition of the upper mantle estimated from the composition of lherzolites. *Contributions to Mineralogy and Petrology*, **63**, 161–173.

MAHLBURG KAY, S. & KAY, R. W. 1985. Role of crystal cumulates and the oceanic crust in the formation of the lower crust of the Aleutian arc. *Geology*, **13**, 461–464.

MCDONOUGH, W. F. 1990. Constraints on the composition of the continental lithospheric mantle. *Earth and Planetary Science Letters*, **101**, 1–18.

MCDONOUGH, W. F. 1992. Chemical and isotopic systematics of continental lithospheric mantle. *In*: MEYER, H. O. A. & LEONARDOS, O. H. (eds) *Proceedings of the 5th International Kimberlite Conference*. CPRM, Brasilia, 478–485.

MCKENZIE, D. & BICKLE, M. J. 1988. The volume and composition of melt generated by extension of the lithosphere. *Journal of Petrology*, **29**, 625–679.

MENZIES, M. A. 1990. Archaean, Proterozoic and Phanerozoic lithospheres. *In*: MENZIES, M. A. (ed.) *Continental Mantle*. Clarendon, Oxford, 67–86.

MILLER, G. H., STOLPER, E. M. & AHRENS, T. J. 1991. The equation of state of molten komatiite, 2: application to komatiite petrogenesis and the Hadean mantle. *Journal of Geophysical Research*, **96**, 11 849–11 864.

NESBITT, R. W., SUN, S. S. & PURVIS, A. C. 1979. Komatiites: geochemistry and genesis. *Canadian Mineralogist*, **17**, 165–186.

NISBET, E. G. & WALKER, D. 1982. Komatiites and the structure of the Archean mantle. *Earth and Planetary Science Letters*, **60**, 105–113.

NISBET, E. G., CHEADLE, M. J., ARNDT, N. T. & BICKLE, M. J. 1993. Constraining the potential temperature of the Archaean mantle: a review of the evidence from komatiites. *Lithos*, **30**, 291–307.

O'HARA, M. J., SAUNDERS, M. J. & MERCY, E. L. P. 1975. Garnet-peridotite, primary ultrabasic magma and eclogite; interpretation of upper mantle processes in kimberlite. *Physics and Chemistry of the Earth*, **9**, 571–604.

OHTANI, E., MORIYAMA, J. & KAWABE, I. 1988. Majorite garnet stability and its implication for genesis of komatiite magmas. *Chemical Geology*, **70**, 147.

O'REILLY, S. Y. & GRIFFIN, W. L. 1996. 4-D lithosphere mapping: a review of the methodology with examples. *Tectonophysics*, **262**, 3–18.

PARMAN, S., GROVE, T. L., DANN, J. & DE WIT, M. J. 1996. Pyroxene compositions in 3.49 Ga komatiite: evidence of variable H_2O content. *EOS Transactions, American Geophysical Union*, **77**, 280.

PEARSON, D. G. 1999. Evolution of cratonic lithospheric mantle: an isotopic perspective. *In*: FEI, Y., BERTKA, C. M. & MYSEN, B. O. (eds) *Mantle Petrology: Field Observations and High-Pressure Experimentation: a Tribute to Francis R. (Joe) Boyd*. Geochemical Society, Special Publications, **6**, 57–78.

POUDJOM-DJOMANI, Y., O'REILLY, S. Y., GRIFFIN, W. L. & MORGAN, P. 2001. The density structure of subcontinental lithosphere through time. *Earth and Planetary Science Letters*, **184**, 605–621.

TAYLOR, S. R. & MCLENNAN, S. M. 1985. *The Continental Crust: its Composition and Evolution*. Blackwell Scientific, Oxford.

VAN DER HILST, R. D. & KARASON, K. 1999. Compositional heterogeneity in the bottom 1000 km of Earth's mantle: towards a hybrid convection model. *Science*, **283**, 1885–1888.

VILJOEN, M. J. & VILJOEN, R. P. 1969. The geology and geochemistry of the lower ultramafic unit of the Onverwacht Group and a proposed new class of igneous rocks. *In: Upper Mantle Project*. Geological Society of South Africa, Special Publication, **2**, 55–85.

WALLACE, P. L. 1998. Water and partial melting in mantle plumes: inference from dissolved H_2O concentrations of Hawaiian basaltic magmas. *Geophysical Research Letters*, **25**, 3636–3642.

WALTER, M. J. 1999. Melting residues of fertile peridotite and the origin of cratonic lithosphere. *In*: FEI, Y., BERTKA, C. M. & MYSEN, B. O. (eds) *Mantle Petrology: Field Observations and High-Pressure Experimentation: a Tribute to Francis R. (Joe) Boyd*. Geochemical Society, Special publications, **6**, 225–240.

WATSON, S. & MCKENZIE, D. 1991. Melt generation by plumes: a study of Hawaiian volcanism. *Journal of Geophysical Research*, **83**, 5989–6004.

ZHANG, H., MENZIES, M., GURNEY, J. J. & ZHOU, X. 2001. Cratonic peridotites and Si-rich melts: diopside–enstatite relationships in polymict xenoliths, Kaapvaal, South Africa. *Geochimica et Cosmochimica Acta*, **65**, 3365–3377.

Pb isotope variations in Archaean time and possible links to the sources of certain Mesozoic–Recent basalts

BÉATRICE LUAIS[1] & CHRIS J. HAWKESWORTH[2]

[1] *Centre de Recherches Pétrographiques et Géochimiques (CRPG)–CNRS UPR 2300, 15, rue Notre-Dame-des-Pauvres, BP 20, 54501 Vandoeuvre-les-Nancy Cédex, France (e-mail: luais@crpg.cnrs-nancy.fr)*
[2] *Department of Earth Sciences, University of Bristol, Wills Memorial Building, Queens Road, Bristol BS8 1RJ, UK*

Abstract: The present-day Pb isotope ratios of mafic and felsic rocks, feldspars and ore samples (galenas) from Archaean cratons ranging in age from 3.8–3.7 Ga to 2.6–2.7 Ga have been used to investigate their initial $^{207}Pb/^{204}Pb$ ratios, and μ_1 ratios (source U/Pb ratios). Two broad trends of initial $^{207}Pb/^{204}Pb$ evolution are observed, with samples from Greenland, Abitibi and SW India tending to have lower initial $^{207}Pb/^{204}Pb$ ratios through time than those from eastern India, Australia and southern Africa (Kaapvaal and Zimbabwe). The calculated μ_1 values are in the range of 7–9 and they tend to be constant for a given craton through time, with Zimbabwe as a notable exception, for both mafic and felsic rock types. It follows that each craton is characterized by specific μ_1 values, or initial $^{207}Pb/^{204}Pb$ ratios; μ_1 for Greenland is 7.37 ± 0.27, and they are higher for Abitibi at 7.57 ± 0.13, which are similar to SW India, where $\mu_1 = 7.57 \pm 0.13$. In Scotland $\mu_1 = 7.94 \pm 0.152$, in Finland μ_1 is 8.03 ± 0.08, in eastern India it is 8.03 ± 0.1, and for Australia $\mu_1 = 8.07 \pm 0.14$, and there tend to be higher values of 8 ± 0.2 and 8.32 ± 0.29 in the Kaapvaal and Zimbabwe, respectively. Overall, there is therefore a strong provinciality of μ_1 values in the Archaean areas, and unexpectedly the high μ_1 Archaean terranes are close to the areas of high μ_1 Mesozoic continental flood basalts (CFB), and the DUPAL anomaly in recent oceanic basalts. The cratons with high μ_1 values are in the southern hemisphere near the 130–190 Ma CFB from the Gondwana super-continent, and the maximum DUPAL anomalies at latitudes 30–45°S south of Africa and 0–15°S south of India. The variations in μ_1 are based on $^{207}Pb/^{204}Pb$ ratios, and so they primarily reflect events that took place early in Earth history. In one model, the high μ_1 values were initially generated at the core–mantle boundary, through processes linked to core formation and late accretion events, and they were mobilized in plumes responsible for at least some of the early Archaean mafic crust. Preservation of domains characterized by different U/Pb ratios is inferred to have been at relatively shallow levels, as they are preserved in both felsic and mafic rocks, and they appear to have survived for long periods of time, despite the effects of mantle convection and plate motion. Recycling of early Archaean lithosphere in the mantle by delamination might perpetuate these anomalies through time in a given area, perhaps even contributing to the modern DUPAL Pb anomalies. Delamination rather than subduction as the process of lithosphere recycling might explain why such regional domains are not observed in the kappa (κ_1: source Th/U) ratios of these Archaean samples.

The U–Th–Pb isotope systems have half-lives ranging from 0.7 to 4.6 Ga, and can therefore be used to investigate variations in time-integrated U/Pb (μ_1) and Th/U (κ_1) ratios, and to constrain models for the evolution of the mantle and crust from early Archaean time to the present (Sinha & Tilton 1973; Stacey & Kramers 1977; Kramers & Tolstikhin 1997). The Pb isotope compilation and model of Kramers & Tolstikhin (1997) highlight the range in the initial $^{207}Pb/^{204}Pb$ ratios (and thus in time-integrated μ_1) in suites of Archaean to early Proterozoic rocks. Inferred differences between high-μ_1 lower crust and low-μ_1 upper crust could not be simply explained by intracrustal processes such as magmatic fractionation, partial melting, or U transport in fluid phases during metamorphism. Instead, their model includes concomitant crustal fractionation and erosion–delamination processes, and hence fluxes between young upper crust and old lower crust. It is perhaps less clear whether the main differences in μ_1 are between different geographical areas or between upper- and lower-crustal rock types. In rocks from 2 Ga to the present there is then a restricted range of the nitial $^{207}Pb/^{204}Pb$ ratios, and a greater range in $^{206}Pb/^{204}Pb$.

From: FOWLER, C. M. R., EBINGER, C. J. & HAWKESWORTH, C. J. (eds) *The Early Earth: Physical, Chemical and Biological Development.* Geological Society, London, Special Publications, **199**, 105–124. 0305-8719/02/$15.00
© The Geological Society of London 2002.

This reflects both the presence of less parental ^{235}U, and the more efficient recycling of U into the mantle under the more oxidizing atmospheric conditions that have prevailed since 2 Ga. This major change in the behaviour of uranium has been reaffirmed in discussions of Th–U–Nb systematics (Collerson & Kamber 1999) and Th/U mantle evolution (Elliott et al. 1999).

The record of ^{207}Pb/^{204}Pb (and μ_1) variability in modern ocean island basalt (OIB) and mid-ocean ridge basalt (MORB) samples defines chemical heterogeneities in the Earth's mantle today that were initially recognized by Manhès et al. (1977) to be as old as 2 Ga. Among them, the globe-encircling DUPAL anomaly (Dupré & Allègre 1983) is one of the largest isotopically enriched mantle reservoirs known, and the debate on its origin remains open. Finally, Sinha (1971) described distinct Pb isotopic signatures in samples with a wide range of ages from the northern and southern hemispheres.

These Pb isotope features raise a number of questions: How were the variable U/Pb ratios generated in the early Earth's mantle and recorded in samples from early Archaean time? How have enriched mantle domains with different U/Pb ratios been preserved through time despite the effects of mantle convection? Is there any link between the presence of high-μ_1 Archaean crust and the origin of the mantle DUPAL anomaly? It might be speculated, for example, that the DUPAL anomaly is an old feature that formed in Archaean times, as first suggested by Hart (1984) and Castillo (1988), and has been preserved in the lower mantle (Kamber & Collerson 1999). Alternatively, it might be a young feature related to oceanic crust and/or sediment recycling (Rehkämper & Hofmann 1997), or shallow-level mobilization of continental mantle lithosphere (Hawkesworth et al. 1986).

To evaluate these questions further, we have looked again at the variations in ^{207}Pb/^{204}Pb ratios (and thus μ_1) of Archaean basalts and granitoids worldwide in the age range 3.7–3.8, 3.5, 3.2, 2.9 to 2.7–2.6 Ga. We show that there are provinces with different ^{207}Pb/^{204}Pb and μ_1 values, and note that some unexpectedly coincide with Mezosoic continental flood basalts and oceanic DUPAL rocks with similar Pb isotope features.

Selection of data

The Pb isotope data used here are from a worldwide compilation of samples from Archaean cratons (These data can be obtained from the Society Library or the British Library Document Supply Centre, Boston Spa, Wetherby, West Yorkshire LS23 7BQ, UK as Supplementary Publication No. SUP 18176 (14 pages)) selected to cover a range in ages from 3.74 to 2.6 Ga, and in the rocks and minerals analysed (mafic and granitoid rocks, K-feldspars and galenas) (Moorbath et al. 1972, 1975, 1977; Baadsgaard 1973; Hawkesworth et al. 1975, 1979; Hawkesworth & O'Nions 1977; Nunes and Pyke 1980; Cattell et al. 1984). They include new data from the Belingwe belt of Zimbabwe. The locations of the selected areas are summarized in Figure 1 and the data sources are given in the figure captions. The samples are from: Greenland (3.7–2.7 Ga); Scotland (3.3–2.9 Ga; Finland (2.9–2.6 Ga); India (3.3–2.7 Ga); southern Africa (3.5–2.7 Ga), with samples from the Zimbabwe and Kaapvaal cratons; Western Australia (3.5–2.7 Ga), with samples from Yilgarn and Pilbara cratons; Canada, Superior Province, in particular the 2.7 Ga Abitibi belt.

The mafic rocks are komatiites, basalts, gabbros and dolerites, and the felsic rocks are both tonalite–trondhjemite–granodiorites (TTG) and granodiorite–granite–monzogranites (GGM). Analyses of galena and K-feldspar provide initial Pb isotopic compositions, as they contain little or no uranium.

Knowledge of the crystallization ages of the samples (i.e. the time t) is critical for the accurate calculation of initial ^{207}Pb/^{204}Pb ratios, and U/Pb (μ_1) and Th/U (κ_1) values. One criterion for selection was therefore that each sample lies on an isochron for which the calculated U–Pb zircon age is in agreement with the one inferred from other decay systems, typically Sm–Nd and T_{DM} ages. This clearly restricts the number of samples that can be used. The second criterion was that the samples selected should not have undergone any of the following: (1) alteration; (2) high-grade metamorphism and deformation, as with the granulites from Scourie in Scotland (Moorbath et al. 1969; Whitehouse 1989a), Greenland and Labrador (Schiotte & Bridgwater 1989), or as in the 3.36 Ga Acasta grey gneisses (Yamashita et al. 2000); (3) crustal contamination of mantle-derived magmas, as indicated by low ε_{Nd} values (e.g. Wyoming: Wooden & Mueller 1988; komatiites from Kambalda: Dupré & Arndt 1990; 2.7–2.6 Ga late Archaean crust of the Slave Province: Davis et al. 1996; Yamashita et al. 2000); or (4) were the products of crustal reworking (Kreissig et al. 2000).

Methods

For each sample, the μ_1 and μ_2 ratios were calculated using the radioactive decay equations for ^{238}U, ^{235}U and ^{232}Th (equations (1) and (2)).

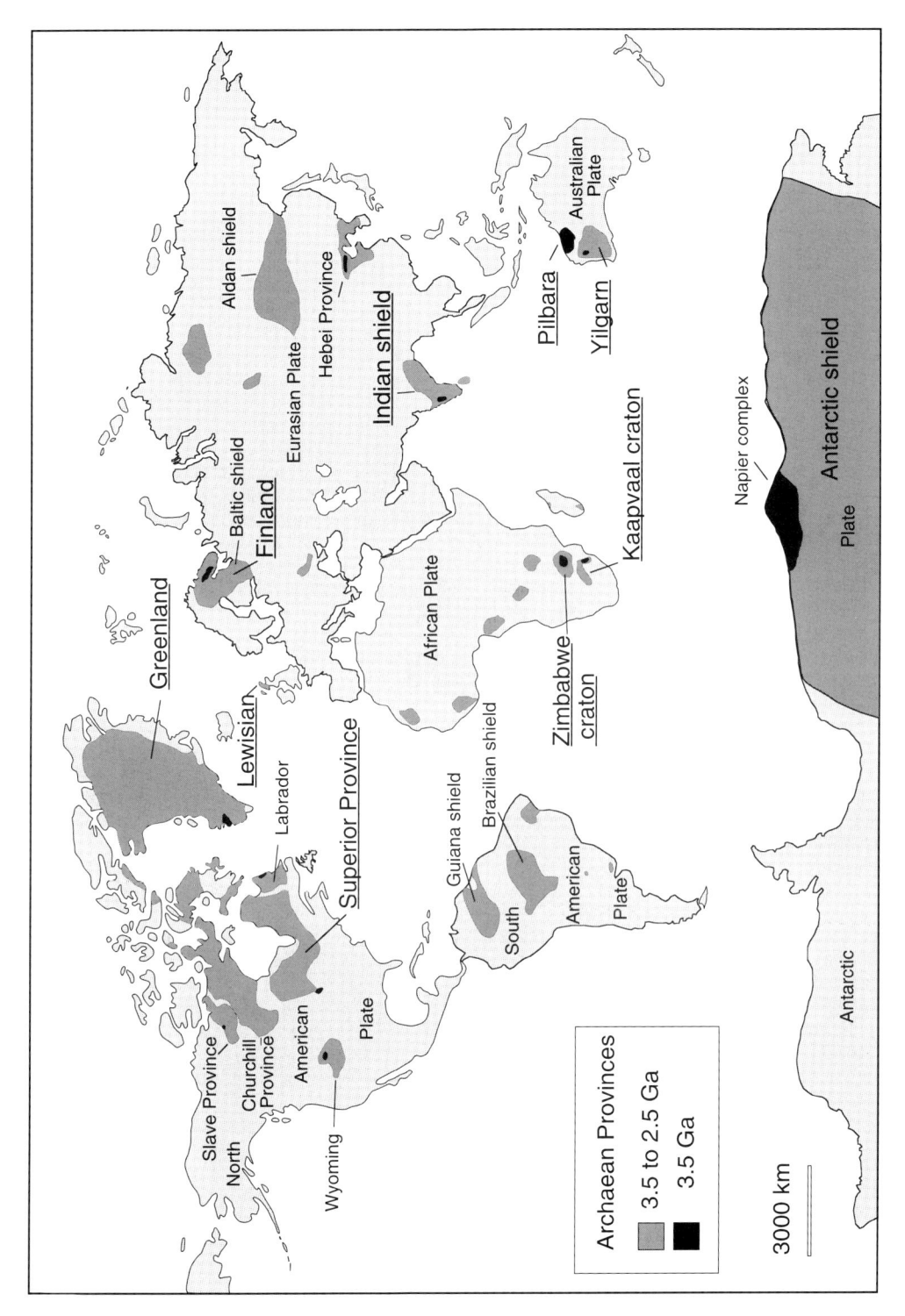

Fig. 1. Map of the world with locations of Archaean provinces (from Condie 1993). Selected areas for this study are underlined.

The initial $^{207}\text{Pb}/^{204}\text{Pb}$ compositions at the time when the rock or mineral crystallized (t) were then calculated:

$$\frac{^{206}\text{Pb}}{^{204}\text{Pb}_{meas}} = \frac{^{206}\text{Pb}}{^{204}\text{Pb}_{T=4.57}} + \mu_1(e^{\lambda T} - e^{\lambda t})$$

$$+ \mu_2(e^{\lambda t} - 1) \qquad (1)$$

$$\frac{^{207}\text{Pb}}{^{204}\text{Pb}_{meas}} = \frac{^{207}\text{Pb}}{^{204}\text{Pb}_{T=4.57}} + \frac{\mu_1}{137.88}(e^{\lambda' T} - e^{\lambda' t})$$

$$+ \frac{\mu_2}{137.88}(e^{\lambda' t} - 1) \qquad (2)$$

$$\frac{^{208}\text{Pb}}{^{204}\text{Pb}_{meas}} = \frac{^{208}\text{Pb}}{^{204}\text{Pb}_{T=4.57}} + \mu_1 \kappa_1(e^{\lambda'' T} - e^{\lambda'' t})$$

$$+ \mu_2 \kappa_2(e^{\lambda'' t} - 1) \qquad (3)$$

where T is the age of the Earth (4.57 Ga); t is the age of formation of the sample; λ', λ'' and λ'' are the decay constants of ^{238}U, ^{235}U and ^{232}Th, respectively; μ_1 and μ_2, κ_1 and κ_2 are the U/Pb and Th/U ratios for the time intervals of $(T - t)$ and (t-present), respectively; $(^{206}\text{Pb}/^{204}\text{Pb})_{4.57}$ and $(^{207}\text{Pb}/^{204}\text{Pb})_{4.57}$ ratios are the primordial terrestrial Pb isotopic compositions for the Earth at 4.57 Ga, assumed to be the same as those of the Canyon Diablo (CD) troilite (Tatsumoto et al. 1973).

The μ_2 and κ_2 values reflect the time-integrated bulk-rock U/Pb and Th/U ratios from the time of formation t to the present, and these may be the original bulk-rock values or those disturbed by subsequent post-emplacement processes. The extent to which secondary processes have influenced the calculated μ_2 and κ_2 values depends on how early in the history of the rock such disturbance took place.

The μ_1 parameter is the time-averaged U/Pb ratio between the age of the Earth T and the formation of the rock at time t. Thus it reflects the trace element characteristics of the source of each sample, assuming that the source had the same U/Pb ratio from $T = 4.57$ Ga to the time of formation of the sample. This is clearly an oversimplification geologically, but μ_1 values remain a powerful tool for accessing time-averaged source characteristics, particularly for old rocks, because ^{235}U has a relatively short half-life compared with other long-lived decay schemes.

The μ_1 values are then used to calculate the initial Pb isotopic ratios for each sample according to equations (1) and (2). The decay constant for the decay of ^{235}U to ^{207}Pb ($\lambda'_{235} = 0.98485 \times 10^{-9} \text{a}^{-1}$) is six times higher than that for the $^{238}\text{U}-^{206}\text{Pb}$ system ($\lambda_{238} = 0.155125 \times 10^{-9} \text{a}^{-1}$) (Jaffey et al. 1971), and so the first 2.5 Ga of the Earth history is characterized by greater variation in $^{207}\text{Pb}/^{204}\text{Pb}$ compared with $^{206}\text{Pb}/^{204}\text{Pb}$. Thus Pb isotope heterogeneities in Archaean time are best investigated by looking at initial μ_1 and $^{207}\text{Pb}/^{204}\text{Pb}$ ratios.

Pb isotope evolution in Archaean time

Temporal evolution of the initial $^{207}Pb/^{204}Pb$ isotopic compositions and μ_1

The initial $^{207}\text{Pb}/^{204}\text{Pb}$ isotope values of the selected 3.74–2.6 Ga samples are plotted against their age of formation in Figure 2. In addition to a general overall increase in the initial $^{207}\text{Pb}/^{204}\text{Pb}$ ratios from c. 13 at 3.7 Ga to c. 15 at 2.6 Ga, different areas do appear to be characterized by distinct initial Pb isotope ratios. Thus, there is: (1) a low initial $^{207}\text{Pb}/^{204}\text{Pb}$ group defined by samples from Greenland and Abitibi, and from SW India; in these rocks the initial $^{207}\text{Pb}/^{204}\text{Pb}$ values range from c. 13 at 3.7 Ga (Greenland) to c. 14.6 at 2.6 Ga; (2) a high initial group defined by samples from Australia (Yilgarn, Pilbara), Zimbabwe, Kaapvaal, east India, Scotland and Finland, with $^{207}\text{Pb}/^{204}\text{Pb}$ values ranging from c. 13.7 at 3.5 Ga to c. 15 at 2.6 Ga.

This distinction into two groups is especially clear for the samples formed at 2.9 and 2.7 Ga, for which there are more extensive datasets. At a given age, there is therefore a large range in initial $^{207}\text{Pb}/^{204}\text{Pb}$ ratios, from 14.3 to 14.8 at 2.9 Ga, and from 14.1 to 15 at 2.7 Ga. Moreover, the higher values tend to be in samples from Australia, Zimbabwe, Kaapvaal, east India, Scotland (Scourie) and Finland rather than in those from Greenland, Abitibi and SW India. Whether the suggestion of two groups will survive collection of more data is unclear, but they reflect the scatter in $^{207}\text{Pb}/^{204}\text{Pb}$ ratios in ores and feldspars of Archaean and Early Proterozoic rock suites observed by Kramers & Tolstikhin (1997).

According to equation (2), initial $^{207}\text{Pb}/^{204}\text{Pb}$ values reflect the μ_1 ratios calculated for the period $(T - t)$, and the age of the sample t. In Figure 3 the calculated μ_1 ratios for each sample are plotted against their formation ages. For a given age, for example at 3.5–3.6 Ga, we again observe a large variation in μ_1 values from seven in the Greenland samples to nine in those from Zimbabwe. This large range reflects the two groups of samples with different initial $^{207}\text{Pb}/^{204}\text{Pb}$ ratios, as discussed for Figure 2: (1) the samples from Abitibi and Greenland, and some of those from SW India, have low μ_1 values from c. seven at 3.6 Ga to c. 7.7 at 2.6 Ga; (2) the samples from Australia, Zimbabwe, Kaapvaal, east India, Scotland (Scourie) and

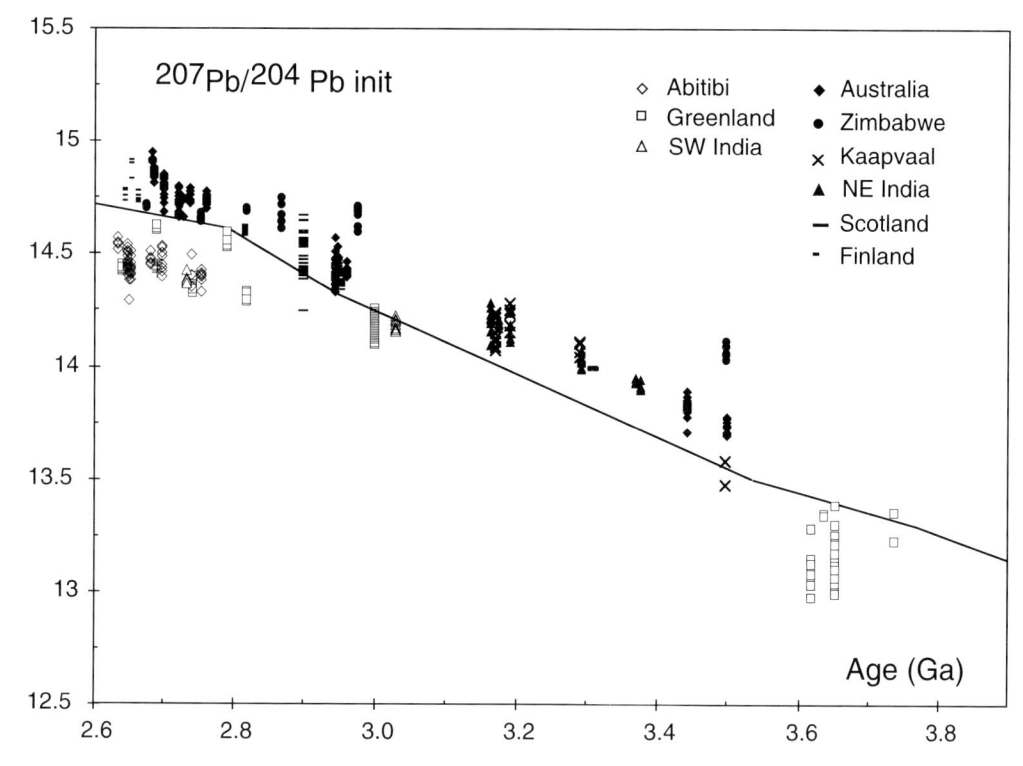

Fig. 2. Initial $^{207}Pb/^{204}Pb$ v. age for the 3.8–2.6 Ga felsic, mafic and ore (galenas, feldspars) samples. Two evolution trends are shown: a high initial $^{207}Pb/^{204}Pb$ trend is defined with samples from East India, Finland, Scotland, Australia, Zimbabwe and Kaapvaal cratons; a low initial $^{207}Pb/^{204}Pb$ trend is defined with samples from Greenland, Superior Province (Abitibi) and SW Indian cratons. An arbitrary separation line between these two groups is drawn. For the 2.6–2.8 Ga felsic samples: data from Greenland (Kalsbeek et al. 1988; Taylor et al. 1992), Superior Province (Sihna & Tilton 1973; Abitibi: Gariépy & Allègre 1985; Tilton & Kwon 1990; Carignan et al. 1993; Vervoort et al. 1993; Stern et al. 1994), Finland (Vidal et al. 1980, Putchel et al. 1998), Zimbabwe (Somabula granite, Taylor et al. 1991), Australia (Yilgarn craton, Sihna & Tilton 1973; Oversby 1975; Bickle et al. 1983b). For 2.6–2.8 Ga mafic samples, data from Superior Province (Tilton 1983; Brévart et al. 1986; Dupré et al. 1984; Abitibi, Vervoort et al. 1993), India (Balakrishnan et al. 1990), Finland (Vidal et al. 1980; Putchel et al. 1998), Australia (Yilgarn, Roddick 1984; Chauvel et al. 1985), Zimbabwe (Belingwe belt, Dupré & Arndt 1990). For the 2.9 Ga samples: data from Greenland (Nûk gneisses, Taylor et al. 1980, 1984a), Scotland (Whitehouse 1989b), India (Taylor et al. 1984b), Zimbabwe (Belingwe belt, Taylor et al. 1991; this work, Table 3), Australia (Pilbara, Korsh & Gulson, 1986; Bickle et al. 1989). For the 3.2 Ga samples: data from India (Taylor et al. 1984b; Moorbath et al. 1986; Sengupta et al. 1991), Scotland (Burton et al. 1994), South Africa (Kaapvaal–Onverwacht: Sihna & Tilton 1973; Wilson & Carlson 1989; Huang et al. 1995; Barberton: Sihna & Tilton 1973), Zimbabwe (Mashaba galenas, Robertson 1973). For the 3.5 Ga samples: data from Greenland (Amitsôq, Black et al. 1971; Gancarz & Wasserburg 1977; Kamber & Moorbath 1998), South Africa (Onverwacht, Dupré & Arndt 1990), Zimbabwe (Belingwe belt, Taylor et al. 1991), Australia (Pilbara, Bickle et al. 1983a, 1993). For the 3.8 Ga samples: data from Isua (Appel et al. 1978). Open symbols correspond to low initial $^{207}Pb/^{204}Pb$; filled symbols to high initial $^{207}Pb/^{204}Pb$ samples (mafic, felsic, galenas, feldspars).

Finland tend to have higher μ_1 values, of 7.5–9 over a similar age range.

Overall, the values of μ_1 remain fairly constant in rocks of different ages from particular Archaean terranes. Thus,

- the Greenland samples have μ_1 values of 7–8 with most around 7.37 ± 0.27 as mainly seen in the 3.74–2.9 and 2.7 Ga samples;
- the Abitibi samples have remarkably constant μ_1 values of 7.57 ± 0.13;
- the Indian samples split into two distinct groups of $\mu_1 = 8.03 \pm 0.1$ for felsic rocks from eastern India at 3.2 Ga, and $\mu_1 = 7.57 \pm 0.08$ for felsic rocks at 2.9 Ga and for mafic rocks at 2.7 Ga from SW India;
- the Scotland samples have homogeneous μ_1 values of 7.94 ± 0.15 at 3.2 and 2.9 Ga;

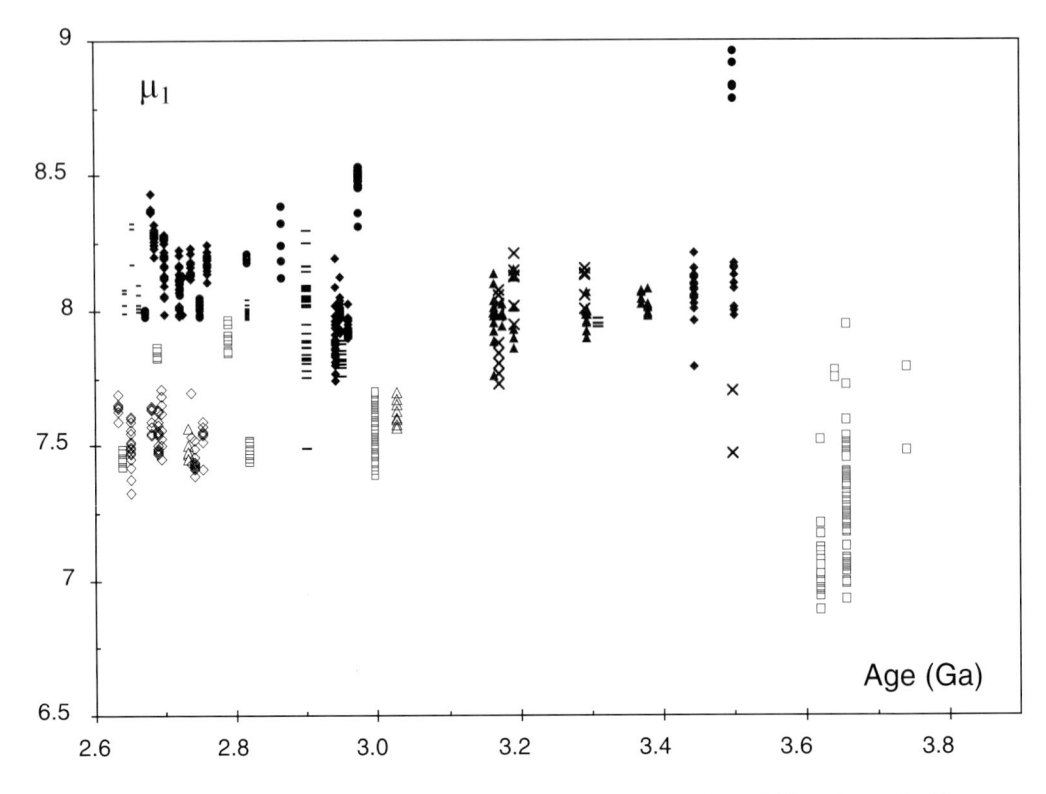

Fig. 3. μ_1 (source U/Pb) v. age for the 3.8–2.6 Ga felsic, mafic and ore (galenas, feldspars) samples. Same symbols, localities and references as in Figure 2.

- the Finland samples have μ_1 values of 7.96 ± 0.02 at 2.8 Ga and 8.09 ± 0.11 at 2.64 Ga;
- the Western Australian samples exhibit a narrow range of μ_1 values at 8.07 ± 0.14 from 3.5 to 2.7 Ga, for both mafic and felsic rocks;
- the Kaapvaal samples have μ_1 values of 8.01 ± 0.14 at 3.2 Ga;
- the Zimbabwe samples have the highest μ_1 values of 8.85 ± 0.06 at 3.5–3.2 Ga, and values of 8.32 ± 0.16 at 2.9 Ga, and 8.00 ± 0.02 at 2.7 Ga, apparently irrespective of whether the rocks are mafic or felsic in composition.

Propagation of the errors on the ages of these samples into the calculation of μ_1 results in small differences in μ_1 compared with the ranges of μ_1 discussed above. Most rocks have U–Pb zircon ages with errors of only few millions of years, and yet an error of 100 Ma on the formation age would propagate to an error of ±0.24 on μ_1 at 3.7 Ga, and of ±0.06 at 2.7 Ga. Such differences are small in the context of the range of μ_1 from a given craton at a given age.

Another issue is how the initial ^{207}Pb/^{204}Pb and μ_1 values of different rock types and minerals vary within different areas. Thus, Figure 4 presents μ_1 frequency plots for four selected cratons where data are available for both mafic and felsic samples of the same age: Abitibi, India, Australia and Zimbabwe. Strikingly, there are no clear differences in the initial ^{207}Pb/^{204}Pb and μ_1 values of the felsic or mafic whole rocks, the potassic feldspars or galenas. This suggests that the mafic and felsic magmas were derived from source regions with similar Pb isotope histories, and either that U/Pb ratios were not fractionated by intracrustal processes or that there were short time periods between the formation of basaltic crust and remelting to generate the more evolved lithologies. Thus, we reiterate that the data show a striking consistency in μ_1 values for given areas, with Greenland μ_1 < Abitibi μ_1 < India μ_1 < Scotland–Finland ≤ Australia μ_1 < Kaapvaal–Zimbabwe μ_1, regardless of the age and the composition of the samples. This highlights the strong provinciality in the μ_1 (U/Pb) values

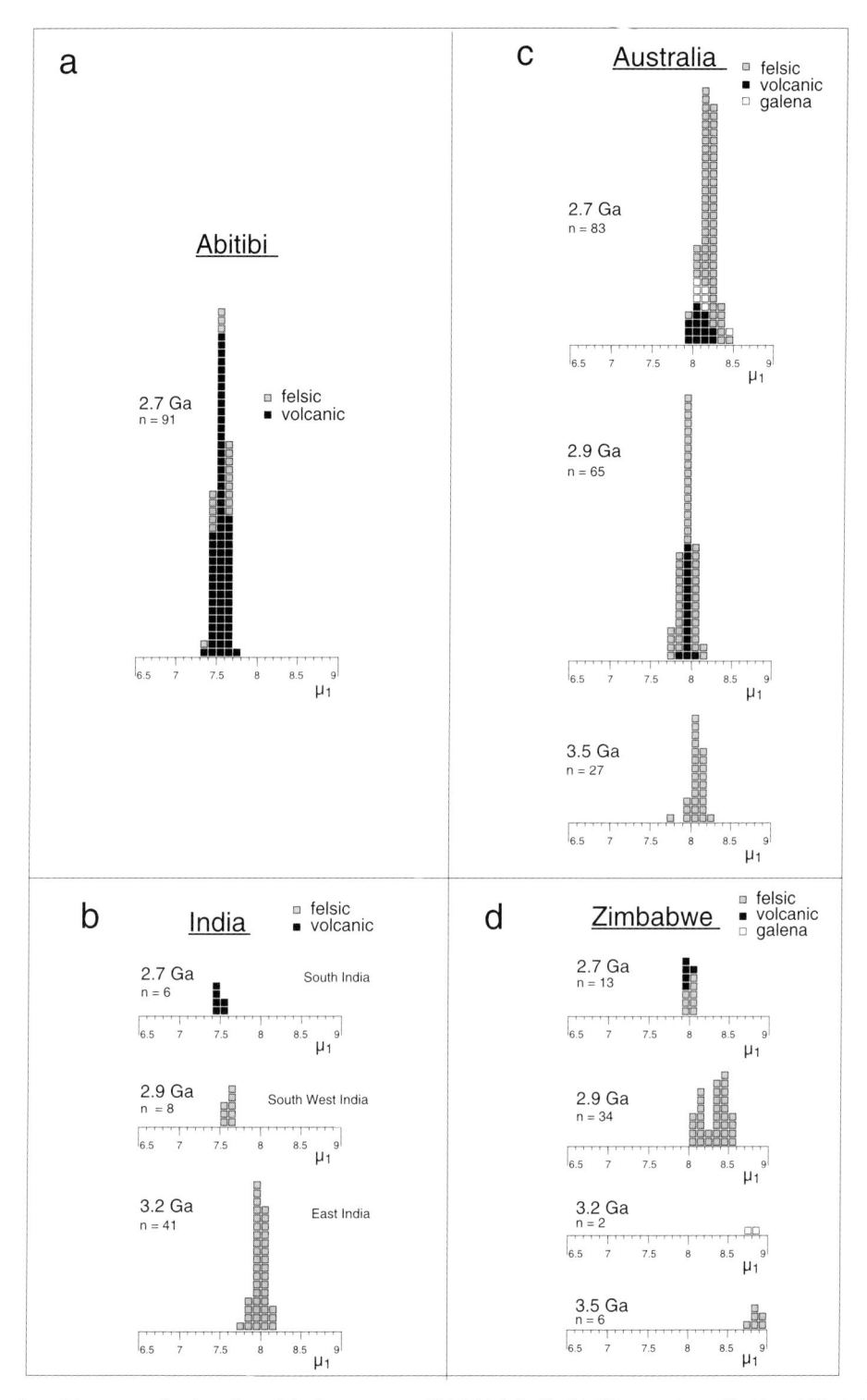

Fig. 4. μ_1 histograms for the selected Archaean areas: Abitibi (**a**), India (**b**), Western Australia (**c**) and Zimbabwe (**d**). *n*, number of samples. References as in Figure 2.

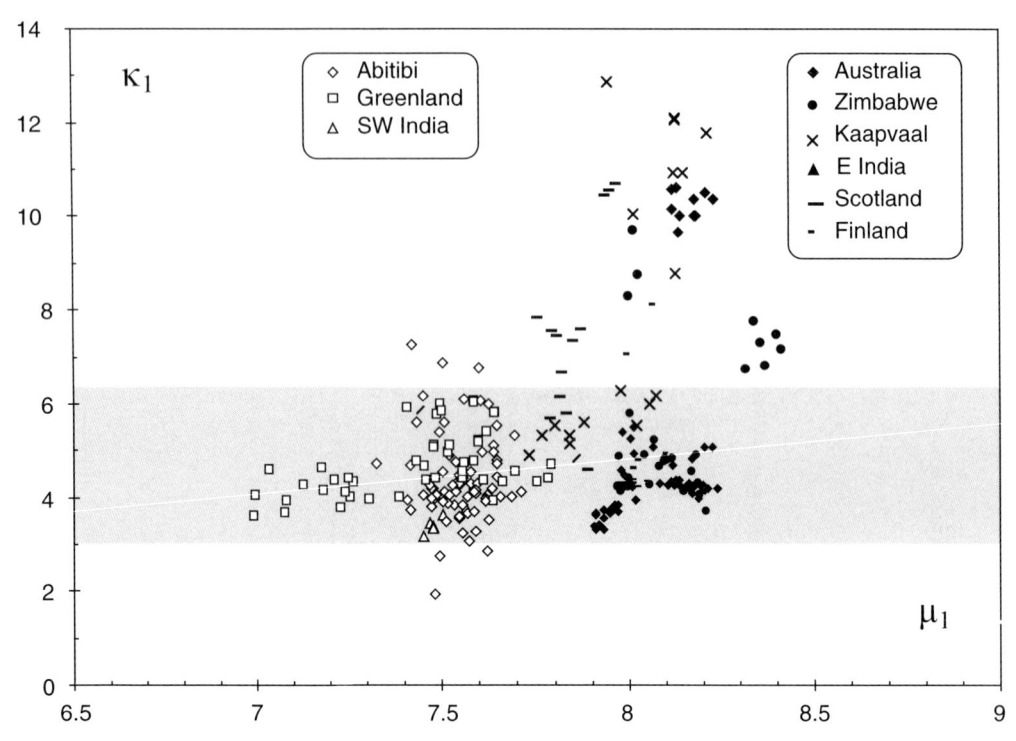

Fig. 5. κ_1 (source Th/U) v. μ_1 (source U/Pb) for the 3.8–2.6 Ga felsic, mafic and ore (galenas, feldspars) Archaean samples. Same symbols, localities and references as in Figure 2. Most of the Archaean samples have κ_1 values of 3–6.5 (grey field). Higher values up to 13 can be found only in high-μ_1 samples.

of the crust in the preserved Archaean terranes. In some areas this may usefully be described in terms of high- and low-μ_1 groups.

Variations in the source Th/U (κ_1)

The source Th/U (κ_1) values calculated for each sample are reported together with μ_1 in Figure 5. The κ_1 values vary from two to 11, with most of the samples having κ_1 values between three and five. There is no pattern of different κ_1 values either within or between cratons, and hence the variations in the source Th/U (κ_1) parameter appear to be decoupled from those of μ_1. One exception is that high κ_1 values of up to 11 are found only in some of the high-μ_1 felsic or mafic rocks from Yilgarn (granodiorite), Kaapvaal (mafic granulite), Zimbabwe (tonalite), Scotland (amphibolite) and Finland (amphibolite).

Discussion

In summary, the available initial ^{207}Pb/^{204}Pb, μ_1 and κ_1 ratios in 3.74–2.6 Ga Archaean samples have shown a number of features: (1) regional differences in ^{207}Pb/^{204}Pb and μ_1, but not κ_1, with the former being lower in Abitibi, Greenland and SW India than in Scotland, Finland, east India, Australia (Yilgarn, Pilbara) and southern Africa (Kaapvaal, Zimbabwe); (2) a remarkable consistency of μ_1 in rocks of different ages from the same cratonic areas; (3) similar initial ^{207}Pb/^{204}Pb and μ_1 values in felsic and mafic rocks, galenas and K-feldspars of similar ages from the same area.

The provinciality of initial $^{207}Pb/^{204}Pb$ and μ_1 ratios

The two groups of initial ^{207}Pb/^{204}Pb values observed in the Archaean rocks ranging in age from 3.74 to 2.6 Ga (Fig. 2) highlight the regional variations in Pb isotopic compositions. Large variations in ^{207}Pb/^{204}Pb ratios in Archaean time have been previously recognized by Kramers & Tolstikhin (1997). They interpreted the low ^{207}Pb/^{204}Pb and hence low U/Pb values as being from the lower crust and the high ^{207}Pb/^{204}Pb and U/Pb values being from the upper crust, for both old and young terranes, as a result of intracrustal fractionation and erosion processes. However, as illustrated in Figure 2 samples from the upper crust can have low initial $^{207}Pb/^{204}Pb$ (14.3–14.65) as in the Abitibi TTG (Vervoort et al. 1994) and Greenland diorites (Taylor et al. 1992), or high initial $^{207}Pb/^{204}Pb$ (14.7) as in the Yilgarn craton (granodiorite from Kambalda; Oversby 1975) and Zimbabwe craton (Somabula tonalite; Taylor et al. 1991). Thus, initial $^{207}Pb/^{204}Pb$ values in the Archaean do not appear to be simply distributed between upper- and lower-crustal rock types, and instead it is suggested that the range in initial $^{207}Pb/^{204}Pb$ largely reflects regional differences between geographically distinct terranes. Moreover, the observation that this provinciality in initial $^{207}Pb/^{204}Pb$, and hence μ_1, is seen in the mafic samples and feldspars strongly suggests that this provinciality is in the mantle, and not generated by intracrustal processes.

Two previous studies also noted regional variations in μ_1 values. Sinha (1971) described similar variations in Pb isotopes in a smaller subset of crustal samples, and suggested that the μ_1 values might be different in the northern and southern hemispheres. He observed that samples from southern Africa, Australia and India had higher μ_1 values (9.1–9.5) than those from Canada, the USA, Japan and Europe in the northern hemisphere. Such differences are similar to those described here (Figs 3 and 4), although there is now at least one exception to the simple pattern of different μ_1 values in the two hemispheres, in that uncontaminated 3.5 Ga samples from the Karelia Craton (high ε_{Nd} of +2.8–+3.4) have high μ_1 ratios of eight (Vidal et al. 1980; Putchel et al. 1998), similar to those of the southern hemisphere (Fig. 2). In contrast, other high-μ_1 signatures in the northern hemisphere appear not to be primary. The 3.5 Ga high-μ_1 Acasta gneisses from the Slave province (Yamashita et al. 2000) reflect U–Pb disturbance during subsequent metamorphism, and those from Hebei province, China, have been modified by crustal contamination (Jahn et al. 1987). More recently, Galer & Goldstein (1996) used early Archaean samples from Greenland (3.81 Ga Isua galenas, and 3.59 Ga feldspars from Amitsôq gneisses), and Australia (3.45 Ga Pilbara galenas) to model the increase in μ_1 values $\bar{\mu}_1$ over an accretion time ΔT (the time between condensation of the solar nebula at 4.566 Ga and the cessation of the Earth growth). Galer & Goldstein (1996) noted that the Isua samples had $\bar{\mu}$ values that were c. two units less than those from the Pilbara, similar to the differences observed here for Greenland and Australian samples of similar ages (Figs 3 and 4).

These differences in μ_1 between Archaean areas raise a number of interesting questions that were highlighted above. First, how were the different U/Pb ratios generated in the Earth's mantle in early Archaean time (before 3.8–3.5 Ga), and is there any reason for their apparent spatial distribution? Is there any evidence that mantle domains of different U/Pb ratios have been preserved since early Archaean time despite the effects of mantle convection? It might, for example, be argued that the geographical variations in μ_1 in Archaean rocks are evidence that the high-μ_1 cratons were close together at the time of formation. Reconstructions of craton assemblages in Archaean time (Rogers 1996), inferred from the basement stabilization ages of 3 Ga, indicate that the high-μ_1 cratons of Kaapvaal, Pilbara and east India (Darhwar) formed an early old continent at that time. The Zimbabwe, Yilgarn and western and southern Indian cratons were then accreted along marginal belts to this continent from 2.8 to 1.5 Ga. Given that the high-μ_1 character is preserved in rocks of very different compositions, and in galenas and feldspars, we infer that it reflects the character of new crustal material generated from the mantle in these areas. The low-μ_1 Archaean areas in Canada, Greenland and Siberia may have finally stabilized much later, certainly by c. 1.8 Ga, but how that links to their initial U/Pb characteristics is unclear.

Generation of large-scale μ_1 heterogeneities in the early Archaean mantle

The high-μ_1 and low-μ_1 cratons, which have distinct initial Pb isotopic signatures, may have been located in distinct areas in early Archaean time. We infer that they were generated by processes involving high- and low-μ_1 mantle source regions, and these are often explained by interactions with core material, and contributions from enriched materials, perhaps linked to lithosphere recycling, and from old depleted mantle (Kramers & Tolstikhin 1997).

The presence of cratons implies that they had thick lithospheres, and Pearson et al. (1995), for example, highlighted that they tend to be as old as the rocks in the overlying crust. Pearson et al. (1995) favoured a model in which the 150–200 km thick lithosphere of the high-μ_1 Kaapvaal Craton was linked to the presence of high-degree melt fractions. Such melts are likely to have involved plume activity and there is an increasing consensus that plumes typically originate in the deep mantle, or at the core–mantle boundary (Campbell et al. 1989; Campbell & Griffiths 1992; Abbott 1996; Polat & Kerrich 2000). This is consistent with seismic models (Zhao 2001) and models of mantle convection (Kellogg et al. 1999; Van Der Hilst & Karason 1999), although the latter often invoke a thick boundary, or hidden zone, in the lower mantle from where larger plumes may be initiated. This zone may be replenished by subducted slabs, which may also reach the D″ core–mantle boundary layer. The D″ layer may therefore contain (1) recycled oceanic crust, and hence be enriched in U and Th relative to Pb (Coltice & Ricard 1999; Ferrachat & Ricard 2001); (2) material from the outer core (Ito et al. 1995; Walker et al. 1995). If the D″ layer contains core material it would be enriched in siderophile elements relative to the rest of the lower mantle. Such an enrichment in the modern D″ layer has been inferred from ^{186}Os–^{187}Os systematics (Walker et al. 1995; Brandon et al. 1998). The D″ layer would then have lower U, Th and K (Kellogg et al. 1999) and higher Pb contents (Widom & Shirey 1996), and therefore low U/Pb.

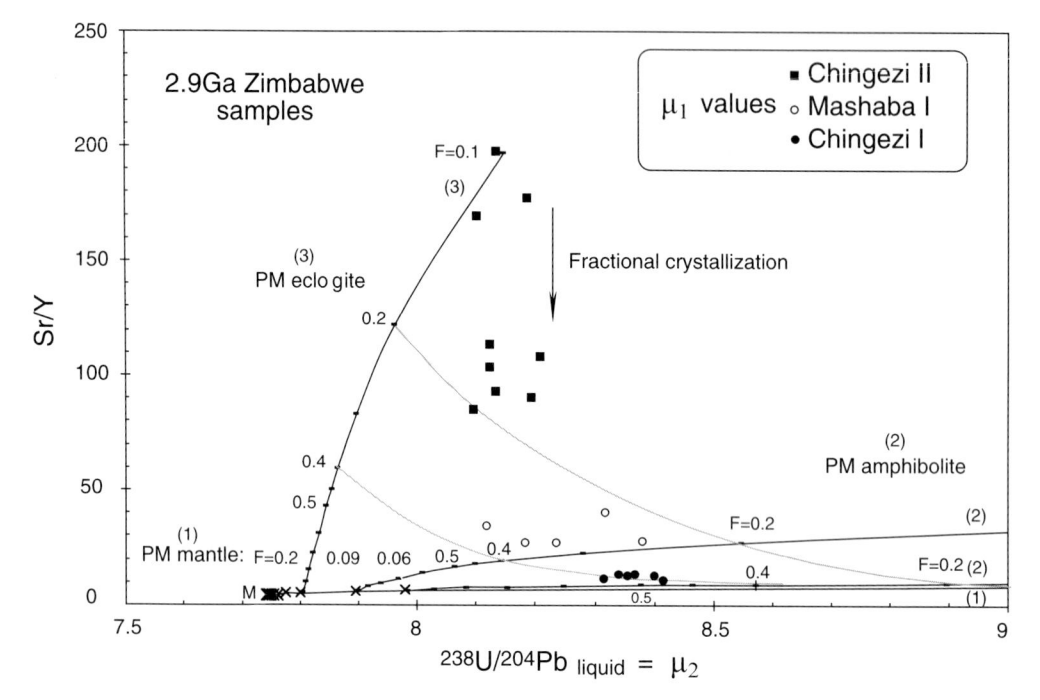

Fig. 6. Modelling U/Pb and Sr/Y fractionation by intracrustal melting of amphibolite and eclogite sources adapted from Luais & Hawkesworth (1994). Sr/Y parameter reflects difference in the ratio Plagioclase/Garnet, i.e. upper-crust v. lower-crust source regions. The residue assemblages are as given by Luais & Hawkesworth (1994). U and Pb partition coefficients for plagioclase are from Dostal & Capedri (1976) and Leeman (1979), Sr and Y partition coefficients as given by Luais & Hawkesworth (1994), and the others as given by Halliday et al. (1995). Curve (1): first stage of melting: partial melting of a primitive mantle (modified from Hofmann 1988) to obtain μ values similar to uncontaminated mafic rocks from the Belingwe belt ($\mu \approx 8$ for location A samples, Table 3; Dupré & Arndt 1990). Second stages of melting: curves (2): partial melting of metamorphosed basalts in amphibolite facies with 45% and 15% plagioclase in the residue; curve (3): partial melting of metamorphosed basalts in eclogite facies. Large degrees of melting ($F = 0.4$) are necessary to keep U/Pb fractionation to less than 10% during melting if plagioclase is present in the residue. Negligible U/Pb fractionation results in the case of an eclogite residue. Lines of equal degree of partial melting ($F = 0.2, 0.4$) are shown in grey. The Zimbabwe samples are from Luais & Hawkesworth (1994). The μ_2 values of Zimbabwe samples (from Table 3, not shown here) are consistent with this model, in that Chingezi I samples have higher μ_2 values (22–36) than Mashaba I samples (6–8.6). This has been modelled by partial melting of an amphibolite residue with 45% plagioclase and 15% plagioclase, respectively, and the lower μ_2 values (4.7–9) for the Chingezi II samples are attributed to partial melting of an eclogite residue. As U/Pb was not strongly fractionated by partial melting, there is a good fit for modelled μ_2 with μ_1 values of the samples reported here. Restricted variations in μ_2 ratios during crustal genesis at the scale of the craton can be achieved for large degree of partial melting ($F = 0.4$–0.45) or melting of an eclogite source.

Can these models be tested geochemically? First, crustal recycling would provide a high-μ_1 source for new crust in Archaean time. However, if this occurred by subduction, entrained sediments might have high Th/U (McLennan & Taylor 1980), and hence result in correlations between μ_1 and κ_1 ratios. As pointed out by Kramers & Tolstikhin (1997), this is inconsistent with the lack of correlation between Th/U and μ_1 (Fig. 5) ($\kappa_1 = 3$–6) both in low-μ_1 (e.g. Greenland, Abitibi) and high-μ_1 regions (Zimbabwe, Australia, India). Moreover, in contrast to slabs in the geologically recent past which dehydrate at shallow levels, and may therefore reach deep levels in the mantle, slabs in Archaean time may have melted and disaggregated in the upper 100–200 km (McCulloch & Woodhead 1993). If that is correct, subducted slabs would be less likely to have contributed to the composition of D″ layer in Archaean time.

Second, delamination of mafic crust may be geologically a more likely process of recycling, but it may result in little fractionation of U/Pb. It can be seen from a case study on 3.5–2.9 Ga Archaean crust from Zimbabwe (Fig. 6) that during partial melting of the mantle, which generates mafic crust, U behaves more incompatibly than Pb and a 20% increase in U/Pb (μ_2) is achieved at 1% partial melting. Critically, the extent of fractionation is drastically reduced to less than 5% at 4% melting, and it is negligible at 10–20% melting, which may be more realistic for the generation of new mafic crust. This highlights that U/Pb is not fractionated during mafic crust generation, and consequently delamination of mafic crust into the mantle is not the process that readily generates high-μ_1 mantle.

Third, the model of an outer core signal in the D″ layer (Walker et al. 1995) faces apparently contradictory data from Pb and Os isotopes in Archaean samples. Os isotopic data on Archaean samples are still scarce, but there appears to be a broad correlation between μ_1 and $^{187}Os/^{188}Os$ isotopic ratios, expressed as γ_{Os}, which is the percentage difference from the chondritic evolution curve. The Abitibi komatiite has chondritic initial γ_{Os} (Shirey 1997; Shirey & Walker 1998) and low μ_1, whereas suprachondritic γ_{Os} values are found in Koatomuska komatiites from the Karelian Craton (Finland) (Putchel et al. 2001) and in the 3.3 Ga greenstone belt in South Africa (Shirey et al. 1998), regions that have high-μ_1 signatures (Fig. 2). Similarly, anomalously high platinum group element (PGE) contents (especially Os) are found in Ni–Fe deposits located in the ultra-mafic complex of the Barberton greenstone belt (Kaapvaal Craton) (Tredoux et al. 1989). In contrast, core material should with time have low μ_1 and elevated Os isotope ratios, and so it is an unlikely cause for the apparent correlation between high-μ_1 and elevated Os isotope signatures and PGE abundances in Archaean rocks. Alternatively, the core material might have low Pb contents of 0.2 ppm (Widom & Shirey 1996), in which case it might influence the Os isotopes but not the Pb isotopes of D″.

In general, models of Earth formation invoke metal–silicate partitioning during core formation, and accretion of a late veneer in the upper mantle, to explain element abundances in the primitive mantle. The relative depletion of Pb in the mantle may be mainly due to partitioning of Pb into the core during core formation (Azbel et al. 1993; Allègre et al. 1995; Galer & Goldstein 1996; Halliday et al. 1996). This is facilitated by having sulphur in the core (Murthy 1991; O'Neill 1991; Allègre et al. 1995; Kilburn & Wood 1997), and this would not be readily achieved under the strongly reducing conditions of core segregation at low pressures. One implication is that it was followed by later accretion of oxidized material to the core (Kilburn & Wood 1997). This might be an Fe–S liquid that would scavenge Pb from the deep mantle, and result in elevated U/Pb ratios there. However, the problem with these global-scale processes is that they offer no explanation for the inferred regional variations in U/Pb in Archaean time.

One way to explain regional geochemical variations is to invoke specific events. Thus, some of the late accreted oxidized material may have been added in a large localized impact, such as the one thought to be responsible for the origin of the Moon (Cameron & Benz 1991). Similarly, it might be that regional variations in U/Pb ratios are developed by recycling of oceanic crust into relatively shallow reservoirs in the mantle.

Survival of μ_1 heterogeneities

The high- and low-μ_1 signatures appear to have been maintained through the history of the different Archaean areas, and it has been argued above that these are source features. Their survival, at least until late Archaean time, implies that they existed in reservoirs protected from the effects of mantle convection (e.g. Davies 1992), and the two most obvious sites are the D″ layer and shallow-level reservoirs such as the lithosphere. The former is often invoked as a source of mantle plumes, but to maintain particular isotope signatures within specific areas may require that the dominant contributions were from shallower levels. Moreover, these distinctive Pb isotope ratios are preserved in the felsic, as well as in the mafic rocks, and in the ore samples (galenas). We now examine the role of mafic lithosphere as the source of felsic crust in Archaean time, and the extent to which U/Pb ratios are fractionated during intracrustal melting.

The Archaean felsic crust is characterized by tonalites and trondhjemites (TTG) that are typically strongly depleted in heavy rare element elements (HREE). Thus, partial melting took place in garnet-amphibolite (Martin 1986) or eclogite facies (Rapp 1991; Luais & Hawkesworth 1994), of mafic material underplated into the lower crust or in subducted slabs. The occurrence of Archaean eclogite, e.g. the 2.9 Ga Re–Os age from Siberian eclogite (Pearson et al. 1995), demonstrates that mafic material occurred at those depths in Archaean time. The questions are the extent to which such melting processes would fractionate U/Pb ratios in the generation of μ_2 and would result in distinct U/Pb signatures between felsic crust and mafic protoliths.

The extent of such fractionation is quantified in the case study from the 3.5–2.9 Ga Archaean crust from the Zimbabwe Craton (Fig. 6). In the first step, the generation of the mafic crust from the mantle involved little U/Pb fractionation. Remelting of that mafic crust in amphibolite facies, with plagioclase in the residue ($Kd_{Pb} = 0.1$; Leeman 1979) will then produce a larger fractionation of U/Pb of up to 43% and 18% fractionation relative to the mafic source, for 45% and 15% plagioclase in the residue, respectively, at 10% melting. However, large degrees of partial melting are probably more reasonable (up to 40% melting to generate tonalites from greenstones; Luais & Hawkesworth (1994) and in the experimental data of Rutter & Wyllie (1988), Rapp et al. (1991) and Rapp & Watson (1995)) and these will strongly reduce the U/Pb fractionation to c. 5% higher than the mafic protolith. In the case of an eclogitic residue, fractionation of U/Pb ratio is even smaller, at 4% for 10% melting and 0.7% for 40% melting (Fig. 6). Thus, the amounts of U/Pb fractionation are small and, at similar degrees of partial melting, U/Pb (or μ_2) are less fractionated when the felsic melts were derived from an eclogitic than amphibolitic source rock. Comparisons of such models with natural rocks relies on little secondary modification of the U/Pb ratios. None the less, the μ_2 values from the Zimbabwe samples, and their variation in Sr/Y, are broadly consistent with such a model, as illustrated in Figure 6.

Thus, U/Pb ratios are not generally fractionated significantly by the crust formation processes widely invoked in the generation of Archaean greenstone belts and the TTG suites (see also Bickle et al. 1989) (Table 1), and thus μ_2 tends to be broadly similar to μ_1 (source). Any variations in μ_1 in the mantle are thus transferred to the crust. Additionally, recycling of unaltered mafic crust to maintain its Pb isotope signature in a segment of mantle would be efficient by delamination according to the model developed by Zegers & van Keken (2001). They suggested that in the case of thick mafic crust, the lower eclogite part will tend to sink back into the depleted mantle, and so there might be a link between the existence of anomalous thick lithospheric mantle and the perpetuation of a high-μ_1 signature. As possible examples, the Kaapvaal and the Pilbara cratons have consistent high-μ_1 signatures through time (Fig. 3) and they preserve geophysical evidence of an anomalously thick mantle lithosphere (Drummond 1988; Vinnik et al. 1996).

Preservation of U/Pb heterogeneities through time: are the spatial links between μ_1 in the Archaean, Mesozoic CFB and DUPAL oceanic basalts just coincidence?

Lithospheric components have been invoked in the generation of certain continental flood basalts (CFB), and basalts with the DUPAL isotope signatures. For the CFB, the debate has focused on the extent to which the 'continental signature' is a feature of their main source regions, as in the mantle lithosphere, or it has been introduced as crustal or mantle contaminants to plume-derived melts (Lassiter & DePaolo 1997). CFB provinces belonging to the Gondwana supercontinent (Parána, Karoo, Tasmania and Ferrar), generated 132–190 Ma ago, are thought to be very largely derived from lithospheric sources (Hawkesworth et al. 1986, 1990; Ellam & Cox 1989; Molzhan et al. 1996; Lassiter & DePaolo 1997; Antonini et al. 1999; Peate et al. 1999). In contrast, other CFB provinces such as the Deccan, Columbia River, Greenland and perhaps Siberia contain at least significant contributions from their respective mantle plumes, and they may well have been generated beneath thinned continental lithosphere (Gill et al. 1992; Hooper & Hawkesworth 1993; Thirlwall et al. 1994; Chesley et al. 1996; Hooper 1997). However, they also include units that contain contributions from the mantle lithosphere and/or crustal contamination (Sharma et al. 1992; Hooper & Hawkesworth 1993; Wooden et al. 1993; Hooper 1997). The Pb isotopic anomalies of DUPAL samples have been interpreted in terms of old anomalous regions in the mantle

(Hart 1984), or the reinjection of sediments (Weaver et al. 1987) or old anomalous Pb material from the mantle lithosphere (Richardson et al. 1982; Hawkesworth et al. 1986) into the convecting mantle.

These CFB provinces and DUPAL regions are illustrated with their respective μ_1 values on the map of low-μ_1 and high-μ_1 Archaean cratons (Fig. 7). There is some sense that the provinciality of μ_1 observed in the Archaean rocks broadly corresponds to that in the CFB provinces with high μ_1 values ($\mu_1 \gg 8$) and even to the areas of DUPAL oceanic basalts. This is most noticeable with the position of the higher-μ_1 Archaean cratons (Zimbabwe, Australia), the CFB provinces of the Gondwana supercontinent with lithospheric source signatures (Parana, Karoo, Ferrar and Tasmania), and the maximum present-day DUPAL anomalies just south of Africa (latitude 30–45°S) (Hart 1984). In view of plate movements since Late Archaean time, this juxtaposition is either a coincidence of no wider meaning or it implies that these large-scale anomalies are transported with plate motion, consistent with their presence in the lithosphere. This case is perhaps strongest for NE India (3.2 Ga Bonaï pluton), where the high μ_1 values of eight for the cratons of India, Zimbabwe and Australia are consistent with their adjacent locations in Archaean time. Displacement of the Indian lithosphere northwards in Mesozoic time may have contributed to the maximum DUPAL signatures being south of India (latitude 0–15°S), and in the Yarlung–Zangpo ophiolites (NE India, Mahoney et al. 1998), which is clearly disjointed from the main elongated present-day DUPAL anomalies at latitude 30–45°S.

It has been argued that the change to a more oxidized atmosphere from c. 2 Ga resulted in a shift to more efficient recycling of U back into the mantle (Kramers & Tolstikhin 1997; Collerson & Kamber, 1999; Elliott et al. 1999). That may imply that any genetic link between high-μ_1 Archaean provinces, high-μ_1 Mesozoic CFB and the DUPAL basalts is less likely to reflect processes that involve water–rock interaction and subduction. Rather it may provide circumstantial evidence that (1) the high-μ_1 signal was linked to delamination from the base of areas of thick high-μ_1 mafic crust, and (2) at least some of the DUPAL signature may reflect recycled lithospheric material, and that it may be of Archaean age. Overall, the low U/Pb ratios of residual mantle, resulting from mantle melting and crust extraction (White 1993) may be buffered by the high U/Pb lithospheric component recycled back and mixed into the depleted mantle.

DUPAL anomalies in the Arctic area?

One exception to the inferred broad geographical coherence of μ_1 ratios since Archaean time is the high μ_1 values of c. eight in samples from the Archaean Baltic shield, which show no crustal contamination (Vidal et al. 1980; Putchel et al. 1998). Moreover, this μ_1 signature is also present in the recent discovery of a DUPAL-like signature in fresh MORB samples from the Arctic Ocean (Mühe et al. 1997), emphasizing again a temporal consistency of high μ_1 values in particular areas. This can be reconciled in models that invoke contributions from mantle lithosphere perhaps as a result of the presence of anomalous thick crust in these areas, which would facilitate delamination processes (Putchel et al. 1998; Hellebrand et al. 2002).

Conclusions

A review of Pb isotope data on Archaean mafic and felsic rocks, feldspars and galenas ranging from 3.7–3.8 Ga to 2.7–2.6 Ga shows the following features.

(1) There is a provinciality of μ_1 values, with higher μ_1 values in some cratons from the southern hemisphere, where Zimbabwe, Kaapvaal, NE India and Australia have μ_1 mean values in the range 8.85–8.00, in contrast to a μ_1 mean of 7.37–7.57 in Greenland, Abitibi and SW India. Moreover, this is irrespective of their age, and the μ_1 values are similar for both the felsic and the mafic rock types, and for feldspars and galenas from a given area.

(2) The μ_1 values in different areas appear to have persisted for long periods of time, and there are currently spatial links with the locations of certain Mesozoic CFB and with oceanic basalts with DUPAL isotope signatures. Such long-lived U/Pb variations may therefore correspond to large-scale mantle anomalies, perhaps at relatively shallow levels in the mantle.

(3) The origins of μ_1 variations in Archaean time are extremely speculative, but greenstone belt volcanic rocks may have been associated with mantle plumes that may have originated in the lower mantle, and possibly at the core–mantle boundary. These μ_1 variations are very old, and perhaps linked to late accretion of oxidized material and core formation. In view of the long-lived nature of the differences in μ_1, it is argued that they were stored at shallow levels in the mantle, where they have been preserved from the effects of mantle convection and homogenization. Felsic crust generated from mafic protoliths also retained the regional μ_1 values.

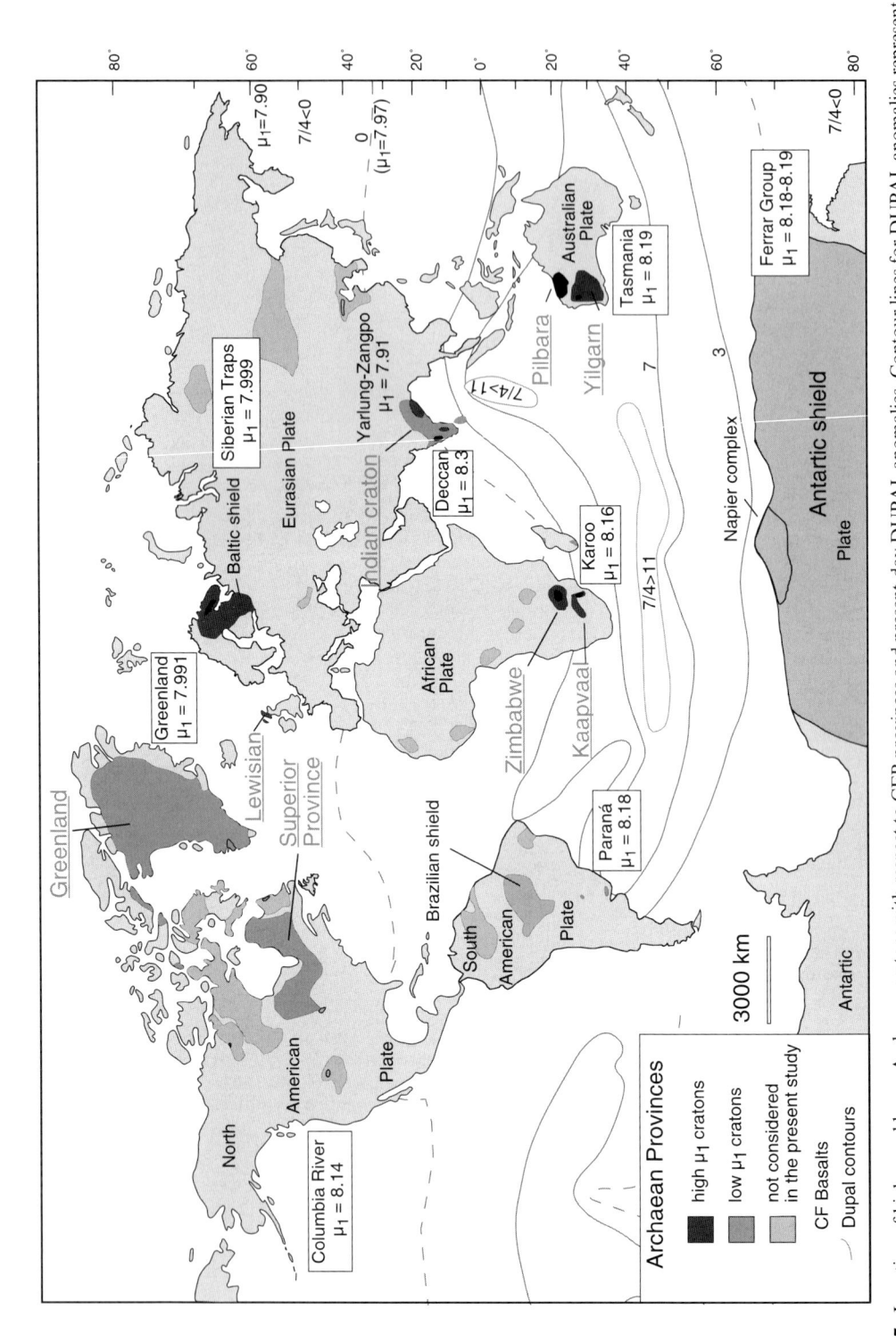

Fig. 7. Location of high-μ and low-μ Archaean cratons with respect to CFB provinces and present-day DUPAL anomalies. Contour lines for DUPAL anomalies represent equal Δ7/4 values (Hart 1984). μ₁ values for CFB provinces are calculated from Pb isotopic ratios (Carlson 1984; Ellam & Cox 1989; Hergt *et al.* 1989, 1991; Sharma *et al.* 1992; Wooden *et al.* 1993; Thirlwall 1994; Molzhan *et al.* 1996; Antonini *et al.* 1999).

(4) The distribution of high-μ_1 Archaean cratons is broadly similar to that of high-μ_1 CFB provinces generated from lithospheric sources, and to that of the present-day DUPAL basalts. This further suggests a common lithospheric origin of these anomalies, in that old high-μ_1 Archaean lithosphere is recycled back to the mantle by delamination and contributed to the geographical and temporal perpetuation of such anomalies. High μ_1 values in Archaean samples from India coincide with the maximum of present-day DUPAL anomalies south of India (Ninetyeast Ridge), and this may represent one of the best examples of survival of μ_1 heterogeneities in the lithosphere, migrating with plate tectonic motion.

A first version of this manuscript has benefited from the comments of F. Albarède. B.L. is grateful to L. Reisberg for fruitful discussions about Os systematics. Comments and constructive criticisms from two anonymous reviewers greatly improved the manuscript. This is a CRPG Contribution 1570.

References

ABBOTT, D. 1996. Plumes and hotspots as sources of greenstone belts. *Lithos*, 37, 113–127.

ALLÈGRE, C., POIRIER, J., HUMLER, E. & HOFMANN, A. 1995. The chemical composition of the Earth. *Earth and Planetary Science Letters*, 134, 515–526.

ANTONINI, P., PICIRILLO, E., PETRINI, R., CIVETTE, L., D'ANTONIO, M. & ORSI, G. 1999. Enriched mantle – Dupal signature in the genesis of the Jurassic Ferrar tholeiites from Prince Albert Mountains, Victoria Land, Antarctica. *Contributions to Mineralogy and Petrology*, 136, 1–19.

APPEL, P. W. U., MOORBATH, S. & TAYLOR, P. N. 1978. Least radiogenic terrestrial lead from Isua, West Greenland. *Nature*, 272, 524–526.

AZBEL, I. Y., TOLSTIKHIN, I., KRAMERS, J., PECHERNIKOVA, G. & VITAZEV, A. 1993. Core growth and siderophile element depletion of the mantle during homogeneous Earth accretion. *Geochimica et Cosmochimica Acta*, 57, 2889–2898.

BAADSGAARD, H. 1973. U–Th–Pb dates on zircons from the Early Precambrian Amitsôq gneisses. *Earth and Planetary Science Letters*, 33, 261–267.

BALAKRIHNAN, S., HANSON, G. N. & RAJAMANI, V. 1990. Pb and Nd isotope constraints on the origin of high-Mg and tholeiitic amphibolites, Kolar schist belt, South India. *Contributions to Mineralogy and Petrology*, 107, 279–292.

BICKLE, M. J., BETTENAY, L. F., BARLEY, R., CHAPMAN, H. J., GROVES, D. J., CAMPBELL, J. H. & DE LAETER, J. R. 1983a. A 3500 Ma plutonic and volcanic calc-alkaline province in the Archaean east Pilbara Block. *Contributions to Mineralogy and Petrology*, 84, 25–35.

BICKLE, M., BETTENAY, L. F., CHAPMAN, H. J., GROVES, D. I., MCNAUGHTON, N. J., CAMPBELL, I. H. & DE LAETER, J. R. 1989. The age and origin of younger granitic plutons of the Shaw Batholith in the Archaean Pilbara Block, Western Australia. *Contributions to Mineralogy and Petrology*, 101, 361–376.

BICKLE, M. J., BETTENAY, L. F., CHAPMAN, H. J., GROVES, D. I., MCNAUGHTON, N. J., CAMPBELL, I. H. & DE LAETER, J. R. 1993. Origin of the 3500–3300 Ma calc-alkaline rocks in the Pilbara Archaean: isotopic and geochemical constraints from the Shaw batholith. *Precambrian Research*, 60, 117–149.

BICKLE, M. J., CHAPMAN, H. J., BETTENAY, L. F., GROVES, D. I. & DE LAETER, J. R. 1983b. Lead ages, reset rubidium–strontium ages and implications for the Archaean crustal evolution of the Diemals area, Central Yilgarn Block, Western Australia. *Geochimica et Cosmochimica Acta*, 47, 907–914.

BLACK, L. P., GALE, N. H., MOORBATH, S., PANKHURST, R. J. & MCGREGOR, V. R. 1971. Isotopic dating of very early Precambrian amphibolite facies gneisses from the Godthaab district, West Greenland. *Earth and Planetary Science Letters*, 12, 245–259.

BRANDON, A. D., WALKER, R. J., MORGAN, J. W., NORMAN, M. D. & PRICHARD, H. M. 1998. Coupled ^{186}Os and ^{187}Os evidence for core–mantle interaction. *Science*, 280, 1570–1573.

BREVART, O., DUPRÉ, B. & ALLÈGRE, C. 1986. Lead–lead age of komatiitic lavas and limitations on the structure and evolution of the Precambrian mantle. *Earth and Planetary Science Letters*, 77, 293–302.

BURTON, N. K. W., COHEN, A. S., O'NIONS, R. K. & O'HARA, M. J. 1994. Archaean crustal development in the Lewisian complex of northwest Scotland. *Nature*, 370, 552–555.

CAMERON, A. G. W. & BENZ, W. 1991. The origin of the Moon and the single impact hypothesis IV. *Icarus*, 92, 204–216.

CAMPBELL, I. H. & GRIFFITHS, R. W. 1992. The changing nature of mantle hotspots through time: implications for the chemical evolution of the mantle. *Journal of Geology*, 92, 497–523.

CAMPBELL, I. H., GRIFFITHS, R. W. & HILL, R. I. 1989. Melting in an Archaean mantle plume: heads it's basalts, tails it's komatiites. *Nature*, 339, 697–699.

CARIGNAN, J., GARIEPY, G., MACHADO, N. & RIVE, M. 1993. Pb isotopic geochemistry of granitoids and gneisses from the late Archaean Pontiac and Abitibi Subprovinces of Canada. *Chemical Geology*, 106, 299–316.

CARLSON, R. 1984. Isotopic constraints on Columbia River flood basalt genesis and the nature of the subcontinental mantle. *Geochimica et Cosmochimica Acta*, 48, 2357–2372.

CASTILLO, P. 1988. The Dupal anomaly as a trace of the upwelling lower mantle. *Nature*, 336, 667–670.

CATTELL, A., KROGH, T. E. & ARNDT, N. T. 1984. Conflicting Sm–Nd whole rock and U–Pb zircon ages for Archaean lavas from Newton Township, Abitibi belt, Ontario. *Earth and Planetary Science Letters*, 70, 280–290.

CHAUVEL, C., DUPRÉ, B. & JENNER, J. A. 1985. The Sm–Nd age of Kambalda volcanics is 500 Ma too old! *Earth and Planetary Science Letters*, **74**, 315–324.

CHESLEY, J., RUIZ, J. & HOOPER, P. 1996. Crust–mantle mixing: implications based on the Re–Os isotope systematics of the Columbia River basalt group. *EOS Trans*, American Geophysical Union, **77**, 832.

COLLERSON, K. D. & KAMBER, B. S. 1999. Evolution of the continents and the atmosphere inferred from Th–U–Nd systematics of the depleted mantle. *Science*, **283**, 1519–1522.

COLTICE, N. & RICARD, Y. 1999. Geochemical observations and one layer mantle convection. *Earth and Planetary Science Letters*, **174**, 125–137.

CONDIE, K. C. 1993. *Plate Tectonics and Crustal Evolution*. Pergamon, New York.

DAVIES, G. 1992. On the emergence of plate tectonics. *Geology*, **20**, 963–966.

DAVIS, W. J., GARIÉPY, C. & VAN BREEMEN, O. 1996. Pb isotopic composition of late Archaean granites and the extent of recycling early Archaean crust in the Slave Province, northwest Canada. *Chemical Geology*, **130**, 244–269.

DOSTAL, J. & CAPEDRI, S. 1976. Uranium in spinel peridotite inclusions in basalts from Sardinia. *Contributions to Mineralogy and Petrology*, **54**, 245–254.

DRUMMOND, B. J. 1988. A review of the crust/upper mantle structure in the Precambrian areas of Australia and implications for Precambrian crustal evolution. *Precambrian Research*, **40**, 101–116.

DUPRÉ, B. & ALLÈGRE, C. 1983. Pb–Sr isotope variation in Indian Ocean basalts and mixing phenomena. *Nature*, **303**, 142–146.

DUPRÉ, B. & ARNDT, N. T. 1990. Pb isotopic compositions of Archaean komatiites and sulfides. *Chemical Geology*, **85**, 35–56.

DUPRÉ, B., CHAUVEL, C. & ARNDT, N. T. 1984. Pb and Nd isotopic study of two Archaean komatiitic flow from Alexo, Ontario. *Geochimica et Cosmochimica Acta*, **48**, 1965–1972.

ELLAM, R. & COX, K. 1989. A Proterozoic lithospheric source for Karoo magmatism: evidence from the Nuanetsi picrites. *Earth and Planetary Science Letters*, **92**, 207–218.

ELLIOTT, T., ZINDLER, A. & BOURDON, B. 1999. Exploring the kappa conundrum: the role of recycling in the lead isotope evolution of the mantle. *Earth and Planetary Science Letters*, **169**, 129–145.

FERRACHAT, S. & RICARD, Y. 2001. Mixing properties in the Earth's mantle: effect of the viscosity stratification and of oceanic crust segregation. *Geochemistry Geophysics Geosystems* **2**, Paper number 2000GC000092.

GALER, S. & GOLDSTEIN, S. 1996. Influence of accretion on lead in the Earth. *In*: BASU, A. & HART, S. (eds) *Earth Processes: Reading the Isotopic Code*. Geophysical Monograph, American Geophysical Union, **95**, 75–98.

GANCARZ, A. J. & WASSERBURG, G. J. 1977. Initial Pb of the Amîtsoq gneiss, West Greenland, and the implications for the age of the Earth. *Geochimica et Cosmochimica Acta*, **41**, 1283–1301.

GARIÉPY, C. & ALLÈGRE, C. J. 1985. The lead isotope geochemistry and geochronology of late-kinematic intrusives from the Abitibi greenstone belt, and the implications for late Archaean crustal evolution. *Geochimica et Cosmochimica Acta*, **49**, 2371–2383.

GILL, R. C. O., PEDERSEN, A. K. & LARSEN, J. G. 1992. Tertiary picrites from West Greenland: melting at the periphery of a plume? *In*: STOREY, B. C., ALABASTER, T. & PANKHURST, R. J. (eds) *Magmatism and the Cause of Continental Breakup*. Geological Society, London, Special Publications, **68**, 335–348.

HALLIDAY, A. N., LEE, D. C., TOMMASINI, S., DAVIES, G. R., PASLICK, C. R., FITTON, J. G. & JAMES, D. E. 1995. Incompatible trace elements in OIB and MORB source enrichment in the sub-oceanic mantle. *Earth and Planetary Science Letters*, **133**, 379–395.

HALLIDAY, A. N., REHKÄMPER, M., LEE, D. C. & YI, W. 1996. Early evolution of the Earth and Moon: new constraints from Hf–W isotope geochemistry. *Earth and Planetary Science Letters*, **142**, 75–89.

HART, S. R. 1984. The DUPAL anomaly: a large scale isotopic mantle anomaly in the Southern Hemisphere. *Nature*, **309**, 753–757.

HAWKESWORTH, C. J. & O'NIONS, R. K. 1977. The petrogenesis of some Archaean volcanic rocks from Southern Africa. *Journal of Petrology*, **18**, 487–520.

HAWKESWORTH, C., MOORBATH, S., O'NIONS, R. & WILSON, J. 1975. Age relationships between greenstone belts and 'granites' in the Rhodesian Archaean craton. *Earth and Planetary Science Letters*, **25**, 251–262.

HAWKESWORTH, C. J., BICKLE, M. J., GLEDHILL, A. R., WILSON, J. F. & ORPEN, J. L. 1979. A 2.9 Ga event in the Rhodesian Archaean. *Earth and Planetary Science Letters*, **43**, 285–297.

HAWKESWORTH, C., KEMPTON, P., ROGERS, S. N., ELLAM, R. & VAN CALSTEREN, P. 1990. Continental mantle lithosphere, and shallow level enrichment processes in the Earth's mantle. *Earth and Planetary Science Letters*, **96**, 256–268.

HAWKESWORTH, C., MANTOVANI, M., TAYLOR, P. & PALACZ, Z. 1986. Evidence from the Parana of south Brazil for a continental contribution to Dupal basalts. *Nature*, **322**, 356–359.

HELLEBRAND, E., SNOW, J. E. & MÜHE, R. 2002. Mantle melting beneath Gakkel Ridge (Arctic Ocean): abyssal peridotite spinel compositions. *Chemical Geology*, **182**, 227–235.

HERGT, J., CHAPPELL, B., MCCULLOCH, M., MCDOUGALL, I. & CHIVAS, A. 1989. Geochemical and isotopic constraints on the origin of the Jurassic dolerites of Tasmania. *Journal of Petrology*, **30**, 841–883.

HERGT, J., PEATE, D. & HAWKESWORTH, C. 1991. The petrogenesis of Mesozoic Gondwana low-Ti flood basalts. *Earth and Planetary Science Letters*, **105**, 134–148.

HOFMANN, A. W. 1988. Chemical differentiation of the Earth: the relationship between mantle, continental crust, and oceanic crust. *Earth and Planetary Science Letters*, **90**, 297–314.

HOOPER, P. 1997. The Columbia River Flood basalt province: current status. *In*: COFFIN, M. F. & MAHONEY, J. J. (eds) *Large Igneous Provinces. Continental, Oceanic, and Planetary Flood Volcanism*. Geophysical Monograph, American Geophysical Union, **100**, 1–27.

HOOPER, P. & HAWKESWORTH, C. 1993. Isotopic and geochemical constraints on the origin and evolution of the Columbia River basalt. *Journal of Petrology*, **346**, 1203–1246.

HUANG, Y., VAN CALSTEREN, P. & HAWKESWORTH, C. 1995. The evolution of the lithosphere in southern Africa: a perspective on the basic granulite xenoliths from the kimberlites in South Africa. *Geochimica et Cosmochimica Acta*, **59**, 4905–4920.

ITO, E., MOROOKA, K., UJIKE, O. & KATSURA, T. 1995. Reactions between molten iron and silicate at high pressure: implications for the chemical evolution of the Earth's core. *Journal of Geophysical Research*, **100**, 5901–5910.

JAFFEY, A. H., FLYNN, K. F., GLENDENIN, L. E., BENTLEY, W. C. & ESLING, A. M. 1971. Precision measurement of the half-lives and specific activities of U^{235} and U^{238}. *Physical Reviews C*, **4**, 1889–1907.

JAHN, B. M., AUVRAY, B., CORNICHET, J., BAI, Y. L., SHEN, Q. H. & LIU, D. Y. 1987. 3.5 Ga old amphibolites from Eastern Hebei province, China: field occurrence, petrography, Sm–Nd isochron age and REE geochemistry. *Precambrian Research*, **34**, 311–346.

KALSBEEK, F., TAYLOR, P. N. & PIDGEON, R. T. 1988. Unreworked Archaean basement and Proterozoic supracrustal rocks from northeastern Disko Bugt, West Greenland: implications for the nature of Proterozoic mobile belts in Greenland. *Canadian Journal of Earth Sciences*, **25**, 773–782.

KAMBER, B. S. & COLLERSON, K. D. 1999. Origin of ocean island basalts: a new model based on lead and helium isotope systematics. *Journal of Geophysical Research*, **104**, 25 479–25 491.

KAMBER, B. S. & MOORBATH, S. 1998. Initial Pb of the Amîtsoq gneiss revisited: implications for the timing of early Archaean crustal evolution in West Greenland. *Chemical Geology*, **150**, 19–41.

KELLOGG, L., HAGER, B. & VAN DER HILST, R. 1999. Compositional stratification in the deep mantle. *Science*, **283**, 1881–1884.

KILBURN, M. & WOOD, B. 1997. Metal–silicate partitioning and the incompatibility of S and Si during core formation. *Earth and Planetary Science Letters*, **152**, 139–148.

KORSCH, M. J. & GULSON, B. L. 1986. Nd and Pb isotopic studies of an Archaean layered mafic–ultramafic complex, Western Australia, and implications for mantle heterogeneity. *Geochimica et Cosmochimica Acta*, **50**, 1–10.

KRAMERS, J. & TOLSTIKHIN, I. 1997. Two terrestrial lead isotope paradoxes, forward transport modelling, core formation and the history of the continental crust. *Chemical Geology*, **139**, 75–110.

KREISSIG, K., NÄGLER, T. F., KRAMERS, J. D., VAN REENEN, D. D. & SMIT, C. A. 2000. An isotopic and geochemical study of the northern Kaapvaal craton and the Southern Marginal Zone of the Limpopo belt: are they juxtaposed terranes? *Lithos*, **50**, 1–25.

LADBURY, R. 1999. Models suggest deep-mantle topography goes with the flow. *Physics Today*, 21–24.

LASSITER, J. & DEPAOLO, D. 1997. Plume/lithosphere interaction in the generation of continents and oceanic flood basalts: chemical and isotopic constraints. *In*: COFFIN, M. F. & MAHONEY, J. J. (eds) *Large Igneous Provinces. Continental, Oceanic, and Planetary Flood Volcanism*. Geophysical Monograph, American Geophysical Union, **100**, 335–355.

LEEMAN, W. P. 1979. Partitioning of Pb between volcanic glasses and coexisting sanidine and plagioclase feldspars. *Geochimica et Cosmochimica Acta*, **43**, 171–175.

LUAIS, B. & HAWKESWORTH, C. 1994. The generation of continental crust: an integrated study of crust-forming processes in the Archaean of Zimbabwe. *Journal of Petrology*, **35**, 43–93.

MAHONEY, J., FREI, R., TEJEDA, M., MO, X., LEAT, P. & NÄGLER, T. 1998. Tracing the Indian ocean mantle domain through time: isotopic results from old West Indian, East Tethyan, and South Pacific seafloor. *Journal of Petrology*, **397**, 1285–1306.

MANHÈS, G., DUPRÉ, B., HAMELIN, B. & ALLÈGRE, C. 1977. Composition isotopique du plomb des roches basaltiques et hétérogénéités du manteau à diverses échelles. *Bulletin de la Societé Géologique de France*, **7**, XIX(6), 1189–1195.

MARTIN, H. 1986. Effect of steeper Archaean geothermal gradient on geochemistry of subduction-zone magmas. *Geology*, **14**, 753–756.

MCCULLOCH, M. & WOODHEAD, J. 1993. Lead isotopes: evidence for deep crustal-scale fluid transport during granite petrogenesis. *Geochimica Cosmochimica Acta*, **57**, 659–674.

MCLENNAN, S. & TAYLOR, S. 1980. Th and U in sedimentary rocks: crustal evolution and sedimentary recycling. *Nature*, **285**, 621–624.

MOLZAHN, M., REISBERG, L. & WÖRNER, R. G. 1996. Os, Sr, Nd, Pb, O isotope and trace element data from the Ferrar flood basalts, Antarctica: evidence for an enriched subcontinental lithospheric source. *Earth and Planetary Science Letters*, **144**, 529–546.

MOORBATH, S., WELKE, H. & GALE, N. H. 1969. The significance of lead isotope studies in ancient, high-grade metamorphic basement complexes, as exemplified by the Lewisian rocks of Northwest Scotland. *Earth and Planetary Science Letters*, **6**, 245–256.

MOORBATH, S., O'NIONS, R., PANKHURST, R. J., GALE, N. H. & MCGREGOR, V. R. 1972. Further rubidium–strontium age determinations on the very early Precambrian rocks of the Godthaab district, West Greenland. *Nature, Physical Science*, **240**, 78–82.

MOORBATH, S., O'NIONS, R. & PANKHURST, R. J. 1975. The evolution of early Precambrian crustal rocks at Isua, West Greenland – geochemical and isotope evidence. *Earth and Planetary Science Letters*, **27**, 229–239.

MOORBATH, S., WILSON, J., GOODWIN, R. & HUMM, M. 1977. Further Rb–Sr age and isotope data on early and late Archaean rocks from the Rhodesia craton. *Precambrian Research*, **5**, 229–239.

MOORBATH, S., TAYLOR, P. N. & JONES, N. W. 1986. Dating the oldest terrestrial rocks – fact and fiction. *Chemical Geology*, **57**, 63–86.

MÜHE, R., BOHRMANN, H., GARBE-SCHÖNBERG, D. & KASSENS, H. 1997. E-MORB glasses from the Gakkel ridge (Arctic Ocean) at 87°N: evidence for the Earth's most northerly volcanic activity. *Earth and Planetary Science Letters*, **152**, 1–9.

MURTHY, V. 1991. Early differentiation of the Earth and the problem of mantle siderophile elements: a new approach. *Science*, **253**, 303–306.

NUNES, P. & PYKE, D. 1980. *Geochronology of the Abitibi Metavolcanic Belt, Timmins Area, Ontario – Progress Report*. Ontario Geological Survey, Miscellaneous Papers, **92**, 34–39.

O'NEILL, H. S. 1991. The origin of the Moon and the early history of the Earth – a chemical model. Part 2: The Earth. *Geochimica et Cosmochimica Acta*, **55**, 1159–1172.

OVERSBY, V. M. 1975. Lead isotopic systematics and ages of Archaean acid intrusives in the Kalgoorlie–Norseman area, Western Australia. *Geochimica et Cosmochimica Acta*, **39**, 1107–1125.

PEARSON, D. G., CARLSON, R. W., SHIREY, S. B., BOYD, F. R. & NIXON, P. H. 1995. Stabilisation of Archaean lithospheric mantle: a Re–Os isotope study of peridotite xenoliths from the Kaapvaal craton. *Earth and Planetary Science Letters*, **134**, 341–357.

PEATE, D. W., HAWKESWORTH, C. J., MANTOVANI, M. M. S., ROGERS, N. W. & TURNER, S. P. 1999. Petrogenesis and stratigraphy of the high-Ti/Y Urubici magma type in the Paraná flood basalt province and implications for the nature of 'Dupal'-type mantle in the South Atlantic region. *Journal of Petrology*, **40**, 451–473.

POLAT, A. & KERRICK, R. 2000. Archaean greenstone belt magmatism and the continental growth–mantle evolution connection: constraints from the Th–U–Nb–LREE systematics of the 2.7 Ga Wawa subprovince, Superior Province, Canada. *Earth and Planetary Science Letters*, **175**, 41–54.

PUTCHEL, I., BRÜGMANN, G. & HOFMANN, A. W. 2001. ^{187}Os-enriched domain in an Archaean mantle plume: evidence from 2.8 Ga komatiites of the Kostomuksha greenstone belt, NW Baltic Shield. *Earth and Planetary Science Letters*, **186**, 513–526.

PUTCHEL, I. S., HOFMANN, A. W., MEZGER, K., JOCHUM, K. P., SHCHIPANSKY, A. A. & SAMSONOV, A. V. 1998. Oceanic plateau model for continental crustal growth in the Archaean: a case study from the Kostomuksha greenstone belt, NW Baltic shield. *Earth and Planetary Science Letters*, **155**, 57–74.

RAPP, R. P. 1991. Origin of Archaean granitoids and continental evolution. *EOS Transactions, American Geophysical Union*, **22**, 225–229.

RAPP, R. P. & WATSON, E. B. 1995. Dehydration melting of metabasalt at 8–32 kbar: implications for continental growth and crust–mantle recycling. *Journal of Petrology*, **36**, 891–931.

RAPP, R. P., WATSON, E. B. & MILLER, C. F. 1991. Partial melting of amphibolite/eclogite and the origin of Archaean trondhjemites and tonalites. *Precambrian Research*, **51**, 1–25.

REHKÄMPER, M. & HOFMANN, A. 1997. Recycled ocean crust and sediment in Indian Ocean MORB. *Earth and Planetary Science Letters*, **147**, 93–106.

RICHARDSON, S., ERLANK, A., DUNCAN, A. & REID, D. 1982. Correlated Nd, Sr and Pb isotopic variation in Walvis Ridge basalts and implications for the evolution of their mantle source. *Earth and Planetary Science Letters*, **59**, 327–342.

ROBERTSON, D. K. 1973. A model discussing the early history of the Earth based on a study of lead isotope ratios from veins in some Archaean cratons of Africa. *Geochimica et Cosmochimica Acta*, **37**, 2099–2124.

RODDICK, J. C. 1984. Emplacement and metamorphism of Archaean mafic volcanics at Kambalda, Western Australia – geochemical and isotopic constraints. *Geochimica et Cosmochimica Acta*, **48**, 1305–1318.

ROGERS, J. 1996. A history of continents in the past three billion years. *Journal of Geology*, **104**, 91–107.

RUTTER, M. J. & WYLLIE, P. J. 1988. Melting of vapour-absent tonalite at 10 kbar to simulate dehydration-melting in the deep crust. *Nature*, **331**, 159–160.

SCHIOTTE, L. & BRIDGWATER, D. 1989. Multi-stage archaean granulite facies metamorphism in northern Labrador, Canada. *In*: VIELZEUF, D. & VIDAL, P. (eds) *Granulites and Crustal Evolution*. Kluwer Academic, Dordrecht, 157–169.

SENGUPTA, S., PAUL, D. K., DE LAETER, J. R., MCNAUGHTON, N. J., BANDOPADHYAY, P. K. & DESMETH, J. B. 1991. Mid-Archaean evolution of the Eastern Indian craton: geochemical and isotopic evidence from the Bonai pluton. *Precambrian Research*, **49**, 23–37.

SHARMA, M., BASU, A. & NESTERENKO, G. 1992. Temporal Sr, Nd and Pb isotopic variations in the Siberian flood basalts: implications for the plume-source characteristics. *Earth and Planetary Science Letters*, **113**, 365–381.

SHIREY, S. B. 1997. Initial Os isotopic composition of Munro Township, Ontario, komatiite revisited: additional evidence for near-chondritic, late-Archaean convecting mantle beneath the Superior Province. *In: 7th Annual V. M. Goldschmidt Conference*, 193.

SHIREY, S. B. & WALKER, R. J. 1998. The Re–Os isotope system in cosmochemistry and high-temperature geochemistry. *Annual Review of Earth and Planetary Sciences*, **26**, 423–500.

SHIREY, S. B., WILSON, A. H. & CARLSON, R. W. 1998. Re–Os isotopic systematics of the 3300 Ma Nondewi greenstone belt, South Africa: implications for komatiite formation at a craton age. *EOS Trans*, American Geophysical Union, **79**, 372.

SINHA, A. K. 1971. Primary differences in U/Pb ratios in two primitive crustal rocks. *Carnegie Institution of Washington Yearbook*, **69**, 405–408.

SINHA, A. K. & TILTON, G. R. 1973. Isotopic evolution of common lead. *Geochimica et Cosmochimica Acta*, **37**, 1823–1849.

STACEY, J. S. & KRAMERS, J. D. 1975. Approximation of terrestrial lead isotope evolution by a two-stage model. *Earth and Planetary Science Letters*, **26**, 207–221.

STERN, R. A., PERCIVAL, J. A. & MORTENSEN, J. K. 1994. Geochemical evolution of the Minto block: a 2.7 Ga continental magmatic arc built on the Superior proto-craton. *Precambrian Research*, **65**, 115–153.

TATSUMOTO, M., KNIGHT, R. J. & ALLÈGRE, C. J. 1973. Time differences in the formation of meteorites as determined from the ratio of lead-207 to lead-206. *Science*, **180**, 1279–1283.

TAYLOR, P. N., CHADWICK, B., MOORBATH, S., RAMAKRISHNAN, M. & VISWANATHA, M. N. 1984b. Petrography, chemistry and isotopic ages of Peninsular gneiss, Dharwar acid volcanic rocks and the Chitradurga granite with special reference to the late Archaean evolution of the Karnataka craton, Southern India. *Precambrian Research*, **23**, 349–375.

TAYLOR, P. N., JONES, N. W. & MOORBATH, S. 1984a. Isotopic assessment of relative contributions from crust and mantle sources to the magma genesis of Precambrian granitoid rocks. *Philosophical Transactions of the Royal Society of London, Series A*, **310**, 605–625.

TAYLOR, P. N., KRAMERS, J. D., MOORBATH, S., WILSON, J. F., ORPEN, J. L. & MARTIN, A. 1991. Pb/Pb, Sm–Nd and Rb–Sr geochronology in the Archaean Craton of Zimbabwe. *Chemical Geology*, **87**, 175–196.

TAYLOR, P. N., LALSBEEK, F. & BRIDGWATER, D. 1992. Discrepancies between neodymium, lead and strontium model ages from the Precambrian of southern East Greenland: evidence for a Proterozoic granulite-facies event affecting Archaean gneisses. *Isotope Geoscience*, **94**, 281–291.

TAYLOR, P. N., MOORBATH, S., GOODWIN, R. & PETRYKOWSKI, A. C. 1980. Crustal contamination as an indicator of the extent of early Archaean continental crust: Pb isotopic evidence from the late Archaean gneisses of West Greenland. *Geochimica et Cosmochimica Acta*, **44**, 1437–1453.

THIRLWALL, M. F., UPTON, B. G. J. & JENKINS, C. 1994. Interaction between continental lithosphere and the Iceland plume – Sr–Nd–Pb isotope geochemistry of Tertiary basalts, NE Greenland. *Journal of Petrology*, **353**, 839–979.

TILTON, G. R. 1983. Evolution of depleted mantle: the lead perspective. *Geochimica et Cosmochimica Acta*, **47**, 1191–1197.

TILTON, G. & KWON, S. 1990. Isotopic evidence for crust–mantle evolution with emphasis on the Canadian shield. *Chemical Geology*, **83**, 149–163.

TREDOUX, M., DE WIT, M. J., HART, R. J., ARMSTRONG, R. A., LINDSAY, M. M. & SELLSCHOP, J. P. F. 1989. Platinum group elements in a 3.5 Ga nickel–iron occurrence: possible evidence of a deep mantle origin. *Journal of Geophysical Research*, **94**, 795–813.

VAN DER HILST, R. & KARASON, H. 1999. Compositional heterogeneity in the bottom 1000 kilometers of Earth's mantle: toward a hybrid convection model. *Science*, **283**, 1885–1887.

VERVOORT, J., WHITE, W. & THORPE, R. 1994. Nd and Pb isotope ratios of the Abitibi greenstone belt: new evidence for very early differentiation of the Earth. *Earth and Planetary Science Letters*, **128**, 215–229.

VERVOORT, J. D., WHITE, W. M., THORPE, R. I. & FRANKLIN, J. M. 1993. Post-magmatic thermal activity in the Abitibi Greenstone belt, Noranda and Matagami districts: evidence from whole-rock Pb isotope data. *Economic Geology*, **88**, 1598–1614.

VIDAL, PH., BLAIS, S., JAHN, B. M., CAPDEVILA, R. & TILTON, G. R. 1980. U–Pb and Rb–Sr systematics of the Suomussalmi Archean greenstone belt, Eastern Finland. *Geochimica et Cosmochimica Acta*, **44**, 2033–2044.

VINNIK, L. P., GREEN, R. W. E. & NICOLAYSEN, L. O. 1996. Seismic constraints of the mantle of the Kaapvaal craton. *Physics of the Earth and Planetary Interiors*, **95**, 139–141.

WALKER, R., MORGAN, J. & HORAN, M. 1995. ^{187}Os enrichment in some plumes: evidence for core–mantle interaction. *Science*, **269**, 819–822.

WEAVER, B., WOOD, D., TARNEY, J. & JORON, J. L. 1987. Geochemistry of ocean island basalts from the South Atlantic: Ascension, Bouvet, St Helena, Gough and Tristan da Cunha. *In*: UPTON, J. F. B. (ed.) *Alkaline Igneous Rocks*. Geological Society, London, Special Publications, **30**, 253–267.

WHITE, W. M. 1993. ^{238}U/^{204}Pb in MORB and open system evolution of the depleted mantle. *Earth and Planetary Science Letters*, **115**, 211–226.

WHITEHOUSE, M. J. 1989a. Pb isotopic evidence for U–Th–Pb behaviour in a prograde amphibolite to granulite facies transition from the Lewisian complex of north-west Scotland: implications for Pb–Pb dating. *Geochimica et Cosmochimica Acta*, **53**, 717–724.

WHITEHOUSE, M. J. 1989b. Sm–Nd evidence for diachronous crustal accretion in the Lewisian complex of northwest Scotland. *Tectonophysics*, **16**, 245–256.

WIDOM, E. & SHIREY, S. B. 1996. Os isotope systematics in the Azores: implications for mantle plume sources. *Earth and Planetary Science Letters*, **142**, 451–465.

WILSON, A. H. & CARLSON, R. W. 1989. A Sm–Nd and Pb isotope study of Archaean greenstone belts in the southern Kaapvaal craton, South Africa. *Earth and Planetary Science Letters*, **96**, 89–105.

WOODEN, J. L. & MUELLER, P. A. 1988. Pb, Sr, and Nd isotopic compositions of a suite of Late Archaean, igneous rocks, eastern Beartooth Mountains: implications for crust–mantle evolution. *Earth and Planetary Science Letters*, **87**, 59–72.

WOODEN, J. L., CZAMANSKE, G. K., FEDORENKO, V. A & 6 OTHERS 1993. Isotopic and trace element constraints on mantle and crustal contributions to Siberian continental flood basalts, Noril'sk area. *Geochimica et Cosmochimica Acta*, **57**, 3677–3704.

YAMASHITA, K., CREASER, R. A., JENSEN, J. E. & HEAMAN, L. M. 2000. Origin and evolution of mid- to late-Archaean crust in the Hanikahimajuk Lake Area, Slave province, Canada; evidence from U–Pb geochronological, geochemical and Nd–Pb isotopic data. *Precambrian Research*, **99**, 197–224.

ZEGERS, T. E. & VAN KEKEN, P. E. 2001. Middle Archaean continent formation by crustal delamination. *Geology*, **29**, 1083–1086.

ZHAO, D. 2001. Seismic structure and origin of hotspots and mantle plume. *Earth and Planetary Science Letters*, **192**, 251–265.

Seismic evidence for a mantle source for mid-Proterozoic anorthosites and implications for models of crustal growth

GEMMA MUSACCHIO[1,2] & WALTER D. MOONEY[1]

[1] *US Geological Survey, 345 Middlefield Road, MS977, Menlo Park, CA 94025, USA*
[2] *Present address: National Institute of Geophysics and Volcanology, 15 v. Bassini, 20133 Milan, Italy (e-mail: gemma@mi.ingv.it)*

Abstract: Voluminous anorthosite intrusions are common in mid-Proterozoic crust. Historically, two end-member models have been proposed for the origin of these anorthosites. In the first model anorthosites derive from fractionation of a mantle source leaving a residue of metagabbro in the lower crust; in the second model anorthosites are the product of partial melting of the lower crust with residual pyroxene and high-grade minerals (i.e. a pyroxenitic and/or metapelitic lower crust). Although a general consensus has developed that the first model provides the best fit to petrological and geochemical constraints, the sparse evidence for mafic and ultramafic counterparts to the anorthosites leaves the issue still unresolved. We use the absolute P-wave velocity and the ratio between P- and S-wave velocities (V_P/V_S) to infer the composition of the lower crust beneath the Marcy Anorthosite (New York State, USA). Seismic refraction data reveal a lower crust 20 km thick, where V_P and V_P/V_S range from top to bottom between 7.0 km s^{-1} and 7.2 ± 0.1 and 1.84 km s^{-1} and 1.81 ± 0.02, respectively. Laboratory measurements on rock samples indicate that these seismic properties are typical of plagioclase-rich rocks. Magmatic underplating of basaltic melts is a mechanism to form plagioclase-rich bulk composition for the Grenville crust. At the bottom of the lower crust, increase of P-wave velocity, slight decrease of V_P/V_S ratios and the presence of a low-reflective seismic Moho are additional observations supporting crust–mantle interactions related to magmatic underplating. High P-wave velocity (8.6 km s^{-1}) in the upper mantle may indicate that the ultramafic portion (e.g. pyroxenites) of the underplated magma has become eclogite. High average P-wave velocity (6.7 km s^{-1}) and V_P/V_S (1.81), and the exceptional abundance of anorthosites–norites–troctolites among the rocks exposed at the surface, indicate that the Grenville Proterozoic crust may have a unique plagioclase-rich bulk composition. We suggest magmatic underplating, occurring either over a wide time span or with separate syn- and post-collisional magmatic pulses, as being a major crust-forming mechanism operating in mid-Proterozoic time.

The origin of voluminous anorthosite bodies is an essential issue in studies of Precambrian crustal evolution. Archaean anorthosites occur as high-Ca, plagioclase-rich bodies and make up a relatively small fraction of the total Archaean crustal volume. In contrast, mid-Proterozoic anorthosites occur as large complexes composed of several plutonic masses (Thomas 1990; Ashwal 1993; Martignole & Calvert 1996). Both Archaean and mid-Proterozoic anorthosites have a much more plagioclase-rich overall bulk composition than layered mafic intrusions, their closest analogue. In this work we focus on mid-Proterozoic anorthosites, as they represent the largest fraction of all Precambrian anorthosites. We provide evidence in favour of a mantle source for the massive anorthosite intrusions in our study area, and speculate on why Archaean anorthosite intrusions differ from mid-Proterozoic complexes.

Twenty per cent of the surface area of the Grenville Province (Fig. 1) consists of meta-anorthosites, which, with associated granites, mangerites and charnockites, form the distinctive anorthosite–mangerite–charnockite–granite (AMCG) Mid-Proterozoic suite (Emslie & Hunt 1990). This makes the Grenville Province the best area to study the genesis of mid-Proterozoic anorthosites. Meta-anorthosites complexes form relatively thin (2–6 km) slabs of broad horizontal extent where mafic and ultramafic rocks are rare or absent (Ashwal 1993). Magmatic pulses at 1.45–1.15 Ga generated the anorthosites, and the Grenville Orogeny (1.1 Ga) was responsible for their amphibolite- to granulite-facies metamorphism (600–800 MPa; 750°C) (Ashwal 1993; Higgins & van Breemen 1996; McLelland et al. 1996; Scoates & Frost 1996; Corrigan & Hanmer 1997).

For many years, a debate on the origin of the mid-Proterozoic meta-anorthosites has revolved

From: FOWLER, C. M. R., EBINGER, C. J. & HAWKESWORTH, C. J. (eds) *The Early Earth: Physical, Chemical and Biological Development.* Geological Society, London, Special Publications, **199**, 125–134. 0305-8719/02/$15.00
© The Geological Society of London 2002.

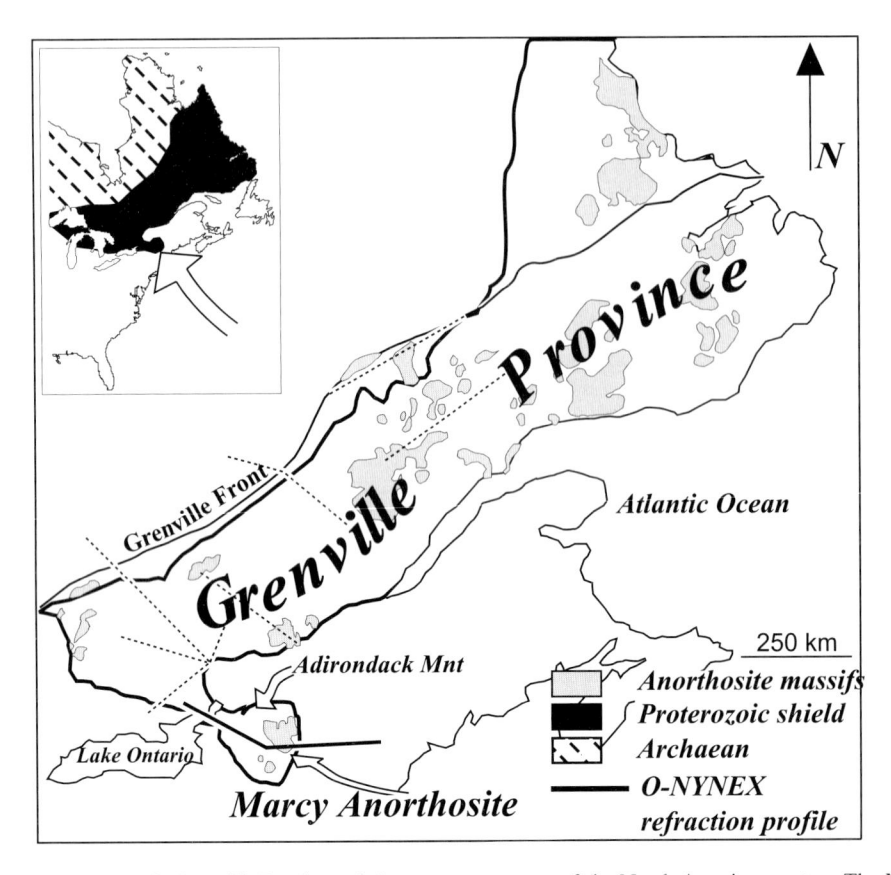

Fig. 1. The Proterozoic Grenville Province of the easternmost part of the North American craton. The Marcy Anorthosite is crossed by the O-NYNEX high-resolution seismic refraction profile. Dashed lines indicate seismic refraction profiles.

around the issue of mantle v. crustal derivation (Ashwal 1993; Corrigan & Hanmer 1997; Duchesne et al. 1999; Glukhovskii & Moralev 1999; Schiellerup et al. 2000). Although today the mantle genetic model is favoured (e.g. Ashwal et al. 1998), the sparse evidence for the required mafic and ultramafic counterpart to the anorthosites represents a critical unsolved issue. Some workers have argued for a lower-crustal source on the basis of a low anorthosite content and Mg number of massif-type anorthosite parent magmas, rare earth element modelling, and erratic isotope signatures (Schiellerup et al. 2000). One possible solution to this apparent problem is the derivation of anorthosites from partial melting of the lower crust.

Whether derived from a mantle or crustal source, the origin of anorthosites must have left a distinct signature in the lower crust. The bulk composition of the lower crust is expected to be of gabbroic composition (metamorphosed in Grenville time) if the anorthosite magma was derived from the fractionation of mantle melt. Thus, either the lower crust represents the remnant of a large magma chamber, or gabbroic–anorthositic sills permeated a pre-existing lower crust. In contrast, if the anorthosite were derived from within the crust, a thick restitic lower crust (pyroxenites and high-grade metapelites) would be present.

Seismic refraction data may constrain crustal composition beneath the anorthosite massifs, and therefore contribute to the debate regarding their origin. Here we focus on the question of whether large volumes of mafic material occupy the lower crust beneath the anorthosite complexes. If the lower crust is largely made of mafic material, we address the question of whether its bulk composition is gabbroic or pyroxenitic and high-grade metapelitic. We use P-wave velocity (V_P) and the ratio between the P- and S-wave velocity (V_P/V_S) to determine the bulk composition of the middle and lower crust. Under the assumption that the Grenville Orogeny did not cause major upper–lower-crust decoupling, we can genetically link the upper-crustal meta-anorthosites to the present-day lower crust beneath it. The assumption is supported by the fact that no regional low-angle faults are detected in seismic reflection imaging (Klemperer et al. 1985) or detailed seismic velocity modelling (Hughes & Luetgert 1991; Zelt & Forsyth 1994; Zelt et al. 1994; Musacchio et al. 1997) for the study area.

V_P and V_S constraints for the composition of the crust

Some parameters of crustal composition can be constrained by V_P/V_S ratios (Kern et al. 1993; Musacchio et al. 1997; Novak et al. 1997; Chevrot & van der Hilst 2000). V_P/V_S ratios are not sensitive to pressure and subsolidus temperatures, but depend on fluid pore pressure, fabric attitude and rock composition, specifically quartz content and plagioclase composition (Christensen 1996). In stable shield areas, high V_P/V_S ratios in the crystalline crust are generally explained in terms of composition. Mafic rocks have V_P/V_S ratios higher than 1.73 as a result of low quartz content and the abundance of mafic minerals (plagioclase, pyroxene, garnet, amphibole and olivine).

Mafic rocks have, in terms of V_P and V_P/V_S ratios, a distinct characteristic depending upon the ratio of plagioclase to other mafic minerals

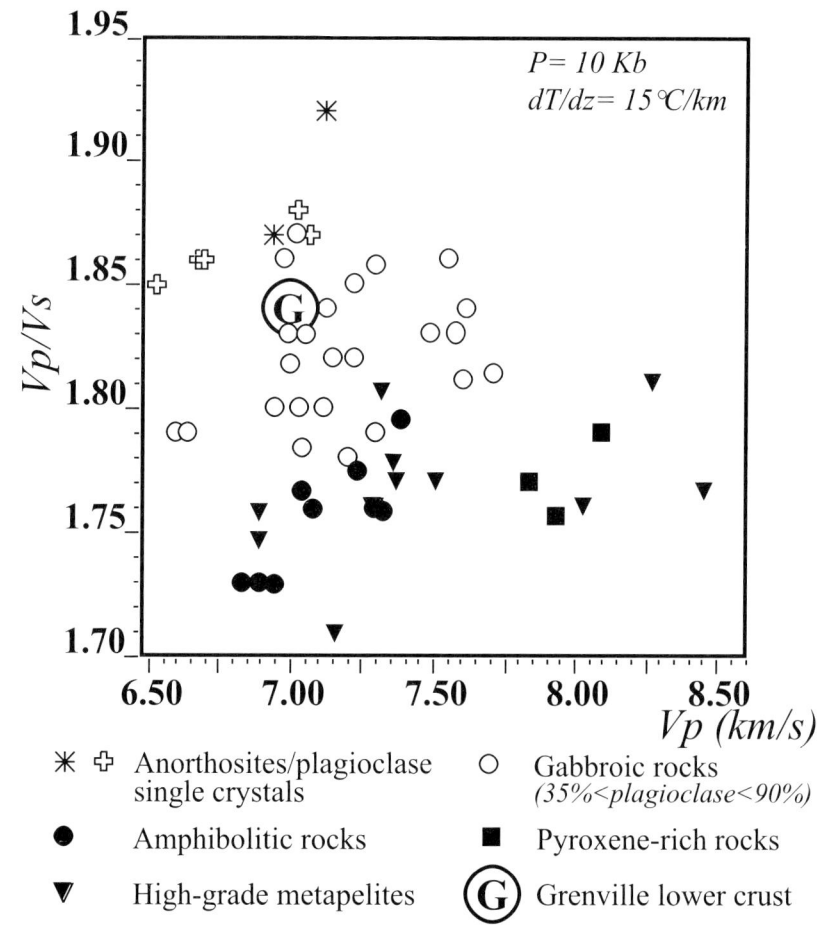

Fig. 2. V_P/V_S v. V_P (km s^{-1}) for laboratory measurements on rock samples (Manghnani & Ramananantoandro 1974; Miller & Christensen 1994; Barruol & Kern, 1996; Christensen 1996; Kern et al. 1996; Musacchio et al. 1997) and for calculated values on single crystals (open crosses; Ryzhova 1964). Plotted rocks are representative of the lower crust and include mafic rocks and high-grade metapelites.

(PG = plagioclase/(pyroxene + garnet + amphibole + olivine)). Rocks with high PG ratios have high V_P/V_S ratios. Plagioclase-rich mafic rocks (35% ≪ plagioclase < 90%) cluster around $6.9 < V_P < 7.3$ km/s and $1.80 < V_P/V_S < 1.85$, whereas anorthosites (plagioclase > 90%) have a V_P/V_S ratio higher than 1.85 (Fig. 2). In mafic rocks, an increase of pyroxene and garnet content correlates with a decrease of the V_P/V_S ratio and an increase of V_P. Amphibole (and/or quartz) minerals are associated with a decrease of V_P/V_S ratio and of V_P (Fig. 2).

The Ontario–New York–New England (O-NYNEX) transect is a high-resolution seismic refraction profile that crosscuts the Proterozoic Grenville Province and the Phanerozoic Appalachian Province (Fig. 1). This transect is of great interest for our study because of the high-resolution V_P and V_P/V_S ratio model (Fig. 3a). In this model, the Proterozoic crust displays an average V_P/V_S ratio that is clearly higher (1.81 ± 0.02) than that of the Phanerozoic crust (1.74 ± 0.02). The average V_P/V_S ratio in the Proterozoic crust is also higher than seismic global average values (1.78; Zandt & Ammon 1995), and crustal averages based on ultrasonic experiments (1.77; Christensen 1996). This high V_P/V_S ratio clearly indicates a mafic composition of the crust and largely reflects the fact that lithologies exposed in the Grenville province are plagioclase dominated. In the seismic cross-section (Fig. 3a), V_P/V_S ratios as high as 1.85 ± 0.02 are associated with V_P ranging from 6.6 to 7.0 ± 0.1 km s^{-1}. These seismic observations are consistent with the presence of anorthositic bodies in the Grenville crust (Musacchio et al. 1997), namely the Marcy Anorthosite, an upper-crustal body that crops out along the profile, and the Tahawus Complex (15–20 km depth), a reflective body identified by Klemperer et al. (1985). The Tahawus Complex may represent an example of a layered meta-anorthosite where contacts between plagioclase-rich (meta-anorthositic) and plagioclase-depleted (metagabbro with plagioclase ≪ 90%) layers are responsible for the reflectivity observed with near-vertical reflection surveys (Klemperer et al. 1985; Musacchio et al. 1997).

The observed V_P (7.0 ± 0.1 km s^{-1}) and the high V_P/V_S ratio (1.84 ± 0.02) for the lower crust beneath the Marcy Anorthosite are consistent with plagioclase-dominated mafic compositions (i.e. gabbros and/or norites). Despite the clear compositional signature, little can be said regarding the metamorphic grade. In fact, the effect of metamorphism on V_P/V_S ratios for gabbroic rocks is significant only when eclogite grade is reached.

The seismic properties of the Grenville lower crust are inconsistent with pyroxene-rich or high-grade metapelites (Fig. 2). This is crucial in determining which source, mantle or crust, is responsible for generating the mid-Proterozoic meta-anorthosites, as we will discuss below.

Although high-resolution V_P/V_S ratios from refraction data are restricted to the Adirondack Mountains transect, estimates from teleseismic data indicate $V_P/V_S = 1.84$ for the entire Grenville Province (Jordan & Frazer 1975; Owens 1987; Zandt & Ammon 1995). These data also confirm that the lower crust is 20 km thick and has a P-wave velocity of about 7 km s^{-1} (Martignole & Calvert 1996).

Despite the sparse available data on V_P/V_S ratios, a worldwide compilation indicates a clear trend of high values for Proterozoic crust (Fig. 4). Specifically, the high V_P/V_S ratio for Proterozoic crust is anomalous compared with average values for the crust of any other age, in which large volumes of plagioclase-rich rocks are inferred to be absent. In other words, large volumes of plagioclase-dominated lithologies seem to be predominantly a Proterozoic phenomenon.

Discussion: mantle derivation of the mid-Proterozoic magma

The high V_P/V_S ratios found for the lower crust in the Grenville Province support the hypothesis of a mantle source for the mid-Proterozoic meta-anorthosites. The seismic constraints are inconsistent with the alternate hypothesis that anorthosites were derived from partial melting of an early Proterozoic lower crust (Taylor et al. 1984; Longhi & Auwera 1996; Schärer et al. 1996), leaving a thick restitic layer (pyroxenites and high-grade metapelites) in the lower-crust. Estimates of lower-crustal composition include: garnetiferous metapelites (Valley & O'Neil 1982; Taylor et al. 1984), mafic granulites (Christensen & Mooney 1995; Rudnick & Fountain 1995;

Fig. 3. (a) Colour contour plot of V_P/V_S ratios along the O-NYNEX seismic profile across the Adirondack Mountains (Fig. 1); (b) schematic representation of the stratified magma chamber analogue. The schematic representation does not refer to actual depths of emplacement and does not include deformations that may have occurred during subsequent orogenesis. Numbers on the cross-section indicate P-wave velocities (in km s^{-1}). GR, Grenville Ramp; CCmz, Cartage–Colton mylonite zone; MA, Marcy Anorthosite; TC, Tahawus Complex.

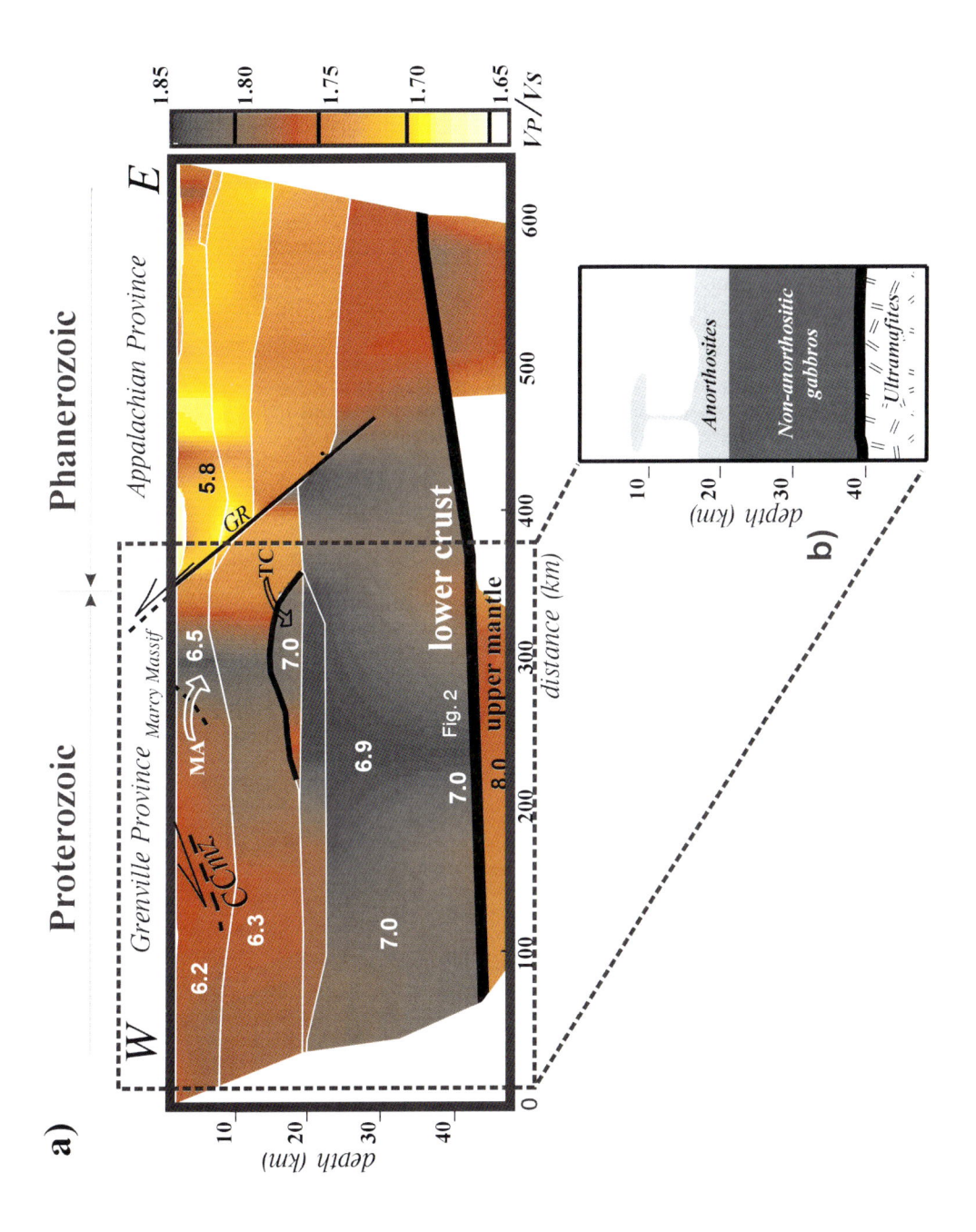

MANTLE SOURCE FOR MID-PROTEROZOIC ANORTHOSITES 129

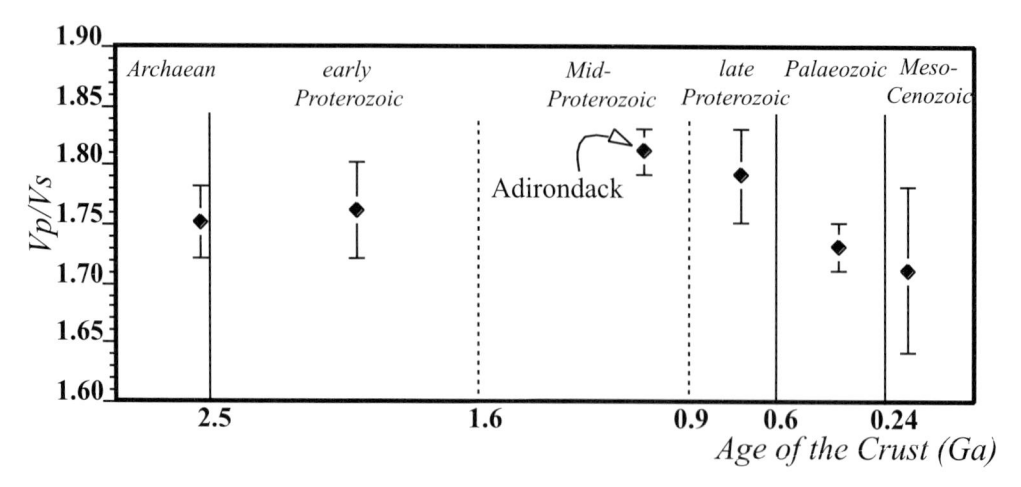

Fig. 4. V_P/V_S v. age for the continental crust. Data are from seismic refraction profiles (Barth & Klemperer 1993; Musacchio 1993; Walther & Flueh 1993; Musacchio et al. 1995, 1997) and teleseismic measurements (Jordan & Frazer 1975; Owens 1987; Zandt & Ammon 1995; Chevrot & van der Hilst 2000) in various areas. Data from some areas (i.e. young island arcs) where composition cannot be reliably inferred from V_P/V_S ratios have been omitted. The high V_P/V_S ratio in the Mid-Proterozoic crust is not found in crust of any other age.

Longhi et al. 1999) and amphibolite (Christensen & Mooney 1995; Rudnick & Fountain 1995). For any of these compositions, partial melting and extraction of the anorthosite suite had to result in a general increase of restitic minerals (pyroxene and garnet). The occurrence of this process at a large scale, as the volume of mid-Proterozoic meta-anorthosites would require, would have produced a restitic bulk composition of the lower crust and a decrease of V_P/V_S ratios to values inconsistent with our observations.

A mantle source requires that the anorthosites represent the upper layers of a stratified magma chamber produced by magmatic underplating at the base of a 30–35 km thick crust during mid-Proterozoic time (Fig. 5). Only the anorthosite portion of the magma chamber rose through the crust (in the form of diapirs) to crystallize at shallow depths, leaving the gabbroic magma chamber in the lower crust and ultramafic residues in the upper mantle. Considering that the rocks that are at present exposed are derived from the middle crust we infer that the crust was thickened by 10–15 km during the Grenville Orogeny (1.1 Ga). Exhumation to the present-day position is associated with extension and thinning of the crust to 40 km.

This model accounts for the high-Na plagioclase (crystallization at 900–1500 MPa; Longhi et al. 1999) and for the presence of high-Al pyroxene megacrysts (crystallization at 1100 MPa and 1250°C; Longhi et al. 1999). If mid-Proterozoic anorthosites differentiated from a broadly basaltic magma (Ashwal 1993) in a manner similar to that in layered mafic intrusions, then the proportion of anorthosite : mafic : ultramafic rocks will be 1:3:1 (e.g. the Stillwater Complex), provided that crustal thickening associated with the Grenville Orogeny did not significantly influence this ratio. The estimated thickness of the mid-Proterozoic meta-anorthosites is at the most 6 km (e.g. Marcy (Thomas 1990; Musacchio et al. 1997); Morin (Thomas 1990; Martignole & Calvert 1996); Bouchette (Martignole & Calvert 1996); Lac Fourier (Thomas 1990)). On the basis of the layered mafic intrusion analogue, the lower-crustal non-anorthositic metagabbro is inferred to be 18–20 km thick, which is consistent with the seismic constraints on thickness of the plagioclase-rich lower crust beneath the anorthosite massifs. V_P/V_S data are also consistent with the lower crust being permeated by plagioclase-rich sills (i.e. the Tahawus Complex). In fact, delamination typically results in low P-wave velocities in the lower crust (e.g. 6.3–6.4 km s^{-1} in the Sierra Nevada (Ruppert et al. 1998)), an observation that is inconsistent with the seismic velocities (6.8–7.2 km s^{-1}) modelled for the Grenville Province.

We note that several recent studies support a crustal source for these magmas. Duchesne et al. (1999) noted that radiogenic isotopes cannot differentiate between juvenile lower-crustal and mantle sources, thus negating the usefulness of this technique to resolve this issue. Coupled with the recent petrological work of Longhi et al. (1999), Duchesne et al. (1999) proposed a crustal gabbronoritic primary source for the parent

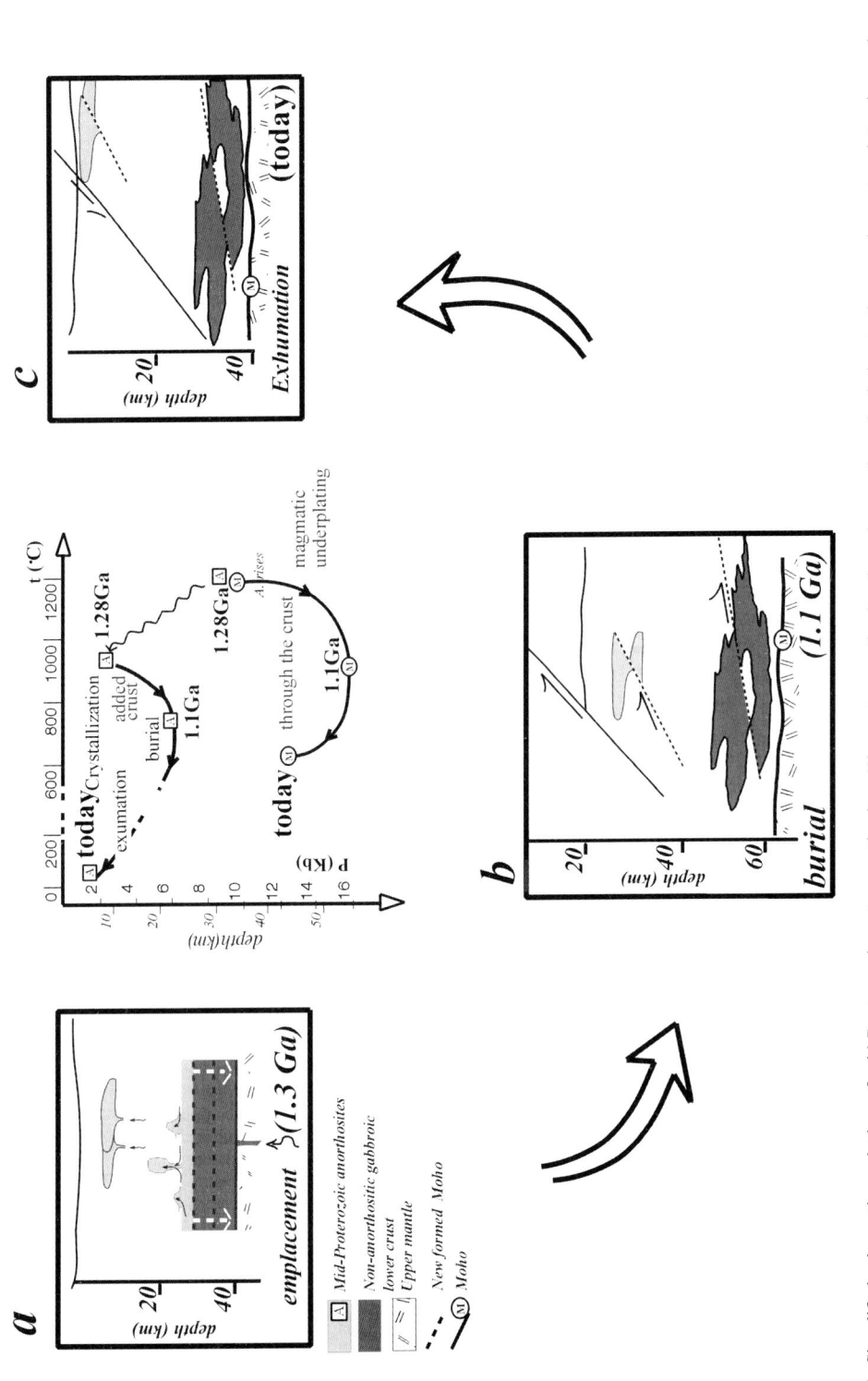

Fig. 5. Simplified schematic evolution of mid-Proterozoic meta-anorthosites. The meta-anorthosites have been enlarged relative to the complementary intrusive rocks (dark grey) for illustration. The model largely refers to the emplacement of the Marcy Anorthosite. Emplacement of the anorthosites follows concentration of plagioclase-rich melts at the upper part of magma chambers that pond at the bottom of a 30–35 km thick Proterozoic crust. Anorthosite diapirs rose through the crust and crystallized at shallow depths and a non-anorthositic gabbroic magma chamber (dark grey) remained in the lower crust. (**a**) A significant amount of mantle-derived magma underplating resulted in crustal growth. (**b**) In late Proterozoic time, the Grenville Orogeny caused crustal thickening, deformation and metamorphism (amphibolite to granulite facies). (**c**) Exhumation of the meta-anorthosites is associated with thinning of the crust to 40 km. In the lower crust, plagioclase-bearing roots are the leftover trace of anorthosite emplacement. The P–T–t path is conceptual.

magma. Dempster et al. (1999) offered a compromise, stating that mantle-derived basic magmas pond at the base of the crust and relatively Al-rich lower-crustal lithologies would result in the precipitation of large volumes of plagioclase. Anorthosite massifs could then be emplaced at higher crustal levels within juvenile Proterozoic crust. A 'mixed' (crust–mantle) source for anorthosites was also suggested by Glukhovskii & Moralev (1999).

Some anorthosite massifs show evidence for a shallow depth of emplacement (e.g. the Marcy Massif; Valley & O'Neil 1982) indicating that the anorthositic fraction of the magma rose through the crust in the form of diapirs (Ashwal 1993). The massifs probably grew to their present dimensions by coalescence of multiple diapirs (Ashwal 1993; McLelland 1994) (Fig. 5a). The ultramafic rocks, which occupy the bottom part of the magma chamber, could be present in the uppermost mantle and be associated with observed upper-mantle seismic reflectors (Hughes & Luetgert 1991).

Conclusion

The seismic properties of the lower crust beneath the mid-Proterozoic meta-anorthosites of the southeastern Grenville Province (Fig. 3) indicate a metagabbroic composition (35% ≪ plagioclase < 90%), representing the mafic counterpart of a mantle-derived basaltic magma. The data strongly support mantle source for the anorthosites occurring in the study area and we speculate that the same mechanism applies to the genesis of worldwide mid-Proterozoic anorthosites. Global P-wave velocity data indicate that the lower crust is generally mafic in composition (amphibolite, mafic granulite and garnet mafic granulite in various percentages; Christensen & Mooney 1995; Rudnick & Fountain 1995). Data on V_P/V_S ratios discussed in this paper allow us to infer that the mid-Proterozoic lower crust is a mafic plagioclase-dominated lithology.

In mid-Proterozoic time, widespread magmatic underplating generated the anorthosite complexes. These results have important implications for the two commonly cited mechanisms for forming continental crust: accretion of subduction complexes or island arcs at continental margins; magmatic underplating of mantle-derived melts (Goodwin 1996; Durrheim & Mooney 1994). Although the subduction-complexes and island-arc model can explain most of the crustal growth in Phanerozoic time, the contribution of magmatic underplating to crustal growth has been difficult to quantify. Given a ratio of 1:3 for anorthosites to non-anorthositic gabbroic rocks, our study demonstrates that significant thickening of the lower crust occurred during the genesis of the mid-Proterozoic meta-anorthosites. This conclusion has two major implications: (1) crustal growth models have to be revised to include more growth in the mid-Proterozoic time; (2) models for the composition of the Proterozoic crust need to account for a plagioclase-dominated lower crust. We note that the evidence we present for crustal underplating will provide a higher rate of crustal growth in Proterozoic than in Phanerozoic time, as extensive anorthosites are largely restricted to mid-Proterozoic time.

We speculate that both mid-Proterozoic and Archaean anorthosites have a mantle source. However, it is necessary to explain why the Archaean anorthosites are characterized by high-Ca plagioclase and megacrystic texture. These characteristics may be related to the composition of parental magma and/or to the temperature of the lithosphere. A komatiitic composition of parental magma is an attractive explanation because it implies a genetic connection between these two rock types, which are known to be almost entirely restricted to Archaean time (Phinney 1982; Ashwal 1993). A warmer Archaean lithosphere would enhance the formation of megacrystic high-Ca plagioclase-rich melts at shallow depths.

We thank R. Frost, G. Fuis, D. Fountain, C. Bacon, K. Favret, S. Detweiler, O. Fitzpatrick, B. Coleman, W. Hamilton, J. Percival, P. Rey, C. M. R. Fowler and an anonymous reviewer for the constructive criticism.

References

ASHWAL, L. 1993. *Anorthosites*. Springer, New York.

ASHWAL, L., HAMILTON, M., MOREL, V. & RAMBELOSON, R. 1998. Geology, petrology and isotope geochemistry of massif-type anorthosites from southwest Madagascar. *Contributions to Mineralogy and Petrology*, **133**, 389–401.

BARRUOL, G. & KERN, H. 1996. Seismic anisotropy and shear-wave splitting in lower crust and upper-mantle rocks from the Ivrea Zone–experimental and calculated data. *Physics of the Earth and Planetary Interiors*, **95**, 175–194.

BARTH, G. A. & KLEMPERER, S. L. 1993. Proterozoic crust of the southern Baltic Shield: shear wave seismic structure and Poisson's ratios from Babel profiles A and B. *In*: MOONEY, W. D. (ed.) *1993 CCSS Workshop*. Mir Publishing, Moscow, 225–236.

CHEVROT, S. & VAN DEN HILST, R. 2000. The Poisson ratio of the Australian crust: geological and geophysical implications. *Earth and Planetary Science Letters*, **183**, 121–132.

CHRISTENSEN, N. I. 1996. Poisson's ratio and crustal seismology. *Journal of Geophysical Research*, **101**, 3139–3156.

CHRISTENSEN, N. I. & MOONEY, W. D. 1995. Seismic velocity structure and composition of the continental crust: a global view, *Journal of Geophysical Research*, **100**, 9761–9788.

CORRIGAN, D. & HANMER, S. 1997. Anorthosites and related granitoids in the Grenville orogen: a product of convective thinning of the lithosphere. *Geology*, **25**, 61–64.

DEMPSTER, T. J., PRESTON, R. J. & BELL, B. R. 1999. The origin of Proterozoic massif-type anorthosites: evidence from interactions between crustal xenoliths and basaltic magma. *Journal of the Geological Society, London*, **156**, 41–46.

DUCHESNE, J. C., LIÉGEOIS, J. P., AUWERA, J. V. & LONGHI, J. 1999. The crustal tongue melting model and the origin of massive anorthosites. *Terra Nova*, **11**, 100–105.

DURRHEIM, R. & MOONEY, W. D. 1994. Evolution of the Precambrian lithosphere: seismological and geochemical constraints. *Journal of Geophysical Research*, **99**, 15 359–15 374.

EMSLIE, R. & HUNT, P. 1990. Ages and petrogenetic significance of igneous mangerite–charnockite suites associated with massif anorthosites, Grenville Province. *Journal of Geology*, **98**, 213–231.

GLUKHOVSKII, M. Z. & MORALEV, V. M. 1999. Evolution of the tectonic setting of anorthosite magmatism in the Aldan Shield. *Geotectonics*, **33**, 425–435.

GOODWIN, A. M. 1996. *Principles of Precambrian Geology*. Academic Press, London.

HIGGINS, M. D. & VAN BREEMEN, O. 1996. Three generations of anorthosites–mangerite–charnockite–granite (AMCG) magmatism, contact metamorphism and tectonism in the Saguenay–Lac-Jean region of the Grenville Province, Canada. *Precambrian Research*, **79**, 327–346.

HUGHES, S. & LUETGERT, J. H. 1991. Crustal structure of the western New England Appalachians and the Adirondack Mountains, *Journal of Geophysical Research*, **96**, 16 471–16 494.

JORDAN, T. H. & FRAZER, L. N. 1975. Crustal and upper mantle structure from Sp phases. *Journal of Geophysical Research*, **80**, 1504–1518.

KERN, H., GAO, S. & LIU, Q. S. 1996. Seismic properties and densities of middle and lower crustal rocks exposed along the North China Geoscience Transect. *Earth and Planetary Science Letters*, **139**, 439–455.

KERN, H., WALTHER, C., FLUEH, E. R. & MARKER, M. 1993. Seismic properties of rocks exposed in the Polar profile region – constraints on the interpretation of refraction data. *Precambrian Res.*, **64**, 169–187.

KLEMPERER, S. L., BROWN, L. D., OLIVER, J. E., ANDO, C. J., CZUCHRA, B. L. & KAUFMAN, S. 1985. Some results of COCORP seismic reflection profiling in the Grenville-age Adirondack Mountains, New York State. *Canadian Journal of Earth Sciences*, **22**, 141–153.

LONGHI, J., AUWERA, J. V., FRAM, J. & DUCHESNE, J. C. 1999. Some phase equilibrium constraints on the origin of Proterozoic (massif) anorthosites and related rocks. *Journal of Petrology*, **40**, 339–362.

MANGHNANI, M. H. & RAMANANTOANDRO, R. 1974. Compressional and shear wave velocities in granulite facies rocks and eclogites to 10 Kb. *Journal of Geophysical Research*, **79**, 5427–5446.

MARTIGNOLE, J. & CALVERT, A. 1996. Crustal scale shortening and extension across the Grenville province in western Quebec. *Tectonics*, **15**, 376–386.

MCLELLAND, J. 1994. Multi-stage polybaric genesis of Proterozoic massif anorthosite: example from the Adirondack mountains, New York. *Abstracts with Programs, Geological Society of America*, **26**, 40.

MCLELLAND, J., DALY, J. S. & MCLELLAND, J. 1996. The Grenville orogenic cycle (*ca.* 1350–1000 Ma): an Adirondack perspective. *Tectonophysics*, **265**, 1–28.

MILLER, D. J. & CHRISTENSEN, N. I. 1994. Seismic signature and geochemistry of island arc: a multi-disciplinary study of the Kohistan accreted terrane, northern Pakistan. *Journal of Geophysical Research*, **99**, 11 623–11 642.

MUSACCHIO, G. 1993. *Modellizzazione sismica della crosta in un'area d'orogenesi recente: l'esempio delle Alpi centro-occidentali*. PhD thesis, University of Milan.

MUSACCHIO, G., MECHIE, J. & CASSINIS, R. 1995. Babel line 1 (Bothnian Gulf) high resolution 3-component wide-angle data processing. *In*: SNYDER, D. B. & MCBRIDE, H. (eds) *9th Workshop Meeting of the CCSS Workshop, Moscow*. Canadian Geological Survey, Ottawa, Open File, 27–38.

MUSACCHIO, G., MOONEY, W. D., LUETGERT, J. H. & CHRISTENSEN, N. I. 1997. Composition of the crust in the Grenville and Appalachian Provinces of North America inferred from V_P/V_S ratios. *Journal of Geophysical Research*, **102**, 15 225–15 241.

NOVAK, O., RITTER, J. R. R., ALTHERR, R. & 6 OTHERS 1997. An integrated model for the deep structure of the Chyulu Hills volcanic field, Kenya. *Tectonophysics*, **278**, 187–209.

OWENS, T. J. 1987. Crustal structure of the Adirondacks determined from broadband teleseismic waveform modeling, *Journal of Geophysical Research*, **92**, 6391–6401.

PHINNEY, W. C. 1982. Petrogenesis of Archean anorthosites. *In*: WALKER, D. & MCCALLUM, I. S. (eds) *Workshop on Magmatic Processes of Early Planetary Crusts: Magma Oceans and Stratiform Layered Intrusions*. Lunar and Planetary Institute Technical Report, **82-01**, 121–124.

RUDNICK, R. & FOUNTAIN, D. 1995. Nature and composition of the continental crust: a lower crustal perspective. *Reviews of Geophysics*, **33**, 267–309.

RUPPERT, S., FLIEDNER, M. M. & ZANDT, G. 1998. Thin crust and active upper mantle beneath the Southern Sierra Nevada in the Western United States. *Tectonophysics*, **286**, 237–252.

RYZHOVA, T. V. 1964. Elastic properties of plagioclase. *Bulletin of the Academy of Sciences of the USSR, Geophysics Series, English Translation*, **7**, 633–635.

SCHÄRER, U., WILMART, E. & DUCHESNE, J. C. 1996. The short duration and anorogenic character of anorthosite magmatism: U–Pb dating of the Rogaland complex, Norway. *Earth and Planetary Science Letters*, **139**, 335–350.

SCHIELLERUP, H., LAMBERT, D., PRESTVIK, T., ROBINS, B., MCBRIDE, J. & LARSEN, R. 2000. Re–Os isotopic evidence for a lower crustal origin of massif-type anorthosites. *Nature*, **405**, 781–784.

SCOATES, J. S. & FROST, C. D. 1996. A strontium and neodymium isotopic investigation of the Laramie anorthosites, Wyoming, USA: implications for magma chamber processes and the evolution of magma conduits in Proterozoic anorthosites. *Geochimica et Cosmochimica Acta*, **60**, 95–107.

TAYLOR, S., CAMPBELL, I. & MCCULLOCH, T. 1984. A lower crustal origin for massif-type anorthosites, *Nature*, **311**, 372–384.

THOMAS, M. D. 1990. Deep structure of the Middle Proterozoic anorthositic intrusions in the eastern Canadian Shield: insights from gravity modeling. *In*: GOWER, C. F., RIVERS, T. & RYAN, B. (eds) *Mid-Proterozoic Laurentia–Baltica*. Geological Association of Canada, Special Papers, **38**, 353–372.

VALLEY, J. & O'NEIL, J. 1982. Oxygen isotope evidence for shallow emplacement of Adirondack anorthosite. *Nature*, **300**, 497–500.

WALTHER, C. & FLUEH, E. R. 1993. The POLAR Profile revisited: combined P- and S-wave interpretation. *Precambrian Research*, **64**, 153–168.

ZANDT, G. & AMMON, C. J. 1995. Continental crust composition constrained by measurements of crustal Poisson's ratio. *Nature*, **374**, 152–154.

ZELT, C. A. & FORSYTH, D. A. 1994. Modeling wide-angle seismic data for crustal structure: southeastern Grenville province. *Journal of Geophysical Research*, **99**, 11 678–11 705.

ZELT, C. A., FORSYTH, D. A., MILKEREIT, B., WHITE, D. J., ASUDEH, I. & EASTON, R. M. 1994. Seismic structure of the Central Metasedimentary Belt, southern Grenville province, *Canadian Journal of Earth Sciences*, **31**, 243–254.

Deflection of mantle plume material by cratonic keels

N. H. SLEEP[1], C. J. EBINGER[2] & J.-M. KENDALL[3]

[1] *Department of Geophysics, Mitchell Building, Stanford University, Stanford, CA 94305, USA (e-mail: norm@pangea.stanford.edu)*
[2] *Department of Geology, Royal Holloway, University of London, Egham TW20 0EX, UK*
[3] *School of Earth Sciences, University of Leeds, Leeds LS2 9JT, UK*

Abstract: Lithosphere that formed in Archaean and possibly early Proterozoic time is thicker, more buoyant, and geochemically distinct from lithosphere that formed after about 2.3 Ga. Mantle xenolith and seismic data indicate that some cratonic roots, or 'keels', extend to depths of $c.\,250$ km, compared with normal continental lithosphere of thickness 150 km or less; yet many cratons have experienced uplift, dyking and kimberlite emplacement, suggesting interactions with hot, rising asthenosphere referred to as mantle plumes. Plumes supply additional heat to the base of the lithospheric plates, whose base can be heated and entrained in the flow (thermal erosion). How have these cratonic keels persisted despite their interactions with mantle plumes? The geometry of cratonic keels during their interactions with mantle plumes is a critical factor controlling keel preservation. To a laterally spreading plume head, cratonic keels appear as major obstacles, and the hot, buoyant plume material ponds beneath thinner lithosphere. Our model simulations show that deep keels deflect mantle plume material and that steep gradients at the lithosphere–asthenosphere boundary between Archaean keels and 'normal' lithosphere will focus flow, leading to localized adiabatic decompression melting. Plume processes can lead to a reduction in the breadth of a cratonic root where the plume rises beneath the craton, regardless of the initial breadth of the craton. Where the plume rises beneath a craton the hot plume material will spread laterally beneath the keel and attain thicknesses of tens of kilometres. This transfers heat to the base of the lithosphere and could generate small volumes of melt at considerable depth, depending on the composition of the lower lithosphere. We have used model simulations of plumes beneath Africa to predict the magnitude and direction of seismic anisotropy caused by lateral flow of hot plume material beneath and around a cratonic keel. The shear-wave splitting in our models is greatest at the edge of the cratonic keel, and its azimuth is parallel to the plume flow direction.

Earth scientists define cratons by their lack of thermal and mechanical deformation. Cratons are largely aseismic, volcanism is restricted to small-volume melts (e.g. kimberlites) derived from depths >150 km (e.g. Boyd *et al.* 1985; Nixon 1987), and deformation occurs in the thinner lithosphere adjacent to the cratons (e.g. Lenardic & Moresi 2000; Petit & Ebinger 2000). Archaean, and possibly early Proterozoic, cratons are compositionally distinct from younger lithosphere: the upper mantle beneath cratons is depleted and hence more buoyant than younger lithosphere, making it difficult to deform in subduction and collisional zones (e.g. Jordan 1981; Doin *et al.* 1996).

Seismic, heat-flow and xenolith data provide constraints on different physical properties, and differences between estimates are expected. Seismic methods determine the depth to a low-velocity zone at the base of higher-velocity lithospheric lid (e.g. James & Fouch 2002; Kendall *et al.* 1998); heat-flow studies estimate the thickness of the conductive lid (e.g. Rudnick & Nyblade 1999); mantle xenolith data provide pressure–temperature constraints on the base of the lithosphere at the time of eruption, or a palaeo-lithospheric thickness (e.g. Pearson *et al.* 2002). Thus, the various methods will be more or less sensitive to the lithosphere–asthenosphere transition zone.

Seismic studies of cratonal lithosphere at both the global and local scale indicate that Archaean cratons are sites of unusually thick, cold lithosphere, which may have mantle roots extending to depths of 200–400 km compared with 'normal' continental lithosphere, which is less than 150 km thick (e.g. van der Lee & Nolet 1997; Ritsema *et al.* 1998; Ritsema & van Heijst 2000; James & Fouch 2002). Heat-flow data place some bounds on lithospheric thickness, but regional comparisons require major assumptions as to the lower-crustal contribution to radioactive heating.

Jaupart & Mareschal (1999) subtracted the effects of radioactive heating to constrain the contribution from mantle heating, and determined a conductive lithospheric thickness of 200–250 km for the large Canadian Shield. Rudnick & Nyblade (1999) and Artemieva and Mooney (2001) used models of heat-flow and xenolith data to show that cratonic lithospheric thickness is probably less than 300 km worldwide.

Geochemical data help to constrain the longevity of cratonic roots and their interactions with mantle flow. Pearson (1999) and Carlson et al. (2002) examined Re–Os patterns in mantle xenoliths and found that the mantle lithosphere to depths of $c.$ 150 km retains Archaean ages comparable with those of crustal rocks ($c.$ 2.5 Ga), implying stability and great depth since craton accretion. The mantle lithosphere to depths of 200–250 km beneath the Kaapvaal Craton shows ages of 2.5 to $c.$ 1.5 Ga, implying mixing and/or modification of the lowermost mantle with asthenospheric melts. Chesley et al. (1999) reported similar results from the smaller Tanzania Craton. Thus, despite its great age and thickness, the base of cratonic lithosphere shows evidence for modification by the general mantle circulation and/or mantle plumes.

Sleep (2002) modelled heat and mass transfer at the lithosphere–asthenosphere boundary, and suggested that the base of the lithosphere coincides with an increase in viscosity, thereby explaining the longevity of cratonic roots within a convecting, cooling Earth. Basal drag on the plate and convection interact, if a nonlinear viscosity is assumed for the lithospheric thermal boundary layer at the base of the lithosphere. His work shows that smaller cratons will have more heating by stirring, and will tend to lose their cratonic roots over long times. Forte & Perry (2000) inverted global plate convection data to constrain the relationship between thermal and chemical buoyancy in the tectosphere. These two factors appear to be in near equilibrium, stabilizing cratons over geological time.

Throughout geological history mantle plumes have impinged episodically on continental lithosphere; they may contribute a large portion ($c.$ 7 mW m^{-2}) of the global average mantle heat flux (e.g. Davies 1988). An Archaean craton may have been affected by mantle plumes many times during its history. Heating occurs both by ponding of plume material and through mixing caused by the lateral flow of plume material before impingement.

The record of larger melt volumes associated with these plume events is clear in cratonic crust and in the orogenic belts surrounding cratons. Cratons worldwide show large dyke swarms, such as the Proterozoic Mackenzie dyke swarm in the Superior Craton (e.g. Heaman & LeCheminant 1988). Several flood basalt provinces are found in the orogenic belts adjacent to Archaean cratons, such as the Late Proterozoic Keweenawan basalt province along the Superior Craton margin, the Karoo flood basalt province along the Kaapvaal Craton margin, and East African flood basalt province along the northeastern Tanzania Craton margin. How have these cratonic roots persisted through geological time in the presence of mantle plumes, and what parts of cratons are susceptible to melting in the presence of plumes?

Additional insights into the modification of cratonic lithosphere by mantle plumes come from upper-mantle anisotropy patterns. Observations of shear-wave splitting routinely are used to study upper-mantle anisotropy (e.g. Silver 1996), but there is always some ambiguity in the interpretation of this splitting as to where the anisotropic region lies. Much of the debate centres on whether or not the crust is coupled to the mantle during orogenic episodes. Silver (1996) referred to the coupled case as vertically coherent deformation and the decoupled case as simple asthenospheric flow. In the vertically coherent case the anisotropy is thought to be frozen into the rigid lithosphere and crust at the time of orogeny. In the simple asthenospheric flow case the anisotropy reflects the direction of mantle flow beneath the lithosphere. In reality there are probably contributions from both regions (e.g. Kay et al. 1999).

Fouch et al. (1999) compared model simulations of asthenospheric flow and seismic observations, and found that a significant proportion of the shear-wave splitting can be explained by the deflection of asthenospheric flow around a cratonic root. In our study we are interested in the extra complication of anisotropy in flowing mantle plume material that lies beneath and around continental keels.

We design our studies of plume–lithosphere interactions to: (1) predict the distribution in space and time of hot, buoyant plume material beneath cratons of various shapes; (2) determine the physical conditions favourable for the lateral distribution of plume material beneath cratonic keels, which may give rise to small-volume melts (e.g. kimberlites); (3) evaluate the longevity of cratonic keels beneath large and small cratons; (4) predict the behaviour of viscous plume material at the edges of the plume in terms of thickness and temperature. As we show below, a significant thickness of plume material flows beneath a cratonic keel only where the plume rises beneath part of the craton, providing a viable mechanism for the emplacement of kimberlites.

We summarize the results of several plume simulations for a small craton, using the approach of Sleep (1997). We then analyse the effect of plume material emplacement beneath cratonic keels and channels of different dimensions. This work presents new simulations of plume flow beneath the African continent that include deep keels beneath Archaean cratons, which were not considered in the simple models of Ebinger & Sleep (1998). Flow velocities and strains predicted from our preferred model allow us to estimate the magnitude and direction of SKS splitting by plume flow around a cratonic keel. These patterns are then compared with SKS splitting patterns from 'normal' mantle flow around a keel (e.g. Fouch et al. 1999).

Model of plume–lithosphere interactions

The past 5 years have seen considerable progress in the development of whole-earth mantle convection models that show excellent correlations with subduction and mid-ocean ridge patterns, allowing us to begin to model the interactions between asthenospheric flow and the heterogeneous plates moving in response to this flow. Doin et al. (1997) examined controls of asthenospheric flow on plate thickness, and considered thick cratonic roots. Their 3D simulations indicate that cratonic roots are stable only if the root is buoyant and more viscous than 'normal' continental lithosphere. Sleep (1994, 1997) has used these 3D models to parameterize a numerical model of plume–lithosphere interactions that allows for movement of the lithosphere over a stationary plume, and tracks thickness changes in the lithospheric lid caused by mantle plume heating.

Hot, low-density material within starting mantle plumes rises diapirically through the mantle and interacts with the base of the lithosphere, forming a broad head that may partially melt to produce flood basalts and radial dykes (e.g. Griffiths & Campbell 1991; Ribe & Christensen 1994). This buoyant material displaces denser, normal asthenosphere and heats the lithosphere, leading to regional uplift, lithospheric thinning and pressure release magmatism. As the plate moves over the hotspot the lithosphere cools and returns to a stable thickness. General models differ in the interactions between plume and lithosphere and the timing of pressure release volcanism (e.g. Richards et al. 1989).

Plume head flow

We consider the case where a large but finite volume of hot, lower-density material impinges on the base of the lithosphere over a brief time period; for simplicity, we ignore the contributions of additional plume tail material (Fig. 1). The hot plume material transfers heat to the base of the lithosphere, leading to an increase in viscosity within the underlying plume material, which slows the flow. Geometrical spreading of the plume material thins the layer to the point that viscous forces greatly retard flow, and the plume material cools like a static layer emplaced

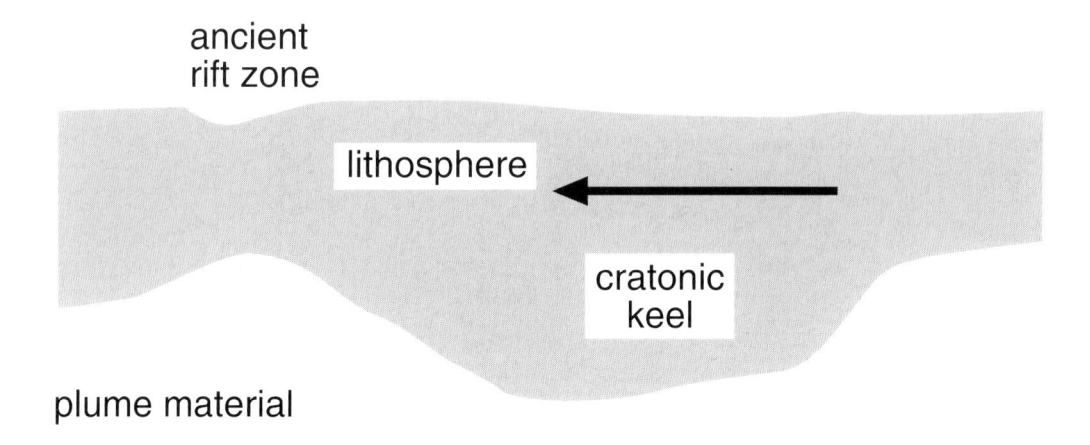

Fig. 1. Schematic diagram illustrating interactions between a mantle plume and a cratonic keel. Hot, buoyant plume material will pond in pre-existing thin zones, and be deflected by cratonic keels.

beneath the lithosphere, analogous to a sill. It is this thickness of sluggish plume material that is of most interest, as it allows one to then calculate the amount of heat transferred to the cratonal lithosphere, expected uplift and geoid anomalies, and variations in seismic velocity. Below we consider the spreading of material along a flat, elongate channel (e.g. thinned lithosphere beneath a rift system), beneath a small, flat-bottomed craton with steep edges (e.g. Tanzania Craton), and a craton that is larger than the plume head (>1000 km, e.g. Superior Craton).

We obtain the final thickness of ponded plume material following Huppert (1982) for four geometrically ideal cases: ponding of radially symmetric plume material beneath a very wide craton, ponding beneath a craton of limited area, 2D ponding where material flows outward along a channel of constant width, and 2D flow from a craton of limited width. The objective is to obtain the dimensional dependence of the thickness of plume material on physical parameters.

As a first guess, a starting plume may be represented by a volume Q of material ponding beneath a flat base of the lithosphere. Initially the material has radius R_0 and central height H_0; that is, dimensionally

$$Q = R_0^2 H_0. \quad (1)$$

To use the analytical (lubrication theory) results of Huppert (1982), we represent the plume material as an equivalent layer of constant viscosity η and constant specific weight contrast $\Delta\rho g$. We ignore viscous forces associated with downward flow of the normal mantle at the edges of the blob. The appropriateness of this assumption is discussed in the Appendix. The thickness of the material is $h(R)$ where R is the radial distance from the centre of the blob. Outward flow of plume material is driven by the local slope at its base. The flux (in m³ s⁻¹ per circumferential length of the flow front) is analytically

$$F = \frac{-h^3 \Delta\rho g}{12\eta} \frac{\partial h}{\partial R} \quad (2)$$

where the constant 12 arises from the assumption that the plume material is much less viscous than the underlying and overlying material (Huppert (1982) modelled the laboratory case where the upper boundary is a free surface and the constant is three), and the minus sign indicates that flow is from thick to thin plume material.

The height of plume material at a distance R may increase or decrease with time depending on whether the volume of material carried per time across a radius $2\pi RF$ increases or decreases with distance; that is, the change in $2\pi RF$ across an annular ring of dR is balanced by a change in the volume $2\pi Rh\, dR$ of plume material in the ring. Mathematically, this is equivalent to taking the divergence of F in cylindrical coordinates to obtain

$$\frac{\partial h}{\partial t} = \frac{\partial}{R \partial R} \frac{Rh^3 \Delta\rho g}{12\eta} \frac{\partial h}{\partial R} \quad (3)$$

where t is time. This equation was solved analytically by Huppert (1982). Rather than repeating this bulky derivation, we obtain the result dimensionally to illustrate how the result arises and then supply the numerical constant obtained rigorously by Huppert (1982).

To evaluate (3) dimensionally, we let the central height h_0 and extent of the plume R_m replace the differentials:

$$\frac{h_0}{t} = R_m^{-2} \frac{h_0^4 \Delta\rho g}{12\eta} \quad (4)$$

where t is the time since the blob started to spread. Noting that the volume is by assumption conserved, $Q = R_m^2 h_0$ and solving yields

$$h_0 = 0.53 t^{-1/4} \left[\frac{12\eta Q}{\Delta\rho g} \right]^{-1/4} \quad (5)$$

where the constant $0.53 = (4\pi)^{-1/4}$ is given in equation (3.4) of Huppert (1982).

The case where the craton has a finite width, allowing the plume material to cascade off its sides, follows similarly. In this case, the plume material thickness remains zero at its edge; $h(R_0) = 0$. Equation (4) becomes

$$\frac{h_0}{t} = R_0^{-2} \frac{h_0^4 \Delta\rho g}{12\eta} \quad (6)$$

and solving yields

$$h_0 = 0.68 t^{-1/3} \left[\frac{12\eta R_0^2}{\Delta\rho g} \right]^{1/3} \quad (7)$$

where the constant was obtained by numerically solving (3).

Two-dimensional linear spreading of plume material may also occur. The first case is analogous to that in (5) in that the material is free to spread indefinitely in one direction. This could represent spreading of an elongate band of plume material beneath flat lithosphere, or the spreading of plume material along a flat elongate channel confined on the sides by thicker lithosphere. The initial half area of plume material in a cross-section in the direction of flow is $q = H_0 X_0$, where X_0 is the initial distance x that the plume material extends from its centre. The flux obtained in (2) still applies, replacing R with x. Equation (3) is replaced by that for linear flow

$$\frac{\partial h}{\partial t} = \frac{\partial}{\partial x} \frac{h^3 \Delta \rho g}{12\eta} \frac{\partial h}{\partial x}. \quad (8)$$

The dimensional equation (4) instead uses the width of the edge of the plume material x_m at time t

$$\frac{h_0}{t} = x_m^{-2} \frac{h_0^4 \Delta \rho g}{12\eta} \quad (9)$$

where t is the time since the blob started to spread. Noting that the volume is by assumption conserved, $q = x_m h_0$, and solving yields

$$h_0 = 0.842 t^{-1/5} \left[\frac{12\eta q^2}{\Delta \rho g} \right]^{1/5} \quad (10)$$

where the constant 0.842 is given in equation (2.12) of Huppert (1982).

The case for the edge of the plume material where $h = 0$ stays constant follows similarly. The thickness of the centre of the plume is

$$h_0 = 0.845 t^{-1/3} \left[\frac{12\eta X_0^2}{\Delta \rho g} \right]^{1/3} \quad (11)$$

where the constant was obtained by numerically solving (8). It should be noted that the maximum thickness obtained from (7) in radial geometry is similar to that obtained for linear flow in (11). That is, the thickness of plume material spreading off a small craton does not strongly depend on the shape of the craton but only on the minimum distance (width or diameter) across it.

Figure 2 shows the thickness of spreading plume material for the cases of linear and radial flow over a finite region. The thicknesses illustrate a general feature also shared by the infinite radial and linear cases studied by Huppert (1982): the thickness is near its maximum expected value very near the edge because the gradient in (2) is large where the thickness is small. In fact, the assumption that the flow is essentially horizontal breaks down near the edge. Huppert (1982) addressed this issue by comparing the lubrication theory with experiments. (The issue of flow in the underlying mantle near the edge is addressed in the Appendix.) In fact, the behaviour near the edge has little effect on the rest of the flow and the theory gives a good representation of observations. This feature applies to a substance with a nonlinear rheology. For example, in the limit of

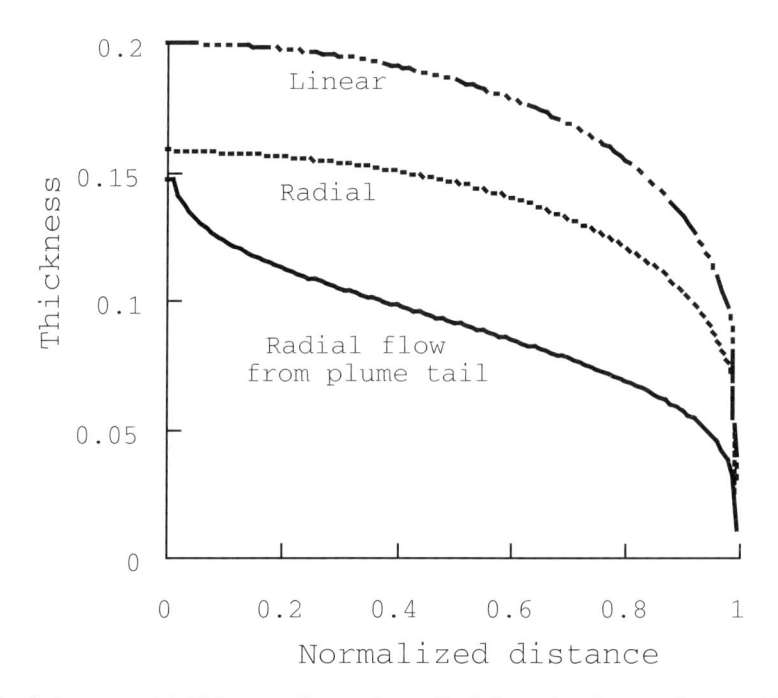

Fig. 2. Graph of plume material thickness v. distance (normalized) from plume's centre for several lithospheric geometries. For linear flow (e.g. along a channel at the LAB below a pre-existing rift zone) and radial flow (e.g. plume arising beneath a flat-keeled craton) material from the plume head is relatively uniform in thickness, except at its edges where viscosity increases associated with cooling retard flow. The shape of the 'radial' curve is similar for topography of limited (small craton) or infinite extent (large craton). Unlike the top two cases, the flow from the plume tail thins more rapidly with distance from the centre (continuous line).

perfect plasticity, the shear traction, dimensionally $\Delta\rho g h(\partial h/\partial x)$, is constant and the thickness h scales to the square root of the distance from the edge. For application to the Earth, we can conclude that ponded plume material beneath a flat craton is near its maximum thickness except right at its edges.

Summarizing, cooling of plume material and consequent increase in viscosity near the edge of the plume has little effect on the rest of the flow. This result implies that plume material maintains a relatively constant thickness beneath a craton, except very near its steep edges.

How far can plume material spread before it cools and becomes sluggish? Initially, the plume material is thick and spreads rapidly. Little cooling occurs away from the boundaries of the plume material and the centre of the ponded layer remains hot and fluid. Eventually, the plume material becomes thin enough that conduction becomes important. The time for a thickness of material to cool is

$$t_\kappa = C_\kappa h_0^2/\kappa \qquad (12)$$

where κ is thermal diffusivity, h_0 is initial plume thickness and C_κ is a constant of order one. The value of C_κ can be constrained to some extent by noting that the viscosity of mantle material is strongly temperature dependent. Cooling of the material by 40–100°C will increase viscosity by a factor of e. This is a fractional cooling of around 15–50% of the difference between plume material and ordinary adiabatic material, and it is a smaller fraction of the difference between plume material and the overlying lithosphere. Cooling needs to penetrate from the top and bottom to the centre of the plume material, with a distance scaling to $h_0/2$ significantly retarding flow. In general, the constant C_κ includes the effects of heat transfer by small-scale convection, the actual temperature variations within the plume material, and the details of the thermal structure of the overlying lithosphere.

The transition from spreading to stagnation by conductive cooling occurs when the time for material to spread to thickness h_0 exceeds the time for the material of that thickness to cool. For example, the spreading time for material spreading radially over a flat surface is

$$t_\eta = (0.53)^4 h_0^{-4} \left[\frac{12\eta Q}{\Delta\rho g} \right] \qquad (13)$$

where Q is volume of starting plume material of constant viscosity η. As the viscosity of plume material is strongly temperature dependent, spreading will become sluggish after a time scaling to t_κ, but be fast before that; that is, the transition from fast to sluggish spreading occurs approximately when $t_\eta = t_\kappa$. Solving yields the thickness of plume material h_0 present at that time for each of the special cases listed above. The thickness for infinite radial flow is

$$h_{\text{pond}} = 0.53^{4/6} \left[\frac{12\kappa\eta Q}{C_\kappa \Delta\rho g} \right]^{1/6}. \qquad (14)$$

The thickness for finite radial flow is

$$h_{\text{pond}} = 0.680^{3/5} \left[\frac{12\eta R_0^2 \kappa}{C_\kappa \Delta\rho g} \right]^{1/5}. \qquad (15)$$

The thickness for infinite linear flow is

$$h_{\text{pond}} = 0.842^{5/7} \left[\frac{12\eta\kappa q_2}{C_\kappa \Delta\rho g} \right]^{1/7}. \qquad (16)$$

The thickness for finite linear flow is

$$h_{\text{pond}} = 0.854^{3/5} \left[\frac{12\eta X_0^2 \kappa}{C_\kappa \Delta\rho g} \right]^{1/5}. \qquad (17)$$

One can conclude from these considerations that 'chilling' of spreading plume material is only weakly dependent on material properties, with an order of magnitude increase in viscosity corresponding to an increase in plume material thickness of 1.39 for finite radial flow. For example, if the viscosity was an order of magnitude higher, the estimated thicknesses would increase by factors of 1.58–1.39. The estimates are weakly dependent on the assumed initial conditions, provided that the initial plume material thickness is somewhat more than the final one, because the initial spreading of the thick plume material is fast, but becomes slow when the plume material is near its final thickness at time t_η. In all these cases the net heat anomaly in a column is dimensionally $h_{\text{pond}} \rho C_p$ where ΔT_p is the initial temperature anomaly of the plume material and ρC is volume specific heat. That is, the equivalent thickness h_{pond} includes heat both in the sluggishly flowing plume material, and heat transferred already to the lithosphere over the time since flow started. An additional unknown multiplicative factor of order one arises from this effect, which we do not explicitly model.

For purposes of computation we used parameters obtained by Ebinger & Sleep (1998) for the East African plume. The material properties assumed are a viscosity, $h = 0.3 \times 10^{18}$ Pa s thermal diffusivity $\kappa = 0.8 \times 10^{-6}$ m^2 s^{-1}, and specific weight contrast 200 N m^{-3}. The scaling constant is $C_\kappa = 1/16$. The starting plume head volume in the infinite radial model is $Q = 0.2 \times 10^{18}$ m^3, which yields $h_{\text{pond}} = 39$ km. The finite widths R_0 and X_0 are assumed to be 250 km. This yields thicknesses of 21 and 25 km, respectively. The area q in the infinite linear model is assumed to be 10^{11} m^2, which yields a thickness of 51 km. From this, we can conclude that the final thickness of plume material is tens of kilometres and not drastically dependent on assumptions of the flow geometry over a flat surface. Conversely, if there is significant local relief on the base of the lithosphere, local closed thin spots will act as traps (e.g. Thompson & Gibson 1991; Bank et al. 1998). Local relief will dominate ponding if the traps are greater than the plume material thickness.

The effective heat flow transferred to the craton from the ponded plume material over geological time can be estimated. For example, if we assume a 40 km equivalent thickness of 200 K excess plume temperature ponds on average at a spot under the craton every 300 Ma and that the volume specific heat is $4 \times 10^6 \text{ J K}^{-1} \text{ m}^{-3}$, then the average additional heat flow is 3.4 mW m^{-2}. This is a modest part of typical mantle heat flows from cratonal regions (e.g. Jaupart et al. 1998).

Plume tail contributions

The steady-state solution for material fed by a plume tail is also obtained following Huppert (1982). For simplicity we centre the plume tail within a flat-bottomed region of radius R_0, which represents a small craton. The steady-state solution of (3) where the plume material thickness is zero at $R = R_0$ is

$$h = h_1 [\ln(R_0/R)]^{1/4} \quad (18)$$

where h_1 is a constant with dimensions of length. This equation holds best near the leading edge of the material. It holds approximately beyond a distance R_p from the centre of the plume where flow becomes horizontal rather than the vertical upward conduit flow at the edge R_0. The flux of material outward from radius R is

$$\chi = -\frac{2\pi R h^3 \Delta \rho g}{12 \eta} \frac{\partial h}{\partial R} = \frac{\pi h_1^4 \Delta \rho g}{24 \eta} \quad (19)$$

which is expected to maintain steady state and is independent of radius R. The thickness of plume material at the distance R_p at the edge of plume conduit is

$$h(R_p) = \left[\frac{24 \chi \eta}{\Delta \rho g \pi}\right]^{1/4} \left[\ln\left[\frac{R_0}{R_p}\right]\right]^{1/4} \quad (20)$$

which is weakly dependent on the ratio of the distances and hence the uncertain definition of R_p. For the parameters used above, the flux $\chi = 200 \text{ m}^3 \text{ s}^{-1}$ for the East Africa plume of Ebinger & Sleep (1998), and $R_p = 50 \text{ km}$, the thickness $h(R_p)$ is 39 km, which is within the range of estimates obtained above. An order of magnitude increase in viscosity would increase the estimate by a factor of 1.78.

Summarizing, we first state two intuitive inferences, which we illustrate below using East Africa as an example.

(1) lithospheric thinning and, subsequently, melt production focus along steep gradients at the LAB (e.g. Fig. 1); compositional differences between cratonic and non-cratonic lithosphere could enhance the focusing effects.

(2) plume processes can effectively erode the margins of cratons, laterally reducing their breadth.

Our analytical models provide inferences on the behaviour of ponded plume material.

(1) Plume material does not pond beneath the cratonic keel unless the centre of the plume is located beneath the cratonic root. These relations hold for large and small cratons. Thus, in most cases the cratonic keel effectively deflects hot plume material to thinner lithosphere, providing a mechanism for the preservation of cratonic keels.

(2) Where a plume arises beneath a craton, plume material will attain thicknesses of tens of kilometres, raising heat transfer to the base of the lithosphere by c. 20% relative to normal input (e.g. Jaupart et al. 1998). Plume material maintains a relatively constant thickness beneath a craton, except very near its steep edges.

(3) Larger cratons will be more susceptible to rifting as a result of basal drag at the LAB. Variations in lithospheric thickness within a large craton may lead to ponding of plume material, generating extensional body forces within the cratonic interior.

Application to East Africa

This paper compares model simulations of mantle plumes placed near and beneath the small (c. 500 km wide) Tanzania Craton (Fig. 3). One or more mantle plumes have impinged on the lithosphere beneath East Africa during the past 45 Ma (e.g. George et al. 1998; Nyblade et al. 2000). Only small volumes of melt have erupted in the Western rift, whereas much larger melt volumes cover large parts of the Eastern and Ethiopian rift systems and surrounding uplifted plateaux (Fig. 3). Of specific interest to this study of plume–cratonic keel interactions, the Western and Eastern rift systems have developed in the Proterozoic orogenic belts ringing the narrow Tanzania Craton since Miocene time (Fig. 3).

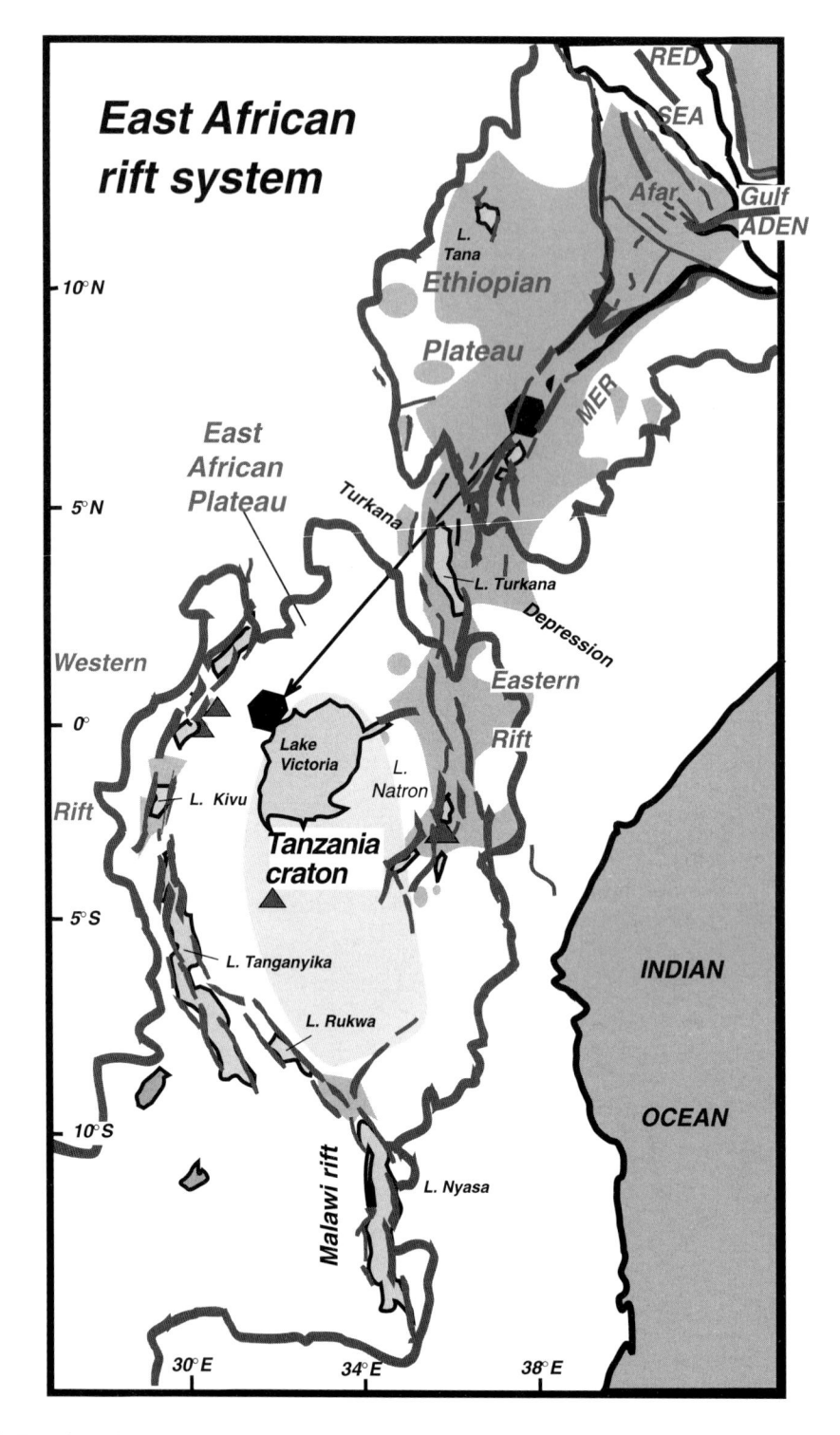

Fig. 3. Location of Tanzania Craton (Archaean), Oligocene–Recent Red Sea and Gulf of Aden rifts, and Miocene–Recent Ethiopian, Western and Eastern rift systems (after Ebinger *et al.* 1999). Dark shading encloses elevations above 1000 m; lighter shading shows Late Eocene–Recent volcanic provinces. Hexagons mark the assumed path of the African plate over a mantle plume (from George *et al.* 1998). ▲, locations of Cenozoic kimberlites and carbonatites. MER, Main Ethiopian rift.

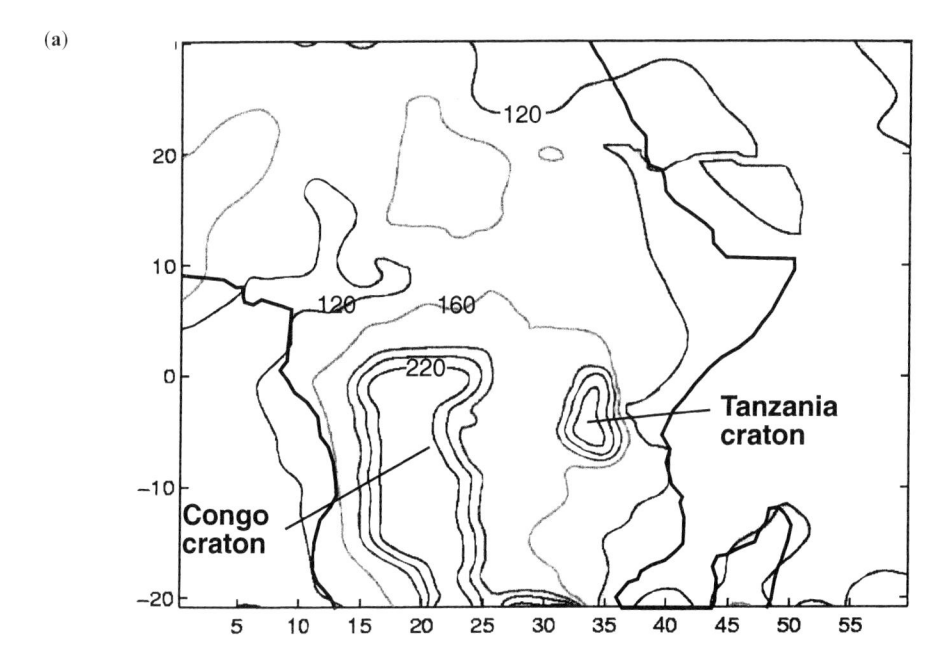

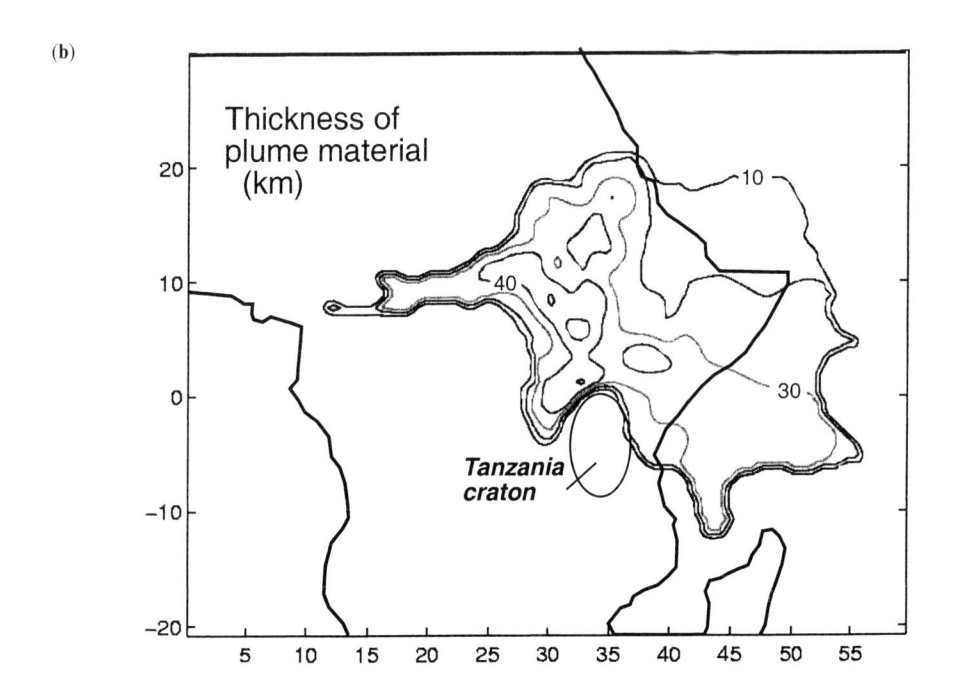

Fig. 4. (a) Model of pre-plume lithospheric thickness at c. 40 Ma (before the separation of Arabia and opening of Red Sea and Gulf of Aden) with present-day outline of African continent overlain. Cratonic thickness is 220 km, with constraints on lithospheric thickness summarized by Ebinger & Sleep (1998). (b) Contours of plume material 40 Ma after plume impingement, or present day, predicted by our model interactions of a mantle plume with the lithospheric lid shown in (a). It should be noted that our models do not include the lithospheric stretching in the Red Sea, Gulf of Aden and East African rift system that occurred as the plume material spread beneath East Africa. This stretching would focus melt towards zones of stretching, enhancing melting beneath the rift systems.

We assume that a single plume impinged on the lithosphere beneath central Ethiopia, and that the African plate has moved slowly NNE over the stationary plume since 40 Ma (George et al. 1998) (Fig. 4). Below we discuss the general results, rather than comparison with observations from East Africa, which have been summarized by Ebinger & Sleep (1998). Specifically, we use these models to evaluate the effects of plume placement and cratonic size on the distribution of plume material susceptible to melting.

Tomographic models show >200 km thick lithosphere beneath the kimberlite-rich Tanzania Craton, with steep gradients at the lithosphere–asthenosphere boundary (LAB) along its margins (Ritsema et al. 1998). We have used seismic, heat-flow and subsidence models of African lithosphere to construct a model of lithospheric thickness before plume impingement, updating the simplified model of Ebinger & Sleep (1998) to include cratonic keels imaged in tomographic models (Ritsema et al. 1998; Ritsema & van Heijst 2000). Cratonic keels beneath the Tanzania and Zaire Cratons would have caused significant variations in the depth to the 1300°C isotherm (base of thermal lithosphere) before onset of a plume in Cenozoic time, yet their effects have not been considered in plume models. In addition, incomplete thermal equilibration of thinned lithosphere beneath Late Cretaceous–Paleogene rifts and Mesozoic passive margins produced topographic relief at the LAB. This topographic relief disrupts the radial spread of buoyant plume material, leading to ponding below thinner lithosphere

We feel that the melting process is too poorly parameterized to do more than make spatial comparisons of small- v. large-volume melting. (The models, which represent the plume material as an equivalent lubrication theory layer, are not suited for studying the details of melting.) Although not calculated in these models, we can predict that melting is enhanced where plume material cascades over steep LAB relief (Sleep 1994; Ebinger & Sleep 1998) (Fig. 4). The steep gradients at the LAB along the Tanzania Craton–mobile belt boundaries focus flow and decompression melting, explaining <30 Ma carbonatitic and kimberlitic magmas derived from small melt volumes at depths <150 km (e.g. Chesley et al. 1999) (Figs 3 and 4).

Our models include the movement of the lithosphere over a stationary hotspot, thereby allowing us to evaluate the effects of plume proximity on the distribution of plume material susceptible to melting. We have considered models with 150 km (Ebinger & Sleep 1998) and 220 km thick cratonic keels, and varied the location of the plume stem relative to the craton. From these tests, another clear result is that plume material does not pond beneath cratonic keels unless the plume stem (or part of the initial plume head) lies beneath that part of the craton. Below we describe the behaviour of the material flowing beneath small and large cratonic keels.

A north–south cross-section through the Tanzania Craton and adjacent younger lithosphere illustrates the effects of plume proximity on thermal thinning of the lithosphere (Fig. 5). Over 45 Ma the plate moves northward over the plume, which now lies beneath the northern margin of the Tanzania Craton (e.g. George et al., 1998). South of the plume stem, we see that the cratonic lithosphere cools and thickens (ignoring frictional heating and secondary convection), whereas north of the craton, the lithosphere thins. More importantly, the steep gradient along the craton margin steepens as the breadth of the keel decreases; the remaining keel serves as a barrier to lateral flow of plume material (Fig. 4b).

We present no new information to resolve the issue of the location of plume heads and/or plume stems beneath the East African rift system (e.g. George et al. 1998; Nyblade et al. 2000); our studies are designed to understand the physics of plume flow beneath cratons. Nyblade et al. (2000) noted a depression of the 410 km discontinuity beneath the eastern side of the Tanzania Craton and the Eastern rift system, and suggested that a >200 km thick plume head underlies this area. Our analytical studies of plume material flow shows that a thin (c. 50 km thick) layer of hot plume material would underlie a craton, but that this layer could be thicker at the craton margin. Conduction of heat from the plume stem over the last c. 45 Ma is expected to have produced a thermal halo of warm mantle. The passage of the starting plume head at c. 45 Ma would leave a broad region of warm mantle in its wake. Both these effects depress the 410 km discontinuity beneath East Africa without requiring a layer of plume material of 200 km or more thickness as suggested by Nyblade et al. (2000).

Upper-mantle anisotropy and shear-wave splitting

The primary mechanism for seismic anisotropy in the mantle is thought to be due to the lattice preferred orientation (LPO) of olivine and other dominant minerals. Numerical predictions of mantle flow-fields can be used to predict LPO as a result of dislocation mechanisms and hence the seismic anisotropy (e.g. Blackman et al. 1996;

Fig. 5. Thermal evolution of the lithosphere along a cross-section of Africa through 34°E. Arrow shows the location of the plume relative to the northward-moving African plate since 45 Ma. It should be noted that lithosphere cools as $\sqrt{t}$ south of plume, as we have placed no restriction on maximum thickness of the continental lithosphere. The long-term effect of the plume heating can be crudely estimated, if we assume that a 40 km equivalent thickness of material that is 200 K hotter than normal mantle, and with specific heat 4×10^6 J K^{-1} m^{-3}, ponds beneath a craton every 300 Ma. The mantle heat flow is increased by 3.4 mW m^{-2}, or 20–25% of typical mantle heat flow from cratonal areas (e.g. Jaupart et al. 1998). (b) Thickness of plume material ponded beneath lithosphere 45 Ma after plume onset.

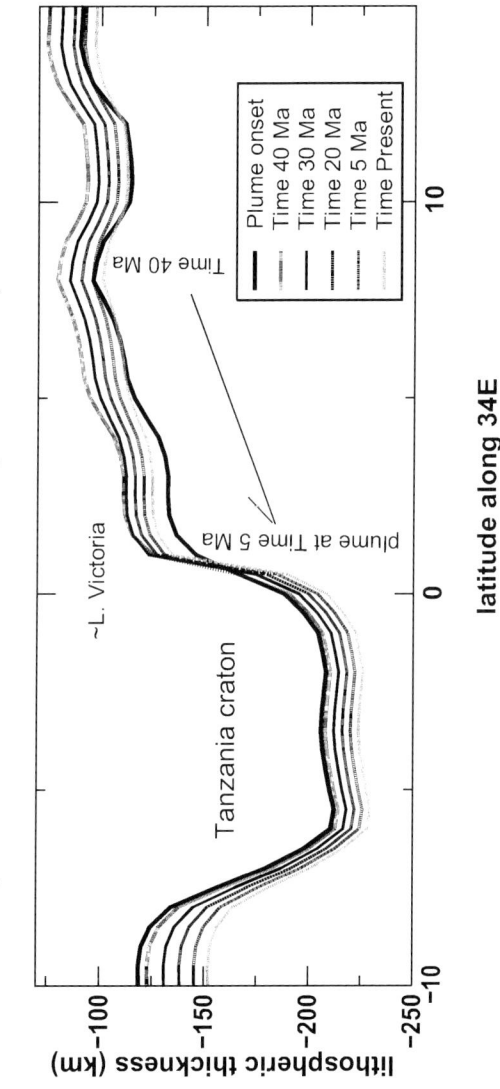

Tommasi 1998). In our study of plume flow around the Tanzania Craton we used the results of Blackman et al. (1996) as constraints on the scaling between the flow vectors and the anisotropy (effective elasticity). This then allowed us to model shear-wave propagation through the model. The anisotropy is dependent on the magnitude and the orientation of the flow, which in our case is plume material deflected by cratonic keels. We use the predicted elastic constants from Blackman et al. (1996) in the scaling. The maximum anisotropy in the Blackman models is tied to the maximum flow magnitudes. Although the aggregate elasticity of the Blackman models is in general triclinic in symmetry, we used the best-fitting hexagonal symmetry (Kendall et al. 1998; Blackman et al. 2002) in our calculations. This simplifies the wave-propagation modelling and does not change the predictions appreciably.

Vertically travelling shear waves (i.e. SKS phases) are then traced through the model using anisotropic ray theory (Kendall & Thomson 1989). The accrued splitting is proportional to the magnitude of the anisotropy (hence flow velocity) and the thickness of the plume material. Figures 6 and 7 show the predicted shear-wave splitting in the plume material and the polarization of the fast shear wave, respectively. The splitting is highest near the edge of the keel and the azimuth of the splitting follows the pattern of plume flow, thus implying that observations of SKS splitting could illuminate the flow patterns around the keel.

We tested the sensitivity of the splitting to the scaling adopted. The orientation of the shear-wave polarizations will remain consistent, but the magnitude of the splitting could vary. What the modelling provides is a prediction of the pattern in splitting that should be observed around the edges of the keel. The results from seismic studies under way in the East African rift system (Nyblade, pers. comm.; Stuart, pers.

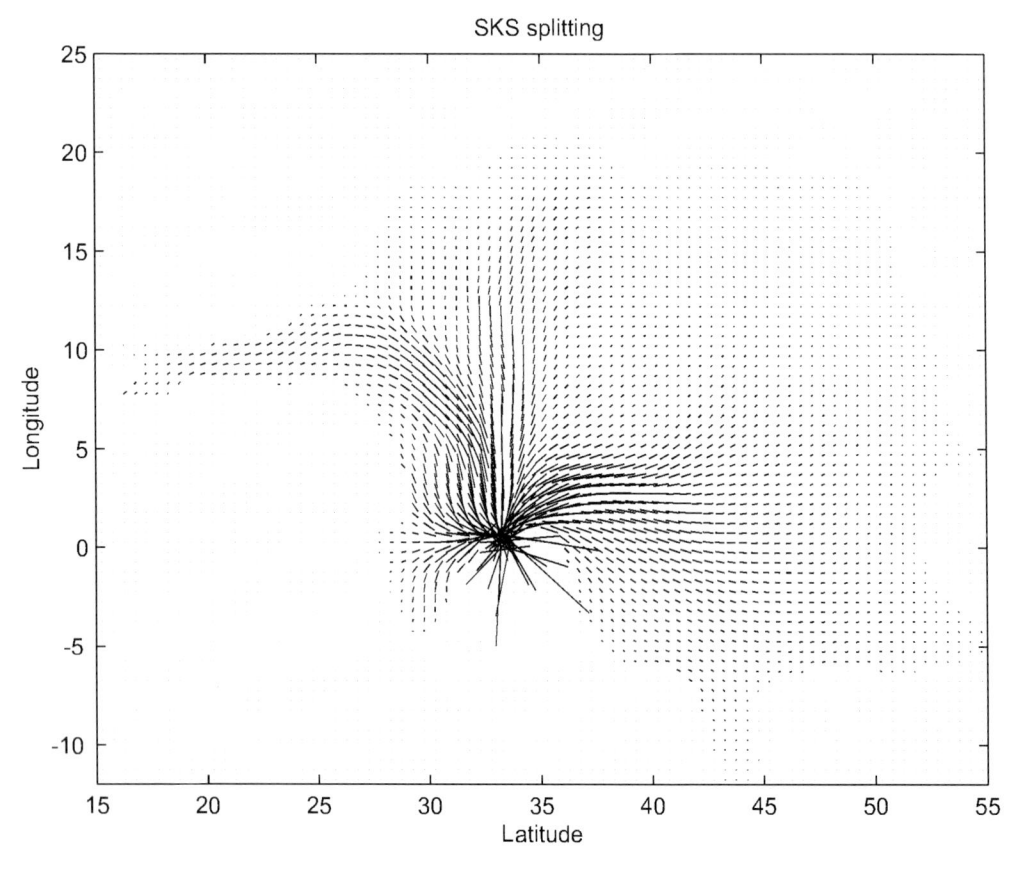

Fig. 6. Predicted SKS splitting as a result of deformation of plume material. The vectors (lines) show the polarization of the fast shear wave. The length of the lines is proportional to the magnitude of the separation between the fast and slow shear wave (see Fig. 7 for scale).

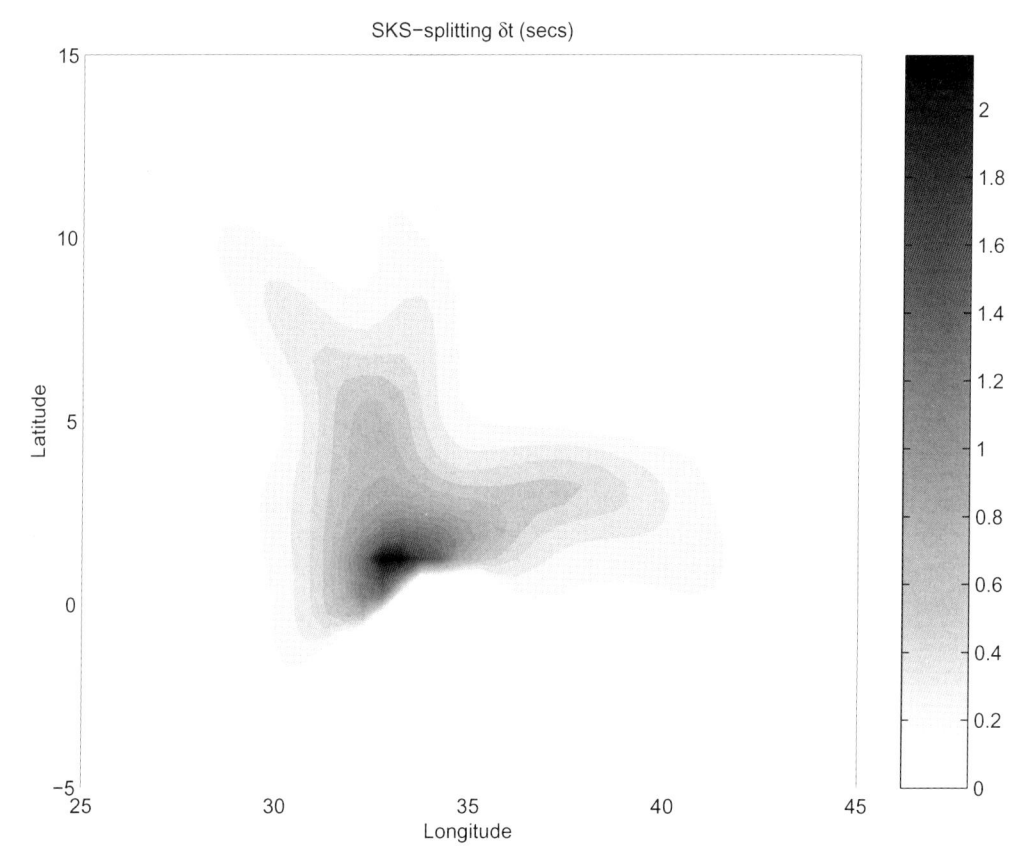

Fig. 7. Predicted magnitude of SKS splitting accrued through deformed plume material. (Note the region of >1 s splitting along northern craton margin.)

comm.) will provide a means for investigating this effect. There is the distinct possibility that there will also be contributions to the splitting from anisotropy in the overlying lithosphere. Local seismic events that originate in the lithosphere could be used to estimate the contribution from lithospheric anisotropy, and remove this effect to observe deep flow.

Conclusions

Numerical and analytical models of the vertical and lateral flow of hot, buoyant plume material beneath variable thickness continental lithosphere show that steep gradients at the lithosphere–asthenosphere boundary (LAB) strongly influence the distribution of plume material, and, consequently, the location and volume of melt. Cratonic keels deflect the plume material, with only minor thinning and heating of the mantle lithosphere beneath the cratons, unless the plume head lies beneath the craton. Clearly, the rate of movement of the plate over the mantle plume will also affect the amount of cratonic keel erosion by plume heating. Using these flow models to predict the preferred orientations of olivine, we predict that shear-wave splitting (SKS) will be greatest along the margins of the cratonic keels. The buoyancy-enhanced deformation in our models effectively generates LPO in olivine within the plume layer. Patterns of SKS splitting that reflect the resulting seismic anisotropy may illuminate high-deformation zones around the keel.

These studies have implications for the stability of cratons and diamond exploration. The steep gradients focus flow and adiabatic decompression melting along cratonic boundaries where extensional body forces arising from the density contrast between normal asthenosphere and hot plume material will be the largest, indirectly focusing tectonism. This in turn further shields the cratonic cores, although lateral erosion of the cratonic root may occur, depending on the location of the mantle plume. Where a plume underlies the deep keel of the craton, a thin sheet of plume material spreads beneath the keel. The lateral dimensions of the cratonic root are not a critical factor. Depending on composition of the overlying mantle lithosphere, the presence of the hot, enriched plume material could lead to small degrees of melting, providing one mechanism for the production of diamondiferous kimberlites.

This work was supported by NERC grant GR9/03469 and NSF grant EAR-0000747. Our studies benefited from discussions with H. Helmstaedt, A. Nyblade and W. Mooney. We thank R. Griffiths for a constructive review and for reanalysis of laboratory results on inviscid blobs, and an anonymous reviewer.

Appendix: Flow of normal mantle at the leading edge

In the lubrication theory results presented in this paper, we ignored viscous forces within the normal mantle that is displaced downward at the leading edge of the blob. Comparison of scaling results from lubrication theory with numerical models of midplate swells (Ribe & Christensen 1999) and on-axis hotspots (Albers & Christensen 2001) gives some confidence in this approach. However, forces within the displaced normal mantle become significant if it is much more viscous than the plume material viscosity. We obtain scaling relationships for when the lubrication theory is applicable to ponded plume material. We consider an equivalent chemical plume and do not attempt to refine the scaling criterion for an actual thermal blob.

We begin with a case considered by Griffiths & Campbell (1991) where the blob is inviscid and all the resistance to flow is within the displaced normal mantle. We modify their derivation to aid comparison with lubrication theory results.

We obtain dimensional results by equating the change in potential energy of the blob with viscous dissipation. The volume of a radial blob Q is again given by (1). Its thickness is essentially h_0 except within a horizontal distance of $c.h_0$ of the leading edge. The potential energy is dimensionally

$$W = \Delta\rho g R_m^2 h_0^2 = \Delta\rho g Q h_0. \quad (A1)$$

The change in gravitation potential with time is

$$\frac{\partial W}{\partial t} = \Delta\rho g Q \frac{\partial h_0}{\partial t} = \Delta\rho g \frac{Q^2}{R_m^3} \frac{\partial R_m}{\partial t} \quad (A2)$$

where the second dimensional equality is obtained using $Q = h_0 R_m^2$. The velocity of the normal mantle near the leading edge scales as the rate of advance of the edge $\partial R_m/\partial t$. The strain rate scales as

$$\varepsilon = \frac{1}{h_0}\frac{\partial R_m}{\partial t} \quad (A3)$$

over a dimensional volume of $h_0^2 R_m$. The viscous dissipation within this volume equals the change of gravitational potential with time in (A2)

$$\frac{\partial W}{\partial t} = \eta_N R_m \left[\frac{\partial R_m}{\partial t}\right]^2 \quad (A4)$$

where η_N is the viscosity of the normal mantle. Solving (A2) and (A4) yields the dimensional position for the leading edge

$$R_m = 0.51\left[\frac{t\Delta\rho g Q^2}{\eta_N}\right]^{1/5} \quad (A5)$$

which is (9a) of Griffiths & Campbell (1991) with the empirical coefficient supplied by Griffiths (pers. comm.). The thickness of the blob is dimensionally

$$h_0 = 1.54 Q^{1/5}\left[\frac{t\Delta\rho g}{\eta_N}\right]^{-2/5} \quad (A6)$$

which is equation (9b) of Griffiths & Campbell (1991) with the empirical coefficient supplied by Griffiths (pers. comm.).

The relationship between volume, maximum thickness and radius gives a measure of the flatness of the blob away from its leading edge. That is, the volume is a constant times $\pi R_m^2 h_0$. The constant is unity for a flat blob with vertical sides, 0.79 for the inviscid blob, and 0.67 for the lubrication theory blob (Huppert 1982). The central region of the inviscid blob is flatter than that of the lubrication theory blob as expected, but the difference is not large.

We are now ready to formulate a criterion for when lubrication theory is applicable. We assume that it is and obtain the total viscous dissipation. The shear traction at the base of the blob is then

$$\tau = \Delta\rho g \frac{h_0^2}{R_m} \quad (A7)$$

and the flow front velocity is dimensionally

$$V \equiv \frac{\partial R_m}{\partial t} = \Delta\rho g \frac{h_0^3}{\eta R_m}. \quad (A8)$$

The ratio of total dissipation in the normal mantle to that within the blob from (A2), (A7) and (A8) is

$$\Pi \equiv \frac{\eta_N V}{\tau R_m} = 1.2 \frac{\eta_N h_0}{\eta R_m} \quad (A9)$$

where the multiplicative constant is obtained by comparing the time derivative of h_0 in with that in (A6) at given h_0 and Q. The dimensionless variable defines three domains. When it is much less than unity, lubrication theory applies. This happens when the blob has spread enough that its short-to-long aspect ratio is less than the ratio of the viscosity of plume material to that of normal mantle. When the ratio is large, the inviscid blob theory applies. This occurs during the initial stages of spreading when the starting plume head impinges on the base of the lithosphere. A transition is implied when the ratio is near unity. Viscous forces within and outside the blob resist flow. Either theory can be used to obtain gross estimates as done in this paper. A quick approximate numerical method is available by computing the time derivatives of maximum blob thickness (or blob radius) for the inviscid blob and the lubrication theory blob at the instantaneous value of thickness and the constant value of the volume. The inverse of the actual time derivative is the sum of the inverses of the two derivatives. The method is not exact, in part, because the shape of an inviscid blob differs from that of a lubrication theory blob.

A similar derivation applies to the linear spreading of a blob. The position of the leading edge is

$$x_m = \left[\frac{t\Delta\rho g q^2}{\eta_N}\right]^{1/3} \quad (A10)$$

where the numerical constant has not been calibrated. The scaling criterion (A8) is obtained as before, but without a known numerical constant. The thickness of the plume material scales as

$$h_0 = q^{1/3}\left[\frac{t\Delta\rho g}{\eta_N}\right]^{-1/3}. \quad (A11)$$

It can be seen from (A5) and (A10) that the behaviour of an inviscid blob is qualitatively similar to that predicted by lubrication theory. The blob spreads quickly at first and slowly thereafter. The qualitative behaviour of a plume of plume material is understood from (A9). Initially, the vertical-to-horizontal aspect ratio of the blob is relatively large and the surrounding mantle resists flow. Eventually the blob becomes thin enough that lubrication theory applies.

In this paper, we are mainly interested in the blob when it is about to pond; that is, when it is thin and it has cooled somewhat so that its viscosity is higher than that of fresh plume material. These are the conditions for which lubrication theory applies. The blob is near this final state for much of its existence as the initial flow is fairly rapid. For somewhat cool plume material, which is a few tens of kilometres thick and has a spread over 1000 km, the criterion that Π in (A9) is less than unity is likely to be satisfied and the scaling results in the text apply.

As the viscosity of the normal mantle and that of plume material are unknown, we present scaling relationships for the spreading time of inviscid material. Truly inviscid material continues to spread forever. Here the plume material is fluid until conductive cooling increases its viscosity. Equating the cooling time in (12) to the spreading time from (A6) yields the ponding thickness for infinite radial flow:

$$h_{pond} = Q^{1/9}\left[\frac{\Delta\rho g}{\kappa\eta_N}\right]^{-2/9} \quad (A12)$$

where a constant analogous to C_κ is omitted. The ponding thickness for infinite linear flow is obtained from (12) and (A11):

$$h_{pond} = q^{1/5}\left[\frac{\Delta\rho g}{\eta_N}\right]^{-1/5}. \quad (A13)$$

The ponding thickness depends weakly on the volume of the blob and weakly on material properties including the viscosity of the normal mantle. This is similar to the behaviour of ponding thickness of lubrication theory blobs in (14) and (16).

References

ALBERS, M. & CHRISTENSEN, U. L. 2001. Channeling of plume flow beneath mid-ocean ridges, *Earth and Planetary Science Letters*, **187**, 207–220.

ARTEMIEVA, I. & MOONEY, W. 2001. Petrological constraints on lithospheric geotherms and implications for Archaean cratons. *Journal of Geophysical Research*, **106**, 16387–16414.

BANK, C.-G., BOSTOCK, M., ELLIS, R., HAJNAL, Z. & VANDECAR, J. 1998. Lithospheric mantle structure beneath the Trans-Hudson orogen and the origin of diamondiferous kimberlites. *Journal of Geophysical Research*, **103**, 10103–10114.

BLACKMAN, D. K., KENDALL, J.-M., DAWSON, P., WENK, H.-R., BOYCE, D. & PHIPPS MORGAN, J. 1996. Teleseismic imaging of subaxial flow at mid-ocean ridges: travel-time effects of anisotropic mineral texture in the mantle. *Geophysical Journal International*, **127**, 415–426.

BLACKMAN, D. K., WENK, H.-R. & KENDALL, J.-M. 2002. Seismic anisotropy of the upper mantle: 1. Factors that affect mineral texture and effective elastic constants. *Geochemistry, Geophysics, Geosystems*, (in press).

BOYD, F. R., GURNEY, J. J. & RICHARDSON, S. 1985. Evidence for a 150–200 km-thick Archaean lithosphere from diamond inclusion thermobarometry. *Nature*, **315**, 387–389.

CARLSON, R., PEARSON, D. G., BOYD, F. R., SHIREY, S. B., IRVINE, G., MENZIES, A. H. & GURNEY, J. J. 2002. Re–Os systematics of lithospheric peridotites: implications for lithosphere formation and preservation. *In*: GURNEY, J. J., GURNEY, J. L., PASCOE, D. & RICHARDSON, S. (eds) *Proceedings of the 7th International Kimberlite Conference*. Redroof Designs, Capetown, 99–108.

CHESLEY, J., RUDNICK, R. & LEE, C.-T. 1999. Re–Os systematics of mantle xenoliths from the East African rift: age structure and history of the Tanzanian craton. *Geochimica et Cosmochimica Acta*, **63**, 1203–1217.

DAVIES, G. F. 1988. Ocean bathymetry and mantle convection, 1, Large-scale flow and hotspots. *Journal of Geophysical Research*, **93**, 10467–10480.

DOIN, M.-P., FLEITOUT, L. & CHRISTENSEN, U. 1997. Mantle convection and stability of depleted and undepleted continental lithosphere. *Journal of Geophysical Research*, **102**, 2771–2787.

DOIN, M.-P., FLEITOUT, L. & MCKENZIE, D. P. 1996. Geoid anomalies and the structure of continental and oceanic lithospheres. *Journal of Geophysical Research*, **101**, 16119–16135.

EBINGER, C. & SLEEP, N. 1998. Cenozoic magmatism throughout east Africa resulting from impact of one large plume. *Nature*, **395**, 788–791.

EBINGER, C., JACKSON, J., FOSTER, A. & HAYWARD, N. 1999. Extentional basin geometry and the mechanical lithosphere. *Philosophical Transactions of the Royal Society, London*, Series A, **357**, 741–762.

FORTE, A. & PERRY, H. C. 2000. Geodynamic evidence for a chemically depleted continental tectosphere. *Science*, **290**, 1941–1944.

FOUCH, M., FISCHER, K., PARMENTIER, E. M., WYSESSION, M. & CLARKE, T. 1999. Shear wave splitting, continental keels, and patterns of mantle flow. *Journal of Geophysical Research*, **103**, 6255–6276.

GEORGE, R., ROGERS, N. W. & KELLEY, S. P. 1998. Two plumes beneath Africa. *Geology*, **26**, 923–926.

GRIFFITHS, R. W. & CAMPBELL, I. H. 1991. Interaction of mantle plume heads with the Earth's surface and onset of small scale convection. *Journal of Geophysical Research*, **96**, 18 295–18 310.

HEAMAN, L. M. & LECHEMINANT, A. 1988. U–Pb baddeleyite ages of the Muskox intrusion and Mackenzie dyke swarms, NWT. *Mineralogical Association of Canada*, **12**, 53.

HUPPERT, H. 1982. The propagation of two-dimensional and axisymmetric viscous gravity currents over a rigid horizontal surface. *Journal of Fluid Mechanics*, **121**, 43–58.

JAMES, D. E. & FOUCH, M. J. 2002, Formation and evolution of Archaean cratons: insights from southern Africa. *In*: FOWLER, C. M. R., EBINGER, C. J. & HAWKESWORTH, C. J. (eds) *The Early Earth: Physical, Chemical and Biological Development*. Geological Society, London, Special Publications, **199**, 1–26.

JAUPART, C. & MARESCHAL, J. C. 1999. The thermal structure and thickness of continental roots. *Lithos*, **48**, 93–114.

JAUPART, C., MARESCHAL, J. C., GUILLOU-FROTTIER, L. & DAVAILLE, A. 1998. Heat flow and thickness of the lithosphere in the Canadian shield. *Journal of Geophysical Research*, **103**, 15 269–15 286.

JORDAN, T. H. 1981. Continents as a chemical boundary layer. *Philosophical Transactions of the Royal Society of London, Series A*, **301**, 359–373.

KAY, I., SOL, S., KENDALL, J.-M. & 5 OTHERS. 1999. Shear wave splitting observations in the Archean craton of western Superior. *Geophysical Research Letters*, **6**, 2669–2672.

KENDALL, J.-M. & THOMSON, C. J. 1989. A comment on the form of the geometrical spreading equations, with numerical examples of seismic ray tracing in inhomogeneous, anisotropic media, *Geophysical Journal International*, **99**, 401–413.

KENDALL, J.-M., RAYMER, D. G. & BLACKMAN, D. K. 1998. Bias due to simple approximations for flow-induced mineral orientation distribution. *EOS Transactions American Geophysical Union*, **79**, S212.

LENARDIC, A. & MORESI, L. 2000. A new class of equilibrium geotherms in the deep thermal lithosphere of the continents. *Earth and Planetary Science Letters*, **176**, 331–338.

NIXON, P. H. 1987. *Mantle Xenoliths*. Wiley, New York.

NYBLADE, A., OWENS, T., GARRULA, H., RITSEMA, J. & LANGSTON, C. 2000. Seismic evidence for a deep upper mantle thermal anomaly beneath east Africa. *Geology*, **28**, 599–602.

PEARSON, D. G. 1999. The age of continental roots. *Lithos*, **48**, 171–194.

PEARSON, D. G., IRVINE, G. J., CARLSON, R. W., KOPYLOVA, M. G. & IONOVA, D. A. 2002. The development of lithospheric keels beneath the earliest continents: time constraints using PGE and Re–Os isotope systematics. *In*: FOWLER, C. M. R., EBINGER, C. J. & HAWKESWORTH, C. J. (eds) *The Early Earth: Physical, Chemical and Biological Development*. Geological Society, London, Special Publications, **199**, 65–90.

PETIT, C. & EBINGER, C. 2000. Flexure and mechanical behaviour of cratonic lithosphere: gravity models of the East African and Baikal rifts. *Journal of Geophysical Research*, **105**, 19 151–19 162.

PRIESTLEY, K. & MCKENZIE, D. 2002. The structure of the upper mantle beneath southern Africa. *In*: FOWLER, C. M. R., EBINGER, C. J. & HAWKESWORTH, C. J. (eds) *The Early Earth: Physical, Chemical and Biological Development*. Geological Society, London, Special Publications, **199**, 45–64.

RIBE, N. & CHRISTENSEN, U. R. 1994. Three-dimensional modelling of plume–lithosphere interaction. *Journal of Geophysical Research*, **99**, 669–682.

RIBE, N. M. & CHRISTENSEN, U. R. 1999. The dynamical origin of Hawaiian volcanism. *Earth and Planetary Science Letters*, **177**, 517–531.

RICHARDS, M., DUNCAN, R. & COURTILLOT, V. 1989. Flood basalts and hotspot tracks: plume heads and tails. *Science*, **246**, 103–107.

RITSEMA, J. & VAN HEIJST, H. 2000. New seismic model of the upper mantle beneath Africa. *Geology*, **28**, 63–66.

RITSEMA, J., NYBLADE, A., OWENS, T., LANGSTON, C. & VANDECAR, J. 1998. Upper mantle seismic velocity structure beneath Tanzania, east Africa. *Journal of Geophysical Research*, **103**, 21 201–21 213.

RUDNICK, R., & NYBLADE, A. 1999. The thickness and heat production of Archean lithosphere: constraints from xenolith thermobarometry and surface heat flow. *In*: FEI, Y., FERTKA, M. & MYSEN, B. (eds) *Mantle Petrology: Field Observations and High Pressure Experimentation*. Chemical Society, Special Publications, **6**, 3–12.

SILVER, P. 1996. Seismic anisotropy beneath the continents: probing the depths of geology. *Annual Review of Earth and Planetary Science*, **24**, 385–432.

SLEEP, N. H. 1994. Lithospheric thinning by midplate mantle plumes and the thermal history of hot plume material ponded at lithospheric depths. *Journal of Geophysical Research*, **99**, 9327–9343.

SLEEP, N. H. 1997. Lateral flow and ponding of starting plume material. *Journal of Geophysical Research*, **102**, 10 001–10 012.

SLEEP, N. H. 2002. Thermal history of cratonal lithosphere. *Journal of Geophysical Research*, in press.

THOMPSON, G. & GIBSON, S. 1991. Subcontinental mantle plumes, hotspots and pre-existing thinspots. *Journal of the Geological Society, London*, **148**, 973–977.

TOMASSI, A. 1998. Forward modelling of the development of seismic anisotropy in the upper mantle. *Earth and Planetary Science Letters*, **160**, 1–13.

VAN DER LEE, S. & NOLET, G. 1997. Upper mantle S velocity structure of North America. *Journal of Geophysical Research*, **102**, 22 815–22 838.

Archaean tectonics: a review, with illustrations from the Slave craton

WOUTER BLEEKER

Continental Geoscience Division, Geological Survey of Canada, 601 Booth Street, Ottawa, Ontario K1A 0E8, Canada (e-mail: WBleeker@NRCan.gc.ca)

Abstract: The tectonic evolution of Archaean granite–greenstone terranes remains controversial. Here this subject is reviewed and illustrated with new data from the Slave craton. These data show that a thick, *c.* 2.7 Ga, pillow basalt sequences extruded across extended sialic basement of the Slave craton at a scale comparable with that of modern large igneous provinces. The pillow basalts do not represent obducted oceanic allochthons. Basement–cover relationships argue for autochthonous to parautochthonous development of the basaltic greenstone belts of the west–central Slave craton, an interpretation that is further supported by geochemical and geochronological data. Similar data exist for several other cratons and granite–greenstone terrains, including the Abitibi greenstone belt of the Superior craton, where stratigraphic and subtle zircon inheritance data are equally incompatible with accretion of oceanic allochthons. Many classical granite–greenstone terrains, including most well-documented komatiite occurrences, thus appear to have formed in extensional environments within or on the margins of older continental crust. Closest modern analogues for such basalt–komatiite–rhyolite-dominated greenstone successions are rifts, marginal basins and volcanic rifted margins. Indeed, these environments have high preservation potential compared with fully oceanic settings. Collapse and structural telescoping of these highly extended volcano-sedimentary basins would allow for the complex structural development seen in granite–greenstone terrains while maintaining broadly autochthonous to parautochthonous tectonostratigraphic relationships. Seismic reflection profiles cannot discriminate between these telescoped autochthonous to parautochthonous settings and truly allochthonous accretionary complexes. Only carefully constructed structural–stratigraphic cross-sections, allowing some degree of palinspastic reconstruction, and underpinned by sufficient U–Pb zircon dating, can address the degree of allochthoneity of greenstone packages. Furthermore, seismic reflection profiles are essentially blind for the steep structures produced by multiple phases of upright folding and buoyant rise of mid- to lower-crustal, composite, granitoid and gneiss domes. Such structures are ubiquitous in granite–greenstone terrains and, indeed, most of these terrains appear to have experienced at least one phase of convective overturn to re-establish a stable density configuration, irrespective of the complexities of the pre-doming structural history. Buoyant rise of mid- to lower-crustal granitoid and gneiss domes can explain the typical size and spacing characteristics of such domes in granite–greenstone terranes, and the coeval deposition of late-kinematic, 'Timiskaming-type' conglomerate–sandstone successions in flanking basins. The extensional and subsequent contractional evolution of granite–greenstone terrains may have occurred in the overall context of a plate tectonic regime (e.g. volcanic rifted margins, back-arc basins) but highly extended, intraplate, rift-like settings seem equally plausible. Explaining the evolution of the latter in terms of Wilson cycles is misguided. Periods of intense rifting and flood volcanism (e.g. 2.73–2.70 Ga) may have been related to increased mantle plume activity or perhaps catastrophic mantle overturn events. Although there is evidence for plate-like lateral movement in late Archaean time (e.g. lateral heterogeneity of cratons, arc-like volcanism, craton-scale deformation patterns, strike-slip faults, etc.), the details of how these plate-like crustal blocks interacted and how they responded to rifting and collision appear to have differed significantly from those in Phanerozoic time. The most productive approach for Archaean research is probably to more fully understand and quantify these differences rather than the common emphasis on the superficial similarities with modern plate tectonics.

In spite of intense research efforts, the detailed nature of Archaean tectonics remains controversial (e.g. de Wit 1998; Hamilton 1998). Why this controversy persists is to some extent clear: the sample of preserved Archaean crust is small, highly fragmented, and invariably complex.

First of all, Archaean crust occupies a mere 7–11 million km^2 (*c.* 5–7.5%; Goodwin 1996; de

Wit 1998) of the exposed continents. Nevertheless, its direct overall mass contribution to continental crust is estimated to be considerably higher (10–20%), as much Archaean crust is covered by platformal successions or buried as lower plates beneath younger collisional orogens. Furthermore, the preserved record of Archaean continental crust is highly fragmented, and scattered around the globe in about 35 sizeable fragments and a less well-defined number of smaller slivers (Bleeker 2002). Except for perhaps some of the very largest fragments (e.g. the Superior craton of the Canadian Shield), few are big enough to preserve complete tectonic systems on the scale of modern collisional orogens and their associated sedimentary basins (e.g. foreland basin, deformed passive margins, a collisional internal zone with or without accreted arcs, hinterland).[1] Therefore, even if modern-style plate tectonics was dominant in Archaean time, tectonic models for the evolution of particular cratons will generally remain underconstrained. Finally, even the best preserved and well-exposed fragments of Archaean crust (e.g. Pilbara Craton, Western Australia; Hickman 1983, 1984; Buick et al. 1995; or the Slave craton, Canada; Padgham & Fyson 1992; Bleeker & Davis 1999) are characterized by complex depositional and volcanic histories that are invariably overprinted by polyphase deformation and multiple granitoid intrusive events. These processes and events need to be carefully resolved and precisely dated to reconstruct the dynamics of the primary tectonic environments. These are not easy tasks.

It is perhaps not surprising therefore that an enthusiastic application of the plate tectonic paradigm to Archaean time has delivered mixed results (e.g. see critique by Hamilton 1993, 1998) and, currently, impedes an equally exciting research avenue into what is different about the Archaean Eon and the 'early Earth', and what this tells us about secular evolution of our planet. A broader view, one naturally promoted by planetary science, is that Earth is just one of several rocky planetary bodies (e.g. Beatty et al. 1999), each with a unique evolution and none of which, other than Earth, shows plate tectonic activity.[2] Of course, the rarity of plate tectonics may relate directly to another unique feature of our planet: it is 'wet'. Earth evolved and maintained a hydrosphere since at least 3.9 Ga, and possibly as early as 4.4 Ga (Wilde et al. 2001). Water, through its significant influence on mantle rheology (weakening of olivine, e.g. Hirth & Kohlstedt 1996) and melting processes (e.g. arc magmas, granite generation), is probably a critical link in promoting plate-like behaviour (Richards et al. 2001) and the generation of buoyant continental crust (Campbell & Taylor 1983).

Nevertheless, even on the modern Earth, the dominant mode of crust–mantle interaction (plate tectonics dominated by a few relatively large plates) appears sensitively tuned to the present thermal regime of the mantle (Davies 1999), thus begging the question whether a significantly hotter Archaean Earth could have operated in a similar way. The general answer has been more ridges (e.g. Hargraves 1986) and smaller, faster plates (the 'permobile regime' of Burke & Dewey 1973; Burke et al. 1976), but there is a fundamental limit to this mode of recycling beyond which small fast plates do not reach negative buoyancy and a fundamentally different heat loss mechanism is required (Davies 1992, 1999).[3] In this hotter Earth, where heat production exceeded plate subduction-mediated heat loss, crust–mantle interaction may have been dominated by episodic, catastrophic, mantle overturn events (Davies 1995), or by heat advection through voluminous basaltic volcanism coupled to a return flow into the mantle of dense eclogite bodies, either by delamination or drip (e.g. Vlaar et al. 1994; Zegers & van Keken 2001). Furthermore, it is likely that throughout Earth history these processes driven by the upper thermal boundary layer of the mantle were complemented by mantle plumes emanating from a deeper boundary layer, probably the core–mantle boundary (Davies 1999; Campbell 2001). Even today, mantle plumes are responsible for Earth's most

---

[1] A craton is defined as a segment of continental crust that has attained and maintained long-term stability, with tectonic reworking being confined to its margins. Although there is no strict age connotation in this definition, e.g. some segment of crust could have attained 'cratonic' stability during Proterozoic time, the term is typically applied to stable segments of Archaean crust. Long-term stability is thought to be a function, in part, of thicker, stronger lithosphere, involving a cool but composition-ally buoyant keel of Fe-depleted upper mantle (i.e. tectosphere; Jordan 1978, 1988).

[2] Relatively recent (<1 Ga) volcanic resurfacing of Venus may have erased earlier evidence for plate tectonics, and there is some discussion on whether Mars may have experienced sea-floor spreading early in its history (e.g. Connerney et al. 1999; see also Stevenson 2001).

[3] In addition to the heat loss problem, another paradox is how 'small, fast plates', with their inherent buoyancy, could have been 'fast' at all given that 'slab pull' is the dominant driving force of modern plates.

active volcanoes (e.g. Hawaii, Iceland), *c.* 10% of the global heat loss, and possibly much of the net continental growth (e.g. Albarède 1998).

It is critical therefore that we go back to the primary rock record and look for signals of processes other than those that may have had their origin in modern-style plate tectonics. In this context I will discuss some important themes in Archaean geology and highlight those features that perhaps appear uniquely Archaean. As part of this discussion I will present new data from the Slave craton, including the first craton-scale cross-section through this well-exposed fragment of Archaean crust.

Earth's secular evolution: what is different about the Archaean Eon?

The doctrine of uniformitarianism has been an important guiding principle in geology since the days of Hutton and Lyell (e.g. Hallam 1992) and has proven particularly useful in understanding Earth's history on a time scale of the last two billion years. At longer time scales, and further back in time to the era of the 'early Earth', it is inevitable to take a complementary view: that Earth's history is a string of singular or 'first' events, most of which were irreversible, catastrophic, or both. These events punctuated longer-term processes many of which were equally irreversible or unidirectional (e.g. heat loss; growth of buoyant, highly evolved felsic crust).

Table 1 lists a selection of these events and processes from the time of accretion to *c.* 2 Ga. This is perhaps a somewhat subjective list of 'events' and only about half relate to the Archaean Eon *sensu stricto* (2.5–*c.* 3.9 Ga), whereas others relate to initial accretion, the Hadean, or the transition to the Proterozoic Eon. Yet other events or processes were diachronous or remain poorly dated.

It should also be stressed that the Archaean to Hadean terrestrial rock record is largely a preservational record rather than an unbiased primary record. For instance, do the 2772 Ma Black Range dykes of the Pilbara craton (Wingate 1999) truly represent the first craton-scale mafic dyke swarm or did others exist earlier but failed to be preserved at a similar scale (e.g. Ameralik dykes, west Greenland; McGregor 1973; Gill & Bridgewater 1976)? There is no straightforward answer to this question: 'absence of evidence does not equal evidence of absence'. None the less, it is important to realize that stable cratonic lithosphere is a prerequisite, first of all, for the formation of the large brittle fracture patterns that accompany the intrusion of large dyke swarms and, subsequently, for providing the stability for their preservation. Once Archaean cratonic lithosphere was generated and buttressed by relatively cool (stiff), but compositionally buoyant tectosphere (Jordan 1978, 1988), its probability for long-term survival was much higher than for average continental crust. The oldest (preserved) cratonized lithosphere is about 3.1–3.0 Ga in age (e.g. Moser *et al.* 2001).

Hence, the rock record perhaps cannot rule out a steady-state evolution of continental crust generation and recycling by plate tectonics since 3.9 or 4.4 Ga, but its fragmented nature certainly does not favour such a scenario. The limited rock record is equally mute on whether the immense Hadean to Archaean time interval, with its steep exponential decline in heat production and average mantle temperatures, saw just one dominant mode of crust–mantle interaction or, perhaps more likely, a progression of different modes. As the record is relatively well populated between 3.0 and 2.6 Ga, the lack of modern-style passive margin sequences and uncontested ophio-lites, in addition to the absence of high-pressure–low-temperature metamorphic rocks and paired metamorphic belts, is particularly troublesome for uncritically projecting modern-style plate tectonics well back into Archaean time. Hamilton (1993, 1998) has repeatedly stressed the lack of well-documented examples of accretionary prisms and mélange belts, although some have countered this with the observation that many younger convergent margins also lack accretionary prisms as a result of subduction erosion (e.g. von Huene & Scholl 1991).

The Archaean rock record

Perhaps the most outstanding feature of the Archaean rock record is its near absence for much of the first half of Archaean time. Supracrustal rocks of 3.7–3.9 Ga age in the Isua greenstone belt, and other nearby supracrustal enclaves, Nain craton, West Greenland (e.g. Nutman *et al.* 1993, 1997), are a singular occurrence: a stochastic oddity that survived what must have been highly efficient crustal destruction and recycling processes of the Hadean and Archaean eons. The next oldest and much better preserved volcano-sedimentary rocks are those of the Coonterunah (3515 ± 2 Ma) and Warrawoona (*c.* 3490–3330 Ma) groups in the Pilbara craton (Buick *et al.* 1995; Nelson *et al.* 1999; Van Kranendonk *et al.* 2002) with possible correlatives in the Barberton greenstone belt of the eastern Kaapvaal craton, South Africa (e.g. de Wit

Table 1. *First-order events and processes in the early Earth*

Event or process	Age, time interval	References
Main accretion and early differentiation	4.56–4.50 Ga	Patterson 1956; Allègre *et al.* 1995; Alexander *et al.* 2001
Giant impacts and Moon-forming event, magma ocean	*c.* 4.51 Ga	e.g. Lee *et al.* 1997; Canup & Asphaug 2001
Core segregation, release of gravitational energy (superheated liquid Fe–Ni ± impurities core?)	Early or delayed until *c.* 4.51 Ga?	Stevenson 1981, 1990; Davies 1990
Accretion of 'late chondritic veneer'	Post-core segregation	e.g. Holzheid *et al.* 2000
Crystallization of primordial crust; episodic magma oceans(?) owing to large impacts		
Impact erosion of early atmosphere and outgassing of secondary atmosphere and hydrosphere (comet delivery of volatiles?)		e.g. Chyba 1990
Oldest preserved detrital zircons; evidence for hydrosphere?	4.41–4.35 Ga	Mojzsis *et al.* 2001; Wilde *et al.* 2001
Exponential decay of heat-producing elements K, U, Th and steep secular cooling; even today, heat loss is *c.* twice production		e.g. Pollack 1997
Emergence of life (seeded from Mars?); primary origin or refuge in deep hydrothermal environments?		e.g. Sleep *et al.* 1989; Nisbet & Sleep 2001
Oldest preserved terrestrial rocks: Acasta Gneiss Complex, metatonalites and gabbros, no supracrustal rocks	4.05–4.03 Ga	Bowring *et al.* 1989; Stern & Bleeker 1998; Bowring & Williams 1999
Lunar cataclysm, late heavy bombardment	*c.* 4.0–3.8 Ga	e.g. Cohen *et al.* 2000; Culler *et al.* 2000
Oldest preserved supracrustal rocks, Isua greenstone belt; stratigraphy includes quartz-rich sedimentary rocks	3.9–3.8 Ga	Nutman *et al.* 1993, 1997
More widespread formation (preservation?) of early basement complexes	3.6–3.5 Ga	e.g. Buick *et al.* 1995; Bleeker & Davis 1999; Horstwood *et al.* 1999
Oldest (preserved?) tectosphere development, from Re–Os dating	*c.* 3.5 Ga	e.g. Pearson 1999
Thick, stacked, flood basalt–komatiite–rhyolite sequences, e.g. Pilbara craton	3.5–3.4 Ga	Hickman 1983, 1984; Van Kranendonk *et al.* 2002
Tailing-off of high impact rates (see Fig. 1)	3.2–3.0 Ga	e.g. Cohen *et al.* 2000; Culler *et al.* 2000
Initiation of inner-core crystallization and growth, and start-up of core dynamo; remnant magnetism in *c.* 2.9 Ga rocks	Pre-3.0 Ga?	but see Labrosse *et al.* 2001
Widespread Archaean quartzite sequences indicating early (transient?) stability of numerous cratonic nuclei	3.1–2.8 Ga	Eriksson & Fedo 1994; Donaldson & de Kemp 1998; Bleeker *et al.* 1999a
Accelerated growth of highly differentiated crust and net flux of heat-producing elements (K, U, Th) to crust		e.g. Collerson & Kamber 1999
Tidal resonances causing spiked heat flow out of liquid outer core into D″, causing sharply increased mantle plume activity	*c.* 3.0–2.7 Ga & *c.* 2.0–1.8 Ga	Greff-Leffk & Lepros 1999
First (preserved?) craton-scale mafic dyke swarm: Black Range swarm, Pilbara craton (but, note *c.* 3.5 Ga Ameralik dykes)	2772 Ma	Wingate 1999
Widespread 2.73–2.70 Ga flood basalt and komatiite volcanism, present in almost every craton (global?)	2.73–2.70 Ga	
Prolific mesothermal Au deposits in sheared and altered host-rocks of granite–greenstone terranes	2.73–2.60 Ga	

Table 1. (*Continued*)

Event or process	Age, time interval	References
Reduced role of komatiite volcanism after 2.68 Ga, with flare-up at 2.0–1.8 Ga but with lower MgO content (18 v. 23 wt%)		
Widespread stabilization of Archaean crust, typically after terminal 'granite bloom' (e.g. 2.60–2.58 Ga granites of Slave craton)	2.65–2.55 Ga	e.g. Davis & Bleeker 1999
First (preserved?) craton-scale strike-slip dominated faults, e.g. Beaulieu River Fault Zone, Slave craton (Fig. 2)	2.68–2.55 Ga	e.g. Sleep 1992; see also Zegers *et al.* 1998 for a *c.* 2.95 Ga fault in the Pilbara
Secular evolution to more evolved (less sodic) crust; Eu anomalies, less Ni, Cr, and higher Th/Sc in average pelitic sediments		e.g. Taylor & McClennan 1985; Condie & Wronkiewicz 1990
Rapid rise of sea-water $^{87}Sr/^{86}Sr$ ratio from mantle dominated to continental weathering dominated	*c.* 2.6–2.5 Ga	Veizer & Compston 1976
Great Dyke intrusion, Zimbabwe craton	2574 ± 2 Ma	e.g. Wingate 2000
First (preserved?) giant, radiating, mafic dyke swarms, e.g. Matachewan dyke swarm, Superior craton	2.48–2.44 Ga	Heaman 1997
First modern-style sediment-rich passive margin sequences, e.g. Huronian Supergroup, southern margin of Superior craton	2.48–2.2 Ga	e.g. Young *et al.* 2001
Giant Superior-type iron formations	2.5–1.9 Ga	e.g. Cloud 1988
Oxyatmospheric conversion	*c.* 2.4–1.9 Ga	e.g. summary in Windley 1995 (see references therein)
First uncontested ophiolite complexes, e.g. Finland, Ungava	*c.* 2.0 Ga	Kontinen 1987; Scott *et al.* 1992

et al. 1992; Zegers *et al.* 1998). Older gneisses (e.g. Black *et al.* 1986; Kinny *et al.* 1988), xenocrystic zircons (e.g. Thorpe *et al.* 1992*b*; Kröner *et al.* 1996; Nelson *et al.* 2000), and detrital zircons (e.g. Liu *et al.* 1992; Mueller *et al.* 1992; Sircombe *et al.* 2001) are relatively widespread but collectively represent a volumetrically insignificant component of the ancient cratons. The Acasta gneisses, at 4.03 Ga the oldest preserved terrestrial rocks, are part of a large gneissic basement complex underlying much of the Slave craton (Bleeker & Davis 1999; Bleeker *et al.* 2000), which to date has yielded >4 Ga rocks from an area only a few square kilometres in size (Bowring *et al.* 1989; Stern & Bleeker 1998; Bowring & Williams 1999).

From about 3.3 Ga, supracrustal rocks become more abundant with examples known from many cratons. After 3.1–3.0 Ga, the record is well populated although in most Archaean cratons the supracrustal record is strongly skewed towards 2.75–2.65 Ga Neoarchaean rocks. Quantifying this record of crustal growth (and preservation) is difficult (e.g. Gurnis & Davies 1986), and attempts using isotopic systematics have yielded a variety of growth models. One interesting way of looking at this problem is to analyse detrital zircon populations of mature, well-mixed, Archaean quartzites. Results of such an analysis for the Slave craton are shown in Figure 1 for an unbiased selection of *c.* 300 grains from five samples of a *c.* 2.8 Ga quartzite unit (Sircombe *et al.* 2001). The detrital zircon record is dominated by younger Archaean ages and thus mimics the global Archaean rock record. The skewed nature of this and similar records could be largely a function of preservation, but another clue is provided by considering them in relation to the violent impact history in the inner Solar System. As shown in Figure 1 the skewed zircon record is to a large degree complementary to that of dated impact-generated glass spherules from lunar soils (Fig. 1; e.g. Culler *et al.* 2000). The gravitational cross-section of the larger and more massive Earth is *c.* 25 times that of the Moon (Chyba 1987) so that the lunar record can only be a weak reflection of the destructive impact record that the Earth must have endured well into Archaean time. Highly destructive impacts and catastrophic mantle overturn events favoured by a hotter mantle are two processes that could have stirred crust back into the mantle, thus providing at least partial explanations for the highly skewed nature of the Archaean rock record.

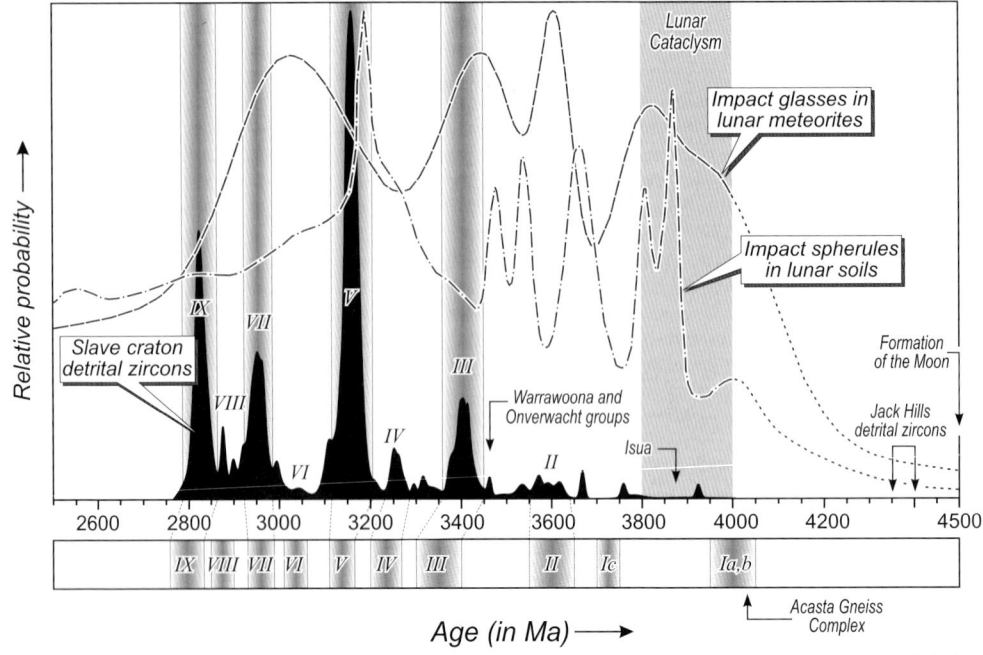

Fig. 1. Relative probability histograms of Slave craton detrital zircons (continuous curve with black infill below; based on data from Sircombe et al. 2001), $^{40}Ar/^{39}Ar$ ages of impact spherules in lunar soil samples (dash–dot curve; after Culler et al. 2000), and $^{40}Ar/^{39}Ar$ ages of impact glasses in lunar meteorites (dashed curve; after Cohen et al. 2000). Time interval spans from 4500 Ma, the approximate age of formation of the Moon, to 2500 Ma, the defined Archaean–Proterozoic boundary. Vertical scales of the three curves are independent. Shaded age bars with roman numerals represent main 'events' in basement of the Slave craton that were initially defined on the basis of individual rock age and their inheritance (see Bleeker & Davis 1999). The detrital zircon data represent c. 300 zircon grains from five widely distributed samples of a c. 2800 Ma quartzite unit overlying the Mesoarchaean to Hadean-age basement complex of the Slave craton. These data represent a least-biased record of pre-2.8 Ga components of the Slave craton. The broad complementarity in the datasets should be noted, with the first major peak in Slave crustal ages (event V: 3100–3200 Ma) immediately following the last major peak in the lunar spherule data. Both lunar soil and meteorite data sets support a 'lunar cataclysm' or 'late heavy bombardment' that appears to have erased or swamped out the pre-4.0 Ga lunar record.

Basement–cover relationships in Archaean cratons

Although Archaean cratons are laterally heterogeneous, they are invariably characterized by tracts of low to medium metamorphic grade volcano-sedimentary rocks, 10–100 km in scale, that are intruded and surrounded by granitoid plutons (de Wit & Ashwal 1997). These tracts are known as 'granite–greenstone terrains' and contrast with more deeply eroded terrains comprising higher-grade gneisses and gneissic granitoids, although distinctions are rarely sharp. In fact, narrow keels of highly metamorphosed greenstones (or enclaves, where the gneissic rocks are younger intrusive tonalites) typically persist in the gneissic terrains (e.g. Windley & Bridgewater 1971). Conversely, many granite–greenstone terrains include older gneissic rocks at their lowest structural level (e.g. Fig. 2).

A significant number of studies, spanning over half a century and many different cratons around the world (e.g. Macgregor 1951; Bickle et al. 1975; Baragar & McGlynn 1976; Henderson 1981, 1985; Chadwick et al. 1985a,b; Srinivasan & Ojakangas 1986; Roscoe & Donaldson 1988; Wilks & Nisbet 1988; Eriksson & Fedo 1994; Hunter et al. 1998; Bleeker et al. 1999a,b), have demonstrated that the critical contacts between these structurally lowest, older, gneissic rocks and overlying volcano-sedimentary sequences are unconformities, complete with weathering horizons and basal conglomerates or quartzites (Fig. 2b). In other granite–greenstone terranes, where gneissic basement has not (yet?) been identified, very similar volcano-sedimentary sequences have nevertheless been shown to unconformably overlie older, subaerially eroded, greenstone successions (e.g. Buick et al. 1995; Donaldson & de Kemp 1998). Hence, many sub-

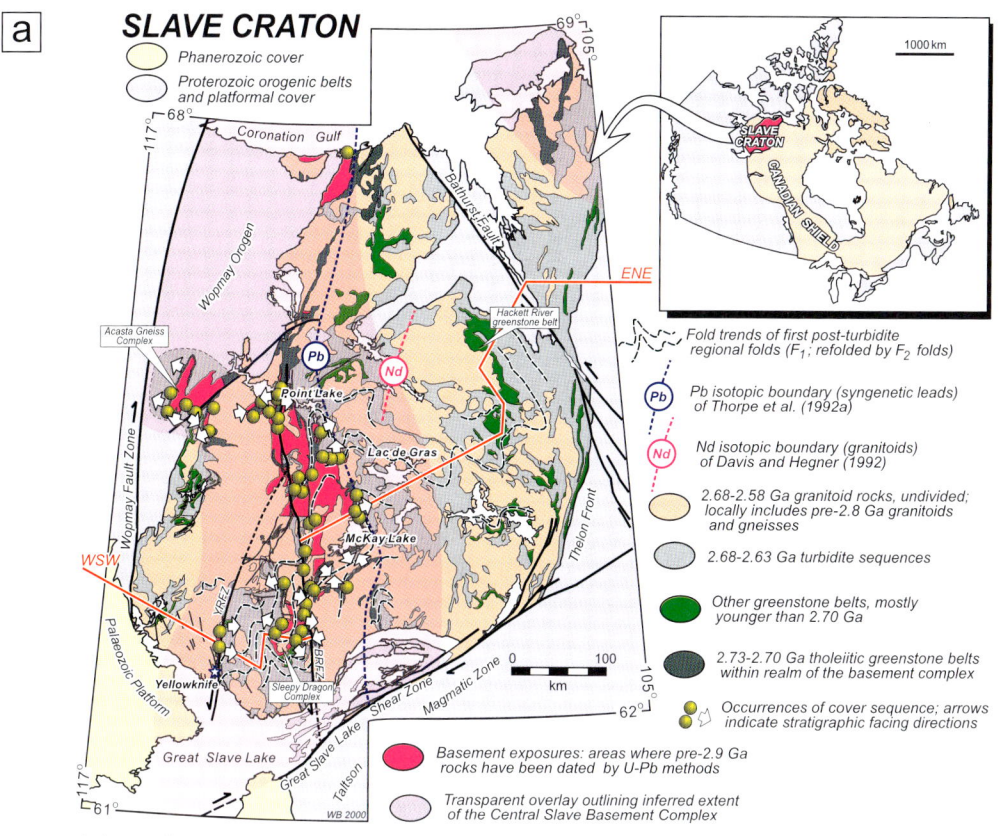

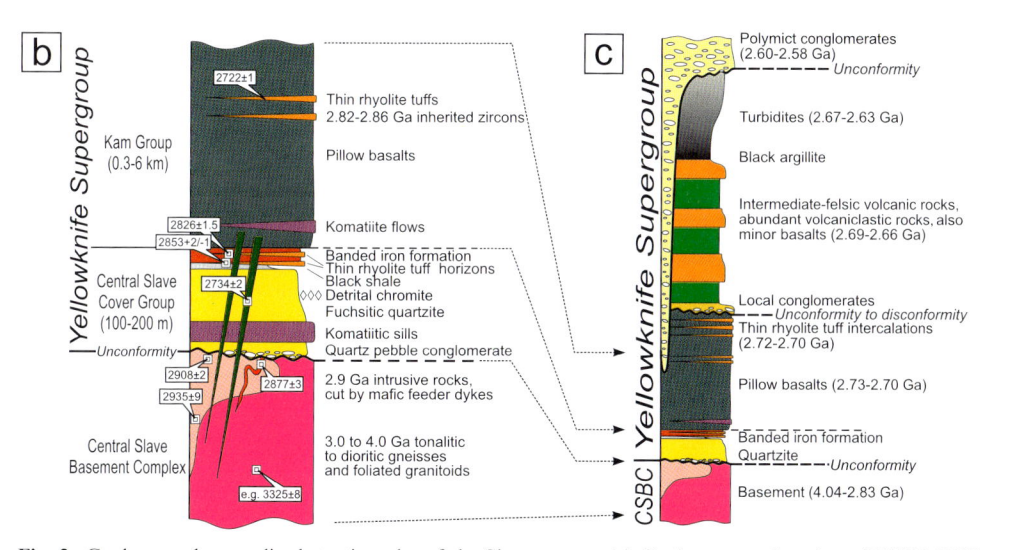

Fig. 2. Geology and generalized stratigraphy of the Slave craton. (**a**) Geology map; location of WSW–ENE cross-section line should be noted (red line; see Fig. 3). Light purple overlay outlines minimum extent of the Mesoarchean to Hadean basement complex of the Slave craton. (**b**) Typical stratigraphy that overlies the basement complex; precise zircon ages from Isachsen & Bowring (1997), Bleeker *et al.* (1999*a*, *b*) and Ketchum & Bleeker (unpubl. data). (**c**) Generalized stratigraphy for the entire craton.

aqueous greenstone sequences and their thin basal sedimentary successions were deposited on pre-existing, emergent, sialic basement. In most cases this view is reinforced by the presence of dense mafic dyke swarms in the underlying basement gneisses, with appropriate relative and absolute ages to have been feeders to the basalts (e.g. Bleeker et al. 1999a, b; Wingate 1999).

In agreement with a rational succession of depositional facies ('Walther's law'), the contact between basal quartzitic sedimentary rocks and overlying basalt–komatiite sequences is typically marked by finer-grained clastic or chemical sedimentary strata such as pelites, banded iron formation (BIF), chert and (or) carbonate. A primary stratigraphic relationship between basal sedimentary rocks and the greenstones can in many places be demonstrated by primary interfingering of the various lithologies, and mapping of this stratigraphy, through various strain states, over distances of tens to hundreds of kilometres (Bickle & Nisbet 1993; Bleeker et al. 1999a, b).

However, greenstone sequences are typically deformed to an extent where primary stratigraphic relationships are complicated by foliation and shear-zone development. Foliation development is commonly strongest across contacts between thick, stiff, members in the stratigraphy, such as basement gneisses or pillow basalt sequences. Fortunately, these deformation patterns are invariably heterogeneous so that with careful and persistent mapping across a variety of scales, ranging from thin section and outcrop to belt and craton scales, one can commonly demonstrate the primary depositional relationship between basement gneiss, basal quartzites, BIFs, basalt and komatiite (e.g. Fig. 2b). This has been done, definitively, in a number of classical granite–greenstone terranes such as the western Dharwar craton, India (Chadwick et al. 1985a, b), the Belingwe greenstone belt in Zimbabwe (Bickle & Nisbet 1993), greenstone belts of the central and western Slave craton (Bleeker et al. 1999a, b), and parts of the western Superior craton (Percival et al. 2001a), among many others and including younger Proterozoic analogues (e.g. Bleeker 1990). The sobering lesson from much of this research is that isolated structural studies, focusing on individual shear zones, or zones of increased foliation development and the local structural complexities they cause, often fail to recognize the regional stratigraphic relationships. Foliation patterns and kinematic indicators are commonly complex and, by themselves, without a detailed geometrical and stratigraphic framework, cannot establish whether a reoriented segment of a shear zone was part of an early thrust, an early extensional fault, or just a localized zone of accommodation as a result of distributed deformation of a heterogeneous volcano-sedimentary terrain with more rigid synvolcanic plutons and basement-cored antiformal structures.

Nevertheless, a number of workers have reinterpreted many of the same or similar greenstone belts as obducted ophiolites lying structurally on thin continental margin sequences (e.g. de Wit et al. 1987; Kusky 1990; Kusky & Kidd 1992; Wilks & Harper 1997; Puchtel et al. 1998; Kusky & Hudleston 1999), generally arguing for a fundamental thrust at the base of the mafic–ultramafic volcanic sequence. If true, this claim would be significant because it would argue for well-preserved Archaean oceanic crust lying on continental basement well inboard of any possible subduction zone. Not only would this be a strong confirmation of plate tectonics in Archaean time but it would also provide critical insights into the thermal regime and chemistry of Archaean oceanic lithosphere and hence a window into the mantle (e.g. Bickle et al. 1994). However, the obduction model is not credible for the areas where it was proposed for the following reasons: (1) it ignores outcrop-scale observations that tie the mafic–ultramafic volcanic rocks to the underlying sedimentary successions, whereas the latter demonstrably overlie sialic basement with erosional contacts; (2) it is inconsistent with the feeder dyke swarms that tie volcanic rocks to basement; (3) it is equally inconsistent with the regional observations of a predictable, conformable stratigraphy at the 10–100 km scale; and (4) although individual contacts may be sheared at any particular outcrop, no consistent structure (i.e. thrust) has been demonstrated at the critical contacts at the same regional scales.

Autochthonous or parautochthonous v. allochthonous greenstones

Clearly, (para)autochthonous stratigraphic development versus the degree of allochthoneity is a central issue in the interpretation of many imperfectly preserved greenstone sequences. Hence, it is worth considering this debate in more detail. Perhaps this discussion should start with structural considerations, before examining relevant geochemical, geochronological and stratigraphic aspects.

Structural geological constraints

Interpretations involving a far-travelled allochthonous origin of greenstones are generally

motivated by indirect geochemical arguments. Timing constraints are commonly such that early thrusting is invoked of thin sheets of 'oceanic' basalts and komatiites. Movement of thin thrust sheets has very predictable consequences, however, and should lead to imbricate structures involving structural stacking of rocks from originally adjacent environments, old over young relationships, uplift, clastic piggyback basins where thrusts become emergent, and a general propagation of structures towards the foreland, before possible re-imbrication by later out-of-sequence thrusts.

Most Archaean terranes show large-scale structural relief as a result of large-wavelength, high-amplitude folds with moderate to steep (or steepened) structural plunges. Hence, structural sections reflecting thrust transport of greenstones should be widely exposed in oblique sections at the surface. It is by this technique of structural–stratigraphic mapping and down-plunge projection of oblique sections that the nappe stack of the Alps was first elucidated and, in general, how fold–thrust belts are reconstructed. Therefore we should be able to demonstrate similar structural sections in the Archaean cratons where large-scale thrusting is invoked. Has this been done in the Zimbabwe craton, for instance, where recent work (e.g. Dirks & Jelsma 1998) invokes craton-scale thrusting? Where do the older gneiss complexes (Horstwood *et al.* 1999) supposedly involved in this thrusting (e.g. Jelsma & Dirks 2002) structurally overlie, on regional scales, younger greenstone sequences? Have the clockwise P–T–t paths and elevated pressure facies series, as one would predict for a structural evolution dominated by thrust stacking, been demonstrated?

In the Yilgarn craton, greenstones of the Eastern Goldfields domain are commonly interpreted as small disparate allochthons overlying a cryptic décollement surface (Swager *et al.* 1997). Where is this décollement surface and overlying imbricate stack exposed in oblique section at the surface? Clearly, the interpretation of the basalt–komatiite sequences of the Eastern Goldfields as oceanic allochthons is contradicted by the presence of ancient zircon xenocrysts (Compston *et al.* 1986; Campbell & Hill 1988), tying at least some of the mafic and ultramafic volcanic rocks to much older continental basement.

In the well-exposed Slave craton, detailed structural–stratigraphic mapping and an ever-expanding dataset of precise U–Pb zircon ages (e.g. Bleeker & Davis 1999) now permit construction of a reasonably constrained craton-scale cross-section (Fig. 3). This cross-section and supporting field mapping show that the central and western part of the craton are underlain by a single contiguous basement complex (Bleeker *et al.* 1999*a*, *b*, 2000), which is overlain by a predictable stratigraphy as shown in Figure 2. Over distances of hundreds of kilometres, a basal quartzite–BIF sequence (Central Slave Cover Group, Fig. 2b) is overlain by a tholeiitic basalt sequence of *c*. 1–6 km thickness (Kam Group; Helmstaedt & Padgham 1986) of massive and pillowed flows. In individual outcrops this critical stratigraphy may be highly deformed or, worse, the basal sedimentary section may be excised entirely by high-strain zones or obliterated by younger granitoid intrusions. Despite these complexities, the autochthonous to parautochthonous nature of the quartzite–BIF–basalt stratigraphy can be established from key outcrops across the basement complex (Bleeker *et al.* 1999*a*, *b*). At the same time, the inherent weakness of the alternative model, structural emplacement of the basalts as a far-travelled allochthonous thrust sheet (Kusky 1990), is well illustrated by the structural section (Fig. 3): the basalts form a thin sheet that is essentially continuous across the realm of the basement complex, without documented imbricate or duplex structures. Hence, structural emplacement of the pillowed and massive lavas as obducted oceanic basalts would require translation of a thin sheet, over hundreds of kilometres, without loss of coherency and piling up into imbricate structures. Furthermore, neither is there evidence for a metamorphic sole at the base of the basalt sequence, nor for imbricated slices of rocks from intermediate palaeogeographical settings (e.g. oceanic islands, slope sediments) that one would expect to be transported with the oceanic allochthon (e.g. see Searle 1985; Robertson 1987; Searle & Cox 1999, for descriptions of the Semail ophiolite and its emplacement history). The verdict is simple: the thick pillow lavas and minor associated komatiites of the Kam Group, like many comparable greenstone successions in other cratons, were deposited on older sialic basement within subsiding volcano-sedimentary basins. The association with dense dyke swarms in the underlying basement indicates that the basement was actively extending at the time of basalt extrusion. Subsequent structural events have locally disturbed the section or even translated parts of the stratigraphy relative to basement, but at the larger scale the stratigraphy is preserved (Bleeker *et al.* 1999*a*, *b*).

Obviously, this conclusion does not mean that other greenstone sequences cannot be allochthonous, but this needs to be proven in each case. Where structural relationships are used as the main evidence for a far-travelled allochthonous origin, this interpretation should be supported by well-constrained structural cross-sections at a variety of scales, not cartoons, that can rationalize where the allochthons originated ('rooted'), how they were transported and imbricated, and, finally, how they were emplaced in their current settings.

Reflection seismic profiles

Seismic reflection profiles have now been acquired across several Archaean granite–greenstone terranes and have added a remarkable dimension to the investigation of these terranes, in some cases imaging what appear to be fossil sutures dipping from the lower crust into the upper mantle (e.g. Calvert et al. 1995). In these cases there can be little doubt that large horizontal shortening movements are involved in the final structural evolution of some Neoarchaean granite–greenstone terranes. It has also been noted that many Archaean reflection profiles do not look fundamentally different from reflection profiles across younger orogens, showing similar images of moderately dipping reflection packages that appear to be imbricated, or truncated by yet younger structures, and generally merging into subhorizontal reflectivity in the lowermost crust (e.g. Cook et al. 1999).

Seismic reflections are caused by elastic waves bouncing off rock packages, individual contacts, or structures across which there is a significant contrast in seismic impedance (product of velocity and density). In general, only larger structures and those with moderate to shallow dips reflect significant energy back to the surface so that seismic sections provide highly 'filtered' images of the crystalline crust. This is particularly important in granite–greenstone terranes, where most of the shallow structures (folded stratigraphy, pluton contacts, faults) are usually steep (e.g. $\geq 70°$) and therefore largely invisible. Hence, despite the spectacular images of the mid- and lower crust, or even upper mantle in some cases, the connection between steep surface structures and moderate- to shallow-dipping reflections at depths of c. 10–15 km or deeper remains unsatisfactory. Too many seismic interpretations appeal to a rapid shallowing of dips in the first few kilometres below the surface. In some cases this interpretation seems justified (e.g. listric faults that sole into a major shallow-dipping detachment fault or thrust), but in many greenstone belts such an interpretation is in contradiction with surface structural data. For example, a seismic reflection profile through the high-amplitude, basement-cored, domes of the central Slave craton (Fig. 3) no doubt would show an intricate image of shallow- to moderately dipping reflectors, whereas the steep flanks of the domes and the near-vertical greenstone stratigraphy would be largely invisible. So what would the reflections represent? In the absence of a detailed structural profile constructed on the basis of closely spaced structural measurements at the surface, and reasonable projection to depth, it seems likely that the reflections would be interpreted as yet another example of shallow-dipping structures along which the greenstone belts of the central Slave craton were imbricated. In fact the southwesternmost part of the Slave craton was imaged by the LITHOPROBE SNORCLE transect. This profile shows two prominent shallow-dipping reflectors in the uppermost 15 km of the crust, just east of the town of Yellowknife (Y1 and Y2, Fig. 3). In the initial interpretation, these reflectors were equated with major shallow-dipping faults (Cook et al. 1999), despite any compelling surface geological features in this area, where all dips are generally steeper than 70°. A more realistic interpretation equates the upper reflection with a thick and shallow-dipping Proterozoic gabbro sill, which is exposed at surface, whereas the lower reflection probably represents the basalt-dominated volcanic stratigraphy between more or less transparent metaturbidites above and mostly felsic basement gneisses below (Fig. 3).

Hence, despite the compelling nature and perhaps simple, but superficial beauty of seismic images of granite–greenstone terranes, many problems remain to be resolved and they are not a close proxy of well-constructed and well-constrained structural cross-sections; rather they are complementary to such sections. Follow-up studies, involving high-resolution seismic profiles, rock property studies, and/or reprocessing are often required to resolve discrepancies between the seismic data and field observations. Additional complications arise from 3D structural effects, which are invariably significant in granite–greenstone terranes, and crooked line geometries. Even subvertically dipping Proterozoic mafic dykes, with trends that intersect the section, can cause bright reflections with apparent moderate dips (Zaleski et al. 1997).

Although the relative ages of reflections can be deduced from apparent truncations, the true age and nature of individual reflections commonly remains uncertain unless they can be tied uniquely to surface structures or features observed in deep drill holes. Finally, it is not clear how the general style of a reflection section can differentiate between a highly shortened collapsed rift or marginal basin on the one hand and a structural collage of truly allochthonous thrust slices on the other.

Geochemical constraints

The geochemistry of greenstone belt volcanic rocks, and associated plutonic rocks, is a complex subject with a voluminous literature. Much of this literature pertains to chemical characterization (e.g. calc-alkalic v. tholeiitic) and comparison with modern analogues using discrimination diagrams. This approach can be criticized because it implicitly assumes that Archaean petrogenetic processes and their relationships to tectonic environments were identical to those in the modern Earth. Although this seems a reasonable assumption it is certain to hide secular trends and differences. With more refined datasets, involving high-quality trace element and isotopic data, interesting differences have started to emerge (e.g. Arndt et al. 1997; Smithies 2000). Even in the modern Earth, more refined datasets show that mixed geochemical signals are commonly present in coherent volcanic successions from a single tectonic environment. Important examples are the primary intercalation of basalts with mid-ocean ridge-like chemistry in breakup related continental flood basalt sequences (e.g. Tegner et al. 1998), the inheritance of arc-signatures in continental flood basalts (e.g. Puffer 2001), the 'leakage' of plume-derived melts into arc sources (e.g. Wendt et al. 1997; Turner & Hawkesworth 1998), the occurrence of calc-alkaline dacites with Nb depletions in Iceland's central volcanoes (Jónasson et al. 1992), and the occurrence of dacites and andesites in extensional environments (e.g. Parson et al. 1989). Hence, given the known complexity of modern processes, the non-uniqueness of many geochemical signatures, and the incomplete preservation of ancient environments, are we not overinterpreting the Archaean record in terms of simple end-members?

Of particular relevance in the context of autochthonous v. allochthonous development of greenstone successions has been the widespread tendency to characterize pillow basalts with flat to light rare earth element (LREE)-depleted patterns or those with positive ε_{Nd} isotopic signatures as 'oceanic' even though careful mapping has demonstrated that many of these basalts (and komatiites) show a primary relationship with continental basement, or are part of complex, cyclical, mafic–felsic stratigraphies. The Kam Group of the central and western Slave craton (Fig. 2) and the Ngezi Group mafic and ultramafic lavas of the Belingwe greenstone belt (Bickle & Nisbet 1993), Zimbabwe craton, are excellent examples of the former, whereas mafic and ultramafic lavas intercalated with felsic volcanic rocks of the Abitibi greenstone belt, Superior craton, are examples of the latter (e.g. Jensen 1985). In these cases the interpretation of these chemical signatures as indicating an oceanic origin is either incorrect or misleading.

In the Slave, Zimbabwe, and many other cratons, depleted upper-mantle sources clearly contributed to lavas extruded on submerged, extended, continental basement. Typically only a small fraction of the magmas feeding such mafic to ultramafic lava sequences interacted sufficiently with extended continental basement to show unequivocal signatures of crustal contamination (e.g. Th and LREE enrichment, relative Nb depletion, variable to negative ε_{Nd} signatures). Rapid ascent, high effusion rates, and the lining of conduits with mantle-derived melts resulted in crustal contamination being a somewhat erratic characteristic. It tends to be stronger towards the top of volcanic cycles, when temporary ponding of magmas allowed for more extensive interaction and assimilation of crust, contributing to the production of felsic magmas. Through a number of detailed studies, crustal contamination has now been conclusively demonstrated in many of these greenstone sequences (e.g. Arndt & Jenner 1986; Chauvel et al. 1993; Cousens 2000; Green et al. 2000).

In the Abitibi greenstone belt of the Superior craton it has long been argued that the volcanic stratigraphy consists, at least in part, of several mafic and ultramafic to felsic cycles. Primary superposition of these cycles has now been demonstrated by zircon inheritance and dating of feeder dykes cutting through underlying packages (Heather et al. 1995; Bleeker et al. 1999c). For instance, rhyolites stratigraphically intercalated with komatiites and basalts of the 2710–2720 Ma Kidd–Munro assemblage contain rare inherited zircons with ages between 2723 and 2735 Ma indicative of older assemblages in the Abitibi (Bleeker et al. 1999c). In general, these important results were obtained during routine, single-zircon dating programmes by isotope dilution thermal ionization mass spectrometry (ID-TIMS), where selection of zircons is biased towards least-magnetic and highest-quality grains to obtain concordant primary crystallization ages. Hence, it seems inevitable that with the application of the sensitive high-resolution ion microprobe (SHRIMP) to the Abitibi a more extensive inheritance history will be uncovered.

Because volcanic and plutonic rocks of the Abitibi greenstone belt show fairly uniform positive ε_{Nd} values (at 2700 Ma), recent interpretations have assumed that the belt represents a collage of intra-oceanic arcs and plateaux (Jackson & Fyon 1991; Desrochers et al. 1993; Jackson et al. 1994) that were accreted to an older terrane that is now exposed along its northern margin. The zircon inheritance data, which show that coherent cyclic stratigraphy does exist in large parts of the Abitibi, now refute this model in which each different package of volcanic rocks is interpreted as yet another allochthon. In a strict sense, the Nd and Hf data (e.g. Corfu & Noble 1992) require only that basement below the volcanic rocks is not significantly older than 2.75–2.70 Ga volcanic rocks at the surface. The resolution and sensitivity of the isotopic tracer methods are insufficient, however, to eliminate weak to moderate interaction with only marginally older crust with ages up to $c.$ 2.85 Ga, or perhaps even 2.95 Ga.

In light of this, it is important to note that several subtle indications for interaction with older crust do indeed exist but have not received widespread attention. (1) In the northeastern Abitibi greenstone belt, Mortensen (1993) dated two samples of $c.$ 2759 Ma felsic volcanic rocks that show prominent inheritance of 2800–2805 Ma zircons. (2) Rocks with such ages are known from the Opatica gneiss belt (Sawyer & Benn 1993), a metaplutonic terrane immediately north of the Abitibi belt. One tonalitic gneiss from this domain contains both 2807 and 2721 Ma zircon populations (Davis et al. 1995). The older age may represent the protolith age, whereas the younger age may reflect plutonic events of the same age as volcanic ages in the adjacent greenstone belt. Other tonalites and diorites of the Opatica belt have been dated at 2820–2830 Ma (Davis et al. 1994). (3) In the western extension of the Abitibi greenstone belt, Moser et al. (1996) described a mid-crustal tonalitic gneiss (Wawa gneiss domain) with an interpreted age of 2935 ± 35 Ma. This same gneiss also contains younger zircons at $c.$ 2823 and 2735 Ma. Although other interpretations are possible (see Moser et al. 1996), it seems likely that this rock provides yet another link between Abitibi volcanic ages and plutonic basement rocks with a complex history extending back to 2.82 and 2.93 Ga. (4) Within the central Abitibi greenstone belt, the large, synvolcanic, $c.$ 2702 Ma Montcalm gabbroic complex shows ε_{Nd} values that vary up-section from +2.8 to +1.0, thus indicating weak interaction with enriched crust (Barrie & Shirey 1991). (5) Finally, in the same area, the Coté Township tonalite intrudes slightly older volcanic rocks of the Kamiskotia Complex. This pluton has been dated at 2694 ± 4 Ma, but also yielded one zircon fraction with a minimum age of 2926 Ma (Barrie & Davis 1990).

Hence, orthogneisses now exposed below the Abitibi greenstone belt (Wawa gneiss domain) and in a structural high to the north of the belt (Opatica belt) are linked to the greenstone belt at the time of volcanism (e.g. 2735 and 2721 Ma), and certainly by the time of voluminous 2696–2693 Ma tonalitic plutonism that stitches the boundary between the Opatica and Abitibi belts (Davis et al. 1995). The south-vergent structural fabric of the main Abitibi seismic section (Calvert et al. 1995; D_2 fabric of Sawyer & Benn 1993) postdates this plutonism and, further west, affects turbiditic sedimentary rocks of the Porcupine Group that contain detrital zircons of all the various volcanic packages (Bleeker & Parrish 1996; Bleeker et al. 1999c). Therefore, the seismic image does not record accretion of oceanic allochthons but rather late-stage south-vergent shortening and imbrication of already assembled stratigraphic units. This implies that many of the stratigraphic elements of the Abitibi greenstone belt are parautochthonous and formed, at least in part, on extended basement of marginally older tonalitic gneisses as now exposed below (Wawa gneiss domain) and to the north of (Opatica belt) the low-grade greenstone domain. This interpretation is compatible with the Nd and Hf isotopic constraints and may help explain the locally significant volume of felsic rocks in parts of the Abitibi greenstone belt.

In summary, lateral accretion of disparate oceanic allochthons is inconsistent with the complex but in part coherent, cyclic, stratigraphy of large parts of the Abitibi and neglects the subtle indications for basement below the volcanic rocks that is probably 2.80–2.85 Ga in age with some components as old as 2.93 Ga. The observed stratigraphic and structural relationships are better explained by a highly extended rift or marginal basin that developed in part on slightly older, $c.$ 2.80–2.93 Ga crust before collapsing during 2.70–2.68 Ga regional shortening, perhaps in response to interaction with a more southerly crustal block such as the Minnesota River Valley terrane (e.g. Hoffman 1989; Percival et al. 2001b).

Geochronological constraints

Some of the most robust constraints on the tectonic provenance of greenstone belt successions come from U–Pb zircon geochronology and several examples have already been discussed. Ideally, ID-TIMS and SHRIMP techniques are applied in parallel, combining the superior precision (e.g. ±1 Ma) of isotope dilution single grain analyses with the high spatial resolution and rapid instrumental, non-destructive, capability of the SHRIMP. Apart from providing basic crystallization ages of volcanic and plutonic units, U–Pb zircon geochronology can conclusively link stratigraphic elements to underlying units and (or) basement through precise dating of inherited xenocrystic zircons, feeder dykes and detrital zircons of intercalated clastic units, and by establishing age links between cover and basement.

Discovery of ancient xenocrystic zircons in the Kambalda volcanic sequence, Yilgarn craton (Compston et al. 1986; Campbell & Hill 1988), clearly showed that these mafic and ultramafic volcanic rocks erupted through older continental basement and hence are not far-travelled oceanic allochthons. Similar data have emerged from the Steep Rock Group (Wilks & Nisbet 1988), a quartzite–BIF–komatiite–basalt sequence overlying c. 3.0 Ga basement tonalites in the southern Wabigoon greenstone belt of the western Superior craton. Komatiitic rocks in this sequence, known as the Dismal Ashrock (Wilks & Nisbet 1988), are developed as lapilli tuffs and breccias that contain accidental clasts of the underlying basement tonalites (Schaefer & Morten 1991). Highly vesicular clasts, cored and composite lapilli, and bedding features indicate explosive komatiitic volcanism and reworking in a shallow-water setting (Schaefer & Morten 1991), and inherited zircons, derived from the entrained tonalite clasts, indicate eruption through basement of the underlying Marmion Complex. Both 2997 and 2780 Ma inherited zircons have been extracted from the Dismal Ashrock (Davis & Tomlinson, pers. comm.), matching known ages of tonalites in the Marmion basement complex. Hence, these komatiitic rocks are not an oceanic allochthon (cf. Kusky & Hudleston 1999) but form part of an autochthonous, shallow-water, sedimentary and volcanic succession deposited during rifting and transgression of an older basement complex sometime after 2780 Ma.

In the basaltic greenstones of the west–central Slave craton (Fig. 2b), intercalations of reworked felsic tuff contain detrital zircons of the same age as, and probably derived from, immediately underlying stratigraphic units (Isachsen & Bowring 1997) that can be tied to basement, thus supporting a coherent autochthonous stratigraphy. Autochthonous development of the basalt stratigraphy is further supported by the presence of a dense mafic dyke swarm, dated at 2734 ± 2 Ma, and of the same age as the approximate onset of volcanism in the overlying basalt sequence (Bleeker et al. 1999b).

In the Bababudan hills of the western Dharwar craton, India, a thick subaerial to subaqueous basalt succession can be mapped across several synformal greenstone belts (Chadwick et al. 1985a, b). Not only do these basalts overlie a similar basal sedimentary succession as in the Slave, but cross-bedded quartzite intercalations also occur higher up in the basalt stratigraphy, suggesting a direct link with sialic basement during basalt eruption. Matching the detrital zircons in these quartzite intercalations to basement sources in the underlying 'Peninsular Gneisses' should be carried out to further test this link. Similar quartzite intercalations are also common in greenstone belts of the Southern Cross domain of the Yilgarn craton (e.g. Griffin 1990).

As already discussed, in the Abitibi greenstone belt of the Superior craton, extensive dating of volcanic units, their inherited zircons, and feeder dykes cutting units lower in the stratigraphy has established coherent stratigraphy where previous workers had proposed a tectonic collage.

Finally, where oceanic allochthons are emplaced on unrelated, thermally cold foreland, there is no *a priori* reason why plutonic events in the basement should match those within the volcanic cover. However, in most well-studied granite–greenstone terrains the opposite observation is made that, with more and higher precision data, various ages in the complex volcanic stratigraphy are echoed by plutonic events at deeper structural level in what are typically polyphase granitoid–gneiss complexes (e.g. Williams & Collins 1990; Thorpe et al. 1992b; Zegers et al. 2001; Van Kranendonk et al. 2002). The more precisely these age links can be established, the stronger the evidence is for more or less autochthonous development.

Stratigraphic constraints

Careful reconstruction of coherent stratigraphic successions, with age constraints from multiple, precise U–Pb zircon dates (e.g. Fig. 2), provides the strongest constraint on the origin of greenstone successions, their primary tectonic setting, and the rates and duration of depositional and basin-forming processes. These stratigraphies should be compared and contrasted with modern analogues and may ultimately provide quantitative information on what was different about the Archaean Earth (e.g. duration and episodicity of volcanism, sedimentation rates, petrogenetic processes, lithospheric thickness, flexural rigidity; e.g. McKenzie et al. 1980).

For those well-studied greenstone successions for which an allochthonous ophiolitic, or more general oceanic origin has been invoked or debated, how do their stratigraphies compare with that of true ophiolites, modern oceanic crust, or theoretical predictions of thicker Archaean oceanic crust? As discussed in this paper, and in detail in an earlier review by Bickle et al. (1994), apart from a superficial resemblance in terms of abundant pillow lavas, the comparison fails in all other critical details. No complete ophiolite stratigraphy (Anonymous 1972), involving (1) harzburgitic tectonites, (2) ultramafic and gabbroic cumulate rocks, (3) a sheeted dyke complex, and (4) a sediment-free massive and pillow lava sequence, has been documented in Archaean time, although claims to the contrary have been made. In the Slave craton, Kusky (1990) interpreted rare ultramafic rocks near the base of the stratigraphic sequence as exotic mantle slices. However, these rocks show cumulate textures and occur as sills within the basal sedimentary section (Fig. 2b; Bleeker et al. 1999a). Although mafic dykes are locally numerous, and may even be sheeted on an outcrop scale (Helmstaedt et al. 1986; but see discussion by Bickle et al. 1994), none are part of an extensive sheeted dyke complex. Instead, they belong to several temporally distinct swarms that can be followed into underlying basement rocks (Lambert et al. 1992; Bleeker et al. 1999a, b).

Perhaps the most compelling argument against an ophiolitic origin is that many greenstone successions show an increase in felsic volcanic and (or) sedimentary rocks towards their top, or for cyclical successions, towards the top of each cycle. These intercalations typically provide evidence that felsic central volcanic complexes became emergent and became a source of volcaniclastic debris. Age dating of these felsic horizons shows that a typical greenstone sequence may span 10–30 (up to 50) Ma of episodic volcanism, a period long enough that even at moderate spreading rates (e.g. $2.5 \, \text{cm a}^{-1}$) a spreading-related volcanic edifice would have drifted hundreds of kilometres off the spreading axis (250 km per 10 Ma).

Are the cyclic, bimodal, mafic to felsic sequences such as those of the Abitibi greenstone belt compatible with an oceanic plateau origin? The zircon inheritance as well as the abundance of felsic volcanic rocks seem incompatible with such an origin. Hence, where are the Ontong–Java plateaux of the past (cf. Kerr et al. 2000)?

Lateral heterogeneity of Archaean cratons

Many Archaean cratons show marked lateral heterogeneity. This heterogeneity can be expressed by many different attributes: ages and distribution of greenstone belts, plutonic and (or) metamorphic age domains, contrasting structural subprovinces, or average Nd crustal extraction ages. Examples are the contrasts between the eastern and west–central parts of the Slave craton (Figs 2 and 3), a similar apparent contrast between the Eastern Goldfields and Southern Cross domains of the Yilgarn craton, and the age contrast between ancient nuclei of the western Superior craton, particularly the $c.\,3.0\,\text{Ga}$ North Caribou terrane and the $c.\,3.0$–$3.6\,\text{Ga}$ Winnipeg River terrane (Thurston et al. 1991; Percival et al. 2001b). Some form of plate tectonics involving lateral 'docking' of contrasting terranes provides an appealing mechanism to explain these contrasts.

In the Slave craton, where the eastern part of the craton has been interpreted as an accreted arc, the contrast is largely defined by isotopic tracer data that suggest that the basement of the eastern Slave is not significantly older than the $c.\,2.7\,\text{Ga}$ volcanic rocks at the surface (Davis & Hegner 1992; Thorpe et al. 1992a). This contrasts with the ages for the Central Slave Basement Complex to the west, which scatter between 2.82 and 4.03 Ga. Recent mapping has shown, however, that the stratigraphic units that were thought to define the eastern Slave or 'Hackett River arc' (Kusky 1989, 1990), including 2690–2670 Ma volcanic rocks and overlying turbidites, overlap the craton from east to west and show no evidence of a suture nor of an exotic arc terrane (Bleeker 2001). Therefore, if the eastern Slave is an accreted arc it must have docked sometime before 2690 Ma, during or immediately following the time that the Central Slave Basement Complex was undergoing flood basalt volcanism and rifting (see Figs 2 and 3). Hence, perhaps a more realistic interpretation is that the eastern Slave domain represents a more highly extended rift or marginal basin developed within or on the edge of the Central Slave Basement Complex. A structural observation that supports this hypothesis is that the sediment-dominated eastern Slave domain exposes, on average, a more shallow structural level (Fig. 3). Less extensive exhumation may be a consequence of thinner, more extended crust. Whether the $c.\,2.7\,\text{Ga}$ rifting event was driven by intraplate processes (e.g. active rifting in response to a mantle plume that contributed to flood basalt volcanism), or by external boundary forces (e.g. passive rifting; or back-arc rifting) is not yet resolved.

As alluded to above, in the western Superior craton a strong case is building for $c.\,2.7\,\text{Ga}$ lateral interaction of distinct crustal nuclei; for example, the North Caribou and Winnipeg River terranes (e.g. Percival et al. 2001b). It is remarkable, however, that even in this more compelling case for lateral accretion we lack many of the elements of modern plate tectonics. Although there are rift-related basalts, neither of the microcontinents involved has an identifiable passive margin sequence. Nor are there true ophiolites, mélange belts or identifiable sutures. Intervening sedimentary belts (e.g. English River and Quetico metasedimentary belts) have been interpreted as accretionary prisms (Percival & Williams 1989; Card 1990), but are younger than, and stratigraphically overlap, adjacent volcanic belts. This stratigraphic development is broadly similar to that of the turbidites in the Slave craton (Fig. 3) and seems inconsistent with a accretionary prism model. There is a detailed seismic profile that supports a generally south-verging imbricate structure (White et al. 2001), but no detailed structural–stratigraphic cross-section. Perhaps this is one of the better places in the world to unravel the differences between modern plate tectonics and late Archaean microcontinent interaction.

Broad domes, narrow synclines

Among the many geological attributes of Archaean cratons that appear distinct, if not unique, two features stand out. These features are intimately related to each other and, ultimately, to significantly higher heat production in both crust and mantle.

The first is stratigraphic in nature and involves the relative abundance, particularly compared with sedimentary rocks, of mafic and ultramafic lavas piled in sequences of 1–10 km thickness, or sometimes multiple sequences, on pre-existing felsic crust. Modern flood basalts, related to mantle plumes, rifts, or both, are the closest analogue but are not nearly as dominant in the modern world relative to sediment-dominated cover sequences and passive margins.

The second is structural in nature and involves the high-amplitude, upright, granitoid or basement-cored domes that are a ubiquitous feature of most cratons, particularly in areas where there is evidence of older sialic basement (e.g. Choukroune et al. 1995; Chardon et al. 1998; Collins et al. 1998). Both flood basalts and high-amplitude structural domes are well exemplified by the cross-section through the central Slave craton (Fig. 3), and have long been known from mapping and satellite imagery of the superbly exposed eastern Pilbara craton (Fig. 4; e.g. Hickman 1983, 1984; Collins 1989; Collins et al. 1998; Van Kranendonk et al. 2002). In his classical paper on the geology of the Zimbabwe craton, Macgregor (1951) drew attention to similar domal features and referred to them as his 'gregarious batholiths'.

In the Slave craton, these broad domes are largely cored by basement, but were further inflated and amplified by multiple phases of granitoid intrusions, some of which are subvolcanic intrusions of similar ages to the volcanic rocks in the cover. Late-kinematic granites (Fig. 3m) are locally large contributors to the volume of the domes. These crustally derived granites, which in the southern Slave craton are referred to as the Morose Suite (c. 2590–2580 Ma, Davis & Bleeker 1999), intruded c. 100 Ma after initial flood basalt volcanism. Perhaps the most remarkable attribute of the domes is their amplitude, with culminations projecting c. 10–15 km above the present erosion surface and troughs projecting 10–15 km into the subsurface. Similar amplitudes have been derived by geophysical modelling for the east Pilbara domes (Wellman 2000). Hence, crest-to-trough structural relief is approximately equal to the present thickness of the crust. The horizontal diameter of the domes is typically of similar magnitude (30–50 km across). An identical picture has emerged from the Zimbabwe

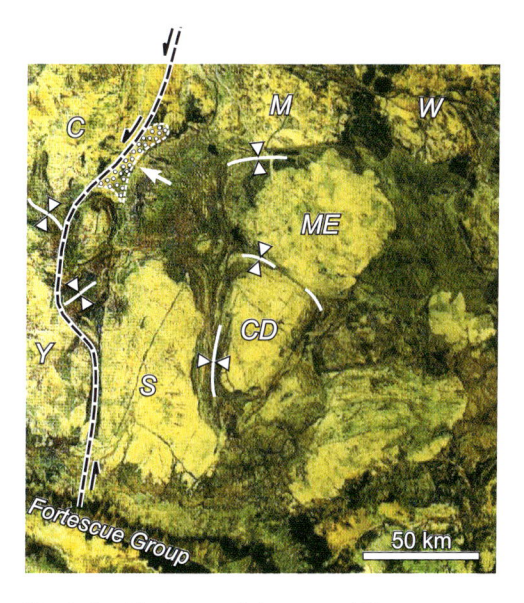

Fig. 4. Satellite image of the eastern Pilbara craton, showing the large granitoid–gneiss domes (light toned, oval structures: C Carlindi dome; CD, Corunna Downs dome; M Muccan dome; ME, Mount Edgar dome; S Shaw dome; W Warrawagine dome; Y Yule dome) and tight intervening greenstone synclines (greenstone in dark tones; some synclines highlighted by white lines). Scale and spacing of the domes should be noted; typical diameters of the domes are c. 40–60 km. Approximate location of the late-kinematic Lalla Rookh clastic basin (at white arrow) and the sinistral Mulgandinnah strike-slip corridor are indicated (see text for references).

craton, where ancient basement gneisses are now known from several of the domes (e.g. Horstwood et al. 1999), whereas others are cored by younger granitoid suites, particularly the late-kinematic, c. 2600 Ma, Chilimanzi Suite (e.g. Wilson et al. 1978, 1995; Jelsma et al. 1996).

Similarities in the dimensions of these domes across several cratons (see, for instance, fig. 7.4 of Cloud 1988, p. 150) suggest that their size and spacings are controlled by the thickness and rheological properties of the crust, almost certainly through growth of Rayleigh–Taylor-type instabilities (e.g. Turcotte & Schubert 1982; Davies 1999) in a hot and therefore weak crust in which a thick basaltic cover overlying felsic basement (or an intrusive tonalite–granodiorite layer) created a density inversion (Macgregor 1951; Ramberg 1972, 1981; Mareschal & West 1980; Dixon & Summers 1983).

In a viscous medium, the initial growth of buoyant instabilities of height h follows an exponential function

$$h(t) = h_0 \, e^{(t/\tau)}$$

in which h_0 is an integration constant that represents the initial height of the perturbation and τ is a time constant that is dependent on width w (horizontal crest-to-trough distance) of the initial perturbation:

$$\tau = \frac{\mu}{g\Delta\rho w}.$$

The value of the time constant is further dependent on the viscosity (μ), gravitational acceleration (g), and the density contrast ($\Delta\rho$) of the inversion; its magnitude gives a measure of the time scale involved in growth. From a range of initial perturbations, those for which the time constant τ is minimized will grow the fastest and will generally dominate the final size spectrum (Fig. 5a and b; but see Schmeling 1987). This condition is met for perturbations that satisfy

$$w \approx D$$

where D is the depth of the buoyant layer (Davies 1999, p. 222). Hence, buoyantly rising lower-crustal material, originating at a depth of c. 20–30 km, would create domes with similar crest-to-trough distances, or c. 40–60 km horizontal spacing between dome centres.[4] In the case of broad domes and narrow intervening synclines this would produce domes with a typical diameter of c. 30–50 km, in agreement with field observations. Interestingly, by significantly reducing the bulk effective viscosity of the buoyant material, granitic partial melt will strongly increase the growth rate but not necessarily change the $w \approx D$ relationship. Granite magmatism therefore may trigger or enhance dome growth but may not affect the spacing and diameter of the domes.

In detail, however, the growth of the domes appears more complicated in many instances. First of all, the rheological behaviour of the crust is considerably more complicated than that of a newtonian viscous fluid. Particularly, the cool upper crust will have a finite yield strength that any buoyant instability will have to overcome before it can rise significantly and break through to the surface. Any structure or deformation event that weakens the upper crust may

---

[4] Assuming a constant viscosity, this approximation arises because the growth of long-wavelength domes is impeded by the increased volume of lateral flow required to maintain their growth. If the mid- to lower-crustal layer that feeds growth of the buoyant dome has a lower viscosity (e.g. higher temperature, or substantial partial melt), this may allow domes for which w is somewhat larger than D (e.g. see Mareschal & West 1980).

Fig. 5. (**a, b**) Schematic diagram illustrating buoyant growth of Rayleigh–Taylor instabilities in a crust in which a density inversion ($\rho_{\text{upper crust}} > \rho_{\text{lower crust}}$) is induced by widespread basaltic volcanism (c. 2.9–3.1 g cm^{-3}) and sedimentation on pre-existing sialic basement (c. 2.7 g cm^{-3}). The buoyancy force of an initial perturbation is a function of its volume. Growth of small perturbations (with crest-to-trough length scale w_s) is therefore slow, whereas growth of long-wavelength perturbations (w_l) is impeded by the high-volume of lateral viscous flow required to feed their growth. Perturbations of intermediate length scale (w_{int}) for which $w \approx D$ (with D being the depth of the buoyantly rising layer) will grow fastest and thus dominate the final size spectrum. Surface uplift above the composite granite and gneiss domes may lead to erosional unroofing of the domes, with coeval sedimentation of late-kinematic conglomerate and sandstone sequences in subsiding flanking basins. (**c**) Typical example of a late-kinematic conglomerate–sandstone sequence, from the centre of the Cross Lake greenstone belt, Superior craton. The layer of granitoid boulder conglomerate within coarse sandstones should be noted. (**d, e**) Late-kinematic, c. 2600–2580 Ma, polymict conglomerates from the Slave craton, occurring in narrow, steeply infolded panels on the flanks of high-amplitude granitoid–gneiss domes: Jackson Lake Formation from the Yellowknife greenstone belt (**d**); Beaulieu Rapids Formation from the centre of the Beaulieu River belt (**e**). Both photographs show vertical outcrop faces with deformed clasts that indicate steeply plunging prolate strain ellipsoids (inset).

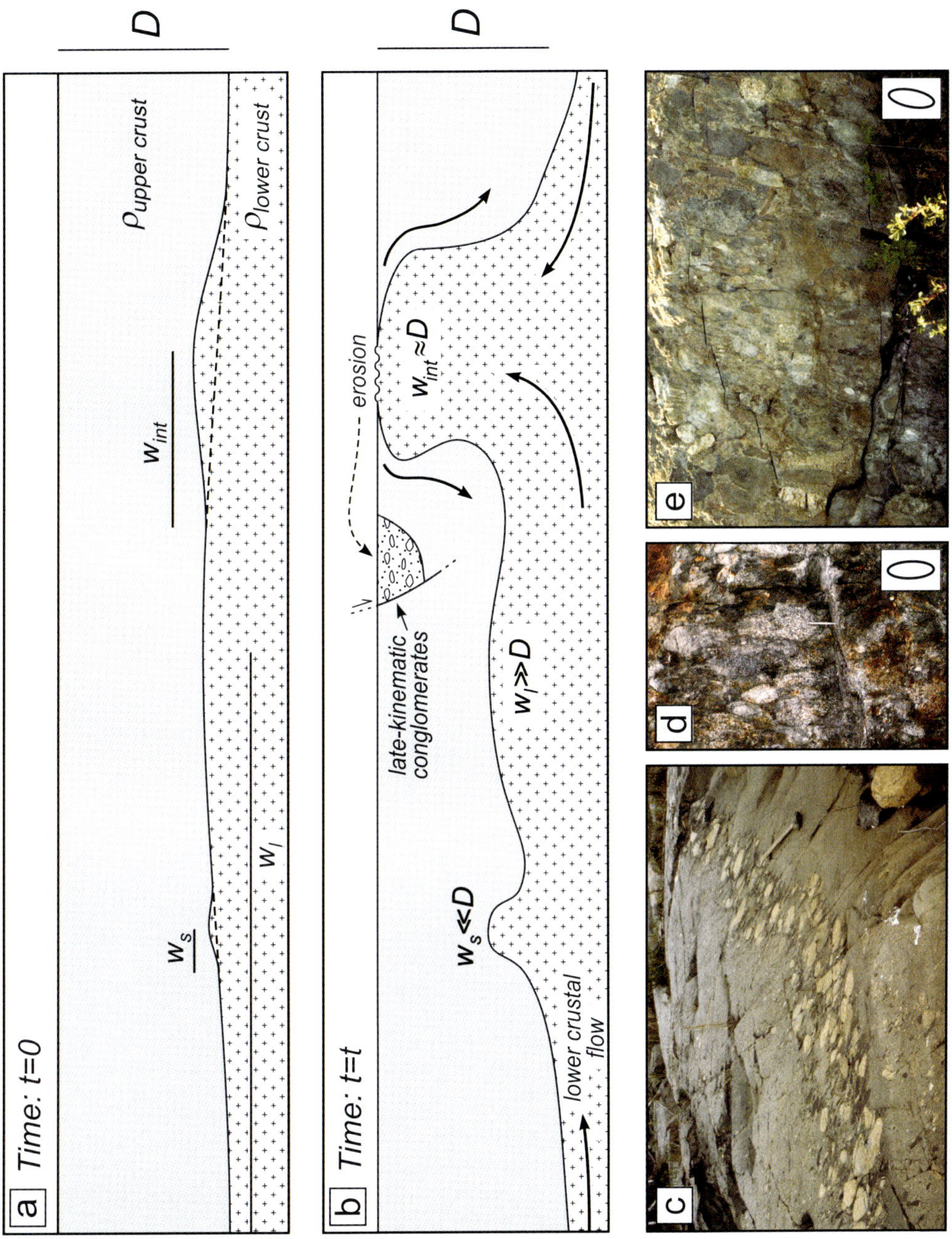
a
Time: t=0
D
ρupper crust
ρlower crust
wint
ws
wl
b
Time: t=t
D
erosion
wint≈D
late-kinematic conglomerates
wl≫D
ws≪D
lower crustal flow
c
d
e

thus localize or trigger dome growth. Regional extension is thought to be particularly effective as it weakens and thins the strong overburden while enhancing initial perturbations in the buoyant layer. Indeed, it plays a critical role in the buoyant rise of modern core complexes (e.g. Martinez *et al.* 2001) and salt diapirs (Jackson & Vendeville 1994), two classes of structures that show interesting analogies with Archaean granitoid–gneiss domes.

The buoyancy force of an initial density instability is a function of its volume. In the absence of extension, a relatively strong overburden may thus be able to resist the growth of all but the very largest of initial instabilities. This could explain why subvolcanic granitoid complexes, which provide high-amplitude primary instabilities, may be favoured to grow into final domes (e.g. Hickman 1984; Collins 1989).

Finally, there is a common association between high-amplitude domes and regional folds. In the Slave craton, the domes are culminations along regional antiforms that are part of a craton-scale F_1 fold belt (Fig. 2a) that developed at *c.* 2635–2640 Ma, at least 60 Ma after flood basalt volcanism but before voluminous granitoid magmatism (Davis & Bleeker 1999). Later, at *c.* 2600 Ma, these antiforms were refolded into mushroom-type interference patterns by a second craton-scale fold set (F_2; Bleeker 1996). During this phase, the basement-cored domes were strongly amplified but maintained a systematic relationship with regional fold patterns. Final amplification occurred during ascent of *c.* 2590–2580 Ma late-kinematic granites, which intruded synkinematically with respect to a third set of folds (F_3). Hence, in the Slave craton dome growth and amplification was episodic and appears to have been triggered or enhanced by both regional deformation and large-scale ascent of lower-crust-derived granitic magmas. These observations argue for a feedback mechanism between buoyancy-driven processes and externally controlled deformation. Initial Rayleigh–Taylor instabilities may have resulted in low-amplitude domes which then controlled the location and wavelength of large-scale antiforms during regional shortening events (F_1). These buoyant domal antiforms then predetermined the loci of further antiformal growth during subsequent deformation (F_2), before undergoing final amplification by large volumes of granitic magma ascending into their cores. A similar interplay between late-stage dome growth and regional deformation was documented by Blewett (2001, 2002) for the Pilbara craton, whereas other workers in the Pilbara have advocated a purely buoyancy-driven process (e.g. Collins 1989; Collins *et al.* 1998).

Irrespective of the details of the dome growth processes, there is a net flow of volcano-sedimentary cover rocks, dominated by thick, dense, basalt sequences, into intervening synclines (Figs 3–5). These synclines are typically tight to isoclinal with nearly vertical greenstone stratigraphy facing away from the basement or granitoid-cored domes. Where the synclines are narrow, between several large domes, the supracrustal stratigraphy is typically highly attenuated and foliated, and locally even partly excised. Asymmetry may develop if one dome rises higher, dragging lower stratigraphic levels higher on one limb of the syncline. The domes commonly develop 'mushroom-like' heads, resulting in overturning of the flanking synclines, an example being the western flank of the Sleepy Dragon Complex dome (Fig. 3), which is locally overturned with layering in the basal sedimentary rocks and overlying basalts dipping 80° into the dome. Finally, late-stage strike-slip faulting, typically along the highly foliated synclinal corridors, may further complicate the structure of the greenstones (e.g. Chen *et al.* 2001). The overall material flow results in a regional structure in which uppermost stratigraphic units, typically characterized by greenschist- or even sub-greenschist-facies metamorphism, are preserved in the cores of synclines or synclinoria and juxtaposed on either side against gneisses or granitoids metamorphosed at mid- to lower-crustal levels. Where the synclines are broad, the Archaean supracrustal rocks may show remarkable preservation with merely open folding and little or no penetrative strain (e.g. Fig. 3j). Where the synclines are tight to isoclinal, stretching lineations are typically steep with highly prolate strains in the constricted areas ('sinks') between domes (e.g. Gorman *et al.* 1978; Choukroune *et al.* 1995; Collins *et al.* 1998; see also Eskola 1949; Brun 1980; Brun *et al.* 1981; Marshak *et al.* 1997, for analogous structures of Palaeoproterozoic age).

Hence, resulting deformation and metamorphic facies tend to be highly domainal, defining patterns that are spatially centred on the domes. Low-pressure facies series dominate and thermal pulses tend to be linked to heat advected by episodic granitoid magmatic events and (or) the ascent of hot gneiss domes. These domainal patterns, without linear, regional-scale, belts of tectonites, no consistent structural vergence, and no clear 'peak of regional metamorphism' (e.g. Warren & Ellis 1996; Bethune *et al.* 1999), are ubiquitous in Archaean granite–greenstone terranes and are unlike the deformational and metamorphic style associated with thrust-dominated orogens. It is possible that early thrusting and accretion of volcanic packages played a role in some greenstone terranes, but the signatures of these early processes must be resolved, uniquely, from the later deformation patterns. This has rarely been done to a satisfactory level. In the Pilbara craton, for instance, Bicke et al. (1980, 1985) documented early recumbent folds and attributed these to an early phase of Alpine-style thrusting. These folds occur in the complex contact zone between volcanic rocks and the subvolcanic tonalite–trondhjemite–granodiorite (TTG) intrusive rocks of the Shaw granitoid complex. It has proven difficult to relate these structures to a regional thrust–fold belt and hence their regional significance remains unclear.

In nearly all granite–greenstone terranes, later structural processes involved at least one phase of partial convective overturn to establish or re-establish a stable density configuration. Widespread granite magmatism played an important role in these later processes by enhancing mobility, thus facilitating or in some cases triggering the overturn. Equally important, the very presence of anatectic granite signals that after rapid deposition of a thick volcano-sedimentary pile the 'thermal incubation' (e.g. Sandiford et al. pers. comm.) of the mid- and lower crust reached its peak and that the crust had become sufficiently hot and weak for buoyancy forces to overcome the yield strength of the brittle upper crust. Differences in composition and thicknesses of crustal profiles (e.g. concentration and vertical distribution of heat-producing elements; thickness of volcano-sedimentary cover) within and between granite–greenstone terranes, and of mantle heat fluxes into the bases of their crust, may explain why these incubation times vary significantly between granite–greenstone terranes. Finally, lower-crustal granites also transfer, irreversibly, minimum melt fractions, aqueous fluids, and heat-producing elements to the upper crust (see also Campbell & Hill 1988) leading to cooling and stiffening of the lower crust and final mechanical coupling with the lithospheric mantle. In most cratons, final stabilization thus occurred shortly after a 'final granite bloom' event as exemplified by the $c.$ 2590–2580 Ma Morose Suite of the Slave craton (Davis & Bleeker 1999) or the similar age Chilimanzi Suite of the Zimbabwe craton (Wilson et al. 1978, 1995; Jelsma et al. 1996).

An important but commonly neglected realization is that whereas regional deformation can be turned on or off depending on external processes, buoyancy-driven processes will operate as long as a significant density inversion remains; these processes can be retarded by low temperature (i.e. high viscosity) but they cannot be turned off, unless the long-term yield strength of the brittle upper crust is too high. A significantly higher heat production in Archaean time ($c.$ 2–4 times modern values; e.g. Richter 1988; Pollack 1997) dictates that buoyancy forces were proportionally more important in Archaean than in Phanerozoic time. The debate whether buoyancy-driven processes ('vertical tectonics') operated in granite–greenstone terranes is therefore not very useful and should be rephrased into 'what else went on besides the buoyant ascent of composite granitoid–gneiss domes'.

Late-kinematic conglomerates

Modelling shows that ascent of large domes is accompanied by significant surface uplift (e.g. 500–1000 m; Mareschal & West 1980) and sinking of denser supracrustal material in the flanking synclines. Increased erosion is thus expected above the domes, possibly linked to clastic sedimentation in flanking basins (Fig. 5b; see also Anhaeusser 1971; Campbell & Hill 1988).

Indeed, restricted, late-kinematic, clastic basins are a ubiquitous feature of granite–greenstone terranes (Fig. 5c–e), typically occurring as tight to isoclinally folded synclinal panels on the flanks of, or between, high-amplitude domes. These basin are filled with locally derived conglomerates and cross-bedded sandstones (e.g. Fig. 5c), whereas prominent granitoid clasts indicate erosional unroofing of nearby plutonic complexes. Although difficult to date, relative age relationships and detrital zircon studies invariably show that these basins formed well after the main phase of volcanic activity in any greenstone terrane, but approximately coeval with voluminous late-stage granitoid magmatism. In the Slave craton, for instance, deposition of the late-kinematic conglomerates and sandstones is bracketed at $c.$ 2600–2580 Ma (Isachsen & Bowring 1994; Sircombe & Bleeker, unpubl. data), i.e. broadly coeval with the main amplification stage of the high-amplitude domes.

Other distinguishing features of these late clastic basins are as follows: (1) they have a restricted volume; (2) some include rare intercalations of alkaline lavas and tuffaceous deposits that, where dated, are of identical age to nearby granite domes or plutons; (3) there is a general lack of cross-cutting granite dykes, indicating their relative young age and near-surface, flanking, location during terminal granitoid pluton emplacement; (4) although late, the tightly folded basins experienced the tail if not the peak of the last thermal pulse of metamorphism; (5) where the conglomerates are highly strained, clasts typically show steeply plunging prolate strain ellipsoids (e.g. Fig. 5d and e). In fact, these clastic basins and their features are so characteristic for granite–greenstone terranes that across the six or seven distinct Archaean cratons of the Canadian Shield they are commonly referred to as 'Timiskaming-type' deposits, after their type locality in the Abitibi greenstone belt of the Superior craton. Their diagnostic features are compatible with a formation on the subsiding flanks of ascending composite granitoid–gneiss domes, late in the evo-lution of granite–greenstone terranes (Fig. 5b). In many ways their subsidence mechanism can be compared with that of peripheral 'withdrawal basins' on the flanks of large salt diapirs (e.g. Trusheim 1960; Jackson & Talbot 1994; Giles & Lawton 1999). The origin of these late-kinematic clastic basins is thus inherently tied to the internal evolution of granite–greenstone terranes, explaining why similar deposits are less typical for post-Archaean settings.

An alternative explanation is that the late clastic basins represent remnants of pull-apart basins formed along late-stage strike-slip fault zones (e.g. Krapez & Barley 1987; Bleeker 1999). Indeed, many of the steeply dipping, late-kinematic, conglomerate panels appear spatially associated with structurally late, regional, fault zones along corridors of highly strained greenstones, but a causal relationship between strike-slip movement and rapid basin subsidence can rarely be proven. Hence, the faults may post-date sedimentation and merely follow the same highly strained greenstone corridors that, earlier, were sites of basin subsidence. The strike-slip model is less successful in explaining the temporal relationship between clastic sedimentation, late-stage granitoid magmatism, unroofing of granite plutons, and the last thermal pulse of metamorphism.

Krapez & Barley (1987) argued that the late clastic Lalla Rookh basin of the eastern Pilbara craton (see Fig. 4 for approximate location) formed as a sinistral pull-apart structure. Indeed, part of the basin is offset by sinistral faults (Van Kranendonk & Collins 1998), but considered on a larger scale (Fig. 4), the Lalla Rookh basin appears to be located along a short NE-trending segment of a larger sinistral, north–south-trending, fault corridor (e.g. Van Kranendonk & Collins 1998; Zegers et al. 1998). Hence, the larger reference frame suggests that the Lalla Rookh basin occurs within a compressional rather than extensional bend of a regional strike-slip fault zone. Equally important, an estimate of granitoid volumes v. age (Van Kranendonk et al. 2002) suggests that several of the nearby granitoid domes contain significant volumes (up to 75% for the Yule dome to the SW) of $c.$ 2940 Ma granitoid material. The poorly dated Lalla Rookh basin is considered to be of a similar age.

In their study of the Eastern Goldfields domain of the Yilgarn craton, Campbell & Hill (1988) concluded that not only late, coarse clastic deposits, but all sedimentation (i.e. including earlier turbiditic rocks) was related to internally generated vertical movements of rising granitoid domes and subsiding greenstones. This proposal merits attention. It is indeed striking that in several well-studied granite–greenstone terrains the onset of regional turbidite sedimentation is coeval with voluminous intrusion of TTG plutons. In the Abitibi greenstone belt, voluminous TTG plutonism peaks at 2696–2690 Ma, coeval with or just before the onset of deposition of the turbiditic Porcupine Group (Bleeker & Parrish 1996; Bleeker et al. 1999c). In the Yellowknife Domain of the southwestern Slave craton, a succession of turbiditic sediments of $c.$ 10 km thickness (Burwash Formation, see Fig. 3a, j and k) was deposited between 2680 and 2658 Ma, coeval with felsic volcanism and voluminous, $c.$ 2687–2660 Ma, TTG magmatism in the cores of adjacent domal granitoid complexes (e.g. Sleepy Dragon Complex, Fig. 3a; Bleeker et al. 1999a, b). Similarly, Kamo & Davis (1994) have shown that in the Barberton greenstone belt, Kaapvaal craton, South Africa, deposition of the Fig Tree and Moodies groups was essentially coeval with intrusion of the large Kaap Valley pluton and related TTG intrusions. In all these terranes, detrital zircons in the turbiditic sediments are reflecting predominantly local sources. Other tests, e.g. an approximate mass balance between turbidites and potential domal sources, or geochemistry, could be applied to this question of internal v. external sediment generation. Turbidites of the Slave craton have been successfully modelled as mixtures of nearby volcanic rocks and a minor granitoid component (Jenner et al. 1981; Yamashita & Creaser 1999), entirely consistent with progressive off-stripping and mixing of the volcanic cover (Bleeker et al. 2001) and incipient unroofing of the basement antiforms (with a large proportion of synvolcanic granitoid plutons). Although much more work is needed on this subject, it is intriguing that the cross-section through the Slave craton (Fig. 3) allows, at least qualitatively, an approximate balance between the volume of the turbidites and that of eroded domes and their volcanic cover.

Discussion and conclusions

The lateral heterogeneity of Archaean cratons, together with their structural and stratigraphic complexity, favours a mobile tectonic regime (e.g. Burke et al. 1976) that, at least in some cases, appears to have involved the lateral interaction or accretion of discrete crustal blocks. Indeed, palaeomagnetic data from the Pilbara (e.g. Strik et al. 2001, and references therein) support motion of this craton in Neoarchaean time. Systematic deformation patterns at the scale of cratons (e.g. F_1 fold belt in the Slave craton, Fig. 2a; or the dominant NNW-trending structural grain of much of the Yilgarn craton) indicate that coherent stress patterns were applied and transmitted over large distances, perhaps most easily explained by collision of semi-rigid lithospheric plates. In the Slave craton, the presence of at least two independent craton-scale deformation patterns at high angle to each other (F_1, F_2), and imposed several tens of million years apart, possibly argues for several discrete collision events along the periphery of the growing craton. However, the colliding crustal blocks do not manifest themselves within the realm of the preserved craton, possibly because of subsequent (i.e. Palaeoproterozoic) rifting and dispersal.

Large-scale strike-slip faults that are a ubiquitous feature of late Archaean cratons also favour collisional interaction of rigid plates (Sleep 1992). These faults probably accommodated strike-slip motion associated with oblique convergence on outboard margins, or lateral escape of tectonic blocks during final collision. It also seems certain that these plate-like interactions involved a return flow of hydrated oceanic crust and minor sediments back into the upper mantle, resulting in hydration and metasomatism of the upper mantle and in arc or back-arc geochemical signatures in contemporaneous volcanic rocks extruded at the surface. Even boninitic signatures, suggesting rapid extension in fore-arc or infant arc settings (Stern & Bloomer 1992), have now been described from several greenstone terranes (e.g. Kerrich et al. 1998).

However, few Archaean granite–greenstone terrains show a fully convincing suite of plate tectonic elements, and widely referenced examples of Archaean ophiolite obduction (Kusky 1990; Kusky & Kidd 1992) have been refuted (Bickle et al. 1994; this study). As discussed in detail in this paper, basalts of the west–central Slave craton are not oceanic in origin but were extruded on pre-existing sialic basement, in response to plume- and/or rift-related volcanism (Bleeker et al. 1999a, b). The critical stratigraphy of the Slave cover sequence (quartzite, BIF, and basalts with minor komatiite and rhyolites) can be mapped across previously proposed terrane boundaries, thus invalidating such terrane divisions. The scale of the basaltic volcanism is similar to that of modern large igneous provinces. Younger volcanic rocks and turbidites overlap much of the craton (Bleeker 2001) and do not define an exotic arc, an accretionary prism, and a foredeep basin (cf. Kusky 1990). Rather than an obducted oceanic plateau, the stratigraphy of the Slave craton (Fig. 2b) invites comparisons with, for instance, the Keweenawan rift (e.g. Ojakangas et al. 2001). In this late Mesoproterozoic rift, an exceedingly thick bimodal basalt ± rhyolite pile overlies rifted basement. The contact between basement and the flood basalt sequence is locally marked by a thin quartz arenite unit, and the large Duluth igneous complex was preferentially intruded along this contact (analogous to the ultramafic sills in the Central Slave Cover Group, Fig. 2b). Basal volcanic rocks are locally pillowed, although most of volcanic pile of the Keweenawan rift extruded subaerially. After the cessation of volcanism, the rift was filled with clastic sediments, some of which are marine. Although the highly extended rift is more than 100 km wide, it never reached an oceanic state. If this analogy is appropriate, it suggests that in the Slave craton, and granite–greenstone terrains in general, extension must have been greater because, after formation of the basal unconformities, conditions became submarine and generally remained so throughout construction of much of the basalt–komatiite lava piles. Typically, only late-stage central dacite–rhyolite volcanoes managed to become emergent, shedding volcaniclastic sediments in surrounding basins.

Generalizing beyond individual studies, it seems clear that the predictive power of the plate tectonic paradigm is much weaker for the Archaean than for the post-Archaean record and simply does not address some of the more characteristic features of Archaean cratons. Against this background, and that of other obvious differences that set the Archaean apart (e.g. higher heat flow; more mafic upper crust, Table 1), a uniformitarian application of plate tectonics to Archaean time seems naive. Archaean lithosphere appears to have been segmented in plate-like entities that were mobile but interacted and reacted in ways that are subtly, if not fundamentally, different from those of modern plates: for instance, less sediment was supplied to continental margins (see also Lowe 1994), probably indicating less relief and less sediment transport by smaller river systems; mature clastic sediments were less well mixed, similarly suggesting restricted continental hinterlands and smaller drainage basins (e.g. Sircombe et al. 2001); although basalt-filled rift basins appear ubiquitous, their flanks were less uplifted and failed to provide abundant coarse-grained, unsorted and immature clastic deposits typical for modern rifts; ophiolites failed to accrete; and if most of the basalt and komatiite sequences discussed in this paper extruded on pre-existing continental crust, even oceanic plateaux failed to accrete (e.g. Saunders *et al.* 1996) or, if they did, remain to be identified; intraplate volcanism and extension were widespread; finally, intraplate reworking of crust in response to density inversions was ubiquitous, resulting in buoyant ascent of composite granitoid–gneiss domes; associated uplift may have contributed to the deposition of post-volcanic sedimentary rocks in granite–greenstone terranes. These examples suggest smaller, hotter and weaker continental plates, with very weak lower crusts and little or no flexural rigidity. Both extension and subsequent shortening events were distributed over large areas, rather than focused in narrow zones, possibly facilitated by decoupling along a hot, weak lower crust (e.g. Royden 1996).

Based on heat-loss arguments (e.g. Bickle 1978, 1986; Davies 1999), it seems inevitable that these proto-continents interacted with a vigorously convecting oceanic realm in which recycling was highly efficient. Locally, flakes of partially hydrated oceanic crust may have piled up, either in intra-oceanic settings or along the margins of the buoyant proto-continents (Hoffman & Ranalli 1988; Davies 1992; de Wit *et al.* 1992). Where thick enough, partial melting of these hydrated basalts produced buoyant tonalite magmas on the one hand and dense eclogitic residues on the other. Delamination of the refractory eclogitic residues ('drip tectonics' of Davies 1992) helps explain the intermediate bulk composition of the continental crust and its depletion in trace elements such as Nb, Ta and Ti (through rutile in eclogites; e.g. Rudnick 1995; Rudnick *et al.* 2000; Zegers & van Keken 2001). Surprisingly though, few of these primitive basalt piles, with ages that predate associated felsic material, have been identified with certainty.

Overall, the spectrum of carefully reconstructed stratigraphies for granite–greenstone terrains around the world is rather restricted, showing only minor variations on a theme that can be exemplified by the stratigraphy for the Slave craton as shown in Figure 2. In other granite–greenstone terrains, the overall age range may be longer, involving perhaps several cycles, but the general lithological associations are remarkably similar. This strongly suggests that granite–greenstone terrains were the product of a more restricted range of processes, with lesser variables, than those that control the enormous variety and complexity of modern plate-tectonic environments and their resulting spectrum of rock assemblages. Most preserved Archaean granite–greenstone terranes appear to represent highly contracted extensional basins with protracted, commonly cylcic, volcanic activity and extensive autochthonous to parautochthonous stratigraphy (e.g. Groves *et al.* 1978). Some partial analogues of these settings do occur in post-Archaean time: for instance, the *c.* 1.1 Ga basalt ± rhyolite and sediment-filled Keweenawan rift of North America (Green 1977; Behrendt *et al.* 1988; Ojakangas *et al.* 2001), the complex volcanic rifted Vöring margin off the coast of Norway (Eldholm *et al.* 1989; Parson *et al.* 1989; Reemst & Cloetingh 2000), and perhaps the collapsed marginal basin of the Cretaceous Rocas Verdes Complex, Chile (e.g. Tarney *et al.* 1976). The preservation potential of these ensialic or marginal settings is high compared with that of intra-oceanic settings and therefore they should indeed be favoured in the rock record. It is not surprising therefore that genetic links between many greenstone sequences and pre-existing continental basement, although commonly subtle, are ubiquitous. Even in the case of the Abitibi greenstone belt, sufficient evidence exists to seriously question, if not rule out, a model of accretion of oceanic allochthons. One of the more recent proposals for an Archaean obducted oceanic plateau is that of Puchtel *et al.* (1998) for the *c.* 2.8 Ga Kostomuksha greenstone belt of the Karelian craton. They have described yet another example of a primitive basalt–komatiite succession, several kilometres thick, that is spatially associated with tonalitic gneisses and a sedimentary sequence consisting of quartzites and BIF. The mafic–ultramafic volcanic pile contains a significant proportion of rhyolites, however, which are characterized by ε_{Nd} values (2.8 Ga) between -1 and -6. These observations, which Puchtel *et al.* (1998) chose to ignore in their interpretation, appear to rule out an oceanic plateau setting and instead favour interaction with significantly older crust. If correct, the *c.* 2.8 Ga Kostomuksha greenstone belt preserves a sequence very similar to that overlying the Central Slave Basement Complex of the Slave craton. Hence, most if not all well-characterized komatiite–basalt sequences discussed in the present paper appear to have been extruded on extended and submerged continental crust (see also Bickle *et al.* 1994; Arndt 1999). The single modern example of partial accretion–obduction of an oceanic basalt–picrite–komatiite sequence, the Caribbean–Columbian plateau (Storey *et al.* 1991), is therefore not a good analogue for Archaean settings.

Some of the extensional basins from which granite–greenstone terranes evolved may well have occurred in an 'intraplate' setting (e.g. analogues to the Keweenawan rift). Attempts to interpret these environments in terms of complete Wilson cycles (Kusky 1989, 1990) are misguided. Other extensional basins may have occurred in broad plate margin settings and their dynamics may have been controlled by plate tectonic-like processes (e.g. back-arc rifting). Even in the latter case, the sutures that mark the active plate boundaries at the time of basin development or later during convergence may well lie outside the realm of a typical granite–greenstone terrane or, worse, outside the preserved extent of small crustal fragments such as the Slave or Zimbabwe cratons. Indeed, scale (e.g. size of an arc and back–arc system, a typical arc–trench gap, or the diameter of an oceanic plateau, etc.) is an important but commonly neglected aspect of the study of ancient tectonic systems and should be incorporated in any rigorous comparison with modern plate-tectonic environments. At a basic level, this requires construction of well-constrained and properly scaled structural–stratigraphic cross-sections. More challenging, it will also require successful correlation of the 35 or so remaining cratonic fragments, to reconstruct the larger landmasses ('supercratons'; Bleeker 2002) and thus more complete tectonic systems of late Archaean time from which the present cratons are merely rifted and drifted fragments.

I wish to thank many of my colleagues at the Geological Survey of Canada for stimulating discussions on Archaean tectonics. My work on the Slave craton has benefited from collaboration with two superb geochronologists, W. Davis (GSC) and J. Ketchum (Royal Ontario Museum). Comments by C. van Staal, W. Hamilton, J. Percival, S. Hanmer, R. Baragar and M. van Kranendonk, and a formal review by M. Bickle, helped to sharpen some of the arguments. This paper is Geological Survey of Canada Contribution 2001235.

References

ALBARÈDE, F. 1998. The growth of continental crust. *Tectonophysics*, **296**, 1–14.
ALEXANDER, C. M. O. D., BOSS, A. P. & CARLSON, R. W. 2001. The early evolution of the inner solar system: a meteoritic perspective. *Science*, **293**, 64–68.
ALLÈGRE, C. J., MANHES, G. & GÖPEL, C. 1995. The age of the Earth. *Geochimica et Cosmochimica Acta*, **59**, 1445–1456.
ANHAEUSSER, C. R. 1971. The Barberton Mountain Land, South Africa; a guide to the understanding of the Archaean geology of Western Australia. *In*: *Symposium on Archaean Rocks*. Geological Society of Australia Special Publication, **3**, 103–111.
ANONYMOUS 1972. Ophiolites. *Geotimes*, **17**, 24–25.
ARNDT, N. 1999. Why was flood volcanism on submerged continental platforms so common in the Precambrian? *Precambrian Research*, **97**, 155–164.

ARNDT, N. T. & JENNER, G. A. 1986. Crustally contaminated komatiites and basalts from Kambalda, Western Australia. *Chemical Geology*, **56**, 229–255.
ARNDT, N. T., ALBARÈDE, F. & NISBET, E. G. 1997. Mafic and ultramafic magmatism. *In*: DE WIT, M. J. & ASHWAL, L. D. (eds) *Greenstone Belts*. Oxford Monographs on Geology and Geophysics, **35**, 223–232.
BARAGAR, W. R. A. & MCGLYNN, J. C. 1976. *Early Archean Basement in the Canadian Shield: a Review of the Evidence*. Geological Survey of Canada Paper, **76-14**, 21.
BARRIE, C. T. & DAVIS, D. W. 1990. Timing of magmatism and deformation in the Kamiskotia–Kidd Creek area, western Abitibi Subprovince, Canada. *Precambrian Research*, **46**, 217–240.
BARRIE, C. T. & SHIREY, S. B. 1991. Nd- and Sr-isotope systematics for the Kamiskotia–Montcalm area; implications for the formation of late Archean crust in the western Abitibi Subprovince, Canada. *Canadian Journal of Earth Sciences*, **28**, 58–76.
BEATTY, J. K., PETERSEN, C. C. & CHAIKIN, A. 1999. *The New Solar System*, 4th edn. Cambridge University Press, Cambridge.
BEHRENDT, J. C., GREEN, A. G., CANNON, W. F. & 5 OTHERS 1988. Crustal structure of the Midcontinent rift system; results from GLIMPCE deep seismic reflection profiles. *Geology*, **16**, 81–85.
BETHUNE, K. M., VILLENEUVE, M. E. & BLEEKER, W. 1999. Laser ^{40}Ar/^{39}Ar thermochronology of Archean rocks in Yellowknife Domain, southwestern Slave Province: insights into the cooling history of an Archean granite–greenstone terrane. *Canadian Journal of Earth Sciences*, **36**, 1189–1206.
BICKLE, M. J. 1978. Heat loss from the Earth; a constraint on Archaean tectonics from the relation between geothermal gradients and the rate of plate production. *Earth and Planetary Science Letters*, **40**, 301–315.
BICKLE, M. J. 1986. Implications of melting for stabilisation of the lithosphere and heat loss in the Archaean. *Earth and Planetary Science Letters*, **80**, 314–324.
BICKLE, M. J. & NISBET, E. G. 1993. *The Geology of the Belingwe Greenstone Belt, Zimbabwe – a Study of the Evolution of Archaean Continental Crust*. Geological Society of Zimbabwe Special Publication, **2**, 239.
BICKLE, M. J., BETTENAY, L. F., BOULTER, C. A., GROVES, D. I. & MORANT, P. 1980. Horizontal tectonic interaction of an Archean gneiss belt and greenstones, Pilbara Block, Western Australia. *Geology*, **8**, 525–529.
BICKLE, M. J., MARTIN, A. & NISBET, E. G. 1975. Basaltic and peridotitic komatiites and stromatolites above a basal unconformity in the Belingwe greenstone belt, Rhodesia. *Earth and Planetary Science Letters*, **27**, 155–162.
BICKLE, M. J., MORANT, P., BETTENAY, L. F., BOULTER, C. A., BLAKE, T. S. & GROVES, D. I. 1985. Archean tectonics of the Shaw Batholith, Pilbara Block, Western Australia: structural and metamorphic tests of the batholith concept. *In*: AYRES, L. D., THURSTON, P. C., CARD, K. D. &

WEBER, W. (eds) *Evolution of Archean Supracrustal Sequences*. Geological Association of Canada Special Paper, **28**, 325–341.

BICKLE, M. J., NISBET, E. G. & MARTIN, A. 1994. Archean greenstone belts are not oceanic crust. *Journal of Geology*, **102**, 121–138.

BLACK, L. P., WILLIAMS, I. S. & COMPSTON, W. 1986. Four zircon ages from one rock: the history of a 3930 Ma-old granulite from Mount Sones, Enderby Land, Antarctica. *Contributions to Mineralogy and Petrology*, **94**, 427–437.

BLEEKER, W. 1990. *Evolution of the Thompson Nickel Belt and its nickel deposits, Manitoba, Canada*. PhD thesis. University of New Brunswick, Fredericton, NB.

BLEEKER, W. 1996. Thematic structural studies in the Slave Province, Northwest Territories: the Sleepy Dragon Complex. Geological Survey of Canada Current Research, **1996-C**, 37–48.

BLEEKER, W. 1999. Structure, stratigraphy, and primary setting of the Kidd Creek volcanogenic massive sulphide deposits: a semiquantitative reconstruction. *In*: HANNINGTON, M. D. & BARRIE, C. T. (eds) *The Giant Kidd Creek Volcanogenic Massive Sulfide Deposit, Western Abitibi Subprovince, Canada*. Economic Geology Monograph, **10**, 71–121.

BLEEKER, W. 2001. *The ca. 2680 Ma Raquette Lake Formation and Correlative Units across the Slave Province, Northwest Territories: Evidence for a Craton-scale Overlap Sequence*. Geological Survey of Canada Current Research, **2001-C7**.

BLEEKER, W. 2002. The late Archean record: a puzzle in ca. 35 pieces. *Lithos*, (in press).

BLEEKER, W. & DAVIS, W. J. 1999. NATMAP Slave Province Project. *Canadian Journal of Earth Sciences*, **36**, 1033–1238.

BLEEKER, W. & PARRISH, R. R. 1996. Stratigraphy and U–Pb zircon geochronology of Kidd Creek: implications for the formation of giant volcanogenic massive sulphide deposits and the tectonic history of the Abitibi greenstone belt. *Canadian Journal of Earth Sciences*, **33**, 1213–1231.

BLEEKER, W., DAVIS, W. J., KETCHUM, J., SIRCOMBE, K. & STERN, R. 2001. Tectonic evolution of the Slave craton, Canada. *In*: CASSIDY, K. F., DUNPHY, J. M. & VAN KRANENDONK, M. J. (eds) *4th International Archaean Symposium, 24–28 September 2001, Extended Abstracts, Perth*. AGSO–Geoscience Australia, Record, **2001/37**, 288–290.

BLEEKER, W., KETCHUM, J. W. F. & DAVIS, W. J. 1999*a*. The Central Slave Basement Complex, Part II: Age and tectonic significance of high-strain zones along the basement–cover contact. *Canadian Journal of Earth Sciences*, **36**, 1111–1130.

BLEEKER, W., KETCHUM, J. W. F., JACKSON, V. A. & VILLENEUVE, M. E. 1999*b*. The Central Slave Basement Complex, Part I: Its structural topology and autochthonous cover. *Canadian Journal of Earth Sciences*, **36**, 1083–1109.

BLEEKER, W., PARRISH, R. R. & SAGER-KINSMAN, A. 1999*c*. High-precision U–Pb geochronology of the late Archean Kidd Creek deposit and the Kidd Volcanic Complex. *In*: HANNINGTON, M. D. & BARRIE, C. T. (eds) *The Giant Kidd Creek Volcanogenic Massive Sulfide Deposit, Western Abitibi Subprovince, Canada*. Economic Geology Monograph, **10**, 43–69.

BLEEKER, W., STERN, R. & SIRCOMBE, K. 2000. *Why the Slave Province, Northwest Territories, Got a Little Bigger*. Geological Survey of Canada Current Research, **2000-C2**.

BLEWETT, R. S. 2001. The tectonic framework and geological evolution of the Archaean Pilbara granite–greenstone terrane: integration of geology and geophysics. *In*: CASSIDY, K. F., DUNPHY, J. M. & VAN KRANENDONK, M. J. (eds) *4th International Archaean Symposium, 24–28 September 2001, Extended Abstracts, Perth*. AGSO–Geoscience Australia, Record, **2001/37**, 291–293.

BLEWETT, R. S. 2002. Archaean tectonic processes: a case for horizontal shortening in the North Pilbara Granite–Greenstone Terrane, Western Australia. *Precambrian Research*, **113**, 87–120.

BOWRING, S. A. & WILLIAMS, I. S. 1999. Priscoan (4.00–4.03 Ga) orthogneisses from northwestern Canada. *Contributions to Mineralogy and Petrology*, **134**, 3–16.

BOWRING, S. A., WILLIAMS, I. S. & COMPSTON, W. 1989. 3.96 Ga gneisses from the Slave Province, Northwest Territories, Canada; with Suppl. Data 89–17. *Geology*, **17**, 971–975.

BRUN, J. P. 1980. The cluster–ridge pattern of mantled gneiss domes in eastern Finland; evidence for large-scale gravitational instability of the Proterozoic crust. *Earth and Planetary Science Letters*, **47**, 441–449.

BRUN, J. P., GAPAIS, D. & LE, T. B. 1981. The mantled gneiss domes of Kuopio (Finland); interfering diapirs. *Tectonophysics*, **74**, 283–304.

BUICK, R., THORNETT, J. R., MCNAUGHTON, N. J., SMITH, J. B., BARLEY, M. E. & SAVAGE, M. 1995. Record of emergent continental crust ~3.5 billion years ago in the Pilbara Craton of Australia. *Nature*, **375**, 574–577.

BURKE, K. & DEWEY, J. F. 1973. An outline of Precambrian plate development. *In*: TARLING, D. H. & RUNCORN, S. K. (eds) *Implications of Continental Drift to the Earth Sciences, Vol. 2*. Academic Press, New York, 1035–1045.

BURKE, K., DEWEY, J. F. & KIDD, W. S. F. 1976. Dominance of horizontal movements, arc and microcontinental collisions during the later permobile regime. *In*: WINDLEY, B. F. (ed.) *The Early History of the Earth*. Wiley, New York, 113–129.

CALVERT, A. J., SAWYER, E. W., DAVIS, W. J. & LUDDEN, J. N. 1995. Archaean subduction inferred from seismic images of a mantle suture in the Superior Province. *Nature*, **375**, 670–674.

CAMPBELL, I. H. 2001. Identification of ancient mantle plumes. *In*: ERNST, R. E. & BUCHAN, K. L. (eds) *Mantle Plumes; their Identification through Time*. Geological Society of America, Special Paper, **352**, 5–21.

CAMPBELL, I. H. & HILL, R. I. 1988. A two-stage model for the formation of the granite–greenstone terranes of the Kalgoorlie–Norseman area, Western Australia. *Earth and Planetary Science Letters*, **90**, 11–25.

CAMPBELL, I. H. & TAYLOR, S. R. 1983. No water, no granites – no oceans, no continents. *Geophysical Research Letters*, **10**, 1061–1064.

CANUP, R. M. & ASPHAUG, E. 2001. Origin of the Moon in a giant impact near the end of the Earth's formation. *Nature*, **412**, 708–712.

CARD, K. D. 1990. A review of the Superior Province of the Canadian Shield, a product of Archean accretion. *Precambrian Research*, **48**, 99–156.

CHADWICK, B., RAMAKRISHNAN, M. & VISWANATHA, M. N. 1985a. Bababudan – a late Archaean intracratonic volcanosedimentary basin, Karnataka, southern India. Part I: stratigraphy and basin development. *Journal of the Geological Society of India*, **26**, 769–801.

CHADWICK, B., RAMAKRISHNAN, M. & VISWANATHA, M. N. 1985b. Bababudan – a late Archaean intracratonic volcanosedimentary basin, Karnataka, southern India. Part II: structure. *Journal of the Geological Society of India*, **26**, 802–821.

CHARDON, D., CHOUKROUNE, P. & JAYANANDA, M. 1998. Sinking of the Dharwar Basin (South India): implications for Archaean tectonics. *Precambrian Research*, **91**, 15–39.

CHAUVEL, C., DUPRE, B. & ARNDT, N. T. 1993. Pb and Nd isotopic correlation in Belingwe komatiites and basalts. *In*: BICKLE, M. J. & NISBET, E. G. (eds) *The Geology of the Belingwe Greenstone Belt, Zimbabwe – a Study of the Evolution of Archaean Continental Crust*. Geological Society of Zimbabwe Special Publication, **2**, 167–174.

CHEN, S. F., LIBBY, J. W., GREENFIELD, J. E., WYCHE, S. & RIGANTI, A. 2001. Geometry and kinematics of large arcuate structures formed by impingement of rigid granitoids into greenstone belts during progressive shortening. *Geology*, **29**, 283–286.

CHOUKROUNE, P., BOUHALLIER, H. & ARNDT, N. T. 1995. Soft lithosphere during periods of Archaean crustal growth or crustal reworking. *In*: COWARD, M. R. & RIES, A. C. (eds) *Early Precambrian Processes*. Geological Society, London, Special Publications, **95**, 67–86.

CHYBA, C. F. 1987. The cometary contribution to the oceans of primitive Earth. *Nature*, **330**, 632–635.

CHYBA, C. F. 1990. Impact delivery and erosion of planetary oceans in the early inner solar system. *Nature*, **343**, 129–133.

CLOUD, P. 1988. *Oasis in Space – Earth History from the Beginning*. Norton, New York.

COHEN, B. A., SWINDLE, T. D. & KRING, D. A. 2000. Support for the lunar cataclysm hypothesis from lunar meteorite impact melt ages. *Science*, **290**, 1754–1756.

COLLERSON, K. D. & KAMBER, B. S. 1999. Evolution of the continents and the atmosphere inferred from Th–U–Nb systematics of the depleted mantle. *Science*, **283**, 1519–1522.

COLLINS, W. J. 1989. Polydiapirism of the Archean Mount Edgar Batholith, Pilbara Block, Western Australia. *Precambrian Research*, **43**, 41–62.

COLLINS, W. J., VAN KRANENDONK, M. J. & TEYSSIER, C. 1998. Partial convective overturn of Archaean crust in the East Pilbara Craton, Western Australia: driving mechanisms and tectonic implications. *Journal of Structural Geology*, **20**, 1405–1424.

COMPSTON, W., WILLIAMS, I. S., CAMPBELL, I. H. & GRESHAM, J. J. 1986. Zircon xenocrysts from the Kambalda volcanics; age constraints and direct evidence for older continental crust below the Kambalda–Norseman greenstones. *Earth and Planetary Science Letters*, **76**, 299–311.

CONDIE, K. C. & WRONKIEWICZ, D. J. 1990. The Cr/Th ratio in Precambrian pelites from the Kaapvaal Craton as an index of craton evolution. *Earth and Planetary Science Letters*, **97**, 256–267.

CONNERNEY, J. E. P., ACUNA, M. H., WASILEWSKI, P. J. & 7 OTHERS 1999. Magnetic lineations in the ancient crust of Mars. *Science*, **284**, 794–798.

COOK, F. A., VAN DER VELDEN, A. J., HALL, K. W. & ROBERTS, B. J. 1999. Frozen subduction in Canada's Northwest Territories; Lithoprobe deep lithospheric reflection profiling of the western Canadian Shield. *Tectonics*, **18**, 1–24.

CORFU, F. & NOBLE, S. R. 1992. Genesis of the southern Abitibi greenstone belt, Superior Province, Canada; evidence from zircon Hf isotope analyses using a single filament technique. *Geochimica et Cosmochimica Acta*, **56**, 2081–2097.

COUSENS, B. L. 2000. Geochemistry of the Archean Kam Group, Yellowknife greenstone belt, Slave Province, Canada. *Journal of Geology*, **108**, 181–197.

CULLER, T. S., BECKER, T. A., MULLER, R. A. & RENNE, P. R. 2000. Lunar impact history from $^{40}Ar/^{39}Ar$ dating of glass spherules. *Science*, **287**, 1785–1788.

DAVIES, G. F. 1990. Heat and mass transport in the early Earth. *In*: NEWSOM, H. E. & JONES, J. H. (eds) *Origin of the Earth*. Oxford University Press, Oxford, 175–194.

DAVIES, G. F. 1992. On the emergence of plate tectonics. *Geology*, **20**, 963–966.

DAVIES, G. F. 1995. Punctuated tectonic evolution of the earth. *Earth and Planetary Science Letters*, **136**, 363–379.

DAVIES, G. F. 1999. *Dynamic Earth*. Cambridge University Press, Cambridge.

DAVIS, W. J. & BLEEKER, W. 1999. Timing of plutonism, deformation, and metamorphism in the Yellowknife Domain, Slave Province, Canada. *Canadian Journal of Earth Sciences*, **36**, 1169–1187.

DAVIS, W. J. & HEGNER, E. 1992. Neodymium isotopic evidence for the tectonic assembly of late Archean crust in the Slave Province, Northwest Canada. *Contributions to Mineralogy and Petrology*, **111**, 493–504.

DAVIS, W. J., GARIEPY, C. & SAWYER, E. W. 1994. Pre-2.8 Ga crust in the Opatica gneiss belt; a potential source of detrital zircons in the Abitibi and Pontiac subprovinces, Superior Province, Canada. *Geology*, **22**, 1111–1114.

DAVIS, W. J., MACHADO, N., GARIEPY, C., SAWYER, E. W. & BENN, K. 1995. U–Pb geochronology of the Opatica tonalite–gneiss belt and its relationship to the Abitibi greenstone belt, Superior Province, Quebec. *Canadian Journal of Earth Sciences*, **32**, 113–127.

DESROCHERS, J. P., HUBERT, C., LUDDEN, J. N. & PILOTE, P. 1993. Accretion of Archean oceanic plateau fragments in the Abitibi greenstone belt, Canada. *Geology*, **21**, 451–454.

DE WIT, M. J. 1998. On Archean granites, greenstones, cratons and tectonics: does the evidence demand a verdict? *Precambrian Research*, **91**, 181–226.

DE WIT, M. J. & ASHWAL, L. D. (eds) 1997. *Greenstone Belts*. Oxford University Press, Oxford.

DE WIT, M. J., HART, R. A. & HART, R. J. 1987. The Jamestown ophiolite complex, Barberton mountain belt: a section through 3.5 Ga oceanic crust. *Journal of African Earth Sciences*, **6**, 681–730.

DE WIT, M. J., ROERING, C., HART, R. J. & 6 OTHERS 1992. Formation of an Archaean continent. *Nature*, **357**, 553–562.

DIRKS, P. H. G. M. & JELSMA, H. A. 1998. Horizontal accretion and stabilization of the Archean Zimbabwe Craton. *Geology*, **26**, 11–14.

DIXON, J. M. & SUMMERS, J. M. 1983. Patterns of total and incremental strain in subsiding troughs: experimental centrifuged models of inter-diapir synclines. *Canadian Journal of Earth Sciences*, **20**, 1843–1861.

DONALDSON, J. A. & DE KEMP, E. A. 1998. Archaean quartz arenites in the Canadian Shield: examples from the Superior and Churchill provinces. *Sedimentary Geology*, **120**, 153–176.

ELDHOLM, O., THIEDE, J., TAYLOR, E. & 25 OTHERS 1989. Evolution of the Voring volcanic margin, Proceedings of the Ocean Drilling Program, Norwegian Sea; covering Leg 104 of the cruises of the Drilling Vessel *JOIDES Resolution*, Bremerhaven, Germany, to St. John's, Newfoundland, Sites 642–644, 19 June 1985–23 August 1985. *In: Proceedings of the Ocean Drilling Program, Scientific Results*, **104**. Ocean Drilling Program, College Station, TX, 1033–1065.

ERIKSSON, K. A. & FEDO, C. M. 1994. Archean synrift and stable-shelf sedimentary successions. *In*: CONDIE, K. C. (ed.) *Archean Crustal Evolution*. Developments in Precambrian Geology, **11**, 171–204.

ESKOLA, P. 1949. The problem of mantled gneiss domes. *Quarterly Journal of the Geological Society, London*, **104**(Part 4), 461–476.

GILES, K. A. & LAWTON, T. F. 1999. Attributes and evolution of an exhumed salt weld, La Popa Basin, northeastern Mexico. *Geology*, **27**, 323–326.

GILL, R. C. O. & BRIDGWATER, D. 1976. The Ameralik dykes of West Greenland, the earliest known basaltic rocks intruding stable continental crust. *Earth and Planetary Science Letters*, **29**, 276–282.

GOODWIN, A. M. 1996. *Principles of Precambrian Geology*. Academic Press, London.

GORMAN, B. E., PEARCE, T. H. & BIRKETT, T. C. 1978. On the structure of Archean greenstone belts. *Precambrian Research*, **6**, 23–41.

GREEN, J. C. 1977. Keweenawan Plateau volcanism in the Lake Superior region. *In*: BARAGAR, W. R. A., COLEMAN, L. C. & HALL, J. M. (eds) *Volcanic Regimes in Canada*. Geological Association of Canada Special Paper, **16**, 407–422.

GREEN, M. G., SYLVESTER, P. J. & BUICK, R. 2000. Growth and recycling of early Archaean continental crust: geochemical evidence from the Coonterunah and Warrawoona groups, Pilbara Craton, Australia. *Tectonophysics*, **322**, 69–88.

GRIFFIN, T. J. 1990. Southern Cross Province. *In: Geology and Mineral Resources of Western Australia*. Geological Survey of Western Australia, Memoir, **3**, 60–77.

GROVES, D. I., ARCHIBALD, N. J., BETTENAY, L. F. & BINNS, R. A. 1978. Greenstone belts as ancient marginal basins or ensialic rift zones. *Nature*, **273**, 460–461.

GURNIS, M. & DAVIES, G. F. 1986. Apparent episodic crustal growth arising from a smoothly evolving mantle. *Geology*, **14**, 396–399.

HALLAM, A. 1992. *Great Geological Controversies*, 2nd edn. Oxford, Oxford University Press.

HAMILTON, W. B. 1993. Evolution of Archean mantle and crust. *In*: REED, J. C., JR, BICKFORD, M. E., HOUSTON, R. S., LINK, P. K., RANKIN, D. W., SIMS, P. K. & VAN SCHMUS, W. R. (eds) *Precambrian: Conterminous, US*. The Geology of North America, **C-2**, 597–614.

HAMILTON, W. B. 1998. Archean magmatism and deformation were not products of plate tectonics. *Precambrian Research*, **91**, 143–179.

HARGRAVES, R. B. 1986. Faster spreading or greater ridge length in the Archean. *Geology*, **14**, 750–752.

HEAMAN, L. M. 1997. Global mafic magmatism at 2.45 Ga; remnants of an ancient large igneous province? *Geology*, **25**, 299–302.

HEATHER, K. B., SHORE, G. T. & VAN BREEMEN, O. 1995. The convoluted 'layer-cake': an old recipe with new ingredients for the Swayze greenstone belt, southern Superior Province, Ontario. Geological Survey of Canada, Current Research, **1995-C**, 1–10.

HELMSTAEDT, H. & PADGHAM, W. A. 1986. A new look at the stratigraphy of the Yellowknife Supergroup at Yellowknife,, N.W.T.; implications for the age of gold-bearing shear zones and Archean basin evolution. *Canadian Journal of Earth Sciences*, **23**, 454–475.

HELMSTAEDT, H., PADGHAM, W. A. & BROPHY, J. A. 1986. Multiple dikes in lower Kam Group, Yellowknife greenstone belt: evidence for Archean seafloor spreading? *Geology*, **14**, 562–566.

HENDERSON, J. B. 1981. Archaean basin evolution in the Slave Province, Canada. *In*: KRÖNER, A. (ed.) *Precambrian Plate Tectonics*. Elsevier, Amsterdam, 213–235.

HENDERSON, J. B. 1985. *Geology of the Yellowknife–Hearne Lake area, District of Mackenzie: a segment across an Archean basin*. Geological Survey of Canada, Memoir, **414**, 135.

HICKMAN, A. H. 1983. Geology of the Pilbara Block and its environs. *Western Australia Geological Survey Bulletin*, **127**, 268.

HICKMAN, A. H. 1984. Archaean diapirism in the Pilbara Block, Western Australia. *In*: KRÖNER, A. & GREILING, R. (eds) *Precambrian Tectonics Illustrated*. Schweizerbart, Stuttgart, 113–127.

HIRTH, G. & KOHLSTEDT, D. L. 1996. Water in the oceanic upper mantle: implications for rheology, melt extraction and the evolution of the lithosphere. *Earth and Planetary Science Letters*, **144**, 93–108.

HOFFMAN, P. F. 1989. Precambrian geology and tectonic history of North America. *In*: BALLY, A. W. & PALMER, A. R. (eds) *The Geology of North America; an Overview.* Geological Society of America, Boulder, CO, 447–512.

HOFFMAN, P. F. & RANALLI, G. 1988. Archean oceanic flake tectonics. *Geophysical Research Letters*, **15**, 1077–1080.

HOLZHEID, A., SYLVESTER, P., O'NEILL, H. ST. C., RUBIE, D. C. & PALME, H. 2000. Evidence for a late chondritic veneer in the Earth's mantle from high-pressure partitioning of palladium and platinum. *Nature*, **406**, 396–399.

HORSTWOOD, M. S. A., NESBITT, R. W., NOBLE, S. R. & WILSON, J. F. 1999. U–Pb zircon evidence for an extensive early Archean craton in Zimbabwe: a reassessment of the timing of craton formation, stabilization, and growth. *Geology*, **27**, 707–710.

HUNTER, M. A., BICKLE, M. J., NISBET, E. G., MARTIN, A. & CHAPMAN, H. J. 1998. Continental extensional setting for the Archean Belingwe greenstone belt, Zimbabwe. *Geology*, **26**, 883–886.

ISACHSEN, C. E. & BOWRING, S. A. 1994. Evolution of the Slave Craton. *Geology*, **22**, 917–920.

ISACHSEN, C. E. & BOWRING, S. A. 1997. The Bell Lake Group and Anton Complex; a basement–cover sequence beneath the Archean Yellowknife greenstone belt revealed and implicated in greenstone belt formation. *Canadian Journal of Earth Sciences*, **34**, 169–189.

JACKSON, M. P. A. & TALBOT, C. J. 1994. Advances in salt tectonics. *In*: HANCOCK, P. L. (ed.) *Continental deformation.* Pergamon, Oxford, 159–179.

JACKSON, M. P. A. & VENDEVILLE, B. C. 1994. Regional extension as a geologic trigger for diapirism; with Suppl. Data 9401. *Geological Society of America Bulletin*, **106**, 57–73.

JACKSON, S. L., FYON, J. A. & CORFU, F. 1994. Review of Archean supracrustal assemblages of the southern Abitibi greenstone belt in Ontario, Canada; products of microplate interaction within a large-scale plate-tectonic setting. *Precambrian Research*, **65**, 183–205.

JACKSON, S. L. & FYON, J. A. 1991. The Western Abitibi subprovince in Ontario. *In*: THURSTON, P. C., WILLIAMS, H. R., SUTCLIFFE, R. H. & STOTT, G. M. (eds) *Geology of Ontario.* Ontario Geological Survey, Special Volume, **4**(Part 1), 405–482.

JELSMA, H. A. & DIRKS, P. H. G. M. 2002. Neoarchaean tectonic evolution in the Zimbabwe Craton. *In*: FOWLER, C. M. R., EBINGER, C. J. & HAWKESWORTH, C. J. (ed.) *The Early Earth: Physical, Chemical and Biological Development.* Geological Society, London, Special Publications, **199**, 183–211.

JELSMA, H. A., VINYU, M. L., VALBRACHT, P. J., DAVIES, G. R., WIJBRANS, J. R. & VERDURMEN, E. A. T. 1996. Constraints on Archaean crustal evolution of the Zimbabwe Craton: a U–Pb zircon, Sm–Nd and Pb–Pb whole-rock isotope study. *Contributions to Mineralogy and Petrology*, **124**, 55–70.

JENNER, G. A., FRYER, B. J. & MCLENNAN, S. M. 1981. Geochemistry of the Archean Yellowknife Supergroup. *Geochimica et Cosmochimica Acta*, **45**, 1111–1129.

JENSEN, L. S. 1985. Stratigraphy and petrogenesis of Archean metavolcanic sequences, southwestern Abitibi Subprovince, Ontario. *In*: AYRES, L. D., THURSTON, P. C., CARD, K. D. & WEBER, W. (eds) *Evolution of Archean Supracrustal Sequences.* Geological Association of Canada, Special Paper, **28**, 65–87.

JÓNASSON, K., HOLM, P. M. & PEDERSEN, A. K. 1992. Petrogenesis of silicic rocks from the Kroksfjordur central volcano, NW Iceland. *Journal of Petrology*, **33**, 1345–1369.

JORDAN, T. H. 1978. Composition and development of the continental tectosphere. *Nature*, **274**, 544–548.

JORDAN, T. H. 1988. Structure and formation of the continental tectosphere. *Journal of Petrology, Special Lithosphere Issue*, 11–37.

KAMO, S. L. & DAVIS, D. W. 1994. Reassessment of Archean crustal development in the Barberton Mountain Land, South Africa, based on U–Pb dating. *Tectonics*, **13**, 165–192.

KERR, A. C., WHITE, R. V. & SAUNDERS, A. D. 2000. LIP reading; recognizing oceanic plateaux in the geological record. *Journal of Petrology*, **41**, 1041–1056.

KERRICH, R., WYMAN, D., FAN, J. & BLEEKER, W. 1998. Boninite series: low Ti-tholeiite associations from the 2.7 Ga Abitibi greenstone belt. *Earth and Planetary Science Letters*, **164**, 303–316.

KINNY, P. D., WILLIAMS, I. S., FROUDE, D. O., IRELAND, T. R. & COMPSTON, W. 1988. Early Archaean zircon ages from orthogneisses and anorthosites at Mount Narryer, Western Australia. *Precambrian Research*, **38**, 325–341.

KONTINEN, A. 1987. An early Proterozoic ophiolite: the Jormua mafic–ultramafic complex, northeastern Finland. *Precambrian Research*, **35**, 313–341.

KRAPEZ, B. & BARLEY, M. E. 1987. Archaean strike-slip faulting and related ensialic basins; evidence from the Pilbara Block, Australia. *Geological Magazine*, **124**, 555–567.

KRÖNER, A., HEGNER, E., WENDT, J. I. & BYERLY, G. R. 1996. The oldest part of the Barberton granitoid–greenstone terrain, South Africa: evidence for crust formation between 3.5 and 3.7 Ga. *Precambrian Research*, **78**, 105–124.

KUSKY, T. M. 1989. Accretion of the Archean Slave Province. *Geology*, **17**, 63–67.

KUSKY, T. M. 1990. Evidence for Archean ocean opening and closing in the southern Slave Province. *Tectonics*, **9**, 1533–1563.

KUSKY, T. M. & HUDLESTON, P. J. 1999. Growth and demise of an Archean carbonate platform, Steep Rock Lake, Ontario, Canada. *Canadian Journal of Earth Sciences*, **36**, 565–584.

KUSKY, T. M. & KIDD, W. S. F. 1992. Remnants of an Archean oceanic plateau, Belingwe greenstone belt, Zimbabwe. *Geology*, **20**, 43–46.

LABROSSE, S., POIRIER, J. P. & LE MOUEL, J. L. 2001. The age of the inner core. *Earth and Planetary Science Letters*, **190**, 111–123.

LAMBERT, M. B., ERNST, R. E. & DUDÁS, F. Ö. L. 1992. Archean mafic dyke swarms near the Cameron

River and Beaulieu River volcanic belts and their implications for tectonic modelling of the Slave Province, Northwest Territories. *Canadian Journal of Earth Sciences*, **29**, 2226–2248.

LEE, D. C., HALLIDAY, A. N., SNYDER, G. A. & TAYLOR, L. A. 1997. Age and origin of the Moon. *Science*, **278**, 1098–1103.

LIU, D. Y., NUTMAN, A. P., COMPSTON, W., WU, J. S. & SHEN, Q. H. 1992. Remnants of >3800 Ma crust in the Chinese part of the Sino-Korean craton. *Geology*, **20**, 339–342.

LOWE, D. R. 1994. Archean greenstone-related sedimentary rocks. *In*: CONDIE, K. C. (ed.) *Archean Crustal Evolution*. Developments in Precambrian Geology, **11**, 121–169.

MACGREGOR, A. M. 1951. Some milestones in the Precambrian of Southern Rhodesia (presidential address). *Transactions and Proceedings of the Geological Society of South Africa*, **54**, XXVII–LXVI.

MARESCHAL, J. C. & WEST, G. F. 1980. A model for Archean tectonism; Part 2, Numerical models of vertical tectonism in greenstone belts. *Canadian Journal of Earth Sciences*, **17**, 60–71.

MARSHAK, S., TINKHAM, D., ALKMIM, F., BRUECKNER, H. & BORNHORST, T. J. 1997. Dome-and-keel provinces formed during Paleoproterozoic orogenic collapse – core complexes, diapirs, or neither? Examples from the Quadrilatero Ferrifero and the Penokean Orogen. *Geology*, **25**, 415–418.

MARTINEZ, F., GOODLIFFE, A. M. & TAYLOR, B. 2001. Metamorphic core complex formation by density inversion and lower-crust extrusion. *Nature*, **411**, 930–934.

MCGREGOR, V. R. 1973. The early Precambrian gneisses of the Godthaab District, West Greenland. *Philosophical Transactions of the Royal Society of London, Series A*, **273**, 343–358.

MCKENZIE, D., NISBET, E. G. & SCLATER, J. G. 1980. Sedimentary basin development in the Archaean. *Earth and Planetary Science Letters*, **48**, 35–41.

MCLENNAN, S. M. & TAYLOR, S. R. 1982. Geochemical constraints on the growth of the continental crust. *Journal of Geology*, **90**, 347–361.

MOJZSIS, S. J., HARRISON, T. M. & PIDGEON, R. T. 2001. Oxygen-isotope evidence from ancient zircons for liquid water at the Earth's surface 4300 Myr ago. *Nature*, **409**, 178–181.

MORTENSEN, J. K. 1993. U–Pb geochronology of the eastern Abitibi Subprovince. Part I: Chibougamau–Matagami–Joutel region. *Canadian Journal of Earth Sciences*, **30**, 11–28.

MOSER, D. E., FLOWERS, R. M. & HART, R. J. 2001. Birth of the Kaapvaal tectosphere 3.08 billion years ago. *Science*, **291**, 465–468.

MOSER, D. E., HEAMAN, L. M., KROGH, T. E. & HANES, J. A. 1996. Intracrustal extension of an Archean orogen revealed using single-grain U–Pb zircon geochronology. *Tectonics*, **15**, 1093–1109.

MUELLER, P. A., WOODEN, J. L. & NUTMAN, A. P. 1992. 3.96 Ga zircons from an Archean quartzite, Beartooth Mountains, Montana. *Geology*, **20**, 327–330.

NELSON, D. R., TRENDALL, A. F. & ALTERMANN, W. 1999. Chronological correlations between the Pilbara and Kaapvaal cratons. *Precambrian Research*, **97**, 165–189.

NELSON, D. R., ROBINSON, B. W. & MYERS, J. S. 2000. Complex geological histories extending for ≥4.0 Ga deciphered from xenocryst zircon microstructures. *Earth and Planetary Science Letters*, **181**, 89–102.

NISBET, E. G. & SLEEP, N. H. 2001. The habitat and nature of early life. *Nature*, **409**, 1083–1091.

NUTMAN, A. P., BENNETT, V. C., FRIEND, C. R. L. & ROSING, M. T. 1997. Approximately 3710 and ≥3790 Ma volcanic sequences in the Isua (Greenland) supracrustal belt: structural and Nd isotope implications. *Chemical Geology*, **141**, 271–287.

NUTMAN, A. P., FRIEND, C. R. L., KINNY, P. D. & MCGREGOR, V. R. 1993. Anatomy of an early Archean gneiss complex; 3900 to 3600 Ma crustal evolution in southern West Greenland. *Geology*, **21**, 415–418.

OJAKANGAS, R. W., MOREY, G. B. & GREEN, J. C. 2001. The Mesoproterozoic Midcontinent Rift System, Lake Superior region, USA. *Sedimentary Geology*, **141–142**, 421–442.

PADGHAM, W. A. & FYSON, W. K. 1992. The Slave Province; a distinct Archean craton. *Canadian Journal of Earth Sciences*, **29**, 2072–2086.

PARSON, L. M., VIERECK, L. G., LOVE, D. A., GIBSON, I. L., MORTON, A. C. & HERTOGEN, J. 1989. The petrology of the lower series volcanics, ODP Site 642. *In*: ELDHOLM, O., THIEDE, J. & TAYLOR, E. (eds) *Proceedings of the Ocean Drilling Program, Scientific Results, 104*. Ocean Drilling Program, College Station, TX, 419–428.

PATTERSON, C. 1956. Age of meteorites and the earth. *Geochimica et Cosmochimica Acta*, **10**, 230–237.

PEARSON, D. G. 1999. The age of continental roots. *Lithos*, **48**, 171–194.

PERCIVAL, J. A., BAILES, A. & MCNICOLL, V. 2001*a*. Mesoarchean Western Margin of the Superior Craton in the Lake Winnipeg Area, Manitoba. Geological Survey of Canada, Current Research, **2001-C16**.

PERCIVAL, J. A., SKULSKI, T., MCNICOLL, V. & 8 OTHERS 2001*b*. Neoarchean assembly of the Superior Province. *In*: CASSIDY, K. F., DUNPHY, J. M. & VAN KRANENDONK, M. J. (eds) *4th International Archaean Symposium, 24–28 September 2001, Extended Abstracts*. AGSO–Geoscience Australia, Record, **2001/37**, 341–343.

PERCIVAL, J. A. & WILLIAMS, H. R. 1989. Late Archean Quetico accretionary complex, Superior Province, Canada. *Geology*, **17**, 23–25.

POLLACK, H. N. 1997. Thermal characteristics of the Archaean. *In*: DE WIT, W. M. J. & ASHWAL, L. D. (eds) *Greenstone Belts*. Oxford Monographs on Geology and Geophysics, **35**, 223–232.

PUCHTEL, I. S., HOFMANN, A. W., MEZGER, K., JOCHUM, K. P., SHCHIPANSKY, A. A. & SAMSONOV, A. V. 1998. Oceanic plateau model for continental crustal growth in the Archaean; a case study from the Kostomuksha greenstone belt, NW Baltic Shield. *Earth and Planetary Science Letters*, **155**, 57–74.

PUFFER, J. H. 2001. Contrasting high field strength element contents of continental flood basalts from plume versus reactivated-arc sources. *Geology*, **29**, 675–678.

RAMBERG, H. 1981. *Gravity, Deformation and the Earth's Crust*, 2nd edn. Academic Press, New York.

RAMBERG, H. 1972. Theoretical models of density stratification and diapirism in the Earth. *Journal of Geophysical Research*, **77**, 877–889.

REEMST, P. & CLOETINGH, S. A. P. L. 2000. Polyphase rift evolution of the Voring Margin (mid-Norway); constraints from forward tectonostratigraphic modeling. *Tectonics*, **19**, 225–240.

RICHARDS, M. A., YANG, W.-S., BAUMGARDNER, J. R. & BUNGE, H.-P. 2001. Role of low-viscosity zone in stabilizing plate tectonics: implications for comparative terrestrial planetology. *Geochemistry Geophysics Geosystems*, **2**, Paper number 2000GC000115.

RICHTER, F. M. 1988. A major change in the thermal state of the Earth at the Archean–Proterozoic boundary: consequences for the nature and preservation of continental lithosphere. *Journal of Petrology, Special Lithosphere Issue*, 39–52.

ROBERTSON, A. 1987. The transition from a passive margin to an Upper Cretaceous foreland basin related to ophiolite emplacement in the Oman Mountains. *Geological Society of America Bulletin*, **99**, 633–653.

ROSCOE, S. M. & DONALDSON, J. A. 1988. Uraniferous pyritic quartz pebble conglomerate and layered ultramafic intrusions in a sequence of quartzite, carbonate, iron formation and basalt of probable Archean age at Lac Sakami, Quebec. *Geological Survey of Canada, Paper*, **88-1C**, 117–121.

ROYDEN, L. 1996. Coupling and decoupling of crust and mantle in convergent orogens; implications for strain partitioning in the crust. *Journal of Geophysical Research, B Solid Earth and Planets*, **101**, 17 679–17 705.

RUDNICK, R. L. 1995. Making continental crust. *Nature*, **378**, 571–578.

RUDNICK, R. L., BARTH, M., HORN, I. & MCDONOUGH, W. F. 2000. Rutile-bearing refractory eclogites; missing link between continents and depleted mantle. *Science*, **287**, 278–281.

SAUNDERS, A. D., TARNEY, J., KERR, A. C. & KENT, R. W. 1996. The formation and fate of large oceanic igneous provinces. *Lithos*, **37**, 81–95.

SAWYER, E. W. & BENN, K. 1993. Structure of the high-grade Opatica Belt and adjacent low-grade Abitibi Subprovince, Canada: an Archean mountain front. *Journal of Structural Geology*, **15**, 1443–1458.

SCHAEFER, S. J. & MORTEN, P. 1991. Two komatiitic pyroclastic units, Superior Province, northwestern Ontario: their geology, petrography, and correlation. *Canadian Journal of Earth Sciences*, **28**, 1455–1470.

SCHMELING, H. 1987. On the relation between initial conditions and late stages of Rayleigh–Taylor instabilities. *Tectonophysics*, **133**, 65–80.

SCOTT, D. J., HELMSTAEDT, H. & BICKLE, M. J. 1992. Purtuniq Ophiolite, Cape Smith Belt, northern Quebec, Canada; a reconstructed section of early Proterozoic oceanic crust. *Geology*, **20**, 173–176.

SEARLE, M. P. 1985. Sequence of thrusting and origin of culminations in the northern and central Oman Mountains. *Journal of Structural Geology*, **7**, 129–143.

SEARLE, M. P. & COX, J. 1999. Tectonic setting, origin, and obduction of the Oman Ophiolite. *Geological Society of America Bulletin*, **111**, 104–122.

SIRCOMBE, K. N., BLEEKER, W. & STERN, R. A. 2001. Detrital zircon geochronology and grain-size analysis of ~2800 Ma Mesoarchean proto-cratonic cover succession, Slave Province, Canada. *Earth and Planetary Science Letters*, **189**, 207–220.

SLEEP, N. H. 1992. Archean plate tectonics: what can be learned from continental geology? *Canadian Journal of Earth Sciences*, **29**, 2066–2071.

SLEEP, N. H., ZAHNLE, K. J., KASTING, J. F. & MOROWITZ, H. J. 1989. Annihilation of ecosystems by large asteroid impacts on the early Earth. *Nature*, **342**, 139–142.

SMITHIES, R. H. 2000. The Archean tonalite–trondhjemite–granodiorite (TTG) series is not an analogue of Cenozoic adakite. *Earth and Planetary Science Letters*, **182**, 115–125.

SRINIVASAN, R. & OJAKANGAS, R. W. 1986. Sedimentology of quartz-pebble conglomerates and quartzites of the Archean Bababudan Group, Dharwar Craton, South India: evidence for early crustal stability. *Journal of Geology*, **94** 199–214.

STERN, R. A. & BLEEKER, W. 1998. Age of the world's oldest rocks refined using Canada's SHRIMP; the Acasta gneiss complex, Northwest Territories, Canada. *Geoscience Canada*, **25**, 27–31.

STERN, R. J. & BLOOMER, S. H. 1992. Subduction zone infancy; examples from the Eocene Izu–Bonin–Mariana and Jurassic California arcs. *Geological Society of America Bulletin*, **104**, 1621–1636.

STEVENSON, D. J. 1981. Models of the Earth's core. *Science*, **214**, 611–619.

STEVENSON, D. J. 1990. Fluid dynamics of core formation. *In*: NEWSOM, H. E. & JONES, J. H. (eds) *Origin of the Earth*. Oxford University Press, Oxford, 231–249.

STEVENSON, D. J. 2001. Mars' core and magnetism. *Nature*, **412**, 214–216.

STOREY, M., MAHONEY, J. J., KROENKE, L. W. & SAUNDERS, A. D. 1991. Are oceanic plateaus the site of komatiite formation? *Geology*, **19**, 376–379.

STRIK, G., BLAKE, T. S. & LANGEREIS, C. G. 2001. The Fortescue and Ventersdorp Groups: a paleomagnetic comparison of two cratons. *In*: CASSIDY, K. F., DUNPHY, J. M. & VAN KRANENDONK, M. J. (eds) *4th International Archaean Symposium, 24–28 September 2001, Extended Abstracts*. AGSO–Geoscience Australia, Record **2001/37**, 532–533.

SWAGER, C. P., GOLEBY, B. R., DRUMMOND, B. J., RATTENBURY, M. S. & WILLIAMS, P. R. 1997. Crustal structure of granite–greenstone terranes in the Eastern Goldfields, Yilgarn Craton, as revealed by seismic reflection profiling. *Precambrian Research*, **83**, 43–56.

TARNEY, J., DALZIEL, I. W. D. & DE WIT, M. J. 1976. Marginal basin 'Rocas Verdes' complex from

S. Chile: a model for Archaean greenstone belt formation. *In*: WINDLEY, B. F. (ed.) *The Early History of the Earth*. Wiley, New York, 131–146.

TAYLOR, S. R. & MCLENNAN, S. M. 1985. *The Continental Crust: its Composition and Evolution*. Blackwell Scientific, Oxford.

TEGNER, C., LESHER, C. E., LARSEN, L. M. & WATT, W. S. 1998. Evidence from the rare-earth-element record of mantle melting for cooling of the Tertiary Iceland Plume. *Nature*, **395**, 591–594.

THORPE, R. I., CUMMING, G. L. & MORTENSEN, J. K. 1992*a*. A significant Pb isotope boundary in the Slave Province and its probable relation to ancient basement in the western Slave Province. *In*: RICHARDSON, D. G. & IRVING, M. (eds) *Project Summaries; Canada–Northwest Territories Mineral Development Subsidiary Agreement 1987–1991*. Geological Survey of Canada, Open-File Report, 179–184.

THORPE, R. I., HICKMAN, A. H., DAVIS, D. W., MORTENSEN, J. K. & TRENDALL, A. F. 1992*b*. U–Pb zircon geochronology of Archaean felsic units in the Marble Bar region, Pilbara Craton, Western Australia. *Precambrian Research*, **56**, 169–189.

THURSTON, P. C., OSMANI, I. A. & STONE, D. 1991. Northwestern Superior Province: review and terrane analysis. *In*: THURSTON, P. C., WILLIAMS, H. R., SUTCLIFFE, R. H. & STOTT, G. M. (eds) *Geology of Ontario*. Ontario Geological Survey, Special Volume, **4**(Part 1), 81–142.

TRUSHEIM, F. 1960. Mechanism of salt migration in northern Germany. *AAPG Bulletin*, **44**, 1519–1540.

TURCOTTE, D. L. & SCHUBERT, G. 1982. *Geodynamics – Applications of Continuum Physics to Geological Problems*. Wiley, New York.

TURNER, S. & HAWKESWORTH, C. 1998. Using geochemistry to map mantle flow beneath the Lau Basin. *Geology*, **26**, 1019–1022.

VAN KRANENDONK, M. J. & COLLINS, W. J. 1998. Timing and tectonic significance of late Archaean, sinistral strike-slip deformation in the central Pilbara structural corridor, Pilbara Craton, Western Australia. *Precambrian Research*, **88**, 207–232.

VAN KRANENDONK, M. J., HICKMAN, A. H., SMITHIES, R. H., NELSON, D. R. & PIKE, G. 2002. Geology and tectonic evolution of the Archaean North Pilbara Terrain, Pilbara Craton, Western Australia. *Economic Geology*, in press.

VEIZER, J. & COMPSTON, W. 1976. $^{87}Sr/^{86}Sr$ in Precambrian carbonates as an index of crustal evolution. *Geochimica et Cosmochimica Acta*, **40**, 905–914.

VLAAR, N. J., VAN KEKEN, P. E. & VAN DEN BERG, A. P. 1994. Cooling of the Earth in the Archaean: consequences of pressure-release melting in a hotter mantle. *Earth and Planetary Science Letters*, **121**, 1–18.

VON HUENE, R. & SCHOLL, D. W. 1991. Observations at convergent margins concerning sediment subduction, subduction erosion, and the growth of continental crust. *Reviews of Geophysics*, **29**, 279–316.

WARREN, R. G. & ELLIS, D. J. 1996. Mantle underplating, granite tectonics, and metamorphic $P-T-t$ paths. *Geology*, **24**, 663–666.

WELLMAN, P. 2000. Upper crust of the Pilbara Craton, Australia; 3D geometry of a granite/greenstone terrain. *Precambrian Research*, **104**, 175–186.

WENDT, J. I., REGELOUS, M., COLLERSON, K. D. & EWART, A. 1997. Evidence for a contribution from two mantle plumes to island-arc lavas from northern Tonga. *Geology*, **25**, 611–614.

WHITE, D. J., MUSSACHIO, G., SOL, S. & LITHOPROBE WESTERN SUPERIOR WORKING GROUP 2001. Evidence for subduction processes and terrane accretion in the Archean Western Superior Province, Canada: results from combined LITHOPROBE deep seismic studies. *In*: CASSIDY, K. F., DUNPHY, J. M. & VAN KRANENDONK, M. J. (eds) *4th International Archaean Symposium, 24–28 September 2001, Extended Abstracts*. AGSO–Geoscience Australia, Record, **2001/37**, 539.

WILDE, S. A., VALLEY, J. W., PECK, W. H. & GRAHAM, C. M. 2001. Evidence from detrital zircons for the existence of continental crust and oceans on the Earth 4.4 Gyr ago. *Nature*, **409**, 175–178.

WILKS, M. E. & HARPER, G. D. 1997. Wind River Range, Wyoming Craton. *In*: DE WIT, M. J. & ASHWAL, L. D. (eds) *Greenstone Belts*. Oxford Monographs on Geology and Geophysics, **35**, 508–516.

WILKS, M. E. & NISBET, E. G. 1988. Stratigraphy of the Steep Rock Group, northwest Ontario: a major Archaean unconformity and Archaean stromatolites. *Canadian Journal of Earth Sciences*, **25**, 370–391.

WILLIAMS, I. S. & COLLINS, W. J. 1990. Granite-greenstone terranes in the Pilbara Block, Australia, as coeval volcano-plutonic complexes; evidence from U–Pb zircon dating of the Mount Edgar Batholith. *Earth and Planetary Science Letters*, **97**, 41–53.

WILSON, J. F., BICKLE, M. J., HAWKESWORTH, C. J., MARTIN, A., NISBET, E. G. & ORPEN, J. L. 1978. Granite–greenstone terrains of the Rhodesian Archaean craton. *Nature*, **271**, 23–27.

WILSON, J. F., NESBITT, R. W. & FANNING, C. M. 1995. Zircon geochronology of Archaean felsic sequences in the Zimbabwe craton: a revision of greenstone stratigraphy and a model for crustal growth. *In*: COWARD, M. P. & RIES, A. C. (eds) *Early Precambrian Processes*. Geological Society, London, Special Publications, **95**, 109–126.

WINDLEY, B. F. 1995. *The Evolving Continents*, 3rd edn. Wiley, New York.

WINDLEY, B. F. & BRIDGWATER, D. 1971. The evolution of Archaean low- and high-grade terranes. *Geological Society of Australia, Special Publication*, **3**, 33–46.

WINGATE, M. T. D. 1999. Ion microprobe baddeleyite and zircon ages for Late Archaean mafic dykes of the Pilbara Craton, Western Australia. *Australian Journal of Earth Sciences*, **46**, 493–500.

WINGATE, M. T. D. 2000. Ion microprobe U–Pb zircon and baddeleyite ages for the Great Dyke and its satellite dykes, Zimbabwe. *South African Journal of Geology – Suid-Afrikaanse Tydskrif vir Geologie*, **103**, 74–80.

YAMASHITA, K. & CREASER, R. A. 1999. Geochemical and Nd isotopic constraints for the origin of late

Archean turbidites from the Yellowknife area, Northwest Territories, Canada. *Geochimica et Cosmochimica Acta*, **63**, 2579–2598.

YOUNG, G. M., LONG, D. G. F., FEDO, C. M. & 6 OTHERS 2001. Paleoproterozoic Huronian basin; product of a Wilson cycle punctuated by glaciations and a meteorite impact. *Sedimentary Geology*, **141–142**, 233–254.

ZALESKI, E., EATON, D. W., MILKEREIT, B., ROBERTS, B., SALISBURY, M. H. & PETRIE, L. 1997. Seismic reflections from subvertical diabase dikes in an Archean terrane. *Geology*, **25**, 707–710.

ZEGERS, T. E., DE KEIJZER, M., PASSCHIER, C. W. & WHITE, S. H. 1998. The Mulgandinnah shear zone; an Archean crustal scale strike-slip zone, eastern Pilbara, Western Australia. *Precambrian Research*, **88**, 233–247.

ZEGERS, T. E. & VAN KEKEN, P. E. 2001. Middle Archean continent formation by crustal delamination. *Geology*, **29**, 1083–1086.

ZEGERS, T. E., NELSON, D. R., WIJBRANS, J. R. & WHITE, S. H. 2001. SHRIMP U–Pb zircon dating of Archean core complex formation and pancratonic strike-slip deformation in the East Pilbara granite–greenstone terrain. *Tectonics*, **20**, 883–908.

ZEGERS, T. E., DE WIT, M. J., DANN, J. & WHITE, S. H. 1998. Vaalbara, Earth's oldest assembled continent? A combined structural, geochronological, and palaeomagnetic test. *Terra Nova*, **10**, 250–259.

Neoarchaean tectonic evolution of the Zimbabwe Craton

HIELKE A. JELSMA[1] & PAUL H. G. M. DIRKS[2]

[1] CIGCES, Department of Geological Sciences, University of Cape Town, Rondebosch 7701, South Africa (e-mail: jelsma@cigces.uct.ac.za)
[2] School of Geosciences, University of the Witwatersrand, Private Bag 3, WITS 2050, South Africa

Abstract: An overview is presented of the field relations, age data and geochemical characteristics of the Neoarchaean granites and greenstones of the Zimbabwe Craton, southern Africa. A major tectono-magmatic event at c. 2.7 Ga produced two distinct greenstone successions. One succession is reminiscent of rift- or back-arc environments and is associated with an old continental fragment. A second succession is indicative of arc magmatism and is associated with juvenile crust. Both were affected by a major accretionary event that, in an apparent sense, swept across the craton between 2.68 and 2.60 Ga. During this 80 Ma time period, concomitant late volcanism, regional deformation, the development of syntectonic sedimentary successions in foreland-type basins, and late syntectonic plutonism took place in selected shear-zone-bounded tectonic domains over limited periods of time (<10–20 Ma). Deformation led to isostatically stable, 30–40 km thick continental crust, without significant exhumation of high-pressure rocks, suggesting that lithospheric shortening was accommodated independently in a rheologically strong upper and weak lower crust. Deformation was followed by pan-cratonic crustal melting and strike-slip shear motions, and led to stabilization of the crust at 2575 Ma, heralded by the emplacement of the Great Dyke.

Archaean crustal fragments are composed of primitive rock assemblages with distinct volcanic, metamorphic and tectonic features as preserved in granite–greenstone terranes, remnants of which are found within continental plates across the globe (de Wit & Ashwal 1997, and references therein). The Archaean crust beneath cratons is typically thin (30–40 km) and unlayered, as demonstrated by seismic studies (e.g. Nguuri et al. 2001). The upper crust is exposed in granite–greenstone terranes and is geochemically composed of 70% TTG (tonalite–trondhjemite–granodiorite) and 30% greenstones. The greenstones contain on average 40% mafic volcanic rocks, 20% intermediate and felsic volcanic rocks, and 40% sedimentary rocks (using estimates from 15 cratons, reported by de Wit & Ashwal 1997). Strain distribution is heterogeneous and mainly partitioned into discrete shear zones. Metamorphic grades range from subgreenschist- to amphibolite-facies conditions. In southern Africa, the mid- and lower crust is exposed in the Vredefort Structure of the Kaapvaal Craton and high-grade gneiss terranes such as the Limpopo belt or Zambezi belt. Compositions are similar to those of the upper crust, with felsic and mafic granulites, paragneisses and metasedimentary rocks, intruded by later granite sheets (e.g. Moser et al. 2001). Strain distribution is homogeneous and the lower-crustal rocks are characterized by a penetrative gneissic fabric, transposition of earlier structures and generally collinear deformation. The Archaean mantle is composed of garnet peridotite and eclogite as shown by Archaean mantle xenoliths in kimberlites (Gurney 1990), and is as old as parts of the overlying crust (Nägler et al. 1997; Carlson et al. 2000).

One of the main differences between the Archaean and Phanerozoic Earth is the rate of radiogenic heat production, which at the end of the Archaean is still 1.6–2.8 times the present-day heat production (Pollack 1997). As a result of the higher heat production the average Archaean lithosphere was probably hotter, less rigid and more buoyant, and initial continental crustal fragments smaller and less homogenized, preventing plate tectonics in the modern sense (Davies 1992; Hamilton 1998). If this reasoning is correct, what were the main processes that formed the Archaean continental crust, which is dominated by TTG? The production of TTG crust requires burial, reheating and partial melting of hydrated basalt at mid- to lower-crustal pressures of 5–10 kbar (Rapp et al. 1991; Wyllie et al. 1997). This may be explained by low-angle subduction and underplating or obduction of oceanic crust (Goodwin & Smith 1980; Abbott &

Table 1. *Geochronological data for Archaean rocks in Zimbabwe*

Area	Stratigraphic unit	Lithology/mineral	Sample no.	Zone	UTM-X	UTM-Y	Age (Ma)	Error (Ma)	Error type	Technique	Reference
Crustal residence ages											
Belingwe belt	Manjeri Formation	Shale	MHZI41a	35	803000	7726400	2965	26	2σ	TDM, Sm–Nd, WR	Hunter 1997
Harare belt	Shamvaian Supergroup	Greywacke	ZIM6, 10–13, 20	36	349500	8090800	3030	37	95% conf.	Pb–Pb, WR	Jelsma et al. 1996
Belingwe belt	Manjeri Formation	Shale	MHZI41b	35	803000	7726400	3099	26	2σ	TDM, Sm–Nd, WR	Hunter 1997
Belingwe belt	Brooklands Formation	Mudstone	MHZ036Bc	36	196800	7727700	3154	242	2σ	TDM, Sm–Nd, WR	Hunter 1997
Belingwe belt	Manjeri Formation	Shale	MHZ317	35	805000	7737900	3206	34	2σ	TDM, Sm–Nd, WR	Hunter 1997
Belingwe belt	Manjeri Formation	Shale	MHZ220B	35	805000	7737900	3239	34	2σ	TDM, Sm–Nd, WR	Hunter 1997
Belingwe belt	Brooklands Formation	Shale	MHZ034	36	196800	7727700	3262	22	2σ	TDM, Sm–Nd, WR	Hunter 1997
Belingwe belt	Brooklands Formation	Shale	33 zircons	36	200500	7732100	3283	×	peak	TIMS, Pb–Pb, zircon	Hunter 1997
Belingwe belt	Manjeri Formation	Mudstone	MHZ037Bb	36	196800	7730300	3312	16	2σ	TDM, Sm–Nd, WR	Hunter 1997
Belingwe belt	Brooklands Formation	Mudstone	MHZ036Ba	36	196800	7727700	3312	34	2σ	TDM, Sm–Nd, WR	Hunter 1997
Belingwe belt	Manjeri Formation	Siltstone	MHZ008Bid	36	196300	7742700	3339	38	2σ	TDM, Sm–Nd, WR	Hunter 1997
Belingwe belt	Manjeri Formation	Ironstone	MHZ007e	36	196300	7742700	3353	62	2σ	TDM, Sm–Nd, WR	Hunter 1997
Belingwe belt	Manjeri Formation	Mudstone	MHZI27b	36	196300	7742700	3362	44	2σ	TDM, Sm–Nd, WR	Hunter 1997
Belingwe belt	Manjeri Formation	Mudstone	MHZ038	36	196800	7730300	3403	18	2σ	TDM, Sm–Nd, WR	Hunter 1997
Belingwe belt	Manjeri Formation	Silicified shale	MHZ098	36	189600	7748700	3437	18	2σ	TDM, Sm–Nd, WR	Hunter 1997
Belingwe belt	Brooklands Formation	Mudstone	MHZ230a	36	196800	7727700	3471	20	2σ	TDM, Sm–Nd, WR	Hunter 1997
Belingwe belt	Manjeri Formation	Black chert	MHZ014	36	196300	7742700	3502	32	2σ	TDM, Sm–Nd, WR	Hunter 1997
Belingwe belt	Manjeri Formation	Shale	MHZI26a	36	196300	7742700	3511	22	2σ	TDM, Sm–Nd, WR	Hunter 1997
Belingwe belt	Manjeri Formation	Mudstone	MHZ002baii	36	196300	7744700	3527	26	2σ	TDM, Sm–Nd, WR	Hunter 1997
Belingwe belt	Manjeri Formation	Mudstone	MHZI90a	36	196800	7727700	3539	54	2σ	TDM, Sm–Nd, WR	Hunter 1997
Belingwe belt	Brooklands Formation	Mudstone	MHZ002bai	36	196300	7742700	3546	36	2σ	TDM, Sm–Nd, WR	Hunter 1997
Belingwe belt	Manjeri Formation	Mudstone	MHZ002baii	36	196300	7744700	3563	24	2σ	TDM, Sm–Nd, WR	Hunter 1997
Belingwe belt	Manjeri Formation	Mudstone	MHZ002b	36	196300	7742700	3589	26	2σ	TDM, Sm–Nd, WR	Hunter 1997
Belingwe belt	Ngezi schist inclusion	Micaschist	MHZI71	36	197700	7726500	3628	28	2σ	TDM, Sm–Nd, WR	Hunter 1997
Belingwe belt	Manjeri Formation	Ironstone	TR43	36	196300	7742700	3630	110	2σ	TDM, Sm–Nd, WR	Hunter 1997
Belingwe belt	Brooklands Formation	Shale	MHZI89	36	196800	7727700	3635	26	2σ	TDM, Sm–Nd, WR	Hunter 1997
Belingwe belt	Manjeri Formation	Shale	MHZ008Bic	36	196300	7742700	3654	24	2σ	TDM, Sm–Nd, WR	Hunter 1997
Belingwe belt	Manjeri Formation	Mudstone	MHZ008Biig	36	196300	7742700	3703	28	2σ	TDM, Sm–Nd, WR	Hunter 1997
Inheritance											
Harare belt	Passaford Formation	Felsic clasts in reworked tuff	ZIM92/31	36	287800	8034800	2700	×		SHRIMP, zircon	Wilson et al. 1995
Harare belt	Maparu Formation	Water-lain tuff	ZIM92/30	36	343200	8097200	2720	×	×	SHRIMP, zircon	Wilson et al. 1995
Midlands belt	What Cheer Formation	Felsic clast, reworked volcanic breccia	ZIM92/2	35	793100	7954300	2780	×	×	SHRIMP, zircon	Wilson et al. 1995
Shurugwi belt	Surprise Formation	Felsic volcanic rock	ZIM92/12	35	814300	7830100	2820	×	×	SHRIMP, zircon	Wilson et al. 1995
Filabusi belt	Eldorado Formation	Foliated felsic volcanic	ZIM92/16	35	754700	7728200	2823	×	×	SHRIMP, zircon	Wilson et al. 1995
Belingwe belt	Chingezi Suite	Granitoid	Zimb95/29	35	796200	7744700	2850	×	×	TIMS, U–Pb, zircon	Horstwood 1998
Filabusi belt	Eldorado Formation	Flaggy felsic gneiss	ZIM92/17	35	753400	7725500	2850	×	×	SHRIMP, zircon	Wilson et al. 1995
Belingwe belt	Koodoovale Formation	Dacite clasts, reworked volcanic breccia	ZIM92/32	35	805400	7737500	2880	×	×	SHRIMP, zircon	Wilson et al. 1995
Bulawayo belt	Avalon Formation	Rhyodacite from vent	ZIM92/13	35	658800	7752200	2895	×	×	SHRIMP, zircon	Wilson et al. 1995
Belingwe belt	Hokonui Formation	Felsic clasts, volcanic breccia	ZIM92/24	35	799100	7738500	2950	×	×	SHRIMP, zircon	Wilson et al. 1995
Tokwe segment	Mashaba tonalite	Weakly foliated tonalite sill cutting 3.5 Ga gneisses	89-Zb-12 core	36	212600	7765000	3250	×	peak	Kober, zircon	Dougherty-Page 1994

Location	Unit	Rock type	Sample								Method	Reference
Belingwe belt	Ngezi tonalite	Tonalite	MHZ143	36	197700	7726600	3267	×	peak	TIMS, Pb-Pb, zircon		Hunter 1997
Masvingo belt	Upper Shamvaian Group	Flow-banded felsite, vent breccia	ZIM92/27	36	297400	7774700	3530	×	×	SHRIMP, zircon		Wilson et al. 1995
Masvingo belt	Upper Shamvaian Group	Flow banded felsite, vent breccia	ZIM92/27	36	297400	7774700	3620	×	×	SHRIMP, zircon		Wilson et al. 1995

Crystallization ages

Location	Unit	Rock type	Sample							Method	Reference
Zambezi belt	Makuti Group	Biotite gneiss	Zim 21	35	761550	8176600	2510	0.4	2σ	Kober, zircon	Dirks et al. 1999
Limpopo NMZ	Razi Suite	Porphyritic granite	L95/8	36	298100	7728800	2517	55	95% conf.	TIMS, U-Pb, zircon	Frei et al. 1999
Zambezi belt	Chilimanzi Suite	Pegmatite	Zim-17	36	413800	8117400	2519	0.4	×	Kober, zircon	Kroner, pers. comm. 1999
Zambezi belt	Mudzi Suite	Gneiss	Zim-43	36	488400	8141100	2526	×	×	Kober, zircon	Kroner, pers. comm. 1999
Masvingo area	Chilimanzi Suite	Porphyritic granite	L95/7	36	279600	7760400	2540	38	95% conf.	TIMS, U-Pb, zircon	Frei et al. 1999
Zambezi belt	Chilimanzi Suite	Porphyritic granite	Zim-19	36	417900	8115400	2559	×	×	Kober, zircon	Kroner, pers. comm. 1999
Zambezi belt	Mudzi Suite	Leucogneiss	ZIM-169	36	396800	8138800	2563	1	×	TIMS, U-Pb, zircon	Vinyu et al. 2001
Zambezi belt	Chilimanzi Suite	Porphyritic granite	J1	36	482200	8090000	2570	0.2	2σ	TIMS, U-Pb, zircon	Hofmann et al. 2002
Limpopo NMZ	Charnockite suite	Charno-enderbite	Renco Mine	36	309000	7718000	2571	5	95% conf.	TIMS, U-Pb, zircon	Blenkinsop & Frei 1997
Zambezi belt	Mudzi Suite	Felsic granulite	Zim-12	36	475300	8123600	2571	×	×	Kober, zircon	Kroner, pers. comm. 1999
impopo NMZ	Razi Suite	Porphyritic granite	L95/6	36	265300	7655700	2576	31	95% conf.	TIMS, U-Pb, zircon	Frei et al. 1999
Limpopo NMZ	Razi Suite	Granite	×	36	205500	7693500	2583	52	2σ	Rb-Sr, WR	Mkweli et al. 1995
Limpopo NMZ	Razi Suite	Charnockite	L95/4	36	203000	7691600	2589	11	95% conf.	TIMS, U-Pb, zircon	Frei et al. 1999
Limpopo NMZ	Razi Suite	Granite	L95/3	36	203000	7691600	2590	7	95% conf.	TIMS, U-Pb, zircon	Frei et al. 1999
Limpopo NMZ	Razi Suite	Porphyritic granite	93,508	36	226000	7690000	2591	4	95% conf.	TIMS, U-Pb, zircon	Kamber et al. 1996
Zambezi belt	Mudzi Suite	Granitic gneiss	Zim-47	36	482000	8092000	2593	0.6	95% conf.	TIMS, U-Pb, zircon	Hofmann et al. 2002
Murehwa batholith	Chilimanzi Suite	Granite	ZIM64	36	374500	8087500	2601	14	2σ	Kober, zircon	Jelsma et al. 1996
Zambezi belt	Mudzi Suite	Granodioritic gneiss	Zim-46	36	482000	8092000	2601	0.6	2σ	Kober, zircon	Hofmann et al. 2002
Limpopo NMZ	Charnockite suite	Charno-enderbite	92/111	36	218100	7672800	2603	64	2σ	TIMS, U-Pb, zircon	Berger et al. 1995
Limpopo NMZ	Razi Suite	Porphyritic granite	L95/5	36	238000	7702000	2604	22	95% conf.	TIMS, U-Pb, zircon	Frei et al. 1999
Zambezi belt	Mudzi Suite	Granite	ZIM-606	36	398900	8148700	2605	11	wt. mean	TIMS, U-Pb, zircon	Vinyu et al. 2001
Harare belt	Internal pluton	Monzodiorite	ZIM89-96	36	319500	8086700	2617	24	95% conf.	Pb-Pb, WR	Jelsma et al. 1996
Harare belt	Internal pluton	Tonalite	ZIM240	36	298000	8087800	2618	6	2σ	TIMS, U-Pb, zircon	Jelsma et al. 1996
Harare belt	Chilimanzi Suite	Pegmatite	CH13	36	339000	8059200	2619	26	2σ	TIMS, U-Pb, apatite	Siegesmund et al. 2001
Zambezi belt	Chilimanzi Suite	Gneiss	Zim-18	36	413800	8117400	2619	0.8	×	Kober, zircon	Kroner, pers. comm. 1999
Zambezi belt	Mudzi Suite	Tonalitic gneiss	ZIM-608	36	397100	8147200	2621	9	wt. mean	TIMS, U-Pb, zircon	Vinyu et al. 2001
Limpopo NMZ	Razi Suite	Microgranite	92/012	36	208000	7693400	2627	7	2σ	TIMS, U-Pb, zircon	Mkweli et al. 1995
Masvingo area	Chilimanzi Suite	Granite	Zimb95/06	36	264700	7824900	2634	17	2σ	TIMS, U-Pb, zircon	Horstwood 1998
Limpopo NMZ	Charnockite suite	Charno-enderbite	92,113	36	228600	7668400	2637	19	2σ	TIMS, U-Pb, zircon	Berger et al. 1995
Harare belt	Shamva porphyries	Andesite porphyry	ZIM1-4, 7-9, 14	36	348100	8087700	2641	65	95% conf.	Pb-Pb, WR	Jelsma et al. 1996
Harare belt	Passaford Formation	Felsic clasts	ZIM92/31	36	287800	8034800	2643	8	95% conf.	SHRIMP, zircon	Wilson et al. 1995
Harare belt	Iron Mask Formation	Rhyolite	ZIM92/29	36	289700	8048100	2645	4	95% conf.	SHRIMP, zircon	Wilson et al. 1995
Harare belt	Internal pluton	Granodiorite	ZIM79	36	278200	8067000	2647	10	2σ	SHRIMP, zircon	Nesbitt et al. 2000
Harare belt	Internal pluton	Granodiorite	ZIM80	36	322000	8088500	2648	6	2σ	SHRIMP, zircon	Nesbitt et al. 2000
Harare belt	Internal pluton	Granodiorite	ZIM80	36	322000	8088500	2649	6	2σ	TIMS, U-Pb, zircon	Jelsma et al. 1996
Harare belt	Arcturus Formation	Rhyolite porphyry	Rh81/23a-f	36	316700	8083700	2659	39	2σ	Pb-Pb, WR	Taylor et al. 1991
Masvingo belt	Upper Shamvaian	Flow banded felsite	ZIM92/27	36	297400	7774700	2661	17	95% conf.	SHRIMP, zircon	Wilson et al. 1995
Harare belt	Internal pluton	Granodiorite	ZIM79	36	278200	8067000	2664	15	2σ	TIMS, U-Pb, zircon	Jelsma et al. 1996

(continued)

Table 1. (*continued*)

Area	Stratigraphic unit	Lithology/mineral	Sample no.	Zone	UTM-X	UTM-Y	Age (Ma)	Error (Ma)	Error type	Technique	Reference
Crystallization ages (continued)											
Harare belt	Lower Shamvaian Group	Tuff	ZIM97-01	36	349000	8083400	2666	36	2σ	SHRIMP, zircon	Nesbitt et al. 2000
Chinamora batholith	Wedza Suite	Granitic gneiss	ZIM60	36	332700	8072600	2667	4	2σ	TIMS, U-Pb, zircon	Jelsma et al. 1996
Midlands belt	Sesombi tonalite	Tonalite	Zimb95/02	35	779300	7925500	2668	17	2σ	TIMS, U-Pb, zircon	Horstwood 1998
Limpopo NMZ	Razi Suite	Granite	92/041	35	766700	7646900	2669	67	2σ	TIMS, U-Pb, zircon	Mkweli et al. 1995
Belingwe belt	Chingezi Suite	Granitoid	Zimb95/29	35	796200	7744700	2672	14	2σ	TIMS, U-Pb, zircon	Horstwood 1998
Harare belt	Black Cat porphyry	Andesitic porphyry	ZIM29	36	349400	8081800	2672	12	2σ	TIMS, U-Pb, zircon	Jelsma et al. 1996
Midlands belt	Sesombi tonalite	Tonalite	Rh73, 176, 178, 179 mixed	35	780000	7923300	2673	10	2σ	Kober, zircon	Dougherty-Page 1994
Midlands belt	Giraffe porphyry	Porphyry	Zimb164	35	780300	7907900	2677	6	2σ	TIMS, U-Pb, zircon	Horstwood 1998
Midlands belt	What Cheer Formation	Felsic clasts	ZIM92/2	35	793100	7954300	2683	8	95% conf.	SHRIMP, zircon	Wilson et al. 1995
Belingwe belt	Reliance Formation	Komatiitic basalt	Z21, 24, 25	36	196700	7730600	2692	9	×	Pb-Pb, WR	Chauvel et al. 1993
Bulawayo belt	Avalon Formation	Rhyodacite	ZIM92/13	35	658800	7752200	2696	9	95% conf.	SHRIMP, zircon	Wilson et al. 1995
Harare belt	Maparu Formation	Water-lain tuff	ZIM92/30	36	343200	8097200	2697	9	95% conf.	SHRIMP, zircon	Wilson et al. 1995
Shurugwi belt	Surprise Formation	Felsic volcanic rock	ZIM92/12	35	814300	7830100	2698	27	95% conf.	SHRIMP, zircon	Wilson et al. 1995
Midlands belt	Maliyami Formation	Andesite	ZIM92/6	35	765400	7915400	2702	9	2σ	SHRIMP, zircon	Wilson et al. 1995
Zambezi belt	Makuti Group basement	Biotite gneiss	Zim 22	35	761550	8176600	2704	0.3	2σ	Kober, zircon	Dirks et al. 1999
Limpopo NMZ	Charnockite suite	Charno-enderbite	90/78G	36	222000	7676600	2710	38	2σ	TIMS, U-Pb, zircon	Berger et al. 1995
Harare belt	Iron Mask Formation	Rhyolite	ZIM74	36	304500	8078000	2715	15	2σ	TIMS, U-Pb, zircon	Jelsma et al. 1996
Mutare belt	internal pluton	Tonalite	×	36	464000	7913200	2741	3	×	TIMS, U-Pb, zircon	Schmidt-Mumm et al. 1994
Mutare belt	internal pluton	Tonalite	×	36	464000	7913200	2742	3	×	TIMS, U-Pb, zircon	Schmidt-Mumm et al. 1994
Shangani batholith	Somabula tonalite	Tonalite	Rh73/31-39	35	783600	7815300	2752	51	2σ	Pb-Pb, WR	Taylor et al. 1991
Limpopo NMZ	Charnockite suite	Charno-enderbite	90/78F	36	222000	7676600	2768	112	2σ	TIMS, U-Pb, zircon	Berger et al. 1995
Shangani batholith	Gwenoro Dam gneiss	Banded migmatite	B80	35	801400	7811800	2769	0.2	2σ	Kober, zircon	Jelsma et al. unpublished data 2002
Filabusi belt	Eldorado Formation	Rhyodacite	ZIM92/16	35	754700	7728200	2788	10	95% conf.	SHRIMP, zircon	Wilson et al. 1995
Filabusi belt	Eldorado Formation	Felsic gneiss	ZIM92/17	35	753400	7725900	2799	9	95% conf.	SHRIMP, zircon	Wilson et al. 1995
Midlands belt	Arizona Formation	Rhyodacite	ZIM92/10	35	774400	7873900	2805	6	95% conf.	SHRIMP, zircon	Wilson et al. 1995
Belingwe belt	Koodoovale Formation	Dacite clasts	ZIM92/32	35	805400	7737500	2831	6	95% conf.	SHRIMP, zircon	Wilson et al. 1995
Shangani batholith	Insukamini	Banded gneiss	B42b	35	772300	7855900	2837	0.3	2σ	Kober, zircon	Jelsma et al. unpublished data 2002
Masvingo belt	'Manjeri Formation'	Limestone	×	36	261100	7772100	2839	33	2σ	Pb-Pb, WR	Moorbath et al. 1987
Shurugwi belt	Chingezi Suite	Migmatite boudin	×	35	823800	7814100	2848	26	2σ	SIMS, Pb-Pb, zircon	Nägler et al. 1997
Shurugwi belt	Chingezi Suite	Migmatite boudin	×	35	823800	7814100	2852	64	2σ	SIMS, Pb-Pb, zircon	Nägler et al. 1997
Belingwe belt	Ngezi tonalite	Tonalite	MHZ143	36	197700	7726600	2875	×	peak	TIMS, Pb-Pb, zircon	Hunter 1997
Midlands belt	Mafic Formation	dacite	Zimb150	35	790600	7913000	2880	8	2σ	TIMS, U-Pb, zircon	Horstwood 1998
Buchwa area	Chipinda gneiss	Gneiss	×	36	216100	7709100	2900	×	×	TIMS, U-Pb, zircon	Fedo et al. 1995
Belingwe belt	Hokonui Formation	Dacite clasts	ZIM92/24	35	799100	7738500	2904	9	95% conf.	SHRIMP, zircon	Wilson et al. 1995
Tokwe segment	Mashaba tonalite	Tonalite	89-Zb-12 rim	36	212600	7765000	2950	100	×	TIMS, U-Pb, zircon	Dougherty-Page 1994
Shurugwi belt	Mont d'Or pluton	Granodiorite	×	35	812500	7815700	3345	55	2σ	Pb-Pb, WR	Taylor et al. 1984
Tokwe segment	Tokwe gneiss	Leucosome	Zimb95/09	35	260200	7704400	3368	9	2σ	TIMS, U-Pb, zircon	Horstwood 1998
Tokwe segment	Mushandike granite	Granite	×	36	249700	7779000	3375	×	×	TIMS, U-Pb, zircon	Horstwood, pers. comm. 1999

Tokwe segment	Tokwe river gneiss	Dioritic gneiss	ZIM-1	36	231700	7779800	3388	28	2σ	TIMS, U–Pb, zircon	Munyanyiwa, in Blenkinsop 1997
Tokwe segment	Tokwe river gneiss	Banded gneiss	Zimb95/14	36	231700	7779800	3455	2	2σ	TIMS, U–Pb, zircon	Horstwood 1998
Rhodesdale batholith	Rhodesdale gneiss	Gneiss	Zimb226	35	797600	7913000	3456	6	2σ	TIMS, U–Pb, zircon	Horstwood 1998
Tokwe segment	Tokwe river gneiss	Tonalitic gneiss	ZIM-2	36	231700	7779800	3554	45	2σ	TIMS, U–Pb, zircon	Munyanyiwa, in Blenkinsop 1997
Midlands belt	Sebakwe river gneiss	Gneiss	Zimb141	35	792600	7912500	3565	21	2σ	TIMS, U–Pb, zircon	Horstwood 1998
Provenance ages											
Makaha belt	Shamvaian Supergroup	Granitoid clast	Zim 6	36	482200	8084000	2613	0.2	2σ	Kober, zircon	Hofmann et al. 2002
Harare belt	Shamvaian Supergroup	Granitoid clast	89-S-19	36	351300	8090700	2650	×	×	Kober, zircon	Dougherty-Page 1994
Harare belt	Shamvaian Supergroup	Granitoid clast	89-S-25	36	351300	8090700	2670	×	×	Kober, zircon	Dougherty-Page 1994
Harare belt	Shamvaian Supergroup	Granitoid clast	89-S-12	36	351300	8090700	2680	×	×	Kober, zircon	Dougherty-Page 1994
Chinhoyi belt	Shamvaian Supergroup	Granitoid clast	89-C-15	36	204000	6079500	2720	12	2σ	Kober, zircon	Dougherty-Page 1994
Chinhoyi belt	Shamvaian Supergroup	Granitoid clast	89-C-20 + 28 rim	36	204000	6079500	2800	40	2σ	Kober, zircon	Dougherty-Page 1994
Harare belt	Shamvaian Supergroup	Granitoid clast	89-S-14 rim	36	351300	8090700	2800	40	2σ	Kober, zircon	Dougherty-Page 1994
Harare belt	Shamvaian Supergroup	Granitoid clast	89-S-23	36	351300	8090700	2800	×	×	Kober, zircon	Dougherty-Page 1994
Belingwe belt	Hokonui Formation	Tonalite clast	89-H-1	35	799700	7741700	2840	×	×	Kober, zircon	Dougherty-Page 1994
Chinhoyi belt	Shamvaian Supergroup	Granitoid clast	89-C-23	36	204000	6079500	2874	6	2σ	Kober, zircon	Dougherty-Page 1994
Chinhoyi belt	Shamvaian Supergroup	Granitoid clast	89-C-21	36	204000	6079500	2875	6	2σ	Kober, zircon	Dougherty-Page 1994
Chinhoyi belt	Shamvaian Supergroup	Granitoid clast	89-C-20 + 28 core	36	204000	6079500	2880	×	×	Kober, zircon	Dougherty-Page 1994
Belingwe belt	Brooklands Formation	Detrital zircons	MHZ142a	36	200500	7732100	2900	×	peak	TIMS, Pb–Pb, zircon	Hunter 1997
Buchwa belt	Buchwa quartzite	Detrital zircons	×	36	221200	7712700	2900	×	peak	TIMS, U–Pb, zircon	Fedo et al. 1995
Harare belt	Shamvaian Supergroup	Granitoid clast	89-S-23	36	351300	8090700	2907	×	×	Kober, zircon	Dougherty-Page 1994
Harare belt	Shamvaian Supergroup	Granitoid clast	89-S-19	36	351300	8090700	2920	60	2σ	Kober, zircon	Dougherty-Page 1994
Harare belt	Shamvaian Supergroup	Granitoid clast	89-S-12	36	351300	8090700	2921	20	2σ	Kober, zircon	Dougherty-Page 1994
Harare belt	Shamvaian Supergroup	Granitoid clast	89-S-25	36	351300	8090700	2924	10	2σ	Kober, zircon	Dougherty-Page 1994
Buchwa belt	Buchwa quartzite	Detrital zircons	×	36	221200	7712700	3050	×	×	TIMS, U–Pb, zircon	Fedo et al. 1995
Harare belt	Shamvaian Supergroup	Granitoid clast	89-S-25 core	36	351300	8090700	3116	×	peak	Kober, zircon	Dougherty-Page 1994
Harare belt	Shamvaian Supergroup	Granitoid clast	89-S-23 core	36	351300	8090700	3182	×	peak	Kober, zircon	Dougherty-Page 1994
Harare belt	Shamvaian Supergroup	Granitoid clast	89-S-14 core	36	351300	8090700	3197	20	2σ	Kober, zircon	Dougherty-Page 1994
Buchwa belt	Buchwa quartzite	Detrital zircons	21 grains	36	221200	7712700	3200	×	peak	SHRIMP, zircon	Dodson et al. 1988
Shurugwi belt	Wanderer Formation	Detrital zircons	3 grains	35	817700	7828700	3200	×	peak	SHRIMP, zircon	Dodson et al. 1988
Shurugwi belt	Shurugwi basal succession	Detrital zircons	8 zircons	35	818300	7822000	3219	×	peak	SIMS, Pb–Pb, zircon	Nägler et al. 1997
Belingwe belt	Brooklands Formation	Detrital zircons	MHZ142a	36	200500	7732100	3275	×	peak	TIMS, Pb–Pb, zircon	Hunter 1997
Belingwe belt	Brooklands Formation	Detrital zircons	MHZ142a	36	200500	7732100	3350	×	peak	TIMS, Pb–Pb, zircon	Hunter 1997
Buchwa belt	Buchwa quartzite	Detrital zircons	15 grains	36	221200	7712700	3350	×	peak	SHRIMP, zircon	Dodson et al. 1988
Shurugwi belt	Wanderer Formation	Detrital zircons	6 grains	35	817700	7828700	3350	×	peak	SHRIMP, zircon	Dodson et al. 1988
Belingwe belt	Brooklands Formation	Detrital zircons	MHZ142a	36	200500	7732100	3425	×	peak	TIMS, Pb–Pb, zircon	Hunter 1997
Buchwa belt	Buchwa quartzite	Detrital zircons	4 grains	36	221200	7712700	3460	×	peak	SHRIMP, zircon	Dodson et al. 1988
Shurugwi belt	Wanderer Formation	Detrital zircons	7 grains	35	817700	7828700	3460	×	peak	SHRIMP, zircon	Dodson et al. 1988
Shurugwi belt	Shurugwi basal succession	Detrital zircons	9 zircons	35	818300	7822000	3477	×	peak	SIMS, Pb–Pb, zircon	Nägler et al. 1997
Belingwe belt	Brooklands Formation	Detrital zircons	MHZ142a	36	200500	7732100	3575	×	peak	TIMS, Pb–Pb, zircon	Hunter 1997
Buchwa belt	Buchwa quartzite	Detrital zircons	6 grains	36	221200	7712700	3600	×	peak	SHRIMP, zircon	Dodson et al. 1988
Shurugwi belt	Wanderer Formation	Detrital zircons	15 grains	35	817700	7828700	3600	×	peak	SHRIMP, zircon	Dodson et al. 1988
Shurugwi belt	Shurugwi basal succession	Detrital zircons	9 zircons	35	818300	7822000	3701	×	peak	SIMS, Pb–Pb, zircon	Nägler et al. 1997
Shurugwi belt	Wanderer Formation	Detrital zircons	8 zircons	35	817700	7828700	3775	×	peak	SHRIMP, zircon	Dodson et al. 1988
Buchwa belt	Buchwa quartzite	Detrital zircons	1 grain	36	221200	7712700	3800	×	peak	SHRIMP, zircon	Dodson et al. 1988

Dating techniques include single-zircon, or multi-zircon (or apatite) U–Pb or Pb–Pb TIMS (thermal ionization mass spectrometry), single-zircon Pb–Pb SHRIMP (sensitive high-resolution ion microprobe), or single-zircon Pb–Pb Kober methods; zircon Pb–Pb SIMS (secondary ion mass spectrometry) method; Pb–Pb, and Sm–Nd whole-rock methods. TDM. Depleted mantle age; UTM, Universal Transverse Mercator projection; WR, whole rock; NMZ, North Marginal Zone; ×, not supplied in reference.

Hoffman 1984; de Wit 1998). For example, TTGs have geochemical characteristics similar to those of modern adakites (e.g. Defant & Kepezhinskas 2001), which are volumetrically rare in modern arc terranes. This may suggest that melting of the subducting slab was common during Archaean times, basically because the subducting lithosphere was, on average, younger, hotter and more buoyant than it is today. Other models relate the origin of TTG crust to gravity-driven convective overturn, following the transformation of thick basaltic crust into eclogite, which descended back into the mantle (Vlaar et al. 1994), or to fractional crystallization of a global shallow-level tholeiitic magma layer (the mantle recycling or MARCY model: Kramers 1988). These models are not mutually exclusive and all may have operated to some extent during the Archaean. Along similar lines, greenstone sequences may reflect remnant oceanic crust or island-arc material that was amalgamated with continental fragments during some form of subduction–accretion (e.g. Card 1990; Hoffman 1991; de Wit et al. 1992; Kusky & Polat 1999), or rifted or overplated sequences related to the emplacement of mantle plumes, with deformation being attributed to vertical tectonic processes (e.g. Bouhallier et al. 1993; Jelsma et al. 1993; Collins et al. 1998).

The enigma is as follows: in many cratons major accretionary events and the emplacement of pan-cratonic granitoid intrusions occur immediately before (<50–100 Ma) to cratonization (e.g. de Wit et al. 1992; Jelsma et al. 2001). Resultant upper-crustal deformation geometries include recumbent folds, thrust zones, and duplex arrays with discrete mylonite zones bounding low-strain domains (e.g. Dirks & Jelsma 1998). These resemble the upper- to mid-crustal imbricate stacks interpreted from seismic reflection profiles for the Superior Province and Yilgarn Craton (Ludden et al. 1993; Wilde et al. 1996) which can be regarded as 'frozen-in' crustal sections. This deformation led to isostatically stable, 30–40 km thick continental crust, but without significant exhumation of medium- to high-pressure rocks, as can be seen in recent orogens. Blueschists, eclogites and paired metamorphic belts are absent (e.g. Hamilton 1998). Lower-crustal rocks are exposed only because of younger orogenic or extraterrestrial events, or because of complex and rare (for the Archaean) 'collisional' events such as in the Limpopo belt.

The explanation of this paradox may lie with the rate of heat production. With the high crustal geotherms prevailing in the Archaean, decoupling of crust and mantle lithosphere is expected, allowing strain to be accommodated independently in both layers, a so-called 'jelly-sandwich' rheological layering (see Brown & Phillips 2000). Decoupled mantle and crust could accommodate independent strain regimes and responses to regional, far-field stresses. This may explain some of the differences between deformed Archaean and Phanerozoic rock assemblages, in particular the lack of crustal overthickening during Archaean 'orogenic' events (Jelsma et al. 2001).

In this paper we test the implications of this model for compatibility with a review of the evolution of the Zimbabwe Craton, using a collation of geochronological, geochemical and field data, focusing on the Neoarchaean (2.72–2.58 Ga). Our aim is to show: (1) that the Zimbabwe Craton comprises greenstone sequences that were formed in different tectonomagmatic environments; (2) that crustal shortening and thickening through tectonic stacking was a prominent deformation mechanism for large parts of the craton; (3) that tectonic styles cannot be entirely reconciled with modern plate tectonic processes. All ages reported in the text are zircon ages, unless otherwise indicated.

Zimbabwe Craton

The Zimbabwe Craton in southern Africa is composed of 26 greenstone belts that constitute about 20% of the craton, and granite–gneiss complexes that make up the remainder (Fig. 1a, Blenkinsop et al. 1997). These granite–greenstone terranes

---

Fig. 1. (a) Simplified regional map showing the position of the Zimbabwe Craton, greenstone belts, major shear zones as visible on Landsat TM images (e.g. RGB741), and orogenic belts. The shear zones may represent terrane boundaries and/or late strike-slip fault zones. Numbers refer to greenstone belt names.
1, Makaha; 2, Mount Darwin–Dindi; 3, Chinhoyi–Guruve; 4, Harare–Shamva; 5, Norton; 6, Chegutu; 7, Midlands; 8, Gweru–Mvuma; 9, Shurugwi; 10, Beatrice; 11, Manesi; 12, Felixburg; 13, Bubi; 14, Bulawayo; 15, Filabusi; 16, Shangani; 17, Gwanda; 18, Antelope–Lower Gwanda; 19, Mweza; 20, Buhwa; 21, Belingwe; 22, Masvingo; 23, Odzi–Mutare–Manica; 24, Francistown; 25, Vumba; 26, Tati. Superimposed on subsequent figures are sample localities of dated rocks, and the approximate current extent of rock units. (b) Crustal growth between 3.55 and 3.0 Ga: formation of the Sebakwe proto-craton and subsequent stable shelf sedimentation. (c) Frequency histograms of detrital zircons from the Belingwe, Buhwa and Shurugwi greenstone belts. (d) Crustal growth between 2.95 and 2.75 Ga.

(a) ZAMBEZI OROGENIC BELT
2600 Ma, 800 Ma, 500 Ma

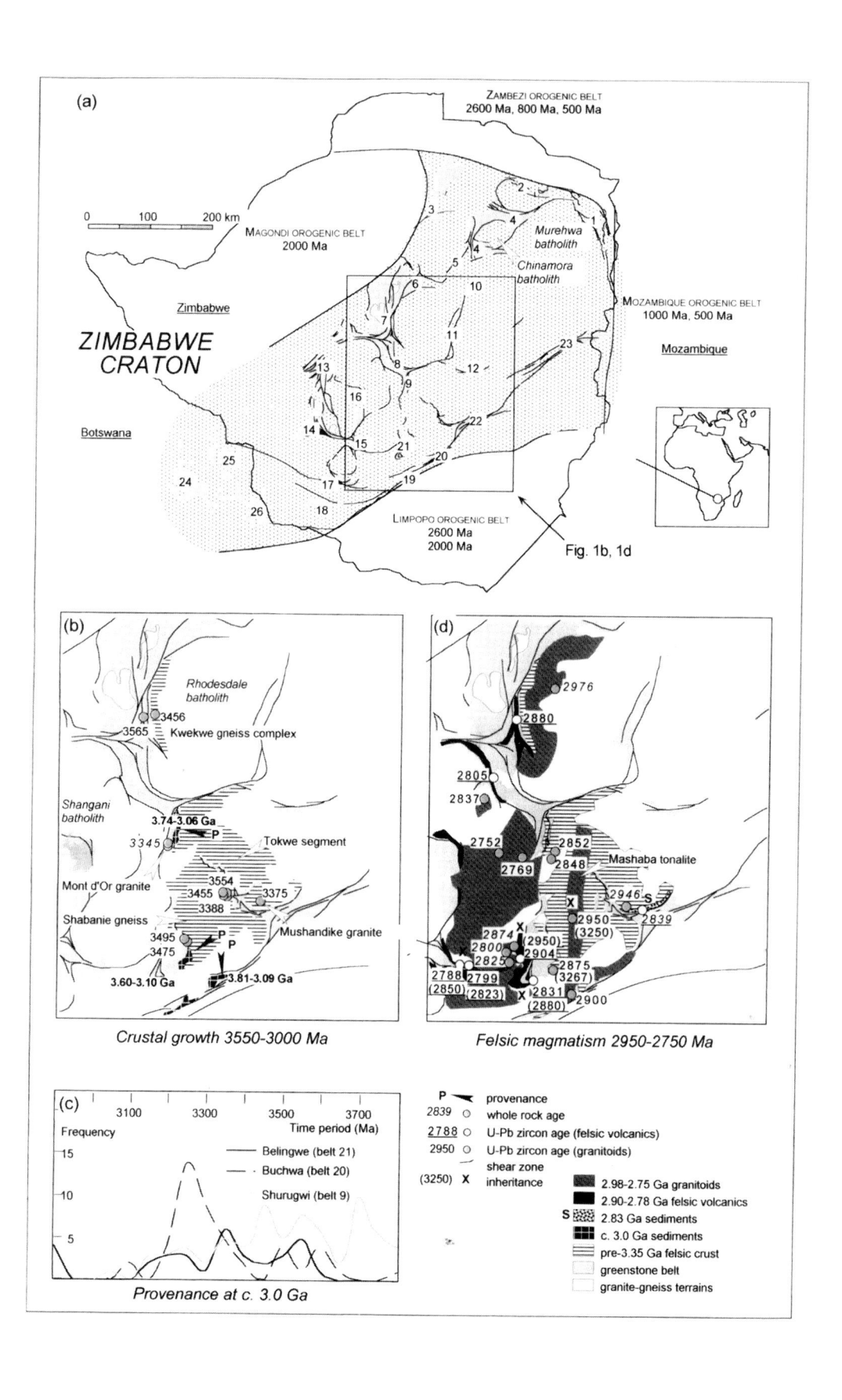

MAGONDI OROGENIC BELT
2000 Ma

Murehwa batholith
Chinamora batholith

Zimbabwe

ZIMBABWE CRATON

MOZAMBIQUE OROGENIC BELT
1000 Ma, 500 Ma

Mozambique

Botswana

LIMPOPO OROGENIC BELT
2600 Ma
2000 Ma

Fig. 1b, 1d

(b)
Rhodesdale batholith
3456
3565 Kwekwe gneiss complex
Shangani batholith
3.74-3.06 Ga
3345
Tokwe segment
Mont d'Or granite
3554
3455   3375
3388
Shabanie gneiss
3495
3475
Mushandike granite
3.60-3.10 Ga   3.81-3.09 Ga

Crustal growth 3550-3000 Ma

(d)
2976
2880
2805
2837
2752   2852
2769   2848   Mashaba tonalite
2946 S
2874 (2950)   2950   2839
2800 2904   (3250)
2825
2788 2799   2875
(2850) (2823) (3267)
2831   2900
(2880)

Felsic magmatism 2950-2750 Ma

(c)
Frequency
Time period (Ma)
3100 3300 3500 3700
15
— Belingwe (belt 21)
--- Buchwa (belt 20)
10
— Shurugwi (belt 9)
5

Provenance at c. 3.0 Ga

P ➤ provenance
2839 ○ whole rock age
2788 ○ U-Pb zircon age (felsic volcanics)
2950 ○ U-Pb zircon age (granitoids)
shear zone
(3250) X inheritance

■ 2.98-2.75 Ga granitoids
■ 2.90-2.78 Ga felsic volcanics
S ▨ 2.83 Ga sediments
▦ c. 3.0 Ga sediments
▤ pre-3.35 Ga felsic crust
□ greenstone belt
□ granite-gneiss terrains

formed during a time span of at least 1000 Ma (3.57–2.57 Ga). To the north and south, the Zimbabwe Craton is bound by the Zambezi and Limpopo orogenic belts, which affected parts of the Zimbabwe Craton during Neoarchaean times (c. 2.6 Ga). The formation of granites and greenstones is commonly linked to the emplacement of mantle plumes below an existing continental crust, causing intracontinental rifting and crustal melting (e.g. Bickle et al. 1975, 1994; Wilson et al. 1995; Silva 1997; Hunter et al. 1998). A number of unconformities have been described at the base of major greenstone sequences overlying older continental crust or older greenstone sequences (e.g. Bickle et al. 1975).

Deformation patterns have been attributed to Neoarchaean horizontal crustal shortening processes such as regional cross-folding (Snowden & Bickle 1976), or the development of nappes (Stowe 1984), or to vertical tectonic processes such as ballooning plutonism (Ramsay 1989) or diapirism (Jelsma et al. 1993; Becker et al. 2000). Some workers have related deformation of the granite–greenstone terranes to far-field stresses associated with collisional processes at plate margins (e.g. indenter tectonics related to the Limpopo belt: Coward et al. 1976; Wilson 1990; Treloar et al. 1992; Blenkinsop & Treloar 2001). Others have presented evidence that tectonic stacking of the stratigraphy played a role in the evolution of the greenstone sequences (Tomschi 1987; Fuchter 1990; Barton et al. 1991; Garson 1995; Kusky & Winsky 1995; Dirks & van der Merwe 1997; Jelsma & Dirks 2000).

Quantitative P–T estimates for the granite–greenstone terranes are scarce. Rocks were generally metamorphosed at low pressures and low to moderate temperatures (Saggerson & Turner 1976). Sub-greenschist-facies rocks are found within the central parts of the craton (e.g. Ngezi Group of the Belingwe greenstone belt and Maliyami Formation in the Midlands greenstone belt), whereas other parts were metamorphosed under greenschist- (central parts of belts) or amphibolite-facies conditions (outer parts of belts, close to batholith contacts). Jelsma (1993) obtained metamorphic estimates of $475° \pm 25°C$ and 1.5–2.0 kbar for the peak assemblage in the central part of the Shamva greenstone belt and $600° \pm 25°C$ and 3.0–4.0 kbar near the contact with the batholiths. Granulite-facies conditions have been recorded for gneisses within the marginal zones of adjacent high-grade gneiss terranes (Zambezi and Limpopo orogenic belts). Here, P–T estimates are in the range of 750–900°C and 5.0–9.0 kbar (e.g. Barton et al. 1991; Tsunogae et al. 1992), indicating that moder-ate to deep sections of Archaean crust are exposed.

3.57–2.75 Ga time period

The oldest rocks of the Zimbabwe Craton are tonalitic to trondhjemitic gneisses that have been dated between 3.57 and 3.37 Ga (Fig. 1b). The outcrop of these gneisses includes the Tokwe Segment (Wilson 1990) and the Kwekwe gneiss complex. The inferred extent of this crustal block was considered by Kusky (1998) and Horstwood et al. (1999) to underlie much of the current Zimbabwe Craton and has been referred to as the (pre-3.35 Ga) Sebakwe proto-craton (Horstwood et al. 1999). However, the eastern extent of these continental crustal fragments is uncertain and the component parts may have been created elsewhere, and assembled at some time later, as the segments are separated by major shear zones (e.g. Campbell & Pitfield 1994) that were active during Neoarchaean times.

The gneisses include banded and isoclinally folded, migmatitic varieties and homogeneous, foliated varieties of tonalitic to granitic composition (TTG: tonalite–trondhjemite–granodiorite). Different phases have been dated between 3565 and 3368 Ma (e.g. Horstwood et al. 1999). Infolded are numerous, up to several kilometres long, narrow greenstone remnants of the Sebakwian Group. The orientation of these inclusions is subvertical, parallel to the banding of the gneisses, and strikes in a NNE direction, giving the proto-craton a pronounced NNE structural trend (Wilson 1968). The gneisses and deformed Sebakwian Group rocks have been intruded by the 3.38 Ga (Horstwood, pers. comm., 1999) Mushandike granite and the 3.35 Ga (Dodson et al. 1988) Mont d'Or 'granite'. Both are granite–granodiorite intrusions that are post-kinematic with respect to the formation of the regional fabric.

Between 3.10 and 2.95 Ga, the western and southern parts of the proto-craton became the sites of stable shelf or rift sedimentation (Fig. 1b). The stratigraphy is characterized by a transgressive assemblage of conglomerates, quartz arenites, siltstones, shales, cherts and ironstones, and associated with (ultra-) mafic lavas (Fedo & Eriksson 1996; Hunter 1997; Nägler et al. 1997). Detrital zircons from sedimentary rocks from the adjacent Buhwa, Belingwe and Shurugwi greenstone belts give ages ranging between 3.81 and 3.06 Ga (Dodson et al. 1988; Hunter 1997; Nägler et al. 1997; Fig. 1c).

Following this initial stabilization of continental crust, felsic magmatism started along the western margin of the proto-craton (Fig. 1d) and perhaps also in other parts. The volcanic rocks form part of the Lower Greenstones (Wilson 1979), which include the 2.90–2.88 Ga Belingwean Supergroup and 2.83–2.79 Ga Lower

Bulawayan Group (Wilson et al. 1995, Horstwood 1998) and comprise calc-alkaline andesites, dacites and rhyolites and their volcaniclastic equivalents. Intercalated with the volcanic rocks are arkoses, conglomerates, cherts and ironstones but also a major development of komatiites, and high Mg- and tholeiitic basalts, interbedded with ironstones and cherts (Bend Formation). TTG-type granitoids of the Chingezi Suite (Wilson et al. 1995) occur as intrusions within the protocraton, and further to the west within the Shangani and Rhodesdale batholiths, associated with the Lower Greenstones (Fig. 1d). These intrusions range in age from 2.98 to 2.75 Ga (Taylor et al. 1991; Dougherty-Page 1994; Fedo et al. 1995; Hunter 1997; Nägler et al. 1997; Horstwood 1998; Bozhko, pers. comm., 1998), and have been interpreted as the plutonic equivalents of the Belingwean and Lower Bulawayan felsic volcanic rocks (Luais & Hawkesworth 1994).

2.75–2.58 Ga time period

Most of the greenstone successions of the Zimbabwe Craton were formed at c. 2.7 Ga and are referred to as the Upper Greenstones (Wilson 1979), or Upper Bulawayan Group (Wilson et al. 1995). Amongst these, two distinctly different greenstone groupings may be recognized in different parts of the craton, which broadly correspond to Wilson's (1979) eastern and western successions (Fig. 2a).

Type 1 greenstone successions: the Belingwe belt as an example of continental rifting

Within the confines of the enlarged Sebakwe proto-craton, a volcano-sedimentary succession was laid down in proximity to older granite–greenstone basement, and has been interpreted as a rift (Bickle et al. 1975, 1994; Wilson et al. 1995; Silva 1997; Hunter et al. 1998), or back-arc (Kusky 1998) succession. This eastern succession can be found in the Belingwe, Filabusi, Shangani, Shurugwi and Gweru greenstone belts (Figs 2a and 3a). It is characterized by a well-preserved stratigraphy with basement–cover unconformities, a basal sedimentary succession and an overlying thick pile of (ultra-) mafic volcanic rocks, which have been dated at c. 2.69 Ga. It comprises on average 10% komatiites (and high-Mg basalts), 63% basalts (and basaltic andesites), 2% andesites and 25% clastic sedimentary rocks. Contemporaneous felsic volcanic rocks are rare or absent.

The type stratigraphy of the Upper Bulawayan Group is in the Belingwe greenstone belt (Wilson 1979) and has been referred to as the Ngezi Group (Nisbet et al. 1993). The Ngezi Group is situated wedged between the c. 3.57–3.35 Ga proto-craton to the east and a c. 2.9–2.8 Ga crustal segment to the west (Fig. 3a and b). It comprises a basal sedimentary succession (Manjeri Formation) overlain by ultramafic and mafic volcanic rocks (Reliance and Zeederbergs Formations), and capped by a sedimentary succession (Cheshire Formation). The base of this succession is an unconformity between the c. 2.7 Ga Manjeri Formation and gneisses of the Tokwe Segment to the east and Lower Greenstones to the SE and west (Bickle et al. 1975).

The Manjeri Formation (Fig. 3b) is a thin (0–120 m thick) clastic sedimentary succession with marked lateral thickness and facies variations. The lower sequence comprises pebble–cobble conglomerates and trough cross-bedded quartz sandstones exhibiting features indicative of soft sediment deformation. These sedimentary rocks grade upward into an upper sequence of arkosic sandstones, siltstones, cherts and ironstones, and rare stromatolitic limestones, capped by a sulphide-facies ironstone horizon (Hunter 1997). Hunter et al. (1998) described the formation as a fluviatile and tidal-dominated delta-plain sequence, overlain by immature alluvial or fan-delta deposits at the margins of a subsiding basin, overlain by a shallow-water exhalative sequence. They interpreted the depositional environment as a continental rift, with changing conditions attributed to increasing topographic relief. The age of the Manjeri Formation is poorly constrained, but younger than 2.83 Ga, which is the age of the (locally) underlying Koodoovale Formation. Crustal residence ages based on Nd isotopic compositions for fine-grained sedimentary rocks (ironstones, shales, mudstones and siltstones) show a variation across the greenstone belt, ranging between 3.71 and 3.32 Ga for the eastern part of the belt, and between 3.24 and 2.97 Ga for the western part (Hunter et al. 1998). The data suggest that the eastern sediments were mainly derived from the Sebakwe proto-craton, whereas the western sediments were derived from a mixed provenance that includes both 3.5–3.3 Ga and 2.9–2.8 Ga components.

The Manjeri Formation is overlain by Reliance Formation (ultra-) mafic volcanic rocks (Fig. 3b). The contact between the two formations is a high-strain zone of 1–5 m width overlain by relatively unstrained pillow lavas. It has been interpreted as (1) conformable but accommodating layer-parallel slip during refolding of the belt (Blenkinsop et al. 1993), or (2) a thrust or detachment surface that accommodated obduction of an oceanic plateau (Kusky & Kidd 1992).

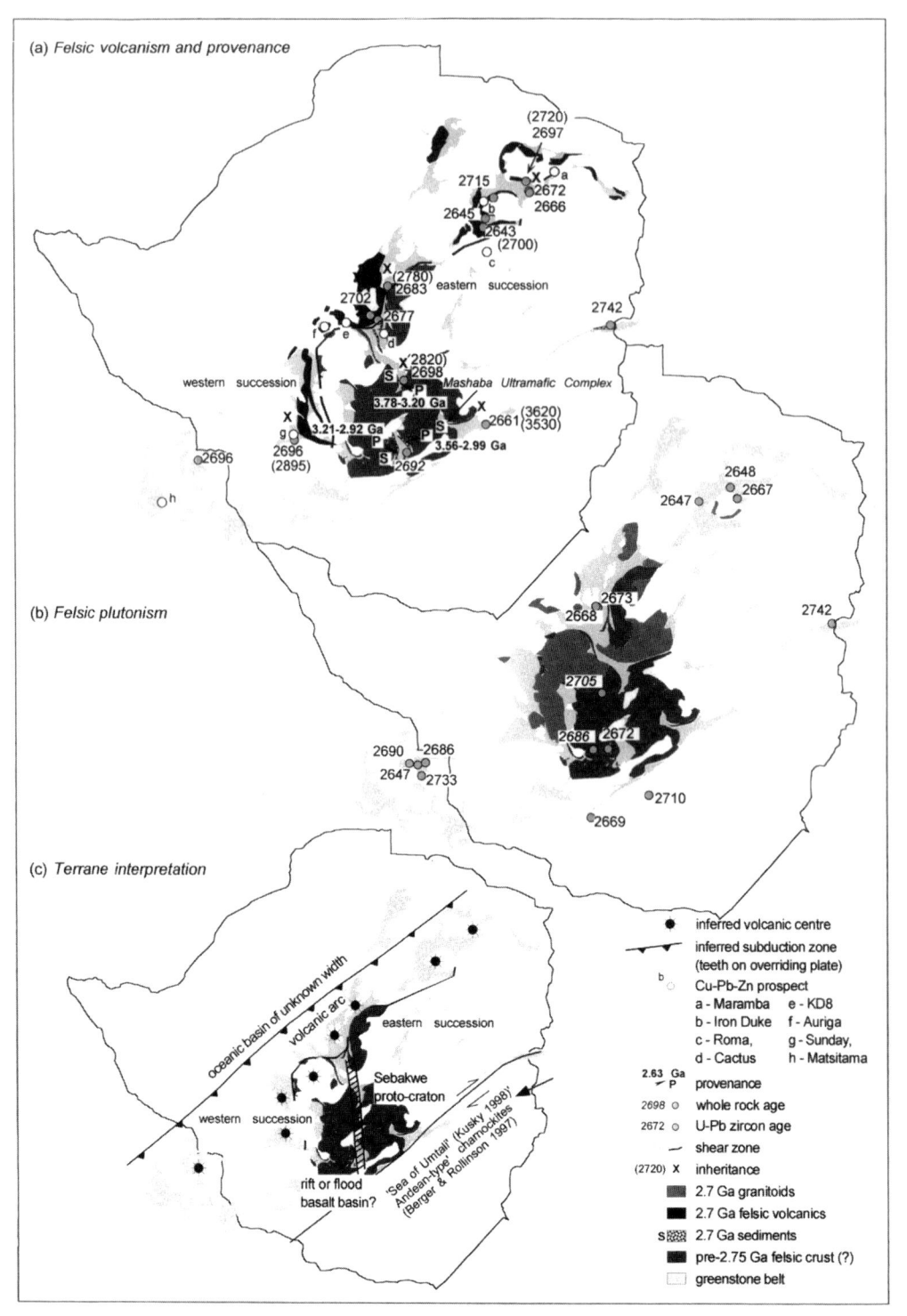

Fig. 2. Crustal growth between 2.72 and 2.64 Ma. (**a**) Volcanic rocks, sedimentary rocks and base metal mineralization. (**b**) Emplacement of Chingezi Suite granitoids. (**c**) Tentative terrane interpretation. Dashed line indicates the position of the boundary between the western and eastern succession of Upper Greenstones of Wilson (1979).

The Reliance Formation consists of komatiites, high-Mg and tholeiitic basalts, and minor tuffaceous intercalations including accretionary lapilli (Nisbet et al. 1977, 1993). The succession has been divided into a lower mafic member (120-160 m thick), central high-Mg basalt and komatiite members (c. 240 m and 0-180 m thick, respectively), and a poorly exposed uppermost volcaniclastic mafic member (200-350 m thick, Nisbet et al. 1977; Scholey 1992). High-Mg basalts from the lower mafic member have been dated at 2692 Ma (Chauvel et al. 1993, Pb-Pb whole-rock age). Nisbet et al. (1993) reported thin beds of quartz grit in drill core intermingled with tuffs of the basal Reliance Formation, suggesting that quartzose clastic sediment was deposited after the onset of volcanic activity.

The upper contact of the Reliance Formation with the overlying Zeederbergs Formation (Fig. 3b) is poorly exposed and placed at the base of a range of hills of basalts, overlying a horizon of volcaniclastic rocks (Nisbet et al. 1993). The Zeederbergs Formation is a c. 2850 m thick sequence of mostly pillowed tholeiitic basalts with minor massive (sills, flows?), hyaloclastic and tuffaceous intercalations, and with an andesitic horizon near the top (Brake 1996). The basalts form flow units up to 200 m thick. Massive basalt is locally gradational to pillow basalt, indicating that the massive variety represents, at least in part, extrusive lava flows. Discordant bodies (<2 m thick dykes, 15-20 m wide feeder vent) of similar material have been interpreted as feeder systems to overlying flows (Brake 1996). Intercalated epiclastic sedimentary rocks include shale and graded units showing partial and complete Bouma sequences indicative of turbidity current deposits (Hofmann et al. 2001a).

The c. 1.3 km thick Cheshire Formation (Fig. 3b) forms the uppermost stratigraphic succession of the Belingwe greenstone belt and is distrbuted along the synclinal axis of the belt (e.g. Martin 1978). It consists predominantly of sedimentary rocks of very low metamorphic grade. The western contact between the Zeederbergs and Cheshire formations is sharp and sheared; the eastern contact is in part unconformable (Hofmann et al. 2001a,b). A detailed analysis of the sedimentology of the formation shows that it comprises a basal carbonate ramp sequence of shoaling-upward, metre-scale carbonate-shale cycles that grade vertically and laterally into deeper-water siliciclastic facies. These are represented by conglomerate, shale and minor sandstone, and mainly formed by high- to low-density turbidity currents in an eastward-deepening, asymmetric basin (Hofmann et al. 2001a).

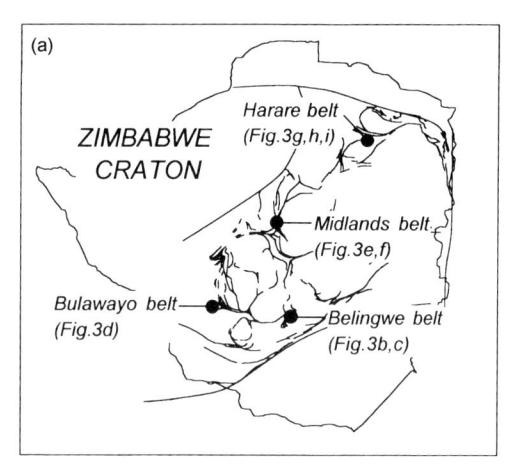

Fig. 3. Geological maps for selected areas in three greenstone belts. (**a**) Overview map. (**b**) Belingwe greenstone belt (modified from Hofmann et al. 2001b). (**c**) Detailed map of the western Zeederbergs-Cheshire Formation contact (modified from Hofmann et al. 2001b; see Fig. 3b for location). (Note the occurrence of ironstone, which completely surrounds the carbonate unit in the eastern part of the map.) (**d**) Detailed map of the Sauerdale thrust stack within the Bulawayo greenstone belt (modified after Garson 1995). (**e**) Sebakwe river area, Midlands greenstone belt (modified after Dirks et al. 2002). (**f**) Detailed map of Sebakwe Poort area (modified after Dirks et al. 2002; see Fig. 3c for location). (**g**) Harare greenstone belt (modified from Jelsma & Dirks 2000). Shown is the generalized (tectono)stratigraphy and division between the various sequences, prominent shear zones and age data. (**h**) Detailed map of the area around Trojan nickel mine, Harare greenstone belt (see Fig. 3f for location), showing the lensoidal distribution pattern of lithological units and anastomosing nature of sedimentary and volcanic horizons within the Arcturus Formation, separated by D_1 shear zones (modified from Dirks & Jelsma 1998). (**i**) Detailed map of the area around Shamva gold mine, Harare greenstone belt (see Fig. 3g for location).

According to Hofmann et al. (2001a), deposition of the Cheshire Formation clastic sediments took place in a foreland-type basin. Accommodation was controlled by tectonic loading of westward-advancing nappes of Zeederbergs-type rocks. Sediment was mainly derived from erosion of the thrust sheets ('molasse-type' conglomerates, Nisbet et al. 1993), but the occurrence of granitic detritus also indicates unroofing of granitoid crust as part of the source terrain. Hofmann et al. (2001a,b) also showed that thrusting gave rise to duplications of the stratigraphy and juxtaposition of Zeederbergs basalts onto Cheshire sediments. This is illustrated in Figure 3c. The

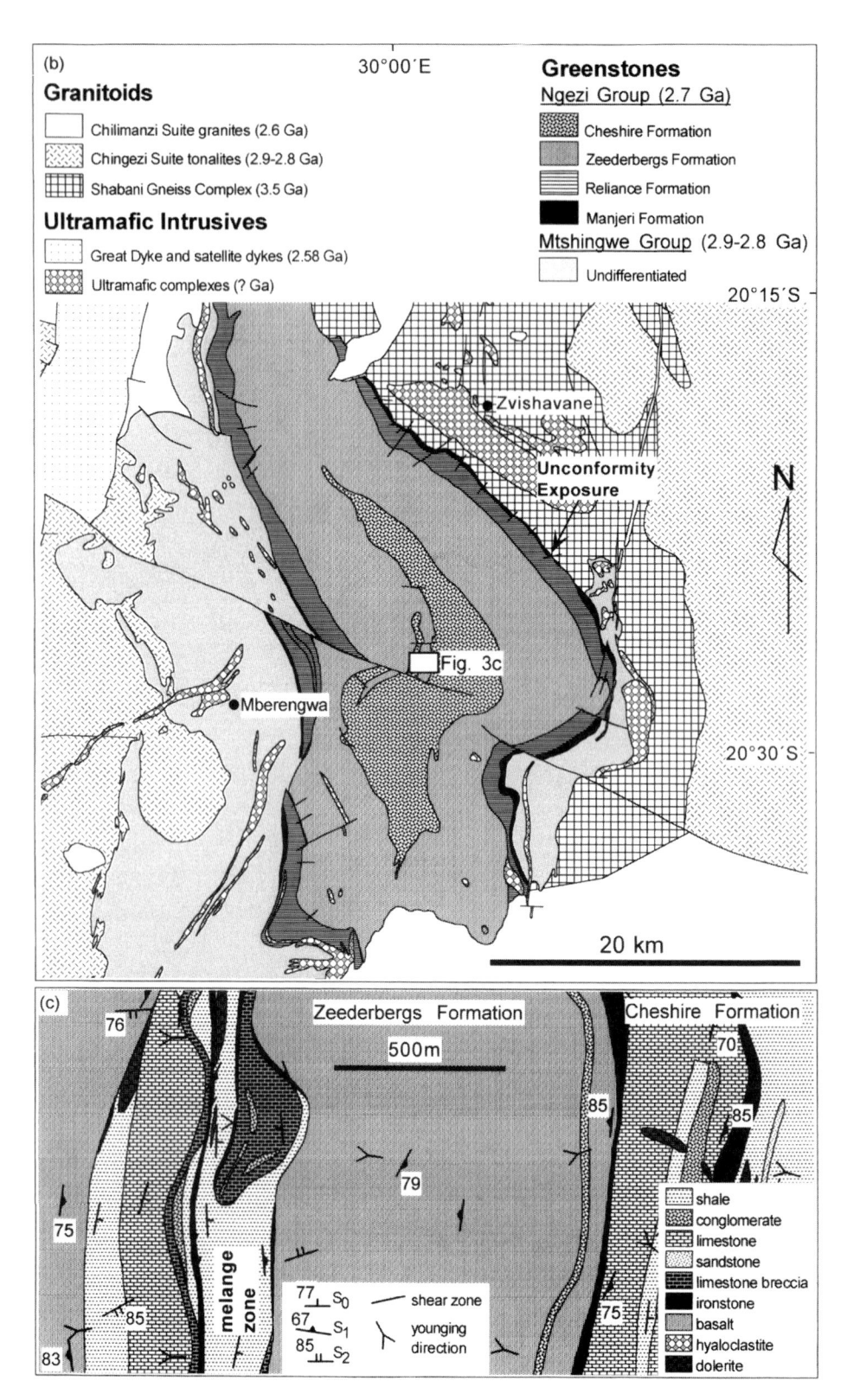

Fig. 3. (*continued*)

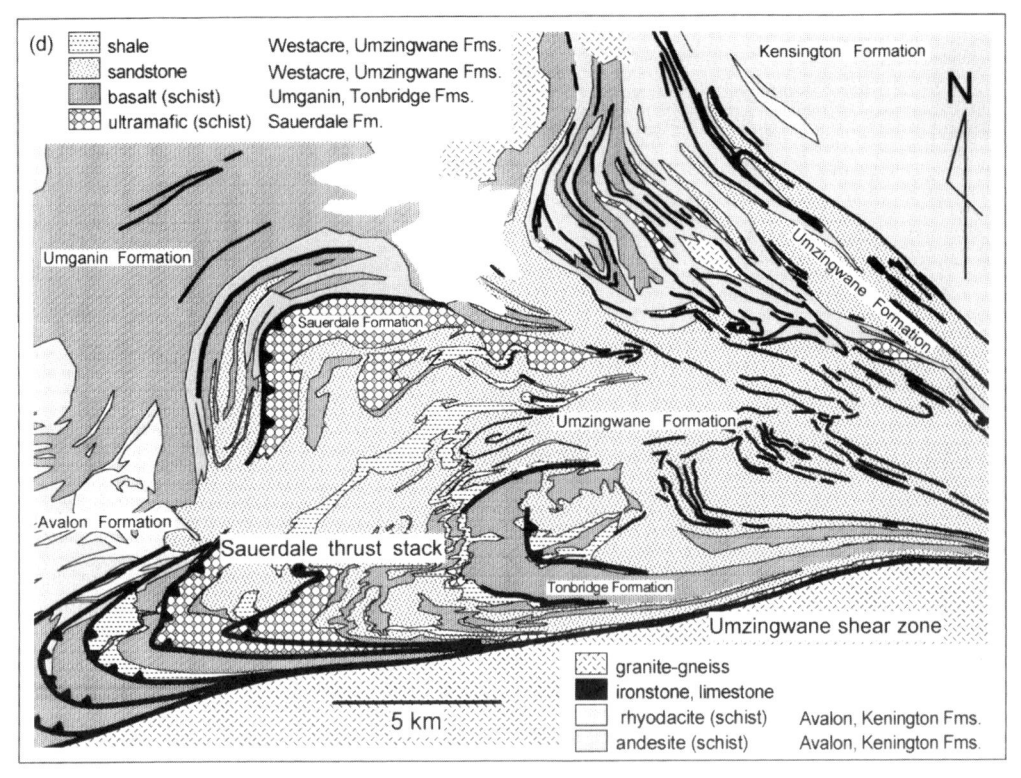

Fig. 3. (*continued*)

map shows the western outcrop of the Cheshire Formation, north of the Ngezi river. Shear zones occur at the contacts between formations and within the Cheshire Formation. These are zones of silicification, ferruginization ('ironstones') and high strain, and include massive, laminated and brecciated varieties, with asymmetric, disharmonic and isoclinal intrafolial folds, and low-angle foliation truncations. These shear zones show cross-cutting relationships with bedding of surrounding sedimentary rocks. The local stratigraphy is characterized by Zeederbergs basalts in the west, overlain by c. 150–400 m of cyclic limestone intercalated with silt- and sandstone and capped by a limestone breccia (Fig. 3c). The carbonate unit is overlain by a folded and strongly sheared siliciclastic unit, which shows features similar to a tectonic mélange (Hofmann et al. 2001a, b). It comprises inclusions of various lithologies wrapped around by a sheared, argillaceous matrix (block-in-matrix structure). This mélange is in structural contact with the siliciclastic unit in the south and is structurally overlain by pillowed basalts of the Zeederbergs Formation to the east (Fig. 3c).

Type 2 greenstone successions: the Bulawayo, Midlands and Harare belts as examples of arc magmatism

To the west and north of the proto-craton, calc-alkaline felsic volcanism is recorded along an almost continuous belt of greenstones, which stretches over a distance of c. 700 km from the Vumba greenstone belt in Botswana in the SW, to the Mount Darwin greenstone belt in Zimbabwe in the NE. This succession can be found, for example, in the Bulawayo, Midlands and Harare greenstone belts (Figs 2a and 3a). It is characterized by a relatively high proportion of felsic and intermediate volcanic rocks (15% rhyodacites and 30% andesites) and associated intrusions, ranging in age from 2715 to 2683 Ma, with a superimposed late phase of magmatism in the northern part at c. 2648–2643 Ma (Harare Sequence, Wilson et al. 1995). These felsic volcanic rocks occur as part of a complex volcano-sedimentary stratigraphy that also includes mafic volcanic rocks (35% basalts, 2% komatiites). This stratigraphy has been affected by a complex deformation history.

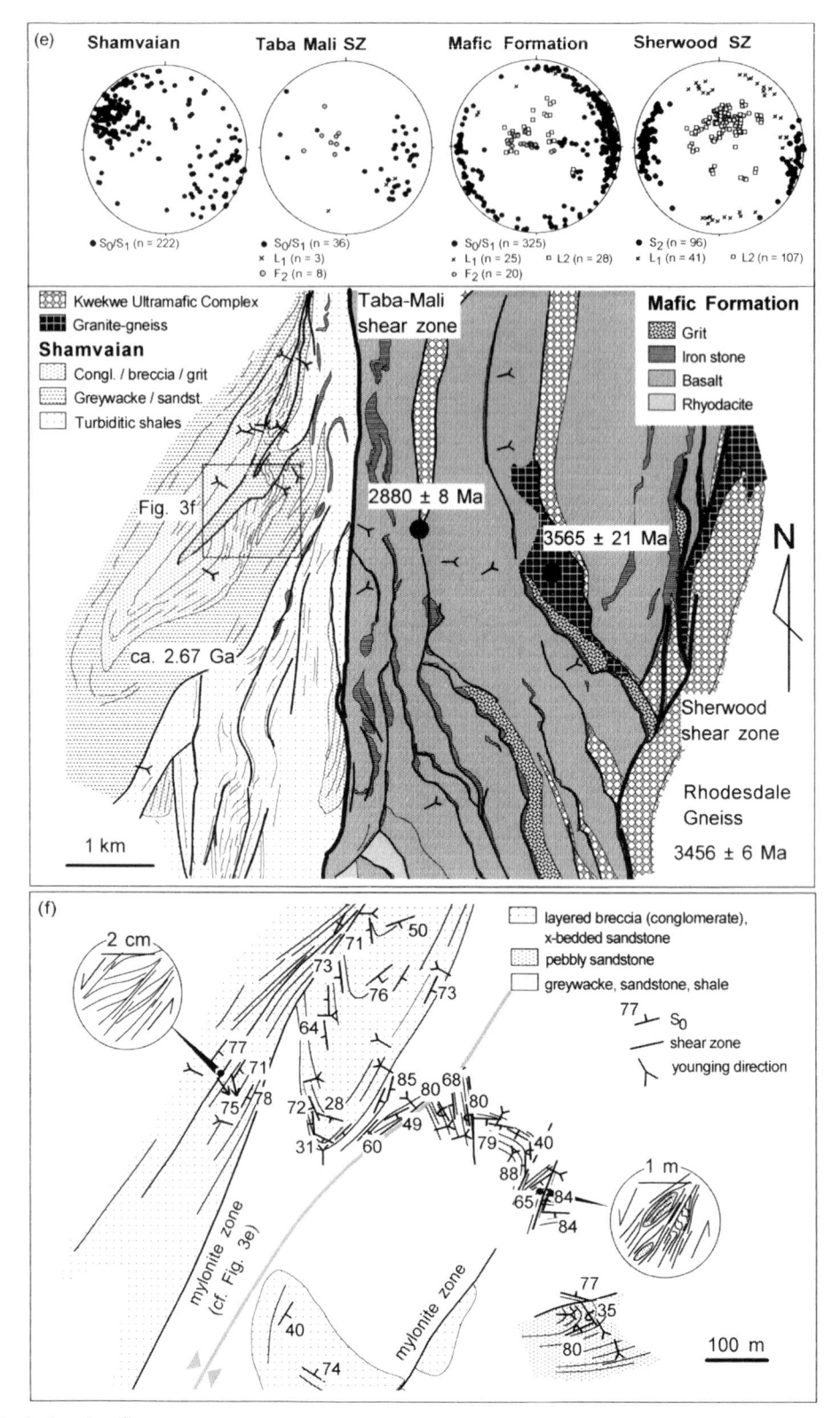

Fig. 3. (*continued*)

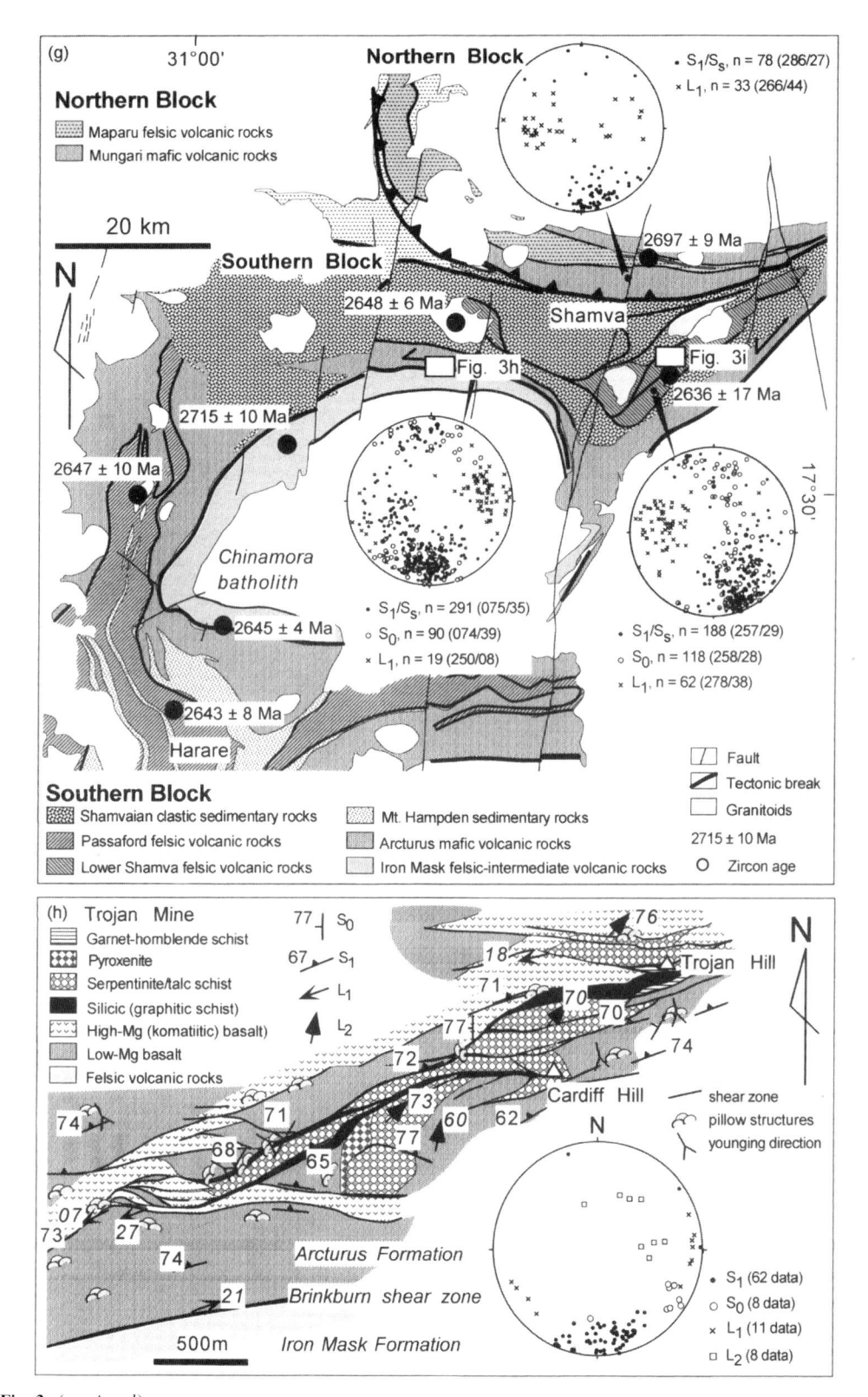

Fig. 3. (*continued*)

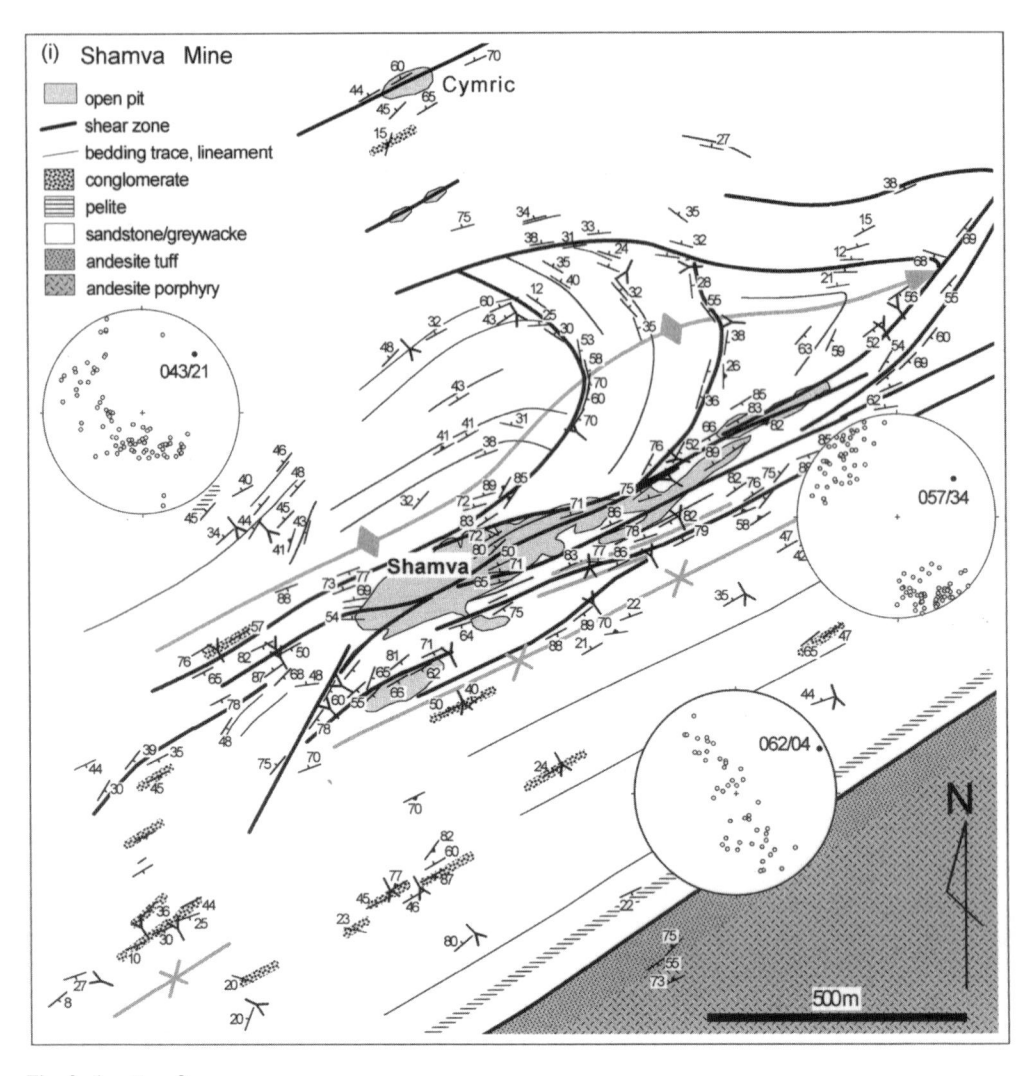

Fig. 3. (*continued*)

In the Bulawayo greenstone belt (Fig. 3a and d), infolded with and lateral to (ultra-) mafic volcanic rocks are calc-alkaline volcanic sequences (Avalon and Kensington formations). These include lava flows and volcanic breccias of basaltic andesite, andesite and rhyodacite composition that have been dated at 2696 Ma (Fig. 2a, Wilson *et al.* 1995). Clast types include andesite, rhyodacite, (ultra-) mafic rocks, greywacke and shale (Garson 1995). The top of the stratigraphic sequence (Umzingwane Formation) consists of an ironstone horizon overlain by a thick, continuous layer of volcanic laharic-type breccia that passes transitionally southeastwards into volcanic breccias and tuffs. Overlying fine-grained sedimentary rocks (sandstones, mudstones, shales and ironstones) resemble volcaniclastic turbidites and overlap or override other formations. Intercalations of rhyodacitic lava provide evidence of continuing calc-alkaline volcanism in parts of the basin. According to Garson (1995), such deposits can be compared with modern back-arc sediments.

In the Midlands greenstone belt (Fig. 3a and e) strongly deformed ultramafic rocks (Kwekwe ultramafic complex) intrude along the Sherwood shear zone into 3456 ± 6 Ma gneiss of the Rhodesdale Complex to the east and tholeiitic basalts and gabbros to the west. The basalts have been dated in different areas at 2683 ± 8 Ma

(What Cheer Formation, Wilson et al. 1995) and 2880 ± 8 Ma (Mafic Formation, Horstwood 1998), respectively. The Mafic Formation comprises a cyclic repetition of basalts, quartz grits and siltstones, topped by laminated jaspilitic cherts. Younging directions are consistently to the west (Fig. 3e). The Mafic Formation is overlain to the west, across a major shear zone (Taba–Mali shear zone), by a unit of clastic sedimentary rocks that has been assigned to the Shamvaian Supergroup (Fig. 3e and f). Further to the west, the nature of the greenstones changes to a thick, monotonous sequence of calc-alkaline andesites and dacites (Maliyami Formation) that have been dated at 2702 ± 6 Ma (Fig. 2a, Wilson et al. 1995). The boundary between the two sequences is sheared and accentuated by elongated bodies of quartz porphyries that were emplaced at 2677 ± 6 Ma (Horstwood 1998). The greenstones were intruded by several large, syn- or inter-tectonic plutons of granodioritic to tonalitic composition assigned to the Sesombi Suite (Wilson 1979) and dated at c. 2670 Ma (Fig. 2b, Dougherty-Page 1994; Horstwood 1998).

In the Harare greenstone belt (Fig. 3a and g) an older, c. 2.7 Ga greenstone sequence and a younger c. 2.65 Ga sequence (the Harare Sequence) have been recognized (Wilson et al. 1995). The basal succession of the older sequence comprises mainly calc-alkaline rhyodacitic rocks with intercalated ironstone and chert horizons (Iron Mask Formation, Jelsma et al. 1996). Rocks near the base and top of the Iron Mask Formation have yielded very different ages, of 2715 ± 15 Ma and 2645 ± 4 Ma (Figs 2a and 3g, Wilson et al. 1995; Jelsma et al. 1996; Nesbitt et al. 2000). They are structurally overlain by a volcanic pile of up to 6 km thick, of pillowed and massive tholeiitic basalts (Arcturus Formation, Fig. 3g and h). Intercalated are komatiitic basalts, ultramafic schists and serpentinites, ironstone, chert and marble, and thin horizons of felsic volcaniclastic and pelitic sediments.

The Arcturus Formation is structurally overlain by felsic volcanic rocks and volcaniclastic sedimentary rocks (Passaford and Lower Shamva formations) and intercalated graphitic argillites (Mount Hampden Formation). Tuffaceous rocks and volcanic breccias occur in close association with porphyry stocks. All are andesitic to dacitic in composition. Associated epiclastic sedimentary rocks are poorly sorted massive, structureless and planar-bedded deposits. Clasts are felsic in composition and include crystal tuff, porphyry and chert, suggesting a very localized provenance. This felsic volcanism has been dated at 2643 ± 8 Ma (Figs 2a and 3g, Wilson et al. 1995) and is associated with the emplacement of late-tectonic tonalites and granodiorites between 2648 ± 6 Ma and 2647 ± 10 Ma (Figs 2b and 3g, Jelsma et al. 1996; Nesbitt et al. 2000). Along the northern margin of the greenstone belt, the stratigraphy comprises a locally preserved basal sliver of calc-alkaline andesitic volcanic rocks and an overlying, up to 6 km thick pile of pillowed and massive tholeiitic basalts (Mungari Formation). The Mungari Formation contains intercalated rhyolites and associated quartz porphyries (Maparu Formation) that have been dated at 2697 ± 9 Ma (Figs 2a and 3g, Wilson et al. 1995).

Geochemistry and magmatic source areas of type 1 and 2 volcanic successions

Geochemical data for volcanic rocks of the type 1 and 2 successions include analyses by Scholey (1992) and Brake (1996) for the Belingwe greenstone belt, analyses from Horstwood (1998) for the Midlands greenstone belt, and analyses from Tomschi (1987) and Jelsma (1993) for the Harare greenstone belt. The volcanic rocks were screened for alteration (using the alkali ratio diagram) and least altered samples have been divided on the basis of their SiO_2 and MgO contents into five categories: (1) high-Mg basalt (MgO contents between 10 and 20 wt%); (2) basalt (SiO_2 contents between 48 and 52 wt%); (3) basaltic andesite (SiO_2 contents between 52 and 56 wt%); (4) andesite (SiO_2 contents between 56 and 63 wt%); (5) rhyodacite (SiO_2 contents >63 wt%). Cumulates, komatiites (with MgO contents >20 wt%) and picrites were excluded from this dataset. The volcanic rocks of the Belingwe greenstone belt (139 analyses) are tholeiitic (or komatiitic) in nature and range in composition from high-Mg basalt to andesite (Fig. 4). The volcanic rocks of the Midlands and Harare greenstone belts (61 analyses) include tholeiitic and calc-alkaline varieties and range from basalt to rhyodacite (Fig. 4). The average MgO contents of the basalts + basaltic andesites and andesites of the Belingwe belt are much higher (8.2 and 6.2 wt%) than those for similar volcanic rocks of the Midlands and Harare belts (6.0 and 4.9 wt%). Rhyodacites of the Midlands and Harare belts have average MgO contents of 1.5 wt% (Fig. 4).

Normalized incompatible trace element diagrams are shown in Figure 4. The basalts of the Belingwe belt are generally characterized by flat incompatible element patterns (1.7–4.6× primitive mantle) with marked enrichment of Th_N (28.3×) and Sr_N (8.2×). The komatiitic basalts show comparable patterns to those of the basalts but with spiked large ion lithophile element

Fig. 4. Geochemistry of the greenstone belt volcanic rocks. Data are from Scholey (1992) and Brake (1996) for the Midlands belt, and from Tomschi (1988) and Jelsma (1993) for the Harare belt, from Horstwood (1998) for the Belingwe belt. (Note the difference in geochemical fingerprint of the basalts, andesites and rhyodacites of the Belingwe, Midlands and Harare belts in terms of MgO contents and primitive mantle-normalized incompatible elements.) Normalizing values are after Wood (1979) and Yb after Holm (1985).

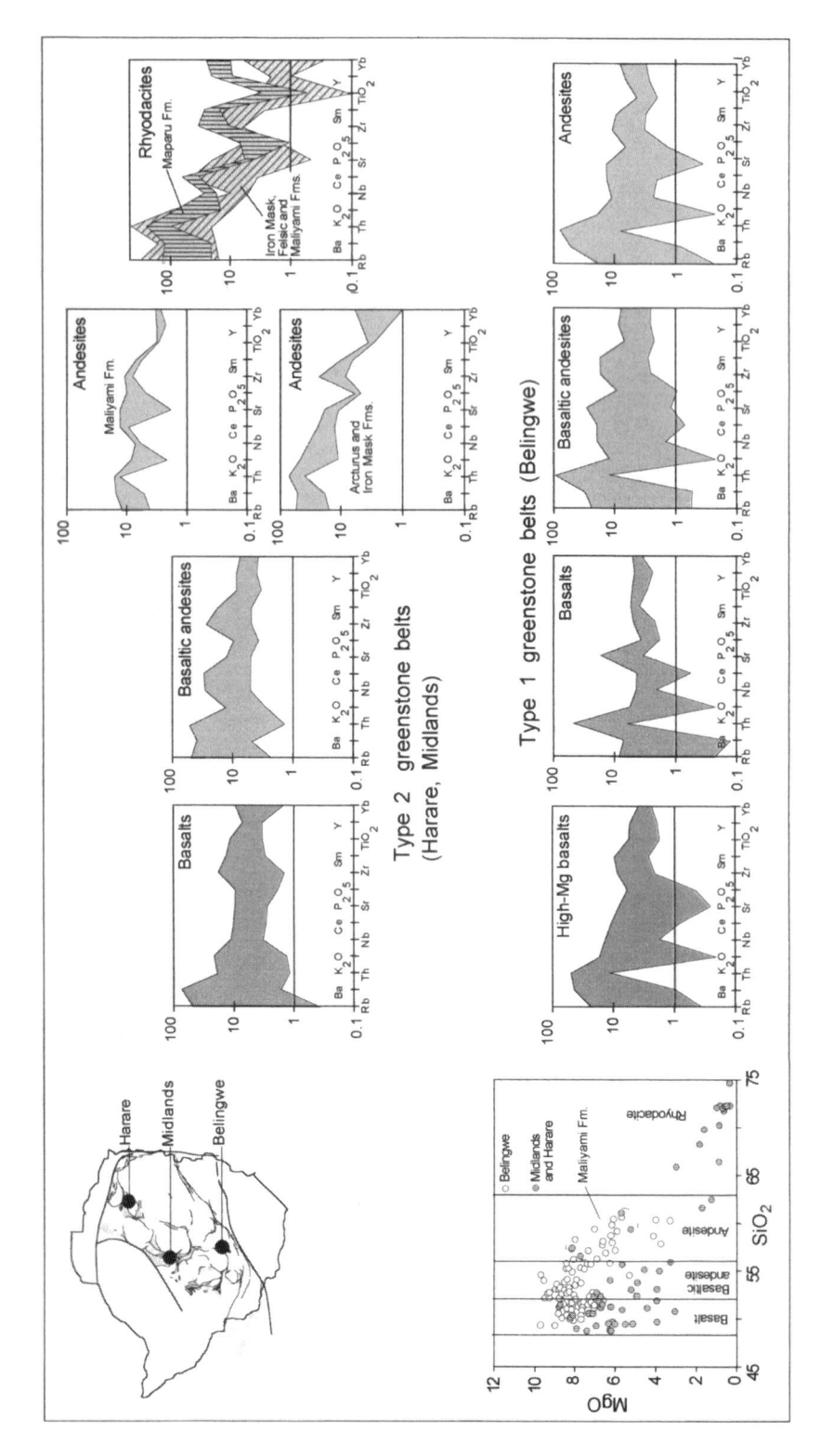

(LILE) values. The basaltic andesites and andesites show flat to gently sloping patterns with slight fractionation from the least to the most incompatible elements ($Rb_N/Yb_N = 1.4$ and $r_N/Yb_N = 1.3$) but also display enrichment of Th_N (37.0×). The volcanic rocks of the Midlands and Harare belts show two different patterns. Pattern 1 can be found in basalts, basaltic andesites and andesites of the Maliyami Formation and is generally flat or gently sloping without marked enrichment of Th_N (4.6× for basalts) or other trace elements ($Rb_N/Yb_N = 2.8$ and $Sr_N/Yb_N = 1.4$). Pattern 2 can be found in evolved andesites and rhyodacites, and is strongly sloping ($Rb_N/Yb_N = 61.7$ and $Sr_N/Yb_N = 5.1$), similar to the patterns found in recent suprasubduction zone settings, with marked depletion of Nb–(Ta), Sr, P and Ti (Fig. 4).

For the Belingwe greenstone belt, trace element modelling and isotopic data suggest that some of the volcanic rocks (high-Mg basalts and Al-depleted basalts) were contaminated by continental crustal material (Scholey 1992; Chauvel *et al.* 1993; Brake 1996). The marked enrichment in Th_N compared with primitive mantle in all Belingwe volcanic rocks may also be indicative of contamination of upper-crustal material. The field and laboratory data lend support to interpretations that the (ultra-) mafic volcanic rocks are autochthonous and that the volcanic rocks were extruded on top of sediments in a continental rift (Hunter *et al.* 1998) or newly opened oceanic basin, or as continental flood basalts on continental basement (Bickle *et al.* 1994; Silva 1997). The rifting model may be further supported by the presence of an array of layered ultramafic intrusions (Mashaba Ultramafic Suite) and mafic dyke swarms (Mashaba–Chibi dykes) that cut and appear restricted to pre-2.75 Ga terranes (Wilson 1979). The excellent preservation of the Mashaba–Chibi dykes within the proto-craton emphasizes the stable nature of this crustal nucleus by *c*. 2.7 Ga (Wilson 1990). The Mashaba Ultramafic Suite was emplaced as sills near the base of the *c*. 2.7 Ga (ultra-) mafic sequence, immediately below the Manjeri Formation. These intrusions can be followed around the Shangani batholith and in the proto-craton, and have been interpreted as the subvolcanic feeders to the (ultra)- mafic volcanic rocks of the Upper Bulawayan Group in this area (Wilson 1979).

For the Bulawayo, Midlands and Harare greenstone belts, inherited zircons in the felsic volcanic rocks yield ages between 2895 and 2650 Ma (Fig. 2a), suggesting that juvenile or mixed continental crustal material was involved in their petrogenesis. Intermediate and felsic volcanic rocks make up more than half of the present-day aerial extent of the volcanic rocks. The felsic volcanic rocks and associated tonalite–granodiorite intrusions of the Sesombi Suite (Fig. 2b) are characterized by relatively unevolved isotope characteristics. Initial Sr ratios range between 0.701 and 0.704 (Baldock & Evans 1988; Vinyu 1994), model μ_1 values between 7.7 and 8.6 (Taylor *et al.* 1991) and ε_{Nd} values are 2.5 ± 0.8 (16 analyses of internal plutons; Jelsma *et al.* 1996; Horstwood 1998) and 2.5 ± 0.4 (eight analyses of andesites and rhyodacites; Horstwood 1998). We interpret this 700 km long NE–SW-trending belt as a calc-alkaline volcanic arc that developed along the northwestern margin of the 3.55–2.75 Ga continent, between 2.72 and 2.68 Ga (Fig. 2c). It is characterized by intermediate–felsic lavas, pyroclastic rocks and related volcaniclastic sedimentary rocks, synvolcanic porphyry intrusions (feeder systems to the volcanic rocks?) and base metal mineralization (e.g. Matsitama, Sunday, Auriga, KD8, Roma, Iron Duke, Maramba prospects, Fig. 2a). A younger phase of volcanism between 2.65 and 2.64 Ga (Harare Sequence, Wilson *et al.* 1995) may be interpreted as a late subduction-related magmatic event and is superimposed on 2.72–2.69 Ga volcanism.

Other workers such as Condie & Harrison (1976) and Nisbet *et al.* (1981) have earlier suggested that rift-controlled magmatism evolved into a magmatic arc (Maliyami Formation andesites) for the Midlands belt. Tomschi (1987) drew analogues between tholeiitic mafic volcanic rocks (Arcturus Formation) and oceanic crust, and between calc-alkaline intermediate and felsic volcanic rocks (Iron Mask and Passaford formations) and arc volcanic rocks for the Harare belt. Kusky (1998) referred to this belt as the Northern Magmatic Belt, a convergent continental margin-type province. A possible second belt of calc-alkaline felsic magmatism can be followed to the SE of the enlarged proto-craton (Fig. 2c) and encompasses the 'Sea of Umtali' of Kusky (1998) and 'Andean-type' charnockites and enderbites of the North Marginal Zone of the Limpopo Belt of Berger & Rollinson (1997).

Clastic sedimentary rocks

Clastic sedimentary rocks cap the stratigraphy in most greenstone belts and have been referred to as the Shamvaian Supergroup (Fig. 5). The central part of the Midlands greenstone belt is occupied by a narrow (1–10 km) NNE-trending fault-bounded unit of clastic sedimentary rocks, interbedded with felsic volcanic rocks (Fig. 3e, Bliss 1970; Harrison 1970; Wilson 1979). A similar unit is found in the Harare belt (Fig. 3g). Both

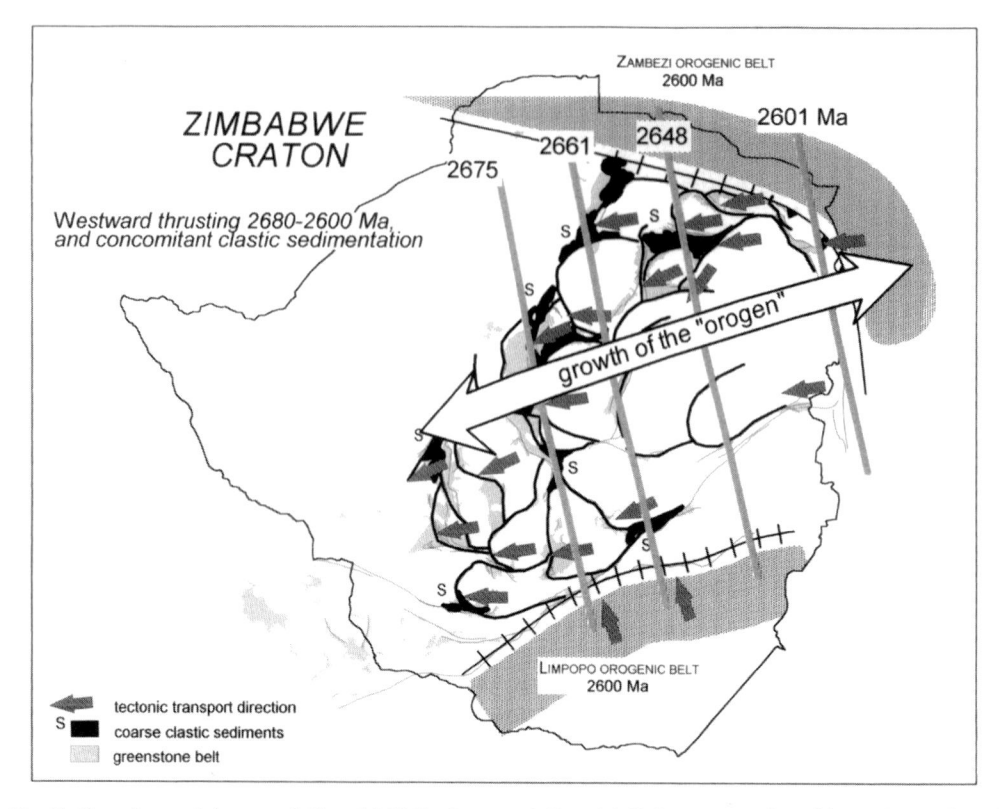

Fig. 5. Crustal growth between 2.68 and 2.58 Ga. Between 2.68 and 2.60 Ga, westward stacking and accretion of domains takes place concomitant with sedimentation in foreland basins. Data from Wilson (1968), Fuchter (1990), Barton *et al.* (1991), Campbell & Pitfield (1994), Fedo *et al.* (1995), Garson (1995), Treloar & Blenkinsop (1995), Dirks & van der Merwe (1997), Jelsma & Dirks (2000), Dirks *et al.* (2001), Hofmann *et al.* (2001a, b, 2002) and Jelsma *et al.* (2002). Lines indicate the position of the deformation front with time.

show a predominantly coarse-grained fluvial–deltaic–turbiditic sedimentary association composed of conglomerates, grits, poorly sorted sand-stones and shales. In places, lenses of intermediate to felsic volcaniclastic sedimentary rocks are present.

In the Sebakwe Poort area of the Midlands belt the conglomerate unit comprises coarsening-upward pebbly sandstones and clast-supported conglomerates (Fig. 3e). Clasts are poorly sorted and angular to subangular, and consist of jaspilite, chert and vein quartz. It is significant that none of the conglomerates near Sebakwe Poort (and in the Battlefields area in the same greenstone belt) contain mafic or intermediate volcanic clasts, in spite of the fact that basalts and intermediate volcanic rocks surround these units. This strongly suggests that the source areas for the conglomerates were removed from the greenstone domains that currently flank the conglomerates and indicates that juxtaposition of the various domains to their current configuration occurred during a later deformational event (Dirks *et al.* 2002). The conglomerate is cut by internal mylonites as well as by the Taba–Mali shear zone (Fig. 3e and f).

The Shamvaian sedimentary rocks have been indirectly dated at 2680 Ma (Midlands belt, Dirks *et al.* 2002), 2645 Ma (Harare belt, Jelsma & Dirks 2000) and 2610 Ma (Makaha belt, Hofmann *et al.* 2002), using ages for clasts, associated felsic volcanic rocks and intruding granitoids. In the Midlands belt, the Shamvaian sedimentary rocks are younger than 2683 ± 8 Ma volcanic rocks (Wilson *et al.* 1995), and have been intruded by quartz porphyries that have been dated at 2677 ± 6 Ma (Horstwood 1998). In the Harare belt, the Shamvaian sedimentary rocks unconformably overlie 2643 ± 8 Ma volcanic rocks (Wilson *et al.* 1995) and have been intruded by a 2648 ± 6 Ma granodiorite intrusion (Nesbitt *et al.* 2000); these ages are within error. In the Makaha belt, the Shamvaian sedimentary rocks are younger than 2613 ± 0.2 Ma (granite clast age) and, at least in part, older than a 2601 ± 0.6 Ma intrusive granodiorite (Hofmann *et al.* 2002).

Accretion with respect to stratigraphic developments

Structural–metamorphic relationships in all studied greenstone belts show that the greenstone sequences cannot be regarded as coherent and non-repetitive, autochthonous successions of (ultra-) mafic and felsic greenstones (Dirks & Jelsma 1998; Jelsma & Dirks 2000).

In the Bulawayo greenstone belt (Fig. 3d), shear-zone geometries, eastward-plunging lineations and shear sense observations indicate a first phase of west-directed thrusting, during which the Umzingwane shear zone acted as the southern lateral ramp to the Sauerdale imbricate thrust stack (Garson 1995). Thrusting formed part of a progressive regional deformational event that was initially associated with the development of large recumbent folds or nappes.

In the Midlands greenstone belt (Fig. 3e), the structural history is characterized by a high degree of strain partitioning, in which numerous anastomosing shear zones envelop lenses of less penetratively deformed, but generally folded domains (e.g. Campbell & Pitfield 1994; Dirks *et al.* 2002). Structural truncations are common, individual stratigraphic units are laterally discontinuous, and most contacts between major stratigraphic units are strongly sheared (e.g. Catchpole 1987; Campbell & Pitfield 1994). Early isoclinal folds and thrusts have been noted in sections of the Midlands greenstone belt (e.g. Nutt 1984; Dirks & van der Merwe 1997) and Dirks *et al.* (2002) identified a number of shear-zone-bounded structural domains, each with a unique structural–metamorphic history (Fig. 3e). These domains include a sliver of 3.57 Ga gneiss as well as clastic sedimentary sequences and mafic and intermediate to felsic volcanic units that have been dated between 2.88 and 2.67 Ga. Concomitant upright folding and 'minor' shearing accommodated strain internal to the domains.

The complexity of this deformation is shown in Figure 3e, where the Mafic Formation is composed of a 10-fold cyclic repetition of a similar lithological package of basalt, quartz grit, siltstone, chert and a tectonic horizon of brecciated chert and foliated ironstones. The cycles are separated by anastomosing shear zones that merge toward the NW, and in the central part of the area a tectonic sliver of 3565 ± 21 Ma gneiss occurs (Fig. 3e). In Fig. 3f a discrete mylonite zone cuts conglomerate beds of the Shamvaian Supergroup at low to moderate angles (Dirks *et al.* 2002). Splays of this shear zone are exposed as 5–15 cm wide ultramylonite horizons. The main mylonite zone is marked by up to 50 cm wide quartz veins and localized sulphidation and can be followed around a synformal fold structure (Fig. 3e and f). To the east, the conglomerate beds have been cut out, and the mylonite zone continues into the surrounding shales, where it occurs as a 10 m wide zone of strongly contorted and folded shales and thinly bedded, cross-laminated fine-grained sandstones and siltstones.

For the Harare greenstone belt (Fig. 3g), Jelsma & Dirks (2000) described an early generation of recumbent folds and layer-parallel shear zones that are characterized by early lineations that plunge at shallow to moderate angles to the west or east. Related kinematic indicators consistently record the hanging-wall block to have moved westward (structural top to the west). Mafic volcanic rocks of the Arcturus Formation structurally overlie gneisses of the Murehwa batholith in the east and felsic–intermediate volcanic rocks of the Iron Mask Formation in the west (Fig. 3g). These contacts have been interpreted as major tectonic breaks (Brinkburn shear zone of Jelsma & Dirks 2000; Tategura shear zone of Blenkinsop *et al.* 2000) considering the intensity of deformation features (planar and linear fabrics, disharmonic and sheath folds, grain shape fabrics, *S–C* fabrics, and truncation of bedding). Low-Mg and high-Mg basalts and ultramafic rocks within the overlying Arcturus Formation are distributed as a series of stacked low-strain lenses with typical sigmoidal outcrop patterns (Fig. 3h) and are separated by locally discordant silicified horizons or schist zones comprising chlorite–sericite schists, (silicified) graphitic schists, 'quartzites', 'ironstones' and 'cherts' (Jelsma & Dirks 2000). These rocks are characterized by a fine grain size and a distinct mylonitic layering.

Another prominent mylonite zone truncates a sequence of tightly folded Shamvaian sedimentary rocks and separates these from basalts of the overlying Mungari Formation, which young to the south (Jelsma & Dirks 2000, Fig. 3g). High-strain fabrics are pervasive, and bedding and foliation are subparallel. Mineral lineations plunge at moderate angles to the west, and the sense of shear, as indicated by asymmetric porphyroclasts and *S–C* fabrics, shows westward movement of the hanging wall. The structural geometry at Shamva mine is characterized by networks of anastomosing shear or fracture zones together making up the Shamva shear zone (Fig. 3i). This shear zone represents the sheared-out limb of a close, anticlinal fold structure to the north and an isoclinal, synclinal fold structure to the south. The rocks of the anticlinal structure has been dissected by low-angle shear or fracture zones and the geometry resembles a folded imbricate thrust stack.

The Dindi and Makaha greenstone belts are positioned along the boundary between the craton and the migmatitic gneiss terrane of the adjacent Zambezi Orogenic Belt (Fig. 5). The boundary between the greenstone belts (the craton) and the high-grade gneiss terrane is gradational in terms of structures and metamorphic grades (Hofmann et al. 2002; Jelsma et al. 2002). In the two domains, similar generations of fabrics can be recognized; linear fabrics such as mineral and stretching lineations, rodding, boudins and collinear fold axes are pervasive and have identical orientations, plunging at shallow angles to the east or west. Differences can be observed in terms of the finite strain and strain distribution. Within the craton, strain is heterogeneous and mainly partitioned into discrete shear zones that accommodated west-directed stacking of the greenstone stratigraphy and intermittent gneiss segments. Near the craton margin strain partitioning becomes less and in the migmatitic gneiss terrane, strain is homogeneous and uniformly high, but still associated with the same lineation direction and shear sense.

In all greenstone belts studied, a general structural pattern emerges of an early generation of tectonic mélanges, recumbent folds, duplex geometries and shear zones bounding low-strain domains (Fig. 3c, e and g). The anastomosing, layer-parallel shear zones are associated with east-west-trending lineations and accommodated west-directed movement. This event can be interpreted as terrane accretion and shortly follows arc magmatism. This event cannot be interpreted as shortening of a basin with a heterogeneous make-up. Distinct shear-zone-bounded tectonostratigraphic domains can be recognized, each with a unique structural-metamorphic history (e.g. Dirks et al. 2002). The shear zones are commonly strongly silicified, sulphidized or carbonated, show mylonitic banding and truncate primary layering at low angles. In the Sebakwe Poort area, the minimum amount of displacement on one of the mylonite zones must have been in the order of several kilometres, to allow for the complete truncation and dismemberment of a coarse-grained conglomerate horizon. Considering that the mylonite horizon is less than 50 cm wide, it is envisaged that this horizon must have been characterized by superplastic conditions and high fluid pressures. Such mylonite horizons are pervasive throughout individual formations accommodating complex internal tectonic stacking (Fig. 3c, f, h and i).

The shear zones in combination with the outcrop pattern of the greenstone sequences appear to underlie semicircular tectonic blocks commonly centred on gneiss units (Fig. 5). The blocks are arranged in an overlapping, fish-scale-like fashion, with the underlying thrusts recording east over west movement. This event also includes the development of syntectonic sedimentary successions (Fig. 5 Shamvaian Supergroup; e.g. Jelsma & Dirks 2000; Hofmann et al. 2001a, b). Each block can be interpreted as a thrust or nappe structure, all of which were piled up between 2.68 and 2.60 Ga (e.g. Jelsma & Dirks 2000). The age of this deformation is diachronous and youngs from west to NE, as indicated by the crystallization age of late syntectonic granitoid plutons, which intruded the various greenstone belt lithologies and cut structural fabrics. The internal plutons have ages from 2673 ± 5 Ma (Midlands belt, Dougherty-Page 1994), to 2648 ± 6 Ma (Harare belt, Nesbitt et al. 2000) and 2601 ± 0.6 Ma (Makaha belt, Hofmann et al. 2002).

Jelsma & Dirks (2000) also recognized a second set of fabrics, strain patterns and contact aureoles that overprint the earlier geometries, and are related to the emplacement of granites in the Chinamora batholith (Fig. 3g). These structures are concentrated in shear zones that are marginal to granite-gneiss domes, and include radial, down-dip lineations and triaxial flattening strains (Jelsma et al. 1993). Kinematic indicators consistently show a dome-side-up sense of movement. This event caused local refolding and steepening of the thrust stacks.

Emplacement of late- to post-tectonic granites

Deformation of the granite-greenstone terranes was accompanied and followed by the pan-cratonic emplacement of granodiorites, granites and monzogranites (Fig. 6a) of the Wedza and Chilimanzi suites. An early generation is characterized by a migmatitic appearance and has been termed Wedza Suite (Wilson et al. 1995). A later, more homogeneous and massive variety is referred to as the Chilimanzi Suite (Razi Suite in Limpopo belt, Frei et al. 1999; Mudzi Suite in Zambezi belt, Barton et al. 1991). The unfoliated or weakly foliated Chilimanzi granites are monzogranitic in composition and are commonly characterized by large K-feldspar megacrysts. With their evolved geochemical (high K, Th, U, ΣREE, Jelsma 1993) and isotope (initial Sr ratios of 0.7040, Taylor et al. 1991; ε_{Nd} values

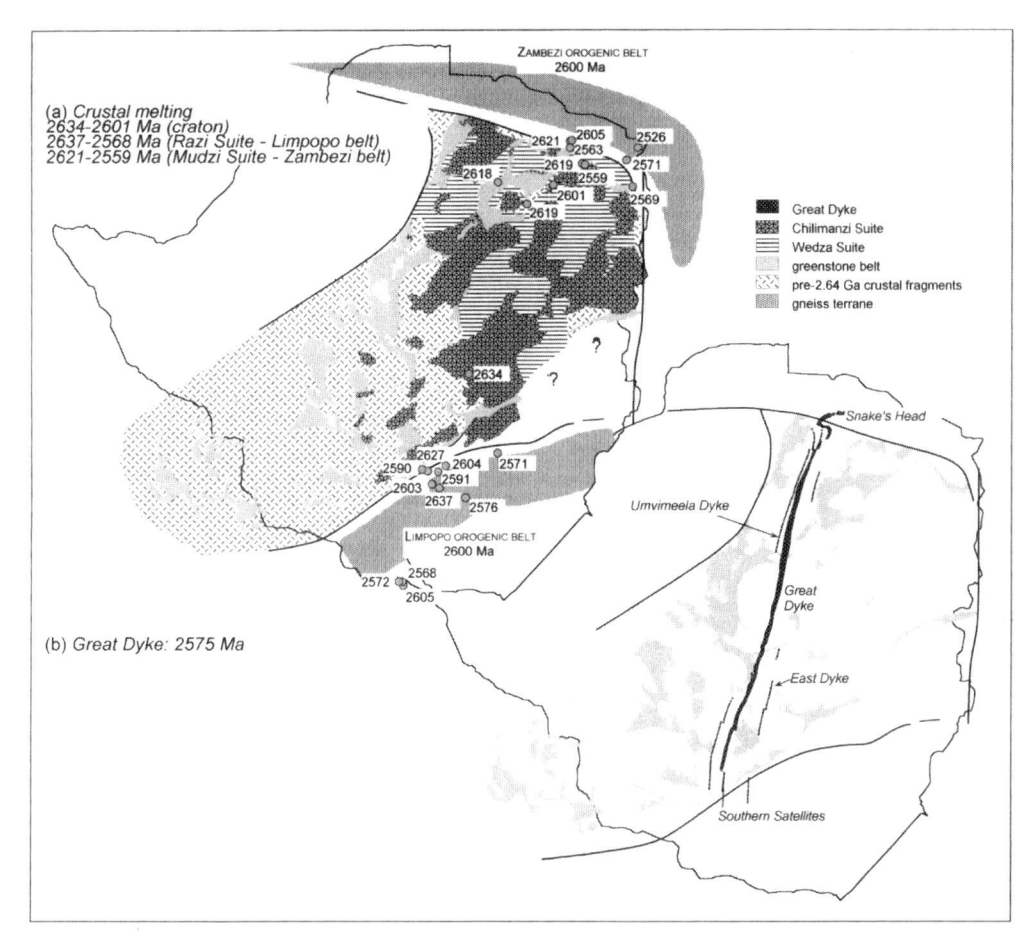

Fig. 6. Crustal growth between 2.68 and 2.58 Ga. (**a**) Emplacement of late-tectonic Wedza and Chilimanzi suites granitoids (2.63–2.58 Ga). (**b**) Emplacement of Great Dyke at 2.58 Ga, following cratonization.

of −1.6, Horstwood 1998) characteristics, they have been interpreted as crustal melts (Jelsma et al. 1996), and may have been a product of loading following accretion. They were probably emplaced as large, thick horizontal sheets, fed by vertical dykes (Blenkinsop & Treloar 2001). Their age of emplacement appears diachronous and is oldest in the west and youngest in the east. The oldest age of 2634 Ma is from the Chilimanzi type area north of the Masvingo belt (Horstwood 1998). Younger ages have been obtained for similar granites east of the Harare belt (2601 Ma, Jelsma et al. 1996) and adjacent to the Dindi and Makaha belts (2559 Ma, 2569 Ma, Hofmann et al. 2002). In the Limpopo belt the Razi Suite shows a spread of ages from 2669 Ma in the west (Mkweli et al. 1995) to 2517 Ma in the east (Frei et al. 1999).

2.58 Ga event: emplacement of Great Dyke

The Great Dyke is a linear, NNE-trending intrusive body of mafic and ultramafic rocks 550 km in length and between 4 and 11 km wide (Fig. 6b). It intruded the stabilized craton at 2575 Ma (Armstrong & Wilson 2000; Wingate 2000). The northern extremity and southern satellites cut the marginal zones of both Zambezi and Limpopo orogenic belts, respectively. The fractures that controlled emplacement of the Great Dyke can be found across the craton. The Popoteke fault set strikes NNE and has a sinistral shear sense. The Mtshingwe fault set strikes ESE and is characterized by an early sinistral and later dextral shear sense. They are generally regarded as a conjugate fault set, related to NNW–SSE crustal shortening during the Limpopo Orogeny (Wilson

1990), but may also be attributed to extensional relaxation following east–west accretion and crustal thickening, along the lines of extensional collapse of modern-day orogens.

Discussion

Neoarchaean crustal growth

On the basis of the above structural–metamorphic, geochemical and geochronological age data, we propose a new model to explain the Neoarchaean crustal growth leading to final stabilization of the Zimbabwe Craton. Initial accretion of greenstone and TTG-type granite material occurred around 3.6–3.3 Ga leading to the formation and stabilization of one or several cratonic nuclei (Fig. 1b). This proto-cratonic continental crust was affected by regional deformation before 3.35 Ga. Crustal stabilization was followed by stable shelf sedimentation at c. 3.0 Ga. Calc-alkaline felsic volcanism took place between 2950 and 2750 Ma along the western margin of the protocraton (Fig. 1d). Volcanism was accompanied at depth by the emplacement of the Chingezi Suite of TTG-type granites.

Most of the greenstone lithologies were formed during 2.72–2.64 Ga events (Fig. 2a). In the south–central part of the craton (e.g. Belingwe area) a rift or back-arc may have opened, separating the ancient c. 3.55–3.35 Ga crustal nucleus (east) from a c. 2.9–2.8 Ga crustal fragment (Fig. 2c). At the same time a NE–SW-striking calc-alkaline volcanic arc (2715 and 2683 Ma) formed on relatively juvenile crust, west of the c. 2.9–2.8 Ga outcrop and stretching over a distance of at least 700 km (Fig. 2c). Volcanism was associated with and followed by the emplacement of the Sesombi Suite of granitoids (Fig. 2b). Along the NE end of this arc calc-alkaline felsic volcanism continued at 2.65–2.64 Ga and was accompanied by the emplacement of late-tectonic tonalites and granodiorites.

The 2.72–2.58 Ga time period saw the formation of present features of the Zimbabwe Craton. During this time period late felsic volcanism and porphyry intrusions, deposition of coarse clastic sediments, complex deformation, and late-tectonic plutonism took place in selected shear-zone-bounded tectonic domains over limited periods of time of between 10 and 20 Ma. In the Midlands belt this time span ranges from 2683 to 2673 Ma, in the Harare greenstone belt from 2647 to 2643 Ma and in the Makaha belt from 2613 to 2601 Ma. In many greenstone belts (Bulawayo, Midlands, Harare, Dindi, Makaha, Belingwe) a similar structural pattern emerges. An early phase of deformation produced large recumbent folds, duplex geometries and shear zones bounding low-strain domains (Fig. 3). These anastomosing, layer-parallel shear zones are associated with shallow, east–west-trending lineations. Thrusting and recumbent folding of distinct tectono-stratigraphic greenstone domains was associated with the development of syntectonic sedimentary successions in foreland-type basins (Fig. 5).

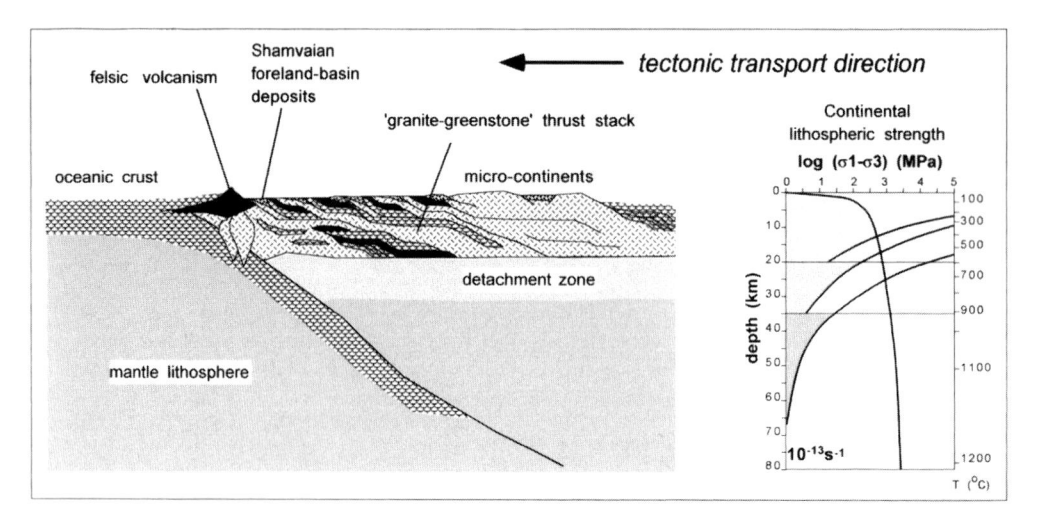

Fig. 7. Tectonic interpretation showing accretion of mafic and ultramafic oceanic sequences, juvenile volcanic arcs, and accretionary prisms against an older crustal nucleus. Inset shows strength envelopes for the lithosphere showing differential stress ($\sigma_1-\sigma_3$, MPa) against depth (km) for granite, granulite and dunite (see Jelsma et al. 2001) and Byerlee's (1978) friction relation.

This stacking was diachronous across the craton. Deformation was oldest in the western part of the craton (Midlands belt: c. 2675 Ma, Horstwood 1998) intermediate in the north–central part (Harare belt: c. 2645 Ma, Wilson et al. 1995; Jelsma et al. 1996; Nesbitt et al. 2000) and youngest in the northeastern part (Dindi–Makaha belts: c. 2605 Ma, Vinyu et al. 2001; Hofmann et al. 2002). This deformation appears to have culminated in an oblique, east–west-directed 'orogenic' event (Fig. 5 Zambezi belt; Vinyu et al. 2001). This could be interpreted as the progressive lateral accretion of tectonic domains between 2.68 and 2.60 Ga (Fig. 7).

The implication of the presence of thrust zones in the stratigraphy and juxtaposition of unrelated sequences is significant. It means that the concept of a craton-wide stratigraphy is incomplete and that stratigraphic thicknesses for some greenstone sequences are incorrect. It also suggests that some greenstone sequences are allochthonous and may have formed in unrelated tectonomagmatic environments. Without precise geochronology and detailed structural studies lithostratigraphic correlations are not possible, given the existence of structural discontinuities, separating similar sequences of widely different ages (e.g. Jelsma & Dirks 2000). The model proposed is at variance with models that relate deformation of the Zimbabwe Craton to Limpopo belt deformation. NNW–SSE shortening and thrusting along the North Marginal Zone of the Limpopo belt probably took place only at a late stage of the evolution of the Zimbabwe Craton, at c. 2.62–2.58 Ga (Mkweli et al. 1995). The east–west accretion of domains documented in the studied greenstone belts started earlier (>2.68 Ga).

Crust–mantle decoupling

The deformation described above led to isostatically and mechanically stable, 30–40 km thick continental crust. Deformation geometries observed in the field but also from seismic reflection profiles available for other cratons, such as the Superior Province and Yilgarn Craton (Ludden et al. 1993; Wilde et al. 1996), are characterized by imbricate stacks or complex duplex arrays of shear zones. This deformation, however, did not cause significant exhumation of high-pressure rocks. P–T estimates for the Zimbabwe Craton vary only between sub-greenschist and upper amphibolite-facies conditions. The high heat flow during the Archaean and the resultant crustal buoyancy must have prevented modern-day-type lithospheric subduction–accretion processes and may have sustained a rheologically stratified Archaean crust, with a brittle upper crust and a highly ductile lower crust (Brown & Phillips 2000). Flow in this weak lower crust may have allowed only limited crustal overthickening and acted as a rheological boundary layer allowing large differential movements between upper crust and mantle lithosphere (Fig. 7). It effectively means that lithospheric shortening was accommodated independently in both layers. It also means that shortening of crust and mantle lithosphere did not necessarily occur along the same boundary zones; i.e. crustal shortening may have occurred above relatively rigid and stable mantle lithosphere that developed its own set of structures independently from the overlying crust.

Crustal melting and stabilization

What may have caused the formation of the pan-cratonic late- to post-tectonic granites (Fig. 6a)? We mentioned above that geochemically these granites are crustal melts. Dirks & Jelsma (1998) suggested that accretion produced a crustal pile of alternating 'cold' and 'hot' crustal slices and a disturbed and stratified crustal geotherm, which would have resulted in crustal melting. We acknowledge that this needs to be addressed more fully. As for the main deformation, emplacement of the granites may have been a diachronous event and oldest in the west and youngest in the east. After the emplacement of the late- to post-tectonic granites the craton cooled and stabilized, with further deformation partitioned into lower-grade, strike-slip shear zones. By 2575 Ma the Zimbabwe Craton was cut by the Great Dyke, its satellites and related fracture patterns (Fig. 6b).

Future directions

The model proposed for Neoarchaean crustal growth leading to final stabilization of the Zimbabwe Craton needs to be tested further with precise age dating across selected greenstone sequences and granite–gneiss terranes and regional studies of the geochemistry of greenstone belt volcanic rocks. Compared with other cratons such as the Superior Province, the Zimbabwe Craton has far fewer good age data. Numerical models of the rheology of the Archaean lithosphere would aid in understanding the concept of crust–mantle decoupling and lower-crustal delamination. Additional studies need to address comparisons of deformational history and geodynamic setting of rocks from the Zimbabwe Craton and those from other cratons.

We thank Stichting Schurmannfonds (grants 1996-2002/13) and the National Research Foundation (NRF) for financially supporting this research. M. Horstwood generously allowed us to use geochemical and geochronological data from his PhD research in the Midlands area, to whom we are grateful. A. Mikhailov kindly provided help with the GIS database. The paper greatly benefited from critical comments by J. Wilson, T. Blenkinsop, P. Treloar, W. Bleeker and H. Rollinson.

References

ABBOTT, D. H. & HOFFMAN, S. E. 1984. Archean plate tectonics revisited 1, Heat flow, spreading rate and the age of subducting lithosphere and their effects on the origin and evolution of continents. *Tectonics*, **3**, 429–448.

ARMSTRONG, R. & WILSON, A. H. 2000. A SHRIMP U–Pb study of zircons from the layered sequence of the Great Dyke, Zimbabwe, and a granitoid anatectic dyke. *Earth and Planetary Science Letters*, **180**, 1–12.

BALDOCK, J. W. & EVANS, J. A. 1988. Constraints on the age of the Bulawayan Group metavolcanic sequence, Harare greenstone belt, Zimbabwe. *Journal of African Earth Sciences*, **7**, 795–804.

BARTON, C. M., CARNEY, J. N., CROW, M. J., DUNKLEY, P. N. & SIMANGO, S. 1991. *The Geology of the Country around Rushinga and Nyamapanda*. Zimbabwe Geological Survey Bulletin, **92**.

BECKER, J. K., SIEGESMUND, S. & JELSMA, H. A. 2000. The Chinamora batholith, Zimbabwe: structure and emplacement-related magnetic rock fabric. *Journal of Structural Geology*, **22**, 1837–1853.

BERGER, M. & ROLLINSON, H. 1997. Isotopic and geochemical evidence for crust–mantle interaction during late Archaean crustal growth. *Geochimica et Cosmochimica Acta*, **61**, 4809–4829.

BERGER, M., KRAMERS, J. D. & NÄGLER, TH. F. 1995. Geochemistry and geochronology of charnoenderbites in the Northern marginal zone of the Limpopo belt, southern Africa, and genetic models. *Schweizerische Mineralogische und Petrographische Mitteilungen*, **75**, 17–42.

BICKLE, M. J., MARTIN, A. & NISBET, E. G. 1975. Basaltic and peridotitic komatiites and stromatolites above a basal unconformity in the Belingwe Greenstone Belt, Rhodesia. *Earth and Planetary Science Letters*, **27**, 155–162.

BICKLE, M. J., NISBET, E. G. & MARTIN, A. 1994. Archaean greenstone belts are not oceanic crust. *Journal of Geology*, **102**, 121–138.

BLENKINSOP, T. G. 1997. *The Limpopo Belt in Zimbabwe: a Field Guide*. Conference on Intraplate Magmatism and Tectonics of Southern Africa (Geological Society of Zimbabwe), Excursion Guide, **1**.

BLENKINSOP, T. G. & FREI, R. 1997. Archaean and Proterozoic mineralization and tectonics at Renco mine (Northern marginal zone, Limpopo belt, Zimbabwe). *Economic Geology*, **91**, 1225–1238.

BLENKINSOP, T. G. & TRELOAR, P. J. 2001. Tabular intrusion and folding of the late Archaean Murehwa granite, Zimbabwe, during regional shortening. *Journal of the Geological Society, London*, **158**, 653–664.

BLENKINSOP, T. G., FEDO, C. M., BICKLE, M. J., ERIKSSON, K. A., MARTIN, A., NISBET, E. G. & WILSON, J. F. 1993. Ensialic origin for the Ngezi Group, Belingwe greenstone belt, Zimbabwe. *Geology*, **21**, 1135–1138.

BLENKINSOP, T. G., MARTIN, A., JELSMA, H. A. & VINYU, M. L. 1997. The Zimbabwe Craton. *In*: DE WIT, M. J. & ASHWAL, L. D. (eds) *Greenstone Belts*. Oxford University Press, Oxford, 567–580.

BLENKINSOP, T. G., OBERTHUR, T. & MAPETO, O. 2000. Gold mineralization in the Mazowe area, Harare–Bindura–Shamva greenstone belt, Zimbabwe: I. Tectonic controls on mineralization. *Mineralium Deposita*, **35**, 126–137.

BLISS, N. W. 1970. *The Geology of the Country around Gatooma*. Rhodesia Geological Survey Bulletin, **64**.

BOUHALLIER, H., CHOUKROUNE, P. & BALLEVRE, M. 1993. Diapirism, bulk homogeneous shortening and transcurrent shearing in the Archaean Dharwar craton, the Holenarsipur area, southern India. *Precambrian Research*, **63**, 43–58.

BRAKE, C. 1996. *Tholeiitic magmatism in the Belingwe greenstone belt, Zimbabwe*. PhD thesis, University of Edinburgh.

BROWN, C. D. & PHILLIPS, R. J. 2000. Crust–mantle decoupling by flexure of continental lithosphere. *Journal of Geophysical Research*, **105**(B6), 13 221–13 237.

BYERLEE, J. D. 1978. Friction of rocks. *Pure and Applied Geophysics*, **116**, 615–626.

CAMPBELL, S. D. G. & PITFIELD, P. E. J. 1994. *Structural Controls of Gold Mineralization in the Zimbabwe Craton – Exploration Guidelines*. Zimbabwe Geological Survey Bulletin, **101**.

CARD, K. D. 1990. A review of the Superior Province of the Canadian Schield, a product of Archaean accretion. *Precambrian Research*, **48**, 99–156.

CARLSON, R. W., BOYD, F. R., SHIREY, S. B. & 14 OTHERS 2000. Continental growth, preservation and modification in southern Africa. *GSA Today*, **10**, 1–7.

CATCHPOLE, S. J. 1987. Gold mineralisation related to shear zones in the Venice Group of mines near Kadoma, Zimbabwe. *In*: DEMPSTER, E. L. & VIEWING, K. A. (eds) *African Mining*, 71–88.

CHAUVEL, C., DUPRE, B. & ARNDT, N. T. 1993. Pb and Nd isotopic correlation in Belingwe komatiites and basalts. *In*: BICKLE, M. J. & NISBET, E. G. (eds) *The Geology of the Belingwe Greenstone Belt, Zimbabwe*. Geological Society of Zimbabwe, Harare, Special Publications, **2**, 167–174.

COLLINS, W. J., VAN KRANENDONK, M. J. & TEYSSIER, C. 1998. Partial convective overturn of Archaean crust in the east Pilbara Craton, Western Australia: driving mechanisms and tectonic implications. *Journal of Structural Geology*, **20**, 1405–1424.

CONDIE, K. C. & HARRISON, N. M. 1976. Geochemistry of the Archaean Bulawayan Group, Midlands greenstone belt, Rhodesia. *Precambrian Research*, **3**, 253–271.

COWARD, M. P., JAMES, P. R. & WRIGHT, L. I. 1976. Northern margin of the Limpopo mobile belt, southern Africa. *Geological Society of America Bulletin*, **87**, 601–611.

DAVIES, G. F. 1992. On the emergence of plate tectonics. *Geology*, **20**, 963–966.

DEFANT, M. J. & KEPEZHINSKAS, P. 2001. Evidence suggests slab melting in arc magmas. *EOS Transactions, American Geophysical Union*, **82**, 65–69.

DE WIT, M. J. 1998. On Archean granites, greenstones, cratons and tectonics: does the evidence demand a verdict? *Precambrian Research*, **91**, 181–226.

DE WIT, M. J. & ASHWAL, L. D. (eds) 1997. *Greenstone Belts*. Oxford University Press, Oxford.

DE WIT, M. J., ROERING, C. HART, R. J. & 6 OTHERS 1992. Formation of an Archaean continent. *Nature*, **357**, 553–562.

DIRKS, P. H. G. M. & JELSMA, H. A. 1998. Horizontal accretion and stabilization of the Zimbabwe Craton. *Geology*, **26**, 11–14.

DIRKS, P. H. G. M. & VAN DER MERWE, I. 1997. Early duplexing in an Archaean greenstone sequence and its control on gold mineralisation. *Journal of African Earth Sciences*, **24**, 603–620.

DIRKS, P. H. G. M., JELSMA, H. A. & HOFMANN, A. 2002. Accretion of an Archean greenstone belt in the Midlands of Zimbabwe. *Journal of Structural Geology*, in press.

DIRKS, P. H. G. M., KRÖNER, A., JELSMA, H. A., SITHOLE, T. A. & VINYU, M. L. 1999. Pb–Pb zircon dates from the Makuti gneisses; evidence for a crustal-scale Pan-African shear zone in the Zambezi belt, NW Zimbabwe. *Journal of African Earth Sciences*, **28**, 427–442.

DODSON, M. H., COMPSTON, W., WILLIAMS, I. S. & WILSON, J. F. 1988. A search for ancient detrital zircons in Zimbabwean sediments. *Journal of the Geological Society, London*, **145**, 977–983.

DOUGHERTY-PAGE, J. S. 1994. *The evolution of the Archaean continental crust of northern Zimbabwe*. PhD thesis, Open University, Milton Keynes.

FEDO, C. M. & ERIKSSON, K. A. 1996. Stratigraphic framework of the c. 3.0 Ga Buhwa greenstone belt: a unique stable-shelf succession in the Zimbabwe Archean craton. *Precambrian Research*, **77**, 161–178.

FEDO, C. M., ERIKSSON, K. A. & BLENKINSOP, T. G. 1995. Geologic history of the Archean Buhwa greenstone belt and surrounding granite–gneiss terrain, Zimbabwe, with implications for the evolution of the Limpopo Belt. *Canadian Journal of Earth Sciences*, **32** 1977–1990.

FREI, R., BLENKINSOP, T. G. & SCHÖNBERG, R. 1999. Geochronology of the late Archaean Razi and Chilimanzi suites of granites in Zimbabwe: implications for the late Archaean tectonics of the Limpopo belt and Zimbabwe craton. *South African Journal of Geology*, **102**, 55–63.

FUCHTER, W. A. H. 1990. *The geology and gold mineralization of the northwestern mining camp, Gwanda greenstone belt, Zimbabwe*. PhD thesis, Queen's University, Kingston.

GARSON, M. S. 1995. *The Geology of the Area around Bulawayo*. Zimbabwe Geological Survey Bulletin, **93**.

GOODWIN, A. M. & SMITH, J. E. M. 1980. Chemical discontinuities in Archaean volcanic terrain and the development of Archaean crust. *Precambrian Research*, **10**, 301–311.

GURNEY, J. J. 1990. The diamondiferous roots of our wandering continents. *Transactions of the Geological Society of South Africa*, **93**, 424–437.

HAMILTON, W. B. 1998. Archean magmatism and deformation were not products of plate tectonics. *Precambrian Research*, **91**, 142–179.

HARRISON, N. M. 1970. *The geology of the Country around Que Que*. Rhodesia Geological Survey Bulletin, **67**.

HOFFMAN, P. F. 1991. On accretion of granite–greenstone terranes. *In*: ROBERT, F., SHEAHAN, P. A. & GREEN, S. B. (eds) *Nuna Conference on Greenstone Gold and Crustal Evolution, Val d'Or*. Geological Association of Canada Special Publications, 32–45.

HOFMANN, A., DIRKS, P. & JELSMA, H. 2001a. Late Archaean foreland basin evolution, Belingwe greenstone belt, Zimbabwe. *Sedimentary Geology*, **141–142**, 131–168.

HOFMANN, A., DIRKS, P. & JELSMA, H. 2001b. Horizontal tectonic deformation geometries in a late Archaean sedimentary sequence, Belingwe greenstone belt, Zimbabwe. *Tectonics*, **20**, 909–932.

HOFMANN, A., YAGOUTZ, O., KRÖNER, A., DIRKS, P. & JELSMA, H. 2002. The Chirwa dome: late Archaean thrusting and granite diapirism along the northeastern margin of the Zimbabwe craton. *South African Journal of Geology*, in press.

HOLM, P. E. 1985. Geochemical fingerprints of different tectonomagmatic environments using hygromagmatophile abundances of tholeiitic basalts and basaltic andesites. *Chemical Geology*, **51**, 303–323.

HORSTWOOD, M. S. A. 1998. *Stratigraphy, geochemistry and zircon geochronology of the Midlands greenstone belt, Zimbabwe*. PhD thesis, University of Southampton.

HORSTWOOD, M. S. A., NESBITT, R. W., NOBLE, S. R. & WILSON, J. F. 1999. U–Pb zircon evidence for an extensive early Archean craton in Zimbabwe: a reassessment of the timing of craton formation, stabilization, and growth. *Geology*, **27**, 707–710.

HUNTER, M. A. 1997. *The tectonic setting of the Belingwe greenstone belt, Zimbabwe*. PhD thesis, University of Cambridge.

HUNTER, M. A., BICKLE, M. J., NISBET, E. G., MARTIN, A. & CHAPMAN, H. J. 1998. Continental extensional setting for the Archean Belingwe Greenstone Belt, Zimbabwe. *Geology*, **26**, 883–886.

JELSMA, H. A. 1993. *Granites and greenstones in northern Zimbabwe, tectono-thermal evolution and source regions*. PhD thesis, Free University of Amsterdam.

JELSMA, H. A. & DIRKS, P. H. G. M. 2000. Tectonic evolution of a greenstone sequence in northern Zimbabwe: sequential early stacking and pluton diapirism. *Tectonics*, **19**, 135–152.

JELSMA, H. A., DIRKS, P. H. G. M & DE WIT, M. J. 2001. Rheological heterogeneity of Archean continental lithosphere: implications for Archean tectonics. *SAGA, Drakensberg, 9–12 October 2001, Extended Abstracts.*

JELSMA, H. A., DIRKS, P. H. G. M & VINYU, M. L. 2002. The margin of an Archean craton: structural relationships between the Zimbabwe craton and the Zambezi orogenic belt. *Precambrian Research*, in press.

JELSMA, H. A., VAN DER BEEK, P. A. & VINYU, M. L. 1993. Tectonic evolution of the Bindura–Shamva greenstone belt (northern Zimbabwe): progressive deformation around diapiric batholiths. *Journal of Structural Geology*, **15**, 163–176.

JELSMA, H. A., VINYU, M. L., VALBRACHT, P. J., DAVIES, G. R., WIJBRANS, J. R. & VERDURMEN, E. A. T. 1996. Constraints on Archaean crustal evolution of the Zimbabwe craton: a U–Pb zircon, Sm–Nd and Pb–Pb whole-rock isotope study. *Contributions to Mineralogy and Petrology*, **124**, 55–70.

KAMBER, B. S., BIINO, G. G., WIJBRANS, J. R., DAVIES, G. R. & VILLA, I. M. 1996. Archaean granulites of the Limpopo belt, Zimbabwe: one slow exhumation or two rapid events? *Tectonics*, **15**, 1414–1430.

KRAMERS, J. D. 1988. An open-system fractional crystallization model for very early continental crust formation. *Precambrian Research*, **38**, 281–295.

KUSKY, T. M. 1998. Tectonic setting and terrane accretion of the Archean Zimbabwe craton. *Geology*, **26**, 163–166.

KUSKY, T. M. & KIDD, W. S. F. 1992. Remnants of an Archaean oceanic plateau, Belingwe Greenstone Belt, Zimbabwe. *Geology*, **20**, 43–46.

KUSKY, T. M. & POLAT, A. 1999. Growth of granite-greenstone terranes at convergent margins, and stabilization of Archean cratons. *Tectonophysics*, **305**, 43–77.

KUSKY, T. M. & WINSKY, T. M. 1995. Structural relationships between a shallow water platform and an oceanic plateau, Zimbabwe. *Tectonics*, **14**, 448–471.

LUAIS, B. & HAWKESWORTH, C. J. 1994. The generation of continental crust: an integrated study of crust-forming processes in the Archaean of Zimbabwe. *Journal of Petrology*, **35**, 43–93.

LUDDEN, J., HUBERT, C., BARNES, C., MILKEREIT, B. & SAWYER, E. 1993. A three dimensional perspective on the evolution of Archaean crust: LITHOPROBE seismic reflection images in the southwestern Superior province. *Lithos*, **30**, 357–372.

MARTIN, A. 1978. *The Geology of the Belingwe–Shabani Schist Belt*. Rhodesia Geological Survey Bulletin, **83**.

MKWELI, S., KAMBER, B. S. & BERGER, M. 1995. A westward continuation of the Zimbabwe craton – northern marginal zone tectonic break and new age constraints on the timing of the thrusting. *Journal of the Geological Society, London*, **152**, 77–83.

MOORBATH, S., TAYLOR, P. N., ORPEN, J. L., TRELOAR, P. & WILSON, J. F. 1987. First direct radiometric dating of Archaean stromatolitic limestone. *Nature*, **326**, 865–867.

MOSER, D. E., FLOWERS, R. M., HART, R. J. 2001. Birth of the Kaapvaal tectosphere 3.08 Ga ago. *Science*, **291**, 465–468.

NÄGLER, T. F., KRAMERS, J. D., KAMBER, B. S., FREI, R. & PRENDERGAST, M. D. A. 1997. Growth of subcontinental lithospheric mantle beneath Zimbabwe started at or before 3.8 Ga: Re–Os study on chromites. *Geology*, **25**, 983–986.

NESBITT, R. W., FANNING, C. M., WILSON, J. F., JELSMA, H. A. & HORSTWOOD, M. S. A. 2000. Zircon U–Pb geochronology of the Zimbabwean Craton; untangling the stratigraphy of greenstone belts. *Geoscience 2000, 17–20 April, Manchester*, abstract.

NGUURI, T. K., GORE, J., JAMES, D. E. & 6 OTHERS 2001. Crustal structure beneath southern Africa and its implications for the formation and evolution of the Kaapvaal and Zimbabwe cratons. *Geophysical Research Letters*, **28**, 2501–2504.

NISBET, E. G., BICKLE, M. J. & MARTIN, A. 1977. The mafic and ultramafic lavas of the Belingwe greenstone belt, Rhodesia. *Journal of Petrology*, **18**, 521–566.

NISBET, E. G., MARTIN, A., BICKLE, M. J. & ORPEN, J. L. 1993. The Ngezi Group: komatiites, basalts and stromatolites on continental crust. *In*: BICKLE, M. J. & NISBET, E. G. (eds) *The Geology of the Belingwe Greenstone Belt, Zimbabwe*. Geological Society of Zimbabwe Special Publications, **2**, 69–86.

NISBET, E. G., WILSON, J. F. & BICKLE, M. J. 1981. The evolution of the Rhodesian craton and adjacent Archaean terrain: tectonic models. *In*: KRÖNER, A. (ed.) *Precambrian Plate Tectonics*. Elsevier, Amsterdam, 161–183.

NUTT, T. H. C. 1984. Archean gold mineralisation in the Nando and Pinkun mines, Kadoma district, Zimbabwe. *In*: FOSTER, R. P. (ed.) *Gold '82*. Balkema, Rotterdam, 268–284.

POLLACK, H. N. 1997. Thermal characteristics of the Archaean. *In*: DE WIT, M. J. & ASHWAL, L. D. (eds) *Greenstone Belts*. Oxford University Press, Oxford, 223–232.

RAMSAY, J. G. 1989. Emplacement kinematics of a granite diapir: the Chindamora batholith, Zimbabwe. *Journal of Structural Geology*, **11** 191–209.

RAPP, R. P., WATSON, E. B. & MILLER, C. F. 1991. Partial melting of amphibolite/eclogite and the origin of Archean trondhjemites and tonalites. *Precambrian Research*, **51**, 1–25.

SAGGERSON, E. P. & TURNER, L. M. 1976. A review of the distribution of metamorphism in the ancient Rhodesian Craton. *Precambrian Research*, **3**, 1–53.

SCHMIDT-MUMM, A., CHENJERAI, K. G., BLENKINSOP, T. G., OBERTHUR, T., VETTER, U. & CHATORA, D. 1994. The Redwing gold deposit, Mutare greenstone belt: geology, mineralogy, geochemistry and fluid inclusion studies. *Geologische Jahrbuch*, **D100**, 423–475.

SCHOLEY, S. P. 1992. *The geology and geochemistry of the Ngezi Group volcanics, Belingwe greenstone belt, Zimbabwe*. PhD thesis, University of Southampton, UK.

SIEGESMUND, S., JELSMA, H. A., BECKER, J. & 5 OTHERS 2001. Constraints on the age of granite emplacement, deformation and metamorphism in the Shamva area, Zimbabwe. *International Journal of Earth Sciences*, **91**, 909–932.

SILVA, K. E. 1997. *Komatiites from the Belingwe greenstone belt, Zimbabwe: constraints on the development of Archaean greenstone belts*. PhD thesis, University of London.

SNOWDEN, P. A. & BICKLE, M. J. 1976. The Chinamora batholith: diapiric intrusion or interference fold? *Journal of the Geological Society, London*, **132**, 131–137.

STOWE, C. W. 1984. The early Archean Selukwe nappe, Zimbabwe. *In*: KRÖNER, A. & GREILING, R. (eds) *Precambrian Tectonics Illustrated*. Nägele und Obermiller, Stuttgart, 41–56.

TAYLOR, P. N., JONES, N. W. & MOORBATH, S. 1984. Isotopic assessment of relative contributions from crust and mantle sources to the magma genesis of Precambrian granitoid rocks. *Philosophical Transactions of the Royal Society of London, Series A*, **310**, 605–625.

TAYLOR, P. N., KRAMERS, J. D., MOORBATH, S., WILSON, J. F., ORPEN, J. L. & MARTIN, A. 1991. Pb/Pb, Sm–Nd and Rb–Sr geochronology in the Archean craton of Zimbabwe. *Chemical Geology (Isotope Geosciences Section)*, **87**, 175–196.

TOMSCHI, H.-P. 1987. *Goldvorkommen im Archaischen Harare–Bindura-Greenstone Belt, Zimbabwe: Zusammenhänge zwischen Lagerstättenbildung und Greenstone Belt Entwicklung*. PhD thesis, Universität of Cologne.

TRELOAR, P. J. & BLENKINSOP, T. G. 1995. Archaean deformation patterns in Zimbabwe: true indicators of Tibetan-style crustal extrusion or not? *In*: COWARD, M. P. & RIES, A. C. (eds) *Early Precambrian Processes*. Geological Society, London, Special Publications, **95**, 87–108.

TRELOAR, P. J., COWARD, M. P. & HARRIS, N. B. W. 1992. Himalayan–Tibetan analogies for the evolution of the Zimbabwe craton and Limpopo belt. *Precambrian Research*, **55**, 571–587.

TSUNOGAE, T., MIYANO, T. & RIDLEY, J. R. 1992. Metamorphic P–T profiles from the Zimbabwe Craton to the Limpopo Belt, Zimbabwe. *Precambrian Research*, **55**, 259–278.

VINYU, M. L. 1994. *The geochronology and geochemistry of post-orogenic granitoids within the Harare–Shamva greenstone belt*. PhD thesis, University of Zimbabwe, Harare.

VINYU, M. L., MARTIN, M. W., BOWRING, S. A., HANSON, R. E., JELSMA, H. A. & DIRKS, P. H. G. M. 2001. First U–Pb zircon ages from a craton-margin Archaean orogenic belt in northern Zimbabwe. *Journal of African Earth Sciences*, **32**, 103–114.

VLAAR, N. J., VAN KEKEN, P. E. & VAN DEN BERG, A. P. 1994. Cooling of the Earth in the Archaean: consequences of pressure-release melting in a hotter mantle. *Earth and Planetary Science Letters*, **121**, 1–18.

WILDE, S. A., MIDDLETON, M. F. & EVANS, B. J. 1996. Terrane accretion in the southwestern Yilgarn Craton: evidence from a deep seismic crustal profile. *Precambrian Research*, **78**, 179–196.

WILSON, J. F. 1968. *The Geology of the Country around Mashaba*. Rhodesia Geological Survey Bulletin, **62**.

WILSON, J. F. 1979. A preliminary reappraisal of the Rhodesian basement complex. *In*: ANHAEUSSER, C. R., FOSTER, R. P. & STRETTON, T. (eds) *A Symposium on Mineral Deposits and Transportation and Deposition of Metals*. Geological Society of South Africa Special Publications, **5**, 1–23.

WILSON, J. F. 1990. A craton and its cracks: some of the behaviour of the Zimbabwe block from the late Archaean to the Mesozoic in response to horizontal movements and the significance of some of its mafic dyke patterns. *Journal of African Earth Sciences*, **10**, 483–501.

WILSON, J. F., NESBITT, R. W. & FANNING, C. M. 1995. Zircon geochronology of Archaean felsic sequences in the Zimbabwe craton: a revision of greenstone stratigraphy and a model for crustal growth. *In*: COWARD, M. P. & RIES, A. C. (eds) *Early Precambrian Processes*. Geological Society, London, Special Publications, **95**, 109–126.

WINGATE, M. T. D. 2000. Ion microprobe U–Pb zircon and baddeleyite ages for the Great Dyke and its satellite dykes, Zimbabwe. *South African Journal of Geology*, **103**, 74–80.

WOOD, D. A. 1979. A variably veined suboceanic upper mantle – genetic significance for mid-ocean ridge basalts from geochemical evidence. *Geology*, **7**, 499–503.

WYLLIE, P. J., WOLF, M. B. & VAN DER LAAN, S. R. 1997. Conditions for formation of tonalites and trondhjemites: magmatic sources and products. *In*: DE WIT, M. J. & ASHWAL, L. D. (eds) *Greenstone belts*. Oxford University Press, Oxford, 256–266.

Formation and early evolution of the atmosphere

BERNARD MARTY[1,2] & NICOLAS DAUPHAS[1]

[1] *Centre de Recherches Pétrographiques et Géochimiques, Rue Notre-Dame des Pauvres, B.P. 20, 54501 Vandoeuvre Lès Nancy Cedex, France (e-mail: bmarty@crpg.cnrs-nancy.fr)*
[2] *Ecole Nationale Supérieure de Géologie, Avenue du Doyen Roubault, 54501 Vandoeuvre Lès Nancy Cedex, France*

Abstract: The tectonic activity of the Earth allowed exchange of volatile elements (H, C, N, rare gases) between the surface of the Earth (atmosphere, crust, sediments, oceans) and the mantle. However, some of these elements still present elemental and isotopic heterogeneities that allow us to reconstruct the volatile composition of the terrestrial mantle. The protosolar nebula supplied a significant fraction of helium and neon, which were presumably trapped during the major phase of the Earth's accretion and were possibly hosted by accreting dust and/or small porous planetesimals. Surprisingly, volatile elements are in chondritic proportion despite their drastic (10^{-3}) depletion in the mantle relative to chondrites, in a way that recalls the case of highly siderophile elements. From stable isotope systematics, we find that the contribution of comets to the volatile inventory of the Earth was very limited. The integrated flux of chondritic-like material necessary to provide water, carbon and nitrogen is consistent with that required for the formation of the lunar craters as well as that necessary to account for the inventory of siderophile elements in the mantle. A consequence of this scenario is that the Earth's surface was oxidized very early. Alternatively, volatile and siderophile elements of the mantle could be the remnant of small patches of chondritic material that did not equilibrate with the core nor drastically degas.

Introduction

The development of life required adequate environmental conditions at the Earth's surface during the Hadean eon. Investigating these conditions and by consequence the origin and early evolution of the atmosphere is necessary to understand which chemical path and metabolism had been favoured during this period. As a result of the Earth's tectonics, the geological record during the Hadean Eon has been completely erased, and most of our knowledge on atmospheric evolution is indirect. Volatile elements in the present-day atmosphere have elemental and isotopic compositions (those of live and extinct radioactivity products) that provide information on the timing of atmospheric evolution. The mantle contains volatile elements that were trapped during Earth's accretion and might have been preserved since then. However, this reservoir exchanged volatile elements with the atmosphere through volcanism and subduction (Fig. 1) and only few elements still keep a record of the volatile component trapped in accreting silicates. Independently, the analysis of extra-terrestrial bodies and of solar wind allows us to infer the composition of potential contributors and to compare it with compositions observed in terrestrial reservoirs. One of the problems of this comparative approach is that the knowledge of these precursors is still limited. For example, the composition of comets is mainly known through remote sensing and in few instances through direct measurements (and possibly through the analysis of interplanetary dust particles (IDPs) but the origin of these objects remains to be established firmly). Other extremely important sources of information are the compositions of planetary atmospheres. The composition of the Jovian, Martian and Venusian atmospheres have been already measured and there is no doubt that missions planned for the next two decades will allow significant advances in this field. In this paper we focus on the case of the Earth and its early environment.

The terrestrial atmosphere cannot derive directly from the protosolar nebula (hereafter labelled PSN): (1) the abundance of atmospheric rare gases (normalized to a non-volatile element such as Si) is 6–10 orders of magnitude lower than solar (Brown 1952); (2) the isotopic compositions of rare gases (e.g. Ozima & Podosek 1983) and nitrogen (Hashizume et al. 2000) are drastically different from those of solar gases. The depletion of rare gases in the Earth resulted in high parent/daughter ratios (where the parent

From: FOWLER, C. M. R., EBINGER, C. J. & HAWKESWORTH, C. J. (eds) *The Early Earth: Physical, Chemical and Biological Development.* Geological Society, London, Special Publications, **199**, 213–229. 0305-8719/02/$15.00
© The Geological Society of London 2002.

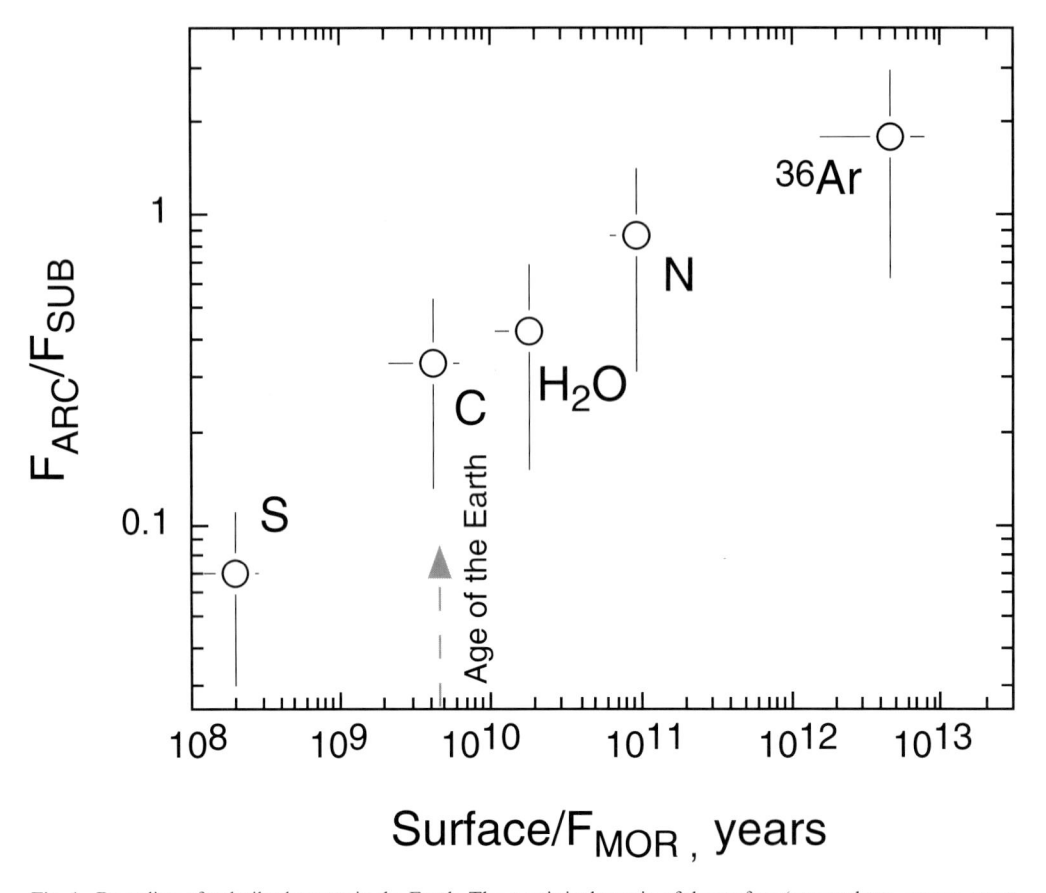

Fig. 1. Recycling of volatile elements in the Earth. The *x*-axis is the ratio of the surface (atmosphere, oceans, crust, sediments) inventory divided by the present-day flux at ridges. The *y*-axis is the amount of volatile elements carried by the oceanic crust and sediments to subduction zones by the volcanic flux at arcs. Neon is not represented because its volcanic flux is not known. Data sources: Craig *et al.* (1975), Matsuo *et al.* (1978), Le Guern (1982), Staudacher & Allègre (1988), Staudigel *et al.* (1989), Allard (1992), Marty (1992, 1995), Rea & Ruff (1996), Sano & Williams (1996), Marty & Tolstikhin (1998), Marty & Zimmermann (1999), and references therein.

element is a non-volatile element and the daughter is a rare gas isotope), allowing us to use natural radioactivity products to quantify the evolution of the terrestrial atmosphere. The first significant observation in the field was by Von Weizsacker, who proposed in 1937 the decay of ^{40}K ($T_{1/2} = 1.25$ Ga) to ^{40}Ar to explain the relatively high content of argon in the atmosphere (0.934%) compared with other rare gases (e.g. 0.0016% Ne). As the major reservoir of terrestrial potassium is the silicate Earth (mantle + crust), this observation implies transfer of radiogenic ^{40}Ar from the K-bearing silicates to the atmosphere through time by magmatism and metamorphism. Rubey (1951) showed that the alteration of rocks at the Earth's surface could not account for the budget of atmospheric gases and, noting that the composition of volatiles present at the Earth's surface (atmosphere + sediments + oceans) resembled to that of volcanic gases, proposed that the atmosphere was formed by volcanic degassing. This view is fully consistent with the potassium–argon budget of the Earth: approximately half of radiogenic argon produced by the decay of ^{40}K is now in the atmosphere. The timing of this transfer was investigated when the analysis of oceanic basalts, presumably derived from the convective mantle, showed that the radioactive/primordial, ^{40}Ar/^{36}Ar ratio of the mantle was much higher than that of the atmosphere. To obtain such a high isotopic ratio in the mantle relative to that of the atmosphere, it is necessary to transfer primordial ^{36}Ar from the mantle to the atmosphere 'before' significant radiogenic in-growth of ^{40}Ar in the mantle from the decay of ^{40}K. Models quantifying this observation led to the concept of 'catastrophic' degassing early in the Earth's history, a major event that was thought to have occurred in the Hadean Eon and might have been linked with the formation of the core (Ozima 1975; Alexander & Schwartzman 1976; Allègre et al. 1983; Ozima & Podosek 1983; Sarda et al. 1985).

Another major advance in the field was the discovery of ^{129}Xe excess (relative to Xe isotopic composition of the atmosphere) in a few CO_2-rich natural gases (Zartman et al. 1961; Phinney et al. 1978) and in mantle-derived, mid-ocean ridge basalts (MORB) (Staudacher & Allègre 1982; Marty 1989; Staudacher et al. 1989). ^{129}Xe is the radioactive daughter of ^{129}I ($T_{1/2} = 17$ Ma), a now extinct radioactive isotope that was synthesized before formation of the Solar System. Assuming that the atmosphere was derived from the mantle, the existence of higher ^{129}Xe/^{130}Xe ratios (where ^{130}Xe is a non-radiogenic Xe isotope used as a reference) in the mantle than in the atmosphere implied that the catastrophic degassing of the Earth occurred before ^{129}I of the mantle was completely decayed, that is, within a few half-lives of ^{129}I. The study of xenon isotopes ('terrestrial xenology') allowed model ages of the atmosphere to be computed within the range of 0.1–0.2 Ga after the start of condensation in the Solar System (Staudacher & Allègre 1982; Allègre et al. 1987; Pepin 1991; Tolstikhin & Marty 1998). It is important to note that the concept of rapid atmospheric formation is still valid even though the starting assumptions have failed to reproduce all rare gas observations.

The first problem with the concept of catastrophic degassing as a source of atmospheric gases came from other radioactive decays that produced Xe isotopes. $^{131-136}$Xe are produced by two radioactivities, the spontaneous fission of still existing ^{238}U ($T_{1/2} = 4.46$ Ga), and that of ^{244}Pu ($T_{1/2} = 82$ Ma). These two decays produce Xe isotopes in proportions that are fairly close, but precise analysis of Xe isotopes of natural gases and of MORB has allowed computation of the fraction of fissiogenic Xe in the mantle produced by plutonium-244 (Ozima et al. 1985; Kunz et al. 1998) and therefore the ^{129}Xe*/^{136}Xe$^*_{Pu}$ ratio (where the asterisk indicates radiogenic or fissiogenic and ^{136}Xe is taken as an example among the $^{131-136}$Xe isotopes) resulting from the decays of ^{129}I and ^{244}Pu in the mantle (Marty 1989). The atmosphere contains also Xe isotopes produced by extinct radioactivities and the ^{129}Xe*/^{136}Xe$^*_{Pu}$ ratio of the atmospheric source could also be computed (Pepin & Phinney 1979;

Igarashi 1986). It turned out that there was more plutonium-derived Xe relative to iodine-derived Xe in the atmosphere than in the mantle (Ozima et al. 1985; Marty 1989). Degassing of the atmosphere after formation of the mantle would have produced the opposite situation, as the half-life of ^{244}Pu is longer than that of ^{129}I. This comparison was at odds with the classical view of simple derivation of atmospheric gases from the mantle reservoir early in the Earth's history: for a common I–Pu geochemical source, the atmosphere as a geochemical reservoir would have been formed before the mantle, which is logically absurd. A way round this problem is to consider that the hypothesis of a common I–Pu geochemical source is not adequate and that two different geochemical sources contributed xenon in the mantle and in the atmosphere, respectively (Marty 1989). This implies either (1) heterogeneous accretion of bodies with contrasted compositions having different I/Pu ratios (Marty 1989) and/or (2) active exchange between atmosphere and mantle while atmospheric xenon was fractionated during escape to space (Pepin 1991; Tolstikhin & Marty 1998). It must be noted, however, that the identification of xenon components in the Earth–atmosphere is not yet definitive, as the isotopic composition of primitive Xe that was presumably trapped during the Earth's formation has not been measured directly in extraterrestrial samples but has been inferred from a statistical treatment of meteoritic data (Pepin and Phinney 1979; Igarashi 1986), making the terrestrial budget of ^{129}Xe* and ^{136}Xe$^*_{Pu}$ model dependent. Moreover, the non-radiogenic isotopic composition of xenon in the mantle, necessary to compute these two values, is not precisely known, although CO_2-well gas studies suggest that there exists a non-atmospheric Xe component in the Earth (Caffee et al. 1999).

The aim of this study is to identify the volatile (H, C, N, rare gases) components present at the Earth's surface and in its interior. We then use such compositions to infer the potential contributing sources and processes that resulted in the surface and the mantle inventory of volatile elements.

Reservoirs and potential contributors

Identification of volatile components in the Earth–atmosphere

As a result of plate tectonics, plume activity and volcanism, volatile elements are continuously exchanged between the surface and the mantle.

The efficiency of these homogenizing processes is illustrated in Figure 1 which compares the 'mean degassing duration' (MDD) of the atmosphere for several volatile elements with the efficiency of recycling at arcs. The former is defined as the surface inventory (the total amount of a given volatile element, whatever its chemical state, in the crust, the sediments, the oceans and the atmosphere) divided by its degassing rate from the mantle, thought to occur mainly at mid-ocean ridges (degassing at plumes is minor compared with that at ridges; see, e.g. Marty & Tolstikhin 1998). MDD lower than the age of the Earth would imply fast recycling whereas MDD greater than the age of the Earth would suggest either decreasing degassing rate with time or the occurrence of a volatile component at the Earth's surface not derived from the mantle, or both. The y-axis of Figure 1 is the ratio F_{ARC}/F_{SUB} between the volatile flux through arc volcanism and the amount of volatiles carried by plates towards subduction zones (F_{ARC}/F_{SUB} values close to unity imply no recycling). The correlation between F_{ARC}/F_{SUB} and MDD strongly suggests a recycling and mantle–atmosphere exchange efficiency decreasing from sulphur (MDD $c.$ 200 Ma) to carbon (MDD $c.$ 4 Ga), H_2O, N and finally to rare gases. The isotopic compositions of these volatiles between the mantle and the surface follow this logic well, as both S and C isotopic ratios are similar in the mantle and in the surface inventory, whereas water (δD $c.$ -80‰ relative to ocean water in the upper mantle), nitrogen ($\delta^{15}N$ $c.$ -4‰ in the upper mantle relative to atmospheric N; e.g. Marty & Humbert 1997) and neon ($\delta^{22}Ne \leq -200$‰ relative to atmospheric N; e.g. Sarda et al. 1988) present isotopic heterogeneities that probably represent different volatile sources that were not fully homogenized. Hence it is possible to distinguish partly a mantle reservoir from the 'atmosphere' (surface inventory) one and to explore the origins of both.

The potential components that might have delivered volatile elements to the Earth are the PSN (the major reservoir in the Solar System) and solid matter bodies such as meteorites and comets. The composition of meteoritic volatiles is thought to have been derived from the PSN through elemental and isotopic fractionation. Contributions from sources outside the Solar System such as pre-solar grains or species affected by interstellar chemistry are attested by the discovery of pre-solar grains in primitive meteorites on one hand, and by the large variation of the D/H ratio in the Solar System on another hand, but their extent is a matter of debate. A comparison of the abundances of He, Ne, Ar, C and N between cosmochemical potential precursors (PSN and chondrites) and terrestrial reservoirs (the atmosphere and the mantle source of MORB) is given in Figure 2. Volatile abundances are normalized to ^{20}Ne and the Sun, which, in this figure, results in a flat pattern for the solar abundance. Neon is used for normalization as its isotopic (non-radiogenic) composition in the mantle is clearly different from that of the atmosphere (see below).

Meteorites

Figure 2 illustrates the large difference in abundance pattern between primitive meteorites and the PSN, which presumably reflects different trapping efficiency during condensation of the Solar System material and further exchange between gas and solid in the forming Solar System. The chondritic excesses of C and N relative to solar may have resulted from preferential trapping of C and N compounds, which were stable in the reducing conditions of the hydrogen-rich PSN. The origin of these compounds is unclear: the large heterogeneity of the nitrogen isotopic composition among Solar System objects (Kerridge & Swindle 1988) is not compatible with a single homogeneous source such as a well-mixed solar nebula, but rather suggests mixing between different components (Hashizume et al. 2000). It is likely that a major fraction of H and N (and C by analogy, although in this case the isotope argument is less straightforward) was incorporated as compounds and that the ultimate origin of these elements could be partly interstellar. Notably, rare gases are also fractionated in meteorites and the origin of such fractionation is a matter of debate (Pepin 1991; Ozima et al. 1998).

Terrestrial mantle

The analysis of rare gases together with C and N of mantle-derived samples allows us to investigate the volatile composition of the mantle. However, data need to be corrected for surficial processes such as partial melting, fractional crystallization, degassing and atmospheric contamination, and we have developed in the past few years analytical and correction techniques that allow reconstruction of the elemental and isotopic composition of the mantle for He, C, N and Ar (Marty & Jambon 1987; Marty 1995; Marty & Humbert 1997; Marty & Zimmermann 1999). The data used in Figure 1 were summarized by

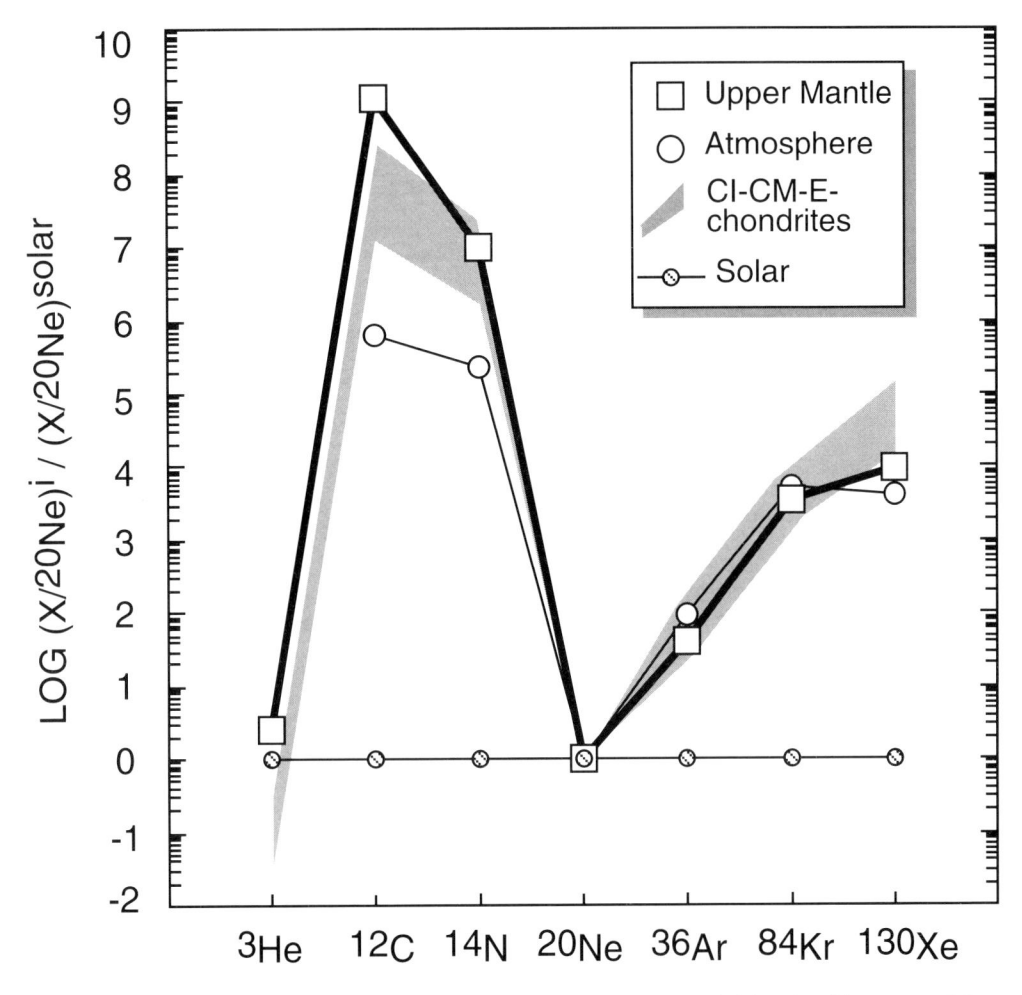

Fig. 2. Comparison of volatile abundance data in protosolar nebula (PSN), primitive chondrites, terrestrial mantle and 'atmosphere' (atmosphere *sensu stricto*, crust, sediments, oceans). Data are normalized to ^{20}Ne and PSN, so that the PSN pattern is flat. The choice of ^{20}Ne as a normalizing isotope is based on the observation that recycling of atmospheric neon in the mantle is limited as indicated by its isotopic composition. Data sources: Mazor *et al.* (1970), Marty & Jambon (1987), Anders & Grevesse (1989), Pepin (1991), Moreira *et al.* (1998), Ozima *et al.* (1998), Marty & Zimmermann (1999).

Marty & Zimmermann (1999) and, for heavy rare gases, we have also used the best available estimates of the upper-mantle composition (Moreira *et al.* 1998).

The mantle displays a chondritic pattern, and this similarity provides important information on the process of delivery, as it suggests that the bulk fraction of these volatiles was delivered with minimal fractionation to the Earth by chondritic-like material. In fact, it is very surprising that such a chondritic pattern could have been preserved despite the severe fractionating processes that are likely to have occurred during formation of the Earth, such as impact degassing, core formation or atmospheric escape. Indeed, the Earth's mantle is extremely depleted in volatile elements, by a factor of 10^{-3} relative to carbonaceous chondrites in the case of nitrogen (1 ± 0.4 ppm N for the mantle, based on K–Ar–N systematics (Marty 1995; Dauphas & Marty 1999), compared with about $(1-3) \times 10^3$ ppm for carbonaceous chondrites). The situation resembles that of highly siderophile elements for which a chondritic pattern is found in the mantle despite drastic depletion (e.g. Righter & Drake 1997). These elements should have been largely partitioned into the core, leaving the mantle much more depleted than is at present observed, and largely fractionated relative to the chondritic pattern. The late addition of a chondritic veneer (after core formation) is frequently advocated for siderophile elements (see 'The case of major volatiles', below). Could it be also the case for volatile elements? We discuss this possibility below for H and possibly C and N but it seems doubtful that rare gases could have been preserved in silicates during late impact shocks. The observation that volatile elements seem to be in chondritic proportion (and isotopic ratios for at least H, N and Ar) despite their drastic depletion is enigmatic at present and will certainly put strong constraints on future models of volatile geochemistry.

There are two notable exceptions to the chondrite analogy. First, carbon is enriched in the mantle relative to other cosmochemical reservoirs, and this difference is likely to reflect preferential recycling of carbon from the surface, as carbon at the Earth's surface is mainly present in sediments as carbonates and organic matter (Javoy *et al.* 1986). This view is consistent with the first-order homogeneity of C isotopes in the mantle and at the Earth's surface, as mentioned above (part of surface nitrogen has probably been recycled also and its fate will be discussed below). Second, the He/Ne ratio of the mantle is similar to the solar ratio within a factor of two (Honda & McDougall 1997; Marty *et al.* 1998; Moreira *et al.* 1998), strongly suggesting that a major fraction of light rare gases in the mantle is derived from the PSN with limited fractionation.

The possibility that light rare gases originated partly from the PSN is confirmed by Ne isotopic ratios in mantle-derived rocks (Fig. 3). In the now classical three-isotope diagram for neon (^{20}Ne/^{22}Ne v. ^{21}Ne/^{22}Ne), the mantle differs from atmospheric neon by excesses of ^{20}Ne and ^{21}Ne relative to ^{22}Ne (Sarda *et al.* 1988; Marty 1989; Staudacher *et al.* 1989; Honda *et al.* 1991, 1993*a,b*; Hiyagon *et al.* 1992; Marty *et al.* 1998; Moreira *et al.* 1998). In this diagram format, mixing between two Ne components translates into a straight line joining dots representing these components. Excesses of ^{21}Ne result from nuclear reactions such as neutron activation of oxygen in the mantle or spallation reactions by cosmic rays and are not a primordial character. In contrast, high ^{20}Ne/^{22}Ne ratios in the mantle relative to the air composition require the presence of a primordial Ne component different from that present in the atmosphere, as no known nuclear process can produce the observed excesses of ^{20}Ne. ^{20}Ne/^{22}Ne ratios observed for some of the mantle-derived rocks analysed so far are higher than those of neon-Q (^{20}Ne/^{22}Ne = 10.7), a meteoritic component thought to characterize best rare gases trapped in chondrites (Wieler *et al.* 1992), or SEP neon (^{20}Ne/^{22}Ne = 11.2) (e.g. Ozima *et al.* 1998), a high-energy component of the solar corpuscular emission. Therefore, mantle samples display Ne isotope variations that are best explained by mixing between atmospheric neon (^{20}Ne/^{22}Ne = 9.80) and a ^{20}Ne-rich component such as neon (^{20}Ne/^{22}Ne = 13.8) as measured in solar-wind irradiated lunar soils (e.g. Ozima *et al.* 1998). The observation that the abundance pattern of volatiles normalized to ^{20}Ne reproduces a chondritic pattern despite a presumably solar origin for Ne is not problematic. Indeed, neon in the mantle seems to include also a meteoritic-like component (Trieloff *et al.* 2000). Trieloff *et al.* remarked that very few, if any, Ne isotope data with good accuracy have ^{20}Ne/^{22}Ne higher than 12.5 and proposed that neon in the mantle was not exactly solar but included a component of the type found in some gas-rich chondrites. In addition, as the solar pattern of rare gases is highly enriched in He and Ne relative to the chondritic pattern, mixing of the two results in He/Ne ratio close to solar together with a near-chondritic-like Ar/N ratio. The fact that the N/Ne ratio is also chondritic is the result of the extreme relative depletion of nitrogen in the solar composition (N/Ne = 0.8) relative to the chondritic pattern (N/Ne *c.* 10^7).

The case of trapping light rare gases in the terrestrial mantle from the PSN with little fractionation requires the presence of the PSN and gas-dust exchange during the early stage of accretion of terrestrial planets. The process of trapping is unclear and could be adsorption on accreting dust. However, such a process would probably result in elemental He-Ne fractionation (although the lack of adsorption data for light rare gases prevents quantitative evaluation of this possibility). PSN gases could also have been occluded in pores of accreting material, a process that would have been able to prevent elemental fractionation. Further compaction would have then resulted in occlusion of pores and subsequent trapping into mantle-forming silicates. Another possibility is gas exchange between the PSN and the Earth in a molten stage; for example, during episodes of magma ocean in the presence of a thick proto-atmosphere derived from the PSN (Abe & Matsui 1986). Equilibrium dissolution of rare gases into silicate melts would have resulted in He-Ne fractionation by a factor of *c.* 2 (e.g. Jambon *et al.* 1986) which is within the uncertainty of the mantle He/Ne ratio.

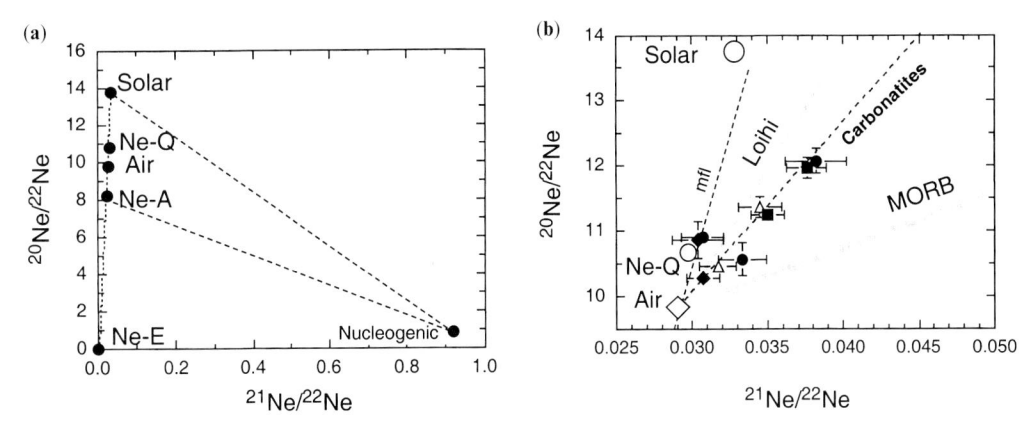

Fig. 3. Diagrams of three neon isotopes. In this format, mixing between end-members is represented by straight lines. (**a**) Overall variations of Ne isotopes in nature. ^{21}Ne and ^{22}Ne are produced significantly by natural nuclear reactions such as neutron activation of oxygen and fluorine, and spallation by cosmic rays. The PSN component is represented by the analysis of solar wind and displays the highest ^{20}Ne/^{22}Ne ratio measured so far. Neon-Q is a ubiquitous rare gas component trapped in primitive meteorites (Wieler et al. 1992); neon-A, often called planetary neon, is another meteoritic end-member but it is not clear if it represents a single, well-individualized component or a mixture between other neon components; neon-E is almost pure ^{22}Ne of nucleosynthetic origin. It should be noted that atmospheric neon presents a ^{20}Ne/^{22}Ne ratio intermediate between solar and Ne-A. Dashed lines between Ne-A, solar and nucleogenic Ne represent the field of values observed in meteorites. (**b**) Ne isotope variations in mantle-derived samples. Mid-ocean ridge basalts (MORB) are thought to derived directly from the convective mantle driving plate tectonics and are represented by an array between atmospheric neon and a mantle component enriched in both ^{20}Ne and ^{21}Ne (Sarda et al. 1988). The latter enrichment is accounted for by production and accumulation of nucleogenic neon in the mantle whereas the ^{20}Ne enrichment is regarded as representing the occurrence of solar-type neon in the mantle. A similar situation is observed for samples linked with mantle plumes such as Hawaii (Loihi Seamount is the youngest volcano of the Hawaiian chain (Honda et al. 1991; Hiyagon et al. 1992)) or carbonates and associated minerals from the Kola Peninsula (data points are represented to illustrate their spread and uncertainties (Marty et al. 1998)). mfl (Mass fractionation line) represents the effect of isotopic mass fractionation on neon isotopes.

Atmosphere and volatile recycling

The atmosphere presents a fractionated rare gas pattern somewhat similar to that of chondrites, but there are also two major differences. First, C and N are underabundant relative to chondritic pattern (Fig. 2). We have advocated recycling through time for carbon and possibly for nitrogen as a net sink for surface C and N. For carbon, this would make the mantle and 'atmosphere' patterns converge towards the shaded zone representing the field of chondrites. A similar effect is also expected for nitrogen. The 'atmosphere' reservoir and the mantle reservoir contain approximately the same amount of nitrogen (Marty 1995), equivalent to 1 ppm when normalized to the mass of the mantle. Contrary to the case of carbon, the two reservoirs differ isotopically by c. 5‰, which is significant for the Earth, but not for extraterrestrial reservoirs. Therefore, we postulate that nitrogen was exchanged actively between the surface and the mantle, but not to a point erasing isotopic differences (the isotopic difference may alternatively result from isotopic fractionation, a possibility that will be discussed in another paper). It must be emphasized that most of nitrogen recycling must have taken place before c. 3 Ga ago, as both mid-ocean-ridge basalts, sampling the present-day upper mantle, and diamonds (Cartigny et al. 1997), which sampled the ancient (for some of them Archaean) subcontinental lithospheric mantle, present a similar distribution of δ^{15}N values centred around -4‰. A further argument for nitrogen recycling is based on the N_2/Ar ratio of the upper mantle (Marty 1995). The N_2/^{40}Ar ratio of this reservoir is comparable with that of the atmosphere and, as ^{40}Ar is produced by ^{40}K, this similarity suggests a similar behaviour of N and K. Nitrogen can substitute for K^+ in the form of NH_4^+, which make these two elements potentially recyclable at similar rates. In contrast, the $N_2/^{36}$Ar ratio of the mantle is two orders of magnitude higher than that of the atmosphere, exactly what is expected if nitrogen is recycled and ^{36}Ar is not.

Apart from the case of recycling as a source of atmosphere alteration, the isotopic compositions of atmospheric neon (and xenon; see Introduction) do not support a mantle origin. An important feature of the atmosphere is its $^{20}Ne/^{22}Ne$ ratio of 9.8, much lower than that of the mantle. Assuming that atmospheric Ne was also solar-like initially, its present-day value requires extensive isotopic fractionation, which might have taken place either in precursors or in the atmosphere itself during early evolution (Zanhle et al. 1988). In the first possibility, late accreting material would have released a low $^{20}Ne/^{22}Ne$ component (Marty 1989). Indeed, $^{20}Ne/^{22}Ne$ ratios as low as eight are observed in bulk chondrites (Mazor et al. 1970) and are thought to result from mixing between a nucleosynthetic ^{22}Ne-rich end-member and the well-characterized trapped Ne-Q component (Wieler et al. 1992) and/or solar Ne. Alternatively, atmospheric neon could have been isotopically fractionated by hydrodynamic escape of a thick proto-atmosphere (Zanhle et al. 1988). Early models required an extremely thick hydrogen-rich atmosphere to start with (Hunten et al. 1987; Sasaki & Nakasawa 1988), which faced a serious problem with conventional models of planetary formation in the terrestrial feeding zone. In this type of scenario, a hydrogen-rich atmosphere, either a remnant of the PSN or the result of strong reduction of water by molten iron, possibly ionized by an enhanced solar UV flux, escapes from the terrestrial gravitational field and entrains other volatile species following a distillation process that greatly fractionates isotopes according to their mass difference. Subsequent works showed the role of other atmospheric constituents such as CO, CO_2 and H_2O in increasing significantly the efficiency of isotopic fractionation during escape (Ozima & Zahnle 1993), which removed the need for a very thick proto-atmosphere. Model calculations show that, depending on the atmospheric composition, Ne can be fractionated from the solar composition to the atmospheric composition with limited Ne loss (Ozima & Zahnle 1993). Further models advocated giant impacts to blow off partially the early atmosphere of the Earth (Pepin 1997), but it is not yet clear if such events would have resulted in extensive isotopic fractionation (Melosh & Vickery 1989). Atmospheric escape could also account for the underabundance of C and N in the atmosphere relative to Ne (Fig. 2), provided that these elements were in the form of species lighter than ^{20}Ne, which could be lost more efficiently than neon. The best candidates for these are methane ($m = 16$) and ammonia ($m = 17$), implying escape when the atmosphere was strongly reducing (a detailed scenario for isotope fractional loss of nitrogen has been developed by Tolstikhin & Marty (1998)).

Atmospheric xenon is also isotopically unique among Xe components in the Solar System (with the possible exception of Martian atmospheric Xe, e.g. Swindle 1995) as it is fractionated by 3% per atomic mass unit (a.m.u.). In addition, atmospheric xenon is depleted relative to the chondritic or solar patterns: the Kr/Xe ratio of air is 25 times the mean ratio of chondrites (Fig. 2). The depletion of xenon is not consistent with its isotopic fractionation because, in the first case, the heavy element (Xe) is depleted relative to the lightest one (Kr) and, in the second case, the light isotopes (e.g. ^{124}Xe) are depleted relative to the heavy ones (e.g. ^{136}Xe). The elemental depletion of Xe could be due to xenon trapping in a terrestrial reservoir, but possibilities exclude ice or sediments, which cannot explain the 25-fold depletion (Podosek et al. 1981; Bernatowicz et al. 1985). Preferential subduction of Xe trapped in sediments could account qualitatively for it, but is not yet documented.

Timing

In this section we review briefly chronological constraints relevant to atmospheric evolution. The onset of condensation and accretion of the Solar System is well dated at 4.556 Ga from the analysis of primitive meteorites (Allègre et al. 1995). Recent accretion models propose that the duration to obtain Mars-sized planets would have been fairly short, less than 10 Ma (see, e.g. Chambers & Wetherhill 1998). The typical time interval indicated by extinct radioactivities to differentiate primitive meteorite parent bodies into metal and silicates is also short, of the order of few million years (e.g. Wetherill 1975; Lee & Halliday 1995). The formation of the giant planets would have also been a relatively short process (<10 Ma), but the process of accretion of terrestrial planets would have slowed down, given the decreasing availability of bodies in the inner Solar System. Recent models postulate a duration of 50–80 Ma to obtain a terrestrial planet of the size of the Earth (Vityazev et al. 1990; Weidenschilling et al. 1997). Some models postulate that, because of the presence of giant planets, gravitational forces would have deflected bodies formed in colder regions of the Solar System and therefore richer in volatiles to the inner Solar System (e.g. Chambers & Wetherhill 1998; Petit et al. 1999). Periodic gravitational forces produced by Jupiter and Saturn could move bodies from the asteroidal belt located in specific narrow zones known as orbital resonances into elongated orbits crossing the terrestrial region. The efficiency of such a process is so high that such zones would have become quickly exhausted and, recently, Vokrouhlicky & Farinella (2000) have proposed that such zones could be continuously refuelled in meteoritic fragments because of the drag exerted by emission of thermal radiation from objects asymmetrically exposed to the Sun. Linking the efficiency of terrestrial bombardment with the solar radiation presents the exciting possibility of modulating the bombardment of the Earth with solar activity.

The life duration of the PSN, the presence of which is required to feed solar-type Ne in the terrestrial matter forming the mantle, is thought to have not exceeded 10–20 Ma (Montmerle 1999). Hence it is conceivable that, if solar-like light rare gases now seen in the mantle were trapped directly from the PSN during accretion or during exchanges between the proto-mantle and a PSN-rich atmosphere, these processes could not have lasted more than a few tens of million years, and that both waning of the PSN and increasing rate of degassing of bodies impacting on the growing Earth would have prevented further delivery of rare gases to the interior of the Earth. The Hf–W chronometer allows proposal of a time interval of 10–60 Ma to form the core (Lee & Halliday 1995), depending on which type of model (e.g. continuous accretion and differentiation v. single-stage differentiation) is adopted. As stated above, the time required to grow the Earth to its present-day mass is ≤100 Ma, and such a duration is in agreement with the lead isotope record of the mantle (Allègre et al. 1995). Notably, the age of the Moon-forming event is of the order of 4.46 Ga (e.g. Heiken et al. 1991), and the use of the coupled $^{146}Sm-^{147}Sm$ system allows us to constrain the initial mantle differentiation yielding the first crustal reservoir to 4.47 Ga (Jacobsen & Harper 1996). Finally, evidence has been reported from zircon analysis that a continental crust existed 4.4 Ga ago (Wilde et al. 2001), and that liquid water was present 4.3 Ga ago (Mojzsis et al. 2001).

We postulate, in view of these constraints, that rare gases were delivered to the Earth and fractionated in the atmosphere mainly within the first 100 Ma, possibly within 50 Ma, when most of the terrestrial accretion took place. During this period, atmosphere and proto-mantle rare gases were actively exchanged and rare gas fractionation took place following a combination of impact degassing, atmospheric escape and mantle–atmosphere exchange during magma ocean stages (Sasaki & Nakasawa 1988; Pepin 1997; Tolstikhin & Marty 1998). Rare gas modelling of mantle–atmosphere differentiation suggests that these reservoirs might have remained 'open', that is, might have continued to exchange their volatiles between reservoirs or with space during a longer period up to 100 Ma, but also predicts that the corresponding fluxes would have decreased dramatically with time (Pepin 1991; Tolstikhin & Marty 1998).

Rare gas models

The fractionation of rare gases in the atmosphere cannot be explained in a straightforward manner and calls for selective fractionation processes. Several sophisticated models have been developed in which each rare gas is affected selectively by a combination of mixing and isotope fractionation during atmospheric escape events. Pepin (1991, 1997) assumed a two-stage hydrodynamic atmospheric escape. In the first stage, a H_2-rich proto-atmosphere containing CO, N_2 and the noble gases in proportions found in primitive chondrites was driven by intense extreme-ultraviolet (EUV) radiation from the young evolving Sun starting at a solar age of c. 50 Ma. This episode was followed by a long (c. 80 Ma) period of quiescence, followed by abrupt degassing of remnant H_2, CO_2 and N_2 from the mantle and of light rare gases trapped deep in the Earth having a solar composition. Hydrodynamic escape resumed with the available H_2 in a waning but still potent EUV flux. Atmospheric volatiles remaining at the end of this second stage, 4.2 Ga ago, formed the bulk of the present-day atmosphere. The role of atmospheric erosion by giant impacts was further considered but did not change the essence of this two-stage scenario (Pepin 1997). Tolstikhin & Marty (1998) attempted to integrate in a global model the formation of the Earth, the differentiation of the mantle and the formation of the atmosphere. They proposed a combination of isotopic fractionation during hydrodynamic atmospheric escape and mantle–atmosphere exchange, with significant differences from Pepin's models. A gradual, single-stage process in which solar-like rare gases and chondritic N were contributed by impacts in the presence of the PSN was postulated. The selective fractionation of each rare gas took place during ocean magma episodes when rare gases were sorted according to their respective solubilities in basaltic melt during vigorous magma convection. The timing of these processes was derived from the I–Pu–Xe systematics and the model predicted that the atmosphere was settled for its rare gases and nitrogen >4.3 Ga ago. The models of Pepin (1991) and Tolstikhin & Marty (1998) agree on several important predictions: (1) both the PSN and chondritic-type material contributed terrestrial volatiles; (2) during the first tens of million years the forming mantle and atmosphere exchanged volatiles; (3) the atmosphere (and the mantle) were closed early in the Earth's history around 4.3 Ga ago and the atmospheric composition (but not its chemistry) experienced little change since then; (4) they require intense yet decreasing EUV during a long period, e.g. 100–200 Ma. This also poses the problem of penetration of the EUV radiation in an initially dense region. These models call for very specific conditions and processes, which will need close examination to confirm or refute. It may also be possible that we still ignore the composition of the contributing sources, and with this respect comets deserve special attention, as the process of rare gas trapping at low temperature in their ice might have resulted in specific fractionation that remains to be documented. The recent discovery that comet C/1995 O1 (Hale Bopp) contains argon in solar proportion (Stern et al. 2002) suggests, however, that fractionation of heavy rare gases in comets is unlikely.

The case of major volatiles

The case of major volatiles deserves special attention, as it was these elements that shaped the early terrestrial environment necessary for the

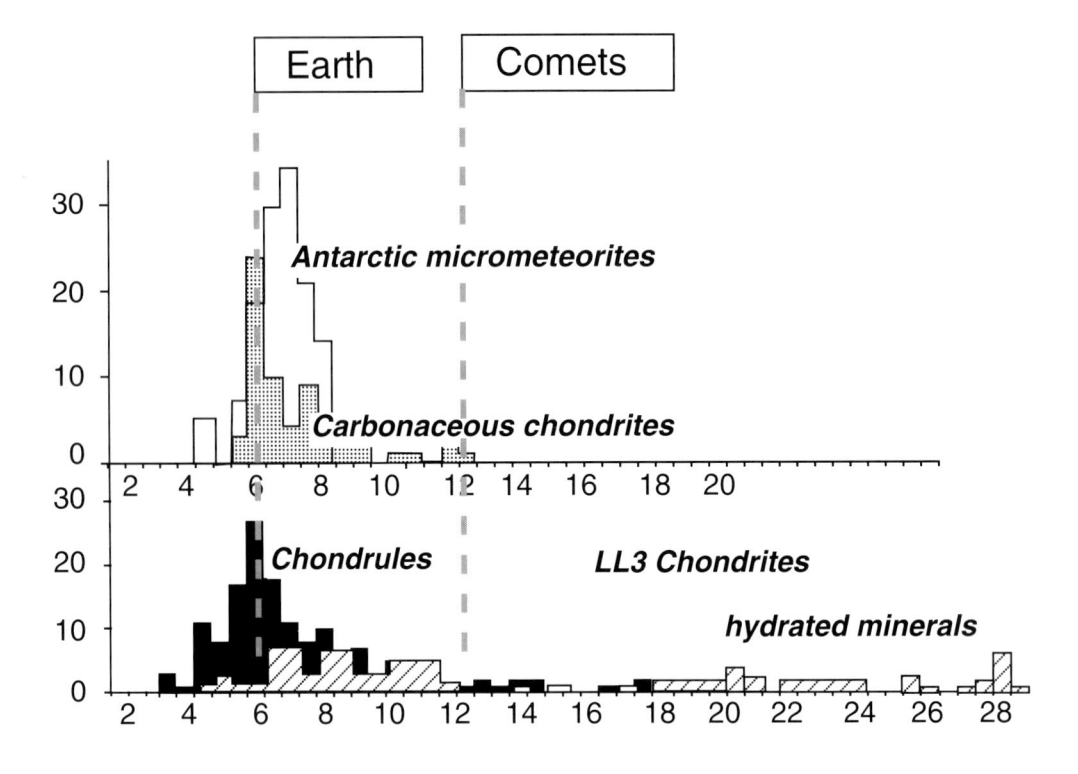

$$f = (D/H)_{H_2O} / (D/H)_{H_2}$$

Fig. 4. Isotopic variations of hydrogen in the Solar System (adapted from Robert et al. 2000). The deuterium/hydrogen ratio of different components is normalized to the D/H ratio of the Sun (as it was before deuterium burning), which is thought to represent H_2 in the protosolar nebula. Numbers along the y-axis represent the numbers of cases. Terrestrial hydrogen is enriched in deuterium by a factor of about six relative to solar. Among Solar System objects analysed so far, carbonaceous chondrites, Antarctic micrometeorites (Engrand & Maurette 1998) and chondrules from LL3 chondrites present a distribution of D/H values that centre around the terrestrial D/H ratio. Notably, comets analysed so far (Halley, Hale Bopp and Hyakutake, references given by Dauphas et al. (2000)) present D/H values about two times higher than the terrestrial value.

development of life. A direct derivation of major volatile elements from the PSN is not supported by abundance data and H, N isotopic ratios, which rather favour a chondritic origin (Fig. 2).

The D/H ratio of Solar System reservoirs presents variations over a factor of 30 (Fig. 4), which are thought to result from mixing of PSN-type hydrogen and D-rich compounds (Geiss & Gloecker 1998; Robert *et al.* 2000). The deuterium enrichment of the latter is attributed to trapping of hydrogen-bearing molecules that experienced low-temperature fractionation in molecular clouds during ion–molecule exchanges. This makes the hydrogen isotopic composition a good diagnostic tracer of sources and mixings among Solar System objects (Robert *et al.* 2000). Hydrogen in water of terrestrial oceans is enriched by a factor of six relative to solar (pre-deuterium burn-ing) hydrogen (Fig. 4) and cannot be derived from the PSN directly. It is possible, however, that hydrodynamic escape of PSN hydrogen would have favoured preferential escape of hydrogen, leaving residual hydrogen rich in deuterium. However, the terrestrial D/H ratio is within the range of values found in primitive meteorites, and this similarity strongly suggests a chondritic origin for terrestrial water. It is also important to note that the terrestrial D/H ratio is clearly different from those of comets as far as the few available data are representative of these objects. Therefore hydrogen isotopic data favour a derivation of terrestrial hydrogen from chondrite-like bodies, with limited, if any, fractionation.

A similar scenario emerges from N isotope systematics. Most planetary objects display $^{15}N/^{14}N$ ratios around the terrestrial atmospheric value within 50‰. These objects include the Earth, Venus (Mars constitutes an exception as its atmosphere is enriched by >600‰ in ^{15}N, presumably as a result of fractional loss of nitrogen, e.g. Yung & DeMore 1999), carbonaceous chondrites and notably enstatite chondrites, which constitute a building material of reference as their oxygen isotopic compositions lie on the terrestrial fractionation line (Javoy *et al.* 1986). Recently, Hashizume *et al.* (2000) proposed that solar wind nitrogen is enriched in ^{15}N relative to the terrestrial atmosphere ($\delta^{15}N \leq 240$‰) based on ion probe analysis of lunar soil grains, and Owen *et al.* (2001) proposed a solar N composition of $\delta^{15}N = -370 \pm 80$‰ from Galileo data for the atmosphere of Jupiter. Thus it is conceivable that nitrogen in the Solar System is the result of mixing between gaseous nitrogen of the PSN and a ^{15}N-rich component carried by solid phases and mostly present in planetary bodies. The incorporation of solar nitrogen would have resulted in ^{15}N-poor nitrogen, which is not seen at the Earth's surface although a remnant of it could be still present in the deep Earth (although the N/Ne ratio of the mantle does not support this possibility). Interestingly, mid-ocean ridge lavas are slightly depleted in ^{15}N by 5‰ relative to atmospheric N. There are, however, other possibilities such as enstatite-type nitrogen, which is depleted by 20–40‰ in ^{15}N relative to atmospheric N (Javoy *et al.* 1986).

Table 1. *Elemental composition of comets*

	Dust (mol g^{-1})	Gas (mol g^{-1})	Bulk (mol g^{-1})
H	0.050 ± 0.010	0.081 ± 0.017	0.067 ± 0.010
C	0.020 ± 0.003	0.009 ± 0.002	0.014 ± 0.002
N	0.0010 ± 0.0003	0.0045 ± 0.0022	0.0029 ± 0.0012
O	0.022 ± 0.002	0.045 ± 0.002	0.034 ± 0.002
Na	0.00025 ± 0.00015	–	0.00012 ± 0.00007
Mg	0.0025 ± 0.0002	–	0.0012 ± 0.0002
Al	0.00017 ± 0.00004	–	0.000079 ± 0.000023
Si	0.0046 ± 0.0005	–	0.0021 ± 0.0004
S	0.0018 ± 0.0006	0.00045 ± 0.00022	0.0011 ± 0.0003
K	0.0000050 ± 0.0000025	–	0.0000023 ± 0.0000012
Ca	0.00016 ± 0.00005	–	0.000073 ± 0.000025
Ti	0.000010 ± 0.000005	–	0.000005 ± 0.0000025
Cr	0.000022 ± 0.000005	–	0.000010 ± 0.000003
Mn	0.000012 ± 0.000005	–	0.0000061 ± 0.0000026
Fe	0.0013 ± 0.0002	–	0.00060 ± 0.00014
Co	0.0000074 ± 0.0000050	–	0.0000034 ± 0.0000024
Ni	0.00010 ± 0.00005	–	0.00005 ± 0.000025

Dust composition is from Jessberger *et al.* (1988), gas composition is from Delsemme (1988). Bulk abundances were calculated assuming that the dust/gas ratio of typical comets is within 0.5–1.3 (Delsemme 1988). Uncertainties were propagated accordingly.

The late bombardment of the Earth

The term late generally refers to addition of material to the growing Earth once its differentiation was completed. It does not quantify any time interval but the term late is used because such addition is thought to have occurred once the Earth had reached a 'significant' size allowing differentiation and internal evolution. Hence material added after the major building period was not in chemical equilibrium with the bulk Earth. Planets formed from the accretion of planetary material, and this process is still going on at present although at a much lower rate than 4.56 Ga ago. The density of lunar craters increases with the age of the exposed surface in a manner indicative of an overall decline with time of the bombardment intensity (e.g. Chyba 1990; Heiken et al. 1991). It is interesting to note that this idea was first stated based on the superposition principle cherished by field geologists 7 years before the first lunar samples were collected and brought back to Earth (Shoemacker & Hackman 1962). It is estimated from the lunar impact record that the mass of asteroids and comets that fell during late bombardment on our planet is within 1×10^{21} to 7×10^{23} kg, with a preferred value of 3.2×10^{21} kg as derived from fitting of the cumulative crater density v. age curve (Chyba 1990).

Additional evidence that remnants of planetary formation struck the Earth at a much higher rate in the Hadean Eon than at present comes from highly siderophile elements (Ru, Rh, Pd, Re, Os, Ir, Pt and Au). When the core segregated from the mantle, highly siderophile element should have been partitioned into the core, leaving the mantle depleted and fractionated. The concentrations of highly siderophile elements in the mantle are much higher that those predicted in the case of equilibrium partitioning between metal and silicate. In addition, the highly siderophile element abundance pattern of the mantle is almost unfractionated relative to potential Earth-forming material (e.g. Righter & Drake 1997). A straightforward interpretation is that a late veneer brought highly siderophile elements into the mantle after the core formed. It has been suggested that highly siderophile elements were not in chondritic proportion in the Earth's mantle and were carried by non-chondritic extraterrestrial material or by the entrainment of core material by plumes. Alard et al. (2000) measured highly siderophile elements in mantle sulphides and concluded that non-chondritic highly siderophile element abundance patterns directly reflect processes occurring in the upper mantle and are not evidence for the addition of core material or exotic meteoritic components. The mass of extraterrestrial matter necessary to account for the budget of siderophile elements in the mantle is $(1-4) \times 10^{22}$ kg (Chyba 1991).

Several observations suggest that our planet was struck by volatile-rich bodies. Carbonaceous asteroids are overwhelmingly the most abundant type in the main belt. Whether pieces of C-type asteroids are present in meteorite collections is debatable but it seems that their reflectance spectra best match that of the carbonaceous chondrites, specifically of the CI-CM types (Gaffey et al. 1993). The extraterrestrial flux to Earth is dominated by carbonaceous micrometeorites (Engrand & Maurette 1998). They lack a strict equivalent among macroscopic samples but their closest analogues are CM carbonaceous chondrites (Kurat et al. 1994). Most xenoliths in meteorite regolithic breccias share affinities with carbonaceous chondrites, which implies that asteroids were embedded in a swarm of carbonaceous material (Anders 1978). The lunar regolith is enriched in trace elements relative to indigenous lunar rocks. The trace element pattern of this enriched compound best matches that of carbonaceous chondrites, implying in turn that the lunar regolith contains 1-2% of carbonaceous debris (Keays et al. 1970).

When an asteroid or a comet strikes the Earth, part of it and of Earth's atmosphere might be ejected back into space. Recent analytical and computational modelling of impact induced erosion led Newman et al. (1999) to conclude that impact events would not remove significant atmospheric gases. As discussed above, the isotopic ratio of hydrogen supports the contribution of carbonaceous chondrites for the origin of water on Earth. Recent observations indicate that comets are enriched in deuterium relative to the oceans by a factor of two, which prevents the possibility that comets contributed a significant fraction of Earth's hydrosphere (<10%) unless one calls on hypothetical comets. It is noteworthy that all the three comets so far measured originate in the Oort cloud, and the scientific community is in need of the determination of the D/H ratio in Kuiper belt objects to ascertain the idea that comets did not significantly contribute to Earth's oceans. Delsemme (1988) calculated the composition of comets and we give an updated version of this compilation where uncertainties were propagated rigorously and the dust composition was modified according to Jessberger et al. (1988). The D/H ratio of the deep mantle is much lower than that of Earth's surface (Deloule et al. 1991). An appealing possibility is that the D/H ratio of the deep mantle is a remnant of the hydrogen isotopic composition of

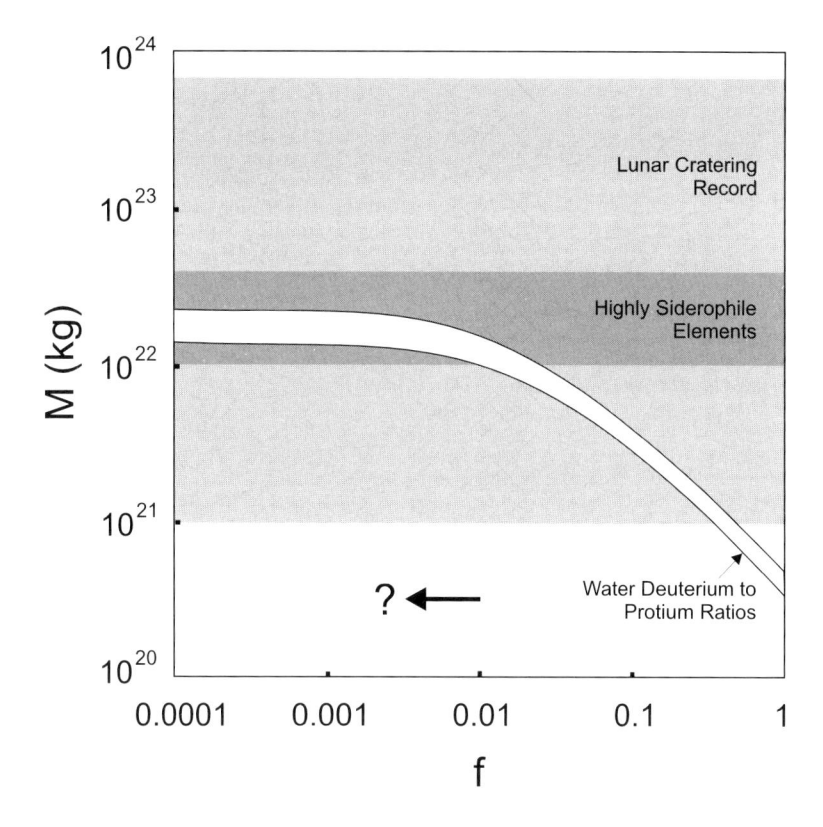

Fig. 5. The mass (M) of asteroids and comets that fell on Earth is reported as a function of the mass fraction (f) of comets among impacting bodies. Lunar cratering record (Chyba 1990), highly siderophile elements (Chyba 1991) and water deuterium to protium ratios (Dauphas *et al.* 2000) are shown. If these three approaches record the late bombardment over the same period of time, then M is $c.\,2 \times 10^{22}$ kg and $f \leq 0.01$. Actually, the mass of asteroids and comets evaluated by the lunar cratering record and highly siderophile elements depends on the mass fraction of comets. Given the level of uncertainty on all these estimates (M and f are reported in a log–log plot), taking this effect into account would not affect our conclusions.

Earth-forming planetesimals, which later evolved as a result of the late accretion of asteroids and comets. Thus, the mass of asteroids and comets incident on Earth since the time of its accretion required to balance the low D/H ratio of the deep mantle is estimated to be 4×10^{20} to 2×10^{22} kg (Dauphas *et al.* 2000). It is worth while to note that fractionation of D/H within the Earth might explain the isotopic heterogeneity of Earth, but we wish to emphasize that the few elements for which there exists an isotopic heterogeneity are those that are not efficiently recycled, those that are prone to record the heterogeneous accretion of the Earth.

It is worth while to note that the estimates based on the lunar cratering record, highly siderophile elements and water deuterium to protium (D/H) ratios integrate the late bombardment of Earth over time scales that are not necessarily identical. In the case of the lunar cratering record, time zero corresponds to the moment when the Moon formed. In the case of highly siderophile elements, time zero corresponds to the moment when the Earth's core formed. In the case of water deuterium to protium ratios, time zero corresponds to the moment when the bulk Earth had a D/H ratio lower than that of the present deep mantle. One might see the remarkably good agreement between the various approaches as merely coincidental. Alternatively, it means that the Moon, highly siderophile elements and water D/H ratios recorded the late heavy bombardment over the same period of time. If so, the mass of extraterrestrial matter that fell on Earth throughout its history must be the same from all point of views. The only locus where the mass of extraterrestrial matter incident on Earth is the same from all point of views is for low mass fractions of comets among impacting bodies (<1%). This inference is highly speculative because it relies on the assumption that all three approaches integrated the late bombardment of Earth over the same period of time, but if it was proven to be right it would have strong implications for cometary dynamics and delivery of organic compounds to Earth.

It was suggested, based on water D/H ratios, that the late veneer consisted predominantly of carbonaceous asteroids and that the mass of extraterrestrial bodies incident on Earth since the time of its accretion was $c.\ 2 \times 10^{22}$ kg. The carbon and nitrogen concentrations of carbonaceous chondrites are 1.5×10^{-3} and 4.2×10^{-5} mol g^{-1}, respectively. If a late veneer supplied Earth with the water required to balance the low D/H of the deep mantle, then it should have contributed to Earth $c.\ 3 \times 10^{22}$ and 8×10^{20} mol of carbon and nitrogen, respectively. The Earth's surface (atmosphere + sediments + igneous crust) has 9×10^{21} and 3.5×10^{20} moles of carbon and nitrogen, respectively. These estimates are in surprisingly good agreement with the inferred contributions of late bombardment for carbon and nitrogen (3×10^{22} mol and 8×10^{20} mol, respectively) if one considers that a significant fraction of carbon and possibly nitrogen has been recycled into the mantle through time. It implies that carbonaceous asteroids might have contributed a significant fraction of hydrogen, carbon and nitrogen on Earth, presumably in the form of prebiotic molecules such as amino acids and heterocycles.

Conclusions

The major conclusion of this work is that there exists a chondritic-like component for volatile elements in the terrestrial mantle, not only for elements presenting a chemical affinity for silicates or metal such as C and N but also for chemically inert gases. A chondritic, rather than solar, origin for volatile elements is consistent with the stable isotope composition of H, C and N and eliminates in the case of H a major contribution from comets. However, a major fraction of He and Ne may be derived from the protosolar nebula and this contribution can only be seen for these two elements because of their high abundance in the PSN. Despite their chondritic chracteristics, the absolute abundances of volatiles in the mantles are three orders of magnitude lower than those of chondrites, as are also strongly siderophile elements. It may well be possible that the growing Earth did not fully equilibrate with the forming proto-atmosphere and the core, and preserved minor (10^{-3}) fractions undifferentiated and later incorporated in the convection of the mantle.

This study was funded by the INSU:CNRS programme 'Interieur de la Terre'. We thank the organizers of the Fermor meeting for the opportunity to present and discuss this work, and numerous colleagues for fruitful discussions, among them F. Albarède, A. Morbidelli, M. Ozima, F. Robert, F. Selsis, I. Tolstikhin and K. Zahlne. The paper benefited from constructive comments by two anonymous reviewers. This paper is CRPG Contribution 1526.

References

ABE, Y. & MATSUI, T. 1986. Early evolution of the Earth: accretion, atmosphere formation, and thermal history. *Proceedings of The 17th Lunar and Planetary Science Conference. Journal of Geophysical Research*, **91**, Supplement, E291–E302.

ALARD, O., GRIFFIN, W. L., LORAND, J. P., JACKSON, S. E. & O'REILLY, S. Y. 2000. Non-chondritic distribution of the siderophile elements in mantle sulphides. *Nature*, **407**, 891–894.

ALEXANDER, E. C. & SCHWARTZMAN, D. W. 1976. Argon isotopic evolution of upper mantle. *Nature*, **259**, 104–108.

ALLARD, P. 1992. Global emissions of helium-3 by subaerial volcanism. *Geophysical Research Letters*, **19**, 1478–1481.

ALLÈGRE, C. J., MANHÈS, G. & GÖPEL, C. 1995. The age of the Earth. *Geochimica et Cosmochimica Acta*, **59**, 1445–1457.

ALLÈGRE, C. J., STAUDACHER, T. & SARDA, P. 1987. Rare gas systematics: formation of the atmosphere, evolution and structure of the Earth's mantle. *Earth and Planetary Science Letters*, **81**, 127–150.

ALLÈGRE, C. J., STAUDACHER, T., SARDA, P. & KURZ, M. 1983. Constraints on evolution of Earth's mantle from rare gas systematics. *Nature*, **303**, 762–766.

ANDERS, E. 1978. Most stony meteorites come from the asteroid belt. *In*: WELLS, D. M. W. C. (ed.) *Asteroids: an Exploration Assessment*. NASA, Houston, Texas, 57–75.

ANDERS, E. & GREVESSE, N. 1989. Abundances of the elements: meteoritic and solar. *Geochimica et Cosmochimica Acta*, **53**, 197–214.

BERNATOWICZ, T. J., KENNEDY, B. M. & PODOSEK, F. A. 1985. Xe in glacial ice and the atmospheric inventory of noble gases. *Geochimica et Cosmochimica Acta*, **49**, 2561–2564.

BROWN, H. 1952. *The Atmospheres of the Earth and Planets*. University of Chicago Press, Chicago, IL.

CAFFEE, M. W., HUDSON, G. B., VELSKO, C., HUSS, G., ALEXANDER, E. C. JR & CHIVAS, A. R. 1999. Primordial noble gases from the Earth's mantle: identification of a primitive volatile component. *Science*, **285**, 2115–2118.

CARTIGNY, P., BOYD, S. R., HARRIS, J. W. & JAVOY, M. 1997. Nitrogen isotopes in peridotitic diamonds from Fuxian, China: the mantle signature. *Terra Nova*, **9**, 175–179.

CHAMBERS, J. E. & WETHERHILL, G. W. 1998. Making the terrestrial planets: N-body integrations of planetary embryos in three dimensions. *Icarus*, **136**, 304–327.

CHYBA, C. 1990. Impact delivery and erosion of planetary oceans in the inner solar system. *Nature*, **343**, 129–133.

CHYBA, C. F. 1991. Terrestrial mantle siderophiles and the lunar impact record. *Icarus*, **92**, 217–233.

CRAIG, H., CLARKE, W. B. & BEG, M. A. 1975. Excess ^{3}He in deep water on the East Pacific Rise. *Earth and Planetary Science Letters*, **26**, 125–132.

DAUPHAS, N. & MARTY, B. 1999. Heavy nitrogen in carbonatites of the Kola Peninsula: a possible signature of the deep mantle. *Science*, **286**, 2488–2490.

DAUPHAS, N., ROBERT, F. & MARTY, B. 2000. The late asteroidal and cometary bombardment of the Earth as recorded in water deuterium to protium ratio. *Icarus*, **148**, 508–512.

DELOULE, E., ALBARÈDE, F. & SHEPPARD, S. M. F. 1991. Hydrogen isotope heterogeneities in the mantle from ion probe analysis of amphiboles from ultramafic rocks. *Earth and Planetary Science Letters*, **105**, 543–553.

DELSEMME, A. H. 1988. The chemistry of comets. *Philosophical Transactions of the Royal Society of London, Series A*, **325**, 509–523.

ENGRAND, C. & MAURETTE, M. 1998. Carbonaceous micrometeorites from Antarctica. *Meteoritics and Planetary Science*, **33**, 565–580.

GAFFEY, M. J., BURBINE, T. H. & BINZEL, R. P. 1993. Asteroid spectroscopy: prospects and perspectives. *Meteoritics*, **28**, 161–187.

GEISS, J. & GLOECKER, G. 1998. Abundances of deuterium and helium in the protosolar cloud. *Space Science Reviews*, **84**, 239–250.

HASHIZUME, K., CHAUSSIDON, M., MARTY, B. & ROBERT, F. 2000. Solar wind record on the Moon: deciphering presolar from planetary nitrogen. *Science*, **290**, 1142–1145.

HEIKEN, G. H., VANIMAN, D. T. & FRENCH, B. V. 1991. *Lunar Sourcebook*. Cambridge University Press, Cambridge.

HIYAGON, H., OZIMA, M., MARTY, B., ZASHU, S. & SAKAI, H. 1992. Noble gases in submarine glasses from mid-oceanic ridge and Loihi Seamount: constraints on early history of the Earth. *Geochimica et Cosmochimica Acta*, **56**, 1301–1316.

HONDA, M. & MCDOUGALL, I. 1997. Primordial helium and neon in the Earth: a speculation on early degassing. *Seventh Annual Goldschmidt Conference*. Lunar and Planetary Institute Contribution, **921**, 98–99.

HONDA, M., MCDOUGALL, I., PATTERSON, D. B., DOULGERIS, A. & CLAGUE, D. A. 1991. Possible solar noble-gas component in Hawaiian basalts. *Nature*, **349**, 149–151.

HONDA, M., MCDOUGALL, I., PATTERSON, D. B., DOULGERIS, A. & CLAGUE, D. A. 1993a. Noble gases in submarine pillow basalt glasses from Loihi and Kilauea, Hawaii: a solar component in the Earth. *Geochimica et Cosmochimica Acta*, **57**, 859–874.

HONDA, M., PATTERSON, D. B., MCDOUGALL, I. & FALLOON, T. J. 1993b. Noble gases in submarine pillow basalt glasses from the Lau Basin: detection of a solar component in back-arc basalts. *Earth and Planetary Science Letters*, **120**, 135–148.

HUNTEN, D. M., PEPIN, R. O. & WALKER, J. C. B. 1987. Mass fractionation in hydrodynamic escape. *Icarus*, **69**, 532–549.

IGARASHI, G. 1986. Components of xenon in carbonaceous chondrites and fission component in the terrestrial atmosphere. *Japan–US Seminar on Terrestrial Noble Gases, Abstracts*, 20–23.

JACOBSEN, S. B. & HARPER, C. L. J. 1996 Accretion and early differentiation history of the Earth based on extinct radionuclides. *In*: BASU, S. H. A. (ed.) *Earth Processes: Reading the Isotopic Code*. Geophysical Monograph, American Geophysical Union, **95**, 47–74.

JAMBON, A., WEBER, H. & BRAUN, O. 1986. Solubilities of He, Ne, Ar, Kr and Xe in a basalts melt in the range 1250–1600°C: geochemical implications. *Geochimica et Cosmochimica Acta*, **50**, 401–408.

JAVOY, M., PINEAU, F. & DELORME, H. 1986. Carbon and nitrogen isotopes in the mantle. *Chemical Geology*, **57**, 41–62.

JESSBERGER, E. K., CHRISTOFORIDIS, A. & KISSEL, J. 1988. Aspects of the major element composition of Halley's dust. *Nature*, **332**, 691–695.

KEAYS, R. R., GANAPATHY, R., LAUL, J. C., ANDERS, E., HERZOG, G. F. & JEFFERY, P. M. 1970. Trace elements and radioactivity in lunar rocks: implications for meteorite infall, solar-wind flux, and formation conditions of Moon. *Science*, **167**, 490–493.

KERRIDGE, J. F. & SWINDLE, T. D. 1988. *Meteorites and the Early Solar System*. University of Arizona Press, Tucson.

KUNZ, J., STAUDACHER, T. & ALLÈGRE, C. J. 1998. Plutonium-fission xenon found in the Earth's mantle. *Science*, **280**, 877–880.

KURAT, G., KOEBERL, C., PRESPER, T., BRANDSTÄTTER, F. & MAURETTE, M. 1994. Petrology and geochemistry of Antarctic micrometeorites. *Geochimica et Cosmochimica Acta*, **58**, 3879–3904.

LEE, D. C. & HALLIDAY, A. (1995) Hafnium–tungsten chronometry and the timing of terrestrial core formation. *Nature*, **378**, 771–774.

LE GUERN, F. 1982. Les débits de CO_2 et de SO_2 volcaniques dans l'atmosphère. *Bulletin of Volcanology*, **45**, 197–202.

MARTY, B. 1989. Neon and xenon isotopes in MORB: implications for the earth–atmosphere evolution. *Earth and Planetary Science Letters*, **94**, 45–56.

MARTY, B. 1992. *Volcanic Fluxes of Volatiles. Preliminary Estimates Based on Rare Gas and Major Volatile Calibration*. International Atomic Energy Agency, Vienna.

MARTY, B. 1995. Nitrogen content of the mantle inferred from N_2–Ar correlation in oceanic basalts. *Nature*, **377**, 326–329.

MARTY, B. & HUMBERT, F. 1997. Nitrogen and argon isotopes in oceanic basalts. *Earth and Planetary Science Letters*, **152**, 101–112.

MARTY, B. & JAMBON, A. 1987. C/^{3}He in volatile fluxes from the solid Earth: implications for carbon geodynamics. *Earth and Planetary Science Letters*, **83**, 16–26.

MARTY, B. & TOLSTIKHIN, I. N. 1998. CO$_2$ fluxes from mid-ocean ridges, arcs and plumes. *Chemical Geology*, **145**, 233–248.

MARTY, B. & ZIMMERMANN, J. L. 1999. Volatiles (He, C, N, Ar) in mid-ocean ridge basalts: assessment of shallow-level fractionation and characterization of source composition. *Geochimica et Cosmochimica Acta*, **63**, 3619–3633.

MARTY, B., TOLSTIKHIN, I. N., KAMENSKY, I. L., NIVIN, V., BALAGANSKAYA, E. & ZIMMERMANN, J. L. 1998. Plume-derived rare gases in 380 Ma carbonatites from the Kola region (Russia) and the argon isotopic composition in the deep mantle. *Earth and Planetary Science Letters*, **164**, 179–192.

MATSUO, S., SUZUKI, M. & MIZUTANI, Y. 1978. Nitrogen to argon ratio in volcanic gases. *In*: ALEXANDER, E. C. J. & OZIMA, M. (eds) *Terrestrial Rare Gases*. Japan Science Society Press, Tokyo, 17–25.

MAZOR, E., HEYMANN, D. & ANDERS, E. 1970. Noble gases in carbonaceous chondrites. *Geochimica et Cosmochimica Acta*, **34**, 781–824.

MELOSH, H. J. & VICKERY, A. M. 1989. Impact erosion of the primordial atmosphere of Mars. *Nature*, **338**, 487–489.

MOJZSIS, S. J., HARRISSON, T. M. & PIDGEON, R. T. 2001. Oxygen-isotope evidence from ancient zircons for liquid water at the Earth's surface 4300 Myr ago. *Nature*, **409**, 178–181.

MONTMERLE, T. (1999) La formation des étoiles de type solaire; des conditions aux limites pour l'origine de la vie? *In*: GARGAUD, M., DESPOIS, D. & PARISOT, J. P. (eds) *L'Environnement de la Terre Primitive et l'Origine de la Vie*. Presses Universitaires de Bordeaux, Bordeaux, **CI-1**, 31–52.

MOREIRA, M., KUNZ, J. & ALLÈGRE, C. J. 1998. Rare gas systematics in Popping Rock: isotopic and elemental compositions in the upper mantle. *Science*, **279**, 1178–1181.

NEWMAN, W. I., SYMBALISTY, E. M. D., AHRENS, T. J. & JONES, E. M. 1999. Impact erosion of planetary atmospheres: some surprising results. *Icarus*, **138**, 224–240.

OWEN, T., MAHAFFY, P. R., NIEMANN, H. B., ATREYA, S. & WONG, M. 2001. Protosolar nitrogen. *Astrophysical Journal*, **553**, L77–L79.

OZIMA, M. 1975. Ar isotopes and Earth–atmosphere evolution models. *Geochimica et Cosmochimica Acta*, **39**, 1127–1140.

OZIMA, M. & PODOSEK, F. A. 1983. *Noble Gas Geochemistry*. Cambridge University Press, Cambridge.

OZIMA, M. & ZAHNLE, K. 1993. Mantle degassing and atmospheric evolution: noble gas view. *Geochemical Journal*, **27**, 185–200.

OZIMA, M., PODOSEK, F. A. & IGARASHI, G. 1985. Terrestrial xenon isotope constraints on the early history of the Earth. *Nature*, **315**, 471–474.

OZIMA, M., WIELER, R., MARTY, B. & PODOSEK, F. A. 1998. Comparative studies of solar, Q-gases and terrestrial noble gases, and implications on the evolution of the solar nebula. *Geochimica et Cosmochimica Acta*, **62**, 301–314.

PEPIN, R. O. 1991. On the origin and early evolution of terrestrial planetary atmospheres and meteoritic volatiles. *Icarus*, **92**, 1–79.

PEPIN, R. O. 1997 Evolution of Earth's noble gases: consequences of assuming hydrodynamic loss driven by giant impact. *Icarus*, **126**, 148–156.

PEPIN, R. O. & PHINNEY, D. 1979. Components of xenon in the Solar System, mythical unpublished ms.

PETIT, J. M., MORBIDELLI, A. & VALSECCHI, G. B. 1999. Large scattered planetesimals and the excitation of the small body belts. *Icarus*, **141**, 367–387.

PHINNEY, D., TENNYSON, J. & FRICK, U. 1978. Xenon in the CO$_2$-well gas revisited. *Journal of Geophysical Research*, **83**, 2313–2319.

PODOSEK, F. A., BERNATOWICZ, T. J. & KRAMER, F. E. 1981. Adsorption of xenon and krypton on shales. *Geochimica et Cosmochimica Acta*, **45**, 2401–2415.

REA, D. K. & RUFF, L. J. 1996. Composition and mass flux of sediments entering the world's subduction zones: implications for global sediment budgets, great earthquakes and volcanism. *Earth and Planetary Science Letters*, **140**, 1–12.

RIGHTER, K. & DRAKE, M. J. 1997 Metal/silicate equilibrium in an homogeneous accreting Earth: new results for Re. *Earth and Planetary Science Letters*, **146**, 541–553.

ROBERT, F., GAUTIER, D. & DUBRULLE, B. 2000. The solar system D/H ratio: observations and theories. *Space Science Reviews*, **92**, 201–224.

RUBEY, W. W. 1951. Geologic history of sea water. *Geological Society of America Bulletin*, **62**, 1111–1148.

SANO, Y. & WILLIAMS, S. N. 1996. Fluxes of mantle and subducted carbon along convergent plate boundaries. *Geophysical Research Letters*, **23**, 2749–2752.

SARDA, P., STAUDACHER, T. & ALLÈGRE, C. J. 1985. ^{40}Ar/^{36}Ar in MORB glasses: constraints on atmosphere and mantle evolution. *Earth and Planetary Science Letters*, **72**, 357–375.

SARDA, P., STAUDACHER, T. & ALLÈGRE, C. J. 1988. Neon isotopes in submarine basalts. *Earth and Planetary Science Letters*, **91**, 73–88.

SASAKI, S. & NAKASAWA, K. 1988. Origin of isotopic fractionation of terrestrial Xe, hydrodynamic fractionation during escape of the primordial H$_2$–He atmosphere. *Earth and Planetary Science Letters*, **89**, 323–334.

SHOEMACKER, E. M. & HACKMAN R. J. 1962. Stratigraphic base for a lunar time scale. *In*: MIKHAILOV, Z. K. Z. K. (ed.) *The Moon*. Academic Press, New York, 289–300.

STAUDACHER, T. & ALLÈGRE, C. J. 1982. Terrestrial xenololy. *Earth and Planetary Science Letters*, **60**, 389–405.

STAUDACHER, T. & ALLÈGRE, C. J. 1988. Recycling of oceanic crust and sediments: the noble gas subduction barrier. *Earth and Planetary Science Letters*, **89**, 173–183.

STAUDACHER, T., SARDA, P., RICHARDSON, S. H., ALLÈGRE, C. J., SAGNA, I. & DMITRIEV, L. V. 1989. Noble gases in basalt glasses from a

Mid-Atlantic Ridge topographic high at 14°N: geodynamic consequences. *Earth and Planetary Science Letters*, **96**, 119–133.

STAUDIGEL, H., HART, S. R., SCHMINCKE, H. U. & SMITH, B. M. 1989. Cretaceous ocean crust at DSDP sites 417 and 418: carbon uptake from weathering versus loss by magmatic outgassing. *Geochimica et Csomochimica Acta*, **53**, 3091–3094.

STERN, S. A., SLATER, D. C., FESTOU, M. C., PARKER, J. W., GLASTONE, G. R., HEARN, M. F. A. & WILKINSON, E. 2002. The discovery of argon in comet C/1995 01 (Hale Bopp). *Astrophysical. Journal*, **544**, L169–L172.

SWINDLE, T. D. 1995. How many Martian noble Zgas reservoirs have we sampled? *In*: FARLEY, K. (ed.) *Volatiles in the Solar System*. American Institute of Physics, **341**, 175–185.

TOLSTIKHIN, I. N. & MARTY, B. 1998. The evolution of terrestrial volatiles: a view from helium, neon, argon and nitrogen isotope modelling. *Chemical Geology*, **147**, 27–52.

TRIELOFF, M., KUNZ, J., CLAGUE, D. A., HARRISSON, C. J. & ALLÈGRE, C. J. 2000. The nature of pristine noble gases in mantle plumes. *Science*, **288**, 1036–1038.

VITYAZEV, A. V., PECHERNIKOVA, G. V. & SAFRONOV, V. S. 1990. *Planets of the Earth's Group*. Nauka, Moscow.

VOKROUHLICKY, D. & FARINELLA, P. 2000. Efficient delivery of meteorites to the Earth from a wide range of asteroid parent bodies. *Nature*, **407**, 606–608.

VON WEIZSÄCKER, C. F. 1937. Ber die Muglochkeiteines dua Len ß-Zerfalls von Kalium. *Phys. Zeitschr.*, **38**, 623–624.

WEIDENSCHILLING, S. J., SPAUTE, D., DAVIES, D. R., MARZARI, F. & OHTSUKI, K. 1997. Accretional evolution of a planetary swarm. *Icarus*, **128**, 429–438.

WETHERILL, G. W. 1975. Radiometric chronology of the early solar system. *Annual Review of Nuclear Science*, **25**, 283–328.

WIELER, R., ANDERS, E., BAUR, H., LEWIS, R. S. & SIGNER, P. 1992. Characterization of Q-gases and other noble gas components in the Murchison meteorite. *Geochimica et Cosmochimica Acta*, **56**, 2907–2921.

WILDE, S. A., VALLEY, J. W., PECK, W. H. & GRAHAM, C. M. 2001. Evidence from detrital zircons for the existence of continental crust and oceans on the Earth 4.4 Gyr ago. *Nature*, **409**, 175–178.

YUNG, Y. L. & DEMORE, W. D. 1999. *In*: *Photochemistry of Planetary Atmospheres*. Oxford University Press, Oxford.

ZANHLE, K. J., KASTING, J. F. & POLLACK, J. B. 1988. Evolution of a steam atmosphere during Earth's accretion. *Icarus*, **74**, 62–97.

ZARTMAN, R. E., WASSERBURG, G. J. & REYNOLDS, J. H. 1961. Helium, argon and carbon in some natural gases. *Journal of Geophysical Research*, **66-1**, 277–306.

Carbon dioxide cycling through the mantle and implications for the climate of ancient Earth

KEVIN ZAHNLE[1] & NORMAN H. SLEEP[2]

[1] *NASA Ames Research Center, Mountain View, CA 94035, USA*
(e-mail: kzahnle@mail.arc.nasa.gov)
[2] *Department of Geophysics, Stanford University, Stanford, CA 94305, USA*

Abstract: The continental cycle of silicate weathering and metamorphism dynamically buffers atmospheric CO_2 and climate. Feedback is provided by the temperature dependence of silicate weathering. Here we argue that hydrothermal alteration of oceanic basalts also dynamically buffers CO_2. The oceanic cycle is linked to the mantle via subduction of carbonatized basalts and degassing of CO_2 at the mid-ocean ridges. Feedback is provided by the dependence of carbonatization on the amount of dissolved carbonate in sea water. Unlike the continental cycle, the oceanic cycle has no thermostat. Hence surface temperatures can become very low if CO_2 is the only greenhouse gas apart from water. Currently the continental cycle is more important, but early in Earth's history the oceanic cycle was probably dominant. We argue that CO_2 greenhouses thick enough to defeat the faint early Sun are implausible and that, if no other greenhouse gases are invoked, very cold climates are expected for much of Proterozoic and Archaean time. We echo current fashion and favour biogenic methane as the chief supplement to CO_2. Fast weathering and probable subduction of abundant impact ejecta would have reduced CO_2 levels still further in Hadean time. Despite its name, the Hadean Eon might have been the coldest era in the history of the Earth.

Introduction

Evidence for relatively mild climates on ancient Earth has been a puzzle in light of the faint early Sun (Ringwood 1961; Sagan & Mullen 1972; Kiehl & Dickinson 1987; Pavlov *et al.* 2000). The usual solution has been to posit massive CO_2 atmospheres (Owen *et al.* 1979; Kasting 1993), although reduced gases (e.g. NH_3 or CH_4) have had their partisans (Sagan & Mullen 1972; Pavlov *et al.* 2002). Evidence against siderite in palaeosols has been used to set a rough upper limit of 30 PAL (present atmospheric levels, where one PAL = 300 ppm) on pCO_2 c. 2.75 Ga (Rye *et al.* 1995), which is less than what is needed to defeat the faint Sun (Rye *et al.* 1995). We present here an independent argument, based on CO_2 fluxes into and out of the mantle, that weighs against high pCO_2 on early Earth, especially during the Archaean and Hadean eons.

There are roughly 5×10^{21} moles of carbon in the crust and more in the mantle (e.g. Zhang & Zindler 1993; Wedepohl 1995). The current CO_2 mantle outgassing rate at the mid-ocean ridges, $(1.5-2.5) \times 10^{21}$ mol a^{-1} (Marty & Jambon 1987), is fast enough to double the crustal inventory in 2–3 Ga. Greater heat flow on early Earth implies faster outgassing, probably reducing the doubling time for crustal carbon to < 1 Ga in Archaean time. The large source implies a large sink to close the mantle cycle. Possibilities include subduction of carbon directly deposited on the ocean floor (pelagic carbonates and organic carbon); subduction of carbon that is scraped off continents and dragged down; and subduction of carbonates formed by sea-water alteration of the oceanic basalt itself. The last depends on the amount of free CO_2 in the ocean, and hence can act as a buffer on atmospheric CO_2 levels.

Here we construct a specific albeit extremely simplified model of the modern CO_2 cycle. We then accelerate the cycle into Archaean time. Then we consider the effects of abundant impact ejecta on the Hadean CO_2 cycle. Throughout we neglect the sometimes different behaviour of reduced carbon, as terms involving reduced carbon are relatively small and the complications and uncertainties introduced by modelling of the oxygen cycle relatively large. This precludes our using carbon isotopes to constrain our models. This paper is intended as an abbreviated and updated version of Sleep & Zahnle (2001). The focus of this paper is on a simpler presentation of model results, although we also clarify the relative insensitivity of our results to the sea-floor spreading rate; the earlier paper placed undue emphasis on this factor. We defer to the earlier paper for detailed discussions of the geological data and weathering rates. We also use this opportunity to correct a typographical error

From: FOWLER, C. M. R., EBINGER, C. J. & HAWKESWORTH, C. J. (eds) *The Early Earth: Physical, Chemical and Biological Development.* Geological Society, London, Special Publications, **199**, 231–257. 0305-8719/02/$15.00
© The Geological Society of London 2002.

that eventually propagated through the earlier paper's discussion of the volume and weathering of impact ejecta during Hadean time.

The current carbonate cycle

On geological time scales, CO_2 cycles between rocks, often by way of the ocean and atmosphere. The rock reservoirs include the mantle, continental carbonates, carbon in reduced form mostly in continental shales, and carbon (mostly carbonate) in or on the sea floor. The small volatile reservoir (ocean plus atmosphere) cycles through carbonate rock in a hundred thousand to a million years. Over longer periods free CO_2 is dynamically controlled by processes that form carbonates at low temperatures and processes that decompose carbonates at high temperatures by (Urey) reactions of the form

$$CO_2 + CaSiO_3 \Leftrightarrow SiO_2 + CaCO_3. \quad (1)$$

Magnesium and, under reducing conditions, iron carbonates can form by analogous reactions. The continental part of this cycle has been extensively studied (e.g. Walker 1977; Holland 1978, 1984; Franck et al. 1999). Weathering of rocks on land dissolves carbonates and releases Ca^{2+} from silicates. Both the dissolved carbonate and calcium are carried by rivers to the oceans, where they react to form $CaCO_3$. Reaction (1) is reversed when carbonates are metamorphosed deep in the crust. Carbonate weathering,

$$H_2O + CO_2 + CaCO_3 \Leftrightarrow Ca^{2+} + 2HCO_3^- \quad (2)$$

is usually regarded as having no direct net effect on carbonate reservoirs on time scales longer than a few thousand years, but silicate weathering, by removing free CO_2 from the atmosphere–ocean system, acts to buffer surface temperatures, because the weathering rate increases with temperature, and temperature rises with atmospheric CO_2 (Walker et al. 1981; Lasaga et al. 1985).

We follow Tajika & Matsui (1992) and apportion carbonate between five significant reservoirs: the atmosphere and ocean, R_{oc}; carbonates lying upon (R_{pel}) or veined within (R_{bas}) oceanic basalt; carbonates on continental platforms, R_{con}; and CO_2 in the mantle, R_{man}. The atmosphere and ocean are tightly coupled on geological time scales. We treat these together as a single reservoir with a current size of $c. 3.3 \times 10^{18}$ moles.

The budget for R_{oc} can be written as

$$\frac{\partial R_{oc}}{\partial t} = F_{meta} + F_{ridge} + F_{arc} + F_{CO_3w}$$
$$- F_{dep} - F_{pel} - F_{SiO_3w} - F_{hydro} - F_{ej}. \quad (3)$$

Sources are the metamorphic flux (F_{meta}), outgassing at mid-ocean ridges (F_{ridge}), outgassing associated with arc volcanoes (F_{arc}), and carbonate weathering (F_{CO_3w}). Sinks are carbonate deposition (F_{dep}) on continental platforms, pelagic carbonate deposition on oceanic crust (F_{pel}), silicate weathering on continents (F_{SiO_3w}), carbonatization of oceanic crust in warm hydrothermal systems (F_{hydro}), and carbonatization of impact ejecta (F_{ej}), respectively. The last is restricted to Hadean time.

The carbonate budget for the oceanic crust $R_{bas} + R_{pel}$ can be written as

$$\frac{\partial(R_{bas} + R_{pel})}{\partial t} = F_{pel} + F_{tect} + F_{hydro}$$
$$+ F_{ej} - F_{sub} - F_{scrape}. \quad (4)$$

Pelagic carbonates, tectonic erosion (F_{tect}, carbonate swept up by the oceanic crust just before subduction), and hydrothermal and ejecta carbonatization add to the inventory of carbonate on or in the oceanic crust. Losses are subduction (F_{sub}) and 'off-scraping' (F_{scrape}, carbonate removed from the oceanic crust just before subduction). By presumption, the subduction flux F_{sub} is proportional to what is on or in the sea floor ($R_{pel} + R_{bas}$), and both F_{sub} and F_{scrape} are inversely proportional to the lifetime of the oceanic crust. The relative importance of subduction and off-scraping depends on whether the carbonate sits on or within the oceanic crust. For example, pelagic carbonates will be more easily scraped off than would carbonatized basalt. In addition to these terms, significant amounts of continentally derived sediments and crust tectonically eroded by the slab are also subducted. We have lumped these together as F_{tect} and accounted them separately from F_{sub}. Neither F_{tect} nor F_{scrape} are well constrained. For modelling purposes we make the arbitrary assumption that the tectonic terms are equal and opposite, so that $F_{tect} = F_{scrape}$.

The continental and mantle budgets are then written as

$$\frac{\partial R_{con}}{\partial t} = F_{SiO_3w} + F_{dep} - F_{CO_3w} - F_{meta} \quad (5)$$

and

$$\frac{\partial R_{man}}{\partial t} = C_{deep}F_{sub} - F_{ridge} \quad (6)$$

respectively. Both reservoirs are large. Representative estimates of their current sizes are $R_{con} = 4 \times 10^{21}$ moles and $R_{man} = (11-18) \times 10^{21}$ moles. At present R_{con} is shrinking because of pelagic carbonate deposition; there are roughly $R_{pel} = 1 \times 10^{21}$ moles of pelagic carbon on the sea floor that have been recently removed from the continents (Wilkinson & Walker 1989). A fraction C_{deep} of the total subduction flux F_{sub} reaches the deep mantle, whereas the balance $(1 - C_{deep})F_{sub} = F_{arc}$ is shunted through arc volcanoes. We think that C_{deep} is not small: we estimate that it is currently $c.\,75\%$ (see below), so that only about a quarter of the subducted carbon fails to reach the mantle.

Continents

The largest fluxes in equations (3)–(6) are those associated with carbonate weathering, which consumes and redeposits carbonates at a rate of $c.\,2 \times 10^{13}$ moles a^{-1}. The associated time scales for recycling R_{con} and R_{oc} are 3×10^8 and 2×10^5 a, respectively. It is sometimes argued that, because carbonate weathering liberates both the anion and the cation (e.g. Ca^{2+}), it serves merely to move calcium carbonate from place to place with no net effect on R_{oc} or the overall carbonate budget. For simplicity we will accept this argument. But at present carbonate weathering and pelagic deposition are moving carbonate from the continents to the sea floor, so that $F_{CO_3w} \approx F_{dep} + F_{pel}$. Thus carbonate weathering currently acts as a net sink on continental carbonate, and to the extent that pelagic carbonate is or will be subducted, it will constitute a net loss of crustal carbonate.

Pelagic carbonates accumulate where the $CaCO_3$ shells made by planktonic organisms fail to dissolve before they reach the ocean floor. Currently F_{pel} is rather large. Catubig et al. (1998) estimated that $F_{pel} \approx (8-9) \times 10^{12}$ moles a^{-1}. But such planktonic organisms precipitated insignificant amounts of carbonate before the middle of Mesozoic time (Sibley & Vogel 1976) and little pelagic $CaCO_3$ is now being subducted compared with the amount being deposited on the sea floor (Plank & Langmuir 1998). When extrapolating the CO_2 cycle into the distant past we set F_{pel} to zero and $F_{CO_3w} = F_{dep}$.

The metamorphic and silicate weathering fluxes are somewhat smaller. Combined, the arc volcanic and metamorphic fluxes are estimated at $(6-6.8) \times 10^{12}$ moles a^{-1} (Brantley & Koepenick 1995; Goddéris & François 1995). Most of this is metamorphic. At current rates the continental carbon inventory is recycled over some 600–800 Ma. For modelling purposes it is usually assumed that silicate weathering and metamorphism balance over time scales of hundreds of millions of years. We will begin our analysis with this case, in which fluxes in and out of the mantle are ignored, to establish context for models that include the mantle.

With the above simplifications equation (3) reduces to an implicit equation for the steady-state value of R_{oc}:

$$F_{SiO_3w} + F_{hydro} + F_{ej} = F_{meta} + F_{ridge} + F_{arc}. \quad (7)$$

In equation (7), the fluxes on the left-hand side are all monotonically increasing functions of R_{oc} (or pCO_2), whereas the terms on the right-hand side are independent of R_{oc}.

Silicate weathering is a strong function of temperature T and a weak function of pCO_2. For simplicity we follow Walker et al. (1981) and use

$$F_{SiO_3w} = C_{SiO_3w} pCO_2^\beta \exp[(T - T_0)/B] \quad (8)$$

where $B = 13.7$ and $\beta = 0.3$ are the weathering parameters suggested by Walker et al. (1981), and C_{SiO_3w} and T_0 are calibration constants.

To obtain pCO_2 histories we need to specify histories of metamorphic recycling and solar luminosity, and we need to parameterize the greenhouse effect. We relate the metamorphic flux to the size of the continental reservoir, $F_{meta} = R_{con}/\tau_{meta}$, where τ_{meta} is the time scale over which continental carbonate is metamorphosed. We assume that the rate of carbonate metamorphism is proportional to the global heat flow Q. We take Q to decrease as $(1 - t/4.5)^\mu$ with $\mu = 0.7$ for high heat-flow models and $\mu = 0.2$ for low heat-flow models. Solar luminosity increases as $L = L_0(1 - 0.07t)$. In both these expressions the time t is in Ga. The heat-flow function is arbitrary; the solar luminosity evolution standard. We use Caldeira & Kasting's (1992) parameterization of the CO_2–water vapour greenhouse effect with an albedo of 31% (so as to reproduce current pCO_2 and T_0). The ice–albedo feedback, which predicts that global ice cover would amplify global cooling, is ignored. Studies of palaeosols constrain pCO_2 in the past 400 Ma to between 1 and 10 PAL (Ekart et al. 1999). We calibrate silicate weathering by assuming a long-term average pCO_2 level of 3.0 PAL and $T_0 = 290$ K.

Illustrative temperature and pCO_2 histories obtained solving equation (8) are shown in Figure 1 for two values of β and (for comparison) for constant pCO_2. The Urey buffer working alone predicts clement ancient climates only if silicate weathering is a weak function ($\beta < 0.2$) of pCO_2. Two observations pertinent to any successful CO_2 climate buffer are that (1) it takes a lot of CO_2 to maintain clement climates and (2) the predicted climates are not necessarily particularly clement. They are usually cooler than

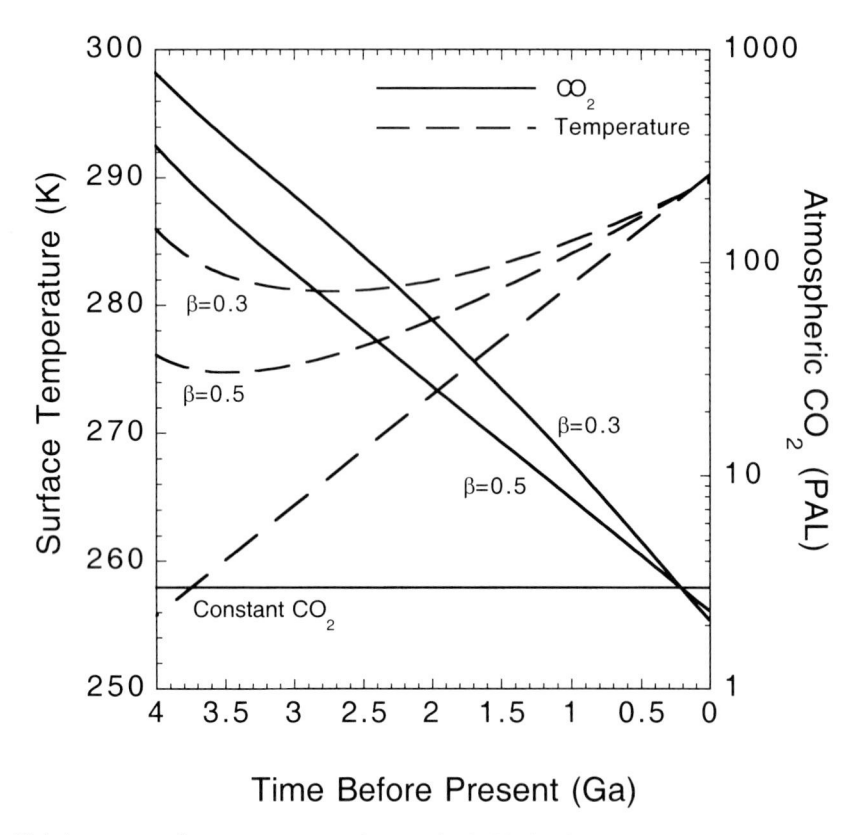

Fig. 1. Global average surface temperatures and atmospheric CO_2 levels predicted by models in which CO_2 is the only greenhouse gas and the temperature dependence of subaerial silicate weathering is the only effective buffer against changing solar luminosity. One PAL indicates one 'present atmospheric level' of 300 ppm. Curves are labelled by the weathering parameter β (equation (8)). These models are inspired by Walker *et al.* (1981). Constant CO_2 is shown for comparison. Unless silicate weathering is nearly independent of pCO_2 (β < 0.2), ancient climates are cool.

the modern climate through most of Precambrian time. An additional effect is required if warmer Precambrian climates are desired. The best explored of these is a hypothesized biotic enhancement of weathering.

Biotic enhancement of weathering means that plants (via roots, respiration and humic acids) increase the effective reaction rate (C_{SiO_3w} in equation (8)) of atmospheric CO_2 with silicates. This can be important if it changes over time. Schwartzman & Volk (1989, 1991) appeared to argue implicitly that C_{SiO_3w} has grown monotonically as the biota became ever better at extracting minerals from the Earth. Their argument is slightly Gaian if it is used to maintain a temperate climate, because it implies that the evolution of biotic weathering has tracked the evolution of the Sun. In later work Schwartzman & McMenamin (1993) went beyond Gaia and argued that biotic weathering has evolved faster, and that despite the brightening Sun, Earth's climate has systematically cooled (the current enhancement in C_{SiO_3w} over that before life would be of order 1000). Schwartzman emphasized the importance of plants stabilizing soils as a means of speeding chemical erosion. This can be questioned. The soils serve not just as micro-environments that might promote chemical weathering but also as armour that can protect the underlying rock. It is not obvious which effect is, on global average, the more important.

Berner (1994, 1997) also emphasized biotic weathering. In his model the advent of vascular land plants and their root systems leads to a step function change in the biotic weathering enhancement beginning in Devonian time. Berner (1994) used more complicated expressions for silicate weathering than we do. These explicitly include topography, runoff, vegetation and climate feedback. When cast in the form of equation (8), Berner's choices are equivalent to taking $B = 11.1$ and $\beta = 0.5$, the latter pertinent before the advent of vascular land plants. Beginning with Devonian time, Berner regarded silicate weathering as dependent on elevated concentrations of CO_2 in the soil: for a given pCO_2, weathering goes some 3–10 times faster with roots than without roots (Berner 1997), but is nearly independent of pCO_2 (roughly equivalent to $\beta \approx 0.14$). Berner relied heavily on data obtained from weathering of recent tillites and recent volcanic rocks. Whether his arguments are appropriate for old soils such as tropical laterites is an open question. To quantify the possible effect of plant roots, we compute 'roots' models (the dotted curves in Fig. 2). In these models we reduce the weathering rate before the advent of roots to 30% of the current rate. For consistency with Berner we use $\beta = 0.5$. The rooted models are especially warm at the end of Precambrian time (see Berner 1997), which may make them a poor match to the climate of the time.

Another factor that can influence the weathering rate is the subaerial extent of the continents. Schwartzman included continental area as a multiplicative factor in his expression for the weathering rate. We think this factor has been overstated. It has been argued that continents grew slowly and only reached their present mass at c. 2 Ga. This view is not universal, and if anything the arguments for constant continental volume seem currently the stronger (Bowring & Housh 1995; Wilde et al. 2001; Mojzcis et al. 2001). Yet if we suppose the continents to have been much smaller in Archaean time, we need also suppose that the continental volcanic and metamorphic CO_2 fluxes were proportionately smaller in Archaean time. The volcanic source cannot simply be held constant while the weatherable surface is shrunk to insignificance.

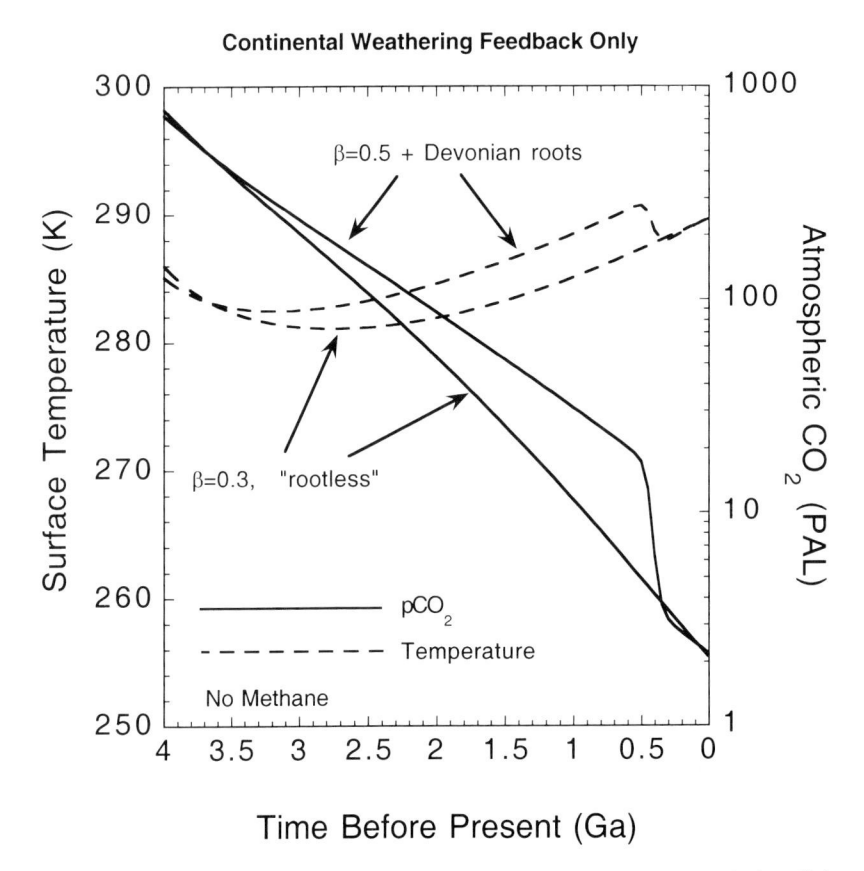

Fig. 2. Comparison of continent-only models with and without an abrupt change in weathering efficiency caused by the advent of vascular land plants in Devonian time. The 'Devonian roots' model uses parameters that follow Berner (1994) as much as possible.

Shrinking the continents reduces their importance as a source as well as a sink of CO_2 vis à vis the sea floor.

The fainter Archaean Sun presents a greater challenge. A warm Archaean climate demands 100–1000 PAL CO_2. Such high CO_2 levels appear to conflict with the absence of siderite in 2.75 Ga palaeosols (Rye et al. 1995; see Figure 7 below). They also imply that ancient oceans were much depleted in Ca^{2+}. The resulting CO_2-rich, Ca-poor sea water would have been much more effective than modern waters at leaching Ca from basalts and much more prone to carbonatize basalts in situ. Extreme Ca depletion of ancient seawaters would seem to be inconsistent with evidence for evaporitic gypsum (Buick & Dunlop 1990).

The mantle

As discussed in the Introduction, CO_2 exchange with the mantle becomes important on Earth's longest time scales. We consider in detail two sources (arc volcanoes and the mid-ocean ridge) and one sink (subduction of carbonitized basalts).

Arc volcanism

Arc volcanism and subduction are related, with arc volcanoes tapping subducted carbonate. We (Sleep & Zahnle 2001) estimated that $F_{arc} = 1.2 \times 10^{12}$ moles a^{-1}. Our estimate is revised downwards from that by Sano & Williams (1996): we assumed an arc volcanism rate of 2 $km^3 a^{-1}$ that is consistent with estimates by Plank & Langmuir (1998), whereas Sano & Williams used 5 $km^3 a^{-1}$. Our estimate for F_{arc} is at the low end of the range estimated by Marty & Tolstikhin (1998); the difference appears to stem from their presuming higher rates of arc volcanism (8 $km^3 a^{-1}$) than we. F_{arc} is a significant fraction of the crustal metamorphic flux ($(6-6.8) \times 10^{12}$ moles a^{-1}) used in other models of the global CO_2 cycle (Brantley & Koepenick 1995; Goddéris & François 1995). Currently, about 75% of the carbon in F_{arc} comes from subducted carbonate, 1/8 from organic matter in subducted sediments, and 1/8 from normal mantle. These fractions imply that some hemipelagic sediments and tectonically eroded crust, which have organic matter as well as pelagic sediments, are subducted and then degassed. The fraction of CO_2 derived from carbonate does not correlate with locations where large amounts of sedimentary carbonate are being subducted (Plank & Langmuir 1998). This indicates that significant amounts of carbonate are within altered basalt as we discuss below. In a detailed study of Kudryavy volcano on the Kurile Islands, Fischer et al. (1998) found that 1/6 of the subducted carbonate and 1/3 of the subducted organic carbon are returned to the surface in arc magmas. We ignore the true mantle source at arc volcanoes as small compared with uncertainties in F_{ridge}.

To calibrate C_{deep}, we take the current flux of pelagic carbonate into subduction zones to be 1.4×10^{12} moles a^{-1}. This is the global extrapolation of the flux reported by Plank & Langmuir (1998). It is currently part of F_{sub}. We take $F'_{sub} = 3.4 \times 10^{12}$ mole a^{-1} as the flux of carbonatized oceanic crust currently entering subduction zones. The latter is Alt & Teagle's (1999) estimate based on carbonate in boreholes in oceanic crust. It is similar to Zhang & Zindler's (1993) estimate of 4×10^{12} moles a^{-1}, which is based on measurements of CO_2 (Staudigel et al. 1989, 1996) in older crust formed at fast ridges in the Pacific that is about to be subducted. Leaving subduction zones we have $F_{arc} = 1.2 \times 10^{12}$ moles a^{-1}. As discussed above, we assume $F_{tect} = F_{scrape}$. This leaves $C_{deep} = 75\%$, consistent with the direct estimate from a Kurile arc volcano of 64–84% (Fischer et al. 1998). High values of C_{deep} are consistent with the thermodynamic stability of carbonates under experimental conditions that simulate those encountered during subduction (Kerrick & Connolly 2001). These experiments indicate that carbonates should mostly survive subduction under most regimes, present or past. In any case, low values of C_{deep} are problematic, as they would imply that CO_2 cannot return to the mantle. The problem would be worse in the past when the mantle was warmer and C_{deep} smaller. If C_{deep} becomes too small then CO_2 is stranded at the surface. When the amount of CO_2 at the surface exceeds the available pool of accessible cations the CO_2 builds up in the atmosphere and the climate becomes insufferably hot. These matters have been discussed in more detail elsewhere (Sleep & Zahnle 2001; Sleep et al. 2001).

Ridge outgassing

The oceanic crust is a source of CO_2 when magma upwelling at mid-ocean ridges degasses and later is a sink when circulating sea water reacts with the sea floor at low temperatures. For simplicity, we assume that at first the basalt degasses completely and later, when the basalt is cool, CO_2 is added to it by reaction with sea water. In reality, some igneous CO_2 is retained by the basalt especially within the dyke complex. Various workers following Alt et al. (1986) have reported net gain rather than total CO_2 in their flux estimates.

The degassing flux is proportional to the global rate of crustal production $\partial A/\partial t$, the degassing depth D_s, and the concentration of CO_2 in the mantle:

$$F_{ridge} = \frac{\partial A}{\partial t} \frac{\rho_s}{\rho_m} \left(\frac{R_{man}}{V_{man}}\right) D_s \qquad (9)$$

where ρ_s is the density of the degassed mantle (3.3 g cm^{-3}), ρ_m is the average mantle density (4.5 g cm^{-3}), and V_{man} is the volume of the mantle (8.9×10^{11} km^3). The current source depth D_s is well constrained independently of volatiles at about 56 km (Langmuir et al. 1992). It is a scale depth for degassing that should not be taken literally as the deepest level where any melt can form. We ignore the undegassed mantle as irrelevant because by definition there is virtually no traffic between it and the surface. If an undegassed mantle were included both R_{man} and V_{man} would be reduced but their ratio unchanged.

The global CO_2 flux is obtained from the CO_2/^{3}He ratio in quenched bubbles, quenched glasses or hot hydrothermal fluids, and the known total flux of ^{3}He to the oceans (DesMarais & Moore 1984). Reliable estimates of the ridge flux F_{ridge} are in the range of $(1.5-2.5) \times 10^{12}$ moles a^{-1} (Marty & Jambon 1987; Gerlach 1991; Zhang & Zindler 1993; Goddéris & François 1995; Marty & Tolstikhin 1998). These imply a present mantle reservoir R_{man} between 11×10^{21} and 18×10^{21} moles. Reliable estimates for the flux from plume-derived hotspot magmas are not available. Marty & Tolstikhin (1998) gave an upper limit of 3×10^{12} mole a^{-1} and a negligible lower limit. We do not consider hotspots separately.

Hydrothermal carbonatization

The oceanic crust is a sink for CO_2 by reaction (1). At low temperatures, circulating sea water reacts with the sea floor to form carbonates. Carbonate veins and disseminated carbonate formed in this way have been sampled from the small number of holes that have been drilled deeply into the oceanic crust (Alt & Teagle 1999). The irregular distribution of carbonate within holes and variations between closely spaced holes hinders extrapolation to the global cycle. An indirect approach using geochemistry and heat flow is more practical.

The amount of carbonate added to the oceanic crust can be constrained by chemical analyses of hydrothermal fluids and by using the fluxes of heat and various elements to calibrate the total volume of flow (e.g. François & Walker 1992; Alt 1995; Caldeira 1995; Kadko et al. 1995; Stein et al. 1995; Brady & Gíslason 1997; Sansone et al. 1998). For this purpose, we follow Schultz & Elderfield (1997, 1999) and partition hydrothermal circulation into three regimes: high temperature, c. 350°C, axial flow; warm, c. 20°C, near-axial flow which extensively reacts with the rock; and cool, c. 5°C, off-axial flow through the shallowest ocean crust. The high-temperature flow moves CO_2 from the hot basalt into the ocean. The cold flow circulates through a relatively small volume of shallow crust and is not believed to be significant to the CO_2 cycle. The warm flow appears to be an important CO_2 sink. For example, dissolved CO_2 has been measured from a warm vent on the flank of the Juan de Fuca ridge and found to be 66% depleted relative to normal sea water (Sansone et al. 1998).

The flux of warm sea water W through the oceanic crust can be estimated from the measured global sea-floor heat-flow anomaly H and the temperature anomaly T_w of the hydrothermal sea water relative to the bottom of the ocean. H is the difference between the heat flow expected from the cooling oceanic crust (which declines with the square root of age) and the heat flow obtained by directly measuring the temperature gradient and conductivity in the sediments. The difference is ascribed to hydrothermal cooling: H amounts to about 20% of the global oceanic heat flow and occurs mostly in young (<1 Ma) crust, for which the thermal gradient is steep enough that circulating water can reach warm rock. For the flow of water we write

$$W = \frac{H}{T_w C_w \rho_w} \qquad (10)$$

where $\rho_w C_w = 4$ MJ m^{-3} is the heat capacity per unit volume of water. H can be expressed in terms of the rate at which warm new crust is created and the degree to which hydrothermal circulation cools it. New crust is created at the rate $\partial A/\partial t$, currently 3 km^2 a^{-1}. The flux of CO_2 into the warm crust is therefore

$$F_{hydro} = R_{oc}\left(\frac{W}{V_{oc}}\right)$$

$$= \frac{H}{\rho_w C_w T_w}\left(\frac{R_{oc}}{V_{oc}}\right) \equiv \frac{\partial A}{\partial t}\frac{R_{oc}}{A_{hydro}} \qquad (11)$$

in which V_{oc} refers to the volume of the oceans and

$$A_{hydro} \equiv \frac{\partial A}{\partial t}\frac{V_{oc}}{W} \equiv \frac{\partial A}{\partial t}\tau_{hydro}$$

is a constant with units of area; τ_{hydro} is the current time scale for the whole ocean to circulate through these hydrothermal systems.

In practice, W is directly inferred from heat flow and chemical mass balances. Warm hydrothermal flow is a major sink for dissolved oceanic Mg^{2+}. The Mg^{2+} in the warm hydrothermal fluid is removed quantitatively by reaction with basalt. Because the flux of Mg^{2+} into the ocean in river waters and the competing sinks (e.g. clays) can be estimated, the known Mg^{2+} concentration in sea water can be used to determine W. This argument is independent of CO_2.

Using both heat flow and Mg, Schultz & Elderfield (1997, 1999) concluded that 1.5×10^{12} W, or about a quarter of the total global hydrothermal heat flow, escapes through warm vents with an assumed average temperature anomaly of 20 K; the corresponding flux of water is $W = 6.4 \times 10^{11}$ m^3 a^{-1}. The latter implies that the entire ocean circulates through warm vents in $\tau_{hydro} = 2$ Ma, for which $A_{hydro} = 6 \times 10^6$ km^2. In the current ocean a hydrothermal flux equivalent to quantitative extraction of CO_2 on a cycle time of 2 Ma is $F_{hydro} = 1.65 \times 10^{12}$ moles a^{-1}. On the other hand, using similar data, Sansone et al. (1998) argue for warmer vents (temperature anomaly 64 K) that carry a somewhat smaller fraction (8–20%) of the global hydrothermal heat flow. Their estimates translate into smaller amounts of water circulating over longer time scales, $8 < \tau_{hydro} < 20$ Ma, for which $2.4 \times 10^7 < A_{hydro} < 6 \times 10^7$ km^2. For comparison, Walker (1985) uses 10 Ma for this cycle. As the actual flux is uncertain, we use Schultz & Elderfield's (1997, 1999) estimate as a lower limit on A_{hydro}. We use the higher of Sansone et al.'s (1998) estimates as our upper limit.

An illustrative example

The simplest case is to consider the steady-state value of the oceanic–atmospheric carbon dioxide reservoir R_{oc} that results from equating mantle outgassing to ingassing by subduction. From equation (7), the condition for steady state is $F_{ridge} = C_{deep}F_{hydro}$. Using equation (9) for F_{ridge} and equation (11) for F_{hydro}, we obtain a steady-state reservoir size of

$$\langle R_{oc}\rangle = \frac{A_{hydro}D_s}{V_{man}C_{deep}}\frac{\rho_s}{\rho_m}R_{man}. \qquad (12)$$

This expression is independent of the rate of crustal overturn: to first approximation the amount of CO_2 in the ocean and atmosphere does not depend on global heat flow. If we set the constant $A_{hydro} = 1 \times 10^7$ km^2, take $11 \times 10^{21} < R_{man} < 18 \times 10^{21}$ moles, take $D_s = 56$ km, and $C_{deep} = 0.75$, the current steady-state value of R_{oc} would be $(6.6$–$11) \times 10^{18}$ moles. This is comparable with the current amount of 3.3×10^{18} moles. Assuming the same partitioning between ocean and atmosphere as today, this steady-state value of R_{oc} is equivalent to $2.0 < pCO_2 < 3.4$ PAL, an amount not differing significantly from what it has been on average over the past hundred million years or so. What is most interesting about this near coincidence is that (1) it is obtained based on geological and geochemical arguments in which CO_2 plays no part and (2) it ignores the continents and the Urey cycle. With A_{hydro} fixed, D_s doubled and C_{deep} halved, equation (12) predicts an Archaean CO_2 level of just 8–14 PAL.

Hydrothermal carbonatization, redux

To extrapolate the sea-floor weathering flux F_{hydro} into the past requires a model of the water–rock reaction. There are two important possibilities, depending on whether the reactable cations are more or less abundant than CO_2. In equation (11) we implicitly assume that CO_2 is quantitatively removed from the sea water. In effect we presume fast reactions with superabundant cations. Walker (1985) and François & Walker (1992) assumed that CO_2 is removed in linear proportion to its concentration. This differs from our picture only in the value of A_{hydro}: in either model it is the delivery of CO_2-rich waters to the oceanic crust by hydrothermal circulation that limits uptake, so that CO_2 is quantitatively depleted in the circulating sea water until available CaO, MgO and FeO in the rock are exhausted.

However, there is some experimental evidence that this reaction is not strongly dependent on CO_2 or pH (Caldeira 1995; Brady & Gíslason 1997). In these experiments reaction rates are kinetically limited by the availability of cations. If this is the case in real hydrothermal systems, then it is possible that the uptake of CO_2 by the sea floor has always been similar to what it is at present, as Caldeira (1995), Goddéris & François (1995) and Brady & Gíslason (1997) assumed. This would make the current rough equality between subduction and ridge degassing rates in equation (12) a coincidence, and leave as an open puzzle how the mantle CO_2 cycle is to be closed over long time scales.

In view of these uncertainties, we will write $F_{hydro} \propto pCO_2^\alpha$, where the parameter α represents the dependence of hydrothermal carbonatization on pCO_2. Quantitative depletion corresponds to $\alpha = 1$. Sansone et al's data imply something short of this. Brady & Gíslason's laboratory data correspond to $\alpha = 0.23$, but of course their data are for reactions running enormously faster than in nature. We can test these different assumptions at a higher CO_2 concentration by comparing reconstructed CO_2 levels with the observed carbonate in Mesozoic ocean cores. As noted, quantitative extraction of CO_2 in the current ocean over $\tau_{hydro} = 2$ Ma corresponds to $F_{hydro} \times 1.65 \times 10^{12}$ moles a^{-1}. This flux is compatible with what is known about the abundance of carbonate from holes drilled into layer 2A of the oceanic crust. Using $\alpha = 1$ and $F_{sub} = 3.4 \times 10^{12}$ mole a^{-1} (the current rate at which carbonatized sea floor is entering subduction zones; see above) implies that the Mesozoic oceans contained about 7×10^{18} moles of CO_2. This is comparable with Lasaga et al.'s (1985) estimate of 10×10^{18} moles of CO_2 (i.e. 3 PAL). On the other hand, Ekart et al.'s (1999) more direct estimate from palaeosol data that pCO_2 was as high as 8 PAL in the Mesozoic is consistent with $\alpha \approx 0.4$.

It is possible that some of the carbonate lost from the circulating hydrothermal fluid has simply precipitated because the sea water has been heated (M. Bickel, pers. comm.). If so, then both the carbonate and the cation are supplied by the sea water. If the ultimate source of the cations is continental weathering, the precipitated carbonate would be equivalent to a pelagic carbonate and not true sea-floor weathering; whereas if the ultimate source of the cation is somewhere else in the sea floor, then for accounting purposes precipitation is equivalent to *in situ* reaction. Solubility of $CaCO_3$ is not a strong enough function of temperature for temperature alone to cause quantitative precipitation of sea water calcium carbonate. The effect of pH is probably greater, but such fluids develop high pH by reaction with the rock. Currently the Ca/bicarbonate ratio in sea water is about four. If the Ca that precipitated had been extracted from the sea floor (as we surmise) this would amount to but one part in five of the calcium. As we go back in time saturation of $CaCO_3$ requires higher levels of pCO_2 that imply lower levels of Ca in seawater. A value of 4 PAL CO_2 reverses the ratio of Ca to bicarbonate. Thus in Mesozoic time (taking $pCO_2 = 4$ PAL for the illustration), quantitative carbonatization of sea floor basalts would have required that 3/4 of the calcium be scavenged from the basalt itself. As discussed above, the Mesozoic sea floor appears to have been well carbonatized. Sea-floor weathering as we have described it is implicated.

Oceanic pH

Oceanic pH controls the partitioning of CO_2 between the ocean and the atmosphere and hence controls pCO_2 given R_{oc}. Over geological times, the pH of the modern ocean is dynamically buffered by low-temperature weathering and sea-floor alteration of silicates (which produce alkaline waters) and high-temperature axial hydrothermal circulation (which produces acid waters) (e.g. Grotzinger & Kasting 1993). The relative importance of high-temperature and low-temperature vents on the early Earth is debated. Kempe & Kazmierczak (1994) noted that impact ejecta and vigorous tectonic activity on the early Earth would have exposed much more rock to low-temperature alteration, and that fluids reacting with Archaean komatiitic lavas are more alkaline than those reacting with modern basalts. They contended that the early ocean was significantly more alkaline than the present ocean. Macloed et al. (1994) and Russell & Hall (1997) in contrast focused on the large amount of hot hydrothermal activity on the early Earth and contended that the early ocean was acid and that low-temperature hydrothermal vents served only to produce local alkaline conditions. Direct evidence to support either position is at best model dependent. We have no strong reason to expect the relative proportions of high- and low-temperature hydrothermal systems to differ between Archaean time and today. In particular, the bulk of both the low- and high-temperature alteration occurs within young crust near the ridge axis, implying that the lifetime of oceanic crust is not a major factor. It is a reasonable compromise for the present to regard pH as fixed in our calculations. We note that the CO_2 levels we predict for the atmosphere depend strongly on pH. A more alkaline ocean, which we expect for Hadean time when impact ejecta are important, would remove more CO_2 from the atmosphere.

Ancient CO_2 cycles

Extrapolating the CO_2 cycle into ancient times requires us first to consider what happens when plate tectonic cycles run more quickly and then, for earlier times, to consider effects that are no longer important on Earth. In order of increasing ancestry and uncertainty, we consider higher heat flows and related issues within the constraints of plate tectonics, and then abundant

Archaean cycle

We solve the time-dependent problem with the coupled equations (4)–(6) using R_{oc}, T and pCO_2 obtained solving equations (7) and (8) in steady state. As above, we let the heat flow evolve as $Q(t) = (1 - t/4.5)^\mu$, with $\mu = 0.7$ or $\mu = 0.2$ for high or low heat flow, respectively. The metamorphic flux F_{meta} is assumed proportional to the product of the heat flow and the continental carbonate reservoir, QR_{con}. We use equation (9) for the mid-ocean degassing flux F_{ridge} and equation (11) for the sea-floor weathering flux F_{hydro}. In our standard models we take the degassing depth to decrease as the mantle cools. We write this as $D_s = 56$ km $(1 + t/3$ Ga$)$ for $t < 3$ Ga and $D_s = 112$ km for $t > 3$ Ga. We also assume that subduction of carbonate to the mantle is less efficient when the mantle is hotter than today. We assume C_{deep} is inversely proportional to D_s. We take the sea-floor creation rate to go as the heat flow squared, $\partial A/\partial t \propto Q^2$, the relationship predicted by plate tectonics. For comparison we also consider what would happen if plate tectonics enforces a more constant heat flow (see below). In these models we hold Q constant, but D_s and C_{deep}, because they depend on mantle temperature, both evolve as in the standard case. In all cases we take $F_{hydro} \propto R_{oc}^\alpha$. Adopted parameters are shown in Figure 3.

We calibrate all models to give the current crustal reservoir R_{con} and an average pCO_2 over the past 400 Ma of 3 PAL. For any model we can do this by adjusting two of four parameters. The four are the hydrothermal circulation constant A_{hydro}; the dependence of sea-floor uptake on pCO_2, represented by α; the calibration of the silicate weathering rate C_{SiO_3w}; and the size of the mantle reservoir R_{man}. The last of these is directly proportional to the mid-ocean ridge outgassing flux F_{ridge}. The mantle degassing time and degassing depth D_s are not free

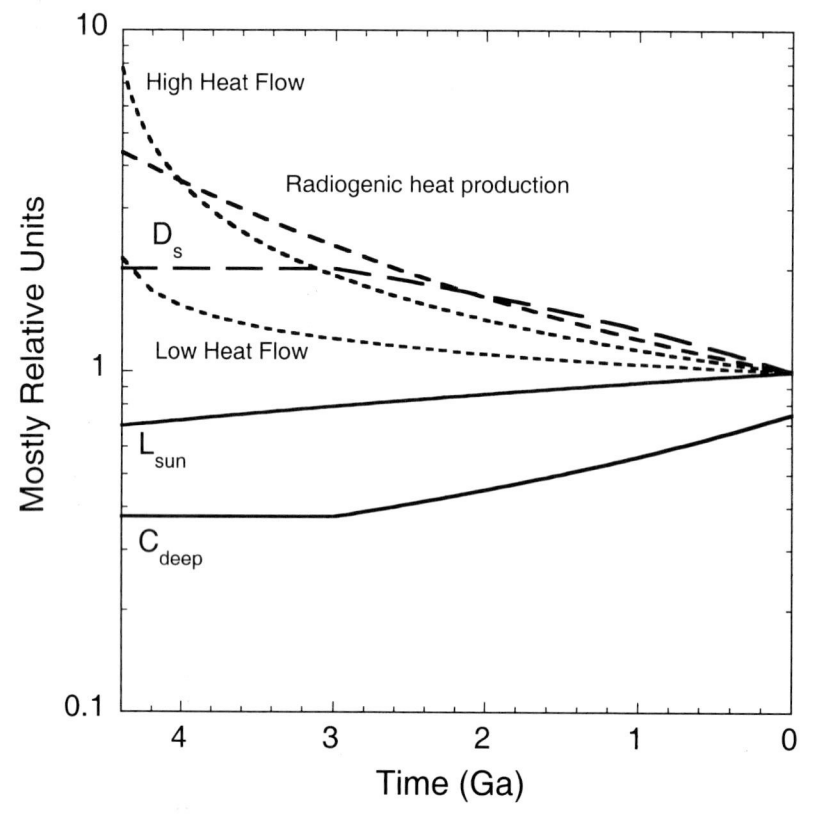

Fig. 3. Relative values of some model parameters through Earth history. Both high and low heat-flow histories are indicated; these are compared with the history of radiogenic heating.

parameters because they are constrained independently of CO_2. Of the four, A_{hydro} and F_{ridge} are observables subject to constraints discussed above. The power α may be near unity but for our purposes it is best regarded as free, as together with A_{hydro} it determines the return flux to the mantle.

Standard models. Figure 4 shows loci of successful standard models as functions of the observables A_{hydro} and F_{ridge}, the warm hydrothermal circulation constant and the mid-ocean ridge outgassing flux, respectively. Heat flow is high ($\mu = 0.7$). The models assume a current continental inventory $F_{con} = 5 \times 10^{21}$ moles. Curves are labelled by α. The fourth parameter, C_{SiO_3w} (not shown), tunes the Urey cycle and needs to be set separately for each model. In successful models F_{SiO_3w} is typically about 10–20% smaller than F_{meta}. Initial conditions are unimportant. Indicated in the figure are two particular models with $F_{ridge} = 2 \times 10^{12}$ moles a^{-1} and different values of A_{hydro} and α. Surface temperature and pCO_2 histories for these models are shown in Figure 5. These compare different dependences of sea-floor weathering on pCO_2 (i.e. different values of α). Lower values of α require faster hydrothermal cycling.

The finite volume of reactable basaltic crust is not explicitly accounted for in these models.

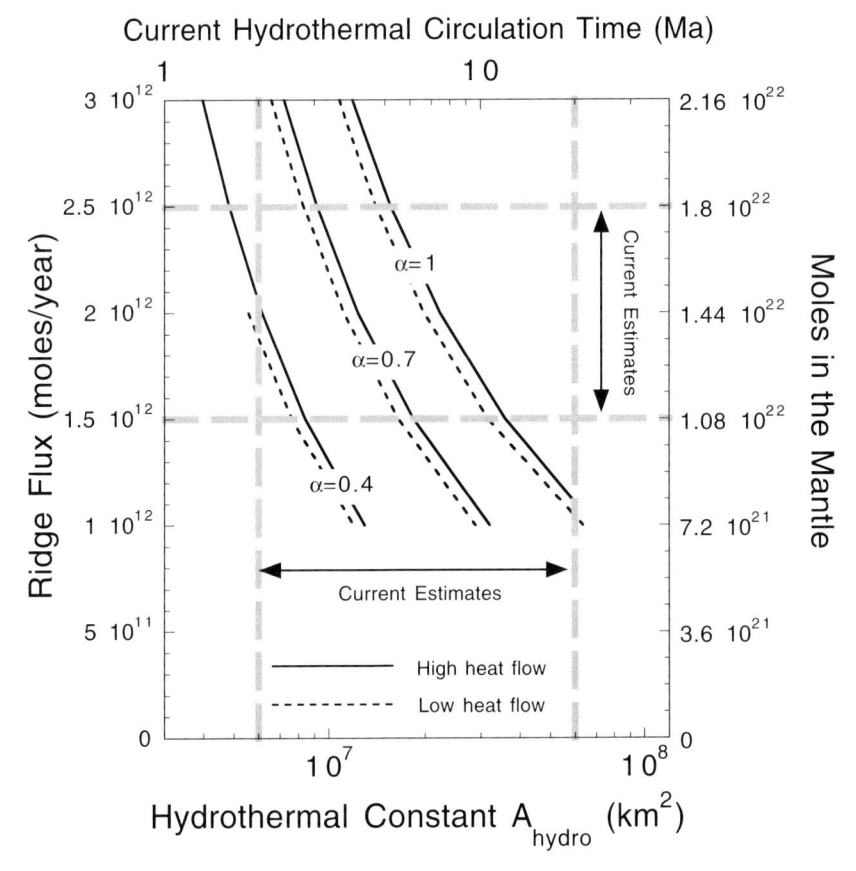

Fig. 4. Parameter choices that cause our model of coupled continental and mantle CO_2 cycles to converge to modern atmospheric and continental CO_2 inventories. We presume standard solar evolution and CO_2 working alone as the greenhouse gas. Quantitative values for the hydrothermal circulation time scale τ_{hydro} and the mid-ocean ridge degassing flux F_{ridge} refer to present values. The mantle inventory presumes whole-mantle convection. The hydrothermal circulation constant A_{hydro} is related to τ_{hydro} and the crustal creation rate $\partial A/\partial t$ by $A_{hydro} = \partial A/\partial t \times \tau_{hydro}$. Best estimates of A_{hydro} and F_{ridge} are discussed in the text. The parameter α represents the dependence of sea-floor carbonatization on the concentration of CO_2 in sea water. The different heat-flow histories are labelled 'high' and 'low' according to heat flow in Archaean time; the low heat-flow models have heat flow that changes relatively little over the past 4 Ga.

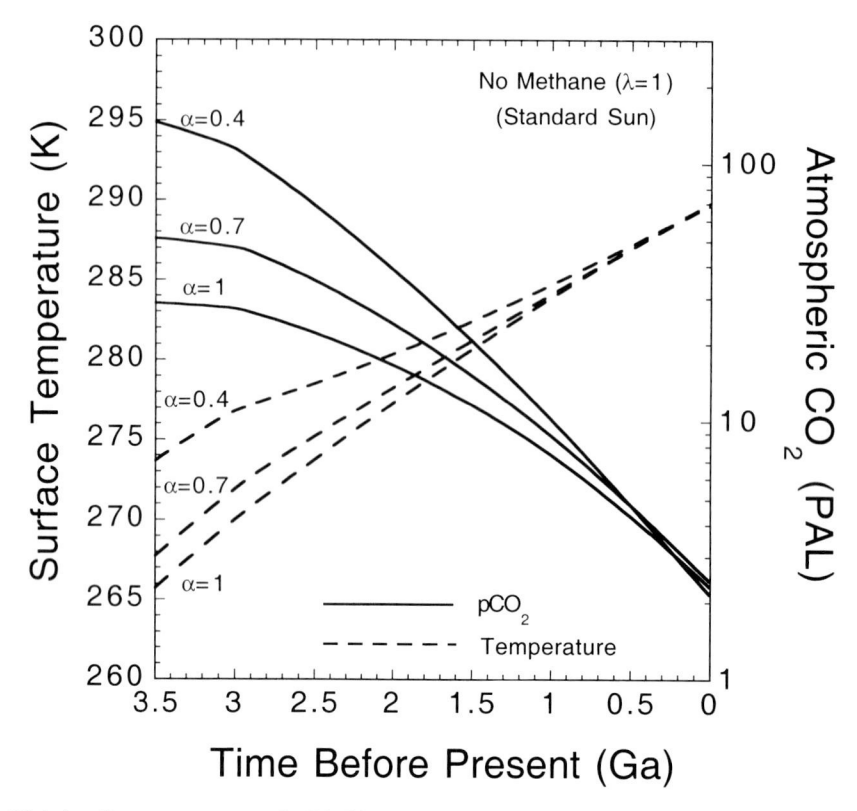

Fig. 5. Global surface temperature and pCO_2 histories for three high heat flow models with different values of α, the CO_2 dependence of the sea-floor weathering rate. The models assume a current mid-ocean ridge flux $F_{ridge} = 2 \times 10^{12}$ moles a^{-1}. Models with higher values of α require higher amounts of CO_2 to obtain the same amount of CO_2 consumption.

A half-kilometre of basalt, a thickness comparable with the modern permeable and reactable layer 2A at fast ridges (it is somewhat thicker at slow ridges, e.g. Hooft et al. (1996)), if fully carbonatized, would yield 300 m of carbonate, or $R_{bas} = 3 \times 10^{12}$ moles. This limit is approached but not exceeded in any of the models discussed here. Our upper limit on the current inventory of carbonatized basalt is equivalent to 40 m of carbonate (i.e. c. 10% carbonatization). A value of α near unity is not unreasonable for carbonatization levels <10% but is untenable if carbonatization levels become much higher. Doubtless the carbonatized layer deepens as CO_2 levels increase, but this requires reaction of increasingly impermeable media, with $\alpha \ll 1$. A related issue is that, in our models, the capacity of the basalt to hold carbonate is exceeded if $C_{deep} < 0.2$ In such conditions CO_2 can build up hugely in the atmosphere, oceans, and on the sea floor, and the surface might become very hot indeed. We have considered such cases in more detail elsewhere (Sleep & Zahnle 2001; Sleep et al. 2001).

Lower heat flow. Our standard higher heat flow assumption is that heat flow Q declined over time somewhat more slowly than radioactive heat generation declined. Although this seems a reasonable assumption it is not especially well supported. In particular, argon degassing of the Earth is consistent with no change in sea-floor spreading rate over 3 Ga (Sleep 1979; Tajika & Matsui 1993). Some theories also suggest that plate tectonics should occur at a relatively constant rate as the mantle cools (F. Nimmo, pers. comm.). Constant heat flow implies that only about half the mantle has cycled through the mid-ocean ridges over the past 3.5 Ga; this slow circulation maintains large geochemically isolated reservoirs without appealing to layered convection. To test the influence of low heat flow we use $Q(t) = (1 - t/4.5)^\mu$ with $\mu = 0.2$ as a lower bound. This parameterization predicts heat flow today that is twice the current rate of radiogenic heating, but c. 3.5 Ga the heat flow is some 10% less than radiogenic heating. The more nearly constant heat flow implies that the hydrothermal sink, subduction, and even continental metamorphism are also nearly constant. Because the mantle has cooled 150–200 K since Archaean time (Abbott et al. 1994), processes that depend directly on the temperature of the mantle (the partial melting depth D_s and the subduction efficiency C_{deep}) evolve as in the standard model.

Figure 4 shows loci of successful constant heat-flow models as functions of the observables A_{hydro} and F_{ridge}. Overall the low and high heat-flow models have similar requirements. Figure 6 which compares two specific models with $\alpha = 0.7$, shows that heat flow has relatively little effect on either temperature or pCO_2. This insensitivity to heat flow was anticipated by the steady-state estimate, equation (12) above, in which the crustal creation rate plays no part. In detail, the low heat-flow models are slightly cooler, owing to a smaller metamorphic source of CO_2. In turn, the smaller value of pCO_2 requires somewhat higher amounts of hydrothermal circulation to maintain the return cycle to the mantle. We will find below that heat flow is important in Hadean time, where impact ejecta compete with new oceanic crust for CO_2.

Effect of another greenhouse gas. The standard models violate the constraint on pCO_2 inferred from palaeosols dated $c.$ 2.75 Ga (Rye et al. 1995). This confrontation with the datum is the subject of Figure 7. Rye et al. showed that Caldeira & Kasting's (1992) climate parameterization for 2.75 Ga implies that siderite would have been stable in soils unless temperatures were below freezing. Iron was present in these soils but siderite was not. There are three ways to resolve this paradox. (1) The palaeosol may represent a local region or time when the temperature was higher than the global average. For example, Ekart et al. (1999) argued that palaeosols record the warmest times of the year, with a typical overestimate of about 10°C for a temperate climate. We therefore also show Rye et al.'s curve offset by 10°C. Doing so eliminates

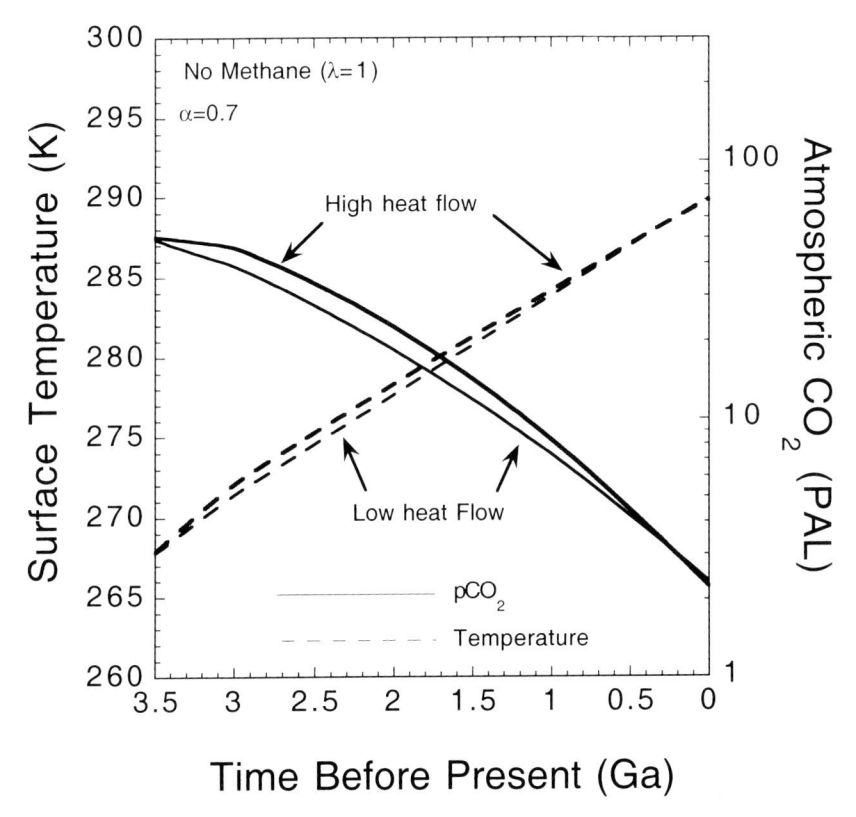

Fig. 6. Global surface temperature and pCO_2 histories for two otherwise identical models with high and low heat flows. The models assume a current mid-ocean ridge flux $F_{ridge} = 2 \times 10^{12}$ moles a^{-1}. The low heat-flow models feature relatively little secular change in the oceanic crustal creation rate and the continental metamorphic recycling rate.

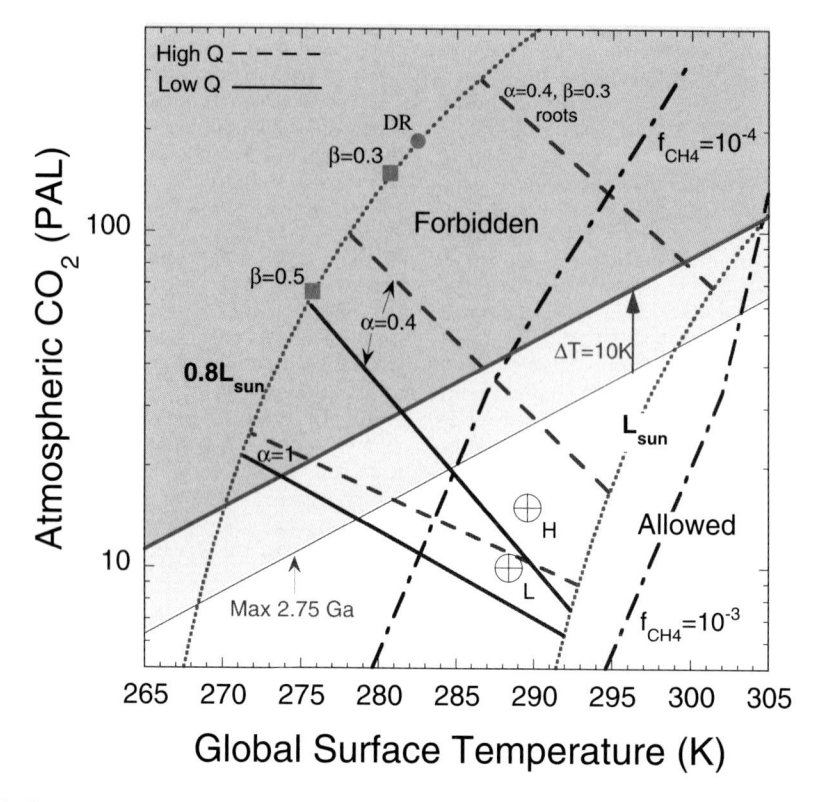

Fig. 7. The figure shows the relationship between pCO_2 and global average surface temperature for 2.75 Ga, and compares this with the siderite stability field (lightly shaded region) deduced by Rye et al. (1995). Siderite is a constraint because it was absent in a 2.75 Ga palaeosol (Rye et al. 1995). The expected seasonal bias in weathering rates (Ekart et al. 1999) relaxes the constraint by some 10 K; the relaxed constraint is indicated by darker shading. The curve labelled $L = 0.8L_{sun}$ shows CO_2's greenhouse effect for nominal insolation (Caldeira & Kasting 1992). Where the models in Figures 1 and 2 plot on this curve is indicated ('DR' denotes 'Devonian roots'). No model on the $L = 0.8L_{sun}$ curve is compatible with the siderite constraint. Either the palaeosol represents a special locality or other greenhouse gases were present. To illustrate the latter, we show CO_2 greenhouses consistent with current insolation ($L = L_{sun}$) and two methane–CO_2 greenhouses consistent with $L = 0.8L_{sun}$ (Pavlov et al. 2000). The latter are labelled by methane mixing ratio. Because the $L = L_{sun}$ and the $f_{CH_4} = 10^{-4}$ curves are nearly parallel, we can vary the effective solar luminosity as a proxy for different greenhouses. The five nearly straight lines connecting the $L = L_{sun}$ and the $L = 0.8L_{sun}$ curves show our model predictions for the climate at 2.75 Ga, treating different amounts of methane by varying $0 < \lambda < 1$ in equation (13). Predicted climates are nearly independent of the mid-ocean ridge degassing flux F_{ridge}. Continuous lines are for high Archaean heat flow; broken lines for low Archaean heat flow. These models are addressed by Figures 4–6 and 8 and 9. The different values of α are meant to span the plausible range of sea-floor weathering rates. The high Q model labelled '$\alpha = 0.4 +$ roots' uses a slower weathering rate before the appearance of land plants with root systems. It maximizes pCO_2 within the context of our models. This model and others like it are addressed by Figures 10 and 11. Finally, the two models labelled '⊕' use $\alpha = 0.7$ and $\lambda = 0.2$ (equivalent to c. 250 ppm CH_4) and the two different heat flow histories. These models are singled out to illustrate the evolution of reservoirs and fluxes in Figures 12–15.

some of the perceived discrepancy. (2) Rye et al. (1995) have the siderite stability field wrong. (3) Other greenhouse gases may have been present, methane in particular.

Here we pursue the more interesting possibility: another greenhouse gas, methane, was impor-tant (Kiehl & Dickinson, 1987; Pavlov et al. 2000). Biologically generated methane is expected to have been present in the atmosphere at levels of hundreds of ppm before the rise of oxygen. Even after oxygen rose to levels of order 0.1 PAL, methane was plausibly present at tens of ppm.

We use the convenient method of modifying the solar luminosity as a proxy for the increased greenhouse effect. We use the arbitrary function

$$L = L_0(1 - 0.07\lambda t) \quad (13)$$

with $\lambda = 0.2$ rather than the standard $\lambda = 1$. This is equivalent to some 250 ppm CH_4 c. 2.75 Ga (see Fig. 7). We will call these 'high methane' rather than 'bright Sun' models, to focus attention on a physically plausible explanation for the temperate climate of ancient Earth. The calculations are performed for all of Earth history even though high levels of methane are inconsistent with the modern atmosphere.

Loci of permitted high methane models are plotted in Figure 8. It should be noted that these models require much higher rates of hydrothermal circulation (lower values of A_{hydro}) than do the standard Sun models. This is required for sea-floor weathering to compete with continental weathering, the rates of which are higher because of the higher surface temperature (recall that sea-floor weathering closes the mantle's CO_2 cycle in our models). Computed temperature and carbon dioxide histories for $\alpha = 0.7$ and different levels of methane are compared with the standard model (also with $\alpha = 0.7$) in Figure 9. As expected, the warmer climates result in less atmospheric CO_2.

Figure 7 compares predictions for 2.75 Ga by two families of models to the local temperature and pCO_2 limits obtained by Rye et al. (1995). The region below the siderite stability field is

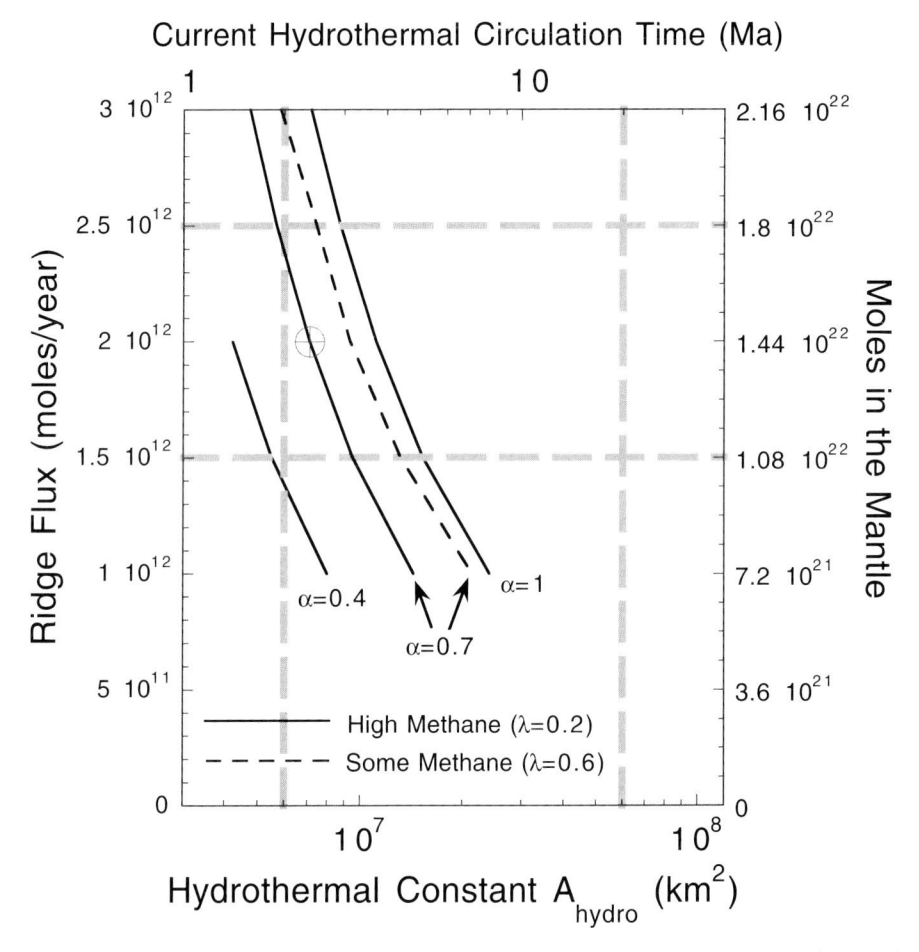

Fig. 8. Parameter choices that converge to modern atmospheric and continental CO_2 inventories for models with methane as an additional greenhouse gas. These models imply relatively low levels of CO_2, which in turn requires that hydrothermal circulation be relatively more efficient if it is to return CO_2 to the mantle. The specific model indicated by the '⊕' symbol is discussed in more detail in Figures 7 and 12–15.

forbidden. We generate surface temperatures and pCO_2 by letting λ range from one to zero. Each model is constrained to evolve to R_{con} = 5 × 10^{21} moles and an average value of pCO_2 = 3 PAL over the past 400 Ma. The standard faint Sun (λ = 1) and the nonstandard bright Sun (λ = 0.2) are shown. Shown for comparison are loci of pCO_2 and T consistent with different amounts of methane (Pavlov et al. 2000). It should be noted that curves of constant methane parallel the curves with λ = 1 and λ = 0.2, thus justifying somewhat λ as a proxy for methane as a greenhouse gas. We indicate a particular pair of models (the '⊕' symbols in Fig. 7) with α = 0.7 and λ = 0.2 (corresponding to c. 250 ppm CH$_4$), but different heat-flow histories, to discuss in more detail below.

Effect of land plants. Land plants with roots have existed only since Devonian time (e.g. Berner, 1997). It might be expected that chemical weathering was sluggish before then (Berner 1997). We use this possibility to illustrate the general conclusion that a clement climate, CO_2 rich, could have existed in Archaean time if both silicate weathering and carbonitization of oceanic crust were sluggish sinks requiring high pCO_2 to operate. It should be noted that these models are mirror images of the high methane models, in that they address the faint early Sun by invoking biological measures to raise CO_2.

To quantify the effect of roots, we compute models based on the situation envisioned by Berner (1997). Our 'roots' models differ from the rootless models by setting the weathering rate

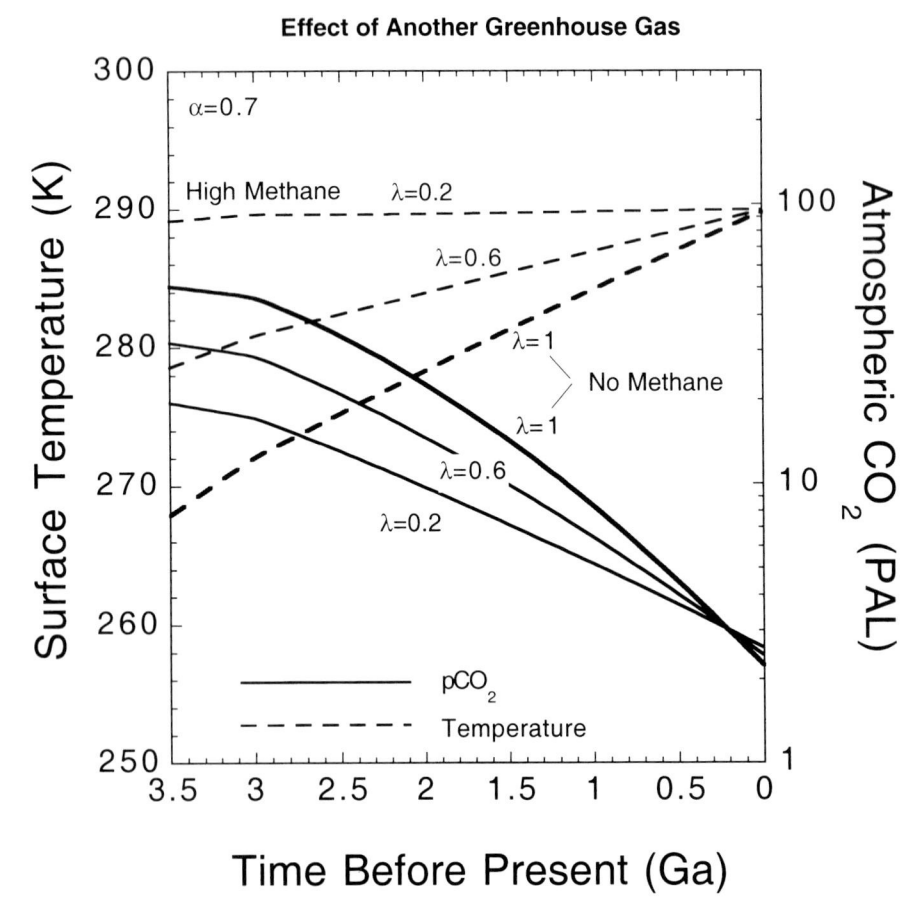

Fig. 9. Global surface temperature and pCO_2 histories for models with different amounts of methane. The models assume a current mid-ocean ridge flux F_{ridge} = 2 × 10^{12} moles a^{-1}. The 'methane' models are constructed by arbitrarily reducing the rate of solar luminosity evolution by the factor λ in equation (13). The added greenhouse effect of the methane produces warmer climates that speed silicate weathering, significantly reducing pCO_2 and thereby indicating that silicate weathering has a significant effect throughout Earth history.

before the advent of roots to 33% of the current rate, and for better consistency with Berner (1994) we consider cases with $\beta = 0.5$ as well as $\beta = 0.3$. Figure 10, the rooted analogue to Figures 4 and 8 shows what parameter values will evolve to modern conditions. Two models with $\alpha = 0.5$ and $F_{ridge} = 2 \times 10^{12}$ moles a^{-1} but different β are paid more attention in Figure 11. Sluggish weathering before the Devonian period results in higher surface temperatures (Fig. 11) and higher pCO_2, as expected. Efficacy of sea-floor carbonatization must then be smaller. With $\alpha = 0.4$ the oceanic sink is smaller and temperate conditions exist throughout Archaean and Proterozoic time, with the warmest climates characterizing early Phanerozoic time.

'Roots' models for $\alpha = 0.4$, $\beta = 0.3$ and different suns ($0 < \lambda < 1$) are compared with Rye et al.'s palaeosol constraint in Figure 7. Rooted models for $\alpha = 1$ (not shown) roughly coincide with $\alpha = 0.4$ rootless models. By construction the rooted models predict higher pCO_2 levels and warmer Earths than do the rootless models. This makes the rooted models harder to reconcile with Rye et al.'s constraint. On the other hand the rooted models are better suited to keeping Earth warm after the rise of oxygen, when a methane greenhouse is less plausible.

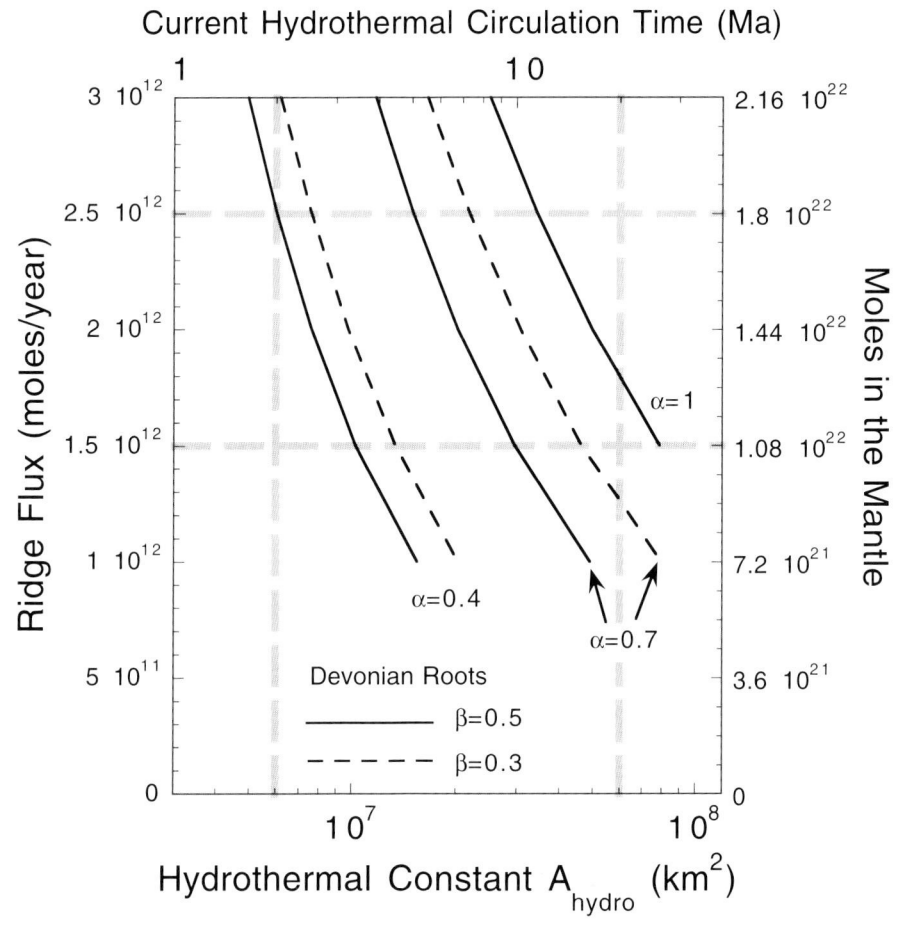

Fig. 10. Parameter choices that converge to modern atmospheric and continental CO_2 inventories for models in which weathering rates increase abruptly upon the advent of roots. These use either $\beta = 0.3$ or $\beta = 0.5$ in equation (8), and reduce the silicate weathering constant (in equation (8)) before the advent of roots to 33% of what it is now. These values are consistent with Berner (1994, 1997)]. These models imply that the ancient atmosphere was fairly CO_2 rich, and so sea-floor weathering needs to have been relatively inefficient or too much CO_2 would have been captured by the mantle.

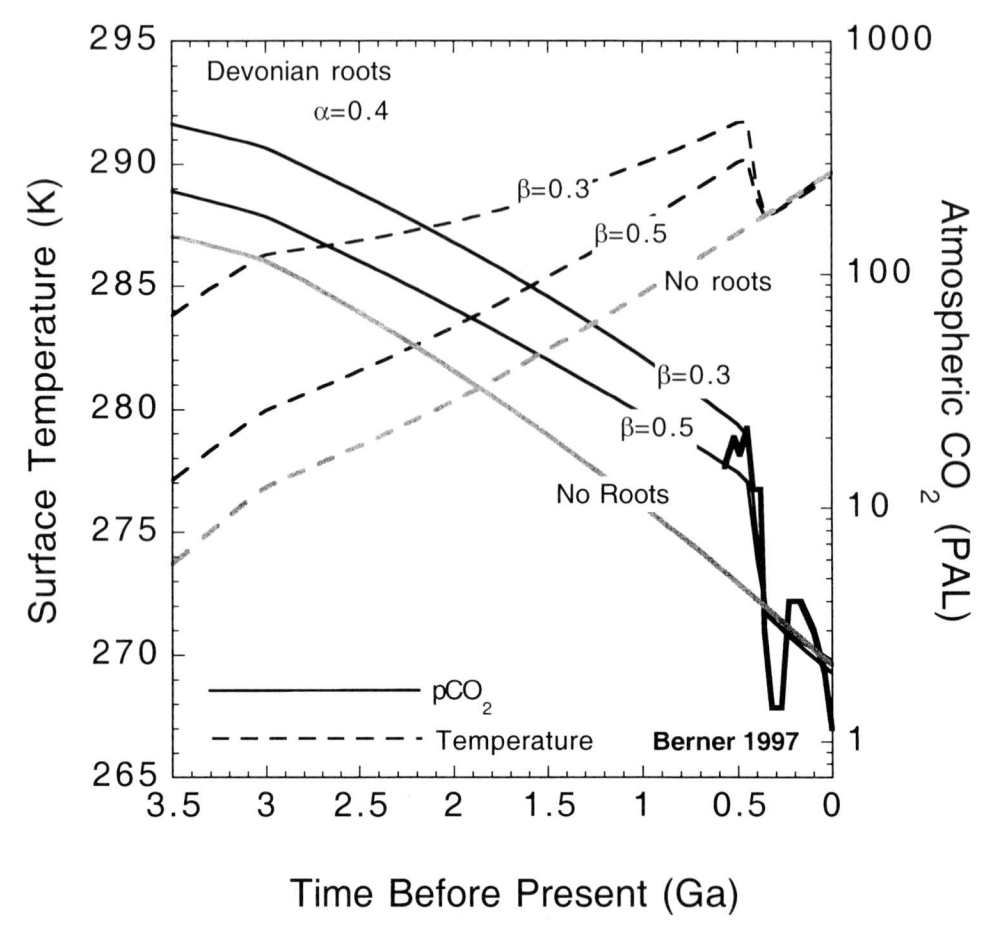

Fig. 11. Surface temperature and $p\text{CO}_2$ histories for some rooted models with current $F_{ridge} = 2 \times 10^{12}$ moles a^{-1}. These are compared with the the standard rootless model (Fig. 6). Also shown for comparison is Berner's (1997) Phanerozoic $p\text{CO}_2$ curve. By choosing favourable parameters ($\alpha = 0.4$, $\beta = 0.3$) we are able to construct a model that maintains a clement Earth using CO_2 alone through most of Earth history. However, such a model badly violates Rye et al.'s constraint on $p\text{CO}_2$ (Fig. 7). It is also worth asking whether such models can be consistent with cold climates at the end of Precambrian time.

The Hadean cycle and impact ejecta

The Hadean eon brought an additional threat to CO_2. During Hadean time, impact ejecta would have been quantitatively important sediments (Koster van Groos 1987). Qualitatively, impact ejecta differ from most other sediments in being pulverized, vitrified or otherwise mechanically damaged, and so relatively easily chemically attacked. Ejecta also tended to be more mafic than conventional sediments, because the most frequent target was oceanic basalt and large impacts excavated material from the mantle. Like other oceanic sediments, weathered ejecta would have been subducted along with the plates on which they accumulated. With more buffering by basic and ultrabasic rocks, and with most of the weathering occurring at relatively low temperatures, the sink for CO_2 would have been greater and the pH may have been higher (Kempe & Degens 1985; Kempe & Kazmierczak 1994). Here we summarize how we parameterized ejecta weathering in our earlier study (Sleep & Zahnle 2001), and in the process correct some errors that marred our first iteration.

The Hadean impact record is best calibrated to the large lunar basins (Zahnle & Sleep 1996; Anbar et al. 2001). To order of magnitude, the mass of the typical basin-forming impactor was c. 10^{20} g although the archtypes (e.g. Orientale) were larger, c. 10^{21} g and the largest that remains clearly visible (S. Pole–Aitken), perhaps

$c.\ 10^{22}$ g. The last of the lunar basins formed $c.\ 3.8$–3.85 Ga, so that to first approximation the average lunar impact rate during Hadean time as a whole (6×10^8 a) was $> 5 \times 10^{13}$ g a^{-1}. At 3.85 Ga the average impact rate was still very high, $> 10^{13}$ g a^{-1}, and possibly as high as 10^{14} g a^{-1}. Taking into account gravitational focusing and Earth's much larger cross-section, these lunar accretion rates correspond to average Hadean terrestrial accretion rates of 10^{15}–10^{16} g a^{-1}. We will be conservative and use 10^{15} g a^{-1} for 3.8–3.9 Ga. The median value is lower and depends on the time scale over which the impacts are sampled. On the time scale of Earth's resurfacing, the median value of the terrestrial accretion rate at 3.85 Ga would have been $c.\ 2 \times 10^{14}$ g a^{-1}. The dependence of the observed accretion rate on the sampling time scale has been discussed in detail by Anbar et al. (2001).

To estimate the mass of reactable ejecta we begin with the total mass of ejecta, as defined by the volume of the crater. According to Schmidt & Housen (1987)

$$m_{ej} = 0.13 m_i (\rho_t/m_i)^{0.217} g^{-0.65} v_i^{1.3} \quad (14)$$

where m_i is the impactor's mass, ρ_t the target density, g gravity, and v_i the impactor's velocity. For $v_i = 15$ km s^{-1} and $\rho_t = 2.86$ g cm^{-3}, the ejected mass is

$$m_{ej} = 2.5 \times 10^{19} (m_i/10^{18})^{0.783} \text{ g} \quad (15)$$

where the impactor mass is also in grams. For a 10^{21} g impactor, $m_{ej} \approx 5.6 m_i$. If we assume a plausible power-law mass distribution for the impactors,

$$N(>m) \approx (m/m_{max})^{-b} \quad (16)$$

where $N(>m)$ is the cumulative number of objects greater than mass m, m_{max} is the mass of the largest object in the distribution and $b \approx 0.8$, the ejecta accumulation rate is

$$\dot{M}_{ej} = \int \left(-\frac{\partial \dot{N}}{\partial m} \right) m_{ej}(m)\, dm$$

$$\approx b \frac{m_{max}^b}{\tau_{Had}} \int m_{ej}(m) m^{-1-b}\, dm. \quad (17)$$

The integral in equation (17) takes an especially simple form for $b = 0.783$:

$$\dot{M}_{ej} \approx \left(\frac{m_{max}}{10^{18}} \right)^{0.783} \frac{2 \times 10^{19}}{\tau_{Had}} \ln \left(\frac{m_{max}}{m_{min}} \right)$$

$$\approx 1 \times 10^{16} \text{ g a}^{-1} \quad (18)$$

evaluated for $m_{min} = 10^{16}$ g and $m_{max} = 10^{23}$ g. For an ocean-covered Earth, m_{min} is equivalent to the smallest impactor that can both penetrate the ocean and excavate pristine basalt. The mass of the largest impact is estimated from the total mass accreted, using the rough approximation $m_{max} \approx (1-b) M_{tot}/b$. We take $\tau_{Had} = 300$ Ma for late Hadean time. The logarithmic dependence makes equation (18) insensitive to specific choices of m_{max} and m_{min}. The logarithmic dependence also indicates that, unlike the iridium accumulation rate (which tracks the mass of the impactors), the ejecta accumulation rate depends only weakly on the sea-floor spreading rate, and the average will not differ significantly from the median. Given radiogenic heating rates at 3.85 Ga $c.\ 3.4$ times higher than today, the heat flow would probably have been some 1.5–4 times higher than today, and thus sea-floor turnover times in the range of 6–45 Ma. The expected standing crop of basaltic and dunitic ejecta is between 70 and 400 m.

As a first approximation, we let all the ejecta react quickly, in effect consuming all the CO_2 for which there are cations (roughly 0.005 moles g^{-1} for basalt and 0.01 moles g^{-1} for ultramafic rocks; these values correct a propagating typo of Sleep & Zahnle (2001)). This yields from equation (18) an average CO_2 sink of $F_{ej} \approx 1 \times 10^{14}$ moles a^{-1}. This is large compared with all geological terms in the modern carbon cycle. Alternatively, we can regard the ejecta simply as additional reactable oceanic crust. New reactable oceanic crust was created at a rate of 1×10^{16} to 7.2×10^{16} g a^{-1}, presuming plate tectonics, global spreading rates of 7–48 km^2 a^{-1} (i.e. 2.3–16 times present), and a reactable depth of 500 m. These set upper limits on the hydrothermal CO_2 sink of 5×10^{13} to 3.6×10^{14} moles a^{-1}. These are comparable with the ejecta sink, but the latter is probably the more easily fully carbonatized, owing to the damaged and more mafic state of the ejecta. Under these assumptions it appears likely that reactions with ejecta will determine pCO_2.

A third, more ambitious, approach is to assume a mass distribution for the ejecta themselves and directly estimate the corresponding weathering rate using measured reaction rates of basaltic glasses. We have done this elsewhere (Sleep & Zahnle 2001); here we quote only the essential results corrected for typos. We described the cumulative mass distribution of the ejecta by a power law $N(>\mu) \propto \mu^{-\gamma}$, where μ is the fragment mass. We found that, for $\gamma < 1$, weathering goes as $pCO_2^{3-3\gamma}$; whereas for $\gamma \geq 1$, weathering is nearly independent of pCO_2. For $\gamma = 0.95$ we obtain

$$F_{ej} \approx 3 \times 10^{13} pCO_2^{0.15} \text{ moles a}^{-1} \quad (19)$$

and for $\gamma \geq 1$ we obtain

$$F_{ej} \approx 5 \times 10^{13} \text{ moles a}^{-1} \quad (20)$$

the latter being, to within a factor two, the same result we obtained above for immediate and complete reaction of ejecta.

Finally, we need an impact history. These are obtained from the lunar record, which itself is controversial. For our purposes it is not particularly important to choose sides in this debate. We arbitrarily assume a linear decline of the Hadean impact flux from an initially high value at 4.4 Ga to zero at 3.55 Ga, which yields a flux at 4.4 Ga closer to the lower end than to the upper end of estimates. There results

$$F_{ej} \approx 2 \times 10^{14} pCO_2^{0.15} \max(0, t - 3.7) \text{ moles a}^{-1} \quad (21)$$

where time t is in Ga and the expression is tuned to give the appropriate impact flux at 3.85 Ga. The steep decline in the impact rate after 3.95 Ga is based on lunar data, although the specific cutoff date of 3.7 Ga is negotiable. Earlier behaviour is debatable. More aggressive extrapolations would imply greater impact flux in the more distant past, less atmospheric CO_2, and colder climates in early Hadean time than in our models.

An exemplary model

Figures 12–15 describe the CO_2 evolution of specific models indicated by the '⊕' symbol in Figures 7 & 8. These models feature a current mantle reservoir R_{man} of 1.44×10^{22} moles and a current ridge degassing flux $F_{ridge} = 2 \times 10^{12}$

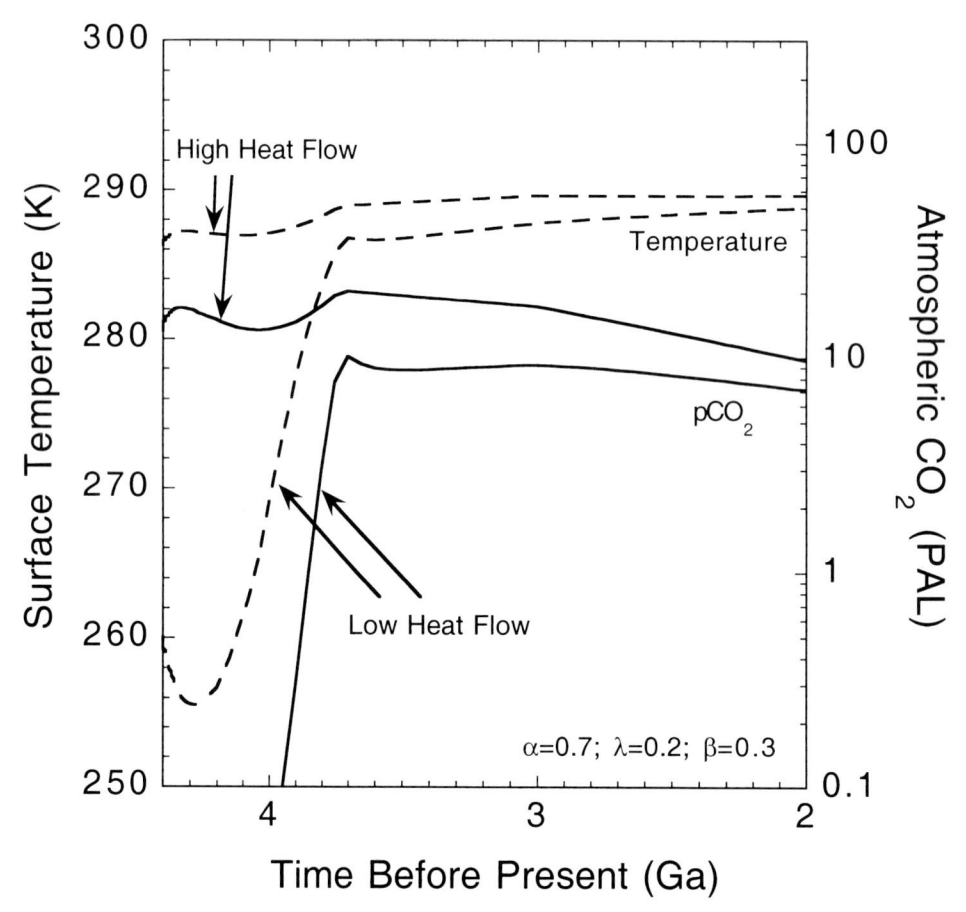

Fig. 12. Temperature and pCO_2 histories for the specific models labelled '⊕' in Figures 7 and 8. These models use $\alpha = 0.7$, and $\beta = 0.3$. The last corresponds to some 250 ppm methane at 2.75 Ga. Results are shown with impact ejecta for low ($\mu = 0.2$) and high ($\mu = 0.7$) heat flow. In Hadean time the heat flow is important because sea floor and ejecta compete for CO_2; high heat flow implies high rates of CO_2 cycling through the upper mantle. With low rates of crustal creation the impact ejecta are an important sink on CO_2, which is driven to negligible levels.

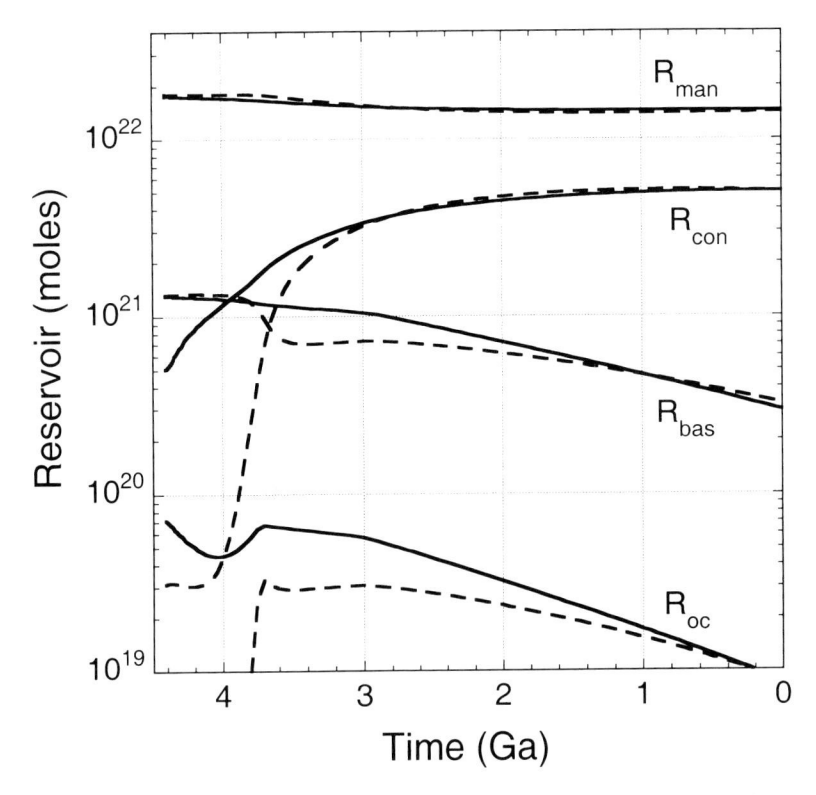

Fig. 13. Four important reservoirs of CO_2 are shown as functions of time for the models in Figure 12. High heat flow is denoted by continuous lines, low heat flow by dashed lines. Here we have chosen models in which the crustal reservoirs are initially constant in time; i.e. we have started from the equilibrium reservoirs. In particular, the equilibrium continental reservoirs are small and so these models begin with very little continental carbonate. The high heat-flow models churn the reservoirs fast enough that if we do not start at equilibrium values, the model quickly evolves to them, but in the low heat-flow models circulation is slow enough that the arbitrary initial conditions are remembered well into Archaean time. In general, the effect of abundant Hadean impact ejecta is to remove CO_2 from the continents and oceans and put it into the mantle.

moles a^{-1}. They use $\alpha = 0.7$, $\beta = 0.3$ and $\lambda = 0.2$ (i.e. 200 ppm CH_4 at 2.75 Ga) to illustrate two possible Earth histories. Continents are held constant. We make no special allowance for biological advances in chemical weathering. The most important nonstandard or nonuniformitarian feature of these models is that they assume a substantial methane greenhouse ($\lambda = 0.2$) all the way back. Without the additional greenhouse gas we expect a hard snowball Earth through most of Hadean time (because of large impact events even a snowball Hadean eon will stochastically enjoy temperate, tropical or hypertropical conditions) and all of Archaean time.

Figure 12 shows surface temperature and pCO_2 histories. These focus on Archaean and Hadean time. Results are shown for the different heat-flow histories. The models with ejecta weathering use equation (19). By construction these are a warm model with relatively modest amounts of atmospheric CO_2. They implicitly depend on a large (biogenic?) methane source to maintain clement conditions. Without methane, ejecta weathering can easily drive surface temperatures below 240 K as we have shown elsewhere (Sleep & Zahnle 2001).

Figure 13 shows how the four major CO_2 reservoirs evolve in the ⊕ models. Because the models address Hadean time, they can be sensitive to the assumed initial conditions. Here we have arbitrarily chosen to present models in which the crustal reservoirs (R_{con} and R_{bas}) are initially in equilibrium with the sources and sinks. The high heat-flow models churn these reservoirs fast enough that the equilibrium sizes are quickly obtained whatever the initial choice is. But in low heat-flow models evolution is slow enough that the arbitrary initial conditions are

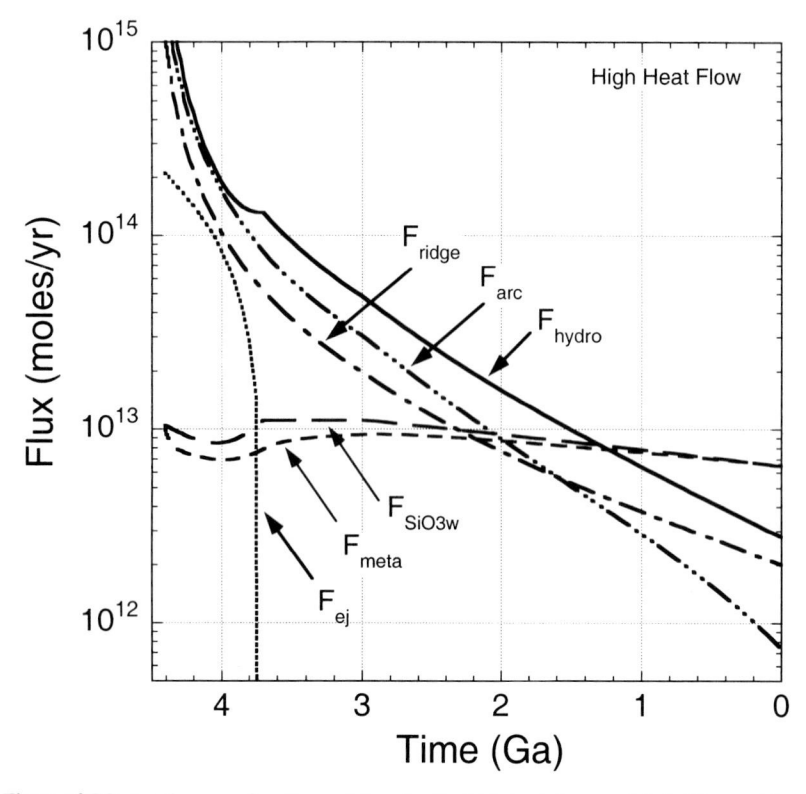

Fig. 14. Fluxes of CO_2 are shown as functions of time for the high heat-flow models in Figures 12 and 13. Impact ejecta are of minor importance compared with the rapid churning of the oceanic crust. In Archaean time, CO_2 is mostly controlled by processes involving the creation and subduction of oceanic crust. Continents become increasingly important through Proterozoic time, with the transition from mantle to continental control occurring at c. 1.4 Ga.

remembered well into Archaean time. In particular, in the low heat-flow ⊕ model shown here (dashed lines in Fig. 13), the equilibrium continental reservoir is initially very small. This is agreeable if one imagines that in the beginning CO_2 was begat by the Earth. But if instead the CO_2 arrived late or owing to very different environments on early Earth began mostly partitioned into surface reservoirs, as we expect would be the case if Earth cooled from a magma ocean (Sleep & Zahnle 2001; Sleep et al. 2001), it is possible that the early continental CO_2 reservoir was large and the low equilibrium value effectively unreachable. Both models predict that early in Earth's history $R_{bas} \approx 1.3 \times 10^{21}$ moles, equivalent to c. 40% carbonatization of the reactable upper 500 m oceanic basalt. The model requires 8% carbonatization today.

Figures 14 and 15 show the fluxes of CO_2 between reservoirs in the ⊕ models. Figure 14 addresses the high heat-flow ($\mu = 0.7$) ⊕ model. In these, the largest fluxes in Hadean time are associated with the oceanic cycle of crustal creation and subduction; weathering of impact ejecta is of secondary importance. Thereafter the oceanic crustal cycle monotonically declines whereas the continental cycle of silicate weathering and metamorphism holds roughly constant with time whereas fluxes to and from the mantle decrease. (The continental cycle increases with time for $\alpha = 1$ (not shown).) The continental cycle becomes increasing important as time passes, as might be expected. The transition from mantle to continental control over the CO_2 cycle (here at c. 1.4 Ga) is a general feature of our models, although the timing varies.

It should be noted that the model predicts arc volcanic fluxes that are currently too low. This occurs because, in our model, the steady-state relation between the mid-ocean ridge flux and F_{arc} is $F_{arc}/F_{ridge} \approx (1 - C_{deep})/C_{deep}$. For $C_{deep} = 3/4$, the model predicts that the current arc volcanic flux should approach $F_{arc} \approx F_{ridge}/3 = 6.7 \times 10^{11}$ moles a^{-1} for $F_{ridge} = 2 \times 10^{12}$ moles a^{-1}. This is only marginally smaller than what we have assumed it would be without pelagic carbonates,

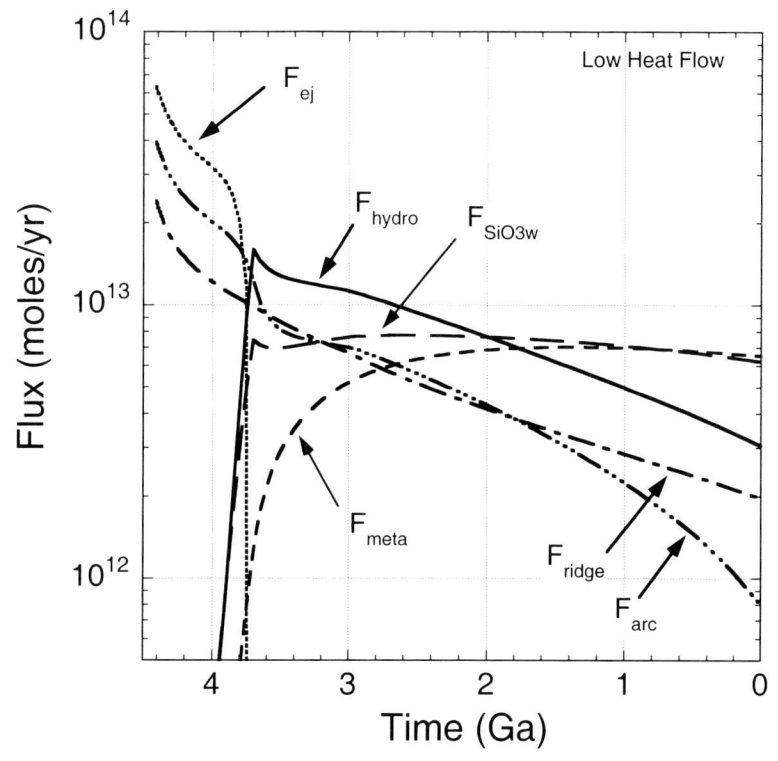

Fig. 15. Fluxes of CO_2 are shown as functions of time for the low heat-flow models shown in Figures 12 and 13. Impact ejecta dominate the CO_2 cycle in Hadean time. The oceanic crustal cycle controls CO_2 in Archaean time. Continents are important throughout Proterozoic time, with the transition from mantle to continental control occurring at $c.\,2.1$ Ga.

$(1 - C_{deep})F'_{sub} = 8.5 \times 10^{11}$ moles a^{-1}, but it is smaller than other estimates in the literature (see Marty & Tolstikhin 1998). To the extent that plumes contribute to the mantle's degassing, our estimate of the steady-state value of F_{arc} is raised proportionately.

Figure 15 addresses the low heat-flow ($\mu = 0.2$) $\oplus$ model. In this model impact ejecta completely dominate the Hadean CO_2 cycle. When impacts cease, control is passed to the oceanic crustal cycle, but continents become important fairly early and by $c.\,2$ Ga the continental cycle is the more important. The transition between oceanic and continental control is driven in part by mantle cooling and in part by decreasing heat flow. Overall, the transition between regimes is clear and its timing reasonable.

Conclusions, caveats and implications

It is well known that the crustal Urey cycle of silicate weathering and metamorphism can function as a dynamic buffer for atmospheric pCO_2 and climate (Walker et al. 1981). Feedback is provided by the strong temperature dependence of silicate weathering. Less appreciated is that hydrothermal alteration of oceanic basalts can also function as a dynamic buffer on CO_2 (Sleep & Zahnle 2001). Feedback is provided by the dependence of carbonatization on dissolved carbonates in sea water. The Urey cycle links atmospheric CO_2 to continental carbonates and silicates; water serves as a catalyst, and the oceans and mantle are otherwise ignored. The oceanic cycle links surficial CO_2 to the mantle via subduction and outgassing at the mid-ocean ridges. Currently the continental cycle is more important, but earlier in Earth's history, especially if heat flow were higher than it is now, a warmer mantle would have made the oceanic cycles dominant. Oceanic control would have been greater still if continents were late to develop. The important feature of the oceanic cycle is that, although it can control CO_2, it does so without a thermostat. We infer that massive CO_2 greenhouses are implausible and that, if there were no other greenhouse gases available to help, globally freezing temperatures are indicated for early Proterozoic and especially Archaean time.

In Hadean time huge amounts of mafic and ultramafic impact ejecta would probably have held atmospheric CO_2 levels strongly in check. This view of the Hadean Eon contrasts starkly with more traditional views in which significant ejecta and basaltic sinks are neglected and massive warm CO_2 atmospheres result (Morse & Mackenzie 1998). Despite its name, the Hadean Eon would have been the coldest era in the history of the Earth. Indeed, an ice-covered Earth has itself been recommended as an excellent environment for the origin of life (Bada et al. 1994).

Obviously models such as ours depend strongly on a wide range of poorly known parameters and debatable assumptions. It would not be fitting to claim any great certainty. To pick an example of how our models could fail, it could prove that Caldeira (1995) and Brady & Gislason (1997) are right, and carbonatization of oceanic basalt proceeds at rates that depend little if at all on the sea-water CO_2 concentration. Readers can probably provide their own examples.

A possibly troubling feature of our models is that we implicitly predict a secular trend in the isotopic composition of carbonates. This would occur because the carbonate subducted would presumably have the isotopic composition of sea water, whereas CO_2 currently outgassed at mid-ocean ridges is isotopically lighter. One way to avoid this problem is to presume that carbonates and reduced carbon are subducted in the same proportions as they are produced. Because the mantle emits some reduced gases and subducts oxidized ferric iron (Lecuyer & Ricard 1999), some reductant must be subducted lest the continents grow secularly more reduced. The natural candidate, perhaps the only viable candidate, is reduced carbon, which is notably refractory and already known to be subducted. This is a reasonable expectation, and it is consistent with what is known of arc volcanoes today (see above), but it is unmotivated by our model.

Taking our conclusions at face value, there are two issues that need to be addressed. One issue is how our model breaks down at early times if carbonate subduction were inhibited by a hotter mantle. In our model the subduction of carbonated oceanic crust closes the mantle's CO_2 cycle. Factors that inhibit carbonate subduction have the effect of stranding CO_2 in surface reservoirs. Although the capacity of the permeable layer of oceanic basalt to store CO_2 is large ($c.\ 3 \times 10^{21}$ moles), it falls well short of the global inventory of $c.\ 2 \times 10^{22}$ moles of CO_2. In our models the sea floor's capacity is exceeded if C_{deep}, the fraction of subducted carbonate that reaches the mantle, becomes smaller than 20% (the modern value is between 64 and 84% (Fischer et al. 1998)). A massive CO_2 atmosphere could result if $C_{deep} < 0.2$. Surface temperatures exceeding 200°C are implied (Kasting & Ackerman 1986). We have addressed in more detail elsewhere (Sleep & Zahnle 2001; Sleep et al. 2001) what is required to make C_{deep} small, and how quickly the Earth might make the transition from uninhabitably hot to uncomfortably cold Hadean climates.

The other issue is the nature of, and implications of, a strong early greenhouse effect in which CO_2 is not the only player. Methane is the most plausible alternative because it is relatively stable and effective, and can easily be produced in the required amounts by modern levels of biogenic activity (Pavlov et al. 2000). But a methane greenhouse comes at an interesting price: large amounts of methane imply large amounts of irreversible hydrogen escape to space (Catling et al. 2001). At the limiting flux (Walker 1977), 100 ppm CH_4 implies an H_2 escape rate of 1.4×10^{12} moles a^{-1} and, if the methanogenesis ultimately derives from oxygenic photosynthesis (as it almost certainly must), a corresponding O_2 production rate of 7×10^{11} moles a^{-1}. This is about a tenth of the current rate at which O_2 is consumed by weathering (Walker 1977), but unlike weathering, global oxidation by hydrogen escape is unbalanced by reduction elsewhere. Eventually a methane greenhouse must oxidize the Earth (Catling et al. 2001).

This study grew slowly from a brief argument made to a sceptical audience into a longer version of the same. We thank especially A. Anbar, K. Arrigo, M. Bickle, R. Buick, D. Catling, R. Dunbar, M. Green, H. Holland, J. Kasting, D. Lowe, B. Marty, C. McKay, E. Nesbit, F. Nimmo, A. Pavlov, T. Plank, F. Sansone, D. Schwartzman, and P. Wyllie. We thank NASA's Exobiology and Astrobiology Programs for support.

References

ABBOTT, D. L., BURGESS, L., LONGHI, J. & SMITH, W. H. F. 1994. An empirical thermal history of the Earth's upper mantle. *Journal of Geophysical Research*, **99**, 13 835–13 850.

ALT, J. C. 1995. Subsurface processes in mid-ocean ridge hydrothermal systems. *In*: HUMPHRIS, S. E. ZIERENBERG, R. A. MULLINEAUX, L. S. & THOMSEN, R. E. (eds) *Seafloor Hydrothermal Systems, Physical, Chemical, Biological, and Geological Interactions*. Geophysical Monograph, American Geophysical Union, **91**, 85–114.

ALT, J. C. & TEAGLE, D. A. H. 1999. The uptake of carbon during alteration of oceanic crust. *Geochimica et Cosmochimica Acta*, **63**, 1527–1535.

ALT, J. C., MUEHLENBACHS, K. & HONNOREZ, J. 1986. An oxygen isotope profile through the upper kilometer of the oceanic crust, DSDP Hole 504B. *Earth and Planetary Science Letters*, **80**, 217–229.

ANBAR, A. D., ARNOLD, G. L., MOJZSIS, S. J. & ZAHNLE, K. J. 2001. Extraterrestrial iridium, sediment accumulation and the habitability of the early earth's surface. *Journal of Geophysical Research*, **106**, 3219–3236.

BADA, J. L., BIGHAM, C. & MILLER, S. L. 1994. Impact melting of frozen oceans on the early Earth Implications for the origin of life. *Proceedings of the National Academy of Sciences of the USA*, **91**, 1248–1250.

BERNER, R. A. 1994. 3GEOCARB II: A revised model of atmospheric CO_2 over Phanerozoic time. *American Journal of Science*, **294**, 56–91.

BERNER, R. A. 1997. The rise of plants and their effect on weathering and atmospheric CO_2. *Science*, **276**, 544–546.

BOWRING, S. A. & HOUSH, T. 1995. The Earth's early evolution. *Science*, **269**, 1535–1540.

BRADY, P. V. & GISLASON, S. R. 1997. Seafloor weathering controls on atmospheric CO_2 and global climate. *Geochimica et Cosmochimica Acta*, **61**, 965–973.

BRANTLEY, S. L. & KOEPENICK, K. W. 1995. Measured carbon dioxide emissions from Oldoiyo Lengai and the skewed distribution of passive volcanic fluxes. *Geology*, **23**, 933–936.

BUICK, R. & DUNLOP, J. S. R. 1990. Evaporitic sediments of early Archean age from the Warrawoona Group, North Pole, Western Australia. *Sedimentology*, **37**, 247–277.

CALDEIRA, K. 1995. Long term control of atmospheric carbon: low-temperature seafloor alteration or terrestrial silicate-rock weathering? *American Journal of Science*, **295**, 1077–1114.

CALDEIRA, K. & KASTING, J. F. 1992. The life span of the biosphere revisited. *Nature*, **360**, 721–723.

CATLING, D. C., ZAHNLE, K. & MCKAY, C. P. 2001. Biogenic methane, hydrogen escape, and the irreversible oxidation of the early Earth. *Science*, **293**, 839–843.

CATUBIG, N. R., ARCHER, D. E., FRANÇOIS, R., DEMENOCAL, P., HOWARD, W. & YU, E.-F. 1998. Global deep-sea burial rate of calcium carbonate during the last glacial maximum, *Paleoceanography*, **13**, 298–310.

DESMARAIS, D. J. & MOORE, J. G. 1984. Carbon and its isotopes in mid-oceanic basaltic glasses. *Earth and Planetary Science Letters*, **69**, 43–57.

EKART, D. D., CERLING, T. E., MONTAÑEZ, I. P. & TABOR, N. J. 1999. A 400 million year carbon isotopic record of pedogenic carbonate: implications for paleoatmospheric carbon dioxide. *American Journal of Science*, **299**, 805–827.

FISCHER, T. P., GIGGENBACH, W. F., SANO, Y. & WILLIAMS, S. N. 1998. Fluxes and source of volatiles dischanged from Kudryavy, a subduction zone volcano, Kurile Islands. *Earth and Planetary Science Letters*, **160**, 81–96.

FRANCK, S., KOSSACKI, K. & BOUNAMA, C. 1999. Modelling the global carbon cycle for the past and future evolution of the earth system. *Chemical Geology*, **159**, 305–317.

FRANÇOIS, L. M. & WALKER, J. C. G. 1992. Modelling the Phanerozoic carbon cycle and climate: constraints from the $^{87}Sr/^{86}Sr$ isotopic ratio of sea water. *American Journal of Science*, **292**, 81–135.

GERLACH, T. M. 1991. Present-day CO_2 emission from volcanoes. *EOS Transactions, American Geophysical Union*, **72**, 249–251.

GODDÉRIS, Y. & FRANÇOIS, L. M. 1995. The Cenozoic evolution of the strontium and carbon cycles: relative importance of continental erosion and mantle exchanging. *Chemical Geology*, **126**, 167–190.

GROTZINGER, J. P. & KASTING, J. F. 1993. New constraints on Precambrian ocean composition. *Journal of Geology*, **101**, 235–243.

HOLLAND, H. D. 1978. *The Chemistry of the Atmosphere and Oceans*. Wiley, New York.

HOLLAND, H. D. 1984. *The Chemical Evolution of the Atmosphere and Ocean*. Princeton University Press, PRINCETON, NJ.

HOOFT, E. E. E., SCHOUTEN, H. & DETRICK, R. S. 1996. Constraining crustal emplacement processes from the variation in seismic layer 2A thickness at the East Pacific Rise. *Earth and Planetary Science Letters*, **142**, 289–309.

KADKO, D., BAROSS, J. & ALT, J. 1995. The magnitude and global implications of hydrothermal flux. *In*: HUMPHRIS, S. E., ZIERENBERG, R. A., MULLINEAUX, L. S. & THOMSEN, R. E. (eds) *Seafloor Hydrothermal Systems, Physical, Chemical, Biological, and Geological Interactions*. Geophysical Monograph, American Geophysical Union, **91**, 446–466.

KASTING, J. F. & ACKERMAN, T. P. 1986. Climatic consequence of very high carbon dioxide levels in the early Earth's atmosphere, *Science*, **234**, 1383–1385.

KEMPE, S. & DEGENS, E. T. 1985. An early soda ocean? *Chemical Geology*, **53**, 95–108.

KEMPE, S. & KAZMIERCZAK, J. 1994. The role of alkalinity in the evolution of ocean chemistry, organization of living systems, and biocalcification processes. *Bulletin de l'Institut océanographique, Monaco, numéro spécial*, **13**, 61–117.

KERRICK, D. M. & CONNOLLY, J. A. D. 2001. Metamorphic devolatization of subducted oceanic metabasalts: implications for seismicity, arc magmatism and volatile recycling. *Earth and Planetary Science Letters*, **189**, 19–29.

KIEHL, J. T. & DICKINSON, R. E. 1987. A study of the radiative effects of enhanced atmospheric CO_2 and CH_4 on early Earth surface temperatures. *Journal of Geophysical Research*, **92**, 2991–2998.

KOSTER VAN GROOS, A. F. 1987. Weathering, the carbon cycle, and differentiation of the continental crust and mantle. *Journal of Geophysical Research*, **93**, 8952–8958.

LANGMUIR, C. H., KLEIN, E. M. & PLANK, T. 1992. Petrological systematics of mid-ocean ridge basalts: constraints on melt generation beneath ridges. *In*: PHIPPS MORGAN, J., BLACKMAN, D. K. & SINTON, J. M. (eds) *Mantle Flow and Melt Generation of Mid-ocean Ridges*. Geophysical Monograph, American Geophysical Union, **71**, 183–280.

LASAGA, A. C., BERNER, R. A. & GARRELS, R. M. 1985. An improved geochemical model of atmospheric CO_2 fluctuations over the past 100 million years. *In*: SUNDQUIST, E. T. & BROECKER, W. S. (eds) *The Carbon Cycle and Atmospheric CO_2: Natural Variations Archean to Present*. Geophysical Monograph, American Geophysical Union, **32**, 397–411.

LÉCUYER, C. & RICARD, Y. 1999. Long-term fluxes and budget of ferric iron: implication for the redox state of the Earth's mantle and atmosphere. *Earth and Planetary Science Letters*, **165**, 197–211.

MACLOED, G., MCKEOWN, C., HALL, A. J. & RUSSELL, M. J. 1994. Hydrothermal and ocean pH conditions of possible origin of life. *Origins of Life and Evolution of the Biosphere*, **24**, 19–41.

MARTY, B. & JAMBON, A. 1987. $C/^3He$ in volatile fluxes from the solid Earth: implications for carbon geodynamics. *Earth and Planetary Science Letters*, **83**, 16–26.

MARTY, B. & TOLSTIKHIN, I. N. 1998. CO_2 fluxes from mid-oceanic ridges, arcs, and plumes. *Chemical Geology*, **145**, 233–248.

MOJCZIS, S. J., HARRISON, T. M. & PIDGEON, R. T. 2001. Oxygen-isotope evidence from ancient zircons for liquid water at the Earth's surface 4300 Myr ago. *Nature*, **409**, 178–181.

MORSE, J. W. & MACKENZIE, F. T. 1998. Hadean ocean carbonate geochemistry. *Aquatic Geochemistry*, **4**, 301–319.

OWEN, T., CASS, R. D. & RAMANATHAN, V. 1979. Enhanced CO_2 greenhouse to compensate for reduced solar luminosity on early Earth. *Nature*, **277**, 640–642.

PAVLOV, A. A., KASTING, J. F., BROWN, L. L., RAGES, K. A. & FREEDMAN, R. 2000. Greenhouse warming by CH_4 in the atmosphere of early Earth, *Journal of Geophysical Research*, **105**, 11 981–11 990.

PLANK, T. & LANGMUIR, C. H. 1998. The chemical composition of subducting sediment and its consequences for the crust and mantle. *Chemical Geology*, **145**, 325–394.

RINGWOOD, A. E. 1961. Changes in solar luminosity and some possible terrestrial consequences. *Geochimica et Cosmochimica Acta*, **21**, 295–296.

RUSSELL, M. J. & HALL, A. J. 1997. The emergence of life from iron monosulphide bubbles at a submarine hydrothermal redox and pH front. *Journal of the Geological Society, London*, **154**, 377–402.

RYE, R., KUO, P. H. & HOLLAND, H. D. 1995. Atmospheric carbon dioxide concentrations before 2.2 billion years ago. *Nature*, **378**, 603–605.

SAGAN, C. & MULLEN, G. 1972. Earth and Mars: evolution of atmospheres and surface temperatures. *Science*, **177**, 52–56.

SANO, Y. & WILLIAMS, S. N. 1996. Fluxes of mantle and subducted carbon along convergent plate boundaries. *Geophysical Research Letters*, **23**, 2749–2752.

SANSONE, F. J., MOTTL, M. J., OLSON, E. J., WHEAT, C. G. & LILLEY, M. D. 1998. CO_2-depleted fluids from mid-ocean ridge-flank hydrothermal springs. *Geochimica et Cosmochimica Acta*, **62**, 2247–2252.

SCHMIDT, R. M. & HOUSEN, K. R. 1987. Some recent advances in the scaling of impact and explosion cratering. *International Journal of Impact Mechanics*, **5**, 543–560.

SCHULTZ, A. & ELDERFIELD, H. 1997. Controls on the physics and chemistry of seafloor hydrothermal circulation. *Philosophical Transactions of the Royal Society of London, Series A*, **355**, 387–425.

SCHULTZ, A. & ELDERFIELD, H. 1999. Controls on the physics and chemistry of seafloor hydrothemal circulation. *In*: CANN, J. R., ELDERFIELD, H. & LAUGHTON, A. (eds) *Mid-ocean Ridges, Dynamics of Processes Associated with Creation of New Oceanic Crust*. Cambridge University Press, Cambridge, 171–209.

SCHWARTZMAN, D. & MCMENAMIN, M. 1993. A much warmer Earth surface for most of geologic time: Implications to biotic weathering. *Chemical Geology*, **107**, 221–223.

SCHWARTZMAN, D. & VOLK, T. 1989. Biotic enhancement of weathering and the habitability of Earth. *Nature*, **340**, 457–460.

SCHWARTZMAN, D. & VOLK, T. 1991. Biotic enhancement of weathering and surface temperatures on earth since the origin of life. *Paleogeography, Paleoclimatology, Paleoecology*, **90**, 357–371.

SIBLEY, D. F. & VOGEL, T. A. 1976. Chemical mass balance of the Earth's crust: the calcium dilemma and the role of pelagic sediments. *Science*, **192**, 551–553.

SLEEP, N. H. 1979. Thermal history and degassing of the earth: some simple calculations. *Journal of Geology*, **87**, 671–686.

SLEEP, N. H. & ZAHNLE, K. 2001. Carbon dioxide cycling and implications for climate on ancient Earth. *Journal of Geophysical Research*, **106**, 1373–1399.

SLEEP, N. H., ZAHNLE, K. & NEUHOFF, P. S. 2001. Initiation of clement surface conditions on the earliest Earth. *Proceedings of the National Academy of Sciences*, **98**, 3666–3672.

STAUDIGEL, H., HART, S. R., SCHMINCKE, H.-U. & SMITH, B. M. 1989. Cretaceous ocean crust at DSDP sites 417 and 418: carbon uptake from weathering versus loss by magmatic outgassing. *Geochimica et Cosmochimica Acta*, **53**, 3091–3094.

STAUDIGEL, H., PLANK, T., WHITE, B. & SCHMINCKE, H.-U. 1996. Geochemical fluxes during seafloor alteration of basaltic upper oceanic crust: DSDP sites 417 and 418. *In*: BEBOUT, G. E., SCHOOL, D. W., KIRBY, S. H. & PLATT, J. P. (eds) *Subduction Top to Bottom*. Geophysical Monograph, American Geophysical Union, **96**, 19–36.

STEIN, C. A., STEIN, S. & PELAYO, A. M. 1995. Heat flow and hydrothermal circulation. *In*: HUMPHRIS, S. E., ZIERENBERG, R. A., MULLINEAUX, L. S. & THOMSEN, R. E. (eds) *Seafloor Hydrothermal*

Systems, Physical, Chemical, Biological, and Geological Interactions. Geophysical Monograph, American Geophysical Union, **91**, 425–445.

TAJIKA, E. & MATSUI, T. 1992. Evolution of terrestrial proto-CO$_2$ atmosphere coupled with thermal history of the Earth. *Earth and Planetary Science Letters*, **113**, 251–266.

TAJIKA, E. & MATSUI, T. 1993. Evolution of seafloor spreading rate based on 40Ar degassing history. *Geophysical Research Letters*, **20**, 251–266.

WALKER, J. C. G. 1977. *Evolution of the Atmosphere*. Macmillan, New York.

WALKER, J. C. G. 1985. Carbon dioxide on the early Earth. *Origins of Life and Evolution of the Biosphere*, **16**, 117–127.

WALKER, J. C. G., HAYS, P. B. & KASTING, J. F. 1981. A negative feedback mechanism for the long-term stabilization of the Earth's surface temperature. *Journal of Geophysical Research*, **86**, 9776–9782.

WEDEPOHL, K. H. 1995. The composition of the continental crust. *Geochimica et Cosmochimica Acta*, **59**, 1217–1232.

WILDE, S. A., VALLEY, J. W., PECK, W. H. & GRAHAM, C. M. 2001. Evidence from detrital zircons for the existence of continental crust and oceans on the Earth 4.4 Gyr ago. *Nature*, **409**, 175–178.

WILKINSON, B. H. & WALKER, J. C. G. 1989. Phanerozoic cycling of sedimentary carbonate. *American Journal of Science*, **289**, 525–548.

ZAHNLE, K. & SLEEP, N. H. 1996. Impacts and the early evolution of life. *In*: THOMAS, P. J., CHYBA, C. F. & MCKAY, C. P. (eds) *Comets and the Origin and Evolution of Life*, 175–208.

ZHANG, Y. & ZINDLER, A. 1993. Distribution and evolution of carbon and nitrogen in Earth. *Earth and Planetary Science Letters*, **117**, 331–345.

Global modelling of continent formation and destruction through geological time and implications for CO₂ drawdown in the Archaean Eon

JAN D. KRAMERS

Isotope Geology Group, Institute of Geological Sciences, University of Bern, Erlachstrasse 9a, 3012 Bern, Switzerland (e-mail: kramers@geo.unibe.ch)

Abstract: The possible drawdown of a massive CO_2 atmosphere in early Earth history is discussed using two working hypotheses: first, that this removal of CO_2 from the atmosphere occurred mainly via silicate weathering; second, that crust-to-mantle recycling rates found from forward modelling of crust–mantle history can be used to estimate rates of this ancient silicate weathering. Previous U–Th–Pb and Sm–Nd forward modelling efforts are reviewed, from which it was concluded that an insignificant amount of continental crust existed at 4.4 Ga, i.e. so-called 'no-growth models' for the continental crust appear untenable. New modelling carried out is based on a crustal growth curve starting with zero mass at 4.2 Ga and reaching 75% of the present crust mass by 2 Ga. It concerns variations in crust-to-mantle recycling rates through geological time. Best fits to isotope data are obtained if it is assumed that erosion rates (mass removal per unit surface) were approximately constant from early Archaean time to the present. From the results it can be estimated that drawdown of a massive CO_2 atmosphere by silicate weathering could have been completed by the end of Archaean time at the earliest, and about 1.5 Ga ago at the latest.

Silicate weathering on land surfaces was first suggested and studied by Ebelmen in 1845 (see Berner & Maasch 1996) as a mechanism that could have removed CO_2 from the atmosphere. Its importance as a major control on atmospheric CO_2 levels was highlighted by Urey (1952). Its effectiveness in lowering atmospheric CO_2 levels has been demonstrated by Berner *et al.* (1983) and Berner (1991) for Phanerozoic time and it is essential in models linking climate change to uplift of mountain belts (Raymo & Ruddiman 1992). The combined carbonate and organogenic carbon in sediments and metasediments amounts to at least 3.6×10^{21} moles (Ronov & Yaroshevskiy 1968; Wedepohl 1995). If all present as CO_2 this would form a $c.$ 30 bar atmosphere, and there can be no doubt that all this carbon was once in the atmo-hydrosphere, although not necessarily all at the same time or all in the form of CO_2. This has important implications for the climate of the early Earth: Kasting & Ackermann (1986) found from 1D radiative–convective modelling that a dense (10–20 bar) CO_2 atmosphere in very early Archaean time could have produced a greenhouse effect that offset the 30% fainter solar irradiation at that time and prevented a wholly frozen Earth. However, Rye *et al.* (1995) concluded from mineralogical observations on palaeosols and thermodynamic data that the partial pressure of CO_2 in the atmosphere between 2.5 and 2.2 Ga was less than $c.$ 0.04 bar. They noted that CH_4 is also an effective greenhouse gas, and may have been abundant in the Archaean atmosphere. Alternatively, CO_2 that had made up a dense atmosphere in early Archaean time could already have been bound by silicate weathering at 2.5 Ga. Evidence of glaciation in the $c.$ 2.9 Ga Mozaan Group, South Africa (Young *et al.* 1998) and in the early Proterozoic Huronian Supergroup, Canada (Nesbitt & Young, 1982) supports the latter view.

Thus estimates of weathering and erosion rates in Archaean time have a bearing on questions of the early evolution of the atmosphere and the climate of the young Earth. To make use of this connection to constrain early CO_2 drawdown potential, it must first be ascertained that chemical weathering was important in Archaean time. Then, Archaean weathering rates should somehow be estimated. This paper therefore consists of two parts. First, CO_2 drawdown is viewed in relation to continental weathering with specific reference to Archaean time. This leads to a minimum estimate for the amount of weathering and erosion needed to remove a $c.$ 30 bar atmosphere. Second, an attempt is made to constrain erosion rates in Archaean time via forward global models of continental crust growth and recycling, which correctly predict present-day Th–U–Pb and Sm–Nd isotope systematics of major terrestrial reservoirs (Kramers & Tolstikhin 1997; Nägler & Kramers 1998). In such modelling, rates of recycling continental crust into the mantle are critical. If the assumption is made that this recycling is a measure of erosion, and that erosion in turn is a measure of chemical weathering, the results of modelling allow us to place constraints on the potential for early CO_2 drawdown.

CO_2 drawdown and chemical weathering

Weathering of silicates occurs through many different hydrolysis reactions whereby CO_2 is converted to HCO_3^- ions and Si is either bound in alteration minerals or liberated as H_4SiO_4. Alkali and alkaline earth elements are progressively removed in solution. Of these, Ca and Mg are largely ultimately bound in carbonate, so that the whole process can be summarized in a generic way by the 'Urey reaction'

$$CO_2 + CaSiO_3 \rightarrow \text{(weathering, transport and carbonate formation)} \rightarrow CaCO_3 + SiO_2 \quad (1)$$

(and analogously for Mg). The silicate weathering + carbonate precipitation process thus removes 1 mole of CO_2 out of the atmosphere per mole of Ca or Mg leached from a weathering silicate rock. A mass balance of organogenic and carbonate carbon in sediments can be made on the basis of their $\delta^{13}C$ values compared with that of juvenile emanations from the mantle (Schidlowski 1988). These data suggest that a mass ratio of carbonate to organogenic carbon of approximately 4:1 has prevailed since about 3.5 Ga ago. As strong C isotope fractionation is a particular characteristic of photosynthesis, it is not possible that the mass balance merely reflects global redox balances. Thus silicate weathering has through geological time drawn down four times more CO_2 than photosynthesis. CO_2 is also bound by carbonate dissolution, but is then released again during the crystallization of new carbonate:

$$CaCO_3 + CO_2 + H_2O \leftrightarrow Ca^{2+} + 2 HCO_3^- \quad (2)$$

Therefore carbonate weathering and subsequent formation of new limestone cause no net removal of CO_2 and have, from the perspective of geological time, a transient effect on atmospheric CO_2 only.

The amount of carbonate rocks present on Earth is thus related to the release of cations by chemical silicate weathering over geological time. In the present-day world, silicate weathering reactions are much enhanced by soil CO_2 partial pressures 10–100 times greater than atmospheric levels, and the presence of humic acids in soils, both of which are due to land plants and associated soil biota (Lovelock & Whitfield 1982; Schwartzman & Volk 1989; Berner 1997). In Archaean time, land vegetation was absent, and correspondingly larger inherent atmospheric concentrations of CO_2 were required for chemical weathering.

Present-day CO_2 drawdown rates can be estimated from the dissolved Ca^{2+} and Mg^{2+} load in rivers (Berner et al. 1983; Gaillardet et al. 1999). For estimating potential rates in the geological past, the counterpart of this dissolved load, i.e. Ca- and Mg-depleted palaeosols and clastic sediments, can be used. Evidence that intense chemical weathering occurred in Archaean and early Proterozoic time is provided by marked depletions in Na, K and Ca in palaeosols when compared with the corresponding unaltered rock (Rainbird et al. 1990), and also in Archaean shales and metapelites compared with average upper continental crust (e.g. Nesbitt & Young 1982; Fedo et al. 1996; Kreissig et al. 2000). In these studies, the immobile trace and major element geochemistry of the metasediments indicates provenance from ordinary Archaean granitoid crust, so that the mobile element depletion can be attributed with some confidence to chemical weathering in the source area. The amount of depletion in mobile elements in palaeosols or pelitic sediments is often expressed by the Chemical Index of Alteration (CIA, Nesbitt & Young 1982):

$$CIA = 100 \times Al_2O_3/(Al_2O_3 + CaO^* + Na_2O + K_2O) \quad (3)$$

where

$$CaO^* = CaO_{(total)} - CaO_{(phosphates)} - CaO_{(carbonates)}.$$

Fresh granitoid rocks have CIA values between 45 and 55. Average Mesozoic to recent shale values are between 70 and 75 (Nesbitt & Young 1982), and Archaean and Proterozoic shales and metapelites have values ranging from about 70 to an extreme of 99 for some shales from the Buhwa Greenstone Belt, Zimbabwe (Fedo et al. 1996). Losses of Ca and Na are mostly approximately equal, whereas K loss may be difficult to assess because of later metasomatism (Fedo et al. 1995). Mg depletion does not enter the CIA value. From metapelite geochemistry it is more difficult to quantify than K loss, because it is highly sensitive to the mafic rock content of the source area, and may also be enriched during diagenetic metasomatism. Nevertheless, Mg is mobile in weathering, as well as being an important constituent of sedimentary carbonates, particularly in Precambrian time. Therefore it will be assumed below that Mg and Ca are released in proportion to their concentrations in chemical weathering of 'average upper continental crust'.

The upper continental crust contains close to 3% Ca and 1.4% Mg (Taylor & McLennan 1995; Wedepohl 1995). If mobilization of Ca and Mg is proportional, then the observed range of CIA values corresponds to a loss of between 1.5 and 3% Ca and between 0.7 and 1.4% Mg. This corresponds to between 650 and 1300 moles of Mg + Ca lost per tonne of rock weathered. Therefore between 650 and 1300 moles CO_2 would ultimately be bound as carbonate per tonne of average crustal silicate rock weathered. The total mass of carbon occurring as carbonate in sediments and metasediments is reliably estimated at 3.5×10^{22} g, or 2.91×10^{21} moles (Ronov & Yaroshevskiy 1968; Wedepohl 1995). Drawing this out of the atmosphere would thus have required chemical weathering of between 2.2×10^{18} and 4.5×10^{18} tonnes of upper-crustal silicate rock, i.e. between 10 and 21% of the present-day mass of the continental crust (2.1×10^{19} tonnes, Taylor & McLennan, 1995). Shales and metapelites, i.e. residua of chemical weathering, constitute at present c. 4% of the continental crust (Ronov & Yaroshevskiy 1968; Wedepohl 1995). The difference is thus between 6 and 17% of the present continental crust mass. From a first-order estimate, this is the amount of such sediments that should have been recycled into the mantle over geological history to account for the present mass of carbonate in sedimentary rocks. This is a pure mass balance reckoning, which is not affected by consideration on the age distribution of sediments (e.g. Veizer & Jansen 1985).

There are, however, two main reasons why this must be a minimum estimate of the amount of sediment recycled into the mantle over geological time. These are, first, reintroduction of CO_2 into the atmosphere via arc volcanism, and second, the fact that purely clastic sediments are ignored in this mass balance. CO_2 is reintroduced into the atmosphere as a result of decarbonation reactions in calc-silicate rocks (the reverse 'Urey reaction'), requiring further weathering and photosynthesis to remove it. Fluxes associated with these processes over Phanerozoic time have been reviewed by Berner et al. (1983) and Berner (1991), who concluded that such addition of CO_2 by (mainly arc) volcanism and its drawdown by silicate weathering have been the major long-term fluxes over this period, and that drawdown has only slightly outstripped supply. In other words, crust-to-mantle recycling that took place since early Phanerozoic time did not contribute significantly to net CO_2 drawdown, but merely helped to maintain a quasi steady state. Concerning the second reason, a significant portion of sediments recycled into the mantle over time was probably produced by mechanical erosion: greywackes are an abundant rock type throughout the geological record, and are often not strongly depleted in Ca and Mg relative to their source rocks.

Therefore it can be concluded that the total mass of sediment that must have been recycled into the mantle over geological time to account for CO_2 removal from the atmosphere is >6–17% of the present continental mass. Drawdown of most CO_2 before 2.5 Ga, as indicated by the Huronian glaciation (Nesbitt & Young 1982; Kasting 1993) indicates vast amounts of chemical weathering before that time. This cannot *a priori* be equated with sediment recycling into the mantle. However, Veizer & Jansen (1979, 1985) and McLennan (1988) have noted a comparative rarity of very ancient sediments and have shown that Archaean clastic sediments are mainly juvenile, i.e. they have model Nd 'crust formation' ages close to their depositional ages. This means, first, that young terrains were being weathered and eroded, and second, that little redeposition of sediments ('cannibalistic recycling') took place in Archaean time. Combined, these observations point to significant sediment recycling into the mantle before c. 2.5 Ga. Below, the possibility of such Archaean recycling is considered from the global modelling point of view.

Models of crustal growth and recycling

Existing models of continental crust growth and recycling offer a perspective on the potential for CO_2 drawdown through geological time. There is a vast number of such models, reviewed, for example, by Reymer & Schubert (1984) and Taylor & McLennan (1995). They range from models that are based strictly on the age distribution of continental rocks observed today and thus imply minimal recycling or none at all (Hurley & Rand 1969; Condie 1990), to 'no-growth' models such as first put forward by Armstrong (1981) based on the principle that continental freeboard should have remained constant through geological time. In such models it is postulated that an amount of continental crust similar to that observed today was generated very early in Earth history (typically, in the first 500 Ma or less) and that since then, newly formed crust replaced old crust recycled into the mantle.

Although the 'no recycling' models appear incompatible with the above requirements for early CO_2 drawdown, the 'no growth' models require a very large amount of continental crust recycling to reproduce the observed crustal age distribution, and thus offer a huge excess capacity for CO_2 drawdown, probably allowing low atmospheric CO_2 levels to be attained in early Archaean time. This range of models thus has important consequences for the development of the Earth's early atmosphere and climate, and needs to be evaluated. One approach to do this has been a search for depleted mantle isotope signatures in the Archaean time, and another has been forward transport balance modelling.

The first approach was based on the observed complementarity with respect to Sm/Nd ratios of continental crust and depleted mantle. With time, the latter reservoir develops radiogenic $^{143}Nd/^{144}Nd$ ratios relative to the undifferentiated Chondritic Uniform Reservoir (CHUR; see Jacobsen & Wasserburg 1984), i.e. positive ε_{Nd} values, where $\varepsilon_{Nd} = [(^{143}Nd/^{144}Nd_{sample})/(^{143}Nd/^{144}Nd_{CHUR}) - 1] \times 10^4$. Positive initial ε_{Nd} values in ancient crustal provinces constitute evidence that a significant mass of continental crust existed before these ancient provinces were formed. Initial ε_{Nd} values were obtained on c. 3.7 Ga rock suites from West Greenland (Bennett et al. 1993) and the c. 4 Ga Acasta Gneisses from Northern Canada (Bowring & Housh 1995) by correcting measured Nd isotope ratios back using zircon U–Pb ages from the same rocks. The results showed large apparent initial ε_{Nd} variations with values up to four which were interpreted by Bowring & Housh (1995) as evidence that large amounts of crust had developed before 4 Ga, thus supporting no-growth models for the continental crust. The possibility that so large an amount of crust was formed very early was argued by Reymer & Schubert (1984) as follows. The melt fluxes required to form the core, and gravitation potential energy released, would be sufficient to generate the present amount of continental crust in 100 Ma or less. After accretion and core formation had ceased, residual and radiogenic heat would support a much lower rate of melting in the mantle. However, the approach used by Bowring & Housh (1995) has been criticized by Moorbath et al. (1997). They showed that correlations of $^{143}Nd/^{144}Nd$ with $^{147}Sm/^{144}Nd$ in the ancient rock suites yield apparent ages much younger than those obtained from zircons (e.g. 3.4 Ga in the case of the 4 Ga Acasta Gneisses), indicating later resetting of Sm–Nd systematics. The apparent variations in initial ε_{Nd} values were reinterpreted by them as mere effects of this resetting, which should therefore not be used as evidence for the pre-existence of continental crust. The debate about this is still continuing. Further, efforts to find Hf isotope effects in zircons from ancient sediments to substantiate the apparent source ε_{Nd} heterogeneity (Amelin et al. 1999) have been unsuccessful.

Transport balance or box models have been used by many workers in the past in efforts to understand the trace element and isotope characteristics of the Earth's major silicate reservoirs, i.e. continental crust, and upper and lower mantle (e.g. Jacobsen & Wasserburg 1979; Zartman & Haines 1988). Although simple mass balance calculations can be applied to present-day trace element concentrations and Pb, Nd and Hf isotope compositions of major reservoirs, e.g. continental crust and depleted mantle, to test the hypothesis that these reservoirs are complementary, transport balance models are needed to test ideas on their evolution in time. The reason is that the isotope ratio variations are the result of time-integrated trace element variations in the reservoir, modified by fluxes between them. Below, recent transport balance models in which the evolution of the continental crust is examined using Th–U–Pb (Kramers & Tolstikhin 1997) and

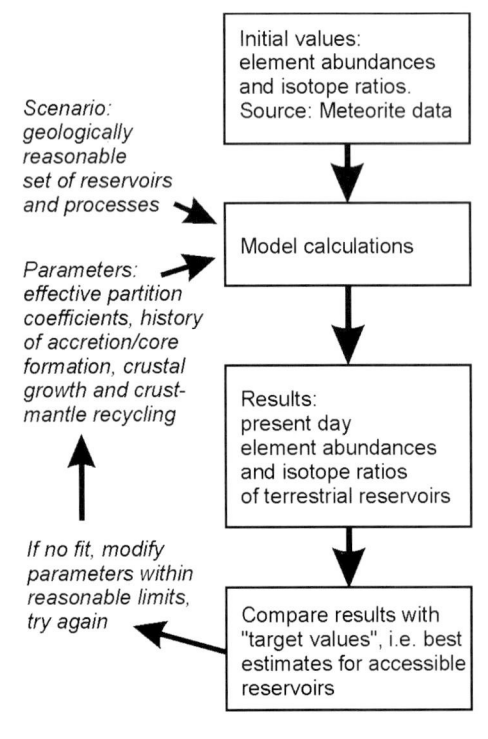

Fig. 1. Principle of forward transport balance modelling used.

Sm–Nd systematics (Nägler & Kramers 1998) are reviewed and then developed further.

Principles and methods of transport balance models; previous work

The principle of the procedure used in the work reviewed and the present study is illustrated in Figure 1. Target values of the modelling are present-day Pb and Nd isotope ratios as well as Th, U, Pb, Sm and Nd concentrations of the depleted mantle, as derived from mid-ocean ridge basalt (MORB) data (Blichert-Toft & Albarède 1994), and of the upper continental crust, for which pelagic sediments, atmospheric dust and river sediments are proxies (Goldstein et al. 1984; Ben Othman et al. 1989). Important sets of target values are summarized in Table 1b. The models include an accretion–core formation stage and therefore chondritic element abundance data are used as a starting point (Table 1a). Forward modelling also allows us to exclude scenarios that do not satisfy other constraints, e.g. accretion–core formation scenarios used were restricted to those that satisfied constraints from siderophile element abundances and Hf–W isotope systematics (Lee & Halliday 1996; Kramers 1998).

Th–U–Pb systematics and Sm–Nd systematics address in part different aspects of Earth differentiation. First, Th/Pb and U/Pb ratios are affected by core formation, whereas the Sm/Nd ratio is not. Second, in the silicate Earth Th/Pb

Table 1. *Input parameters and target values used in modelling*
(a) Input parameters and masses of reservoirs

Parameters, dimensions	Value
Initial reservoir: proto-terrestrial material	
Initial Th (ppm)	0.0607
Initial atomic $^{235}U/^{238}U$	0.31628
Initial atomic $^{232}Th/^{238}U$	2.4115
Initial atomic $^{238}U/^{204}Pb$	3.7628
$^{206}Pb/^{204}Pb$	9.308
$^{207}Pb/^{204}Pb$	10.294
$^{208}Pb/^{204}Pb$	29.440
Nd (ppm)	0.6
Initial $^{143}Nd/^{144}Nd$ ratio	0.506753
Initial atomic $^{147}Sm/^{144}Nd$ ratio	0.20264
Decay constants (Ga^{-1})	
$\lambda(^{238}U)$	0.155125
$\lambda(^{235}U)$	0.98485
$\lambda(^{232}Th)$	0.049475
$\lambda(^{147}Sm)$	0.00654
Masses of present-day reservoirs, 10^{25} g	
Continental crust	2.09
Depleted mantle	130

(b) Target values

Parameters	Depleted mantle	Upper continental crust (ucc)	Recent sediments; proxy for ucc
Th ppm	0.019 ± 0.004	10.5 ± 0.5	
U ppm	0.007 ± 0.002	2.6 ± 0.3	
$^{232}Th/^{238}U$	2.62 ± 0.14		
Pb ppm	0.05 ± 0.15	19 ± 2	
$^{206}Pb/^{204}Pb$	18.46 ± 0.41		18.80 ± 0.24
$^{207}Pb/^{204}Pb$	15.49 ± 0.04		15.64 ± 0.06
$^{208}Pb/^{204}Pb$	38.02 ± 0.37		38.68 ± 0.3
Sm ppm	0.28 ± 0.06	4.6 ± 0.1	
Nd ppm	0.77 ± 0.15	26 ± 2	
$^{147}Sm/^{144}Nd$	0.22 ± 0.02	0.107 ± 0.009	
ε_{Nd}	10.5 ± 0.5		−11.4 ± 4

See Kramers & Tolstikhin (1997) and Nägler & Kramers (1998) for a discussion and sources.

and U/Pb fractionation occurs in intracrustal processes rather than crust formation, because all three elements are so incompatible in mantle melting that their element ratios in the crust are *a priori* not very different from those in the mantle source. In contrast, Sm/Nd is strongly fractionated in mantle melting and thus in crust formation, and much less in intracrustal processes. The two sets of systematics have in common that parent and daughter species are both incompatible, strongly enriched in the continental crust and therefore Pb and Nd isotope ratios in the depleted mantle are highly sensitive to crust–mantle recycling.

The set of reservoirs used in the previous and present work is shown in Figure 2. Between reservoirs mass fluxes operate, to which concentrations of trace elements are assigned that can differ from the concentration in the 'source' of the flux depending on magmatic fractionation or weathering processes. Although the number of reservoirs must be kept to a minimum to limit the number of free parameters in the model, the division of the continental crust reservoir into four subreservoirs (younger and older as well as upper and lower) is needed to portray age heterogeneity of the continental crust as well as chemical heterogeneity. The mass of each of these four subreservoirs is kept at one-quarter of the (variable) mass of the total continental crust. Whereas upper–lower crustal differences are brought about by magmatic fractionation, the distinct character of the older and younger crust reservoirs results from two factors. First, any continental crust newly produced is mixed into the younger part of the crust, and an equalizing flux transfers mass and trace species from the younger to the older reservoir to keep their masses equal. Second, an 'erosion law' (Allègre & Rousseau 1984; Jacobsen 1988) operates whenever continental mass is recycled by erosion: erosion rates for the subreservoir labelled 'old

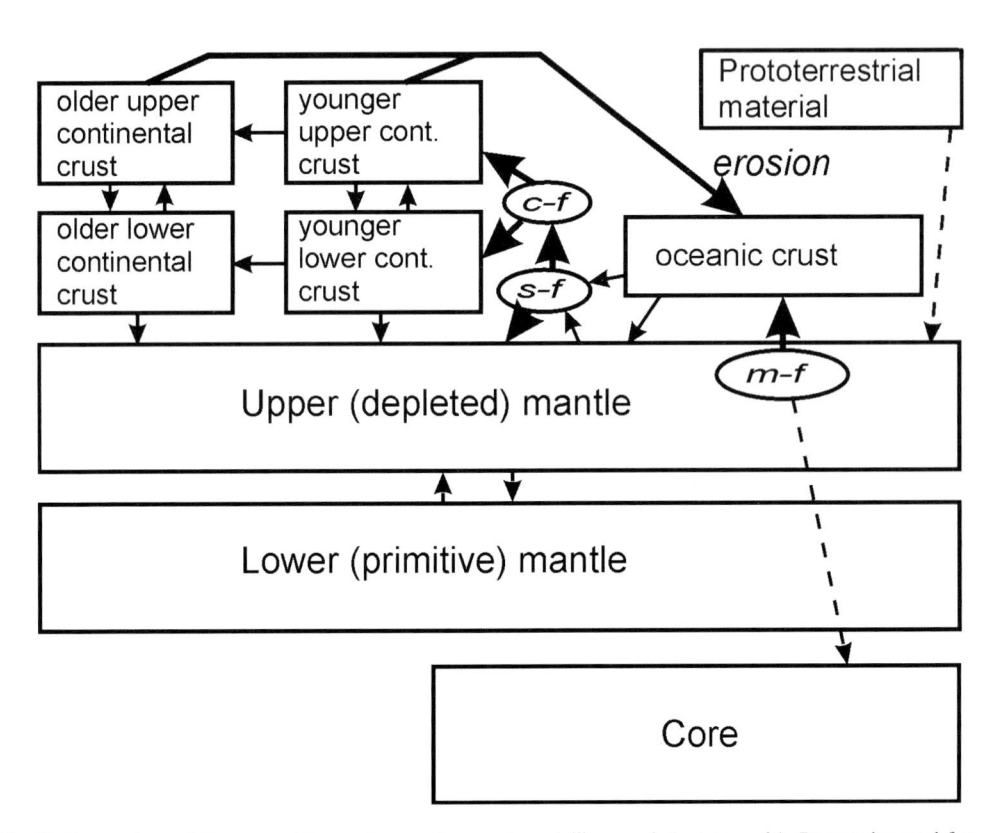

Fig. 2. Reservoirs and fluxes used in previous and present modelling work (not to scale). Rectangles used for reservoirs; ellipses for loci of fractionation by melt processes. *m-f*, depleted mantle melting, oceanic crust (MORB) formation; *s-f*, subduction zone melting leading to continental crust formation; *c-f*, intracrustal fractionation leading to upper and lower crust formation. Bold arrows, fluxes involving trace element fractionation; fine arrows: fluxes without trace element fractionation; stippled arrows, fluxes operating only during accretion and core formation.

upper crust' are set lower than those for that labelled 'young upper crust'. The erosion law factor K describes the disproportionality. If the amounts of older and younger crust are equal, then

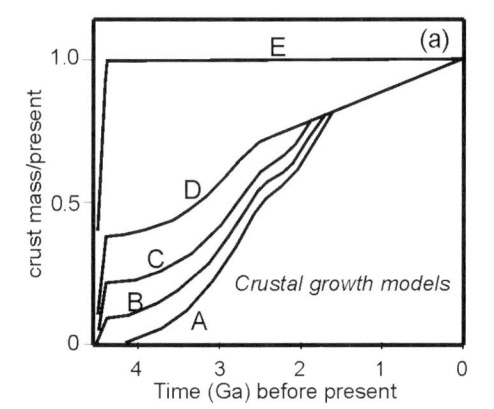

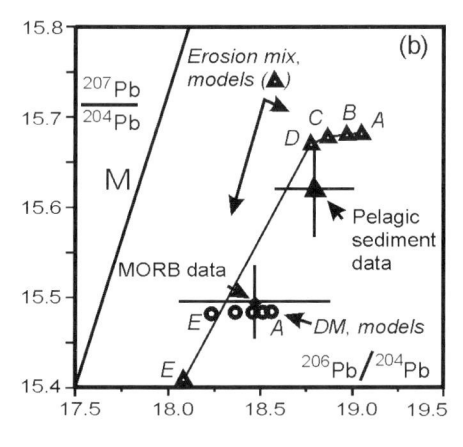

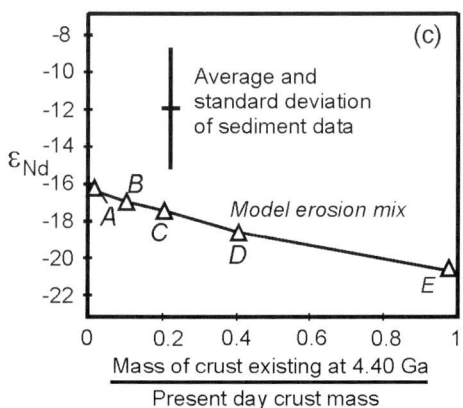

$$K = \Phi_{erosion}(young)/\Phi_{erosion}(old) \quad (4)$$

where $\Phi_{erosion}$ is the erosion mass flux. In forward modelling the application of this erosion law becomes a self-fulfilling prophecy, as the residence time of mass and species in the reservoir labelled 'older' becomes greater than that in the part labelled 'younger' because of the smaller fluxes into and out of the 'older' part. The division into younger and older crust thus exists throughout Earth history, and is alway relative. Erosion affects the upper crust and as it proceeds, fluxes are needed to transfer mass and trace species from the 'lower' to the 'upper' crust to keep their masses equal. Apart from erosion, destruction of continent can also occur by delamination of the lower crust, a process to which a 'delamination law' analogous to the erosion law can apply. Equations for the derivation of the mass fluxes between the reservoirs and their associated trace species fluxes have been given by Kramers & Tolstikhin (1997) and Nägler & Kramers (1998).

In both Th–U–Pb and Sm–Nd global modelling as outlined above, the crustal growth history has turned out to be a critical parameter set for producing fits to target values. With a given present-day age distribution for the continental crust that corresponds closely to that of Taylor & McLennan (1995), the main parameter varied by Kramers & Tolstikhin (1997) and Nägler & Kramers (1998) was the amount of crust generated during accretion and core formation, and surviving after those turbulent processes.

Fig. 3. Some results of previous modelling. (a) Crustal growth scenarios used (curves give crust mass existing at times in the past relative to present-day continental crust mass of 2.1×10^{15} g (Taylor & McLennan 1995). (b) $^{207}Pb/^{204}Pb$ v. $^{206}Pb/^{204}Pb$ plot showing modelling results for the depleted mantle (○) and average sediments (erosion mix) from the continental crust (△) for these scenarios, compared with data averages and standard deviations for MORB (proxy for depleted mantle) and for pelagic sediments (proxy for eroded continental crust). Data are from multiple sources quoted by Kramers & Tolstikhin (1997; see this reference also for details of modelling methods). M, 'Meteoritic' 4.55 Ga Pb/Pb isochron, drawn in for reference. Clearly no fit is achieved at all for the no-growth scenario E. (c) ε_{Nd} value of model erosion mix for these scenarios shown dependent on same parameter (Nägler & Kramers 1998), compared with data on recent sediments (Goldstein *et al.* 1984). In A, B, C and D models are forced to produce ε_{Nd} values of 10.5 ± 1 for the depleted mantle (Blichert-Toft & Albarède 1994), whereas in scenario E this value for the mantle could not be reached. In all this modelling $K = 2$ was used. Fits for ε_{Nd} in scenarios A, B and C (but not D and E) improve if $K = 3$. It should be noted that scenario A (no continental crust present at 4.4 Ga) gives the best fit to data in both cases.

Some results of these tests are summarized in Figure 3. From Figure 3b it is clear that no Pb isotope fit to pelagic sediment data is possible with a no-growth model (E in Fig. 3a). Figure 3c shows that fits for the erosion mix from the crust become progressively better as the amount of continental crust present at the end of the accretion and core formation period (c. 150 Ma after chondrite formation; 4.4 Ga ago) is reduced (models D, C, B and A). Both sets of results can be explained via the large amount of crust-to-mantle recycling needed in no-growth models or similar scenarios. Regarding Pb isotopes, the upper crust has a high U/Pb ratio, and has developed a radiogenic Pb signature whereby the ^{207}Pb/^{204}Pb ratio became particularly enhanced in early Earth history when ^{235}U was still abundant. This causes the observed Pb isotope difference between upper continental crust and depleted mantle. The explanation for the misfit of Pb isotope data in no-growth models is that the excessive crust-to-mantle recycling inherent in them brings large amounts of radiogenic Pb from the upper crust into the mantle, and therefore no Pb isotope difference similar to the observed one is predicted. With regard to Nd isotopes, in no-growth models large amounts of unradiogenic Nd from the crust are introduced into the mantle over time, so that the present-day ε_{Nd} value of the depleted mantle is only about three (Nägler & Kramers 1998) instead of the observed present-day value of around 10.5 (Blichert-Toft & Albarède 1994). This lower ε_{Nd} value for the mantle causes also the younger crust to have a lower average ε_{Nd} value, leading to the observed misfit.

Although the Pb isotope ratios on their own seem to allow scenarios B, C and D the fit for another parameter, the Th/U ratio of the mantle, worsens dramatically from scenario A to D (Kramers & Tolstikhin 1997). The conclusion, reached independently from Th–U–Pb and Sm–Nd modelling, was that only insignificant amounts of crust could have survived from the time of Earth accretion to become incorporated into the growing continental crust. This does not mean that no protocontinental crust could have been generated during Earth accretion and core formation (see Wilde et al. 2001) but this should have been destroyed rapidly in terms of geological time. Further, good fits for both Th–U–Pb and Sm–Nd systematics have resulted from models in which the average net growth of continental crust before 2 Ga was about twice that prevailing afterwards, so that the mass of crust in existence at 2 Ga was about 70% of the present mass.

These main results of previous forward global modelling are in broad agreement with the estimate of continental growth by Taylor & McLennan (1995) and also with the mass balance obtained from Th/Nb ratios of mantle–derived volcanic rocks through time (Collerson & Kamber 1999). Further, if the lower crust is considered to be characterized by low Th/Pb and U/Pb ratios and unradiogenic Pb, a further result from the modelling to date is that crust destruction by lower-crustal delamination is insignificant compared with erosion of the upper crust. This result stems from the relatively radiogenic average Pb isotope composition of the upper mantle, and is fairly robust (Kramers & Tolstikhin 1997). Essentially, crust–mantle recycling thus depends on surface erosion. However, weathering and erosion rates (and therefore

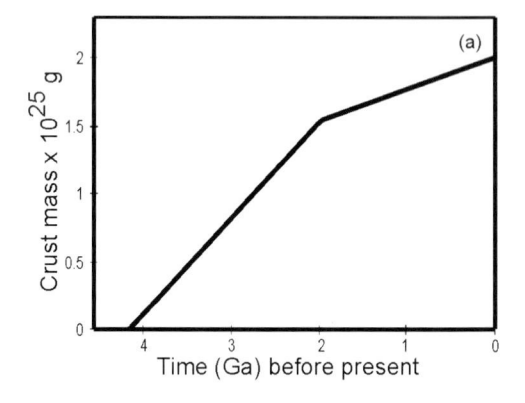

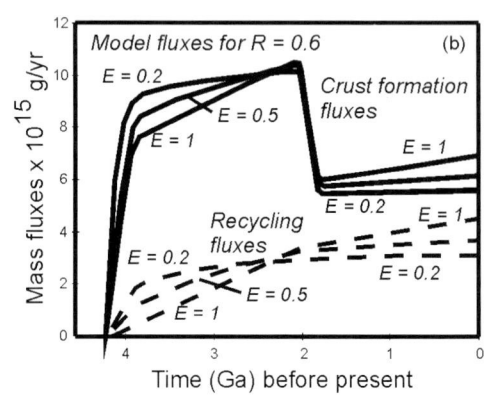

Fig. 4. (a) Crust mass v. time curve used in the present study, a simplified version of curve A in Figure 3a. (b) Bulk mass fluxes in and out of continental crust over time for three values of the geometrical parameter E and for $R = 0.6$ (see text). Continuous lines, fluxes into continental crust; dashed lines, fluxes out of continental crust. Pairs of curves for each value of E balance out to the crust mass v. time curve in (a).

potential CO$_2$ drawdown) in Archaean time cannot be assessed from net crust mass v. time curves alone: any curve giving the mass of crust in existence at any time in the past allows an infinite number of crust formation and crust destruction scenarios, whereby larger rates of crust formation can be compensated by larger rates of erosion to produce the same net crust mass v. time curve, as illustrated in Figure 4. Whereas the previous forward global modelling discussed above (Kramers & Tolstikhin 1997; Nägler & Kramers 1998) has examined the effect of varying very early continental crust mass, various scenarios of crustal growth and recycling rates based on a single crust mass v. time curve (Fig. 4a) are studied in the following section.

Variations on a theme: a single crust mass v. time curve

The single crust mass v. time curve in Figure 4a results from best fits of previous work (see Fig. 3), and the modelling described below explores different combinations of rates of crust formation and recycling, which all balance to this same net curve. To limit the number of unconstrained variables, the assumption is made that recycling is through erosion, with no lower-crustal delamination, as indicated by previous modelling. Two parameters, R and E (explained below), are introduced which, together with the erosion law factor K (see equation (4)), define crustal growth and recycling scenarios based on the curve of Figure 4a.

First, the total amount of crust-to-mantle recycling is defined by R, the time-integrated recycling mass flux over Earth history expressed as a fraction of present-day crust mass:

$$R = \frac{\int_{t=-4.55}^{today} \Phi_{recycling}\, dt}{M_{crust}(today)}. \quad (5)$$

Second, it is assumed that the recycling mass flux and its ancillary trace species fluxes are at any time dependent on the amount of crust in existence at that time. This dependence can take the form of various functions, and a variable geometry parameter E is introduced, which describes the crust to mantle recycling flux $\Phi(t)_{recycling}$ as a function of crust mass in existence.

$$\Phi(t)_{recycling} = f M_{crust}(t)^E \quad (6)$$

where f is a constant proportionality factor with dimension mass$^{(1-E)}$time^{-1}. Its value is calculated for given values of R and E by integrating the flux over time (equation (5)) and adjusting.

The meaning of E is illustrated by the following examples. (1) If it is assumed that, through Earth history, the worldwide crust recycling flux is proportional to the land surface, and the thickness of continental crust is constant, then recycling is also at any time directly proportional to the crust mass in existence. In this case, $E = 1$. (2) If the assumption is that erosion takes place predominantly at continental margins, then the recycling flux is proportional to the length of these margins. If a smaller amount of continental mass in the geological past implies a larger margin to surface (or margin to mass) ratio, then the worldwide recycling flux through time would more probably be proportional to the square root of the continental mass: thus $E = 0.5$. Further, as total crust mass increases with time, $E < 1$ means more intense erosion in early Earth history, and $E > 1$ the opposite. Figure 4b illustrates the variation of bulk mass fluxes into and out of the continental crust as a function of E for the crust mass v. time curve used in this study.

For any given crust mass v. time curve, the two parameters R and E determine the bulk crust formation and the crust recycling (erosion) fluxes through time, as

$$\Phi_{crust\,formation}(t) - \Phi_{recycling}(t) = \frac{dM_{crust}}{dt}(t). \quad (7)$$

Further, the erosion law factor K, defined in equation (4), regulates the age distribution of eroded material relative to the age structure of the crust. Thus for a given crust mass v. time curve, the three parameters E, R and K fully describe the crust formation and erosion history. In the present modelling E was varied between 0.2 and 1, R from 0.3 to 0.7, and K from 1.5 to 3.5. The approach used and the effective partition coefficients, as well as initial (meteoritic) concentrations of the trace species are the same as those used by Kramers & Tolstikhin (1997) and Nägler & Kramers (1998).

The results of numerical modelling with the above range of parameter variations are summarized in Figures 5 and 6 and Table 2. Sm and Nd concentrations of the model reservoirs closely approximate target values (Table 1a). The same is true for Th, U and Pb concentrations in the depleted mantle. Model Th and U concentrations for the upper continental crust are somewhat low, and those for Pb are significantly lower than observed values. This problem, also encountered by Kramers & Tolstikhin (1997), may not be real, as the upper-crust Pb concentration estimates of Taylor & McLennan (1995) and Wedepohl (1995) are too high to be consistent with the average ^{238}U/^{204}Pb ratios of c. 10

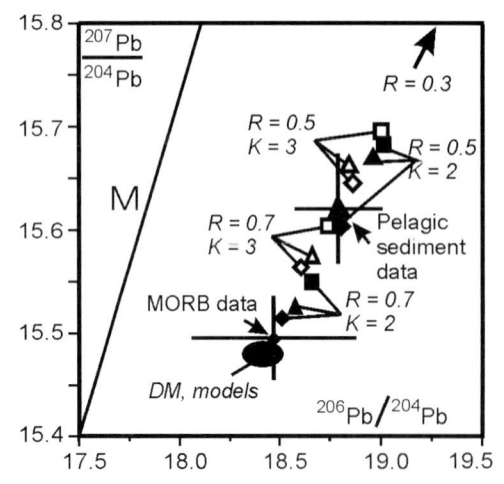

Fig. 5. Pb isotope results of present modelling compared with MORB and pelagic sediment data in ^{207}Pb/^{204}Pb v. ^{206}Pb/^{204}Pb diagram (for data source and 'M'; see caption to Fig. 3b). Filled ellipse below MORB data point is locus of depleted mantle Pb isotope compositions predicted by modelling. Also shown are present-day model isotope compositions for 'erosion mix'. Diamonds, $E=1$; triangles, $E=0.5$; squares, $E=0.2$. Closed symbols, $K=2$; open symbols, $K=3$. Results shown for $R=0.5$ and 0.7. Values from $R=0.3$ are outside figure frame (arrow indicates direction) and clearly do not represent a fit. (Discussion in the text.)

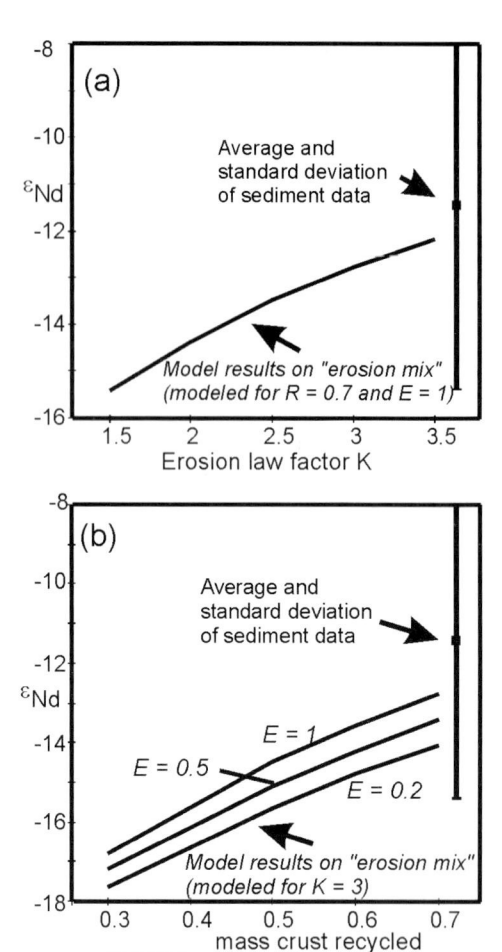

Fig. 6. ε_{Nd} values of model erosion mix resulting from varying parameters R, E and K compared with recent sediment data of Goldstein et al. (1984). (a) Effect of K, modelled with $R=0.7$ and $E=1$. (b) Effect of R and E, modelled with $K=3$. It should be noted that ε_{Nd} value for no-growth crust model are c. -21.5 (not shown; see Fig. 3c).

needed to generate observed average upper-crust Pb isotope ratios. Bearing in mind that between 1960 and 1985 the anthropogenic emission flux of Pb exceeded the crust–mantle recycling flux for Pb by more than an order of magnitude, it is not improbable that published estimates of upper-crust Pb concentrations are biased by contamination.

Regarding Th–U–Pb systematics, depleted mantle Th/U ratios around 2.9 and ^{208}Pb/^{204}Pb ratios around 37.8 result in all cases (Table 2). The latter is a near-perfect fit to data. The Th/U ratio is slightly too high compared with the best estimate of 2.62 ± 0.14 (O'Nions & McKenzie 1993) but nevertheless a fair fit. The relevant results from U–Pb isotope systematics modelling are summarized in Figure 5. Depleted mantle values for ^{207}Pb/^{204}Pb and ^{206}Pb/^{204}Pb ratios are closely reproduced, as in previous modelling, and those for the erosion mix all plot on a relatively narrow band through the average of pelagic sediment data. A measure of the goodnes of fit is how close the modelled erosion mix ratio results plot to the average of pelagic sediment data. Here a strong dependence on total recycling (parameter R) is apparent, with low total recycling resulting in a strongly radiogenic erosion mix, and vice versa. The reason is twofold. First, a larger amount of recycling leads to more lower crust material becoming incorporated in the upper crust over time, and second, average eroded crust is younger if the recycling rate is high, and therefore less radiogenic. Although the value of the exponential parameter E is surprisingly uncritical for this Pb isotope modelling (the early and late effects of variations in this parameter counteract each other), the erosion law factor K is clearly important. With a value of $K=2$ the Pb isotope difference between

Table 2. *Ranges of element concentrations and isotope ratios for reservoirs, resulting from present modelling*

Result	Depleted mantle	Upper crust (young)	Upper crust (old)	Erosion mix
Age (Ga)		1.1–1.6	2.7–2.9	
U (ppm)	0.006–0.0066	0.95–1.05	1.45–2.0	1.95–2.1
Th (ppm)	0.023–0.026	6.4–7.2	7.4–11	7.4–7.6
Pb (ppm)	0.036–0.04	6.3–6.9	9.4–11.1	7.52
Sm (ppm)	0.266	4.4–4.5	4.3–4.5	4.3–4.44
Nd (ppm)	0.677	21.8	24–25.5	22–23
$^{232}Th/^{238}U$	2.85–2.99	4.8–5.1	4.0–4.5	2.65–2.85
$^{238}U/^{204}Pb$	10.1–10.6	9.6–10.0	10–12	16.5–18
$^{206}Pb/^{204}Pb$	18.3–18.5	18.5–19.0	18.6–20.4	Fig. 6
$^{207}Pb/^{204}Pb$	18.65–18.8	15.5–15.7	15.51–16.2	Fig. 6
$^{208}Pb/^{204}Pb$	37.6–37.9	38.6–39.2	38.7–40.4	38.5–39.7
$^{147}Sm/^{144}Nd$	0.2375	0.1203	0.1085	0.1172
ε_{Nd}	9.5–11.5	−11 to −6.6	−34 to −30	Fig. 7

If no range is given for concentrations and element ratios, the variation is <1%.

the erosion mix and the depleted mantle almost disappears for high total recycling ($R = 0.7$). If $K = 3$ this difference is still discernible. As young crust is less radiogenic than old crust, this result is counter-intuitive and it is explained by enhanced retention of ^{207}Pb in the early continental crust if the K value is high. From Figure 5 it can be judged that a number of combinations of parameters would be possible. For instance, if $E = 1$ and $K = 2$ a good fit results for R c. 0.5. If $K = 3$ a value of R c. 0.6 gives a best fit.

In Sm–Nd isotope systematics the ε_{Nd} value of the erosion mix is used as a measure of the quality of the fits obtained (Fig. 6a and b). For all model variations, Sm and Nd concentrations are in the range of accepted values, and ε_{Nd} values between +10 and +11 were obtained for the depleted mantle (Table 2). Figure 6a shows a clear improvement of the fit with increasing K factor, as also noted by Nägler & Kramers (1998). This is intuitive, as a greater K factor means a greater proportion of younger crust in the erosion mix. Further, the quality of fit of erosion mix ε_{Nd} becomes progressively better as the values of E and/or R are increased (Fig. 6b). The effect of R is that (with constant E) greater time-integrated recycling reduces the average age of the present-day crust and therefore renders its ε_{Nd} value less negative. The effect of E is more subtle: for a fixed value of R, a lower value of E leads to a greater average age of the continental crust because erosion fluxes in the second half of geological history (see Fig. 4b) are reduced. The erosion law ultimately affects the average crust age more strongly if applied to late erosion than to early erosion, because the spread in ages in the crust becomes greater with time. Figure 6a and b shows that no perfect fit for model ε_{Nd} values of the erosion mix to recent sediment data is achieved with the range of values of K, E and R investigated, although a reasonably close match is found if all three parameters are set high. Values of R lower than c. 0.5 and values of E lower than c. 0.5 appear highly unlikely.

It is useful to combine the results of U–Pb and Sm–Nd modelling, because they mutually constrain each other and together permit a narrow set of scenarios. If the lower limit of c. 0.5 for exponential parameter E imposed by Sm–Nd modelling is applied to U–Pb systematics, it is seen (Fig. 5) that values for R (total crust recycling) larger than c. 0.6 lead to model erosion mix Pb isotope compositions which are not compatible with data from pelagic sediments, even if $K = 3$. Thus c. 0.6 is an upper limit for parameter R from U–Pb systematics. As the lower limit for this parameter from Sm–Nd systematics is c. 0.5, this yields a rather well-constrained estimate for total crust recycling. $R \approx 0.6$ combined with $E \approx 1$ and $K \approx 3$ yields the best possible approximation of model erosion mix isotope compositions to sediment data for the combined sets of systematics.

Following this result, the total amount of crust recycled by erosion into the mantle would be about 60% of the present crust mass ($R \approx 0.6$). This is not in conflict with any known geochemical data. An apparent contradiction exists with a maximum value of 30% derived from the mass balance of Ar in the atmosphere and K in the continental crust by Coltice et al. (2000). The argument used by these workers was based on the assumption that the amount of ^{40}Ar in the atmosphere and the continental crust combined is equal to or greater than that generated over geological time by the amount of K that has ever been in the continental crust by (equation (3) of Coltice et al. 2000). As they found an excess amount of ^{40}Ar in the atmosphere + crust of only 30% over what could have been produced by the equivalent amount of K at present in the crust, they concluded that not more than 30% of the continental crust mass can have been recycled into the mantle. In the basic assumption of this approach it is implied that the portions of mantle that melted to form continental crust are pristine and undegassed since the origin of the Earth. However, by far the most important degassing of Ar from the mantle occurs at mid-ocean ridges, and oceanic crust and mantle providing K in crust formation are already largely degassed (e.g. Tolstikhin & Marty 1998). Therefore the amount of Ar in the atmosphere and crust together cannot be used to place constraints on the amount of K that was added to the continental crust over time (and, by proxy, the total amount of continental crust formed over time).

The best fit result $E \approx 1$ means that the rate of recycling at any time in the past would have been approximately proportional to the mass of crust in existence at that time. If Archaean continental crust had the same thickness as today's, this implies an approximately constant rate of continent denudation per unit surface through geological time. It is interesting to compare predictions from this modelling with present-day denudation rates and oceanic sediment data. The combination $E = 1$, $R = 0.6$ yields a present-day erosion flux of 4.3×10^{15} g a^{-1} (see Fig. 4b). Averaged over today's continental surface of $c. 1.5 \times 10^8$ km^2, and assuming an average density of 2.7 for the eroded material, this yields an average denudation rate of $c.$ 10 mm ka^{-1}. Using cosmogenic nuclide abundances in river sediments, Schaller et al. (1999) have determined denudation rates over the last 10–20 ka^{-1} of between 20 and 100 mm ka^{-1} in Western European areas outside the Alps. Direct river load gauging data yield somewhat lower values, but have even more a snapshot nature. As much of the material eroded from continents is resedimented on continental shelves and in deltas, measured denudation rates must be higher than those predicted by crust recycling models, which is true in this case. The mass flux of $c. 4.3 \times 10^{15}$ g a^{-1} predicted by modelling can also be compared with estimated rates of oceanic sediment accumulation and sediment subduction. The Cenozoic (pre-anthropogenic) terrigenous oceanic sedimentation flux is estimated at $c. 10^{16}$ g a^{-1} (Milliman & Syvitski 1992), whereas the flux of terrigenous pelagic sediments into subduction zones is at present 1.1×10^{15} g a^{-1} (Rea & Ruff 1996). The difference between these estimates illustrates the difficulty with estimating crust–mantle recycling over large time scales, as pointed out by Rea & Ruff (1996). For instance, a large portion of today's sediment flux to the ocean floor is represented by the Bengal Fan, which is an extremely short-lived phenomenon in Earth history. On the whole, predictions from the present modelling are comfortably within the range defined by estimates based on very different observations.

Implications for CO_2 drawdown in the Archaean time

By linking crust-to-mantle recycling via erosion and weathering to CO_2 drawdown, these modelling results can be used to provide a semi-quantitative estimate on net Archaean CO_2 drawdown rates and the length of time that might have been required to remove a massive CO_2 atmosphere. Such a venture can only be tentative. The Archaean crust–mantle recycling fluxes estimated from modelling provide minimum values for the total erosion rates, as resedimentation on continents is documented for Archaean time (Fedo et al. 1996; Hunter et al. 1998) and may have been significant. On the other hand, the relative importance of purely mechanical erosion leading to graywacke-type sedimentation is not well known for Archaean time, so that total erosion rates impose only a maximum limit on chemical weathering rates. Further, although the extensive loss of mobile cations from Archaean fine clastic sediments and metasediments (e.g. Eriksson 1995; Fedo et al. 1996) indicates that Archaean weathering and erosion were linked to CO_2 drawdown, the rate of reintroduction of CO_2 into the atmosphere by metamorphic decarbonation and (arc) volcanism in Archaean time are unknown. On the one hand, the freeboard criteria indicate that Archaean oceans were probably shallower than todays', increasing the likelihood of carbonate subduction. On the other hand, the calcite compensation depth (CCD) level must also have been shallower in Archaean time as a result of the high atmospheric CO_2 pressure, and this would make carbonate subduction less likely. Following this reasoning, chemical weathering rates can thus provide only a maximum estimate of net CO_2 removal from the atmo-hydrosphere. From this combination of maxima and minima it follows that no firm constraints on Archaean CO_2 drawdown can be imposed from this work.

However, the results do provide a perspective, which is summarized in Figure 7. Integrating the crust–mantle recycling fluxes of Figure 4b over

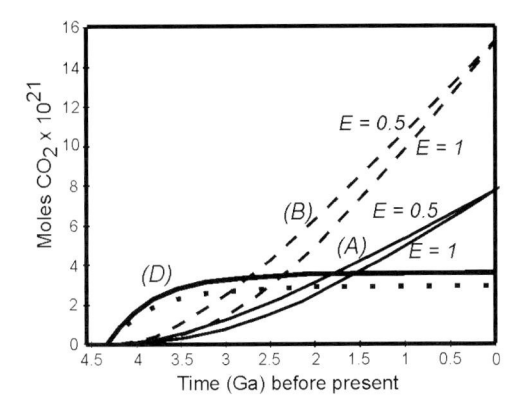

Fig. 7. Potential cumulative CO_2 drawdown capacity by weathering (curves A and B) compared with cumulative mantle CO_2 degassing (curve D) over geological time. A and B are derived from time integration of recycling curves in Figure 4b by assuming that erosion flux equals recycling flux, and drawdown of 650 moles of CO_2 per tonne of rock weathered (continuous curves, A) or 1300 moles (dashed curves, B). D is derived from upper-mantle melting flux of model of Kramers & Tolstikhin (1997), assuming MORB is fully degassed and no carbon goes back into the mantle, and is scaled to total carbon in sedimentary reservoirs (continuous curve) or to the portion in carbonate (dotted curve). D therefore depicts accumulation of carbon in atmo-hydrosphere and in sedimentary rocks. (Discussion in the text.)

time yields the total amount of crust recycled as functions of geological time. Such functions can be converted into cumulative CO_2 drawdown curves (indicating the total amount of CO_2 removed from the atmosphere, as a function of time) by assuming, first, that erosion equals crust–mantle recycling, and second, an amount of CO_2 removed from the atmosphere per unit mass of rock weathered. Curves A and B in Figure 7 are based in this way on recycling fluxes obtained for $R=0.6$ (Fig. 4b). Curves for $E=1$ represent the best-fit scenario for Th–U–Pb and Sm–Nd systematics, and those for $E=0.5$ are added for comparison. Curves A give the cumulative CO_2 drawdown capacity for 650 moles CO_2 ton^{-1} of rock weathered, and curves B for 1300 moles CO_2 tonne^{-1}.

Curve D is one possible description of the amount of CO_2 accumulated in the atmo-hydrosphere and sediments as a function of time. It describes cumulative degassing of the upper mantle and is scaled to the amount of C that is in these reservoirs today (the stippled variant corresponds to the amount of C in carbonates). This curve is derived from the mantle melting flux (i.e. MORB production flux) inherent in the modelling of Kramers & Tolstikhin (1997) and Nägler & Kramers (1998) using the assumptions that volatiles including CO_2 are completely lost from MORB, and that subducted oceanic crust returned to the mantle is free of CO_2. The late start of atmosphere accumulation in this model reflects a probable major atmosphere loss between 0.2 and 0.3 Ga as indicated by ^{129}Xe(I) and ^{136}Xe(Pu) systematics (Azbel & Tolstikhin 1993). This scenario for mantle degassing is slow compared with most others. For instance, Tolstikhin & Marty (1998) have modelled Ne, Ar and N_2 development of the atmosphere using a much greater degassing flux in early Earth history. For the purpose of estimating Archaean atmospheric conditions the differences are, however, insignificant, as even the slow degassing model used here leads to almost complete upper-mantle degassing by 3 Ga ago.

Following Figure 7 from the origin of the Earth towards the present, curve D is first above the cumulative drawdown curves A and B. The vertical difference between the sets of curves would roughly indicate the amount of free CO_2 (or more reduced species that can be oxidized to CO_2) in the atmosphere. This reaches between 1.8×10^{21} and 2.5×10^{21} moles, or 16 and 22 bar. At the intersection of curve D with curves A or B this large amount of CO_2 in the atmosphere could have been converted to carbonate. Later in time, when curves A or B are above curve D it would be expected that CO_2 levels will be kept low continuously by silicate weathering.

The question of which cumulative mantle degassing curve D is really relevant in this context is difficult to answer and depends on the relative rates of drawdown by photosynthesis and silicate weathering–carbonate formation in Archaean time. From the carbon isotope record, these are mainly considered to have remained constant (Schidlowski 1988), in which case the stippled lower curve D would be more appropriate. However, for the timing of drawdown derived from Figure 7 the choice of mantle accumulation curve is unimportant, given the tentative character of this whole consideration.

Following curves A and B for the best-fit scenario ($E=1$), a massive CO_2 atmosphere could have been removed by about c. 2.5 Ga ago at the earliest, and by 1.5 Ga ago at the latest, provided that reintroduction of CO_2 by arc volcanism was of minor importance compared with drawdown. Bearing in mind that curves A and B represent minimum drawdown potential (resedimentation on continents must be added) and that CIA values for Archaean shales can be very high (Fedo et al. 1996) the earlier estimate is probably closer to the truth. This result would be consistent with the low atmospheric CO_2

content derived by Rye et al. (1995) for the age range 2.5–2.2 Ga, and also with the observed Huronian glaciation (Nesbitt & Young 1982). Given the low time resolution of the method, it is also not in serious conflict with the glaciation observed in the 2.9 Ga Mozaan Group (Young et al. 1998).

Notwithstanding the large uncertainties mentioned earlier, it can be concluded from the above argumentation that net surplus CO_2 drawdown capacity by continental weathering probably existed since early Proterozoic time. This leads to a scenario of the evolution of atmospheric CO_2 content in a three-stage history. In Phanerozoic time, silicate weathering takes place, in spite of very low atmospheric CO_2 contents, under CO_2 partial pressures enhanced by biological activity in soils (Lovelock & Whitfield 1982; Schwartzman & Volk 1989; Berner 1997). In Archaean time, it took place under an inherently high CO_2 pressure. Between these two regimes, during most of Proterozoic time, a lower limit on atmospheric CO_2 partial pressure was probably set in the first instance by the reaction equilibria associated with silicate weathering, i.e between 0.01 and 0.1 bar (Rye et al. 1995). The CO_2 levels of these three stages and the timing of changes between them coincide very closely with the atmospheric CO_2 history proposed by Kasting (1993) based on the increase of solar luminosity with time, and the variations in greenhouse effect this requires. The transitions between the three types of climate regime would have been relatively sharp: 'Snowball Earth' events (Hoffman et al. 1998) could have occurred in early Proterozoic time and, as recorded, in very late Proterozoic time, but not in Archaean or mid-Proterozoic time, as solar luminosity increased gradually, whereas atmospheric CO_2 levels decreased in a more stepwise manner.

Although global transport balance modelling thus provides some perspective on the rate at which CO_2 could have been removed from the atmosphere in early Earth history, the resolution of the method is inherently poor and it can provide only broad outlines. The large uncertainties and the problems encountered highlight the need to find independent constraints on weathering and erosion rates in Archaean time to understand the early evolution of the Earth's atmosphere, climate and life.

Isotope research at Bern University is supported by the Swiss Nationalfonds (Grant 20-53865.98). I. Kleinhanns and F. von Blanckenburg provided helpful criticism of an earlier version of this manuscript. M. Bickle and an anonymous reviewer are thanked for their constructive reviews.

References

ALLÈGRE, C. J. & ROUSSEAU, D. 1984. The growth of continents through geological time studied by Nd isotope analysis of shales. *Earth and Planetary Science Letters*, **67** 19–34.

AMELIN, Y., LEE, D.-C., HALLIDAY, A. N. & PIDGEON, R. T. 1999. Nature of the Earth's earliest crust from hafnium isotopes in single detrital zircons. *Nature*, **399**, 252–255.

ARMSTRONG, R. L. 1981. Radiogenic isotopes: the case for crustal recycling on a near-steady-state no-continental-growth Earth. *Philosophical Transactions of the Royal Society of London*, **301**, 443–472.

AZBEL, I. YA. & TOLSTIKHIN, I. N. 1993. Accretion and early degassing of the Earth: constraints from Pu–U–I–Xe isotopic systematics. *Meteoritics*, **28**, 609–621.

BENNETT, V. C., NUTMAN, A. P. & MCCULLOCH, M. T. 1993. Nd isotopic evidence for transient, highly depleted mantle reservoirs in the early history of the Earth. *Earth and Planetary Science Letters*, **119**, 299–317.

BEN OTHMAN, D., WHITE, W. M. & PATCHETT, J. 1989. The geochemistry of marine sediments, island arc magma genesis, and crust–mantle recycling. *Earth and Planetary Science Letters*, **94**, 1–21.

BERNER, R. A. 1991. A model for atmospheric CO_2 over Phanerozoic time. *American Journal of Science*, **291**, 339–376.

BERNER, R. A. 1997. The rise of land plants and their effect on weathering and atmospheric CO_2. *Science*, **276**, 544–546.

BERNER, R. A. & MAASCH, K. A. 1996. Chemical weathering and controls on atmospheric O_2 and CO_2; fundamental principles were enunciated by J. J. Ebelmen in 1845. *Geochimica et Cosmochimica Acta*, **60**, 1633–1637.

BERNER, R. A., LASAGA, A. C. & GARRELS, R. M. 1983. The carbonate–silicate geochemical cycle and its effect on atmospheric carbon dioxide over the past 100 million years. *American Journal of Science*, **283**, 641–683.

BLICHERT-TOFT, J. & ALBARÈDE, F. 1994. Short-lived chemical heterogeneities in the Archaean mantle with implications for mantle convection. *Science*, **263**, 1593–1596.

BOWRING, S. A. & HOUSH, T. 1995. The Earth's early evolution. *Science*, **269**, 1535–1540.

COLLERSON, K. D. & KAMBER, B. S. 1999. Evolution of the continents and the atmosphere inferred from Th–U–Nb systematics of the depleted mantle. *Science*, **283**, 1519–1522.

COLTICE, N., ALBARÈDE, F. & GILLET, P. 2000. ^{40}K–^{40}Ar constraints on recycling continental crust into the mantle. *Science*, **288**, 845–847.

CONDIE, K. C. 1990. Growth and accretion of the continental crust: inferences based on Laurentia. *Chemical Geology*, **83**, 183–194.

ERIKSSON, K. A. 1995. Crustal growth, surface processes, and atmospheric evolution on the early Earth. *In*: COWARD, M.-P. & RIES, A. C. (eds) *Early Precambrian Processes*. Geological Society, London, Special Publications, **95**, 11–25.

FEDO, C. M., ERIKSSON, K. A. & KROGSTAD, E. J. 1996. Geochemistry of shales from the Archaean (~3.0 Ga) Buhwa Greenstone Belt, Zimbabwe: implications for provenance and source-area weathering. *Geochimica et Cosmochimica Acta*, **60**, 1751–1763.

FEDO, C. M., NESBITT, H. W. & YOUNG, G. M. 1995. Unraveling the effect of potassium metasomatism in sedimentary rocks and paleosols, with implications for paleoweathering conditions and provenance. *Geology*, **23**, 921–924.

GAILLARDET, J., DUPRÉ, B., LOUVAT, P. & ALLÈGRE, C. J. 1999. Global silicate weathering and CO_2 consumption rates deduced from the chemistry of large rivers. *Chemical Geology*, **159**, 3–30.

GOLDSTEIN, S. L., O'NIONS, R. K. & HAMILTON, P. J. 1984. A Sm–Nd isotopic study of atmospheric dust and particulates from major river systems. *Earth and Planetary Science Letters*, **70**, 221–236.

HOFFMAN, P. F., KAUFMAN, A. J., HALVERSON, G. P. & SCHRAG, D. P. 1998. A Neoproterozoic snowball earth. *Science*, **281**, 1342–1346.

HUNTER, M. A., BICKLE, M. J., NISBET, E. G., MARTIN, A. & CHAPMAN, H. J. 1998. Continental extensional setting for the Archaean Belingwe Greenstone Belt, Zimbabwe. *Geology*, **26**, 883–886.

HURLEY, P. M. & RAND, J. R. 1969. Pre-drift continental nuclei. *Science*, **164**, 1229–1242.

JACOBSEN, S. B. 1988. Isotopic constraints on crustal growth and recycling. *Earth and Planetary Science Letters*, **90**, 315–329.

JACOBSEN, S. B. & WASSERBURG, G. J. 1979. The mean age of mantle and crustal reservoirs. *Journal of Geophysical Research*, **84**, 7411–7427.

JACOBSEN, S. B. & WASSERBURG, G. J. 1984. Sm–Nd isotopic evolution of chondrites and achondrites, II. *Earth and Planetary Science Letters*, **67**, 137–150.

KASTING, J. F. 1993. Earth's early atmosphere. *Science*, **259**, 920–926.

KASTING, J. F. & ACKERMAN, T. P. 1986. Climatic consequences of very high carbon dioxide levels in the Earth's early atmosphere. *Science*, **234**, 1383–1385.

KRAMERS, J. D. 1998. Reconciling siderophile element data in the Earth and Moon, W isotopes and the upper lunar age limit in a simple model of homogeneous accretion. *Chemical Geology*, **145**, 461–478.

KRAMERS, J. D. & TOLSTIKHIN, I. N. 1997. Two major terrestrial Pb isotope paradoxes, forward transport modeling, core formation and the history of the continental crust. *Chemical Geology*, **139**, 75–110.

KREISSIG, K., NÄGLER, TH. F., KRAMERS, J. D., VAN REENEN, D. D. & SMIT, C. A. 2000. An isotopic and geochemical study of the northern Kaapvaal Craton and the Southern Marginal Zone of the Limpopo Belt: are they juxtaposed terranes? *Lithos*, **50**, 1–25.

LOVELOCK, J. E. & WHITFIELD, M. 1982. Life span of the Biosphere. *Nature*, **296**, 561–563.

LEE, D.-CH. & HALLIDAY, A. N. 1996. Hafnium–tungsten chronometry and the timing of terrestrial core formation. *Nature*, **378**, 771–774.

MCLENNAN, S. M. 1988. Recycling of the continental crust. *Pure and Applied Geophysics*, **128**, 683–724.

MILLIMAN, J. D. & SYVITSKI, J. P. M. 1992. Geomorphic–tectonic control of sediment discharge to the ocean: the importance of small mountainous rivers. *Journal of Geology*, **100**, 525–544.

MOORBATH, S., WHITEHOUSE, M. J. & KAMBER, B. S. 1997. Extreme Nd-isotope heterogeneity in the early Archaean – fact or fiction? Case histories from northern Canada and West Greenland. *Chemical Geology*, **135**, 213–231.

NÄGLER, TH. F. & KRAMERS, J. D. 1998. Nd isotopic evolution of the upper mantle during the Precambrian: models, data and the uncertainty of both. *Precambrian Research*, **91**, 233–252.

NESBITT, H. W. & YOUNG, G. M. 1982. Early Proterozoic climates and plate motions inferred from major element geochemistry of lutites. *Nature*, **299**, 715–717.

O'NIONS, R. K. & MCKENZIE, D. 1993. Estimates of mantle thorium/uranium ratios from Th, U and Pb isotope abundances in basaltic melts. *Philosophical Transactions of the Royal Society of London, Series A*, **342**, 65–77.

RAINBIRD, R. H., NESBITT, H. W. & DONALDSON, J. A. 1990. Formation and diagenesis of a sub-Huronian saprolith: comparison with a modern weathering profile. *Journal of Geology*, **98**, 801–822.

RAYMO, M. E. & RUDDIMAN, W. F. 1992. Tectonic forcing of late Cenozoic climate. *Nature*, **359**, 117–122.

REA, D. K. & RUFF, L. J. 1996. Composition and mass flux of sediment entering the world's subduction zones: implications for global sediment budgets, great earthquakes, and volcanism. *Earth and Planetary Science Letters*, **149**, 1–12.

REYMER, A. & SCHUBERT, G. 1984. Phanerozoic addition rates to the continental crust and crustal growth. *Tectonics*, **3**, 63–70.

RONOV, A. B. & YAROSHEVSKIY, A. A. 1968. Chemical structure of the Earth's crust. *Geochemistry International*, **5**, 1041–1066.

RYE, R., KUO, P. H. & HOLLAND, H. D. 1995. Atmospheric carbon dioxide concentrations before 2.2 billion years ago. *Nature*, **378**, 603–605.

SCHALLER, M., VON BLANCKENBURG, F., KUBIK, P. & KRAMERS, J. D. 1999. Catchment-wide denudation rates from cosmogenic nuclides in river sediment. *Journal of Conference Abstracts*, **4**, 444.

SCHIDLOWSKI, M. 1988. A 3800-million-year isotopic record of life from carbon in sedimentary rocks. *Nature*, **333**, 313–318.

SCHWARTZMAN, D. W. & VOLK, T. 1989. Biotic enhancement of weathering and the habitability of Earth. *Nature*, **340**, 457–460.

TAYLOR, S. R. & MCLENNAN, S. M. 1995. The geochemical evolution of the continental crust. *Reviews in Geophysics*, **33**, 241–265.

TOLSTIKHIN, I. N. & MARTY, B. 1998. The evolution of terrestrial volatiles: a view from helium, neon, argon and nitrogen isotope modeling. *Chemical Geology*, **147**, 27–52.

UREY, H. C. 1952. *The Planets, their Origin and Development*. Yale University Press, New Haven, CT.

VEIZER, J. & JANSEN, S. L. 1979. Basement and sedimentary recycling and continental evolution. *Journal of Geology*, **87**, 341–370.

VEIZER, J. & JANSEN, S. L. 1985. Basement and sedimentary recycling – 2; time dimension to global tectonics. *Journal of Geology*, **93**, 625–643.

WEDEPOHL, K. H. 1995. The composition of the continental crust. *Geochimica et Cosmochimica Acta*, **59**, 1217–1232.

WILDE, S. A., VALLEY, J. W., PECK, W. H. & GRAHAM, C. M. 2001. Evidence from detrital zircons for the existence of continental crust and oceans on the Earth 4.4 Ga ago. *Nature*, **401**, 175–178.

YOUNG, G. M., VON BRUNN, V., GOLD, D. J. C. & MINTER, W. E. L. 1998. Earth's oldest reported glaciation: physical and chemical evidence from the Archaean Mozaan Group (~2.9 Ga) of South Africa. *Journal of Geology*, **106**, 523–528.

ZARTMAN, R. E. & HAINES, S. 1988. The plumbotectonics model for Pb isotopic systematics among major terrestrial reservoirs – a case for bi-directional transport. *Geochimica et Cosmochimica Acta*, **52**, 1327–1339.

Fermor lecture: The influence of life on the face of the Earth: garnets and moving continents

E. G. NISBET

Department of Geology, Royal Holloway, University of London, Egham TW20 0EX, UK
(e-mail: e.nisbet@gl.rhul.ac.uk)

Abstract: The Hadean Earth (before c. 4 Ga) was abiotic, possibly steering a bumpy course between brief periods of hot inferno after meteorite impacts, and long episodes of Norse ice-hell. The earliest Archaean life would probably not have been planet-altering, but restricted to particular habitats. One of the first may have been hot regions around hydrothermal systems where redox contrasts between ocean water and magmatic fluids could be exploited. Molecular evidence suggests that with the evolution of anoxygenic photosynthesis, life became able to occupy wider regions, although focused in the vicinity of hydrothermal systems. Oxygenic photosynthesis by cyanobacteria allowed life fully to occupy the planet, not only forming coastal microbial mats but also possibly inhabiting the broad oceans with abundant photosynthetic bacterial picoplankton, underlain by deeper archaeal picoplankton. In the Belingwe belt, Zimbabwe, textural and isotopic evidence suggests that a complex microbial ecology existed in the late Archaean (2.7 Ga), which was essentially modern in its biochemical abilities and which sequestered into the biosphere the same fraction of primitive carbon emitted from mantle as today. To do this, by the late Archaean the biological productivity must have been significant; not necessarily as large as today, but capable of managing the global carbon budget. When this began is unknown, possibly earlier than 3.5 Ga ago. The controls on the oxidation state of the late Archaean atmosphere–ocean system are not self-evident. Although inorganic controls dominate the long-term balance, short-term biological management of the air may have been crucial. Methane may have played a major role in the pre-metazoan biosphere. The modern atmosphere is a biological construct: oxygen and its reverse, carbon dioxide, are managed by rubisco; nitrogen, its oxides and hydrides mainly by nitrifying and denitrifying bacteria, with a small input from lightning in an oxygen-rich atmosphere; and water (itself the most important greenhouse gas) by its complex interdependence with other greenhouse gases and albedo, including clouds. Earth's air is highly improbable. In controlling surface temperature a subtle interplay between organic and inorganic controls has operated, perhaps to the extent that it is invalid to ask which was the dominant factor. But there is a reasonable uniformitarian argument that life has constructed the air in the past as now, and that, within the broad constraints of the physical setting, this biologically shaped atmosphere has been the dominant control on the planet's surface temperature. In turn, the surface temperature has been one of the various controlling factors on the tectonic evolution of the planet. Thus to a significant extent life has helped shape the physical evolution of the planet.

Half-a-century ago, Sir Lewis Fermor was returning from India, where he had led the superb work of the Indian Geological Survey. Like his distinguished son, Fermor was a traveller. He took the long route, through Gondwana. While du Toit in the interwar years had gone from South Africa to the USA up South American Gondwana, Fermor returned from the land of the Gonds via Africa. In Harare, Fermor addressed one of Africa's oldest scientific societies, the Zimbabwe (then Rhodesian) Scientific Association. His topic was Garnets and moving continents (Fermor 1949). Today we would rephrase that as 'density contrast drives slab movement'.

Fermor's host was the distinguished local geologist, A. M. Macgregor, who spoke on 'The influence of life on the face of the Earth' (Macgregor 1949). Although given far away from the intellectual centres, and published in an obscure journal, both lectures were influential. Neither lecture was polished, neither offers especially penetrating insights or solutions, but the questions they asked addressed new fields. Little more than a decade after Fermor's lecture, Harare became a focus of the palaeomagnetic work that was central to the reconstruction of Gondwana and laid some of the foundations for the acceptance of continental drift and plate tectonics. Macgregor's lecture also bore fruit: the distinguished US biogeologist Preston Cloud once commented that Macgregor had been the greatest single influence on his career.

From: FOWLER, C. M. R., EBINGER, C. J. & HAWKESWORTH, C. J. (eds) *The Early Earth: Physical, Chemical and Biological Development*. Geological Society, London, Special Publications, **199**, 275–307. 0305-8719/02/$15.00
© The Geological Society of London 2002.

Recently, in a review derived from a Macgregor Lecture to the Geological Society of Zimbabwe, the links between the Earth's physical evolution and the habitat of early life were explored (Nisbet & Sleep 2001): that discussion considered 'How has Earth influenced life?' For the converse Fermor Lecture it is appropriate to ask 'How has life influenced the physical development of the Earth?' Following Fermor, this review will seek to ask general questions, although not necessarily to provide answers.

This review will first examine the controls on Earth's modern atmosphere, to give a uniformitarian basis; then consider parallel worlds. Next, models of Earth evolution are built, on uniformitarian grounds; by looking at the early abiotic Earth; and by considering biogeochemical processes in the Archaean. To provide experimental constraint, the Belingwe area is examined in detail. The operation of the Archaean atmosphere is discussed; then the links between air and ocean, and the links with the mantle and interior. Lastly, the overall controls are considered.

The modern atmosphere

The starting point is the modern atmosphere (Walker & Drever 1988; Wayne 1992). Although the present may be the key to the past, it is not the same as the past. Nevertheless, it is good uniformitarian geological practice to consider the present before looking backward. The question is simple, although the answer may not be. Is the modern atmosphere a biological construct whose composition is maintained by a cybernetic set of biological feedback loops (Lovelock 1979, 1988)? If the answer is yes, was ancient air also made by biology? If so, when did biological control begin?

Dinitrogen

Nitrogen is erupted by volcanoes in a variety of oxidation states. Nitrogen oxides eventually rain out, but dinitrogen remains in the air. Dinitrogen is also emitted by a variety of nitrate reduction processes by denitrifying bacteria and also by anaerobic oxidation of ammonium with nitrate and nitrite, by planctomycetes. Thus it collects in the air. The key to the N cycle is nitrogenase, an ancient enzyme, made of two subunits, one with Fe–S and the other with Fe–Mo and S at its heart. On a massive scale, bacteria and some archaea use nitrogenase to capture and split atmospheric dinitrogen. They partition the N into all the possible oxidation states, from ammonia to nitric acid, as 'fixed' nitrogen. Then the denitrifying bacteria capture the fixed N and return it to the air as dinitrogen. Both parts of the cycle are essential to the modern balance: if denitrifying bacteria became extinct, N would accumulate as soluble N in the soil or oceans or muds, eventually depleting the air, or, oppositely, if the nitrifying bacteria died out, N would be progressively sequestered as dinitrogen in air. Thus management of dinitrogen in the modern air is mainly a biological function.

Mainly, but not entirely; planetary atmospheres tend toward maximum entropy. Vigorous weather systems mix the air, and strong temperature gradients, pole to equator contrast, and water in liquid, vapour and probably ice phases generate lightning. In an oxygen-rich atmosphere, lightning (which is closely related to the amount of water in the air and hence the surface temperature) heats N_2 so much that it is split and then oxidized. If oxygen is absent, but CO_2 abundant, the CO_2 can supply O for N-oxides. In contrast, over a wholly abiological, anoxic Earth with an atmosphere mainly N_2, N-fixation by lightning would be less; once present, N_2 would be stable for billions of years.

In the modern oxygen-rich air, lightning N-fixation occurs on a much smaller scale than biological nitrogen fixation (today the largest source is not bacteria but humanity). Nevertheless, lightning fixation is large enough that if resupply did not occur, in a few tens to hundreds of millions of years much inorganic dinitrogen would be lost. Captured N enters the sea floor as organic nitrogen in sediment. Ammonia, in large part made by bacteria, is introduced into new oceanic plate as ammonium minerals during spilitization of basalt by hydrothermal systems (e.g. Hall 1989) and is also introduced into continental rocks by hydrothermal activity in granites (Hall & Alderton 1994). The ammonium minerals in oceanic crust return reduced N via subduction zones to the interior: it re-emerges as dinitrogen and nitrogen oxides at volcanoes. Even if denitrifying bacteria suddenly became extinct, some N_2 would, however, be reconstituted in hydrothermal circulation around volcanoes (Kasting et al. 1993) and returned to air.

Carbon dioxide

Oxygen and carbon dioxide are two sides of the same coin. In the modern air, the seasonal rise and fall of carbon dioxide is almost exactly reversely matched by the opposite fall and rise of the O_2/N_2 ratio: 'almost' matched, but not quite (there are very interesting second-order differences). The relationship between O_2 and CO_2 through photosynthesis and respiration (and fire) is very clear. The biological productivity of the modern Earth, if measured in terms of carbon capture, is mostly on land. In each northern spring, the onset of plant growth draws down carbon from the air; each autumn, the carbon is released again. This cycle dominates the variation in the global distribution of carbon dioxide.

Before the arrival of humanity, the longer-term controls were burial of carbon as carbonate, as methane hydrate in sediment, as gas, as coal, or as reduced organic matter (including charcoal). Before the Devonian, fire would have been impossible, except perhaps on lightning-hit microbial peat bogs after drought.

Carbon isotope lever rule

$\delta^{13}C$ -30   0

CO₂ in air

rubisco fractionation

organic matter

methanogens
-40 to -60

carbonate in sediment
CO₂ from methanotrophs

carbonate

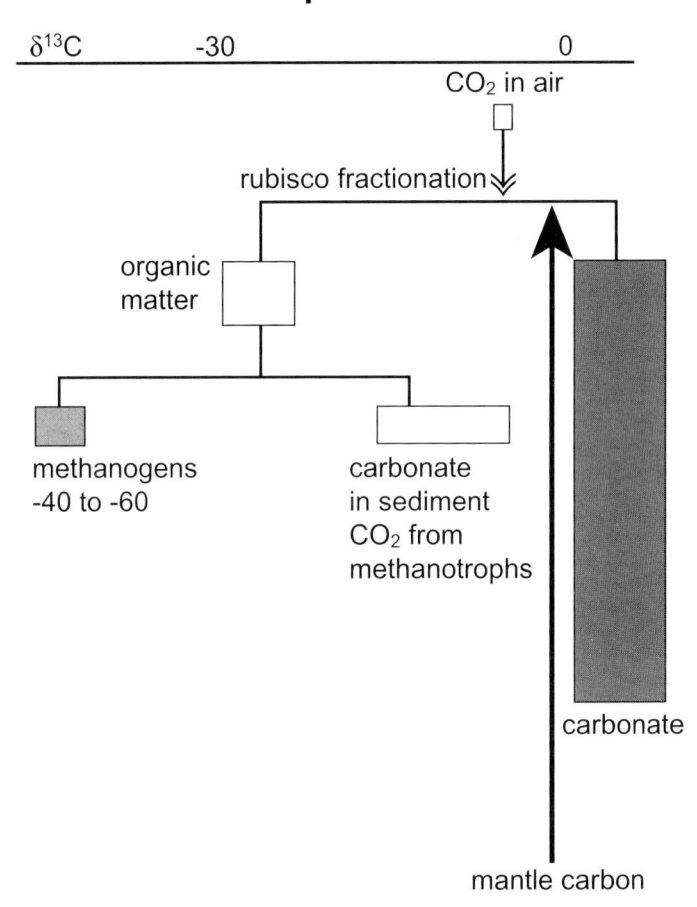

mantle carbon

Fig. 1. Isotopic controls on carbon. The per mil scale for carbon isotope ratio ($\delta^{13}C‰$) is arbitrarily based on fossil carbonate made from sea water. Carbon from the mantle is slightly richer in carbon-12. Roughly a quarter to a fifth of the carbon is processed via rubisco, which preferentially selects carbon-12. This selection by rubisco means organic matter has $\delta^{13}C$ around −25 to −30‰. The remaining carbon is consequently enriched in carbon-13 and precipitates as carbonate. If it is assumed that carbon emitted from the mantle source has had a ratio of about −5 to −7‰ throughout the last 4 Ga, then the dominance in the record of carbonate around 0‰ (apart from rare excursions) is circumstantial evidence that throughout this time organic matter, using rubisco, has captured a fifth to a quarter of the primary carbon coming out of the mantle, as today. Rubisco is characteristically used by bacteria in aerobic or micro-aerobic settings. The main function of rubisco is in oxygenic photosynthesis, although it is also used by purple bacteria. Thus the presence of carbonate at 0‰ for at least the past 3 Ga (own results with P. Abell from Steep Rock) and probably earlier (Schidlowski & Aharon 1992) suggests global oxygenic photosynthesis. However, this must be qualified: other microbial processes, such as carbonate precipitation via a methanotrophic path, can also produce 0‰ values in special circumstances.

Carbon emitted from the mantle has probably had $\delta^{13}C$ roughly between −5 and −7‰ over time, whereas carbonate carbon in sediment has been 0‰ ever since the Archaean, and average organic matter is about −25 to −30‰ (Fig. 1; Schidlowski 2002; see also Grassineau *et al.* 2002). Applying the lever rule, with the fulcrum at the −5 to −7‰ source, this implies that over most of the time since the Archaean (there are exceptions), for every five carbon atoms emitted to the atmosphere from the Earth's mantle, roughly four have been precipitated as carbonate, and one as organic matter (Schidlowski *et al.* 1979; Schidlowski 1988; Schidlowski & Aharon 1992). This isotopic split is the signature, written on almost every carbon compound on the planet's surface (Fig. 1), of the enzyme ribulose bisphosphate carboxylase–oxygenase, or rubisco (Lorimer & Andrews 1973; Lorimer 1981). Rubisco mediates carbon capture in microaerobic or aerobic settings, in purple bacteria and cyanobacteria, and in plants. It is today the dominant link between carbon in the air and biological carbon. Rubisco preferentially accepts 'light' ^{12}C atoms from carbon dioxide and incorporates it into phosphoglyceric acid, the first step on the route to incorporation into carbohydrate: given that the carbon dioxide is abundantly available, rubisco in most plants thus selects ^{12}C sharply. The left-over 'heavy' carbon goes into carbonate.

Rubisco is an apparently 'inefficient' enzyme. In effect, it works both ways (Lorimer & Andrews 1973; Lorimer 1981; Tolbert 1994). If carbon dioxide is at high enough concentration (as it is at present), so that it is easily accessible to rubisco, then reduction power is captured photosynthetically, and carbon is added, via the Calvin cycle, to the organic carbon of the organism. On the other hand, if the carbon dioxide content of the air is low, rubisco works the opposite way, and photorespiration removes carbon from the organism and adds it to the air. The cross-over level depends on several variables (Tolbert 1994). At present atmospheric oxygen levels (where oxygen is about 21% of the air), the balance is at about 40–70 ppm carbon dioxide. The air has about 370 ppm CO_2 at present (and rising), so photosynthesis dominates. In glacial times, air had as little as 190 ppm CO_2, and photosynthesis, although still dominant, would have been less favoured above photorespiration. Total global plant productivity would have fallen sharply. Had the air dropped below 70 ppm CO_2, rubisco-mediated photosynthetic carbon capture would have been greatly restricted.

To escape this problem, some modern plants concentrate carbon dioxide before it is presented to the rubisco, so that they can suppress photorespiration even if CO_2 falls as low as 5 ppm. These are the C4 plants, which appear to be adapted for a low-CO_2 glacial world. Because of the preconcentration, the 'Rayleigh yield' of the carbon capture is much higher and thus in C4 plants the isotopic fractionation is much lower (typically $\delta^{13}C$ around −12‰) than in normal photosynthesis. This adaptation appears to be limited to eucarya and only to have appeared recently, under low-CO_2 conditions during Tertiary glaciation. It is not proven from earlier glaciations, for example in the Permian. However, C4 preconcentration seems to have evolved independently in several lines of plants; thus it must be an 'easy' modification for eukaryotes in times of low ambient CO_2. Could C4-like processes have operated in the Archaean at times of low CO_2? It is an open question. Bacteria do not appear to be able to build the complex physical architecture of the concentration system: thus it is assumed that C4 fractionation was absent in earlier aeons. However, the possibility that something analogous to C4 metabolism emerged in complex microbial consortia must be considered, and would be hard to exclude in the absence of well-preserved organic structural remains.

The 'inefficiency' of rubisco determines that some carbon dioxide must remain in the air, and, conversely, that all the oxygen is not consumed. Overall, in modern organisms, the subtly switched cybernetic balance between the power-generation needs of the mitochondria in plants and animals, burning reduced carbon in oxygen to provide energy, and the photosynthetic restorative capacity of the chloroplasts, may set the daily account of planetary inputs and outputs of carbon dioxide and oxygen (Joshi & Tabita 1996).

Ozone, the cold trap, and the hydrogen budget

The presence in air of free dioxygen liberated by photosynthesis, over a water ocean, and with CO_2 present, has many consequences (Warneck 1988; Wayne 1992). There is also a small volcanic source of O_2. In the modern dioxygen-rich air, in the upper atmosphere, some singlet O is formed, which combines with dioxygen to form ozone. Singlet O drifts into the tropical lower atmosphere, where it meets water vapour, and forms hydroxyl, OH, the 'policeman' of the air (Crutzen 1987). OH is the main species responsible for attacking reduced chemicals in the air and oxidizing them. Nitrogenous gases, sulphur gases and reduced carbon gases are all removed this way, as are methane and organic molecules. Once oxidized, they typically are converted to soluble species which are rained out into the oceans. Thus OH keeps the air 'clean' (where 'clean' is judged from the perspective of an oxygen-dependent organism).

Sunlight is absorbed when it hits ozone. This occurs on a large scale a few tens of kilometres above the Earth's surface (depending on latitude). The energy input makes the ozone warm, and it warms the rest of the air at this level. This effect, and also the various effects of the radiative budget in the air, especially the upper air, and the albedo of the surface, give the Earth's atmosphere a very odd thermal structure (Lewis & Prinn 1984) (Fig. 2).

This curious thermal structure of modern air has remarkable consequences. Liquid water, water vapour and ice coexist on the surface, in complex interchange. There is a top bound to this zone. At, say, 20 km altitude (less in the Arctic, more near the equator), the air is cold and thus very dry: air at $-60°C$ can hold virtually no water, which all rains out before any air rises above this level. This is the cold trap. Above it there is little water (except the small amount that passes through the trap, plus a roughly equal amount made *in situ* high up by oxidation of methane).

If the Earth's ocean surface were to be kept hot for a sustained period, things would be very different. A hot surface would probably mean that water vapour was more abundant in the high atmosphere: the stratosphere would be moist. Without an efficient cold trap, the oceans would imperceptibly be lost as water reached the higher levels of the atmosphere, was broken by photolysis, and produced H. Any H at the top of the atmosphere is vulnerable to ejection to space, leaving matching oxidant that will eventually find its way into reaction with a surface rock.

Although little water manages to pass through the cold trap, there are other sources of hydrogen in the upper air. As mentioned above, methane emissions from the surface mix upwards from the surface, as methane has no cold trap, and in the upper air photolysis leads eventually to release of H. In addition, there is a small emission of H_2 from the surface, some of which will reach the upper part of the atmosphere however, Earth also sweeps up H and H_2 from space. Over time, net hydrogen loss must have been limited: we have kept the oceans.

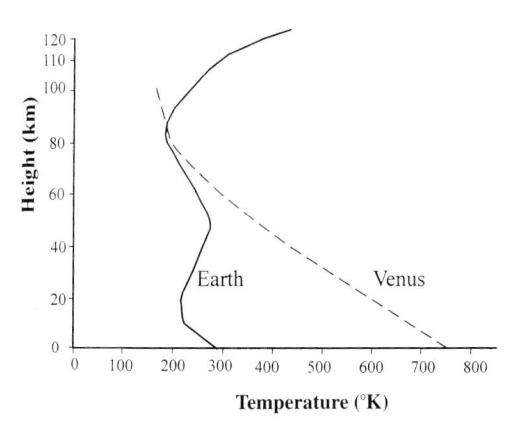

Fig. 2. Height–temperature plots for the atmospheres of modern Venus and modern Earth. On Venus, with an abiogenic atmosphere, the temperature has a straightforward lapse rate upwards until the outer bounds of the atmosphere where energetic radiation is encountered. On Earth, with an atmosphere that is biologically managed, the surface is cool: close to surface in the bottom 10–30 km (troposphere) the lapse rate upward of temperature is adiabatic; above that, the air is warmer from radiation trapped by ozone; higher yet, the air is cooler upwards; in the upmost air, interception of hard incident radiation has a strong heating effect. The cold layer in the Earth's high troposphere is here shown on average at *c*. 20 km above the surface, but its altitude varies with latitude from 10 km (poles) to 30 km (tropics). This layer prevents water rising higher and entering the stratosphere – hence the air above this level, the 'cold trap', is dry. Only minor amounts of water occur in the stratosphere, from the small amount that can pass the cold trap, and from methane oxidation in the higher air. From various sources (see Lewis & Prinn 1984).

Trace gases

The most important traces gases in the air are water and carbon dioxide, both closely linked to oxygen. Next is methane. Natural methane sources are almost entirely biological nowadays. In the pre-Holocene world, methane sources included methanogenic archaea (Woese 1987) living in wetland muds, in ruminants and in termites, and also vegetation fires that released rubisco-captured carbon back to the air as CH_4, CO and CO_2. It should be noted that in the longer run fires deplete the atmosphere–biosphere carbon supply by adding resistant charcoal to the long-term sediment store. The ocean surface biota release relatively minor amounts of methane to air in the modern world, although certain parts of the tropical ocean are significant sources (e.g. in the eastern tropical North Pacific; Sansone *et al.* 2001). Some methane may be released from organic matter in the photic zone, probably much of which is immediately oxidized *in situ* to carbon dioxide. Marine plankton do, however, release large quantities of dimethyl sulphide to the air. This is broken apart in the air, where it adds not only to the supply of methane-related species, but also to the return of sulphur to the land in rain (Lovelock 1979, 1988).

There is a small non-biological source of methane, from serpentinization reactions in hydrothermal systems around volcanoes, especially mid-ocean ridge systems (Holm & Charlou 2001; Kelley *et al.* 2001). Early in the Earth's history this flux may have been greater, though still very small in comparison with modern biogenic fluxes.

Archaea recycle carbon in organic matter, reacting it with hydrogen either from other microbial sources or from volcanic processes, and the waste product is methane. On the modern Earth, possibly 1% of the plant material formed each year is remineralized via methane (Thauer 1998). This methane is isotopically light because the carbon has been twice selected, first by rubisco in photosynthesis, then in methanogenesis. The methane builds up in sediment as gas in solution, as free gas, or in hydrate. Methane is also made by thermal processes, inorganically. This methane comes from the maturation and geothermal heating of detritus rich in residual organic reduced carbon and hence is isotopically heavy. Much of the methane, from both types of source process, migrates upward to be stored in overlying sediment. In effect, the sequestration of methane in sediment much increases the leverage of available P, N, S and other biologically useful elements in managing the carbon gases of the greenhouse. The huge sedimentary stores of methane on continental shelves and slopes, which occur because the Earth is active and has density contrasts that move continents, have a major role in partitioning oxidation power into a highly oxidizing atmosphere and, the other side of the coin, a vast store of reduction power in the sediment. The present-day methane store in offshore sediment contains perhaps 10^{19} g of carbon, or more at best estimate (Kvenvolden 1988; Harvey & Huang 1995). This is tens to thousands of times as much carbon as there is in the active biosphere. These huge stores of methane are a part of the reciprocal crustal store of reduction power that matches the oxidation power stored in the O_2 of the air.

Today there is little direct CH_4 release to atmosphere from the sediment store, except in unusual events such as submarine landslips or pockmark bursts. In an oxygen-rich atmosphere, methane is removed by OH, and to a minor extent by soil methanotrophs. In the sea, unless they are large, bubbles of methane released from sediment are oxidized by methanotrophic bacteria to carbon dioxide as the bubbles rise through the oxygenated column of sea water. In the Archaean, if the air–ocean system had been less oxidized and methanotrophs more restricted, emissions from marine sediment would have been important.

Methane in the present atmosphere, molecule for molecule, is a much more powerful greenhouse gas than carbon dioxide (Hansen & Sato 2001). Today small changes in methane can have large climatic results. In a high-methane atmosphere, however, the incremental impact of emissions would be much less.

Of the other trace gases (e.g. N_2O, dimethyl sulphide, CH_3Cl, etc.) many are, like methane, of dominantly biological origin and are transient disequilibrium components of the air eventually oxidized, whereas others are of volcanic origin, mostly eventually rained out in some form or other.

Argon

Argon is in part a neutral basis for comparison when considering the management of the air. Atmospheric argon is not a biological product. It comes from decay of radioactive potassium in the crust and the subsequent release of the argon by erosion. Moreover, it is always a gas, whatever the temperature. Thus it is at first thought completely unconnected with the overall evolution of the air and ocean. However, erosion depends largely on water, so the argon release rate is dependent on surface temperature and water availability. The major source of K-rich material is subduction-related granitoid magmatism, which depends heavily on melting triggered by water release from subducting slabs. Moreover, the subducting slab may in some circumstances also contribute K to the melt. And on what does subduction and the formation of continental granitoids depend?: water (Campbell & Taylor 1983). Without water the K and hence the argon may have been more deeply trapped in the Earth's mantle, and less substantially released to air.

Parallel worlds

Venus

Venus has about as much carbon dioxide near its surface as Earth, but the carbon dioxide is held as gas, not bound in carbonate, so there is a massive greenhouse effect (Lewis & Prinn

1984). The thermal structure of its atmosphere is thus satisfyingly plain (Fig. 2). Light entering the atmosphere from the Sun is progressively absorbed by the carbon dioxide as it moves down to the surface, which is a gloomy place, with light winds, under perpetual clouds. The lower air, at very roughly 500°C, 90 bars, is in approximate equilibrium with the rocky ground. A rough proxy for this is the reaction between Carbon Dioxide + Wollastonite (i.e. pyroxene-like minerals) to/from Calcite + Quartz. This reaction's equilibrium lies close to the conditions on the Venusian surface. Perhaps changes in carbon dioxide precipitate a calcite snow. Upwards, the air temperature drops in the massive carbon dioxide greenhouse. Were the planet stripped of its greenhouse, it would have more or less the same temperature as a similarly greenhouse-removed Earth (Lewis & Prinn 1984): in some ways Venus seen from space is the cooler of the two planets. With its air, and the resulting greenhouse heating blanket, the surface of Venus would appear to our eyes as a dark cherry-red, as hot as a ring on a kitchen cooker.

The consequences of the thermal regime of the atmosphere permeate the planet to its core. With a surface at roughly 500°C, there is no liquid water and hence no hydrothermal cooling of newly erupted lava. There is no cold trap, and although the planet once had deep oceans it now has virtually no exterior hydrogen, except in sulphuric acid droplets in high clouds (Lewis & Prinn 1984). The high surface temperature means the transfer of heat from the interior to space is relatively inefficient compared with Earth. Compared with the Rolls-Royce radiator of sea-cooled Earth, air-cooled Venus is an old VW beetle. There can be no low-temperature metamorphic rocks. Aphroditotherms are hot, and unlikely to allow diamonds or the garnets in eclogites that so interested Fermor. Perhaps, over the aeons, the hot surface has shaped the interior regime to the extent that formation of an inner core has been hindered. Venus is not 'past-Earth': our planet has never been like this, although it may become like Venus in future. Venus is a mature, evolved stable system in equilibrium with itself: an ideal 'sustainable' world.

Mars

Mars may be an alternate 'future-Earth', but perhaps it can also be seen as a 'might-have-been' Earth. Far from the Sun, with a thin oxidized atmosphere, its surface environment is controlled by equilibrium, but also by the slow kinetics of cold (Carr 1996).

One of the most profound questions in studying the Earth's igneous and metamorphic petrology is 'why doesn't the water simply sink into the surface of the planet?' That is what happens to drizzle on a back garden, and it appears to have happened on Mars. There the original water inventory, of a few hundred metres depth if spread globally, may exist as brine aquifers in the crust, in addition to the polar ice stores. The planet is a permafrost desert. Water is present but largely inactive geologically (and of course biologically), except in rare catastrophic episodes (Carr 1996; Baker 2001).

Earth

On Earth the total water inventory, although minute compared with the planet as a whole, is much larger than on Mars, and it constantly interacts with the mantle (Walker & Drever 1988). Mid-ocean ridge crust is hydrated; carried down subduction zones, and returned. On continents, the near-surface sediments are hydrated and groundwater reservoirs are large. Volcanism at the leading edges of plates heats and metamorphoses buried or stacked sediment, so the water is returned by fluxing upwards. In effect, continents above subduction zones are 'saturated' with water. The lithospheric mantle is also effectively 'saturated', when seen on an aeon-long time scale.

The tectonic water cycle in Earth is today heavily dependent on the presence of liquid water above mid-ocean ridges. The height of the continents is controlled by the depth of the oceans (Hess 1962). If oceans froze, erosion would slow. Over time, if andesitic volcanism and continental collision continued, continents would become somewhat thicker and of smaller area, much more rugged.

Water, in liquid oceans, is what makes Earth's tectonic history different from Venus and Mars. On Earth oceanic crust and hence plate is cooled quickly by water, because the surface is close to 0°C, not 500°C as on Venus. This cooled plate thickens and becomes dense more quickly than it would on Venus. Plates fall into the asthenosphere as a steady regular process. Andesite volcanism is fluxed by water given off by subducted oceanic crust. Mars is so cold that reintroduction of water to the interior does not occur. The only volcanism in the past billion years on Mars appears to have been from deep-rooted plumes.

Neither Mars nor Venus has well-defined smoothly moving plates, thousands of kilometres long from ridge to subduction zone. Plate tectonics needs water to help cool the new plate.

Moreover, water exerts a geochemical control. Continents are built by magmatism that depends on subduction. 'No water, no granites; No oceans, no continents' (Campbell & Taylor 1983). Earth's surface geology and the geochemistry of its upper mantle depend on the presence of water that both fluxes plate geochemistry and physically controls the pressure–temperature regime of the upper mantle and crust. Without water, cool ocean floor would not exist, nor eclogitic garnets, nor the density contrasts that move continents (Fermor 1949). It is the atmosphere that makes the difference.

Apart from the initial inventory of water, many of the geological differences between Earth, Venus and Mars are a matter of the surface temperature, set by the incoming sunlight and the air's composition. Earth is the Goldilocks just-right planet: melting occurs through a subtle interaction of volatile fluxing, heat transfer from depth, and radiogenic build-up. The outward flux of volatiles carried from the mantle by rising melts, and the inward flux down subduction zones are just right, and kept that way by the maintenance of the surface thermostat by the natural greenhouse increment of its air.

A further important factor is the secular change in the Sun, which has slowly brightened over time. The hot modern sunlight needs to be managed by an oxygen-rich, low carbon dioxide atmosphere, but early Archaean sunlight was significantly fainter. If the Archaean surface was as hot as today's surface, then the Archaean atmosphere must have been better at trapping outgoing infrared radiation. Earth must have a self-adjusting thermostat (Lovelock 1979, 1988, and work by Lovelock and Watson discussed therein).

The uniformitarian argument

The surface temperature and the presence of liquid water on Earth are today consequences of the atmospheric thermal control and filtering of light that is exerted by an anomalous, dominantly biogenic, atmosphere. The uniformitarian argument would thus be: 'the present is the key to the past, hence in the past the stability of liquid water on Earth was, as today, biologically maintained – life sustains the oceans'. To this argument could come the further addendum: 'liquid water enables plate tectonics to function, subduction to operate, and andesite volcanoes to erupt'. Hence if water is sustained by life, then plate tectonics and the maintenance of continents are sustained by life. But this is only unsupported conjecture. As an alternative hypothesis it is possible that during Earth's early history, a set of inorganic feedback loops sustaining the presence of liquid water may have accidentally been established on Earth, with consequent development of plate tectonics.

Is such thinking taking uniformitarianism too far? Perhaps; obviously the distant past was different from the present day, so the actualistic instant uniformity model (Fig. 3) does not completely apply: Earth has changed greatly over time, and the past was different. The 'coffee-pot' model has more appeal: the Earth has surely evolved through its own internal constraints, for example by dissipating heat. The cafetière (plunger) model allows periodic catastrophes, and may more nearly describe the Earth's history: part uniformitarian, part-catastrophic.

The appearance of liquid water oceans was probably inevitable early in Earth's history: but was the sustained long-term stabilization of oceans preordained from the moment the Earth system was created? Or is the existence of stable oceans over 4 Ga unexpected? – Anything from mildly unlikely to quite improbable, sustained only by an unpredictable event, such as the onset of life?

Do models say anything? Is Lyellian uniformity worth considering? It is valid to assume that through the later Phanerozoic, life has controlled the composition of the air. Most likely, the air and hence the climate has been set by life at least since the appearance of rain forest in the late Devonian. Probably this has been the case since long before, ever since the dominance of oxygen over carbon dioxide, which perhaps but not certainly began in the early Proterozoic. This is only half of Earth history, but perhaps long enough, if the system is mismanaged, to lose an ocean in a greenhouse runaway, or to freeze the planet permanently. Perhaps such excursions have nearly happened, in snowballs and possibly in hot flushes also. And before the oxygen atmosphere? What of the Archaean and the Hadean?

Abiotic Earth: the Hadean

Life on this planet began possibly around 4.2–3.8 Ga ago, although there is a chance it began earlier elsewhere (see Nisbet & Sleep 2001, 2002). Before the start of life, the abiotic Earth may have undergone dramatic rapid fluctuations in its surface environment, from long periods of ice-oceans to brief moments of heat, sometimes enough to produce a rock-vapour atmosphere, after major meteorite impacts (Sleep et al. 2001). If life had not begun, what sort of Earth would have developed?

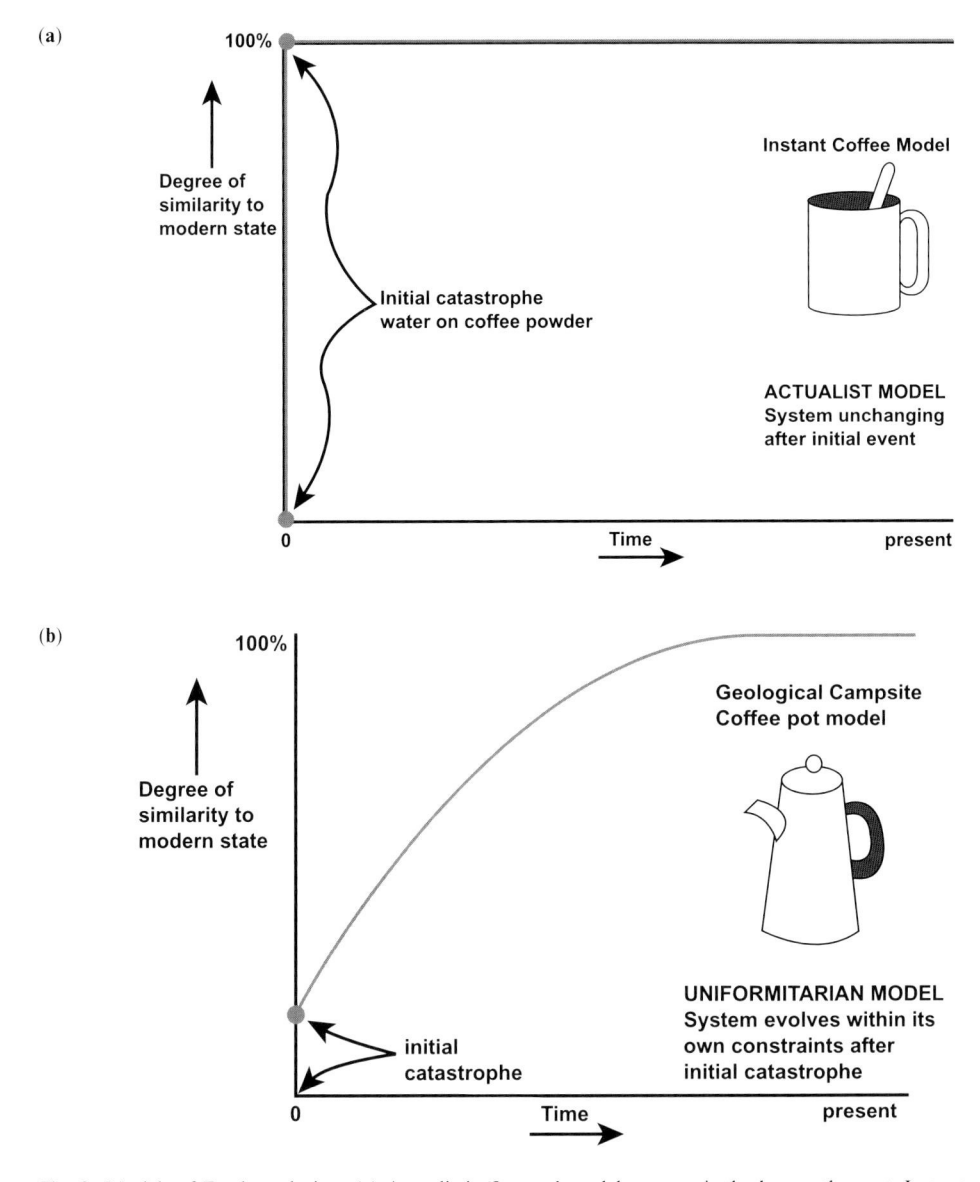

Fig. 3. Models of Earth evolution. (**a**) Actualistic 'Instant' model: present is the key to the past. Instant coffee, once made, changes little (apart from slow cooling). The initial event, in which hot water is poured on powder, is a catastrophe, where catastrophe is defined as sudden downwards change. It should be noted that in the view of an observer viewing from after the catastrophe, what may have been a downturn for the prior system (e.g. the dinosaurs, or King George III) is perhaps an upturn from the perspective of the later observer (e.g. a mammal, or George Bush II): an anastasis. (**b**) Modified uniformitarian or Zimbabwean field coffee model: once set up, the system evolves within its own constraints, as there is no external intervention once the initial catastrophe is over. Pour water into the pot, on coffee grounds, and the coffee evolves according to the internal dynamics of its system, without further external intervention. (**c**) Plunger or Seattle model (catastrophist model): the system evolves within its internal constraints, but is also subject to assorted catastrophes. External catastrophes include both the initial event and later the depression of the plunger. Internal catastrophes occur because of properties inherent in the system as it is created, and without external intervention. In this model, (and indeed, hidden in model (**b**)) there is also an internal catastrophe. When the coffee grounds become denser than the coffee-water, and turbulence reduces, the coffee grounds suddenly sink. The segregation of the core may have been an internal catastrophe; the arrival of life may also have been an inevitable internal catastrophe for abiosis, although an anastasis from our perspective.

(c)

Degree of similarity to modern state (y-axis, 0 to 100%)

1. Initial catastrophe: water on grounds
- decay of initial turbulence
2. settling of grounds
3. depression of plunger

Cafetière

CATASTROPHIST MODEL
Catastrophes 1 & 3 externally driven
Catastrophe 2 is internally constrained

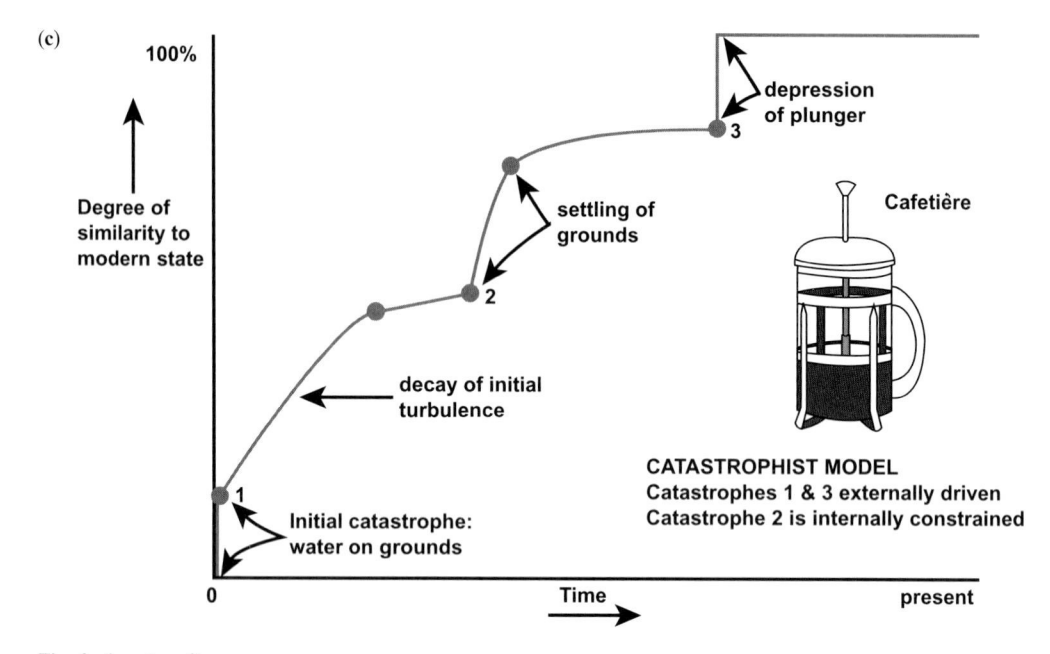

Time → present

Fig. 3. (*continued*)

The answer is not clear. Even an abiotic planet is chaotic, unpredictable: there are too many variables. Some guesses can be made. One possibility is a Venus-like Earth. But with the faint young Sun (Sagan & Chyba 1997), an ice-ocean with a fractured frozen surface over liquid deep water (Sleep *et al.* 2001) might have persisted. If so, despite a cold surface, plate tectonics would have operated. Degassing of carbon dioxide would have been matched by carbon dioxide uptake into spilitized basalt and thence via hydrothermal water around mid-ocean ridges subducted back into the mantle. But in a snowball ice planet, with little precipitation, clastic sediment and the supply of cations by erosion would have been limited. The continents would been more rugged. Possibly the carbon dioxide would have built up in the air, causing eventually a sudden transition not to a cool ocean but instead directly to a hot moist greenhouse when the icecap melted, ocean albedo suddenly changed, and the warming water, rich in carbon dioxide, degassed. The resulting greenhouse, in extreme cases, could have led to a steam atmosphere.

As an alternate hypothesis, it is possible that an early icehouse setting could have been sustained for aeons, if the return of carbon dioxide to the mantle by carbonated oceanic crust (from ridge crest calcite, precipitated by hydrothermal alteration around vigorous volcanoes, in carbon dioxide-rich liquid seas under ice) exceeded the degassing. On rugged polar land surfaces, carbon dioxide could have accumulated as ice. If so, an ice-capped ocean could have persisted over the aeons, perhaps eventually freezing solid under an atmosphere largely of dinitrogen. Volcanic N_2O is a strong greenhouse gas but it would have frozen out or been taken up by pools of water around volcanoes. This is a Mars-like scenario. The lack of erosion would have produced a breccia-laden ancient icy surface, broken up by meteorite impacts and occasionally resurfaced by rare meltwater events. A significant part of the ocean could have been taken back into the cold crust and upper mantle. The Earth's surface water inventory is much greater than that of Mars (Carr 1996), so the entire ocean would be unlikely to disappear, but a considerable amount could have gone.

Unlike Mars the mantle of Earth is hot enough to produce copious melt if a small amount of water is present; water fluxes melting and is driven out with the melt, to collect in the ocean–atmosphere system, where it is the most important greenhouse gas. The process is self-managing to an extent. If for some reason the surface became very cool, as time went on the cold dense surface plate would founder. This would carry down water. If so, the water in the mantle would promote melting, increasing the rate of volcanism and carbon dioxide and water return to the surface, and dust and ash deposition on ice caps, until a degree of transient equilibrium was restored. It should be noted that volcanic heat only makes a negligible contribution to surface temperature.

As the planet ages and becomes geriatric, so much heat is conducted to the cold surface that the lithosphere becomes thicker and the upper-mantle temperature falls. The degree of melting in the mantle becomes less. Near the surface itself progressively more water can be accommodated in the fractured crust and serpentinized uppermost mantle. The oceans will thin and sink into the planet's cooling mantle. Over 4 Ga, all these factors, taken together, could have led towards a Mars-like permafrost surface over a thick cool lithosphere and a still convective deeper upper mantle, occasionally erupting plumes.

Thus Earth could have suffered an icy fate, covered by frozen carbon dioxide and water, reflecting sunlight, and slowly conducting internal heat. Perhaps, but for our planet this so far must have been unlikely except as a transient state under an early faint Sun (Sleep *et al.* 2001). The planet is still too large and too young to follow Mars and the Moon into permanent senescence. Massive meteorite impacts would have vaporized the ice, suddenly altering the albedo. Moreover, although perhaps half the heat in the planet is from the accretionary age, radioactive decay proceeds and the trapped heat would come out. Huge plume eruptions could cover the planet in dark dust and cause it to warm up and escape ice-death.

Whatever the average temperature of the planet, in the Hadean during the likely 'hot' excursions after very early major meteorite impacts Earth would have incurred substantial loss of H to space. During these events, there would have been oxidation of the atmosphere–ocean system relative to the mantle. If the surface were hot, the atmosphere's lapse rate means that water would have been present relatively high in the air. Here, after photolysis, hydrogen would be lost to space at a rate sufficient to dehydrate the planet eventually. Moreover, throughout Earth history, any episodes of methane-rich atmosphere would also have caused H loss when the high atmosphere lost hydrogen derived from methane. The D/H ratio of the modern ocean may be a record of such events (e.g. Yung *et al.* 1989). Possibly Earth has cumulatively lost the H from 1–2 km of ocean water, leaving an excess of oxygen. This excess would have been equivalent in total over a few hundred million years to 100–200 bars of oxygen (although at any moment the actual oxygen burden of the air from H loss would be miniscule). As oxygen was released from water it would have been sequestered into the mantle by plate subduction (e.g. see Kasting 2001). The air is never in equilibrium. These small supplies of oxidation power, present as transient species in the air, may have been vital in sustaining the first living organisms, by permitting a contrast between relatively reduced hydrothermal fluids in exchange with mantle-derived lava, and ocean water in contact with slightly more oxidized air.

Did 'abiotic-Earth' steer a successful course between Hot-Scylla and Ice-Charybdis? Possibly, but it may have been a very bumpy course compared with the narrow middle route maintained today by the fine-scale management of a small greenhouse increment. Moreover, in the crosscurrents of a steadily brightening Sun and a diminishing radiogenic heat flow and hence volcanic gas input, this narrow course would have needed constant correction. Was the planet steered by blind Chance and inorganic feedbacks of chemical buffers? Or did a biological 'Odysseus', guided by self-adjusting biological feed-backs, lead it? Without self-stabilizing cybernetic biological controls, could liquid oceans have been stable over a time span as long as 4 Ga?

Biogeochemical processes in the Archaean

The early settings

Life did begin: the world did not evolve inorganically. Molecular palaeontology (Zuckerkandl & Pauling 1965), which infers phylogeny from nucleic acid sequences in extant modern microbial life (Fig. 4a, b) as interpreted in the light of the geological record, permits some reconstruction of the Archaean biosphere. Evidence in the rocks, including isotopic fractionation in sedimentary minerals, and relict molecules such as kerogens and oils, allows modelling of the microbial ecosystem (Brocks *et al.* 1999; Rasmussen 2000; Rasmussen & Buick 2000; Nisbet & Sleep 2001). Microbial sulphate reduction in particular is probably of great antiquity (Shen *et al.* 2001), implying that a surface supply (not necessarily in equilibrium) of oxidation power was present, to be reacted against the more reduced fluids in equilibrium with mantle-derived rock. By the late Archaean, and perhaps much earlier, this biosphere was complex enough and chemically sophisticated enough that it had the potential power to control the composition of the air. Whether it did or not is unknown: the tests are isotopic, but the controls are not well enough understood yet to be sure.

(a)

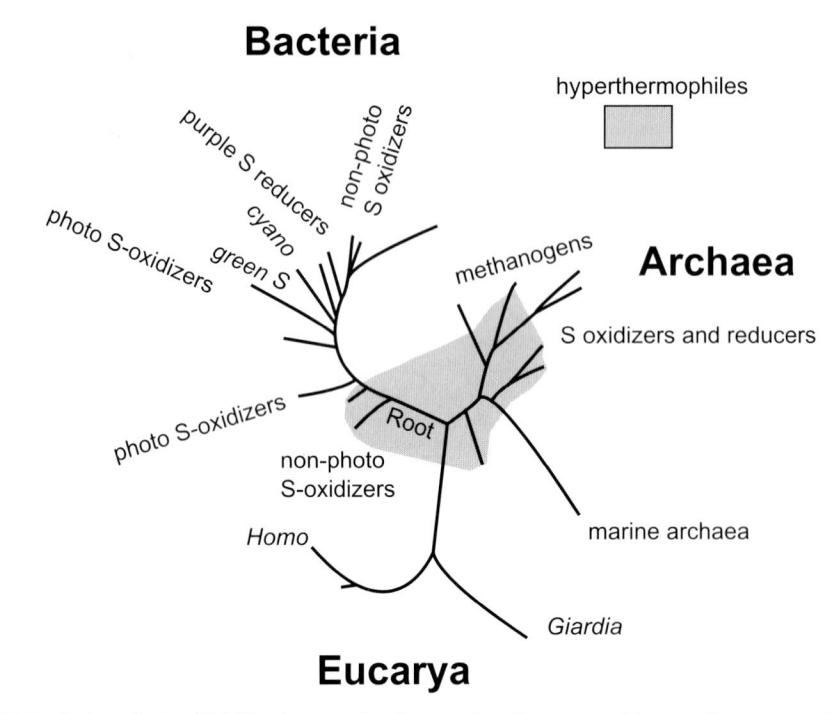

Bacteria

[Phylogenetic tree diagram with branches labeled: purple S reducers, non-photo S oxidizers, cyano, green S, photo S-oxidizers, hyperthermophiles, methanogens, **Archaea**, S oxidizers and reducers, photo S-oxidizers, Root, non-photo S-oxidizers, Homo, marine archaea, Giardia]

Eucarya

Fig. 4. (**a**) Evolution of microbial life: descent of major mat-forming groups. From various sources, including Barnes *et al.* (1996) and Pace (1997). The 'standard view' of the evolutionary tree (Woese 1987; but see Doolittle 1999), based on rRNA. There are three main branches to the tree: bacteria, archaea and eucarya. The most deeply rooted organisms (close to 'root') are exclusively hyperthermophile (shaded area). The implication is that earliest Archaean organisms were exclusively hyperthermophiles, and that the branching of the tree took place in a high-temperature setting (see also Hartman & Fedorov (2002) on the antiquity of the eukaryote branch). The implication of the standard tree is that the microbial S cycle is of great antiquity. It should be noted, however, that this standard phylogenetic tree is much disputed (see discussion by Nisbet & Sleep 2001), and there is interesting evidence to suggest that mesophiles like the modern Planctomycetales were the earliest branch (Brochier & Philippe 2002). If so, the last common ancestor may not have been hyperthermophile. In the 'standard' phylogeny the first bacterial branch to escape the hyperthermophile setting includes the green non-sulphur bacteria, carrying out anoxygenic photosynthesis. Rubisco and micro-aerobic habitats may have been present at this early stage. Green S bacteria may have emerged later, and then cyanobacteria, possibly from a symbiotic combination of photosystems from purple and green bacteria, to give oxygenic photosynthesis. In the archaeal line, the evolution of mesothermophiles may have followed the start of bacterial photosynthesis, with the spread of methanogenic recyclers into mesotherm settings. In the oceans marine archaea may have appeared early, or when overlying waters first hosted photosynthetic bacterial picoplankton. (**b**) Diversification of microbial life in the Archaean, based on the 'standard model' (Woese 1987). In the earliest Archaean, the living community may have been restricted to hyperthermophile habitats. By the late Archaean a diverse ecology was present in a wide variety of habitats, and capable of forming a global biosphere. This diagram assumes the last common ancestor was hyperthermophile. If, however, it was a mesophile planktonic form (e.g. Brochier & Philippe 2002), then (**b**) will refer to the evolution of only a hyperthermophile community and its descendants, whereas the eucarya line may have been mesophile.

The habitat of the last common ancestor of life is still much disputed, but there is much support from studies of rRNA for the argument that the last common ancestor was hyperthermophile (Stetter 1996), perhaps a survivor from a major impact event that heated the ocean (Gogarten-Boekels *et al.* 1995). However, the 'hyperthermophile last common ancestor' hypothesis is not proven (Forterre 1996; Galtier *et al.* 1999). An interesting alternative hypothesis is that the last common ancestor may have been a non-hyperthermophile planktonic bacterium. This hypothesis comes from a reassessment of bacterial rRNA phylogeny using an essentially conservative approach, using the most conserved positions in rRNA (Brochier & Philippe 2002).

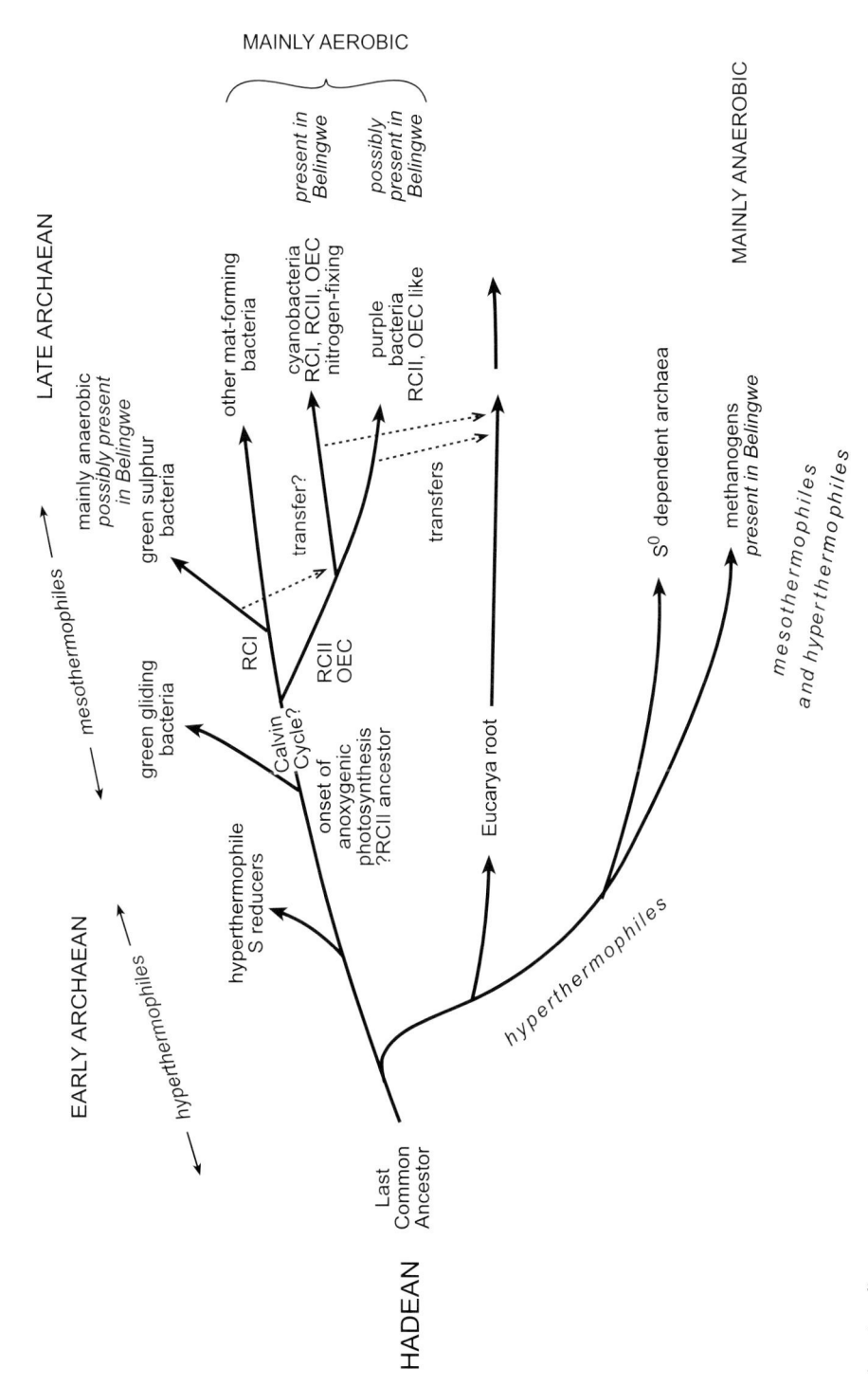

Fig 4. (continued)

The results are that the Planctomycetales emerge at the base of the bacteria as the first branching group. This is a remarkably interesting finding, if correct. Many Planctomycetales are, as their name implies, free-floating mesophiles. In stalked or budding bacteria like the Planctomycetales, cell division occurs by the formation of a daughter cell, with the mother cell retaining her identity after the division is complete. In other words, the cells are mortal, unlike most bacteria, which multiply by binary fission and are hence immortal. Moreover, these organisms, as they have polar growth, have a potential for morphological evolution that is not available to other bacteria. In many ways, these bacteria are reminiscent of eucarya, and just possibly may lie near the most ancient root of the eucarya. Among these budding bacteria, *Pirella* is chemoorganotrophic, hinting at antique habitat.

There are many but not proven arguments to suggest that, whatever the original home of life, the substantial ancestral bacterial diversification took place in a hydrothermal setting (Stetter 1996; Nisbet & Fowler 1996a, b). On the modern world, even on the surface of an active volcano the hydrothermal habitat is limited. Roughly 10 km^3 of new oceanic crust have been produced annually on the Phanerozoic planet. Although in the early Archaean volcanic resurfacing must have been much more extensive than today, and new crust production may have been as much as 100 km^3 per year, it is questionable whether the total thermodynamic power from the heat of the global hydrothermal habitat could have sustained a biosphere productive enough to alter the planet's surface environment as a whole.

Possibly the earliest habitats were thin biofilms of bacteria and archaea that existed by processing redox contrast between hydrothermal products and the external environment (sea water and atmosphere). The productivity of these early microbial mats would have been severely limited by the inorganic sources of redox power from below. Volcanic and hydrothermal processes would have ensured a small but steady supply of H_2, H_2S, CH_4 and possibly HCN from below. Nitrate (from dissolved NO_2), and sulphate are crucial. Some sulphate would have come from dissolved magmatically exhaled SO_3. Some would have been supplied by disproportionation of sulphite in seawater, the sulphite having come from magmatic SO_2. Some sulphate would have been made photochemically in air.

Microbial sulphate reduction is of great antiquity (Shen *et al*. 2001), as may be sulphur reduction by hydrogen. Sulphate chemistry gives many microbial possibilities; for example, an extreme option is

$$CH_4 + SO_4^{2-} \rightarrow HCO_3^- + HS^- + H_2O.$$

Much, perhaps most early microbial ecology around hydrothermal systems may have been sustained by the thermodynamic contrast between sulphate supply from above (water–atmosphere) and reduction power from below (rocks and rock-interacting fluids). This primary thermodynamic power would, however, have been eked out by many recycling steps, in which reduced organic matter was multiply reprocessed. Respiration and fermentation, like burning, run the thermodynamic arrow forwards, but by recycling allow the microbial system to exploit, to the limit, the possible productivity allowed by the inorganic redox bounds of the habitat. The system is like a cuckoo clock. Light quanta wind it up, then as the pendulum's weight falls, it turns innumerable microbial cogwheels, some forwards, some backwards. Complex recycling is possible in microbial mats of sulphur bacteria (Fig. 5). Rubisco must pre-date oxygenic photosynthesis, and is probably of the greatest antiquity. It is used by aerobic or microaerobic organisms, not reducing organisms, suggesting it evolved in a setting where sulphate was present.

The knallgas reaction ($\frac{1}{2}O_2$(aqueous) + H_2(aq.) = H_2O) is used by both bacteria and archaea, some of which (Aquifecales) appear to be very deeply rooted (although see the contrary view of Brochier & Philippe 2002), suggesting perhaps that hydrothermal waters carrying out-of-equilibrium O_2 (possibly from the atmosphere) and H_2 (ultimately volcanogenic) could have been important in early microbial communities. The apparent antiquity of the microbial lines using the reaction is arguably a pointer, although weak, that suggests that O_2 was widely available very early on, and that at least micro-aerobic habitats existed.

Remaking the environment

Photosynthesis (Pierson 1994) uses light to reverse the normal flow of terrestrial thermodynamics. It may have begun in purple bacteria, using bacteriochlorophyll (Xiong *et al*. 2000; Blankenship 2001). Light winds up the spring, so that an endless cycle can operate as the photosynthesizers capture reducing power, and then the other organisms, like cogs, run it down again. Oxygenic photosynthesis, aided by the start of the nitrogen cycle (which may have been nearly simultaneous), broke the bounds by capturing a new source of ordering power, and inflated the

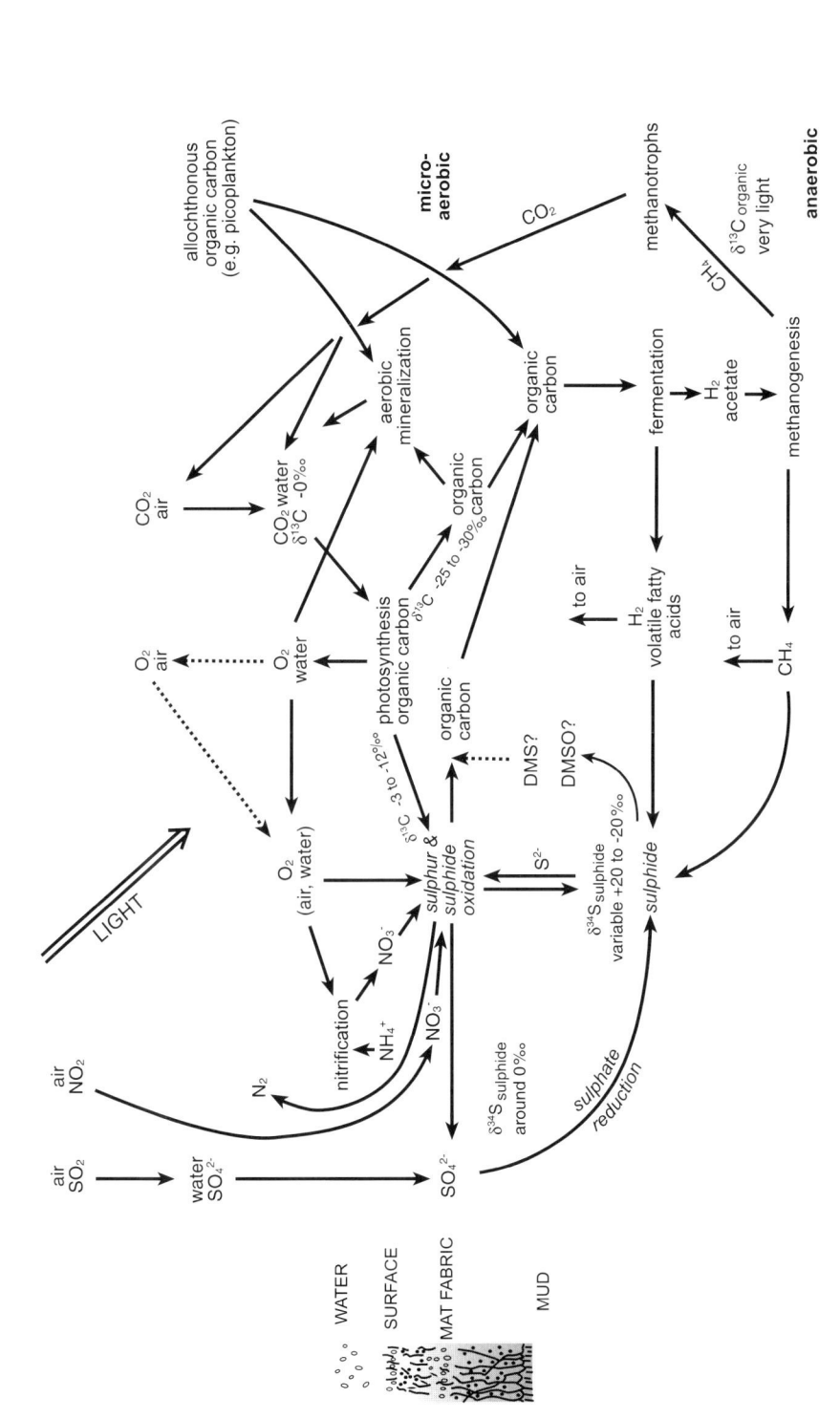

Fig. 5. Inferred biochemical pathways in an Archaean microbial mat built by sulphur bacteria. Model is based on modern mats (Fenchel & Bernard 1995). Many of these processes were probably operating in the Belingwe biota (Grassineau et al. 2002). Although there is direct evidence for only a few processes, for these to have operated the supporting reactions were probably present also. Thus, for example, isotopic evidence for methanotrophs suggests that fermenting bacteria were producing H_2, a conclusion that fits with rRNA trees.

biosphere, until more subtle cation and anion supply limits were confronted.

The early photosynthetic bacteria were limited in the places they could occupy (see Nisbet & Fowler 1999). There is molecular evidence (see summaries by Nisbet et al. 1995; Nisbet & Sleep 2001) that the first photosynthesis was anoxygenic, using bacteriochlorophylls not chlorophyll (Xiong et al. 2000). Most probably such life would still have been confined to the near vicinity of hydrothermal systems where there was adequate redox contrast. Supply of electron donors such as H_2, H_2S and CH_4, produced by hydrothermal processes, is necessary to sustain metabolism. It is also possible that there were enough reduced species in the global ocean to sustain anoxygenic photosynthesis in water plumes that spread far from hydrothermal sources before losing their relatively reduced state. Thus, with the initiation of anoxygenic photosynthesis, organisms may have been able to spread somewhat beyond the early hyperthermophile habitats into lower-temperature mesophile settings (Nisbet 1995; Nisbet & Fowler 1996a), subject to the availability of P and usable N.

The first organisms living by photosynthesis probably colonized the uppermost layers in microbial mats near hydrothermal systems (Nisbet & Fowler 1996b). Below them would be descendants of the earlier respirers and fermenters. The earliest possible evidence for abundant life is from Isua, Greenland, in 3.7 Ga rocks (Rosing 1999), where carbon particles in meta-turbidites appear to record biological isotopic fractionation: if so, they may be debris from plankton. It is tempting to suppose that such plankton, if they existed, were in some way distantly related to the Planctomycetales (Brochier & Philippe 2002). Possibly they were living by anoxygenic photosynthesis. In Earth's modern oceans non-photosynthetic archaeal picoplankton are widely abundant, occupying the levels in the lower photic zone (Beja 2002). Some Isua plankton may have been archaeal also. Today, planktonic archaea live in settings where oxidation and reduction power is controlled by overlying oxygenic photosynthesis by bacteria. In Isua-time, close to submarine volcanic centres, early archaeal plankton could instead have been dependent on chemical contrasts between surface waters and deeper waters influenced by hydrothermal plume sources. Possibly even archaeal methanogens reacted hydrogen (derived from hydrogen in the early air) with dissolved carbon dioxide, controlling the abundance of hydrogen in the air.

Once the planet was populated by organisms carrying out anoxygenic photosynthesis as well as others recycling the available nutrients by respiration and fermentation, life would have begun to have a significant impact on the redox distribution of the sediment–ocean–atmosphere system. Redox segregation would here begin on a grand, planetary scale, partitioning the surface into oxidized and reduced reservoirs. Sulphate and nitrate would be ubiquitously reduced before entering the sediment; hydrogen, methane and hydrogen sulphide would be oxidized. Moreover, the increasing sophistication of life, and the evolution of siderophores, would extend to its ability to capture essential elements, such as Fe, Cu, Mo, Zn, and particularly P, sharply increasing productivity.

At some point in the mid-Archaean oxygenic photosynthesis began (Buick 1992; Summons et al. 1999), allowing the evolution of complex cyanobacterial mats (Fig. 6). Nature has only once succeeded in this, in the creation of the photosystem II water-oxidizing complex. It is possible that the first O_2-evolving photoreaction centre originated in green non-sulphur bacteria, and that this was later incorporated into cyanobacteria (Dismukes et al. 2001) Cyanobacterial life depends on a complex partnership between both bacterial photosystems (Jordan et al. 2001; Kuhlbrandt 2001). Virtually all modern life depends directly or indirectly on oxygenic photosynthesis. Even in modern mid-ocean ridge hydrothermal systems, much of the life depends on sulphate, which in turn depends on the oxygen-rich ocean water created by oxygenic photosynthesis, which ensures an adequate supply of sulphate to the deep water.

An oxygenic photosynthetic ecology depends on water, carbon dioxide and light, and hence has global scope provided it can find enough usable nitrogen, phosphorus, iron, copper, etc. The moment the first oxygenic photosynthesizer evolved, a global ecosystem would have developed.

After the first appearance of cyanobacteria, presumably in microbial mats, the arrival of unicellular cyanobacterial plankton must surely have been rapid. In the modern warm tropical and subtropical oceans, cyanobacterial picoplankton are ubiquitous (Capone et al. 1997), supporting complex microbial consortia, and in the Archaean they could have formed the upper 100 m layer of an open ocean biological community, which may have had great diversity (Karl 2002). Archaeal plankton would have been outcompeted for occupancy of the topmost levels, but could have occupied a now more productive underlying lower layer, 100–300 m thick, dependent on the redox debris (including dissolved chemical species) from the overlying oxygenic photosynthesizers. The immediate results,

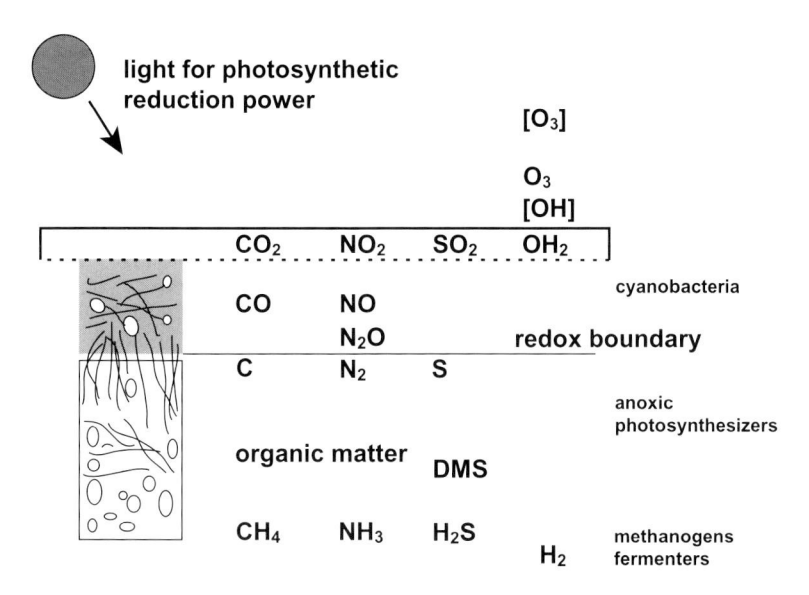

Fig. 6. Sequestration of redox power in microbial mats. An atmosphere–ocean system containing CO_2, NO_2 and SO_2 would supply oxidation power to the water (e.g as NO_3^-, SO_4^{2-}). Microbial mats finely focus and control the redox boundary, so that it becomes a sharp plane rather than a diffuse zone. Above the boundary, rubisco-using bacteria, both photosynthetic and non-photosynthetic, provide reduced organic matter. Below the redox transition, anaerobic microbes recycle the oxidation power, returning chemical species such as CH_4, H_2S and NH_3 to the upper levels. The deeper levels of the mud become a long-term store of reduction power (e.g. methane-rich sediment), sequestered from the atmosphere and inaccessible to it until recycled by large-scale geological processes such as subduction.

possibly as an event around 3.5–3.0 Ga ago, would have been a global crisis. They would have included massive release of dioxygen, massive fixation of nitrogen, and biological control of the carbon and sulphur gases in the air.

Cyanobacterial plankton have a major role in modern nitrogen fixation (Capone et al. 1997; Zehr et al. 2001), a role they presumably also had in the late Archaean (Nisbet & Sleep 2001). Without this, nitrogen supply would have presented a problem (Kasting & Siefert 2001; Navarro-Gonzalez et al. 2001). In air rich in carbon dioxide, lightning would have fixed nitrogen and supplied NO_x to early life. If enhanced organic productivity meant a decline in carbon dioxide content and, from methanogenic recycling, an increase in methane content of the air, there could have been a marked reduction in supply of fixed nitrogen. Life would rapidly consume its resource base of accessible N. Biological nitrogen fixation may have been an evolutionary response. This is a plausible hypothesis, but the crisis may have happened much earlier. Nitrogen fixation may be very ancient, pre-dating the last common ancestor (Line 2002). Possibly nitrogenase may be of great antiquity, and may have first evolved in a microbial mat to handle redox excesses. Perhaps, being available, it was then pressed into service to solve a N-shortage crisis.

There is another knock-on problem with oxygenic photosynthesis. Nitrogenase is very sensitive to dioxygen. One solution is a physical partition, to create a small space where an 'older' pre-oxygen condition can be preserved, despite being in an aerobic habitat. A key part of the apparatus in many cyanobacteria is the heterocyst, a walled-off microhabitat where very oxygen-sensitive nitrogenase can be protected.

The waste disposal from the Mn–oxygen evolving complex, in supporting the nitrogen cycle, is also worth considering. As planktonic cyanobacteria bloomed, the MnO_2 detritus would have become a global carpet on the ocean floor. In modern marine muds, with an infall of MnO_2, anoxic nitrate production during Mn reduction allows coupled sedimentary nitrification–denitrification cycles to occur. These 'nitrificata' link the nitrogen, manganese and sulphur cycles (Hulth et al. 1999), and would have been much promoted when cyanobacteria using Mn–oxygen evolving complexes first became ubiquitous in the oceans and first began to supply abundant MnO_2 debris to the sea bed.

The Belingwe case study

The 2.7 Ga old Belingwe Greenstone Belt, Zimbabwe, gives an insight into the state of life in the late Archaean (Nisbet & Sleep 2001; Grassineau et al. 2002). Several sedimentary facies contain clear evidence of microbial activity, and these have been studied using high-resolution small sample isotopic techniques (Grassineau et al. 2001, 2002). The results give a picture of a complex and interlinked microbial biosphere. Three distinct microbial bio-facies types can be distinguished in the sediment.

(1) The carbonate horizons of the Cheshire and Manjeri Formations contain extensive well-preserved stromatolites. These were formed in shallow or intertidal waters, as demonstrated by interbedded ripple-marked and mud-cracked silts. Texturally, the limestones show many indications of organic activity and gas release structures (Martin et al. 1980). The simplest uniformitarian explanation is that the structures were built by cyanobacteria. Isotopically, carbon from kerogen in the stromatolites typically has $\delta^{13}C = -25$ to $-30‰$. This implies but does not prove fractionation by rubisco of carbon captured from the atmosphere–ocean system. Carbonate in Cheshire limestones is typically close to 0‰, suggesting that carbon in the atmosphere–ocean system was modulated by rubisco and dominated by oxygenic photosynthesis on a planetary scale (Fig. 1) at least by this date. It should be noted that the c. 3 Ga Steep Rock stromatolites are similar (work by Abell, Grassineau and Nisbet).

(2) Shales tell a rather different story (Grassineau et al. 2001, 2002). The Manjeri Formation shales, which are very well-preserved intertidal ripple-marked deposits and shallow sub-tidal deposits, include sulphide stringers that show great S isotopic fractionation (at least for the Archaean). Rare carbonate samples from Cheshire shales show $\delta^{13}C$ significantly lighter than 0‰. Organic carbon in these rocks ranges widely. The simplest, although not the only interpretation of the facies analysis and C and S isotopic results is that anoxygenic photosynthesis was occurring, that sulphuretum-cycling was operating, and in addition, methanotrophy recycled carbon and, presumably to support the methanogens, fermenting bacteria were producing hydrogen. A community of anoxygenic photosynthesizers, respirers and fermenters may have been using light to operate a complex cycle involving sulphate reduction to sulphide and eventually H_2S, producing reduced organic matter that was eventually recycled by methanogens and methanotrophs in a microbial mat. However, it should be noted that this is speculation; in particular, there is no clear evidence for anoxygenic photosynthesis, only inference.

(3) Rocks from the Jimmy Member of the Manjeri Formation record a different facies. They may have been laid down below wave base and include very finely laminated carbon–sulphide layers (Grassineau et al. 2001). They are in very close stratigraphic proximity to submarine volcanic rocks, including pillow lavas of komatiitic basalt. There is strong rare earth element (REE) geochemical evidence for proximal hydrothermal influence. The carbon isotopes in the C–S layers show either rubisco fractionation or in many cases more extreme fractionation, presumably by methanogens. The S isotopes are also highly fractionated, both heavy and light, implying complex S cycling processes. As the rocks are thought to have formed below the wave base (some preserve finely laminated fragments), they may record a deep-water microbial community around a hydrothermal system. This microbial community may have cycled C and S between oxidation states, using sulphate from sea water and possibly reduced (rubisco-fractionated?) organic debris from planktonic life above.

Collectively (Fig. 7), the exceptionally well-preserved Belingwe material shows in its various bio-facies evidence for a complex interlinked biosphere at 2.7 Ga. In coastal waters, oxygenic photosynthesis by cyanobacterial stromatolites occurred. If cyanobacteria existed in stromatolites they surely existed also as free-living plankton, on a global scale. The evidence for anoxygenic photosynthesizers is weak: permissive, not proof. It is consistent with most models of microbial evolution: if cyanobacteria were present, then anoxygenic photosynthesizers probably preceded them. Collectively, if the cyanobacteria were productive enough to influence the global C isotope budget (as is implied by the $\delta^{13}C$ of 0‰ in carbonate), they would have necessarily produced dioxygen on a globally significant scale. Methanogens operated too in stromatolites, recycling organic debris, and probably archaeal plankton lived in deeper waters.

It is possible that mats existing by sulphur-based redox cycling were global in distribution, both in shallow-water muds and in deeper water, driven by the availability of sulphate (in part derived from oxidation power from cyanobacteria), organic carbon, and in shallow settings, light. The Jimmy Member may tell a deep-water story. Here the ecology may have been in part dependent on the abundant hydrothermal nutrients from the underlying volcanic rocks, and in part on the debris and the sulphate that came

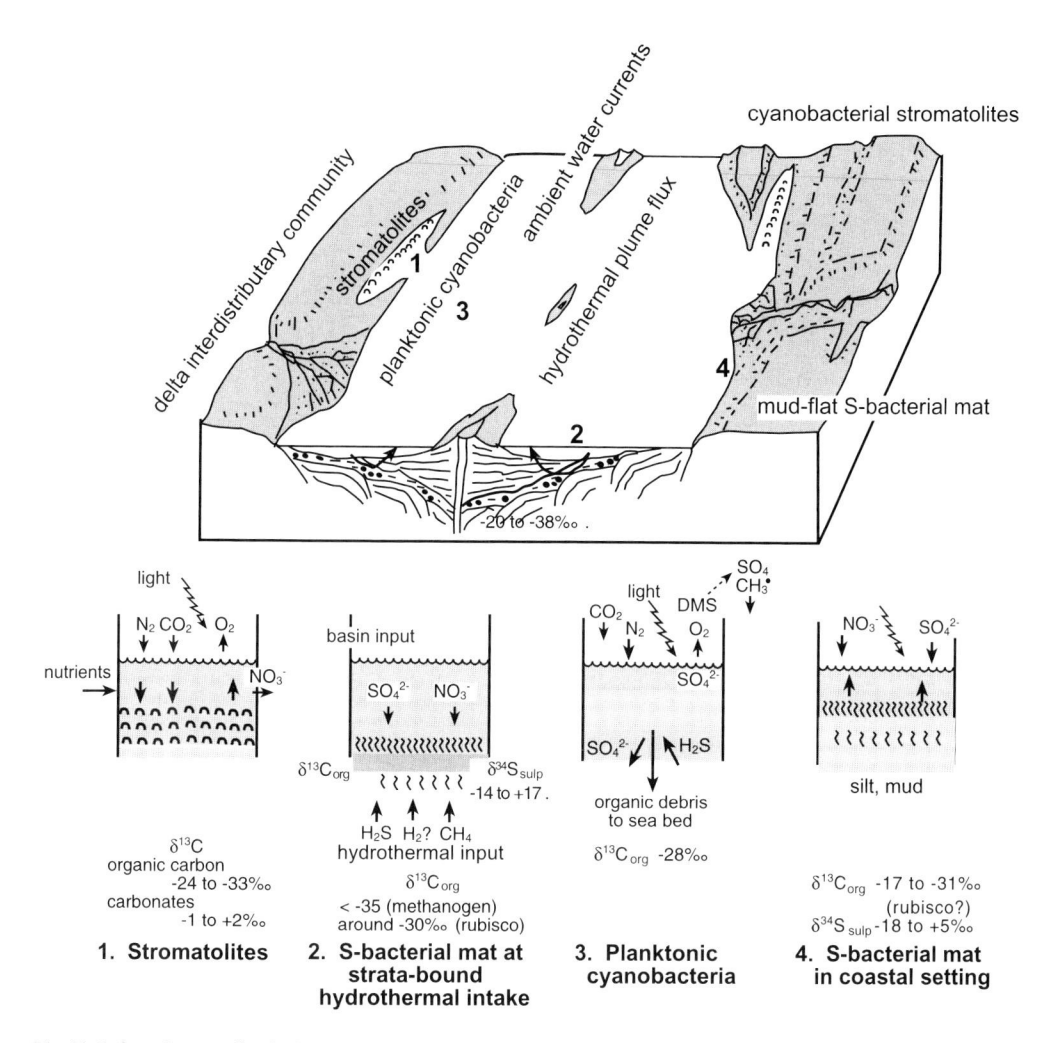

Fig. 7. Inferred generalized photosynthetic microbial mat consortia in an Archaean rift sea floor, inferred from carbon and sulphur isotopic results in the Spring Valley, Shavi, and Jimmy Members of the Manjeri Formation, and the Cheshire Formation (see Grassineau et al. 2002). General palaeoenvironment is reconstructed from sedimentary facies in typical greenstone belts (see summary by Nisbet 1987). Model assumes metabolic pathways from isotopic results and assumes that metabolic processes inferred from rRNA metabolic phylogeny (Woese 1987; Pace 1997) were extant. See also Nisbet & Sleep (2001). It should be noted that the palaeogeographical reconstruction shown here is a generalization, not specific to Belingwe, but drawing from the array of settings studied.

in from the light-driven planktonic communities in the upper layers of water above.

The various facies record microbial ecologies that have close parallels today. A plant can be seen as a community of chloroplasts and mitochondria; a microbial mat in a purpose-built house. The chloroplast is a cyanobacterium, whereas a mitochondrion is a purple bacterium. A land animal is an assemblage of mitochondria able to leave the seas in a walk-around space suit. Animal digestions include archaeal recyclers: to a methanogen a cow or a termite is simply a handy life support system.

Both cyanobacteria and purple bacteria appear to have been present as the basic drivers of the Belingwe ecology, and archaeal recyclers were probably present also. It thus would be logical to expect many similarities between Belingwean and modern biochemical cycles. The most obvious difference is the absence of environment-altering multicelled eukaryotes (e.g. worms) capable of digging into mud, churning sediment, and pumping water and fluids through redox boundaries. This would have gravely limited the productivity of the sea-floor mud biome as recycling would have been restricted. Many primary nutrients, such as accessible metals and phosphorus, would have been buried permanently after one pass through the biosphere, whereas today with eukaryote recycling they may be reused many times before being lost to the tectonic cycle.

Controls on the Late Archaean atmosphere

The implication of the isotopic evidence from late Archaean successions such as Belingwe supports the view from molecular biology that if cyanobacteria existed, a very diverse array of organisms had already evolved (Fig. 4). This diverse community would have necessitated a complex global microbial ecology (Nisbet 1995), and if so, it was probably managing the redox budget and the C, S and N (and probably Fe^{2+}) budgets of the ocean–atmosphere system on a global scale by the late Archaean. Just possibly, carbon and sulphur were managed isotopically on a global scale as early as mid-Archaean. Given the evidence for microbial control of C such as the carbonates with $\delta^{13}C$ c. 0‰, the atmosphere was probably a biological construct in the late Archaean. The further implication is that by the late Archaean, the global surface temperature was modulated by biological processes.

The air is not a simple linear system. There are many feedbacks, giving an atmosphere that appears stable when viewed over the time scale, say, of human evolution. However, even though the air has stabilities, it may have longer-term instabilities, or else bistable or multistable states. How was redox power managed? Models can help.

The bathtub model

The atmosphere is the global reservoir of redox power. The level of oxidant in the air is not set by the various fluxes, but by the switches that control input and output gates. Consider a bathtub. Within wide limits, in a bathtub, the level of water is not set by the flow of water out of the tap, but by whether the plug is in or not. If the plug is in, even a dripping tap will eventually fill the tub so that it spills from the overflow. There is no obvious relationship between tap flow and height of water in the tub. Instead, the level of the reservoir is set by an external physical factor, the height of the overflow hole. If the plug is not in, even the flow from a tap full-on will put only a minimal amount of water in the tub.

The bathtub analogy can be used to consider many aspects of sequestration on the Earth. The most obvious problem, the storage of water in the oceans rather than the mantle, has already been discussed above. A similar problem is the sequestration of oxygen in the air. In the late Archaean Earth, the photosynthetic cause of the flow of oxygen production through the 'tap' was presumably in cyanobacterial mats and global cyanobacterial plankton. This view is based on the evidence for the presence of cyanobacteria (Buick 1992). Bacteria spread very rapidly and once cyanobacteria had evolved they would have occupied every available niche very quickly. Just as rabbits filled Australia within a few decades, so cyanobacteria would spread globally within a very few years of their first evolution. If they are present in one place, they will be present planet-wide.

There is no particular *a priori* reason to suspect that oxygen production was any different from that in the modern ocean, except that there may have been constraints imposed by different availabilities of key nutrients such as phosphorus (Bjerrum & Canfield 2002). Today, the availability of fixed nitrogen may constrain productivity on geological time scales (Falkowski 1997), but in the Archaean phosphorus removal by adsorption on iron oxides could have reduced P availability, significantly reducing productivity compared with today.

Even today cyanobacterial picoplankton are ubiquitous in the ocean. In the late Archaean, resources nowadays used by multicelled eukaryotes would be available for microbial plankton. In the later Archaean, when continental land masses were present, abundant aeolian dust would have provided nutrient for ocean plankton blooms. There may also have been some land production of oxygen from photosynthetic microbial mats growing in damp or swampy areas. In the Archaean, the 'plugholes' that removed dioxygen were dominantly microbial oxidation of reduced organic debris and photorespiration, and, on a longer timescale, flow precipitation of oxidized minerals. This oxygen removal and carbon dioxide return would have matched the photosynthetic oxygen production and net carbon capture (after allowing for methane emission), or the system would have quickly sequestered all its carbon. Broadly, for both carbon dioxide and oxygen, obverse and reverse, the two fluxes of production and removal must be equal, if the oxygen and carbon dioxide levels in the air are stable. Wrapped around this rapid biological cycling of carbon, on a slower time scale, is an outer nested geological carbon cycle, so that any slow carbon burial and crustal storage must be matched by carbon return from volcanoes.

But this applies to the fluxes. The size of the oxidation reservoir in the air (as opposed to the flux) is not necessarily coupled to the flux. In some cases, linkage occurs. For example, if the concentration build-up promotes the oxygen removal process by planktonic oxygen-consuming light harvesting (e.g. see Karl 2002), then controls of chemical equilibrium and kinetics may operate. However, many biological processes are very insensitive to such crude chemical controls, as the cell wall is a powerful barrier. Thus heterocysts protect nitrogenase, which is very oxygen-sensitive, in cyanobacteria. Nitrogen fixation is decoupled from external oxidation state and even in a modern oxygen-rich atmosphere, nitrogenase operates very successfully.

The speed of biological control and recycling of critical nutrients is much faster than the slow action of erosional supply and removal of cations. Thus the dioxygen balance level is not set by an inorganic buffering reaction, nor by a flux rate, but is a plane in O_2–CO_2–T space defined by the competing modes of action of rubisco (see Lorimer & Andrews 1973; Lorimer 1981; Tolbert 1994; Joshi & Tabita 1996), and by biological management and sequestration of crucial elements such as phosphorus.

Oxidation of the surface by methane loss and carbon subduction

Photosynthetically gathered reducing power, collected by bacteria, accumulates as reduced organic matter in sediments. From this, methane and other reduced products are generated by archaea after various processes of microbial fermentation. There are many steps, but finally the fermentation products such as acetate and ethanol are themselves utilized until two ends of anoxic fermentation are reached: isotopically light methane, and isotopically heavy bicarbonate or carbon dioxide. Methane is thus one end member of the full photosynthetic process: $CO_2 + 2H_2O \rightarrow CH_4 + 2O_2$. Other reduced remains include dihydrogen, hydrogen sulphide and ammonia. In the modern world (for example, in the cover of a landfill), these reduced products in turn are reacted by bacteria such as methanotrophs and microbial sulphate reducers, against the opposite waste product, the oxidant stored in the air.

Cyanobacterial mats in modern hypersaline environments also release significant quantities of CH_4. This flux is probably related to H_2-based alteration of the redox potential within the mats (Hoehler et al. 2001). In the Archaean the flux of CH_4 from mats and emissions from muds where methanotrophs were recycling organic matter from plankton would have supplied methane to the air, and hence to the upper air. However, marine microbial consortia of archaea and sulphate-reducing bacteria would probably have been active, carrying out anaerobic oxidation of methane by sulphate reduction. (e.g. Boetius et al. 2000).

In a biologically active planet, microbial activity focuses on redox boundaries (Fig. 6). This enhances the effectiveness of the interaction between oxidant and reductant, and leads to massive sequestration of reducing power as buried reduced biomass. On the surface and near-surface, reservoirs of oxidant and reductant can be sequestered very efficiently. This sequestration is short term (plate-tectonic lifetime of around 100–200 Ma) compared with the history of the Earth, but it is a continuous process and so in effect the sequestration is permanent.

On the modern Earth, near-surface stores of reduction power include coal and oil, gas, and vast reserves of methane in and below hydrate caps, as well as disseminated kerogen and charcoal. The total store of reduced carbon near the surface is many orders of magnitude larger than the total held in the air and living biosphere. Even in Archaean rocks, carbon-rich strata are common. Some Belingwe rocks, for instance, are extremely rich in organic carbon and leave oily sheens on the saw-water when cut. Archaean oil is known to survive (Dutkeiwicz et al. 1998; Summons et al. 1999; Rasmussen & Buick 2000). Equilibrium was clearly not attained: indeed, the biology may have acted to increase the disequilibrium, sequestering redox power into separate reservoirs and hence deepening the bathtub by moving the overflow upwards.

In the modern atmosphere, free methane is oxidized within about a decade by OH, derived from water vapour attacked by O which in turn is derived from ozone and dioxygen. If the Archaean atmosphere were reducing and if it had lower O_2, biologically released methane would have lasted far longer in the air. If abundant, atmospheric methane would have been a powerful greenhouse gas.

There is no cold trap for methane, and thus it would have been abundant in the high Archaean atmosphere. Here it would have been decomposed by UV radiation, releasing hydrogen. Some of this hydrogen would have diffused to the top of the atmosphere and would have been lost to space. There would also have been significant microbial H_2 in the air and hence H loss to space (e.g. Catling et al. 2001). Over time, the Earth could have lost substantial hydrogen and hence substantial water (the original source of the hydrogen in the methane), leaving residual oxygen.

A parallel geological sequestration may have been going on. There is strong evidence that early microbial communities lived in hydrothermal systems (Rasmussen 2000). Such microbial communities would have lived on mid-ocean ridge systems. The reduced carbon in the organisms would have been buried by lavas and eventually subducted. Much of this carbon would be returned to the surface as carbon dioxide within the vigorous Archaean plate cycle in, say, 100 Ma to 1 Ga. However, in part the carbon would have been converted to graphite and then into diamond (Nisbet et al. 1994) when the ocean floor was subducted. Some of this diamond would have been underplated under continent, or else permanently removed from the Earth's surface (the 'exosphere') by return to the deep mantle. Loss of reduced carbon to the depths is, in effect, supply of oxygen to the surface because the C was ultimately derived by splitting CO_2. Similarly, subduction of reduced sulphur (Alt & Shanks 1998) effectively oxidizes the surface atmosphere–ocean system.

Cumulatively, over an aeon, the total impact of H loss to space and C return to the mantle would be relative oxidation of the surface and reduction of the interior. However, if iron oxides were being precipitated in sediment, and then subducted, this process would return oxidation power to the interior. To accomplish both carbon return to the interior and iron oxide return, subduction deep into the mantle is necessary, rather than shallow subduction of relatively hot plate. Thus, to return to Fermor's point, the density contrast that drives subduction, which comes from cooling by liquid water, plays a role in maintaining the oxidation regime of the surface.

Walkerworld

In 'normal' times the air is more oxidized than fresh volcanic rock, and also more oxidized than the sediment column, which is enriched in reduced organic matter. Before the onset of global photosynthesis this contrast with the sediment may have been very different (Walker 1987), especially if methane were abundantly released (Hayes 1994).

In 'Walkerworld' conditions, an 'upside-down' biosphere occurs. The waste oxidant is held in the sediment and the air contains the reducing power. To sustain a Walkerworld, emissions into air of methane, hydrogen, etc. from decay must exceed burial of reductant and hence must exceed photosynthetic emissions of waste oxidant and any inorganic production of oxidant by hydrogen loss to space. These events, if they occurred at all, must be abnormal because the primary wastes of oxygenic photosynthesis are gaseous dioxygen, which tends to escape upwards, and the primary reduced waste, organic matter, which tends to be buried downwards, ensuring an up–down partition, and thus the main steady flux to the air has been oxidant ever since planktonic oxygenic photosynthesis began. Even in a biosphere depending on anoxygenic photosynthesis, the buried organic waste is reduced.

But there is no obvious intrinsic reason why the air cannot for a brief period (geologically speaking) become so rich in methane that the system is inverted, with a reduced atmosphere and a relatively oxidized sedimentary mass. Today a vast reducing biosphere exists in sediment under oxidized skies yet with trace methane in the air, whose lifetime is determined by kinetics. Then the reverse could have been true, especially before the onset of oxygenic photosynthesis, when the output of gaseous oxidation power to the air was less. The air could have been methane rich, with trace oxygen.

If huge quantities of methane are suddenly released from geological stores (for a modern analogue, one can imagine a giant slump releasing several tens of gigatons of methane from hydrate in sediment of a large delta, or a sudden large plume-head intrusion into the giant permafrost-hosted Siberian gas fields), then the onset of methano-dominant events could occur. The microbial ecosystems would invert, and methane release would dominate the air, with carbon dioxide, not methane, being the gas stored in the sediment pile. The geological record may bear witness to such events in major C-isotopic excursions.

Given the absence of eukaryote reworking of muds, it would perhaps have been easier to sustain an inverted ecology in the Archaean than today. In some ways, the deep-water environment of the modern Black Sea is today a Walkerworld, with ambient reduced species in the water above more oxidized sediments (and the chilling possibility that one day the water body will overturn, releasing its gases to air). It is possible that in the Precambrian major events took place with methane as the dominant carbon gas (Hayes 1994). Inversion is, ultimately, arguably unstable as the reduced organic matter settles downwards, so the natural order is reduced below, oxidized above, but Walkerworld events may have been sustained for long periods.

Pavlov et al. (2001) have constructed a plausible history to explain carbon isotope excursions and then major changes in the early Proterozoic atmosphere as the product of a Walkerworld. In this view, the pre-biotic atmosphere was kept warm by high CO_2, with which a significant amount of H_2 coexisted. With the onset of life, methanogens rapidly began to use the H_2 to produce CH_4. In this view, the early C isotope record does not reflect oxygenic photosynthesis but rather methanogenesis. By c. 2.8 Ga ago, however, Pavlov et al. (2001) have suggested that oxygenic photosynthesis began and increased the supply of organic matter and hence of CH_4 and an associated hydrocarbon haze. At this stage, sulphate-reducing bacteria, exploiting the increase in oceanic sulphate after the onset of oxygenic photosynthesis, either outcompeted for the H_2 supply or converted CH_4 back to CO_2 (e.g. see Boetius et al. 2000), causing a decrease in atmospheric methane. In turn, the loss of the greenhouse warming at the collapse of the Walkerworld initiated a massive glaciation at c. 2.3 Ga. Eventually, the response led to the oxygen-rich air.

Building a biosphere

A possible sequence of events can be deduced.

Life before photosynthesis

From the origin of life to the evolution of the first cell capable of anoxygenic photosynthesis (say from between 4–3.8 Ga and 3.8–3.5 Ga), oxidation power was provided by bicarbonate and by oxidized sulphur and nitrogen compounds, especially sulphate and nitrate. Elemental sulphur, as sulphur vapour, may also have been important (Kasting et al. 1989), acting as a shield in the atmosphere to protect against UV radiation and as a supply for sulphur-metabolizing microbial life. On the sea floor, life would have sequestered reduced matter in sediment mainly by burial of organic matter and sulphide precipitation. Organic carbon would have been further recycled by archaeal generation of methane in the deeper parts of microbial mats. Part of this methane would in turn have been reoxidized by methanotrophic bacteria, but some would have escaped to air.

Today, much of the methane that is produced by methanogens is stored in the muds and clays of the mature sediments of the continental slope, where massive reserves build up. In the early Archaean, perhaps with far less continental surface, mature sediments would have been much less common. If methane escape from sediment were easy, Walkerworld conditions (methane, not carbon dioxide, in a 'negative' bathtub) could have been favoured, with the build-up of a reduced atmosphere and a relatively oxidized sediment, possibly rich in precipitated elemental sulphur and sulphate. If methane were more abundant than carbon dioxide it may have created a methane-rich smog that was analogous in function to the modern ozone layer (Lovelock 1988).

Anoxygenic photosynthesis

After the evolution of anoxygenic and before the arrival of oxygenic photosynthesis (say 3.8–3.5 Ga to 3.5–3.3 Ga) (see also Rosing 1999; Brasier et al. 2002), molecular oxygen production from inorganic sources and from biological emissions would have been small, but the added redox power and productivity of microbial mats would have been nevertheless considerable. Consequently, large amounts of reduced organic matter would have built up in sediment, and, though methane emissions from sediment would have increased, so would have the bulk sequestration of reduction power out of the ocean–atmosphere and into the crust. Possibly methane emissions from methanogens were large enough to be a crucial component of the atmospheric greenhouse, able to sustain a warm surface (Pavlov et al. 2000)

Oxygenic photosynthesis

When oxygenic photosynthesis began, whenever this was, the spread of uniquely advantaged cyanobacteria blooms across the planet would have been explosive: an 'Andromeda event' (to recall the Fred Hoyle fiction in which genetically manipulated microbial life suddenly challenged the atmosphere), instantaneously altering the air. Cyanobacterial mats are potent physical separators of oxidant and reductant. Reduced muds would be separated from oxidized surface by the leathery septa of microbial mats. Release of waste oxygen would have been a sudden consequence of the arrival of cyanobacteria, catastrophic to the previous surface ecology. The oxygen level in the bathtub (Fig. 8) would have deepened instantly. Biological sequestration of redox power would have become global, wherever fluid flow occurred into and out of the sediment. But, as argued above, it is not the flux that controls the depth. In the bathtub of a microbial world it is not obvious where or how the overflow level was set.

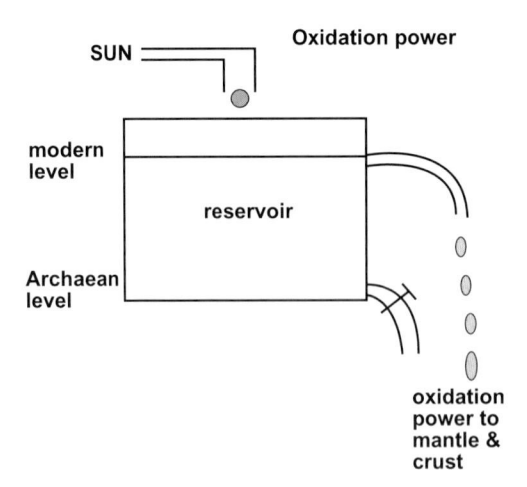

Fig. 8. The bathtub model. The level of the water in the bath is not controlled by the flow from the tap (unless it is huge) but rather by the levels of the output pipes. With the plug in, even an inflow that is just dripping will eventually fill the bath to the overflow level; if the plug is out the bath will start to fill only when the flow from the tap is greater than the capacity of the plughole. If the atmosphere is compared with the bathtub, with the water content analogous to the atmospheric burden of, say, oxygen or carbon dioxide, then the short-term inflow and outflow gates are set by the subtle response of rubisco to CO_2 and O_2 mixing ratios (Lorimer & Andrews, 1973; Lorimer 1981), as managed by chloroplasts and mitochondria and by similar cyanobacteria and purple bacteria in the Archaean. There may be bistable modes, either with low CO_2 and high O_2 as now, with rubisco acting one way, or with high CO_2 and low O_2 in the Archaean. These immediate molecular controls act in a world where the dominant long-term controls are the large-scale inorganic redox stores and geochemical controls (e.g. Lasaga & Ohmoto 2002), for example, iron oxidation, and there are also medium-term organic controls (e.g. wildfire on the modern Earth; microbial blooms and subsequent rapid sinking of reduced organic material in the pre-metazoan planet). Which time scale of control is responsible for setting the atmospheric burden is not obvious, but short-term biological controls, by their ability to sequester redox power (see Figs 5 and 6), may be dominant.

The rise of oxygen

The oxygen burden of the Archaean air (the 'level in the bathtub' Fig. 8) remains an unsettled controversy. Much opinion (e.g. Kasting 1993; Holland 1999) holds that oxygen levels rose sharply in a sudden event that took place in a short time, somewhen around 1.8–2.4 Ga ago. Much evidence, such as the need for reduced settings to transport iron, implies that free dioxygen was virtually absent (Holland 1999), except as a short-lived biologically sustained trace gas (just as 1.8 ppm methane is present in today's oxygen-rich air, with a lifetime around a decade).

Kump et al. (2001) accepted that the initial rise in O_2 levels occurred at some time between 1.8 and 2.4 Ga ago. They suggested that this rise was abrupt, and followed a major period of magmatism, produced by mantle overturn and/or mantle plume activity. In contrast, Catling et al. (2001) have suggested that the transition from low to high dioxygen burden in the atmosphere was a consequence of hydrogen escape into space as a result of high methane concentrations in the Late Archaean–early Proterozoic air. The methane could have been produced biologically (Pavlov et al. 2000) by methanogens acting on carbon captured by photosynthesis, to give atmospheric methane concentrations of 0.1% or more. In the high atmosphere, photolysis of methane would produce hydrogen, leading to steady escape of H. This escape of H to space would slowly oxidize the whole Earth, building up oxygen inventories and increasingly enhancing the kinetic stability of atmospheric O_2, while also slowly reducing the volume of water in the oceans. Atmospheric O_2 would be scavenged in the early reducing environment, by oxidation of crust, but eventually the net photosynthetic production of oxygen would exceed the input of reduced gases to the air, and this would cause irreversible oxygenation of the air.

Kasting (2001) pointed out that serpentinization of new sea floor also produces reduced species, such as H_2 when water is the oxidant, and methane when CO_2 is present. In this process, Fe_3O_4 is produced. Possibly these two processes together sequestered enough oxygen that through the Archaean aeon the atmospheric oxygen was suppressed. Then, when the net photosynthetic flux of oxygen became larger than the flux of reduced gases, the oxygen level suddenly rose. In terms of the bathtub analogy (Fig. 8), the plug would be put in, and the level in the tub would suddenly rise until the new rubisco-controlled overflow operated.

Farquhar et al. (2000) investigated mass-independent fractionation in sulphur isotopes from across the period when oxygen levels are thought to have risen. They found a major change somewhere between 2090 and 2450 Ma ago. In rocks older than this, gas-phase atmospheric reactions may have influenced the sulphur cycle, playing a role in determining the oxidation state of sulphur. This would imply that atmospheric oxygen had low partial pressures, and microbial oxidation and reduction of sulphur were minimal. These results are of great interest, but have been disputed (see questions by Ohmoto et al., and the response by Farquhar et al. (Ohmoto et al. 2001)). It is also not clear how the work is reconciled with the implications of the large fractionation in ^{34}S observed by Grassineau et al. (2001, 2002), in 2.7 Ga material.

With the rise in oxygen would come an increase in ozone and a fall in UV irradiation of the surface (Cockell 2000). Possibly this more benign setting played a role in the onset of eukaryote and metazoan diversification.

The opposite view has been championed by Ohmoto (1997) on the basis of isotopic evidence, interpreted as suggesting relatively high oxygen levels. There is a reasonable case that oxygen levels in the Archaean were higher than thought in orthodox opinion.

Phillips et al. (2001) discussed the central evidence on which the 'reduced' model of the Archaean atmosphere is based. This evidence includes the composition of detrital gold grains, inferred detrital uraninite, inferred detrital pyrite, and inferred palaeosols, especially in the Witwatersrand succession. In a careful detailed review, Philips et al. concluded that the geological evidence for a reducing atmosphere remains ambiguous. In particular, post-depositional processes may need far more examination. Phillips et al. pointed out that some of the mineralogical and field evidence can be interpreted as supporting an oxidized Archaean atmosphere.

There are many pitfalls here. For example (own observations), in the Steep Rock succession of NW Ontario, which is old with abundant stromatolites (Wilks & Nisbet 1985), mining operations about 1100 ft (c. 300–350 m) underground produced 'buckshot ore', a pisolitic laterite iron ore that clearly implied oxygen-rich conditions. However, A. de Barros Machado (pers. comm.) pointed out termite-made features in the ore, and further investigation in the open pit (E.G.N. and R. Bernatchez) produced a small outcrop of the ore, rich in woody fragments. Obviously, wood and termites are not Archaean. Most likely, the ore had been made by Cretaceous or Tertiary termites that had tunnelled down 500 m from the pre-glacial palaeosurface to the contemporary water table, bringing down wood and oxygen. The location 1100 ft. underground in 3 Ga rocks had led to the wrong but obvious assumption that the oxidation of the ore was Archaean; it was not.

Nevertheless, although some arguments for Archaean redbeds can be questioned, Phillips et al. (2001) have a powerful case. The question 'was the Archaean atmosphere reducing' must be regarded as still unsettled.

The evolution of the eucarya

The eucarya may be of very great antiquity (Hartman & Fedorov 2002) even though their geological record is only as far back as possibly 2.7 Ga (Brocks et al. 1999). The possible antiquity of the budding bacteria such as Planctomycetales (Brochier & Philippe 2001) is of great interest here, as it immediately suggests a root stock for the eucarya. The date of incorporation of chloroplasts into the eukaryote cell is not clear, but may have been late Archaean. Whether or not the arrival of the eukaryotes was involved in the rise of oxygen is an open and fascinating question.

Eukaryotes can be regarded as 'microbial-mats-in-a-single-cell'. They incorporate both oxygen production from the chloroplasts (included cyanobacteria) and oxygen removal in the mitochondria (included purple bacteria). The earliest eukaryotes would not necessarily have had any more impact than their individual component parts, as it would have been likely that cyanobacteria and purple bacteria lived in close symbiosis before the arrival of the eukaryotes. Yet the dinoflagellate is a fearsome whip with which to beat the planet: there may have been large changes. There appears to be an intracell switch controlling oxygen levels in the eukaryote cell (Joshi & Tabita 1996), and this switch would probably have had global impact.

Biological controls, acting rapidly, and sequestering redox power into reservoirs, can override the background inorganic chemical controls. For example, Fe^{2+} can be carried in biological organisms, which have captured it from hydrothermal water reducing settings, into oxidized environments, isolated by biological membranes. The debate about the oxidation state of the sea remains open, and it is possible that, like the modern Black Sea, the top of the Archaean ocean may have been oxidizing as a result of widespread cyanobacteria picoplankton, whereas deeper levels were reducing because of the high rates of settling of reduced organic debris. With the arrival of multicelled organisms in the Proterozoic it became possible to pump more water through microbial mats: worms move fluids and rearrange reservoirs of reduction power. The sequestration became intense.

Links between the Archaean atmosphere and the ocean: biology and temperature

The atmosphere controls the planet. The effective temperature of the modern Earth (the temperature without the greenhouse warming of the air) is about $-18°C$. The greenhouse warming of $+33°C$ ensures an actual average temperature around $15°C$ and thus oceans are liquid. The main greenhouse gas is water vapour. The CO_2 and CH_4 have a multiplier effect. If the carbon greenhouse gases are absent, not only is there less water vapour in the air but more of the polar region is white from ice; thus more sunlight is reflected and the water-caused greenhouse declines; which makes the planet cooler yet, and so on: there is a risk of global glaciation. In the early Archaean the Sun was somewhat fainter (Sagan & Chyba 1997), and atmospheric oxygen was probably insignificant. The fainter Sun means that to sustain liquid oceans a more powerful greenhouse may have operated.

Biology-absent planet

What would have controlled surface temperature? Imagine first the snowball risk to an Earth-like but biology-absent planet in the Archaean: the risk that the ocean surface would freeze. Carbon dioxide might eventually freeze at the poles, and an icebox runaway would follow. In the long term, the situation would right itself if the planet were positioned in the Sun's habitable zone. Eventually either a huge volcanic eruption or a meteorite impact would generate enough dust to change albedo. Carbon gases and nitrogen gases would have progressively degassed from volcanoes, and a sudden warming could follow.

The opposite risk is an 'oven', a greenhouse runaway, in which a primary carbon dioxide greenhouse is added to by other gases: indeed, this risk is very present as a planet rights itself from a snowball. Imagine a very large plume eruption, that contributes huge amounts of carbon, sulphur and nitrogen gases to the air and injects huge amounts of dust to the stratosphere. The ocean initially cools under the dust, but soon warms as the dust settles. The carbon dioxide can only be removed on an abiological planet as dry ice or by precipitation; this demands the presence of adequate Ca^{2+}. But if the plume emissions are very large, the weathering that supplies Ca may be too slow in supplying cations, so that precipitation cannot keep up with emissions, despite the acidity of the rain. In this case huge quantities of water can be evaporated, creating a steam atmosphere. In the absence of a temperature inversion at the cold trap, water reaches the high atmosphere and hydrogen is lost to space. Eventually, the Ca supply from weathering precipitates the carbon dioxide, which can then return to the mantle by subduction and normality returns. But there is a risk that such events, repeated too often in the Archaean, could dehydrate the planet.

Biology-present planet

Consider now a biological planet. Imagine it entering a snowball phase for some reason. Methane would accumulate as the photosynthetic biosphere died and sediment was filled with organic debris. Methanogens would create vast unstable reservoirs of methane held in clathrates. Eventually this would emerge to air, probably catastrophically (in terms of geological time); there it would be a powerful greenhouse gas, and as warming occurred methane would be released by hydrates and carbon dioxide would be released massively from the oceans: one hypothesis for the escape from the snowball. The biological planet could probably 'right' itself earlier, after a less prolonged snowball, than an abiological planet.

On a biologically active planet, the feedback controls are faster, to counteract both freezing and heating. In an ice age, biology responds by reducing carbon consumption and then recycling carbon as methane and carbon dioxide. In a hothouse, biology runs rampant, removing carbon by organic capture and precipitation of carbon. This too depends on essential elements, such as Fe, Cu, Zn, Mo and especially P, but the supply is regulated by weathering rates, themselves controlled by biological carbon cycling. Biology is a multiplier, using recycling to extract more carbon management power from the available store of cations than inorganic chemistry. The presence of biological control speeds the return of the planet to 'normality'. Yet the dehydration risk remains, this time via methane partition into the top of the air and subsequent loss of H.

The state of the mantle and crust

Assume a planet in which the upper mantle has about 0.2% water, and the composition is broadly peridotitic, with heat production as on Earth. The geotherm depends on the vertical distribution of the heat production in the interior and on the surface temperature, and on the heat transfer controls. The surface temperature is maintained by the radiative balance of the atmosphere and any greenhouse increment, and is only indirectly dependent on interior processes. If the state of the atmosphere means that the surface is cold, say $-100°C$ to $-200°C$, then the lithosphere will be thick. The top of the mantle adiabat will be forced to cool. The deeper mantle will thus store heat until melting extracts it as plumes. Conduction will carry the heat away over the aeons (depending on how large the planet is), but more quickly than if the surface had been kept warm by the air's greenhouse blanket. If the planet is small the cold and moderately wet mantle will freeze into inactivity.

In the opposite case, consider a Venus-like planet with surface temperature forced by greenhouse blanketing to remain at 500°C. The geotherm will intersect the mantle solidus at shallow depth. Melt will be very efficiently extracted from the topmost levels of the mantle, and with melt will come heat production elements and volatiles, especially water. The crust will be dry, especially deeper than a few kilometres, and will largely be pyroxene-dominated granulite. There will be no return of water to the interior as there is no cold dense wet crust: no muds, no clays, no eclogite, no mud-derived garnets in moving continents. Instead, hydrogen from water photolysis in the high atmosphere will be lost to space.

The extent of the crust will depend on whether a buried mantle ocean, in which dunite floats over denser ultramafic magma, develops in the lower part of the upper mantle. If a mantle magma shell is present, water will be retained in the mantle melt. If a magma shell is not present, melt will rise carrying water, so that the mantle is eventually entirely dehydrated as water is extracted and lost to space. Even if a magma ocean is present, the long-spaced catastrophic overturns of cold dense surface and resulting massive eruptions will lose water to space. The surface and interior will be strongly oxidized by residual oxygen. Although the mantle is so hot, it will be dry after a few billion years.

On a Goldilocks planet the surface temperature is just right. Where volcanism occurs, water is ejected from the interior, but is held under the cold trap until it can be returned to new crust via hydrothermal systems. Old plate is dense and falls in; the water returns to the interior. Much of this water fluxes melt and is returned at andesitic volcanoes. A small amount is retained by plate entering the deep mantle. This intake is controlled by the ambient temperature and in effect the system becomes self-fluxing. If it is locally too hot, abundant melt production removes water and heat. As the planet cools, andesitic melt production falls and the cooler deep mantle accepts more water, to allow melt production to continue at mid-ocean ridges. All this depends on a surface temperature that allows rehydration and cooling of the new plate.

The Earth lies in the habitable zone, the right distance from its star (Kasting *et al.* 1993). But the Sun's output both varies and is steadily brightening. How wide is the allowable range of the Goldilocks temperature window, that allows liquid oceans to be sustained for thousands of millions of years? The upper limit is probably around 70–100°C. Above that, the ocean goes. The lower limit of the window is perhaps a little below 0°C as an equatorial temperature. Below this, there would be less rehydration of new plate, except locally around active volcanism where heat would melt ice and permit hydrothermal systems. The ice-laden planet would be white, and would cool. If it became very cool and carbon dioxide precipitated on high polar continents, the temperature would drop further. Only methane emission could allow an escape from this, but on an abiological planet there would be no massive source.

Thus the Goldilocks window is probably between 0°C and 100°C, or less. This surface temperature also may have influence on the permissible depth of the oceans. Kasting & Holm (1992) have argued persuasively that the depth and hence volume of the oceans is actually controlled by the depth at which the critical pressure for sea water occurs. If mid-ocean ridge axes are at this depth, this optimizes convective heat transport and thus hydrothemal circulation into the new crust. Consequently, the crust is hydrated efficiently, and water is subducted and fluxes the plate system, in a self-stabilizing regime (Kasting & Holm 1992). If this is correct, a surface at, say, 15°C, would imply oceans in which mid-ocean ridges were 2.5 km deep (250 bars pressure) and hence about 3.5 km deep on average in a plate system where ocean depth is determined by the cooling of the plate. Although brief excursions beyond these bounds are possible, indeed likely, the planet must stay in this regime over aeons, despite a brightening Sun, and despite the stochastic occurrence of 'accidents' such as plume eruption and meteorite impacts.

Recapitulation: managing the surface temperature

We exist: therefore the planet has never been dehydrated and there has been liquid water somewhere on the planet's surface ever since life began.

Two questions can be asked:
(1) Could inorganic chemical feedbacks have succeeded, alone, in maintaining planetary temperature?
(2) Did Archaean biology manage the planetary surface temperature?

Before addressing these questions, it should be pointed out that biological and inorganic processes are not independent. They act on the same planet: thus both must contribute jointly to planetary maintenance today, and presumably acted jointly in the past.

The Solar System is not stable, neither is our planet. Both are chaotic, in which luck may allow a certain degree of stasis. Thus orbital collisions occur until there are few bodies left, and interactions become so infrequent that the large planets seem to have attained stability in orbit. On a planet, geochemistry and heat production sort themselves out until the structure of the planet and its tectonic behaviour seem stable. But neither orbits nor tectonics are truly stable. Catastrophic events can and will occur that can upset the system. Planets collide. The Earth has precipitated a core, probably frozen a magma ocean, and will eventually freeze and be still. To stabilize the surface of a planet such that life can exist for 4 Ga surely needs restorative feedbacks, and perhaps luck also.

The inorganic model

Powerful inorganic feedbacks do exist. If a volcanic event degasses huge amounts of carbon dioxide (e.g. eruption of a major plume head), then the air will become warmer and hence wetter, and more acid. Weathering will become more intense; bicarbonate will flow off the land and eventually carbonate will be deposited in sediment and in new oceanic plate, to return the carbon dioxide to the deep continental crust and to the mantle. But this process depends on the availability of exposed land to be weathered, on the rate of erosion, and on the rate of plate motion. Weathering reaches only a certain depth: after the surface and the rock accessible to groundwater have been attacked, the flow of cations is limited. It is possible to imagine a huge plume eruption in which a massive carbon dioxide greenhouse builds up, such that the oceans degas and a global steam greenhouse would exist. Eventually this would subside as carbonate precipitated, but the event could be long-lived.

Conversely, if too much carbonate is withdrawn by tectonics, by subduction or precipitation, then the ocean freezes. Weathering ceases. Carbon dioxide precipitation stops and the gas builds up in the air until part of the ocean melts and degasses more carbon dioxide, and the world switches from total glaciation to partial glaciation. Indeed, intuitively an inorganic world might be expected to see-saw between these two states, with rare excursions into a plume-driven moist greenhouse state.

Despite the abundant evidence for huge plume events, such as the immense Bushveld event, the anthropic principle implies that the whole Earth has not experienced $\gg 100°C$ greenhouse conditions as long as life has existed. Since then, there have always been parts of the Earth below about 100°C, and since the evolution of mesophiles there have been parts of the Earth below about 40°C. Possibly the plume events have simply been too small to disrupt global climate beyond habitability. Or possibly life itself has played a hand.

The organic model

The powerful organic feedbacks allow a much more sophisticated greenhouse control. More

---

Fig. 9. Controls on the Archaean atmosphere. Top: major components of the atmosphere–biosphere–ocean system in the Archaean. The major inputs come from below (magma and volatiles) and above (light energy, meteorite infall, and removal of H to space). The various biomes and inorganic processes on the surface move and sequester the inputs, adding them to the various inventories that constitute the memory of the system. The ocean is assumed to be at least as deep as at present and possibly deeper; the atmospheric pressure may been comparable to that today. The possibility of a deeper ancient ocean arises because today there is probably a net water loss by deep subduction that is slightly greater than the net gain from deep magmatism, as volatile intake acts to flux mantle melt in compensation for secular cooling of the planet The assumption of a comparable old atmospheric pressure (i.e c. 1 bar) is very weakly founded, based on the assumption that nitrification–denitrification by bacteria was broadly subject to the same dynamics as today. The CO_2 partial pressure of the Archaean air is difficult to assess intuitively, as greenhouse constraints (the sedimentary record implies mostly unfrozen oceans) include methane (e.g. Pavlov *et al.* 2000; Catling *et al.* 2001). Bottom: chemical exchanges between parts of the system. The magnitudes of organic fluxes depend on limiting constraints (e.g. availability of fixed N and especially supply of available phosphorus; Bjerrum & Canfield 2002). The surface of the ocean (top 100 m) may have been oxygen-rich, from the waste oxygen released by cyanobacterial photosynthesis in picoplankton. Deeper levels of the ocean, like the modern Black Sea, may have been mildly to strongly reducing, and kept so by the influx of reduced organic debris from the photosynthetic layer (but note there would also have been a transient frequent influx of sulphate and MnO_2). At the bottom of the ocean, sulphide deposits probably formed by microbial reduction of sulphate (e.g. see Grassineau *et al.* 2002). Deeper water would have contained reduced iron species in solution, from hydrothermal vents. Where deep water rich in Fe^{2+} reached the shallow levels in upwellings, blooms of photoferrotrophs would be likely.

THE INFLUENCE OF LIFE ON THE FACE OF THE EARTH

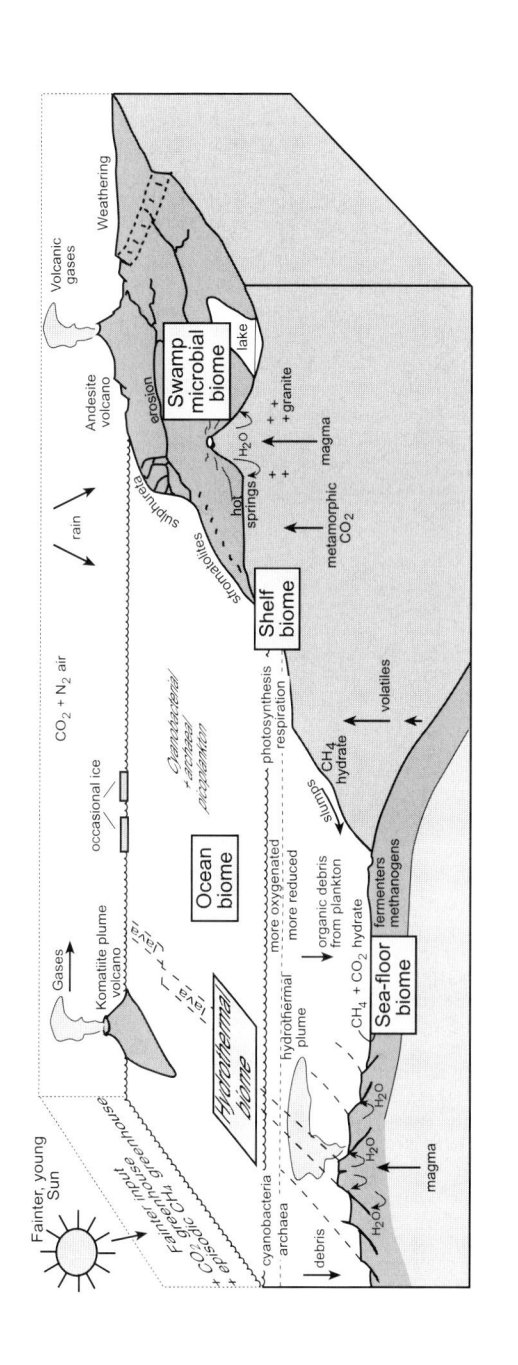

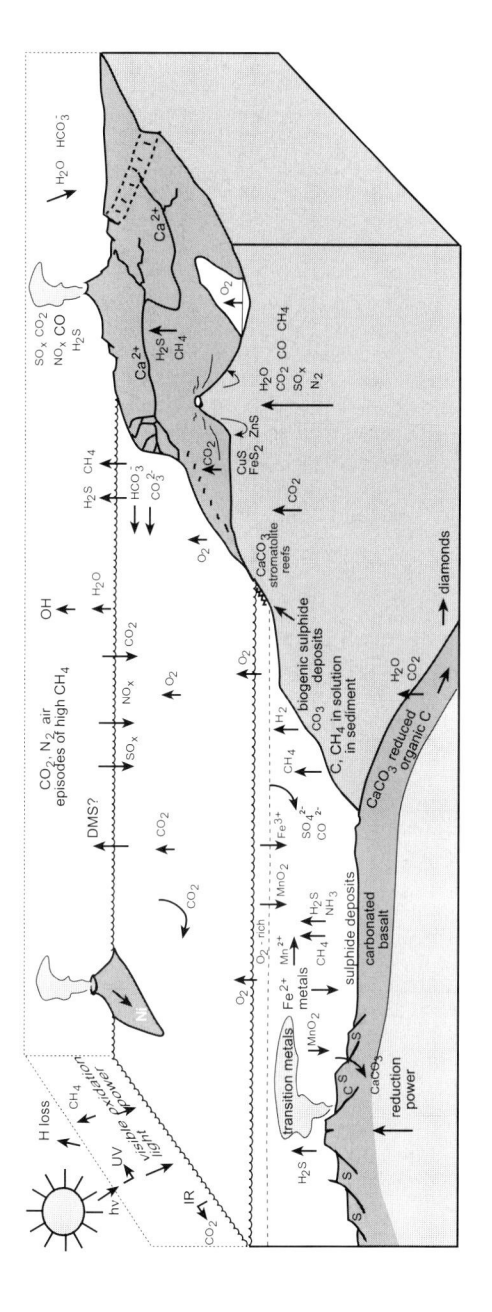

levers are available to control atmospheric water (the main greenhouse gas): in addition to carbon dioxide, methane, nitrous oxide, and albedo change can also be influential. Carbon dioxide management still depends on chemistry, but in addition to calcium, carbon can be deposited as organic matter, which in effect is controlled by access to phosphorus and iron. Thus the potential for greenhouse reduction by carbon removal can be greater, especially after a major outgassing, which will also increase P and Fe availability, increasing biological productivity.

Carbon removed by organic processes is not necessarily shipped to the mantle. Much is deposited as organic matter in sediment, where it is converted in part to methane by methanogens. This is, in effect, a delayed positive feedback: archaea in sediment convert organic matter to methane, and emitted methane can drive sharp strong greenhouse events. Thus any over-removal of carbon by life is ultimately self-correcting.

The real world

Both the inorganic model and the organic model are thought-experiments. The real world is neither. Life, garnets and moving continents are all interdependent. For 4 Ga both have been operating subtly in tandem. Life has shaped the planet; so have inorganic processes. In stabilizing not ice nor steam, but liquid water, on the surface, the combination of life and inorganic processes (Fig. 9) in the Archaean together shaped the planet. The two factors together have sustained plate tectonics by sustaining a cool surface and allowing the reintroduction of water to the interior. Over the aeons they have set the temperature of the top of the mantle adiabat and hence the overall temperature and melting state of the mantle, and thereby its composition. Over 4 Ga it is even arguable that their influence reaches to the precipitation of the inner core, via the adiabat.

Sir Lewis Fermor, half a century ago, considered garnets. In many ways they symbolize crustal material; they also form the richness of the mantle. He rightly recognized the importance of density contrast in driving continental drift, and wondered if garnet formation could help drive movement. He was correct in thinking of density contrast, and of downward recycling of crust, although we now know that the main density drive of subduction is the cooling of the entire lithospheric slab, not just the mafic crust. This cooling would not occur but for the maintenance of the cool surface by the ocean–atmosphere system. Moreover, the recycling of volatiles from the crustal part of the slab as it converts to eclogite is vital in sustaining the continents. Many crustal garnets, those formed from metamorphism of muddy sediment, could exist only on a planet that has continents and erosion. The whole Earth system, garnets, moving continents, and even the mantle adiabat and the modern core temperature, all depends on the sustained existence of the cool wet surface. Is this the result of the influence of life on the face of the Earth? Or accident?

It is probably not a quantifiable question to ask which factor, life or inorganic chance, has been in the driving seat. Both operate. But if any one factor has been central it is the agent that controls the air. Is that agent life?

Much helpful criticism (including sharp and very helpful disagreement) came from T. Lenton, D. Catling, N. Sleep and others, all much appreciated. In particular I thank K. Zahnle, R. Buick and J. Kasting for their great help and comment both wise and very detailed, some of which I have not been diligent enough to follow, and for well-founded, detailed, and kindly but sternly contrary criticism, pointing out error and omission, and clearly demonstrating the weaknesses of these arguments. I thank C. Hawkesworth and C. Ebinger for suggesting the discourse and for comments, Sir Crispin Tickell for drawing my attention to work I had not seen, and the Geological Society of London for asking me to give the Fermor Lecture. Some diagrams were prepared by J. McGraw. For the others, and for much other help and patience, especial thanks go to C. M. R. Fowler. Parts of the work were supported by the Leverhulme Trust and by NERC.

References

ALT, J. C. & SHANKS, W. C. 1998. Sulfur in serpentinized oceanic peridotites: serpentinization processes and microbial sulfate reduction. *Journal of Geophysical Research*, **103**, 9917–9929.

BAKER, V. R. 2001. Water and the martian landscape. *Nature*, **412**, 228–236.

BARNES, S. M., DELWICHE, C. F., PALMER, J. D. & PACE, N. R. 1996. Perspectives on archaeal diversity, thermophily and monophyly from environmental rRNA sequences. *Proceedings of the National Academy of Sciences of the USA*, **93**, 9188–9193.

BEJA, O., SUZUKI, M. T., HEIDELBERG, J. F & 6 OTHERS 2002. Unsuspected diversity among marine aerobic anoxygenic phototrophs. *Nature*, **415**, 630–632.

BJERRUM, C. J. & CANFIELD, D. E. 2002. Ocean productivity before about 1.9 Gyr ago limited by phosphorus adsorption onto iron oxides. *Nature*, **417**, 159–163.

BLANKENSHIP, R. E. 2001. Molecular evidence for the evolution of photosynthesis. *Trends in Plant Science*, **6**, 4–6.

BOETIUS, A., RAVENSCHLAG, K., SCHUBERT, C. & 7 OTHERS 2000. A marine microbial consortium apparently mediating anaerobic oxidation of methane. *Nature*, **407**, 623–626.

BRASIER, M., GREEN, O. R., JEPHCOAT, A. P. & 5 OTHERS 2002. Questioning the evidence for Earth's oldest fossils. *Nature*, **416**, 76–81.

BROCHIER, C. & PHILIPPE, H. 2002. A non-hyperthermophilic ancestor for Bacteria. *Nature*, **417**, 244.

BROCKS, J. J, LOGAN, G. A., BUICK, R. & SUMMONS, R. E. 1999. Archean molecular fossils and the early rise of eukaryotes. *Science*, **285**, 1033–1036.

BUICK, R. 1992. The antiquity of oxygenic photosynthesis: evidence from stromatolites in sulphate-deficient Archaean lakes. *Science*, **255**, 74.

CAMPBELL, I. H. & TAYLOR, S. R. 1983. No water, no granites, no oceans, no continents. *Geophysical Research Letters*, **10**, 1061–1064.

CAPONE, D. G., ZEHR, J. P., PAERL, H. W., BERGMAN, B. & CARPENTER, E. J. 1997. Trichodesmium, a globally significant marine cyanobacterium. *Science*, **276**, 1221–1229.

CARR, M. 1996. *Water on Mars*. Cambridge University Press, Cambridge.

CATLING, D. C., ZAHNLE, K. J. & MCKAY, C. P. 2001. Biogenic methane, hydrogen escape, and the irreversible oxidation of the Earth. *Science*, **293**, 839–843.

COCKELL, C. S. 2000. The ultraviolet history of the terrestrial planets – implications for biological evolution. *Planetary and Space Science*, **48**, 203–214.

CRUTZEN, P. J. 1987. Role of the tropics in atmospheric chemistry. *In*: DICKINSON, R. E. (ed.) *The Geophysiology of Amazonia*. Wiley, New York, 107–132.

DISMUKES, G. C., KLIMOV, V. V, BARANOV, S., KOZLOV, YU. N., DASGUPTA, J. & TRYSHKIN, A. 2001. The origin of atmospheric oxygen on Earth: the innovation of oxygenic photosynthesis. *Proceedings of the National Academy of Sciences of the USA*, **98**, 2170–2175.

DOOLITTLE, W. F. 1999. Phylogenetic classification and the universal tree. *Science*, **284**, 2124–2128.

DUTKEIWICZ, A., RASMUSSEN, B. & BUICK, R. 1998. Oil preserved in fluid inclusions in Archaean sandstone. *Nature*, **395**, 885–888.

FALKOWSKI, P. G. 1997, Evolution of the nitrogen cycle and its influence on the biological sequestration of CO_2 in the ocean. *Nature*, **387**, 272–275.

FARQUHAR, J., BAO, H. & THIEMANS, M. 2000. Atmospheric influence of Earth's earliest sulfur cycle. *Science*, **289**, 756–758.

FENCHEL, T. & BERNARD, C. 1995. Mats of colourless sulphur bacteria. I: Major microbial processes. *Marine Ecology Progress Series*, **128**, 161–170.

FERMOR, SIR, L. L. 1949. Garnets and moving continents. *Rhodesia Scientific Association: Proceedings and Transactions*, **XLII**, 11–18.

FORTERRE, P. 1996. A hot topic: the origin of hyperthermophiles. *Cell*, **85**, 789–792.

GALTIER, N., TOURASSE, N. & GOUY, M. 1999. A non-hyperthermophile common ancestor to extant life forms. *Science*, **283**, 220–221.

GOGARTEN-BOEKELS, M., HILARIO, E. & GOGARTEN, J. P. 1995. The effects of heavy meteorite bombardment on the early evolution – the emergence of the three domains of Life. *Origins of Life and Evolution of the Biosphere*, **25**, 251–264.

GRASSINEAU, N. V., NISBET, E. G., BICKLE, M. J. & 5 OTHERS 2001. Antiquity of the biological sulphur cycle: evidence from sulphur and carbon isotopes in 2700 million year old rocks of the Belingwe belt, Zimbabwe. *Proceedings of the Royal Society of London, Series B*, **268**, 113–119.

GRASSINEAU, N. V., NISBET, E. G., FOWLER, C. M. R. & 7 OTHERS 2002. Stable isotopes in the Archaean Belingwe Belt, Zimbabwe: evidence for a diverse prokaryotic mat ecology. *In*: FOWLER, C. M. R., EBINGER, C. J. & HAWKESWORTH, C. J. (eds) *The Early Earth: Physical, Chemical and Biological Development*. Geological Society, London, Special Publications, **199**, 309–328.

HALL, A. 1989. Ammonium in spilitized basalts of southwest England and its implications for the recycling of nitrogen. *Geochemical Journal*, **23**, 19–23.

HALL, A. & ALDERTON, D. H. M. 1994. Ammonium enrichment associated with hydrothermal activity in the granites of south-west England. *Proceedings of the Ussher Society*, **8**, 242–247.

HANSEN, J. E. & SATO, M. 2001. Trends of measured climate forcing agents. *Proceedings of the National Academy of Sciences of the USA*, **98**, 14778–14783.

HARTMAN, H. & FEDOROV, A. 2002. The origin of the eukaryotic cell: a genomic investigation. *Proceedings of the National Academy of Sciences of the USA*, **99**, 1420–1425.

HARVEY, L. D. D. & HUANG, Z. 1995. Evaluation of the potential impact of methane clathrate destabilisation on future global warming. *Journal of Geophysical Research*, **100**, 2905–2926.

HAYES, J. M. 1994. Global methanotrophy at the Archaean–Proterozoic transition. *In*: BENGTSON, S. (ed.) *Early Life on Earth*. Columbia University Press, New York, 220–236.

HESS, H. H. 1962. History of ocean basins. *In*: ENGEL, A. E. J., JAMES, H. L. & LEONARD, B. F. (eds) *Petrological Studies: a Volume in Honour of A. F. Buddington*. Geological Society of America, Boulder, CO, 599–620.

HOEHLER, T. M., BEBOUT, B. M. & DES MARAIS, D. J. 2001. The role of microbial mats in the production of reduced gases on the early Earth. *Nature*, **412**, 324–327.

HOLLAND, H. D. 1999. When did the Earth's atmosphere become oxic? A reply. *Geochemical News*, **100**, 20–22.

HOLM, N. G. & CHARLOU, J. L. 2001. Initial indications of abiotic formation of hydrocarbons in the Rainbow ultramafic hydrothermal system, Mid-Atlantic Ridge. *Earth and Planetary Science Letters*, **191**, 1–8.

HULTH, S., ALLER, R. C. & GILBERT, F. 1999. Coupled anoxic nitrification/manganese reduction in marine sediments. *Geochimica et Cosmochimica Acta*, **63**, 49–66.

JORDAN, P., FROMME, P., TOBIAS WITT, H., KLUKAS, O., SAENGER, W. & KRAUSS, N. 2001. Three dimensional structure of cyanobacterial photosystem I at 2.5 Å resolution. *Nature*, **411**, 909–917.

JOSHI, H. M. & TABITA, F. R. 1996. A global two-way component signal transduction system that integrates the control of photosynthesis, carbon dioxide assimilation and nitrogen fixation. *Proceedings of the National Academy of Sciences of the USA*, **93**, 14 515–14 520.

KARL, D. M. 2002. Hidden in a sea of microbes. *Nature*, **415**, 590–591.

KASTING, ?. ?. 1993. Earth's early atmosphere. *Science*, **259**, 920–926.

KASTING, J. F. 2001. The rise of atmospheric oxygen. *Science*, **293**, 819–820.

KASTING, J. F. & HOLM, N. G. 1992. What determines the volume of the oceans? *Earth and Planetary Science Letters*, **109**, 507–515.

KASTING, J. F. & SIEFERT, J. L. 2001. The nitrogen 'fix'. *Nature*, **412**, 26–27.

KASTING, J. F., WHITMIRE, D. P. & REYNOLDS, R. T. 1993. Habitable zones around main sequences stars. *Icarus*, **101**, 108–128.

KASTING, J. F., ZAHNLE, K. J., PINTO, J. P. & YOUNG, A. T. 1989. Sulfur, ultraviolet radiation, and the early evolution of life. *Origins of Life and Evolution of the Biosphere*, **19**, 95–108.

KELLEY, D., KARSON, J. A., BLACKMAN, D. K. & 9 OTHERS 2001. An off-axis hydrothermal vent field near the Mid-Atlantic ridge at 30°N *Nature*, **412**, 145.

KUHLBRANDT, W. 2001. Chlorophylls galore. *Nature*, **411**, 896–898.

KUMP, L. R., KASTING, J. F. & BARLEY, M. E. 2001. Rise of atmospheric oxygen and the 'upside-down' Archaean mantle. G^3 *Geochemistry, Geophysics, Geosystems*, **2**, paper 2000GC000114.

KVENVOLDEN, K. 1988. Methane hydrate – a major reservoir of carbon in the shallow geosphere? *Chemical Geology*, **29**, 159–162.

LASAGA, A. C. & OHMOTO, H. 2002. The oxygen geochemical cycle: dynamics and stability. *Geochimica et Cosmochimica Acta*, **66**, 361–381.

LEWIS, J. S. & PRINN, R. G. 1984. *Planets and their Atmospheres*. Academic Press, Orlando, FL.

LINE, M. A. 2002. The enigma of the origin of life and its timing. *Microbiology*, **148**, 21–27.

LORIMER, G. H. 1981. The carboxylation and oxygenation of ribulose 1,5-bisphosphate: the primary events in photosynthesis and photorespiration. *Annual Review of Plant Physiology*, **32**, 349–383.

LORIMER, G. H. & ANDREWS, T. J. 1973. Plant photorespiration – an inevitable consequence of the existence of atmospheric oxygen. *Nature*, **243**, 359.

LOVELOCK, J. E 1979. *Gaia*. Oxford University Press, Oxford.

LOVELOCK, J. E. 1988. *Ages of Gaia*. London, Norton.

MACGREGOR, A. M. 1949. The influence of life on the face of the Earth. Presidential Address. *Rhodesia Scientific Association: Proceedings and Transactions*, **XLII**, 5–11.

MARTIN, A. NISBET, E. G. & BICKLE, M. J. 1980. Archaean stromatolites of the Belingwe greenstone belt, Zimbabwe (Rhodesia). *Precambrian Research*, **13**, 337–362.

NAVARRO-GONZALEZ, R., MCKAY, C. P. & MVONDO, D. N. 2001. A possible nitrogen crisis for Archaean life due to reduced nitrogen fixation by lightning. *Nature*, **412**, 61–64.

NISBET, E. G. 1987. *The Young Earth*. Allen and Unwin, London.

NISBET, E. G. 1995. Archaean ecology: a review of evidence for the early development of bacterial biomes, and speculations on the development of a global scale biosphere. *In*: COWARD, M. P. & RIES, A. C. (eds) *Early Precambrian Processes*. Geological Society, London, Special Publications **95**, 27–51.

NISBET, E. G. & FOWLER, C. M. R. 1996a. Some liked it hot. *Nature*, **382**, 404–405.

NISBET, E. G. & FOWLER, C. M. R. 1996b. The hydrothermal imprint on life: did heat shock proteins, metalloproteins and photosynthesis begin around hydrothermal vents? *In*: MACLEOD, C. J., TYLER, P. A. & WALKER, C. L. (eds) *Tectonic, Magmatic, Hydrothermal and Biological Segmentation of Mid-ocean Ridges*. Geological Society, London, Special Publications, **118**, 239–251.

NISBET, E. G. & FOWLER, C. M. R 1999. Archaean metabolic evolution of microbial mats. *Proceedings of the Royal Society of London, Series B*, **266**, 2375–2382.

NISBET, E. G. & SLEEP, N. H. 2001. The habitat and nature of early life. *Nature*, **409**, 1083–1091.

NISBET, E. G. & SLEEP, N. H. 2002. The early earth: the physical setting for early life. *In*: LISTER, A. & ROTHSCHILD, L. (eds) *Evolution on Planet Earth: The Impact of the Physical Environment*. Linnean Society, London, Special Publication, in press.

NISBET, E. G., MATTEY, D. P. & LOWRY, D. 1994. Can diamonds be dead bacteria? *Nature*, **367**, 694.

NISBET, E. G., CANN, J. R., & VAN DOVER, C. L. 1995. Origins of photosynthesis. *Nature*, **373**, 479–480.

OHMOTO, H. 1997. When did the Earth's atmosphere become oxic? *Geochemical News*, **93**, 12–13.

OHMOTO, H., YAMAGUCHI, K. E. & ONO, S. 2001. Questions regarding Precambrian sulfur fractionation, and Response by Farquhar *et al. Science*, **292**, 1959a.

PACE, N. R. 1997. A molecular view of biodiversity and the biosphere. *Science*, **276**, 734–740.

PAVLOV, A. A., KASTING, J. F., BROWN, L. L., RAGES, K. A. & FREEDMAN, R. 2000. Greenhouse warming by CH_4 in the atmosphere of early Earth. *Journal of Geophysical Research*, **105**, 11 981–11 990.

PAVLOV, A. A., KASTING, J. F., EIGENBRODE, J. L. & FREEMAN, K. H. 2001. Organic haze in Earth's early atmosphere: source of low ^{13}C kerogens? *Geology*, **29**, 1003–1006.

PHILLIPS, G. N., LAW, J. D. M. & MYERS, R. E. 2001. Is the redox state of the Archaean atmosphere constrained? *SEG Newsletter, Society of Economic Geologists*, No. 47, 1–19.

PIERSON, B. K. 1994. The emergence, diversification, and role of photosynthetic eubacteria. *In*: BENGTSON, S. (ed.) *Early life on Earth*. Nobel Symposium, **84**, 161–180.

RASMUSSEN, R. 2000. Filamentous microfossils in a 3235 million year old volcanogenic massive sulphide deposit, *Nature*, **405**, 676–679.
RASMUSSEN, R. & BUICK, R. 2000. Oily old ores, evidence for hydrothermal petroleum generation in an Archean volcanogenic massive sulpide deposit. *Geology*, **27**, 115–118.
ROSING, M. T. 1999. ^{13}C-depleted carbon in >3700 Ma seafloor sedimentary rocks from West Greenland. *Science*, **283**, 674–676.
SAGAN, C. & CHYBA, C. 1997. The early Sun paradox: organic shielding of ultraviolet-labile greenhouse gases. *Science*, **276**, 1217–1221.
SANSONE, F. J., POPP, B. N., GASC, A., GRAHAM, A. W. & RUST, T. M. 2001. Highly elevated methane in the eastern tropical North Pacific and associated isotopically enriched fluxes to the atmosphere. *Geophysical Research Letters*, **28**, 4567–4570.
SCHIDLOWSKI, M. 1988. A 3800 million year record of life from carbon in sedimentary rocks. *Nature*, **333**, 313–318.
SCHIDLOWSKI, M. 2002. Sedimentary carbon isotope archives as recorders of early life: implications for extraterrestrial scenarios. *In*: PALYI, G., ZUCCHI, C. & CAGLIOTI, L. (eds) *Fundamentals of Life*. Elsevier, Amsterdam, 307–329.
SCHIDLOWSKI, M. & AHARON, P. 1992. Carbon cycle and carbon isotopic record: geochemical impact of life over 3.8 Ga of Earth history. *In*: SCHIDLOWSKI, M., GOLUBIC, S., KIMBERLEY, M. M., MCKIRDY, D. M. & TRUDINGER, P. A. (eds) *Early Organic Evolution: Implications for Mineral and Energy Resources*. Springer, Berlin 147–175.
SCHIDLOWSKI, M., APPEL, P. W. U., EICHMANN, R. & JUNGE, C. E. 1979. Carbon isotope geochemistry of the 3.7×10^9 yr old Isua sediments, West Greenland: implications for the Archaean carbon and oxygen cycles. *Geochimica et Cosmochimica Acta*, **43**, 189–199.
SHEN, Y., BUICK, R. & CANFIELD, D. E. 2001. Isotopic evidence for microbial sulphate reduction in the early Archaean era. *Nature*, **410**, 77–81.
SLEEP, N. H., ZAHNLE, K. & NEUHOFF, P. S. 2001. Initiation of clement surface conditions on the earliest Earth. *Proceedings of the National Academy of Sciences of the USA*, **98**, 3666–3672.
STETTER, K. O. 1996. Hyperthermophils in the history of life. *In*: BOCK, G. R. & GOODE, J. A. (eds) *Evolution of Hydrothermal Systems on Earth (and Mars?)*. Ciba Foundation Symposium, **202**, 1–10.
SUMMONS, R. E., JAHNKE, L. L., HOPE, J. M. & LOGAN, G. A. 1999. 2-Methylhopanoids as biomarkers for cyanobacterial oxygenic photosynthesis. *Nature*, **400**, 554–557.
THAUER, R. K. 1998. Biochemistry of methanogenesis: a tribute to Marjory Stephenson. *Microbiology*, **144**, 2377–2406.
TOLBERT, N. E. 1994. Role of photosynthesis and photorespiration in regulating atmospheric CO_2 and O_2. *In*: TOLBERT, N. E. & PREISS, J. (eds) *Regulation of Atmospheric CO_2 and O_2 by Photosynthetic Carbon Metabolism*. Oxford University Press, Oxford, 8–33.
WALKER, J. C. G. 1987. Was the Archaean biosphere upside down? *Nature*, **329**, 710–712.
WALKER, J. C. G. & DREVER, J. I. 1988. Geochemical cycles of atmospheric gases. *In*: GREGOR, C. B., GARRELS, R. M., MACKENZIE, F. T. & MAYNARD, J. B. (eds) *Chemical Cycles in the Evolution of the Earth*, 2. Wiley, New York, 55–75.
WARNECK, P. 1988. *Chemistry of the Natural Atmosphere*. Academic Press, San Diego, CA.
WAYNE, R. P. 1992. Atmospheric chemistry: the evolution of our atmosphere. *Journal of Photochemistry and Photobiology A: Chemistry*, **62**, 379–396.
WILKS, M. E., & NISBET, E. G. 1985. Archaean stromatolites from the Steep Rock Group, N.W. Ontario. *Canadian Journal of Earth Sciences*, **22**, 792–799.
WOESE, C. R. 1987. Bacterial evolution. *Microbiological Reviews*, **51**, 221–271.
XIONG, J., FISCHER, W. M., INOUE, K., NAKAHARA, M. & BAUER, C. E. 2000. Molecular evidence for the early evolution of photosynthesis. *Science*, **289**, 1724–1730.
YUNG, Y., WEN, J.-S., MOSES, J. I., LANDRY, B. M. & ALLEN, M. 1989. Hydrogen and deuterium loss from the terrestrial atmosphere: a quantitative assessment of non-thermal escape fluxes. *Journal of Geophysical Research*, **94**, 14 971–14 989.
ZEHR, J. P., WATERBURY, J. B., TURNER, P. J. & 5 OTHERS 2001. Unicellular cyanobacteria fix N_2 in the subtropical North Pacific Ocean. *Nature*, **412**, 635–638.
ZUCKERKANDL, E. & PAULING, L. 1965. Molecules as documents of evolutionary history. *Journal of Theoretical Biology*, **8**, 357–366.

Stable isotopes in the Archaean Belingwe belt, Zimbabwe: evidence for a diverse microbial mat ecology

N. V. GRASSINEAU[1], E. G. NISBET[1], C. M. R. FOWLER[1], M. J. BICKLE[2], D. LOWRY[1], H. J. CHAPMAN[2], D. P. MATTEY[1], P. ABELL[3], J. YONG[4] & A. MARTIN[5]

[1] *Department of Geology, Royal Holloway, University of London, Egham TW20 0EX, UK (e-mail: e.nisbet@gl.rhul.ac.uk)*
[2] *Department of Earth Sciences, Cambridge University, Downing Street, Cambridge CB2 3EQ, UK*
[3] *Department of Chemistry, University of Rhode Island, Kingston, RI 02881, USA*
[4] *Department of Chemistry, University of Buea, Buea, SW Cameroon, Cameroon*
[5] *6 Autumn Close, Greendale, Harare, Zimbabwe*

Abstract: Sulphide-rich sediments, stromatolitic limestones and tidal-flat deposits in the late Archaean (2.7 Ga) Manjeri and Cheshire Formations, Belingwe greenstone belt, Zimbabwe show evidence for complex and extensive prokaryotic mat communities, including (1) shallow-water coastal sulphur mats; (2) mats, probably in somewhat deeper water; (3) nearby stromatolites that lived by oxygenic photosynthesis in shallow coastal settings. Petrological and geochemical (rare earth element; REE) evidence, coupled with high-resolution stable isotope results, identifies several complex interdependent metabolic consortia of bacteria and archaea. These microbial consortia would have exchanged nutrients and products both locally within prokaryotic mats and more widely via the waters of the Belingwe basin. This isotopic, sedimentological and REE evidence for a complex ecology of bacteria and archaea is consistent with metabolic inferences from rRNA phylogeny and is direct evidence that a diverse prokaryotic community, managing carbon on a global scale, had evolved by the late Archaean.

Molecular palaeontology, based on rRNA studies of modern prokaryotes, has given profound insight into the phylogenetic history of the earliest organisms (Pace 1997). Geological calibration of the rRNA results is slight, however (Nisbet & Fowler 1999). In the Archaean geological record deducing biota and phylogeny from the degraded remnants of organic life is difficult. Textural comparison, inferring the presence of organisms such as cyanobacteria by comparison with modern parallels, provides support for biogenicity, but is often controversial and provides only the most general insight into the nature of the organisms responsible.

In contrast, stable isotopic study offers more powerful insight from the geological record. Biochemical processes leave isotopic fingerprints to calibrate the evolutionary history of metabolic processes inferred from molecular work. Conventional stable isotope study has been used to infer the global role of major processes such as photosynthesis (Schidlowski & Aharon 1992) but analytical limitations make it difficult to study the geological record of microbial consortia.

In this study, rapid high-resolution stable isotope techniques (Grassineau *et al.* 2001a) are used to identify in detail the isotopic fingerprints of metabolic processes in some of the best preserved of all Archaean biogenic rocks, from the 2.7 Ga Belingwe belt, Zimbabwe. The results map out consortia of bacteria and archaea, and show that by late Archaean time very diverse microbial ecology existed, both photosynthetic and sulphur-based, consistent with the implications of the rRNA phylogeny (Nisbet & Fowler 1999; Nisbet & Sleep 2001).

Geological setting

Stratigraphy

The Manjeri Formation is the lowest part of the Ngezi Group in the Belingwe belt (Bickle & Nisbet 1993), Zimbabwe, one of the least deformed and metamorphosed of all Archaean greenstone successions (Fig. 1). It was laid down in a continental basin (Bickle *et al.* 1975; Bickle & Nisbet 1993; Hunter *et al.* 1998) upon a subsiding floor of much older eroded tonalitic gneiss. The contact at the base of the Manjeri Fm is well exposed (Nisbet *et al.* 1993) at many sites, especially at the Unconformity National Monument locality. At this locality (Nercmar drill site) the basal Spring Valley Member (Hunter *et al.* 1998, taking the name from Spring Valley farm nearby) consists of varied shallow-water sediments. On Rupemba mountain, on strike and 10 km south, well-preserved stromatolites are present (Martin *et al.* 1980; Abell *et al.* 1985) on a stratigraphic level equivalent to the Spring Valley Member. Limestones with a mat fabric occur in the Spring Valley Member along strike between the unconformity locality and Rupemba (E.G.N., unpublished mapping).

Above the Spring Valley Member we here introduce a new Shavi Member of the Manjeri Formation, named after the nearby river. This is defined in the Nercmar drillcore (see below), from 128.6 to 143.4 m (Fig. 2). The Shavi Member contains an assortment of shales, often carbon-rich, silts, sulphide stringers, cherts and banded ironstone, and pebble beds. The results previously reported (Grassineau *et al.* 2001*a*) from the Spring Valley Member include samples from this newly defined Member. The reason for introducing it is to remove ambiguity between field

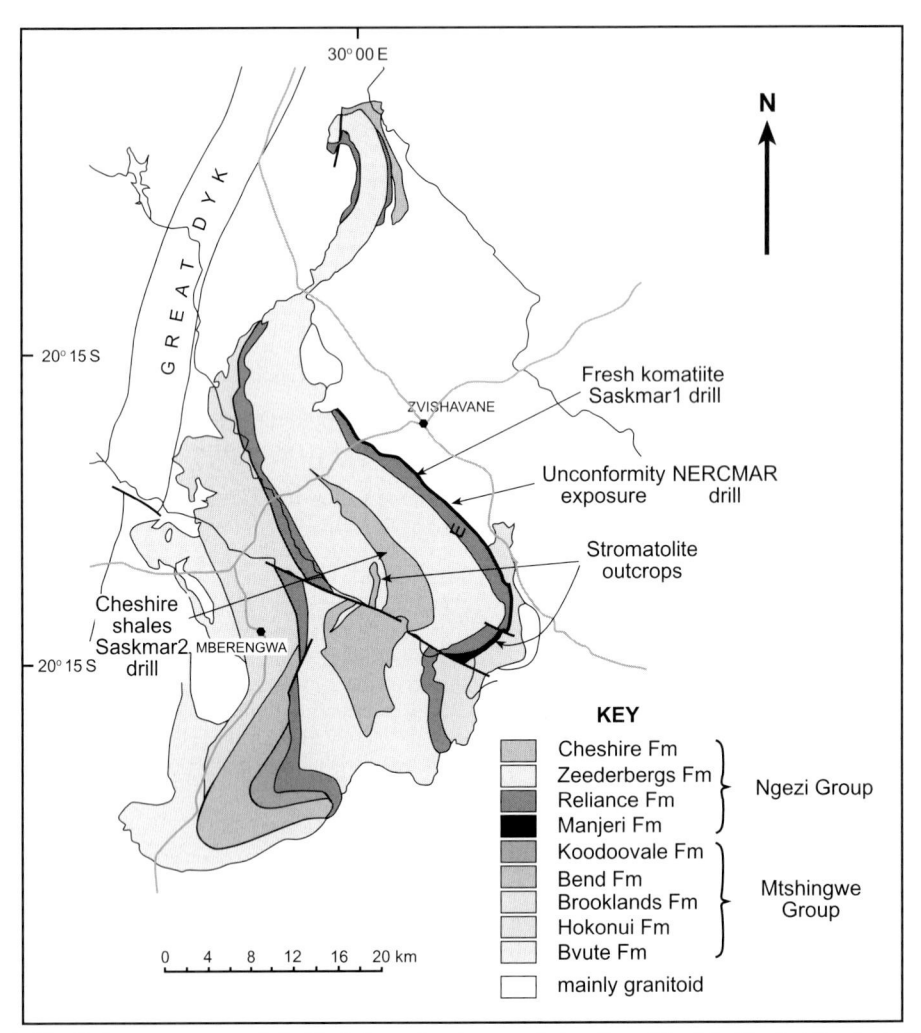

Fig. 1. The Belingwe greenstone belt, Zimbabwe, showing sample locations. The thin Manjeri Fm (including Spring Valley, Shavi, Rubweruchena and Jimmy Members, not shown here) outcrops along the base of the main (Upper Greenstones) syncline.

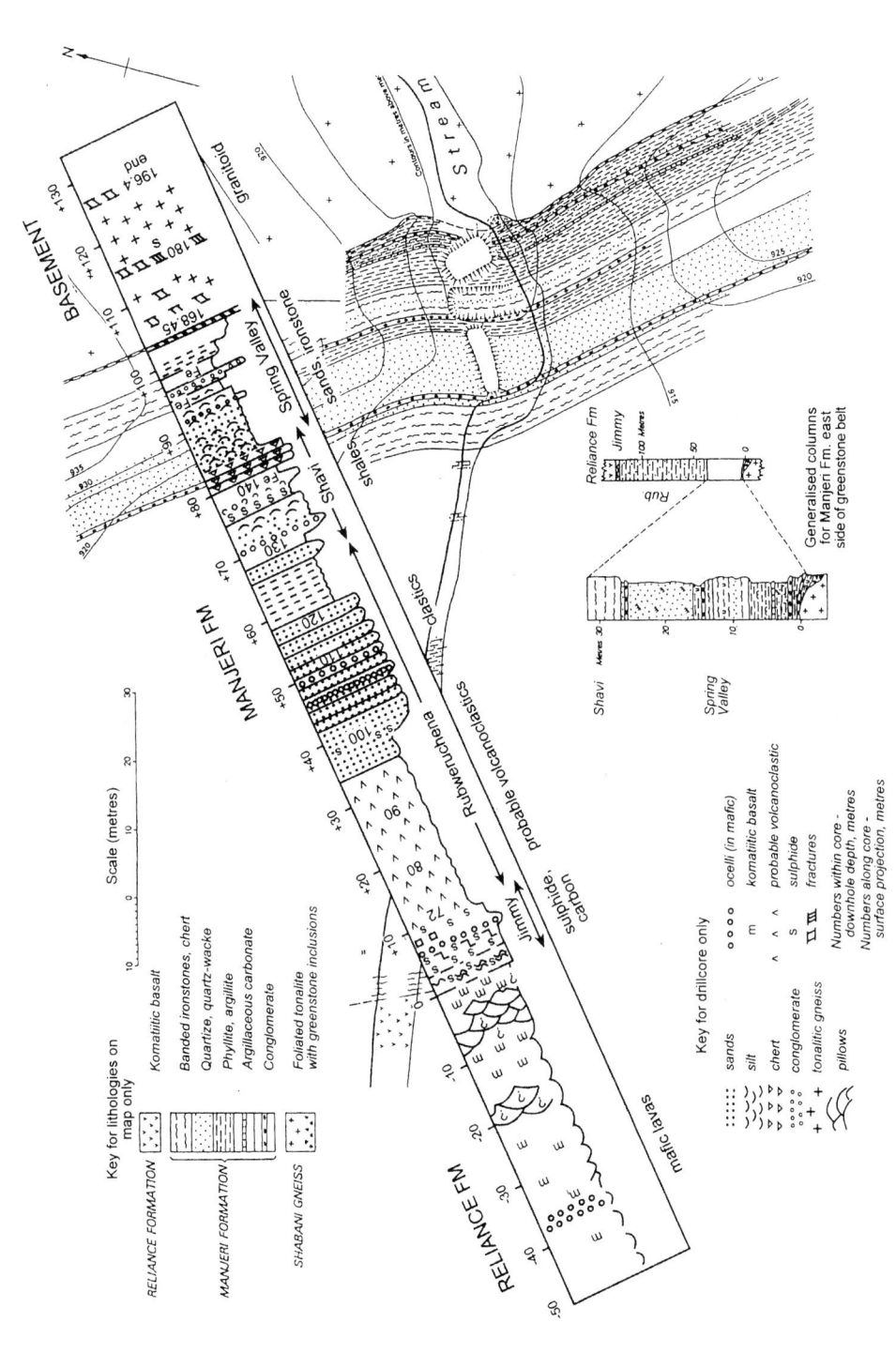

Fig. 2. The Neremar drillcore, at the Unconformity National Monument. (For locality, see Fig. 1.) Drillcore log is projected up to surface outcrop. Numbers on drillcore are down-hole metre depth. Numbers along core are surface projection in metres, set at zero on the top of the Manjeri Fm. Background geological map is from Martin (1978), with north corrected.

and drillcore positioning of the boundary between the Rubweruchena and the Spring Valley Members.

The central Rubweruchena Member of the formation ('white pebbles', named after a nearby hill) contains poorly sorted clastic deposits. The uppermost part of the Manjeri Fm is the sulphide-rich Jimmy Member (Hunter et al. 1998, taking the name from a nearby claim), exposed as a deformed stratabound sulphide–iron gossan with chert bands, complexly folded, very persistent laterally at the contact between the Manjeri Fm and the overlying Reliance Fm, a komatiitic shield volcano, containing komatiitic basalt and komatiite flows and pillows, and tuffs (Nisbet et al. 1993). Above the Reliance Fm are thick mafic pillows and flows of the Zeederbergs Fm and the Cheshire Fm, mainly shallow-water sediment (Nisbet et al. 1993). In the Cheshire Fm, stromatolites of the Macgregor Member (Martin et al. 1980; Abell et al. 1985), are extensive and locally superbly exposed; they are arguably some of the best-preserved Archaean stromatolites extant. Associated shallow-water Cheshire shales locally bear apparently organic carbon.

Nercmar drill

The Nercmar drillhole at the National Monument locality (Fig. 2) provides a nearly complete section downwards through the basal Reliance Fm, the whole Manjeri Fm, and into the underlying tonalitic gneiss that was exposed and eroded in Archaean time. The underlying gneiss is extensively fractured, and the top few metres of gneiss show what appears to be penecontemporaneous Archaean weathering. Above the basal Spring Valley beach, in the Shavi Member (tidal or subtidal), fine sulphide and bands rich in organic carbon occur in mudstones and cherts, underlain by silts. Sulphide stringers occur (although rarely) in the clastic Rubweruchena Member. Core return from the Jimmy Member in the Nercmar hole was variable: this was a very low-budget project with old equipment. In some sections of the Jimmy Member, return was good, but in parts it was moderate to poor (below 50%). In better sections the drilling produced fresh cores made of sulphide, quartz and apparently organic carbon.

Metamorphic grade

The metamorphic grade of the Ngezi Group varies, but is typically low. The petrography and isotopic evidence from the stromatolites of the Macgregor Member, Cheshire Fm, have previously been interpreted as indicating an extremely gentle thermal history with very low-grade metamorphism (Abell et al. 1985). The Rupemba stromatolites, although their thermal history has been warmer, have also had low-grade metamorphism (Abell et al. 1985). This is consistent with the record of the komatiites, which are close by geographically, and stratigraphically lie between the two stromatolite horizons. Some of the Reliance lavas are highly altered, but other localities preserve very fresh material. Exposed in a section about 6 km laterally along strike from the Nercmar locality and immediately above it stratigraphically are some of the freshest Archaean lavas ever found (Nisbet et al. 1987), preserving glass inclusions and abundant fresh olivine. This supports the interpretation that, though local aureoles and shear zones exist, the regional metamorphism has been only to very low grade, and that parts of the sedimentary succession are likely to have near-pristine mineralogy.

Age and structural setting

The Reliance Fm is 2.7 Ga old (Bickle & Nisbet 1993). The sediment in the Manjeri Fm is locally derived, with model ages in the Spring Valley Member around 3.4 Ga (local basement derivation) and varying from 2.7 to 3.1 Ga in the material immediately under the Jimmy Member (Hunter et al. 1998). Our interpretation that the Reliance Fm was directly laid down on the Manjeri Fm is controversial. The alternative hypothesis (Kusky & Kidd 1992; Kusky & Winsky 1995), which we reject (Blenkinsop et al. 1993; Bickle et al. 1994), is that the komatiites and basalts of the Ngezi Group represent a large oceanic plateau, many kilometres thick, which was thrust into place on the Jimmy Member, which includes a high-temperature mylonite: in our contrary view, the Jimmy Member is a low-grade metamorphic rock, preserving much of its original fabric. It has probably undergone synsedimentary slumping and folding, and also shear as the overlying lava sequence was constructed and folded, but the succession is coherent.

Sampling: Manjeri and Cheshire Formations

Field samples come from various sources. These include: a large suite of hand specimens (collections of E.G.N., M.J.B. and A.M.); core from the 200 m Nercmar hole (E.G.N., M.J.B.) through the Manjeri Fm (Fig. 2); core from the 50 m Saskmar 2 hole in the Cheshire Fm shales (P.A., E.G.N.); and hand-drill samples from the

Cheshire Fm and Manjeri Fm stromatolites (collection of P.A.). Because of the importance of the stromatolite exposures, hand specimens were taken from loose boulders on the outcrop, not by hammering.

Spring Valley and Shavi Members, Manjeri Fm. Thin horizons of sulphide–carbon–chert rocks from the Shavi Member in the drillcore are set in a predominantly tidal to shallow-water clastic sequence associated with banded ironstones, underlain by and interlain with silts and muds, and overlain by the basal coarse clastic deposits of the Rubweruchena Member. Deposition may have taken place on the quiet margin of a small delta, in a protected interdistributary bay, before topographic inversion and coarse clastic deposition (Hunter *et al.* 1998). Sulphides are varied and include pyrite and minor sphalerite and arsenopyrite, as well as rare small grains of millerite, and pentlandite.

Rupemba stromatolites, Manjeri Fm. The Rupemba stromatolites are well developed in limestone about 10 km SE of the unconformity locality, along strike in the Manjeri Fm (Bickle *et al.* 1975; Abell *et al.* 1985) Limestones do not occur in the sequence at the unconformity locality itself, although carbonate-rich bands are present; a persistent non-stromatolitic thin limestone band (with probable mat textures) is present along strike to the SE. Texturally they resemble modern photosynthetic stromatolites (Martin *et al.* 1980), although some may be anoxygenic (Schidlowski & Aharon 1992).

Jimmy Member. The Jimmy Member is deeply altered by recent weathering in outcrop. It has been sampled in drillcore, and is fresh in intersections below groundwater level. The unit includes finely (sub-millimetre scale) laminated rock, with folded laminae of quartz, sulphide and carbon presumed to be of organic origin. In addition, some rock is made of laminae that show apparently penecontemporaneous brecciation, and the drillcore also includes massive brecciated sulphide-rich rock. Sulphide is both pyrite and pyrrhotite. In the electron probe, laminae appear to be particularly carbon rich around tiny grains of possibly detrital material. Phosphorus-rich spots occur, possibly tiny (micron-size) apatites. Zircon, rutile and ilmenite are present (possibly wind blown). Some samples contain chlorite-rich phyllite. The depth of deposition of the Jimmy Member is not well constrained, although the fine lamination suggests that deposition is likely to have been below wave base. The lower Reliance Fm immediately above the Jimmy Member is submarine pillow lava, but along strike 5–10 km to the south contains lapilli tuffs with lapilli up to 2 cm, and fiamme, implying subaerial exposure of a volcanic centre. The proximity of the Jimmy Member to these overlying volcanic rocks and to relatively shallow-water facies sediments that are stratigraphically below (Bickle & Nisbet 1993) suggests that the Jimmy Member, although probably below wave base, was not laid down in very deep water.

Macgregor Member stromatolites, Cheshire Fm, Ngezi Group. Stromatolites in the Macgregor Member of the Cheshire Fm are spectacularly developed in reefs, although only locally profusely stromatolitic. Limestone bands up to several hundred metres thick occur, some of which can be traced along strike for up to 10 km (Bickle *et al.* 1975; Martin *et al.* 1980; Abell *et al.* 1985). The Macgregor stromatolite reef (Martin *et al.* 1980) shows a wide variety of resemblances to modern analogues.

Cheshire shales, Cheshire Fm, Ngezi Group. The Cheshire shales were collected from a drillsite (Saskmar 2) in the Cheshire Fm. They are shallow-water silts, with low carbonate and organic carbon contents.

Analytical methods

In any study of Archaean life, modern or Phanerozoic contamination is a problem. This is particularly true of C isotopes in sulphide-rich drillcore, where modern bacteria may have colonized wet core after drilling. Core was wet during drilling and immediately thereafter, during cutting, and in logging. However, visually the deterioration of the cut surface was small. Samples were carefully taken from fresh sulphide and black shale samples broken from the core, to exclude altered margins, and it is unlikely that microbial growth and C fixation from modern CO_2 could have introduced enough carbon significantly to contaminate C isotopic ratios, especially in the more carbon-rich samples. S gain or loss from fresh sample is unlikely. Hand-specimen samples were taken from fresh clean cuts.

The techniques used in carbon isotope studies by Abell and Yong were described by Abell *et al.* (1985). Carbonates (around 120 samples) were analysed by conventional reaction with phosphoric acid. Kerogen (around 250 samples) was extracted with concentrated HCl, rinsed with deionized water and a 5:2 mixture of concentrated HF–HCl (Yong, unpubl. data).

New stable isotope studies (Grassineau, Lowry, and further analyses by Abell) were carried out in the Royal Holloway laboratory, which is equipped with a combustion–continuous flow Fisons Isochrom-EA system, on line to an Optima mass spectrometer (CF-IRMS) (Grassineau et al. 2001a).

Whereas conventional methods for analysing sulphur isotope compositions need >10 mg of physically separated sample and long analytical times, in contrast, the EA-IRMS technique uses small samples (1–2 mg) and has short analysis times. This achieves the high physical resolution and large sample populations needed for adequate study of biological variation in Archaean sulphide-rich organic sediments.

The CF-IRMS analytical method for sulphur is based on rapid oxidation of the samples by flash combustion at 1800°C. The released gases pass through a catalytic-oxidation–reduction column reactor; from this they are chromatographically separated in a 0.8 m packed PTFE column, and then finally analysed in the mass spectrometer. The analytical cycle time is c. 430 s. The size of the samples analysed is optimized when possible to avoid potential linearity effects. One standard is analysed after every five samples, and a blank is run routinely after 10 analyses as a check.

Calibration is by NBS (122, 123, 127) and three laboratory standards that have been calibrated at the Scottish Universities Research Reactor Centre (SURRC). Standards range from $\delta^{34}S$ -17.3 to $+20.3$‰. This wide range permits analysis of samples with wide variations in a single suite. On pyrite, required sample weights are 1.0–1.2 mg; on pyrrhotite 1.6–1.8 mg. The average $\delta^{34}S$ reproducibility of the standards is better than ±0.1‰, and around ±0.15‰ for sulphide mineral samples.

For carbon isotope analysis, individual c. 20 mg fragments rich in organic carbon were used for $\delta^{13}C$ (mainly from black shales). In stromatolitic limestone samples, carbon of probable organic origin was extracted from c. 10 g of rock. In the Royal Holloway work (Grassineau) the samples were treated with HCl for 4 h at 100°C to remove carbonate. On pure carbon samples, $\delta^{13}C$ can be determined using about 0.1 mg. The small sample sizes for sulphur and carbon analysis and large number of analyses give the high resolution needed to study samples lamina by lamina.

The analyses of REE were by isotope dilution at Cambridge University (H.C., M.J.B., M.H.), and major elements by X-ray fluorescence at Royal Holloway (M.H., E.G.N.).

Results: rare earth elements

Rare earth element (REE) analyses are listed in Table 1 and shown in Figure 3. Some of these results were briefly noted by Hunter et al. (1998) and are discussed in more detail here.

In the Spring Valley and Shavi Members, sedimentological facies (Bickle et al. 1975; Martin et al. 1980; Nisbet et al. 1993; Hunter et al. 1998) vary from beach to an assortment of shallow-water settings. REE patterns from oxide-facies ironstones have strong Eu anomalies, but no Ce anomalies. The Spring Valley and Shavi REE are most simply modelled as sediments deposited from water which included: (1) ambient sea water of modern aspect but without a Ce anomaly, as well as (2) an admixed hydrothermal (modern black smoker or white smoker) component. The source of the REE may have been distant.

REE in sulphides from the Jimmy Member (TR44, TR45; Fig. 3) also have positive Eu anomalies, and no Ce anomaly. REE patterns can be simply modelled as a mix of two source components: one component derived from high-temperature hydrothermal fluids, mixed with a component of ambient water (assumed similar to modern sea water without a Ce anomaly). The absence of a Ce anomaly in a modern deposit would suggest water that is not strongly oxidized, but it should be noted that this is not a strong constraint on redox conditions as some modern rocks laid down in oxic conditions do not show Ce anomalies. TR45, from near the top of the Jimmy Member, has ε_{Nd} -1.4. This is less than associated volcanic rocks (ε_{Nd} averaging $+2$, and ranging from $+0.4$ to $+3.2$), but substantially more than contemporary continental crust, or average Manjeri Fm clastic sediment (ε_{Nd} -6) in the east of the Belingwe belt.

The sulphides probably incorporate some crustal-signature detrital REE and perhaps also oxidized sea water REE, but the Eu suggests that the bulk of the REE signature may be from sea water enriched by a hydrothermal plume (see also results of Alibert & McCulloch 1993). Interestingly, oxide-facies cherts have higher positive Eu anomalies, indicating more hydrothermally influenced REE, but more negative ε_{Nd} (i.e. crustal input). Al content tracks continentality. The hydrothermal source may have been proximal. However, it can be argued that sea-water REE and Fe transport could have been over considerable distances if the whole Archaean ocean were reduced. If the REE content of Archaean ocean water were dominated globally by the hydrothermal signal, the Manjeri Fm cherts need not have been linked directly to nearby volcanism.

The REE patterns indicate, in addition to hydrothermal REE, a component from water buffered by crustal inputs of REE (and presumably iron) similar to the input to the modern ocean. The increase in Eu anomaly from the

Table 1. *Manjeri Fm samples*

	SiO$_2$	Al$_2$O$_3$	Fe$_2$O$_3$	MgO	CaO	Na$_2$O	K$_2$O	TiO$_2$	MnO	P$_2$O$_5$	Total	Core depth
TR40	65.87	0.12	30.93	1.44	0.62	0.00	0.00	0.01	0.18	0.04	99.19	S.V. 146 m
TR41	88.37	0.07	10.15	0.41	0.22	0.00	0.00	0.01	0.13	0.03	99.38	Shavi 140.8 m
TR42	80.95	0.08	15.50	1.00	0.95	0.00	0.00	0.01	0.39	0.04	98.90	Shavi 140.2 m
TR43	77.58	0.12	21.50	0.26	0.12	0.03	0.00	0.01	0.03	0.03	99.68	S.V. 146 m

	Sm	1σ	Nd	1σ	Sm/Nd	1σ	^{147}Sm/^{144}Nd	1σ	^{143}Nd/^{144}Nd	1σ	
TR40	0.271	0.001	1.159	0.004	0.23382	0.00117	0.14132	0.00071	0.511470	10	
TR41	0.164	0.000	0.638	0.000	0.25705	0.00051	0.15537	0.00031	0.511702	28	
TR42	0.180	0.000	0.663	0.001	0.27149	0.00054	0.16410	0.00033	0.511871	18	
TR45	0.770	0.001	4.144	0.001	0.18581	0.00037	0.11229	0.00022	0.511064	9	Jimmy 65.8m

	T$_{CHUR}$ (Ma)	1σ	T$_{DM}$ (Ma)	1σ	$\varepsilon_{(2700\,Ma)}$	1σ
TR40	3197	49	3514	40	-3.6	0.3 (0.508952)
TR41	3432	105	3757	74	-4.0	0.6 (0.508934)
TR42	3567	90	3902	60	-3.7	0.4 (0.508948)
TR45	2828	18	3117	15	-1.4	0.2 (0.509064)

	TR40	TR41	TR42	TR43	TR44	TR45
REE, chondrite-normalized [35]						
La	5.796	2.128	1.734	3.929	1.303	15.891
Ce	2.483	1.371	1.258	2.362	1.006	11.641
Nd	1.665	0.959	0.957	1.519	0.639	6.326
Sm	1.153	0.754	0.794	1.035	0.430	3.604
Eu	1.893	1.849	2.118	1.560	0.500	4.060
Gd	1.086	0.932	0.981	1.014	0.356	2.972
Dy	0.863	0.926	0.947	0.775	0.258	2.630
Er	0.915	0.964	1.143	0.784	0.255	3.119
Yb	0.937	0.940	1.151	0.762	0.269	3.572
Lu	0.980	0.972	1.194	0.806	0.267	4.063

Samples TR44 (from 70.5 m) and 45 are Jimmy Member sulphide minerals; others are cherts from oxide-dominated Spring Valley (SV) core. Analyses: isotope dilution, Cambridge group (REE, isotopes); XRF, Royal Holloway, University of London (major elements). REE by standard methods (Greaves *et al.* 1989) on sample solutions containing *c.* 5 ng Nd obtained by dissolution at *c.* 180°C in Teflon-lined bombs containing three stages (HF + HNO$_3$, HNO$_3$, and HCl). DM = 0.509345 at 2700 Ma; CHUR = 0.509136 at 2700 Ma.

sulphide-facies Jimmy Member sample (TR44, 45) to the oxide-facies samples is analogous to the increase seen from high-T black smokers to white smokers and may reflect the partition away from Eu as iron minerals precipitate. The more negative ε_{Nd} values of the oxide samples (-3.6 to -4) compared with -1.4 for the sulphide sample TR45 (Table 1) imply that the sulphide sample contained a smaller proportion of older crustal Nd (ε_{Nd} *c.* -6 in the clastic sediments) and more hydrothermal end member (ε_{Nd} *c.* 0.4 to $+3.2$ in the overlying volcanic rocks).

Eu anomalies are characteristic of sediments from fluids of hydrothermal systems that involve the breakdown of plagioclase. The ubiquitous positive Eu anomaly in Archaean and early Proterozoic banded ironstones is consistent with the chemical models (Drever 1974; Holland 1984), which envisage the ocean predominantly containing relatively reduced iron-rich, sulphate-poor water overlain by a more oxidized surface layer from which the oxide-facies banded ironstone could be precipitated. The REE compositions in the Jimmy Member, and also to a lesser extent in the Spring Valley and Shavi Members, indicate that much of the REE, and by implication much of the iron, is derived from hydrothermal input. High dissolved Fe is incompatible with significant dissolved sulphide, as iron and sulphide together precipitate as pyrite (Drever 1974). The extensive iron deposition from Archaean oceans therefore requires that the water was iron rich. This contrasts with a possible modern parallel, the Black or Euxine Sea. In the Black Sea, much of which is several kilometres deep, high surface productivity

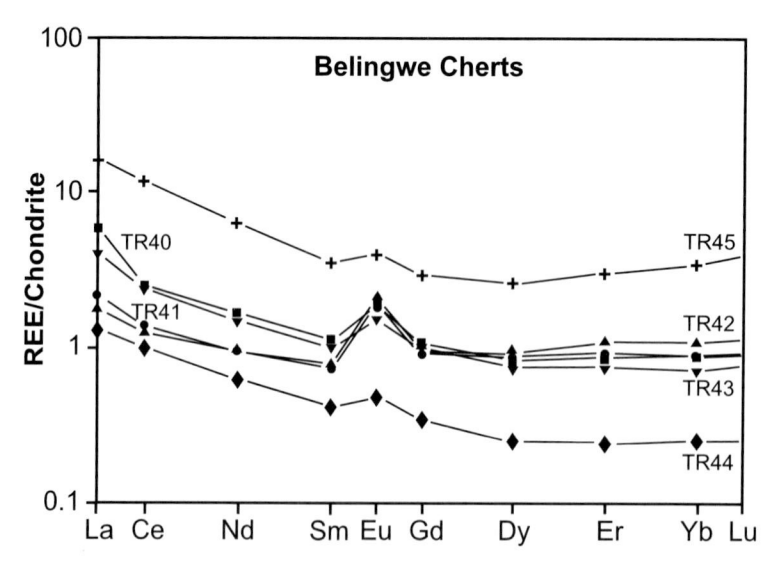

Fig. 3. Isotope-dilution REE analyses (normalized to chondrite) of oxide-facies iron-rich cherts (TR40, 41, 42, 43) and sulphide-facies cherts (TR44, 45) (Table 1). Positive Eu anomalies (typical of modern black smoker fluids) and absence of Ce anomalies (implying low O_2 in modern analogues, although not necessarily in the Archaean) should be noted.

derived from input of sulphate from the Bosporus (Mediterranean), has provided a supply of organic debris to deep levels. Only the top 100 m are oxic, and H_2S-rich anoxic water exists below about 100 m depth.

Hydrothermal fluids with Fe/S mole ratios of c. 2 would be suitable buffers for the reduced part of the Archaean ocean. In such Fe-rich water, sulphide would be removed as pyrite close to hydrothermal vents in a manner reciprocal to deposition of iron in plumes in the modern oceans. The precise oxidation state of the deep water is controversial. It may have been sulphide- or sulphate-bearing (the dissolved sulphide–sulphate boundary lies within the pyrite stability field in fO_2–pH space). Many modern hydrothermal fluids discharging to the sea floor are buffered to fO_2 slightly higher than equilibrium with basalt (py–mag–anhydrite). The oxidation state of water below the photic zone in the Archaean ocean may have been controlled by the relative inputs of reduced hydrothermal fluid, mixing with more oxidized surface water, and settling of organic debris from photosynthetic surface life. Sulphate deposits imply that Archaean surface water was sulphate bearing in equilibrium with iron oxides (Ohmoto 1992; Ohmoto et al. 1993). Like the air, the chemistry of the ocean is not necessarily in equilibrium. It is possible that much of the Archaean ocean was sulphate bearing, the hydrothermal input being more than buffered by mixing with sulphate-bearing surface water.

Deposition of the oxide-facies ironstones requires a direct source of oxygen. Mixing the hypothetical upper oxidized layer with deep reduced water would produce water of intermediate oxygen fugacity and intermediate iron solubility with little iron precipitated. Precipitation of the iron hydroxide or siderite precursors of the oxide-facies ironstones would require addition of oxygen, most plausibly from photosynthetic organisms such as the cyanobacteria in the stromatolites and presumably as cyanobacterial picoplankton.

What are the implications of redox conclusions drawn from REE for the genesis of the Manjeri Formation iron minerals? Sulphide-rich units such as the Jimmy Member could be derived either inorganically, from nearby hydrothermal plumes, or by the action of sulphate-reducing bacteria. Although the setting is inferred to be below wave base, the Jimmy Member is unlikely to have been laid down in very deep water (below 1000 m). Lapilli tuffs are found not far above stratigraphically, suggesting exposure, and shallow-water sediments occur below the Jimmy Member (see Fig. 2). This setting, the very uniform thickness and extent of the Jimmy Member (with a possible original surface extent of the order of 1000 km^2), generally low Cu and Zn, combined with its carbon content and carbon

Table 2. $\delta^{13}C$ and $\delta^{34}S$ values in Belingwe belt material

Rock unit	Sample	$\delta^{13}C$‰	$\delta^{34}S_{Sulphide}$‰	Comment
Cheshire shales (Saskmar 2 drillcore)	Reduced C Carbonate	−44 to −33 −8 to −9	−6	Methanogenic and methane oxidizers? sulphur mat
Stromatolites				
Cheshire Fm (Macgregor Mbr)	Kerogen Carbonate	−33 to −27 c. 0		Cyanobacterial mat community
Manjeri Fm (Rupemba)	Kerogen Carbonate	−24 to −6 (−35 pure kerogen) −0.9 to +1.5		Cyanobacterial mat community
Jimmy Member (Manjeri Fm)	Carbon sulphide	−38 to −20	−14.9 to +16.7 most around 0	Sulphur mat including methanogens
Rubweruchena Mbr (Manjeri Fm)			−1.2 to −0.1	
Shavi and Spring Valley Mbr (Manjeri Fm)		−31 to −17	Sulphur mat −17.6 to +5.4	?Photosynthetic

All samples duplicated. Previous work reported elsewhere (Strauss & Moore 1992) on the cherts collected by us from the Spring Valley Member gave $\delta^{13}C$ in kerogens in the range −32.2 to −8.8‰ (nine samples of cherts) and $\delta^{13}C$ carbonate in the range −5.7 to −7.2‰ (three samples). Kerogens contained 0.08–2.50 mg C g^{-1} TOC (total organic carbon). H/C and N/C ratios have been determined on one sample collected by us from the Spring Valley Member. This gave H/C 0.15 and N/C 0.005 in a sample with $\delta^{13}C$ −33.0‰ (Strauss & Moore 1992).

isotope compositions (see below), make an origin as volcanogenic massive sulphide unlikely.

The alternative to a volcanogenic massive sulphide origin is that the sulphide was precipitated by sulphate-reducing bacteria consuming organic debris sinking from the surface waters. In modern euxinic basins much of the pyrite is precipitated in the water column immediately below the oxic–anoxic boundary (Canfield et al. 1996). The oxide-facies ironstones (Spring Valley) would have been precipitated within the oxic surface waters, oxygen supplied by photosynthesis.

Isotopic evidence

Stable isotope results are shown in Table 2 and Figures 4 and 5 and as a profile through the Nercmar core in the Manjeri Formation at the Unconformity National Monument outcrop (Fig. 6).

Spring Valley and Shavi Members

The sedimentological facies is unambiguously shallow water, in the photic zone close to the unconformity beach. Carbon isotopes (Fig. 4) in sulphide–carbon–chert vary widely in samples with c. 1% carbon (varying up to 3% carbon). Carbon isotopes are moderately resistant against low-grade metamorphism; however, any metamorphic process could have homogenized carbon and thus reduced the range. Application of an isotopic shift (Hayes 1994) in organic carbon of about 7.5–8‰ calculated for an H/C ratio of 0.2, which is slightly higher than that seen in Manjeri chert kerogen (Hayes et al. 1983), would suggest an original isotopic range in organic carbon of about −39‰ to −25‰. However, this shift probably substantially overexaggerates any post-depositional fractionation, which was probably less than 0.2–0.3‰ (Watanabe et al. 1997).

$\delta^{34}S$ in sulphide minerals is very variable (Figs 5a and 6). Millimetre-scale isotopic heterogeneity occurs within units with marked variation even to the millimetre scale in small samples. The S isotopic range is wider than typically reported in most Archaean suites (e.g. Goodwin et al. 1976; Ohmoto 1992; Habicht & Canfield 1996), many of which have a much smaller spread of $\delta^{34}S$, ranging around 0 ± 5‰. Moreover, the original range in the Shavi rocks would have been even greater, as metamorphic homogenization would be expected to have reduced the spread.

Rupemba stromatolites, Manjeri Fm

The content of apparently organic carbon is variable in the Rupemba stromatolites, which, from the sedimentological facies, were formed in a photic zone setting laterally equivalent to the Shavi cherts. Typically $\delta^{13}C$ ranges from −23‰ to −22‰, in material with about 1–2% of organic

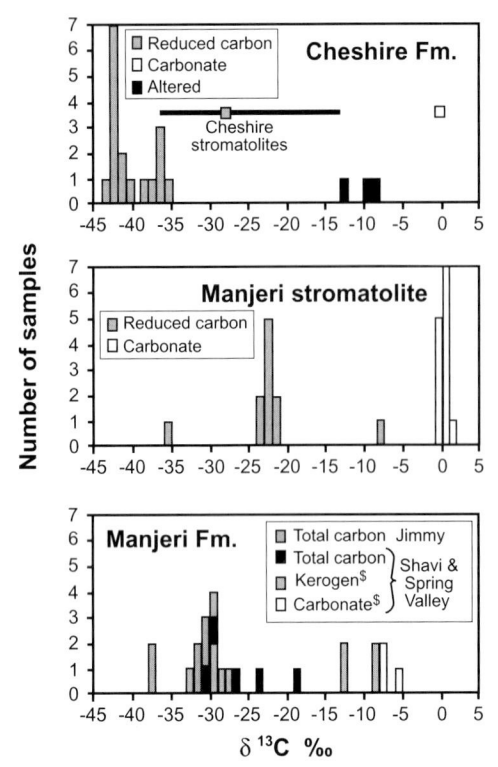

Fig. 4. $\delta^{13}C$ in carbonates and organic carbon from the Manjeri and Cheshire Formations. Data marked $ are from Strauss & Moore, (1992); Cheshire stromatolite data (small boxes) are from Abell *et al.* (1985) on carbonates (average $\delta^{13}C$ 0.22 ± 0.24‰) and on organic carbon samples (unpublished data of J. Yong: around 250 kerogen measurements, average $\delta^{13}C$ -28.8 ± 2.6‰). Other Shavi and Spring Valley Member cherts and Jimmy Member sulphide horizon, this study; Manjeri 'stromatolite kerogen and carbonate' from Rupemba, Cheshire shales kerogen data are from this study.

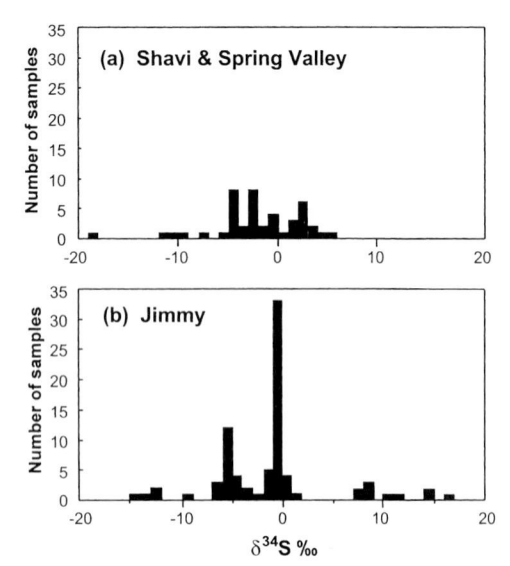

Fig. 5. Stable isotope histograms from Belingwe. (**a**) $\delta^{34}S$ in Spring Valley and Shavi Members; (**b**) $\delta^{34}S$ in Jimmy Member sulphide–carbon–chert samples (data from Grassineau *et al.* 2001 and Grassineau, unpubl. data).

carbon. The Rupemba stromatolites are in a more deformed setting than that of the unconformity outcrop of the Spring Valley and Shavi Members, and metamorphism may have homogenized their isotopic ratios slightly (Abell *et al.* 1985), reducing the isotopic range. Applying a metamorphic shift of about 2–3‰ would imply original $\delta^{13}C$ values around −25‰. Notably, one small clump of dark material (36% organic carbon) had $\delta^{13}C$ of −35‰. Carbonate $\delta^{13}C$ in the Rupemba rocks (Abell *et al.* 1985) is around 0‰.

Jimmy Member

A Nercmar core sample containing laminated sulphide, quartz and carbon-rich bands from the upper part of the Jimmy Member, below the volcanic rocks of the overlying Reliance Fm, is locally as rich as 6.5–11% in organic carbon, but is very variable. The $\delta^{13}C$ in the Jimmy Member samples (Fig. 4) ranges from −38.4‰ to −22.8‰. The carbon content of these samples ranges from low (*c.* 0.5–0.6% in the sample with very light carbon) to 10.8% (in a −29‰ sample). The drill core is heterogeneous in carbon content and in isotopic composition on a fine scale. It is possible that the carbon is secondary, derived from a bitumen after oil migrated into the rock (note that other rocks nearby do have oil of apparent Archaean age); however, the C abundance, isotopic heterogeneity, and the texture and mineralogy of the samples all imply the carbon is syn- or early post-depositional and reflects biological activity in the rock.

The sulphur isotopic ratios (Fig. 5b) have a wide range, with a sharp $\delta^{34}S$ peak around 0‰. As with the Shavi samples, some values are very negative (e.g. −14.9‰; compare Goodwin *et al.* 1976; Hayes *et al.* 1983) considering the age of the rock (Habicht & Canfield 1996).

Cheshire stromatolites

Carbon isotope compositions of organic carbon in the stromatolites of the Macgregor Member of the Cheshire Formation are variable, but

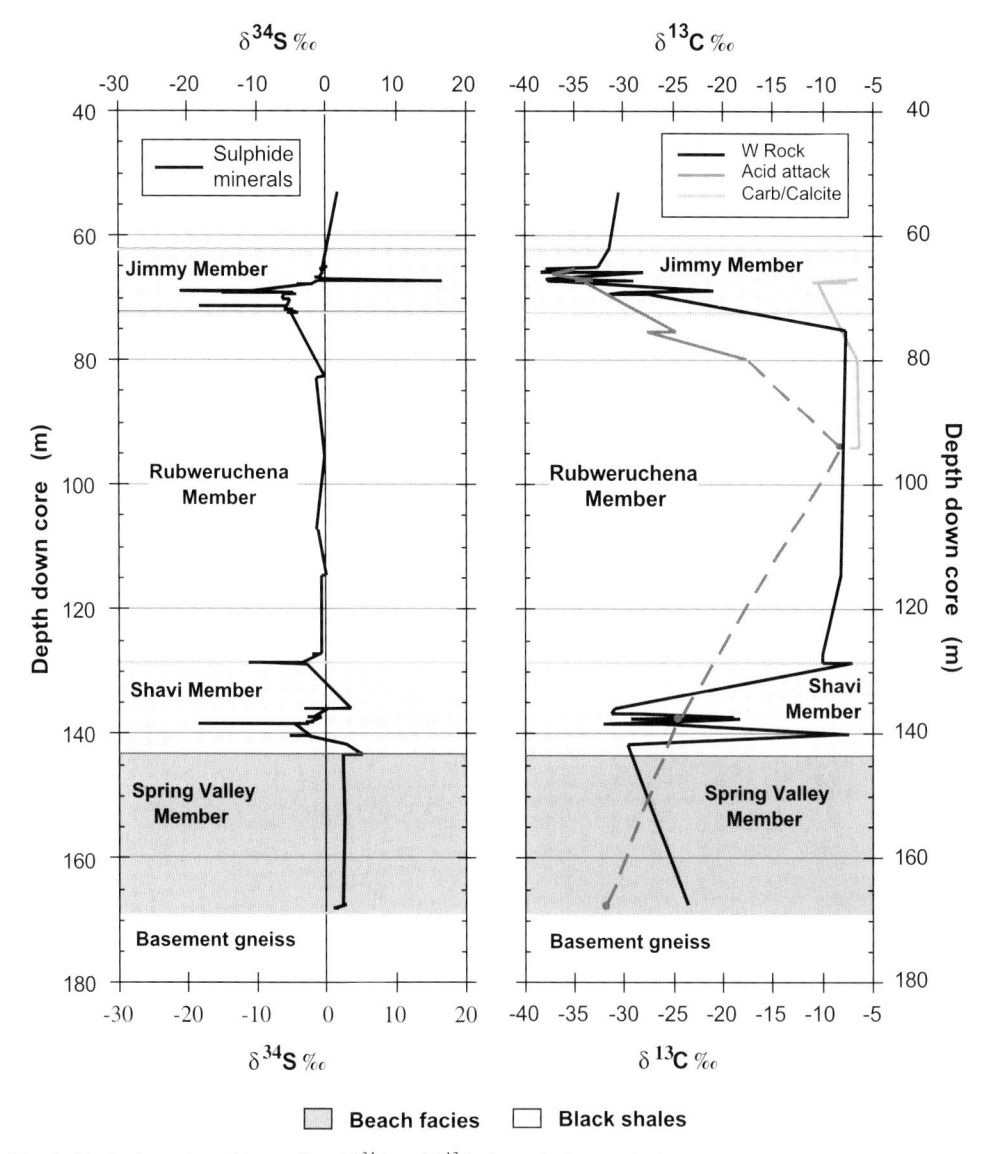

Fig. 6. Vertical stratigraphic profile of $\delta^{34}S$ and $\delta^{13}C$ through the Manjeri Fm.

with $\delta^{13}C$ mainly in the range $-33‰$ to $-27‰$ (Fig. 4). Carbon isotopes in Cheshire stromatolitic carbonate are around 0‰ as in the Rupemba stromatolites (Abell *et al.* 1985).

Cheshire shales

Calcite from the shales has $\delta^{13}C$ around $-8‰$, implying that it was crystallized using carbon from an isotopically light source. Although magmatic gas is a possible source, the shales do not seem to have had proximal volcanism and it is more probable that the carbonate formed from CO_2 derived from organic reactions in the sediment pile. Organic carbon in the Cheshire shales has $\delta^{13}C$ from $-43‰$ to $-33‰$ (perhaps lighter before post-depositional shift).

It is possible that endolithic microbial life persisted in the sediments for some time after deposition and before induration and expulsion of water, but this endolithic life would have been Archaean, before the intrusion of surrounding younger granites and then the Great Dyke. On burial by the lavas of the Reliance and Zeederbergs Fms, and subsequent induration, water flow would have ceased in the Spring Valley, Shavi and Jimmy Members of the Manjeri Fm and continuing endolithic life would have been extremely limited thermodynamically. In the Cheshire Fm limestones, cementation was probably fairly rapid; however, post-depositional endolithic life may have persisted in the Cheshire shales for some time in the Archaean. Later ingress of oil (e.g. in Karroo times) cannot be excluded. There is, however, no evidence for this, or source: the calcite and organic carbon, and any residual oil in the shales are almost certainly of Archaean age.

Discussion

Carbon isotopes in shallow-water sediments: Spring Valley, Shavi and Cheshire stromatolites and black shales

The range of carbon isotopic composition (Fig. 4) in the shallow-water deposits is wide (from 0‰ in the rock, to −44‰ in reduced carbon, possibly as light as −50‰ on original deposition). How much of this records rubisco, and how much other processes such as methanogenesis?

Rubisco is the chief but not the only mediator between atmospheric carbon and carbon in life. It is used by aerobic bacteria: purple bacteria and cyanobacteria. Oxygenic photosynthesis by cyanobacteria produces a fractionation between organic carbon and carbonate of about −29‰ (Pierson 1994), a result typical of much of the organic carbon in the Manjeri Fm rocks. In these very well-preserved and little-metamorphosed rocks, post-depositional processes may have shifted the carbon isotopes slightly by depleting ^{12}C. The many Belingwe C isotope results in the −32‰ to −23‰ range (allowing for the possibility of some post-depositional shift, and the presence of light carbon dioxide, as shown by some carbonates) are thus most simply interpreted as evidence for dominant fractionation by rubisco during oxygenic photosynthesis.

Textural evidence suggests that anaerobic green photosynthetic bacteria may also have been present (Schidlowski & Aharon 1992). Can the isotopic results test this suggestion? Certainly some of the results are not inconsistent with the actions of anoxygenic photosynthesizers. Fractionation from anoxygenic photosynthesis in mats by anaerobic photosynthetic green sulphur bacteria using the reductive citric acid cycle ranges from −12‰ to −3.5‰ (Pierson 1994). The C isotope results in this range (and allowing for possible post-depositional shift of a few per mil) thus support the textural deduction (Schidlowski & Aharon 1992).

In addition, the many more negative C isotope results, especially samples now lighter than $\delta^{13}C$ −33‰, and probably lighter than $\delta^{13}C$ −35‰ before metamorphic shift, imply that carbon reprocessing also occurred in the sediment. In the Rupemba and Cheshire stromatolites, the samples showing this more extreme fractionation suggest the operation of methanogens releasing very fractionated methane from organic debris at the base of a microbial mat, and, shallower in the mat, methanotrophs existing on the methane emitted from below.

Whether marine plankton could have contributed is a moot question. Modern marine phytoplankton in the Gulf of Mexico have an isotopic signature of around $\delta^{13}C$ −21‰ (Jasper & Gagosian 1993). Modern phytoplankton are of course very different from Archaean, although there are reasonable grounds to infer that cyanobacteria, and in addition, eukaryotes were probably present even 2.7 Ga ago (Brocks *et al.* 1999). Just possibly, phytoplankton debris may account for some of the carbon isotope values.

Ambient carbonate in the Cheshire stromatolites shows that precipitation in contemporary sea water had $\delta^{13}C$ of 0‰. This immediately supports the inference that at 2.7 Ga, carbon management by oxygenic photosynthesis took place on a global scale (Schidlowski & Aharon 1992), with mantle derived carbon ($\delta^{13}C$ −7‰ to −5‰) being partitioned into organic ($\delta^{13}C$ around −28‰) and inorganic ($\delta^{13}C$ near 0‰) reservoirs.

In contrast, the carbonate in the shales ($\delta^{13}C$ −8‰), could have formed calcite using ^{13}C-poor carbon dioxide from methanotrophs. This interpretation is supported by organic carbon in the Cheshire shales. which is very light and may record the presence of methanogens.

The full range of carbon isotopic compositions seen in the Cheshire and Manjeri Fm stromatolites is thus collectively interpreted as evidence for abundant rubisco-based oxygenic photosynthetic cyanobacteria, possibly associated with anoxygenic photosynthesizers, and supporting a community of other microbial organisms (as in modern stromatolites; Pierson 1994).

To summarize, the most probable explanation of the C isotope spread in the shallow-water facies of the Manjeri Formation (Spring Valley and Shavi sediments and Rupemba (Spring Valley) stromatolites), and in the Cheshire Formation stromatolites and black shales is thus that carbon isotope fractionation records a diverse ecology that existed based primarily on oxygenic photosynthesis, but with associated organisms exploiting a wide variety of other metabolic pathways. The stromatolites would have grown in favoured sites where other sedimentation was restricted and the currents and geological substrate were suitable.

Moreover, by implication, if coastal cyanobacteria existed 2.7 Ga ago, a very probable speculation is that cyanobacterial picoplankton would also have evolved (see descent trees of Woese 1987; Pace 1997). They were probably abundant at shallow photic levels in open water, although not preserved in the record.

Sulphur isotopes in shallow-water sediments

The diversity of S isotope results in the Shavi Member (Fig. 5) is clear evidence for biogenicity. Both the Shavi–Spring Valley and Jimmy Member histograms show values distributed around 0‰. The strong 0‰ $\delta^{34}S$ peak in the Jimmy Member (see below) is clearly a record of inorganic hydrothermal processes, which may also be reflected in the Shavi–Spring Valley histogram. But much of the diversity is of probable organic origin.

The existence in the Archaean of sulphate-reducing bacteria, producing a spread in S isotopes around $\delta^{34}S = 0 \pm 5$‰ has long been accepted (e.g. Goodwin et al. 1976; Shen et al. 2001). Sulphate would have come from ambient water, with a component supply of metals and other nutrients from hydrothermal sources, as shown by the REE (Table 1). In modern microbial mats, sulphate-reducing bacteria extract sulphur from sea water, fractionating it by an amount that depends on the efficiency and rate of extraction. Conversely, sulphide-oxidizers reverse the process.

The narrow range of S fractionation (0 ± 5‰) seen in many Precambrian biogenic pyrites has been interpreted as showing that sulphate concentrations in Archaean sea water were low (Habicht & Canfield 1996), implying that sulphate reducers, although present, were of limited productivity. However, the counterview (Ohmoto 1992) is that Archaean sea water was sulphate rich, and that the isotopic distribution seen reflects the closed- or partly closed-system fractionation in the absence of large bottom dwellers. Given the absence of metazoan burrowing, the microbial Archaean sea-bottom biome may have been productive, but thinner than today. Thus any heterogeneity would occur on a small scale. To detect this, sample size and number are very important in S isotope studies. To avoid the homogenizing effect of merging fine-scale diversity in large-mass samples, and to map out the diversity of the Archaean community, many small samples are needed, on a millimetre scale. The high-resolution, high sample number techniques used here should show detail not seen in conventional S isotope work.

The total range in $\delta^{34}S$ reported here is −18‰ to +17‰ overall in the Manjeri Fm., with 30‰ fractionation variation in the Jimmy Member, and 23‰ variation in the Shavi Member. The S isotopic variation found in Belingwe rocks is much wider than previously found in most Archaean suites. This is strong evidence for diverse sulphur processors in both settings (Grassineau et al. 2001b). The isotopic diversity suggests (but does not prove) dissimilatory as well as assimilatory reduction. Dissimilatory sulphur and sulphite reduction is probably of the greatest antiquity (Schauder & Kroger 1993; Molitor et al. 1998), but not unambiguously recorded in Archaean sulphur isotopes.

The range shown in these Belingwe results is supported by an earlier study, using conventional techniques but with a large sample population, of biogenic pyrites from the 2.75 Ga Michipicoten ironstones (Goodwin et al. 1976). These ironstones are lithologically similar to the sampled horizons in the Jimmy and Shavi Members. The Michipicoten samples had a $\delta^{34}S$ range of over 20‰ (−10‰ to +10‰). This is less than found in Belingwe, but still much greater than the 0 ± 5‰ conventionally assumed to be typical of the Archaean. Moreover, these pyrites were accompanied by abundant organic carbon ($\delta^{13}C$ from −27.7‰ to −20.4‰), supporting the view that bacterial communities were capable of significant fractionation of S. In modern microbial mats (including sulphate-reducing and also photosynthetic and non-photosynthetic sulphide-oxidizing bacteria and archaea), biological activity, multiply cycling sulphur through its oxidation states, can produce extreme fractionations (Shanks et al. 1995; Canfield & Teske 1996; Habicht & Canfield 1996; Raiswell 1997), with primary sulphate from sea water around +20‰, whereas biologically fractionated sulphide may be −40‰ to −25‰, or in some cases much lighter after a cascade of recycling processes in the sediment. A simple one-way Rayleigh fractionation process, in which sulphur was progressively extracted from seawater sulphate, would give a strongly asymmetric frequency distribution, with a peak at the negative end of the distribution, lighter than sea water, a major population of analysed values somewhat more negative in $\delta^{34}S$ than ambient contemporary sea water, and a long tail of values towards the positive side (Ohmoto 1992; Ohmoto et al. 1993). To achieve a more broadly spread distribution, a rather different model is needed.

A possible model of an Archaean mat community is one in which sea water rich in SO_4^{2-} is reduced in a system that is only occasionally open to ambient water inflow. In such a setting, partly open-system fractionation occurs, giving a moderately symmetric isotopic distribution in sulphides (fig. 4E of Ohmoto 1992). The $\delta^{34}S$ dataset from the Spring Valley and Shavi Members, both in shape and range, is thus here interpreted as probable evidence for a nearly closed-system fractionation in a bacterial mat flushed by sea water that was relatively rich in sulphate, with water sulphate possibly around 3–5‰. However, Archaean seawater $\delta^{34}S$ remains unknown. The ratio is thought to have been not far from 0‰, but this inference is lightly based on a few sulphate deposits that may not be primary: the ratio could have been as heavy as 10–15‰ (Kakegawa et al. 1998).

The assumption is made that the ambient water provided sulphate; which would be expected, given the evidence for volcanism and oxygenic photosynthesis. However, it is possible that in part the system was sulphate poor. The spread of S isotopes in the Spring Valley and Shavi Members (skewed to the light side) perhaps suggests the hypothesis that in the shallow-water facies sulphate reduction was not the only process involved, but also phases of anaerobic bacterial oxidation of S (fig. 4C of Ohmoto 1992). This would imply cycling process similar to those in modern sulphur cycles, although more muted, either because early bacterial diversity was poor, or from post-depositional homogenization and recrystallization. In the Shavi Member, both sulphate reduction and photosynthetic oxidation may have taken place, forming a bacterial mat deposit that was thin compared with modern examples (where benthic eukarya help in the mixing) but nevertheless complex and productive.

The C isotope evidence for rubisco implies aerobic conditions in part of the mat, and thus aerobic sulphide oxidation may have occurred, if organisms capable of this were present. Elsewhere, the evidence for methanogenesis implies reducing settings. At the top of the microbial mats, in shallow water, oxidation power would have been enhanced locally, by oxygenic photosynthesis within the mat, by the ambient water rich in oxidized species and possibly O_2 from laterally equivalent nearby cyanobacterial stromatolites, as recorded in the Rupemba samples. Moreover, if cyanobacterial picoplankton were present, they would have contributed oxidized species in sea water.

Collectively, the sedimentology and facies of the rocks (Bickle et al. 1975; Nisbet et al. 1993; Hunter et al. 1998) and the isotopic evidence from the sulphide-rich parts of the shallow-water rocks can be interpreted to suggest a diverse prokaryotic mat community (Fig. 7), made of both photosynthetic (oxygenic and possibly also anoxygenic) (Cohen et al. 1975) and non-photosynthetic organisms, living in quiet lagoonal waters, protected from clastic incursion, rather like modern sulphur bacteria mats (Shanks et al. 1995), common in coastal settings, although the abundance of aerobic sulphur-oxidizers comparable with modern *Beggiatoa* may have been limited by available oxygen supply.

Jimmy Member

Carbon isotopes. In the Jimmy Member, the C isotopic fractionation, allowing for a possible small metamorphic shift, shows what appears to be the characteristic -30 to -25‰ $\delta^{13}C$ signature of rubisco (i.e probably aerobic bacteria). The sedimentology of the Jimmy Member implies that it may have been laid down below storm base (order of c. 100 m or more), below the photic zone, but not at great depth, given the stratigraphic proximity of shallow-water sediments (below) and subaerial volcanic rocks (stratigraphically above). A possible explanation of this is that cyanobacterial picoplankton existed in the photic zone of the overlying water column, and organic matter produced by photosynthesis settled out to the bottom.

However, this explanation is not the only possibility. The C isotope record of the Jimmy Member is complex. In addition to the probable evidence for rubisco, some organic carbon shows more extreme fractionation (especially if post-depositional shift is allowed for), implying also the presence of methanogens and reduced settings. This suggests three possible hypotheses: (1) the sea floor received organic debris from overlying photosynthetic cyanobacterial plankton, which used rubisco, and on the sea floor methanogens recycled the organic detritus; (2) that the apparent evidence for rubisco is simply a mixing signature from methanogenic and non-methanogenic sea-bottom microbial populations; (3) there is also an alternative possibility that the Calvin cycle operated in a non-photosynthetic sulphate-dependent microbial flora; a bacterial mat cycling oxidation states, but with aerobic or microaerobic bottom waters and a sea-floor population of purple bacteria.

Sulphur isotopes. Sulphur isotopes in the Jimmy Member show a wide range in $\delta^{34}S$ (-14.9 to $+16.7$‰) with $\delta^{34}S$ symmetrically distributed

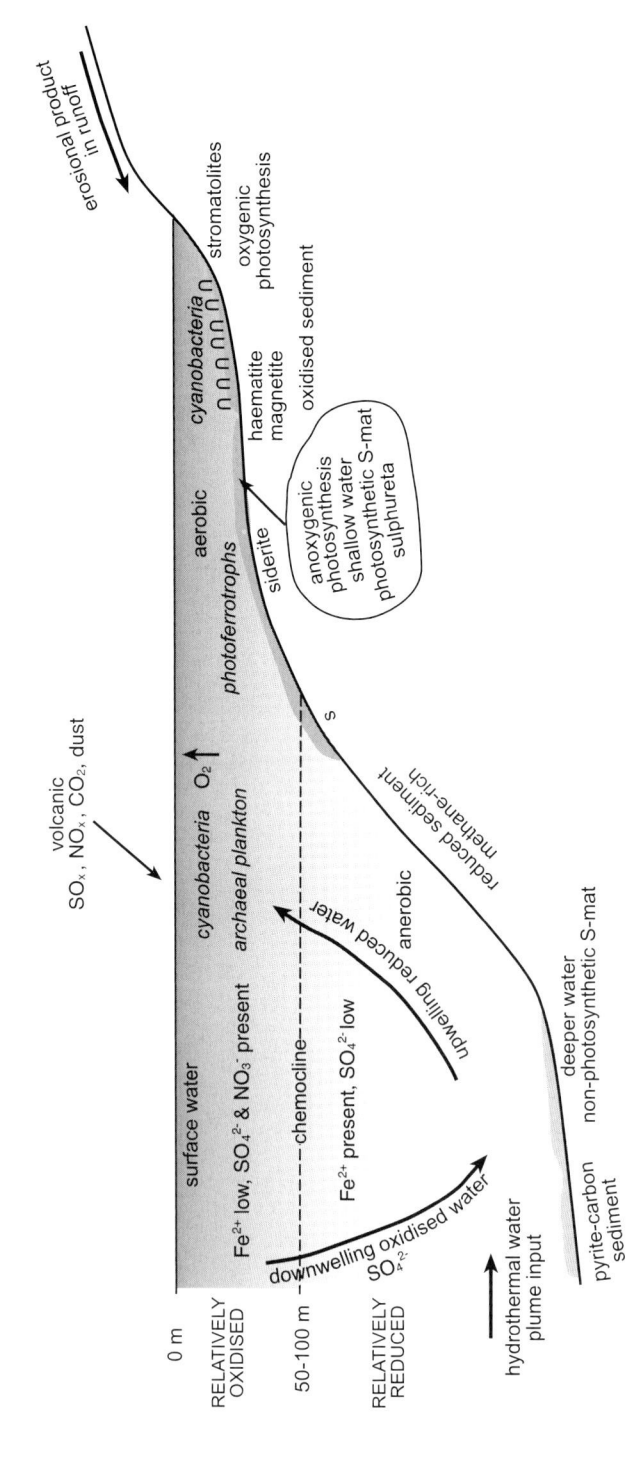

Fig. 7. Model section of the shelf, slope and deeper waters of an Archaean ocean.

around 0‰. The strong peak at 0‰ suggests the strong influence of hydrothermal processes, depositing inorganic sulphide, and is consistent with the REE evidence. The stratigraphic evidence for nearby shallow-water facies (i.e. although below storm base, the water column was not very deep), the lateral persistence of the strata-bound deposit, and its sedimentology, and richness in biologically reduced organic carbon (as shown by the C isotopes) all imply that the deposit must also have strong biogenic aspects.

A possible explanation of the $\delta^{34}S$ variation either side of the 0‰ peak is the hypothesis that both microbial sulphate reduction and sulphide oxidation were proceeding, with a relatively oxidized upper layer and a relatively reduced lower layer to the mat community. Analysis of large mass conventional samples might not have detected wide variation, only apparent from many analyses of small samples. It is worth noting that although the pronounced peak at 0‰ is presumably nearly all a result of hydrothermal processes, in part it may reflect post-depositional homogenization of an original deposit that had fine-scale heterogeneity, either because sample sizes are larger than microbial consortia that fractionated S or, given the shear that is apparent in the top of the unit, by S transport and recrystallization during deformation, so that only a part of the original isotopic diversity remains.

To summarize, given the consistency of the sedimentological and textural evidence (Bickle et al. 1975; Nisbet et al. 1993; Hunter et al. 1998) and the REE and isotopic evidence presented here, together with the amount of carbon in the sample, the most likely origin of the Jimmy Member's carbon-rich sulphide–chert layers is that they represent a bacterial mat deposit, growing in a setting subject to the frequent influence of hydrothermal vent fluids. In addition to the major hydrothermal peak, the S and C isotopic results show unambiguous biological distributions, surviving despite the effects of any later isotopic homogenization by metamorphism or by shear. It is possible that the sedimentology and C and S isotopes suggest that the Jimmy Member may record an diverse ecology (Fenchel & Bernard 1995) of sulphate, sulphite and possibly sulphur reducers (some dissimilatory), possibly sulphide oxidizers, fermenters, methanogens and methanotrophs.

Palaeoenvironmental setting

The broader palaeoenvironmental setting can be reconstructed from the sedimentology, isotopic interpretation, and geochemistry. REE data are most simply interpreted as showing the influence of both hydrothermally derived water and ambient modern-like sea water on the depositional environment. Oxygen release from stromatolitic and perhaps planktonic cyanobacteria in the Belingwe basin would have been proximally supplied to the coeval hydrothermal community; conversely, the nutrients from volcanic hydrothermal plumes would have been available to supply coastal shallow-water cyanobacteria, which would have carried out photosynthesis and probably nitrogen fixing. Evolutionary logic would suggest that if cyanobacteria were present as mat-formers, they were almost certainly also present as oceanic picoplankton. It is a very small step from a cyanobacterial habitat in a mat to a free-floating form. Today, cyanobacterial picoplankton are ubiquitous in tropical waters and form an important part of global plankton productivity.

Sea-water compositions may have alternated between incursions of volcanically influenced plumes, reduced and sulphide bearing, with episodes of more oxidized sulphate bearing ambient water from shallower levels. In this setting, sulphate and nitrate in relatively oxidized upper-level incoming sea water, periodically strongly influenced by the chemistry of vent plumes, is reacted against more reduced chemical species from within the mud, and from deeper-level water.

In part, some of the precipitation of the varied metal sulphides may originally have been biologically mediated. The sulphides indicate the availability of metals, notably Fe but also other essential metals such as Ni (used in urease and hydrogenase), possibly from early alteration of coeval komatiite lava flows. The REE evidence supports the inference of supply of many nutrients from distal hydrothermal sources. Iron transport may in part have been from hydrothermal sources, and partly from runoff from land. Photoferrotrophy may have been important. Even if the water of the Belingwe basin were moderately oxidized from the activity of the stromatolite reefs, iron may have been transported biologically, with initial biological capture of iron in reduced water in the hydrothermal vicinity, or by organic reduction of Fe(III), and transport by magnetite-containing picoplanktonic bacteria (similar to *Aquaspirillum*), which used magnetotaxis for diurnal movement.

The model of Fig. 7 is derived in part from the sedimentological facies (Nisbet et al. 1993; Hunter et al. 1998) and in part from the depositional chemistry of banded ironstones (e.g. Drever 1974). In a world where total biological productivity may have been roughly comparable with the modern level (as implied by the 0‰ carbonate) but with atmospheric pO_2 much lower than today, the entire ocean below the chemocline would be anaerobic, whereas the surface mixed zone would be relatively oxidizing. This occurs in the modern Black Sea, caused by surface oversupply of sulphate and organic matter, and hence sulphide to depth; in the Archaean, the supply of Fe^{2+} at depth from hydrothermal plumes and incoming sediment would have been abundant. Below the Archaean chemocline, in generally anaerobic ambient conditions, downwelling sulphate would support pyrite formation in bacterial mats; above the thermocline, upwelling dissolved Fe^{2+} would be oxidized to goethite, forming oxide ironstones. In shallow settings, photosynthetic mats would form: cyanobacterial stromatolites where protected, or mats of sulphur bacteria on silts. Where organic debris was abundant enough to maintain anaerobic conditions at the sediment–water interface, siderite would form (e.g. close to the boundary between oxidized and reduced water).

Scenario: a microbial ecosystem

Identified communities in the Belingwe belt thus include: (1) sub-tidal or tidal microbial mats in mudflats that formed the shales of the Shavi Member; (2) oxygenic photosynthetic cyanobacterial stromatolites in the Rupemba limestones of the Spring Valley Member and the Macgregor Member of the Cheshire Fm; (3) a sulphate-reducing bacterial mat, possibly non-photosynthetic, in the Jimmy Member; (4) processing of organic matter in Cheshire Fm shales.

Reconstructing the set of prokaryote communities, it is possible to imagine a speculative scenario for an interacting ecosystem (Fig. 7).

(1) In shallow-water reefs, photosynthetic cyanobacteria were active, producing oxygen and capturing carbon. Locally in the Belingwe basin (which may have been confined), the life and death of coastal and planktonic cyanobacteria would have provided a supply of oxidized water, oxidation power to form sulphate. There was probably also production of fixed nitrogen and a supply of organic detritus from cyanobacterial picoplankton. Speculation suggests that archaeal picoplankton were probably also at least as abundant as today. The record of carbonate with $\delta^{13}C$ of 0‰ suggests that burial of organic carbon would have been on a modern scale, and hence atmospheric CO_2 was biologically managed. The modern carbon cycle was already in global operation (Schidlowski & Aharon 1992), implying that oxygenic photosynthesis was widespread. The output of oxidized products (sulphate, O_2, Fe^{3+}) would also have been comparable with modern rates. For microbial life to have an impact on this global scale, cyanobacteria would surely have been present not only in coastal settings but also as picoplankton in open water, as the implications of rRNA trees suggest.

(2) Along depositional strike from the shallow-water reefs, in quiet backwaters between delta distributaries, microbial mat communities exploited the niche between relatively oxidized waters and reducing organic-rich muds, which included debris from local microbes, and, if present, picoplanktonic cyanobacteria and possibly magnetotactic bacteria. Oxygen release from cyanobacterial photosynthesis would have provided oxidation power in the water above the sulphur bacteria mats.

The Archaean atmosphere–ocean system was probably less oxic than today, but supply of oxidant would have occurred nevertheless as, in an Archaean CO_2-based atmosphere, volcanic sulphur and nitrogen sources would have provided SO_x and NO_y, or precursors that would have been oxidized in the atmosphere to SO_x and NO_y. Together with nitrogen-fixing by lightning, this would have supplied sulphate and nitrate for use by microbes living in mud that contained picoplankton debris, to sustain mats, possibly of colourless sulphur bacteria. The REE record a distant hydrothermal input, which, depending on water currents, supplied metals and perhaps episodes of reduction.

Within this setting, in shallow coastal water conditions, consortia of bacteria set up microbial mat columns to exploit the supply of sulphate and nitrate from water. In the muds below the mats, methanogens were active, and above them methane-oxidizing bacteria. The waters were enriched chemically by contributions from hydrothermal water plumes, either from the laterally equivalent beginnings of Reliance Fm volcanism elsewhere in the basin, or from more distant oceanic sources. Photosynthetic green sulphur bacteria may have oxidized H_2S to $S°$, whereas sulphate and sulphur reducers operated in the reverse direction.

(3) In the Jimmy Member, in a strata-bound setting at the boundary above the clastic Manjeri sediment and before the eruption of the volcanic rocks of the Reliance Fm, the isotopic and textural evidence and C abundance implies that a non-photosynthetic bacterial mat grew, supported by a complex sulphur cycle, and by fermentation of organic debris, methanogenesis and methanotrophy. It should, however, be stressed that the evidence for depth below the

photic zone is not conclusive. The setting was probably a confined basin, and the Jimmy Member may be diachronous. The REE evidence demonstrates hydrothermal input, and the high organically fractionated carbon content of some samples indicates a productive mat ecology.

To summarize, the isotopic and textural evidence collectively implies the activity in the Belingwe belt of a variety of prokaryotic processes: (1) sulphate reduction and possibly photosynthetic sulphide oxidation; (2) operation of rubisco both in cyanobacterial stromatolites (as expected) but also possibly in non-photosynthetic sulphur–bacterial mats; (3) oxygenic photosynthesis (in stromatolites); (4) methanogenesis and methane oxidation. Most probably, other sulphur-based metabolic reactions (e.g. dissimilatory sulphate reduction) were also taking place. This complexity is consistent with the relative timing of the metabolic phylogeny deduced from rRNA studies (Woese 1987; Pace 1997).

Conclusion

This evidence supports the view (Nisbet & Sleep 2001, 2002) that by the late Archaean, biological management of carbon and sulphur was on a sufficient scale that the modern carbon cycle was already long established. The absence in other studies (Habicht & Canfield 1996) of evidence to identify the sulphuretum cycle may be in part a consequence of metamorphic isotopic homogenization, or of the difficulty caused by mechanical differences between an Archaean microbial ecology, with thin but efficient mats, and latest Proterozoic and Phanerozoic bioturbated muds with eucarya, including metazoa, producing thicker zones of sea-bed biological redox management. It may be that only high-resolution isotopic studies can identify relicts of the Archaean mats: homogenization in analysing conventional large samples may make it difficult to see more highly fractionated clumps.

Fluxes of carbon, oxygen, sulphur and nitrogen may have been less than today (assuming that the land surface was barren except for cyanobacteria and swamp bacteria), but the isotopic ratios in the carbonates show that organic cycling of carbon was already on a scale to partner inorganic management of carbon isotopes. Net burial of organic carbon, relative to carbonate carbon, was in the same proportion as today, and keeping up with mantle degassing. This also implies that oxygen production, though perhaps much less than today, was nevertheless significant (although the standing crop of O_2 was low), and that the sulphur cycle was also established, as production of sulphate by oxidation would have occurred.

If N, O and C fluxes in the biosphere were all biologically managed then, excepting the noble gases, the late Archaean atmosphere, as today, was a biological construct.

We thank J. F. Wilson and T. Blenkinsop for their sustained interest, and the late Preston Cloud and J. Sutton for their support in the field. A. and C. Rauch of Manjeri Ranch and Zimasco Zimbabwe made the work possible. A. Knoll, M. Walter and N. Pace are thanked for helpful comments on an earlier draft, and D. Des Marais and an anonymous reviewer for very detailed and helpful comments on a later version. R. Raiswell and D. P. Kelly are thanked for advice. This work was supported at varying stages by the University of Zimbabwe, NSERC Canada, NERC-UK, and the Leverhulme Trust.

References

ABELL, P. I., MCCLORY, J., MARTIN, A. & NISBET, E. G. 1985. Archaean stromatolites from the Ngesi Group, Belingwe Greenstone Belt, Zimbabwe. Preservation and stable isotopes – preliminary results. *Precambrian Research*, **27**, 357–383.

ALIBERT, C. & MCCULLOCH, M. T. 1993. Rare earth element and neodymium isotopic compositions of the banded iron-formations and associated shales from Hamersley, Western Australia. *Geochimica et Cosmochimica Acta*, **57**, 187–204.

BICKLE, M. J. & NISBET, E. G. 1993. *The Geology of the Belingwe Greenstone Belt: a Study of the Evolution of Archaean Continental Crust*. Geological Society of Zimbabwe Special Publication, **2**.

BICKLE, M. J., MARTIN, A. & NISBET, E. G. 1975. Basaltic and peridotitic komatiites and stromatolites above a basal unconformity in the Belingwe greenstone belt, Rhodesia. *Earth and Planetary Science Letters*, **27**, 155–162.

BICKLE, M. J., NISBET, E. G. & MARTIN, A. 1994. Archaean greenstone belts are not oceanic crust. *Journal of Geology*, **102**, 121–138.

BLENKINSOP, T. G., FEDO, C. M., BICKLE, M. J., ERIKSSON, K. A., MARTIN, A., NISBET, E. G. & WILSON, J. F. 1993. Ensialic origin for the Ngezi Group, Belingwe greenstone belt, Zimbabwe. *Geology*, **21**, 1135–1138.

BROCKS, J. J., LOGAN, G. A., BUICK, R. & SUMMONS, R. E. 1999. Archaean molecular fossils and the early rise of the eukaryotes. *Science*, **285**, 1033–1037.

CANFIELD, D. E. & TESKE, A. 1996. Late Proterozoic rise in atmospheric oxygen concentration inferred from phylogenetic and sulphur-isotope studies. *Nature*, **382**, 127–132.

CANFIELD, D. E., LYONS, T. W. & RAISWELL, R. 1996. A model for iron deposition to euxinic Black Sea sediments. *Earth and Planetary Science Letters*, **296**, 818–834.

COHEN, Y., JORGENSEN, B. B., PADAN, E. & SHILO, M. 1975. Sulphide-dependent anoxygenic photosynthesis in the cyanobacterium *Oscillatoria limnetica*. *Nature*, **257**, 489–491.

DREVER, J. I. 1974. Geochemical model for the origin of Precambrian banded iron formations. *Geological Society of America Bulletin*, **85**, 1099–1106.

FENCHEL, T. & BERNARD, C. 1995. Mats of colourless sulphur bacteria. I Major microbial processes. *Marine Ecology Progress Series*, **128**, 161–170

GOODWIN, A. M., MONSTER. J. & THODE, H. G. 1976. Carbon and sulphur isotope abundances in ironformations and early Precambrian life. *Economic Geology*, **71**, 870–891.

GRASSINEAU, N. V., MATTEY, D. P. & LOWRY, D. 2001a. Rapid sulphur isotopic analyses of sulphide and sulphate minerals by continuous flow-isotope ratio mass spectrometry (CF-IRMS) *Analytical Chemistry*, **73**(2), 220–225.

GRASSINEAU, N. V., NISBET, E. G., BICKLE, M. J. & 5 OTHERS 2001b. Antiquity of the biological sulphur cycle: evidence from S and C isotopes in 2.7 Ga rocks of the Belingwe Belt, Zimbabwe. *Proceedings of the Royal Society of London, Series B*, **268**, 113–119.

GREAVES, M., ELDERFIELD, H. & WHITFIELD, M. 1989. The determination of the rare-earth elements in natural waters by isotope dilution mass spectrometry. *Analytica Chimica Acta*, **218**, 265–280.

HABICHT, K. & CANFIELD, D. E. 1996. Sulphur isotope fractionation in modern microbial mats and the evolution of the sulphur cycle. *Nature*, **382**, 342–343.

HAYES, J. M. 1994. Global methanotrophy at the Archaean–Proterozoic transition. *In*: BENGTSON, S. (ed.) *Early Life on Earth. Nobel Symposium 84*. Columbia University Press, New York, 220–236.

HAYES, J. M, KAPLAN, I. R. & WEDEKING, K. W. 1983. Precambrian organic geochemistry, preservation of the record. *In*: SCHOPF, W. (ed.) *Earth's Earliest Biosphere*. Princeton University Press, Princeton, NJ, 93–158.

HOLLAND, H. D. 1984. *The Chemistry of the Atmosphere and Oceans*. Princeton University Press. Princeton, NJ.

HUNTER, M. A. H., BICKLE, M. J., NISBET, E. G., MARTIN, A. & CHAPMAN, H. J. 1998. A continental extension setting for the Archaean Belingwe greenstone belt, Zimbabwe. *Geology*, **26**, 883–886.

JASPER, J. P. & GAGOSIAN, R. B. 1993. The relationship between sedimentary organic carbon isotopic composition and organic biomarker compound concentration. *Geochimica et Cosmochimica Acta*, **57**, 167–186.

KAKEGAWA, T., KAWAI, H. & OHMOTO, H. 1998. Origins of pyrite in the ~2.5 Ga Mt. McRae shale, the Hamersley district, Western Australia. *Geochimica et Cosmochimica Acta*, **62**, 3205–3220.

KUSKY, T. M. & KIDD, W. S. F. 1992. Remnants of an Archaean oceanic plateau, Belingwe Greenstone Belt, Zimbabwe. *Geology*, **20**, 43–46.

KUSKY, T. M. & WINSKY, P. A. 1995. Structural relationships along a greenstone/shallow water shelf contact, Belingwe Greenstone Belt, Zimbabwe. *Tectonics*, **14**, 448–471.

MARTIN, A. 1978. *The Geology of the Belingwe–Shabani Schist Belt*. Rhodesia Geological Survey Bulletin, **83**.

MARTIN, A., NISBET, E G. & BICKLE, M. J. 1980. Archaean stromatolites of the Belingwe Greenstone Belt. *Precambrian Research*, **13**, 337–362.

MOLITOR, M., DAHL, C., MOLITOR, I. & 5 OTHERS 1998. A dissimilatory siro-haem-sulfite-reductase-type protein from the hyperthermophile archaeon. *Pyrobaculum islandicum*. *Microbiology*, **144**, 529–541.

NISBET, E. G. & FOWLER, C. M. R. 1999. Archaean metabolic evolution of microbial mats, *Proceedings of the Royal Society of London, Series B*, **266**, 2375–2382.

NISBET, E. G. & SLEEP., N. H. 2001. The habitat and nature of early life. *Nature*, **409**, 1083–1091.

NISBET, E. G. & SLEEP, N. H. 2002. The early Earth: the physical setting for life. *In*: ROTHSCHILD, L. & LISTER, A. (eds) *Evolution on Planet Earth: the Impact of the Physical Environment*. Academic Press–Linnean Society, London, in press.

NISBET, E. G., ARNDT, N. T., BICKLE, M. J. & 8 OTHERS 1987. Uniquely fresh 2.7 Ga old komatiites from the Belingwe greenstone belt, Zimbabwe. *Geology*, **15**, 1147–1150.

NISBET, E. G., MARTIN, A., BICKLE, M. J. & ORPEN, J. L. 1993. The Ngezi Group. Komatiites, basalts and stromatolites on continental crust. *In*: BICKLE, M. J. & NISBET, E. G. (eds) *The Geology of the Belingwe, Greenstone Belt*. Geological Society of Zimbabwe Special Pubublication, **2**, 121–166.

OHMOTO, H. 1992. Biogeochemistry of sulfur and the mechanisms of sulfur–sulfate mineralization in Archean oceans. *In*: SCHIDLOWSKI, M., GOLUBIC, S., KIMBERLEY, M. M., MCKIRDY, D. M. & TRUDINGER, P. A. (eds) *Early Organic Evolution: Implications for Mineral and Energy Resources*. Springer, Berlin, 378–397.

OHMOTO, H., KAKEGAWA, T. & LOWE, D. R. 1993. 3.4 billion-year-old biogenic pyrites from Barberton, South Africa: Sulfur isotope evidence. *Science*, **262**, 555–557.

PACE, N. R. 1997. A molecular view of microbial diversity and the biosphere. *Science*, **276**, 734–740.

PIERSON, B. K. 1994. The emergence, diversification and role of photosynthetic eubacteria. *In*: BENGTSON, S. (ed.) *Early Life on Earth. Nobel Symposium 84*. Columbia University Press, New York, 161–180.

RAISWELL, R. 1997. A geochemical framework for the application of stable sulphur isotopes to fossil pyritisation. *Journal of the Geological Society, London*, **154**, 343–356.

SCHAUDER, R. & KROGER, A. 1993. Bacterial sulphur respiration. *Archives of Microbiology*, **159**, 491–497.

SCHIDLOWSKI, M. & AHARON, P. 1992. Carbon cycle and carbon isotope record: geochemical impact of life over 3.8 Ga of Earth history. *In*: SCHIDLOWSKI, M., GOLUBIC, S., KIMBERLEY, M. M., MCKIRDY, D. M. & TRUDINGER, P. A. (eds)

Early Organic Evolution: Implications for Mineral and Energy Resources. Springer, Berlin, 147–175.

SHANKS, W. C., BOHLKE, J. K & SEAL, R. R. 1995. Stable isotopes in mid-ocean ridge hydrothermal systems: interaction between fluids, minerals and organisms. *In*: HUMPHRIS, S. E., LUPTON, J., MULLINEAUX, L. & ZIERENBERG, R. (eds) *Seafloor Hydrothermal Systems: Physical, Chemical and Biological Interactions.* Geophysical Monograph, American Geophysical Union, **91**, 194–221.

SHEN, Y., BUICK, R. & CANFIELD, D. E. 2001. Isotopic evidence for microbial sulphate reduction in the early Archaean era. *Nature*, **410**, 77–81.

STRAUSS, H. & MOORE, T. B. 1992. Abundances and isotopic compositions of carbon and sulfur species in whole rock and kerogen samples. *In*: SCHOPF, J. W. & KLEIN, C. (eds) *The Proterozoic Biosphere.* Cambridge University Press, Cambridge, 711–798

TAYLOR, S. R. & MCLENNAN, S. M. 1985. *The Continental Crust: its Composition and Evolution.* Blackwell, Oxford.

WATANABE, Y., NARAOKA, H., WRONKIEWICZ, D. J., CONDIE, K. C. & OHMOTO, H. 1997. Carbon, nitrogen and sulfur geochemistry of Archean and Proterozoic shales from the Kaapvaal craton, South Africa. *Geochimica et Cosmochimica Acta*, **61**, 3441–3460.

WOESE, C. R. 1987. Bacterial evolution. *Microbiological Reviews*, **51**, 221–271.

The metamorphic history of the Isua Greenstone Belt, West Greenland

HUGH ROLLINSON

GEMRU, University of Gloucestershire, Francis Close Hall Campus, Swindon Road, Cheltenham GL50 4AZ, UK (e-mail: hrollinson@chelt.ac.uk)
Present address: Department of Earth Sciences, Sultan Qaboos University, Oman

Abstract: New geological investigations in the $c.$ 3.7–3.8 Ga Isua Greenstone Belt in West Greenland have revealed that the belt comprises a number of separate structural domains. Five such domains have been identified on the basis of lithological and structural differences. This study uses the morphology and compositional zoning of garnet porphyroblasts in pelites to investigate the extent to which the various domains within the greenstone belt preserve contrasting deformational and metamorphic histories. Up to three episodes of garnet growth have been identified in a single domain and significant differences in garnet growth history are noted between domains. A distinction is drawn between the relatively simple metamorphic history of a low-strain zone in the NE of the greenstone belt and other domains where more complex histories are preserved. Combining this result with existing geochronology suggests that in the south and west, the greenstone belt was metamorphosed twice, at $c.$ 3.74 Ga and at $c.$ 2.8 Ga, whereas in the NE there was a single event at 3.69 Ga. Preliminary garnet-rim thermometry indicates that some rocks experienced an early metamorphism in which temperatures exceeded 610°C. Kyanite is thought to have been in equilibrium with these assemblages, indicating pressures of at least 6 kbar. A later prograde metamorphic event shows a temperature rise from 480 to 550°C. The high pressures indicate a crustal thickness of at least 20 km at 3.7 Ga.

The Isua Greenstone Belt in West Greenland (Fig. 1) is between 3.7 and 3.8 Ga old and is the oldest known sequence of well-preserved sedimentary and volcanic rocks (Appel *et al.* 1998). The Greenstone Belt is deformed into an arcuate shape about 40 km long and is intruded by the 3.65–3.8 Ga tonalites of the Amîtsoq gneisses (Nutman *et al.* 1996). The Greenstone Belt comprises a thick succession of basaltic pillow lavas and volcanic breccias interleaved with chert and banded iron formation. In addition, there are intrusive ultramafic rocks and minor layers of pelite, conglomerate and felsic igneous rocks of uncertain origin (Fedo *et al.* 2001; Myers 2001). There are no reports of rocks from the Isua succession that are unequivocally 'continental' in origin.

A very early age was established for the Isua Greenstone Belt by Moorbath *et al.* (1973), who calculated a Pb–Pb whole-rock isochron age for the Isua iron formation of 3760 ± 70 Ma. Subsequent studies using Sm–Nd, Rb–Sr and Pb–Pb whole-rock isochrons, and conventional and single-zircon U–Pb ages all confirm an age >3.7 Ga (Moorbath *et al.* 1977, 1986, 1997; Nutman *et al.* 1997; Frei *et al.* 1999) and some parts of the greenstone belt may be as old as 3.9 Ga (Frei & Rosing 2001).

The rocks of the Isua Greenstone Belt have been considerably deformed and are now metamorphosed to amphibolite facies. Typically, metabasaltic assemblages are amphibolites, ultramafic rocks are tremolite–chlorite schists and pelitic rocks are garnet–biotite schists. In places, there is evidence of extensive carbonate metasomatism, postdating the metamorphism (Rose *et al.* 1996; Rosing *et al.* 1996). However, despite the intense deformation and amphibolite-facies metamorphism, low-strain areas have been preserved that are free of metasomatic effects. Areas showing very well-preserved pillow lavas and volcanic breccias have recently been described by Appel *et al.* (1998) and give confidence that in places original lithologies and lithological relationships can still be recognized in this greenstone belt.

Two recent developments in our understanding of the evolution of the Isua Greenstone Belt set the context for this particular study. First, the remapping of the greenstone belt by Myers (2001) has shown that the original stratigraphy and synclinal structure proposed for the greenstone belt by Nutman (1986) is probably incorrect. Rather, the greenstone belt comprises a number of tectonic slices separated by faults, each of which may have had its own independent history before assembly (Fedo *et al.* 2001).

Second, it has been claimed that the rocks of the Isua Greenstone Belt preserve one of the earliest records of life on Earth. Carbon isotope

From: FOWLER, C. M. R., EBINGER, C. J. & HAWKESWORTH, C. J. (eds) *The Early Earth: Physical, Chemical and Biological Development.* Geological Society, London, Special Publications, **199**, 329–350. 0305-8719/02/$15.00
© The Geological Society of London 2002.

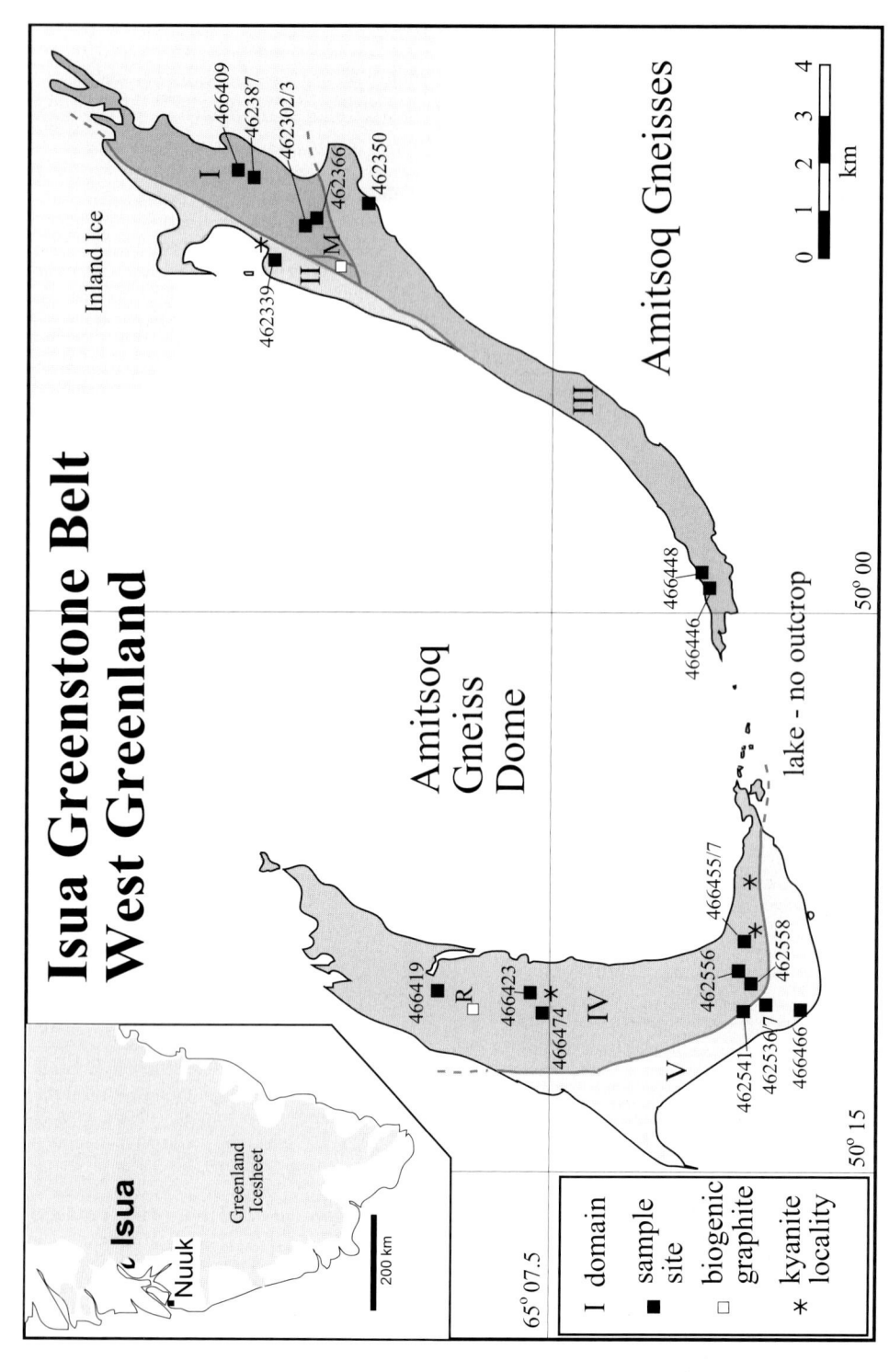

Fig. 1. Geological map of the Isua Greenstone Belt, showing the five structural domains identified in this study (I–V) and the location of the samples examined in this paper. Also shown are localities where kyanite has been recorded and the sites where biogenic graphic is reported (M, Mojzsis et al. 1996; R, Rosing 1999). The inset map shows the location of Isua in West Greenland.

studies initially carried out by Schidlowski (1988) suggested that some of the carbon in Isua sediments may be biogenic in origin. Two more recent studies tend to support this view (Mojzsis et al. 1996; Rosing 1999) and show that graphite from Isua sediments has $\delta^{13}C$ values in the range -20 to -35‰, values that are in the same range as biological debris from sediments 2.7–3.5 Ga old.

The purpose of this paper, therefore, is to present a detailed description of the metamorphic history of the Isua Greenstone Belt. The approach adopted is to establish a relative metamorphic chronology in each of five structural domains of the greenstone belt, using the metamorphic history preserved in garnet porphyroblasts. A particular objective is to identify areas in the Isua Greenstone Belt where the metamorphic imprint is weakest, to maximize the chances of finding primary (that is, pre-metamorphic) geochemical signatures in the sediments.

Previous investigations into the metamorphic history of the Isua Greenstone Belt

The first detailed and systematic study of the metamorphism of the Isua Greenstone Belt was made by Boak & Dymek (1982). These workers calculated the $P-T$ conditions of metamorphism from garnet–biotite pairs in pelites, using the Fe–Mg exchange thermometry of Ferry & Spear (1978). They also noted the presence of kyanite in a small number of samples and on this basis proposed temperatures and pressures for the main stage of metamorphism of c. 550°C, 5 kbar. This estimate was based upon garnet–biotite temperatures for mineral cores. Lower temperatures (c. 460°C) were obtained for garnet–biotite rims, which were thought to have equilibrated during subsequent retrogression. The timing of the metamorphism was not well constrained in the study by Boak & Dymek (1982), but they argued on geological grounds that it was in early Archaean time and occurred before 3600 Ma. These data were used as the basis for a model in which the early Archaean crust at Isua was thickened by the intrusion of the Amîtsoq gneiss tonalites.

A subsequent study of metamorphosed ultramafic rocks in the Isua succession (Dymek et al. 1988) extended the earlier work. Olivine–spinel pairs gave equilibration temperatures of $561 \pm 18°C$, indistinguishable from the results of garnet–biotite thermometry of $541 \pm 43°C$. Mineral reactions plotted in the $CMS-CO_2-H_2O$ system indicate high X_{CO_2} conditions. Subsequent retrogression was at $T < 450°C$ and at lower X_{CO_2}. One locality, however, contains serpentinites in which orthopyroxene overgrows olivine. Here the Al content of the orthopyroxene coexisting with spinel and olivine suggests equilibration temperatures of 650–700°C, indicating the possibility of an earlier, higher-temperature thermal event.

More recently, Rose et al. (1996) and Rosing et al. (1996) showed that the calcsilicate formations mapped by Nutman (1986) are metasomatized ultramafic rocks, and not metamorphosed carbonate-rich sediments as previously supposed. This work has demonstrated the importance of metasomatic processes within the greenstone belt and indicates that the relatively simple stratigraphy and structure proposed for the belt by Nutman (1986), is probably an oversimplification. Komiya et al. (1999), in a study of metabasic rocks from the northeastern part of the greenstone belt, reported a prograde Barrovian-style zonation with increasing metamorphic grade from greenschist facies in the NE to amphibolite facies in the SW part of the belt. No such pattern has been observed in this study.

Domains within the Isua Greenstone Belt

Previous studies of the Isua Greenstone Belt have recognized contrasting 'stratigraphies' in different parts of the belt, implying that it is an assembly of several distinct structural packages (Nutman 1986; Rosing et al. 1996). This idea has been strengthened through recent geochronological studies, which have shown that different parts of the greenstone belt contain primary zircons with crystallization ages that differ by 100 Ma (Nutman et al. 1997). New geological mapping in the greenstone belt (Appel et al. 1998; Myers 2001) has extended this view and identified a number of tectonic domains separated from one another by shear zones. Thus, in this study the Isua Greenstone Belt is subdivided into five tectonic domains (Domains I–V, Fig. 1) on the basis of lithological, structural, geochronological and geochemical differences. These are briefly described below.

Domain I, located in the NE of the greenstone belt was identified as a low-strain domain by Appel et al. (1998). This domain is dominated by mafic volcanic rocks, cherts and banded iron formation, and contains well-preserved, primary igneous and sedimentary features. Measured isotopic ages are between 3742 ± 49 Ma (Sm–Nd whole-rock, chloritic schists) and 3697 ± 70 Ma (Pb–Pb whole-rock, banded iron formation) (Moorbath & Kamber 1998; Frei et al, 1999). Moorbath & Kamber (1998) suggested that their Sm–Nd isochron may represent the depositional age of this part of the Isua succession, whereas Frei et al. (1999) proposed that their Pb–Pb isotope age determination records the timing of amphibolite-facies metamorphism.

Domain II was recognized by Nutman (1986) to be lithologically distinct from the rest of the greenstone belt. The recent geological mapping of Myers (2001) has shown that it is separated from Domain I by a major shear zone. The principal lithologies are pelites, amphibolites and felsic igneous rocks. The felsic igneous rocks have been interpreted as agglomerates but may be deformed tonalites intrusive into the greenstone belt (Rosing et al. 1996). Zircon from these rocks was dated by Nutman et al. (1997) at 3710 ± 4 Ma. Of importance to this study of metamorphism at Isua is the report of kyanite from this part of the greenstone belt (Boak & Dymek 1982). This domain is also where Dymek et al. (1988) recorded anomalously high metamorphic temperatures in ultramafic rocks.

Domain III, the southern part of the eastern limb of the greenstone belt, is the narrowest part of the belt. Observations made during fieldwork for this study indicate that primary igneous features are not preserved in this domain, suggesting that it is more intensely deformed than other domains in the greenstone belt. This unit contains amphibolites, ultramafic schists and pelites, and is intruded by tonalite sheets. Some rocks within this unit are extensively altered to carbonate. Zircons in carbonated felsic igneous rocks have an age of 3806 ± 4 Ma (Nutman et al. 1997), but a plagioclase–hornblende pair from an amphibolite yielded an Sm–Nd mineral isochron of 2849 ± 116 Ma (Grau et al. 1996), implying a late Archaean metamorphic event.

Domain IV is in the eastern part of the western limb of the greenstone belt and is dominated by a thick sequence of metamorphosed mafic pillow lavas, traditionally referred to as the Garbenschiefer Unit. There are also smaller amounts of pelite and iron formation. The domain is generally highly deformed although isolated low-strain lacunae are present. Kyanite has been identified in this domain both in this study and by previous workers (Nutman et al. 1997). Isotopic studies suggest that both late Archaean and early Archaean metamorphic events are recorded in this domain. An Sm–Nd whole-rock isochron for pelites and amphibolites from the Garbenschiefer Unit yielded an age of 3779 ± 81 Ma (Rosing 1999), and zircon ages of 3707 ± 5 Ma and 3711 ± 6 Ma were obtained on iron formation and kyanite schist, respectively (Nutman et al. 1997). Late Archaean ages have also been obtained from this domain on carbonate rocks (2742 ± 25 Ma, Sm–Nd whole-rock; Shimizu et al. 1990) and on magnetite from iron formation (2840 ± 49 Ma, Pb–Pb, step leaching; Frei et al. 1999). Kyanite from this domain yielded a Pb–Pb step-leaching age of 2847 ± 26 Ma (Rosing & Frei 1999), although this is now thought to reflect the age of allanite inclusions from within the kyanite. It has been suggested that the kyanite is of early Archaean age, but that the Pb-isotopic composition of the allanite inclusions was reset at c. 2.85 Ga (Frei et al. 2002).

Domain V, in the SW of the belt, is thought to be the most ancient part of the greenstone belt. This domain was described by Frei & Rosing (2001) and Frei et al. (2002) as the 'amphibolite unit'. Evidence for the great antiquity of this domain is two-fold. Nutman et al. (1997) described a 'quartzite' unit from this area with zircons in the age range 3800–3900 Ma. More recently, Frei & Rosing (2001) have presented Pb isotope evidence for a 3.9 Ga age for a pillow lava unit in this domain. The minimum age of the domain is 3800 Ma and is given by U–Pb zircon ages and Pb isotopic analyses of galena from an intrusive tonalite sheet (Nutman et al. 1997; Frei & Rosing 2001). Domain V is separated from Domain IV by a shear zone (Myers 2001) containing mylonitized ultramafic rocks, pelites and felsic igneous rocks. Most rock types in Domain V are mafic and ultramafic igneous rocks. Polat (pers. comm. 2000) has argued that the trace element chemistry of these metabasalts is significantly different from that of metabasalts elsewhere in the belt. There is also a small amount of pelitic material present. Felsic igneous rocks, previously thought to be volcanic in origin, are now regarded as intrusive sheets of tonalite from the enclosing Amîtsoq gneisses (Myers 2001). In many places the lithological assemblage in Domain V has been extensively metasomatized and several fluid-infiltration events have been recognized. Rosing et al. (1996) drew attention to an important carbonate metasomatic event, and more recently Frei et al. (2002) proposed that the 3740 Ma metamorphism in this domain was closely associated with a fluid infiltration event in which fluids rich in light rare earth elements (LREE), Th and U were emplaced.

A total of 29 samples from 17 sites were studied from the five structural domains. The sample numbering is that of the Geological Survey of Denmark and Greenland. The mineral assemblages for each sample and their global positioning system (GPS) locations (datum WGS 84) are presented in the Appendix. The emphasis in this study is on quartz–biotite–garnet schists. These samples are thought to be metamorphosed argillaceous sediments and are termed pelites in this study. Some pelites contain hornblende and a few contain very calcic plagioclase ($>An_{90}$), reflecting the Ca-rich, Na-poor composition of their source. Some Isua 'pelites' therefore have a more magnesium- and calcium-rich bulk composition than pelites derived from typical continental crust.

Description of garnet morphology and zoning from each domain

The primary aim of this paper is to identify differences in metamorphic history between the domains of the Isua Greenstone Belt using the growth chronology preserved in garnet porphyroblasts in pelites. Central to this approach is the identification of different garnet generations in each of the domains. Separate generations of garnet growth have been identified using the following criteria.

(1) Garnet morphology and optical zoning. This approach utilizes the distribution of inclusions within garnet and the shape and size of the garnet grains.

(2) Compositional zoning in garnet. Here compositional zoning is expressed as changes in the mole fraction of the divalent cations (X_{Fe}, X_{Ca}, X_{Mn}, X_{Mg}) and in the Fe/(Fe^{2+} + Mg) ratio (the fe-number). This latter parameter is a useful monitor of the temperature of equilibration and increases with falling temperature (Spear 1993). The chemical data are presented as compositional profiles across individual grains and are shown on graphs of composition v. distance. Most garnets examined in this study are rich in the almandine component, and Fe^{3+} was estimated using the charge balance equation of Droop (1987).

(3) Plots of the mole fractions of Ca and Mn against the fe-number for individual garnets, or for individual rocks. Here bivariate plots are presented for $\log(X_{Ca})$ v. log(fe-number) and $\log(X_{Mn})$ v. log(fe-number). Garnets with a simple metamorphic growth history tend to show linear arrays on such plots, even if the data show curved or irregular composition–space profiles. In contrast, garnets in which there are discordant overgrowths show either a break in the trend, or a subparallel trend. Given that the fe-number is a simple function of temperature, and that P–T conditions depend upon the log of the equilibrium constant, itself made up of X terms, systematic differences on log(X) plots will reflect differing P–T conditions during garnet growth.

A particular feature encountered in this study was the abundance of elliptical garnets. Although these are well known in metamorphic geology, their origin is something of a puzzle. Experimental studies reveal that garnet does not deform by crystal plasticity or dislocation creep below about 900°C (Ji & Martingnole 1994, 1996), temperatures well above those attained during the metamorphism of the Isua Greenstone Belt. This means that garnets with an elliptical form have attained their present shape through (1) preferential growth within the plane of the foliation (Sakai et al. 1985); this may be assisted by the multiple nucleation in garnets as described by Daniel & Spear (1998); (2) grain boundary diffusional creep (e.g. preferential dissolution normal to the plane of the foliation) (Den Brok & Kruhl 1996; Azor et al. 1997); (3) fracture-controlled dismemberment of initially equant grains (Gregg 1978); (4) in the case of poikilitic grains containing abundant quartz inclusions, cataclastic particulate flow of the garnet component, followed by sintering. Where elliptical garnets are described below, their particular mode of formation will be discussed.

Domain I

Domain I contains regions where primary igneous and sedimentary features are well preserved, and it has been argued that this domain represents the least deformed part of the greenstone belt (Appel et al. 1998). The principal metamorphic features of Domain I are illustrated here by two samples. Pelite 462303 (Fig. 2a) is the matrix from a micro-pillow lava and is described as a 'normal pelite' to distinguish it from a more Mn-rich pelite also examined. It has the mineral assemblage quartz–biotite–garnet–hornblende–plagioclase (An_3 and An_{21-27})–calcite–ilmenite. Pelite 462387 (Fig. 2b) is an Mn-rich pelitic layer from within the amphibolitic pillow breccia described by Appel et al. (2001) and has the mineral assemblage quartz–biotite–garnet–ilmenite, with late chlorite–calcite–muscovite–clinozoisite–allanite–tourmaline. Garnets are up to 5 mm in diameter and are subhedral to euhedral in form. Garnet growth in Domain I is thought to be restricted to a single generation. Only rarely is there evidence for a second garnet growth event.

Garnet morphology. Garnet in pelite 462303 (Fig. 2a) shows a sigmoid trail of quartz and ilmenite inclusions, in which the inclusions close to the margin of the grain are rotated into parallelism with the garnet grain boundary. In the manganiferous pelite 462387 the garnets have two forms. In biotite-rich layers, euhedral garnets are up to 6 mm across and grow across the tectonic foliation (Fig. 2b). These garnets tend to be relatively free of inclusions apart from a few rounded inclusions of quartz, which in some

Domain I

(a) 'normal' pelite

(b) manganiferous pelite

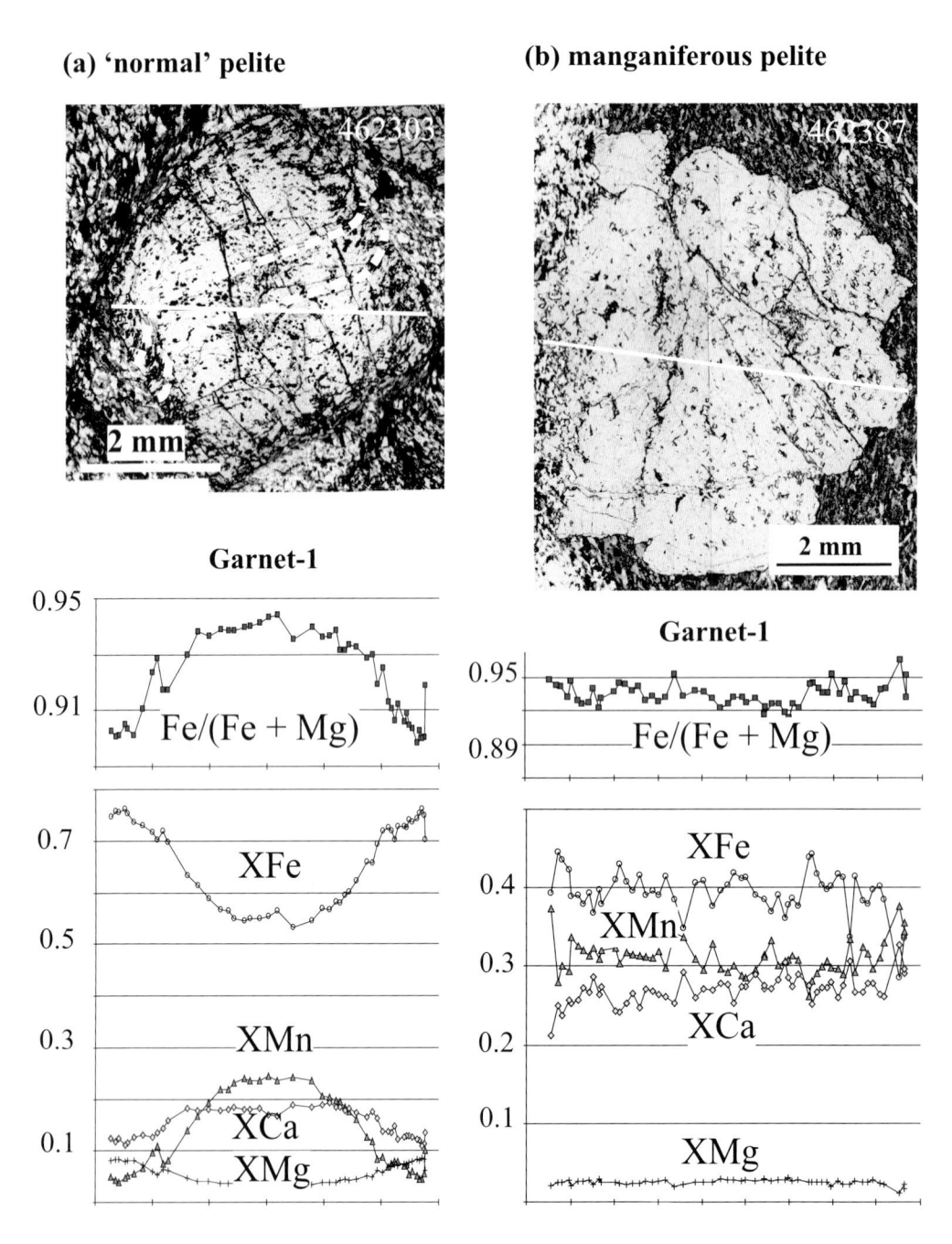

Fig. 2. Garnet zoning profiles from Domain I. (**a**) 'Normal' pelite 462303 shows a single generation of garnet growth and a sigmoidal inclusion trail. (**b**) The manganiferous pelite 462387 records a single generation of garnet growth.

cases define a relict tectonic foliation, and a few inclusions of calcite, biotite, allanite, and tourmaline. These grains also contain a rim of fine dusty inclusions about 50–100 µm from the grain boundary. The garnet rim is embayed by biotite, suggesting some resorption of the garnet at its margins. The more quartz-rich layers in this same sample contain smaller (0.5–3 mm in diameter), strongly poikilitic garnets, which contain abundant inclusions of quartz.

Compositional zoning. Garnet in pelite 462303 (Fig. 2a) shows a classical 'bell-shaped' compositional zoning profile for Mn. Cores are rich in Mn and Ca and depleted in Fe and Mg, and from core to rim there is a systematic increase in X_{Fe} (0.56–0.75) and X_{Mg} (0.05–0.09) and decrease in X_{Mn} (0.24–0.04) and X_{Ca} (0.18–0.10). This systematic change is frequently interpreted as 'growth zoning', and is explained in terms of Rayleigh fractionation of the divalent cations during mineral growth. It is thought that elements that show a strong preference for garnet (such as Mn) are preferentially concentrated in the mineral early in its growth, such that later garnet growth is from a reservoir that is Mn depleted. The change in fe-number from 0.935 at the core to 0.90 at the rim indicates that mineral growth took place during conditions of rising temperature (Spear 1993). There is a narrow outer zone, coincident with rotated inclusions in which the fe-number is relatively low and through which there is minimal compositional zoning.

Garnets from the more Mn-rich pelite 462387 show minimal compositional zoning. X_{Ca} is asymmetrically distributed across the grain. The high value of the fe-number (0.92–0.93 at the core and 0.93–0.95 at the rim) equates with garnet core compositions in other pelites from this domain. For this reason they are thought to have formed during the early stage of garnet growth.

Both types of garnet preserve examples of retrograde diffusional rims. In the case of the Mn-rich pelite this is coincident with the narrow rim separated from the main part of the grain by a surface of very fine inclusions.

Garnet mole fraction plots. Log(X_{Ca}) and log(X_{Mn}) plotted against log(fe-number) show a linear trend for pelites 462303 and 462387. The data for 462303 are illustrated in Figure 3 (Domain I), which shows that both the bell-shaped cores and flat margins of the compositional profiles lie on the same linear trend. This suggests that they represent a single episode of garnet growth. These observations imply that the rotation of inclusions at the margin of garnets in pelite 462303 (Fig. 2a) was synchronous with the main episode of garnet growth.

A pelite collected from the margin of Domain I (462366) shows inclusion-rich cores and relatively wide inclusion-free rims. These garnets show an inflection on a log(X_{Mn}) v. log(fe-number) plot, implying two stages of growth. This is the only clear evidence for a second generation of garnet in Domain I, but as this is found only in a sample close to the domain boundary, the growth of a second generation of garnet is not regarded as typical of this domain.

Domain II

Nutman (1986) was the first to suggest that the rocks of Domain II belong to a different 'lithological package' from the rest of the greenstone belt, a view that has been confirmed by recent mapping (Myers 2001). Boak & Dymek (1982) identified kyanite from this domain in quartz–plagioclase–muscovite–biotite schists. Garnet zonation in Domain II is illustrated from pelite 462339 (Fig. 4). This pelite contains the mineral assemblage quartz–biotite–garnet–grunerite–ilmenite, with later chlorite–calcite–tourmaline. The garnets are up to 10 mm in diameter, are strongly zoned and preserve two distinct generations of garnet growth. They are very different both in appearance and composition from those in Domain I.

Garnet morphology. There are two principal forms of garnet in Domain II. Large garnet grains have inclusion-rich cores and post-foliation, inclusion-free margins. Small garnets form inclusion-free post-foliation grains (Fig. 4). In the large grains, the inclusions in the cores are dominated by quartz, although in Fe-rich samples grunerite is also present. In some garnets, the quartz inclusions are oriented in parallel trails, and in the larger grains show a 'snowball garnet' texture. The inclusion trails probably represent a relict tectonic foliation, although now, even in the most garnet-rich samples, this tectonic banding cannot be traced from one garnet grain into another.

The rims of large garnets are commonly free of inclusions, although in one sample there are ilmenite inclusions in the rim in contrast to quartz inclusions in the core. In another sample there is a strong, late, chlorite schistosity developed, which wraps around the garnet grains. Here garnet rims are incomplete and are found only in 'pressure shadow' regions, where the schistosity does not wrap completely around the garnet grain. The incomplete development of

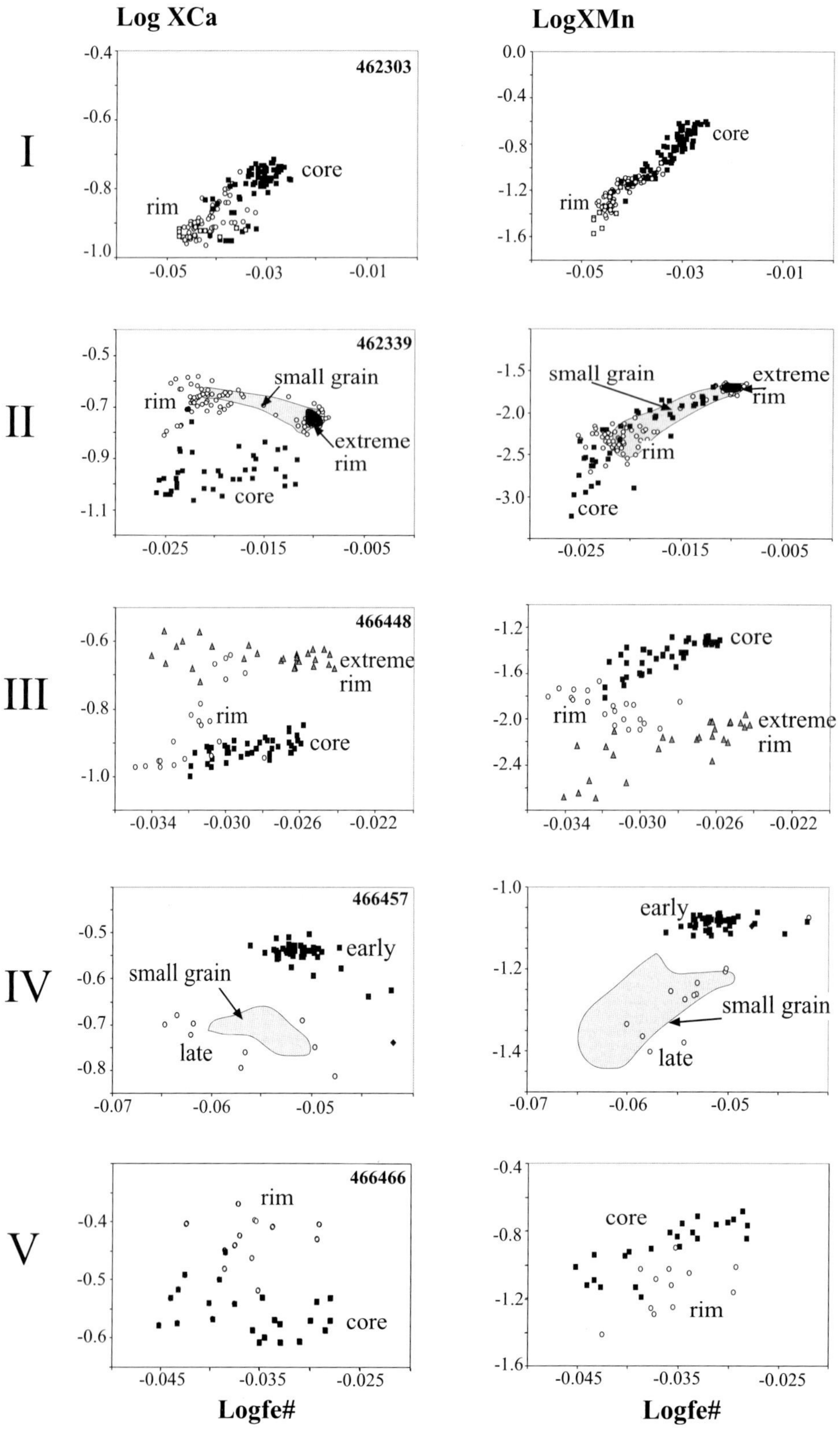

Log XCa
LogXMn
I
462303
core
rim
-0.4
-0.6
-0.8
-1.0
-0.05
-0.03
-0.01
0.0
-0.4
-0.8
-1.2
-1.6
II
462339
small grain
rim
extreme
rim
core
-0.5
-0.7
-0.9
-1.1
-0.025
-0.015
-0.005
-1.5
-2.0
-2.5
-3.0
III
466448
extreme
rim
rim
core
-0.6
-0.8
-1.0
-0.034
-0.030
-0.026
-0.022
-1.2
-1.6
-2.0
-2.4
IV
466457
early
small grain
late
-0.5
-0.6
-0.7
-0.8
-0.07
-0.06
-0.05
-1.0
-1.2
-1.4
V
466466
rim
core
-0.4
-0.5
-0.6
-0.045
-0.035
-0.025
-0.4
-0.8
-1.2
-1.6
Logfe#
Logfe#

the garnet rim gives these grains an elliptical form. In this case it is suggested that the elliptical form of the garnet is due to garnet dissolution associated with chlorite growth during later metamorphism.

Compositional zoning. A compositional profile across a large garnet grain in pelite 462339 (Fig. 4) shows that there are major chemical differences between core and rim. The inclusion-rich cores show a flat-topped bell-shaped profile, with decreasing X_{Fe} and increasing X_{Ca} relative to the central part of the core. In contrast, inclusion-free garnet rims show an increase in X_{Ca}, followed by a decrease in X_{Ca} and then a further increase relative to the margin of the inclusion-rich core. Fe shows the opposite trend. In the outermost part of the rim there is a narrow (c. 100 µm) zone, which is compositionally distinct from the rest of the rim, with high Fe and Mn, and low Mg and Ca. A backscattered-electron image of a small grain (Fig. 4b) shows that garnet with this composition is also found along fractures penetrating the grain.

Garnet mole fraction plots. Log(X_{Ca}) plotted against log(fe-number) shows two compositionally distinct fields for the inclusion-rich cores and the inclusion-free rims (Fig. 3, Domain II). This is less clear on a log(X_{Mn}) v. log(fe-number) plot, although the concentrations of Mn are extremely low. Small, inclusion-free grains plot in the same compositional field as the inclusion-free rims of larger grains, indicating that they represent post-tectonic grains that nucleated independently of the earlier garnet. The composition of the narrow outermost zone of garnet rims plots on the log(X_{Ca}) v. log(fe-number) diagram with the garnet rim compositions suggesting that it is part of this same growth stage in the garnet (Fig. 3, Domain II, extreme rim).

The compositional and morphological features of the garnets in Domain II suggest that there were two distinct episodes of garnet growth. Garnet-1 now forms the inclusion-rich cores of large grains. Garnet-2 forms post-tectonic rims on large garnets and smaller garnet grains which nucleated during Garnet-2 growth. A backscattered-electron image of one of the small garnets that nucleated during Garnet-2 growth shows three distinct compositions (Fig. 4b), indicating that in detail Garnet-2 growth was complex.

Domain III

The garnet chronology of Domain III in the southern part of the greenstone belt is based upon the detailed study of two samples collected about 50 m apart. Sample 466446 (Fig. 5a) is a calcic pelite and contains the mineral assemblage quartz + biotite + garnet + abundant plagioclase (An_{84-92}) + minor clinozoisite, muscovite, chlorite and ilmenite. This pelite has a weak fabric, which is overgrown by subhedral to euhedral garnets, up to 8 mm in diameter. Sample 466448 (Fig. 5b) is a layered Fe-rich pelite (quartz + biotite + garnet + hornblende + clinozoisite + plagioclase (An_{88-91}) + minor ilmenite, muscovite and calcite), with hornblende–biotite-rich and quartz-rich bands. Garnets vary in shape from strongly elliptical (2 mm × 20 mm), boudined grains, to rounded grains (12 mm across), with some euhedral facets. Two generations of garnet growth are recognized in calcic pelite 466446 and three generations in Fe-rich pelite 466448 (Fig. 5). There is geochronological evidence for early and late Archaean metamorphic events in this domain.

Garnet morphology. In Fe-rich pelite 466448 there are three main types of garnet.

(a) Strongly elliptical, boudined grains containing abundant ilmenite inclusions, very similar in form to elongate garnet fragments produced by shear displacement described by Gregg (1978).

(b) Elliptical grains, oriented parallel to the foliation, with euhedral facets (Fig. 5c). These grains are free from inclusions, apart from a few narrow, well-defined zones, rich in ilmenite inclusions, located near grain boundaries. The inclusion-rich zones mimic the euhedral facets of the grain and the distance between these zones is small in the direction normal to the banding but is much greater along the fabric, suggesting that garnet growth took place preferentially along the fabric.

(c) Composite grains with inclusion-rich cores and inclusion-free rims (Fig. 5b). Quartz and platy grains of ilmenite are the main inclusion types in the garnet cores. The orientation of the quartz inclusions is subparallel to that of the fabric of the matrix. Garnet rims contain a few larger, rounded (up to c. 200 µm) quartz inclusions and a few plagioclase inclusions (An_{92}).

In calcic pelite 466446 garnets have dark inclusion-rich cores and lighter-coloured rims

Fig. 3. Log(X_{Ca}) v. log(fe-number) and log(X_{Mn}) v. log(fe-number) plots for garnets in each of Domains I–V. ■, garnet cores; ○, garnet rims. In Domain III the outer rim of garnets (extreme rim) is shown as grey triangles. In Domains II and IV, small garnets with the same composition as the rims of large composite grains are shown by the grey shaded fields. Outer rims in Domain II are shown as the black shaded field (extreme rim).

338 H. ROLLINSON

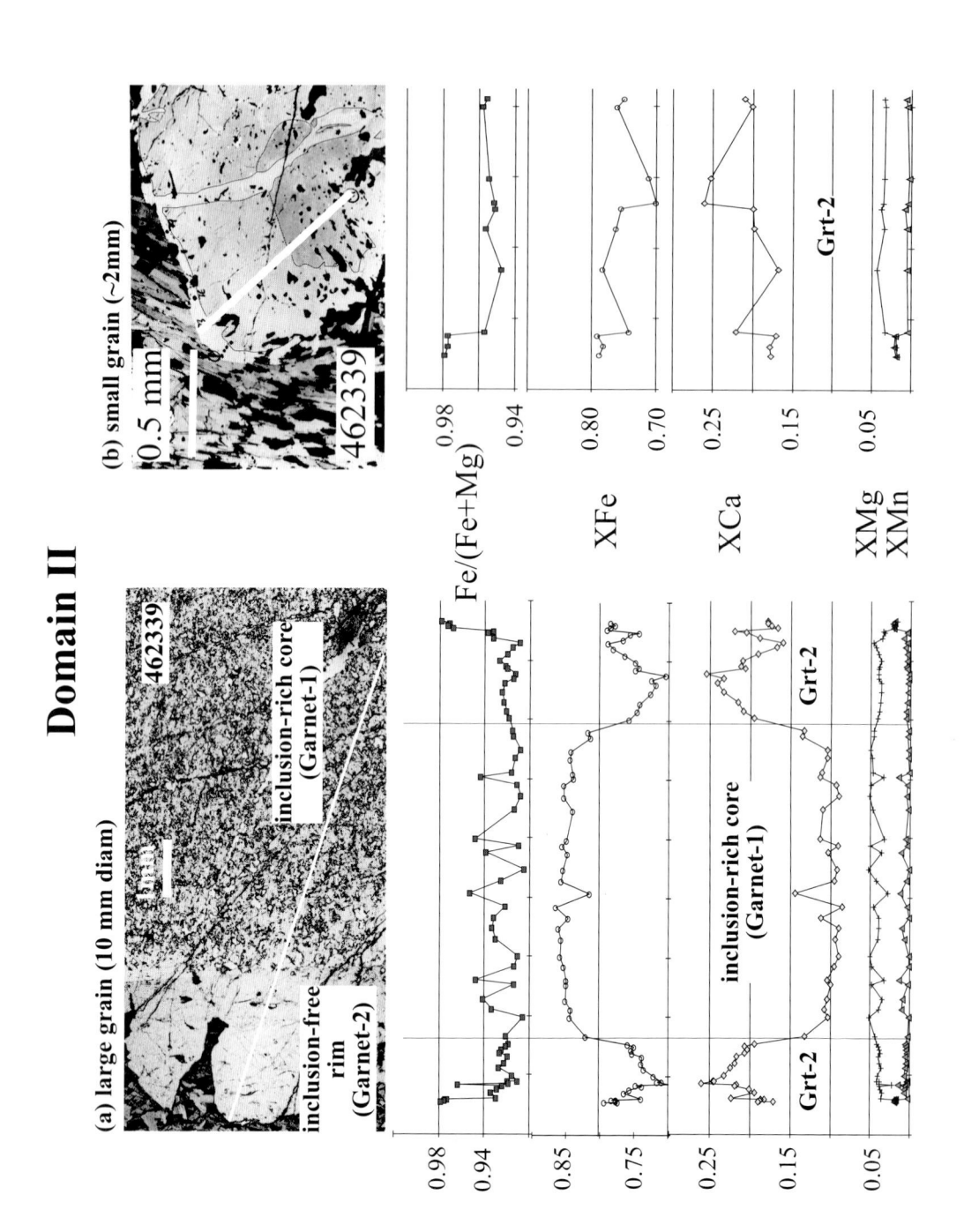

Domain II

(Fig. 5a). The cores are dominated by abundant, very fine-grained silicate inclusions ($c.\ 5\text{--}10\,\mu m$) and a few larger quartz and ilmenite inclusions.

Compositional zoning. A compositional profile across a zoned garnet in calcic pelite 466446 shows complex zoning in the inclusion-rich core and a smoother zoning pattern in the inclusion-free rim (Fig. 5a). This garnet is Ca poor and Fe rich in the centre of the core, but then shows oscillatory zoning with two Ca spikes. Rims show a smoother pattern of decreasing X_{Ca} and increasing X_{Fe} until the extreme rim, where X_{Fe} decreases and X_{Ca} increases, then decreases.

Zoning in Fe-rich pelite 466448 is less extreme but shows a pattern similar to that in the calcic pelite (Fig. 5b). The inclusion-rich core shows a progressive increase in X_{Fe} and decrease in X_{Mn} towards the rim. X_{Ca} also decreases from the centre to edge of the core, but with two small spikes. The inclusion-free rim shows a dramatic decrease in X_{Fe} and increase in X_{Ca}. Compositional zoning at the margin of an inclusion-free elliptical grain in 466448 (Fig. 5c), measured along the length of the grain parallel to the foliation, shows areas that are strongly enriched in Ca and depleted in Fe. This part of the grain also shows oscillatory zoning.

Garnet mole fraction plots. $\text{Log}(X_{Ca})$ and $\log(X_{Mn})$ v. log(fe-number) plots for garnets from Fe-rich pelite 466448 show three groupings of data points (Fig. 3, Domain III). Data points for inclusion-rich cores define a trend that lies at an angle to data points for the inclusion-free rims. Garnet compositions from the margin of an inclusion-free elliptical grain define a third trend (plotted in Fig. 3 (Domain III) as the extreme rim), subparallel to that of garnet cores. Similar plots (not shown) for calcic pelite 466446 define two compositional fields, which correspond to the inclusion-rich core and inclusion-free rims of garnet grains.

The compositional and morphological features of garnets from Domain III suggest that there are three generations of garnet preserved in Fe-rich pelite 466448 and two generations in 466446. Garnet-1 is represented by garnet type (a) (the boudined inclusion-rich garnets in 466448) and the inclusion-rich cores in both samples. Garnet-2 is represented by garnet type (b) (the elliptical inclusion-free garnets in 446448), inclusion-free rims in both samples and grains from which an inclusion-rich core is absent. Garnet-3 is geochemically distinctive (high X_{Ca}, low X_{Fe}) and forms a euhedral outer rim to elliptical Garnet-2 grains in 446448. Zonation preserved as mineral inclusion surfaces suggests that Garnet-3 grew preferentially along the tectonic banding.

Domain IV

Pelitic rocks in Domain IV are strongly banded on a decimetre scale. Field observations show that some bands contain euhedral garnets, whereas in adjacent zones the garnets are elliptical. This is seen particularly clearly in the southern part of this domain, where the most complete chronology of garnet growth in Domain IV is preserved. This chronology is described below and may be used to interpret garnets from other parts of Domain IV. Two samples from a single outcrop collected about 4 m apart show contrasting garnet morphologies. Fe-rich, biotite-rich pelite 466455 has the mineral assemblage quartz–biotite–garnet–plagioclase (An_{28} and An_{88-93}), with later chlorite, clinozoisite and sericite, and contains large garnets, up to about 8 mm in diameter growing across the matrix foliation. In contrast, sample 466457 is a more strongly deformed calcic pelite (quartz–biotite–garnet–hornblende–plagioclase (An_{58}), with later epidote–chlorite–tourmaline), which contains large elliptical garnets, up to about 10 mm in length. These grains are both aligned with and discordant to the matrix foliation and are flanked by pressure shadows of quartz and calcite. In detail, both types of garnet preserve complex internal morphologies and show compositional zoning, indicating three phases of garnet growth (Fig. 6). There is geochronological evidence for early and late Archaean metamorphic events in this domain.

Garnet morphology. Garnets in pelite 466455 possess an inclusion-rich core and have rims that contain fewer, larger inclusions and that have euhedral facets (Fig. 6a). The inclusion-free rims may be developed asymmetrically as in Figure 6a. Another grain in this sample has a zoned inclusion-bearing core. In the centre there is a sigmoidal trail of fine elongate quartz inclusions, surrounded in the outer part of the core by larger, more rounded quartz inclusions. The outermost part of the core is marked by large

---

Fig. 4. Garnet zoning profiles from Domain II. (**a**) Inclusion-free garnet rim (Garnet-2) and the inclusion-rich core (Garnet-1) of a large composite grain. (**b**) Backscattered-electron image of newly nucleated Garnet-2 in the biotite-rich rock fabric, showing compositional differences in Garnet-2.

Domain III

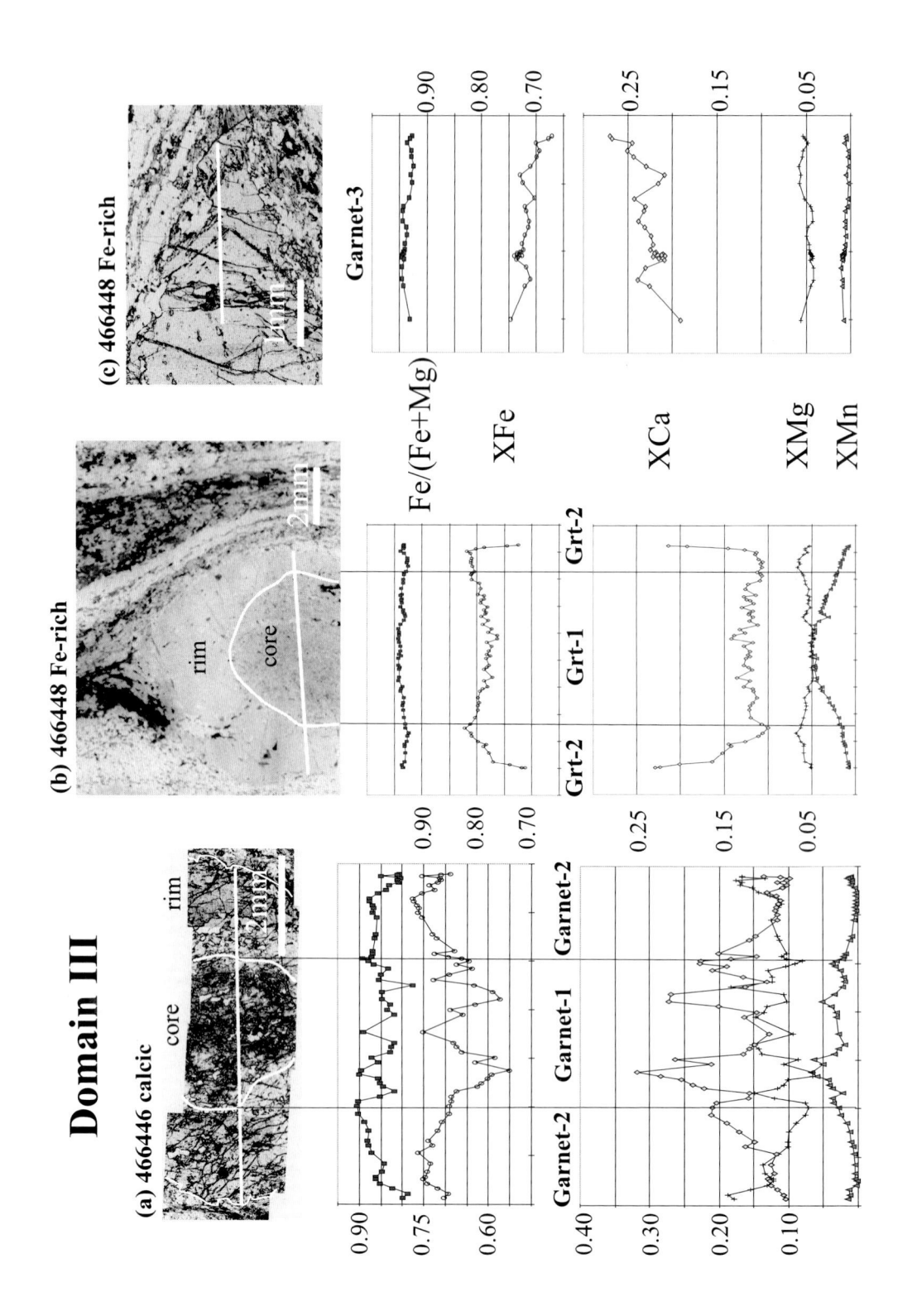

quartz inclusions and these represent the boundary between inclusion-bearing and inclusion-free garnet.

In pelite 466457 a large garnet comprises an inner zone that is relatively free of inclusions surrounded by a later generation of garnet with a crescent shape making an incomplete outer rim. This outer rim is strongly fractured and is partially altered to biotite. The innermost part of the inner zone contains abundant small quartz and calcite inclusions (Fig. 6b).

Compositional zoning. Garnet in pelite 466455 has an asymmetrical compositional profile. X_{Mn} shows a flat-topped, bell-shaped profile, coinciding with the inclusion-rich core, but has lower concentrations in the inclusion-free rim. X_{Ca} shows a similar, but more rounded profile and X_{Mg} and X_{Fe} show the converse, increasing in concentration from core to rim.

In the calcic pelite 466457, the inclusion-free zone has a broad, flat-topped profile, with concentrations rapidly changing over about 100 μm at the outer margin of this zone. The rim has elevated X_{Fe} and X_{Mg} and lower X_{Ca} and X_{Mn} relative to the inclusion-free zone.

In both samples, the presence of a flat-topped Mn profile may reflect the diffusional flattening of an initially bell-shaped compositional profile during high-temperature metamorphism. Volume diffusion of this type is thought to commence in the upper part of the amphibolite facies (Spear 1993).

Garnet mole fraction plots. Log(X_{Ca}) and log(X_{Mn}) v. log(fe-number) plots for garnets from calcic pelite 466457 show two groups of data points (Fig. 3, Domain IV). Garnet compositions for the inclusion-free zone have high Ca and Mn relative to the composition of the outer garnet rim. This is in contrast to Domains II and III, where early formed garnet is less calcic than later garnet growth. A similar bimodal distribution of data points is seen for garnet cores and rims in Fe-rich pelite 466455, although, in addition, the outermost 100 μm of garnet rims in 466455 have compositions that, in places, are distinctly more calcic than the inner part of the rim. This is thought to represent grain boundary diffusion.

These morphological and compositional data for garnets in Domain IV can be used to identify three generations of garnet. The earliest identified garnet is the inclusion-free zone in pelite 466457 (Fig. 6b), although it is possible that the inclusion-rich inner zone is an even earlier generation of garnet. Garnet-2 is represented by the Fe-rich outer rims to garnets in pelite 466457. Small, elliptical, poikilitic garnets, oriented parallel to the fabric in the rock matrix in the same sample, have the same composition as Garnet-2 (Fig. 3, Domain IV), indicating that some garnet nucleated newly during Garnet-2 growth. In pelite 466455 there are two generations of garnet: inclusion-rich cores, enriched in Mn and Ca, and inclusion-free rims with euhedral facets. The inclusion-rich cores are correlated with Garnet-2 in 466457 and the rims identified as Garnet-3. This correlation is made on the basis of the similarity in fe-numbers (Fig. 6) and their similar garnet–biotite K_D^{Fe-Mg}, a parameter that is independent of pelite bulk composition. Garnet-3 growth postdates the formation of the strong biotite fabric found in the rock matrix in these pelites.

Sample 466474 (not figured) is a staurolite–kyanite pelite from 5 km north of the outcrop described here. This pelite contains small (*c.* 1 mm) euhedral garnets, which are zoned. The garnet cores are rich in quartz inclusions, which form sigmoid inclusion trails. The rims are inclusion free, and have euhedral facets. There is a strong biotite fabric, which wraps around the staurolite and kyanite. In contrast, the garnet rims overgrow both the biotite fabric and the staurolite, suggesting that the inclusion-rich garnet cores, the staurolite and the kyanite are synchronous. Garnet cores in 466474 have fe-number between 0.87 and 0.88 and garnet–biotite K_D^{Fe-Mg} values between 0.12 and 0.15, rims have corresponding values of 0.85–0.90 and 0.11–0.16, respectively, and are thought to equate to the compositions of Garnets-1 and -2 in Domain IV, suggesting that the garnet–kyanite–staurolite assemblage was relatively early in the metamorphic history.

Domain V

Domain V is the most ancient part of the Isua Greenstone Belt and is located in the SW of the study area. Two samples have been investigated, from sites about 1 km apart. Sample

---

Fig. 5. Garnet zoning profiles from Domain III. (**a**) Composite grain in 466446 showing the difference in appearance and composition between Garnet-1 and Garnet-2. (**b**) Garnet porphyroblast from a highly strained Fe-rich pelite (466448) showing an inclusion-rich core (Garnet-1), and an inclusion-free rim (Garnet-2). (**c**) Zoned grain margin from an elongate garnet in Fe-rich pelite 466448. This grain does not have an inclusion-rich core and shows Garnet-3 growth parallel to the rock fabric.

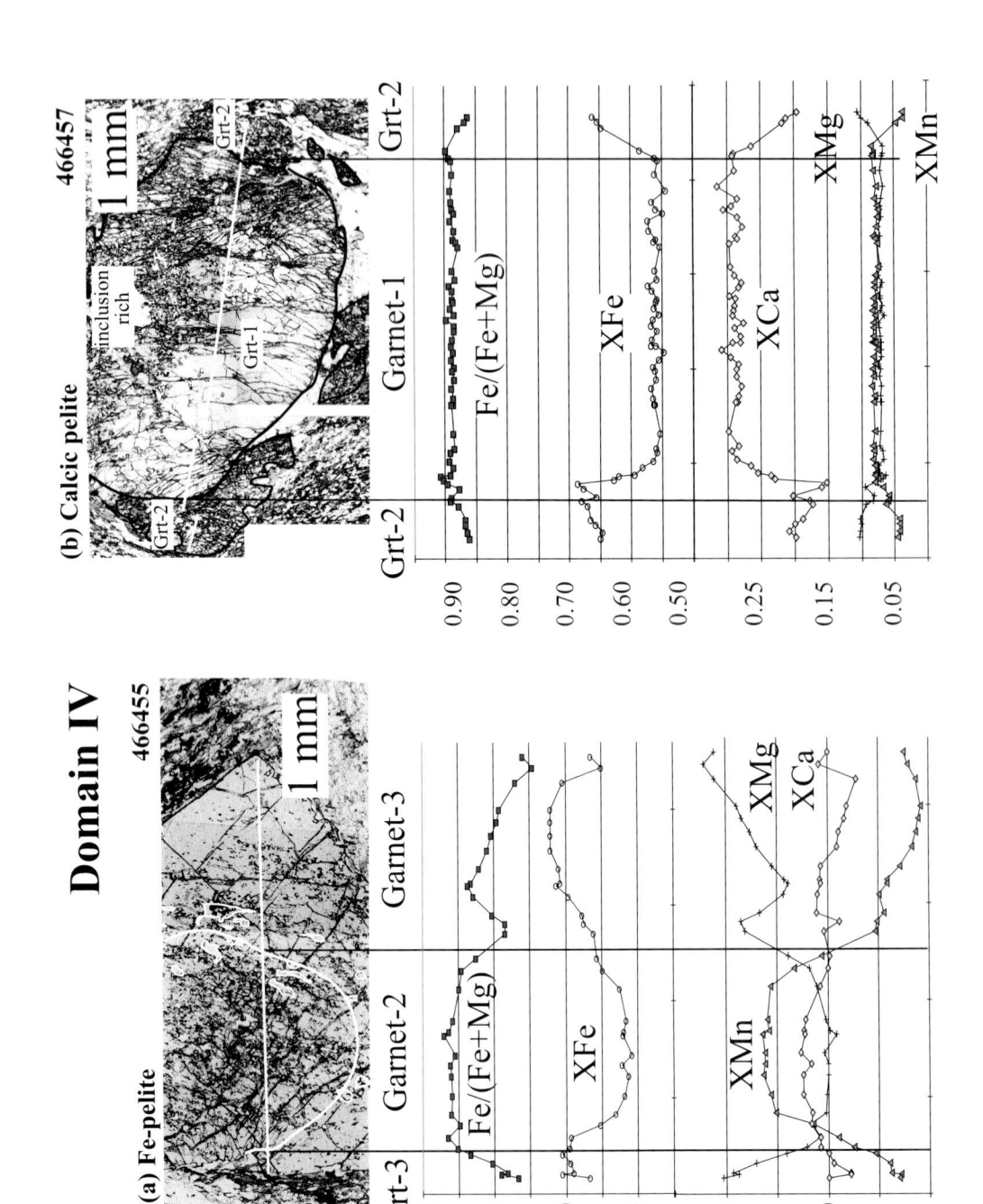

Domain IV

(a) Fe-pelite
466455

(b) Calcic pelite
466457

462536 is a relatively unstrained calcic pelite and is banded, with biotite- and hornblende-rich bands. Biotite-rich bands contain the mineral assemblage quartz–biotite–garnet–plagioclase (An$_{94-96}$) + minor hornblende, clinozoisite, muscovite, chlorite and tourmaline. Hornblende-rich bands contain the mineral assemblage quartz–garnet–plagioclase (An$_{93-97}$)–hornblende + minor clinozoisite, muscovite, chlorite, tourmaline and opaque minerals. Sample 466466 is a more highly strained calcic pelite with biotite-rich and hornblende-rich bands containing the mineral assemblage quartz–biotite–garnet–hornblende –plagioclase (An$_{92-97}$) with minor clinozoisite, calcite and opaque minerals.

Garnet morphology. Garnets in calcic pelite 462536 are similar in form in both the amphibole-rich and biotite-rich bands. Grains are up to 5 mm in diameter and are characterized by a broad inclusion-rich core and an inclusion-poor rim, with euhedral facets. The inclusions in the core are principally quartz and ilmenite and vary in size from a few micrometres to >100 μm. Smaller (c. 2 mm) euhedral garnets preserve only a small zone of inclusions in their core. The garnet depicted in Figure 7a is from the biotite-rich domain in calcic pelite 462536.

Garnets in calcic pelite 466466 are up to about 5 mm in diameter. Some are elliptical in form and are aligned parallel to the fabric in the rock matrix. The larger grains typically have cores rich in quartz inclusions and rims that have fewer inclusions. Unlike sample 462536, these rims do not have euhedral facets. The garnet shown in Figure 7b comprises an altered core, containing a few large quartz inclusions, and an asymmetrically developed rim, which contains smaller quartz inclusions. Part of the garnet rim and garnet interior are altered to calcite.

Pelite 462537 (not figured) contains staurolite. Large elliptical, poikilitic garnets contain abundant staurolite inclusions, suggesting that the garnet has grown as a result of a staurolite breakdown reaction. At garnet grain boundaries there are narrow, inclusion-free zones with euhedral facets.

Compositional zoning. Compositional profiles in both 462536 and 466466 are asymmetric. The altered domain in the core of the garnet in sample 466466 (Fig. 7b) has an Mn-rich bell-shaped profile surrounded by an asymmetric rim that is lower in Mn and richer in Ca. There is an outer zone, which is also developed asymmetrically and shows an initial increase in Mn followed by a decrease in Mn. This zone shows increasing Ca concentrations (Fig. 7b). Garnet in pelite 462536 is more Fe rich and Ca poor than in 466466. The grain depicted in Figure 7a shows similar Mn zoning to that found in the core of garnet in 466466, but the Ca zoning in 462536 is much more irregular than in 466466.

Garnet mole fraction plots. Log(X_{Ca}) and log(X_{Mn}) v. log(fe-number) plots for the garnet from pelite 466466 (Fig. 7b) define two slightly overlapping fields for the altered core and calcic rim (Fig. 3, Domain V). A similar pattern is found for the log(X_{Mn}) v. log(fe-number) plot, for garnet in 462536, although this is not the case for the log(X_{Ca}) v. log(fe-number) plot, on which the samples follow a linear trend.

Taken together these data define two generations of garnet growth. The similarity of the Mn-enriched bell-shaped zoning profile with a flat tail, found in the altered core of 466466 and in the garnets in 462536, suggests that these represent a distinct episode of garnet growth, designated Garnet-1. Garnet-2 is represented by the relatively inclusion-free, low-Mn rims.

The metamorphic history of the Isua Greenstone Belt

Synthesis of garnet zoning

A summary of the chronology of garnet zoning in each of the five domains of the Isua Greenstone Belt is given in Figure 8. The diagram shows the variation in fe-number across garnets from each of the five domains. In each case, garnet zoning profiles are shown and the relative garnet-growth chronology is identified. A few general conclusions are drawn from this synthesis of the metamorphic history of the greenstone belt as follows.

(1) The garnets in Domain I appear to be different from those in all other domains, inasmuch as they preserve a relatively simple growth history.

(2) The garnets in Domains III and IV appear to preserve the most complex history, recording three episodes of garnet growth. These domains both preserve well-documented evidence for

---

Fig. 6. Garnet zoning profiles from Domain IV. (**a**) Euhedral garnet in Fe-pelite 466455, showing inclusion-rich core (Garnet-2) separated by a zone of large quartz inclusions (outlined in white) from a relatively inclusion-free rim (Garnet-3). (**b**) Garnet porphyroblast from calcic pelite 466457 showing an early generation of garnet with an inclusion-bearing core and an inclusion-free rim (Garnet-1) and Garnet-2 growth on the rim.

Domain V

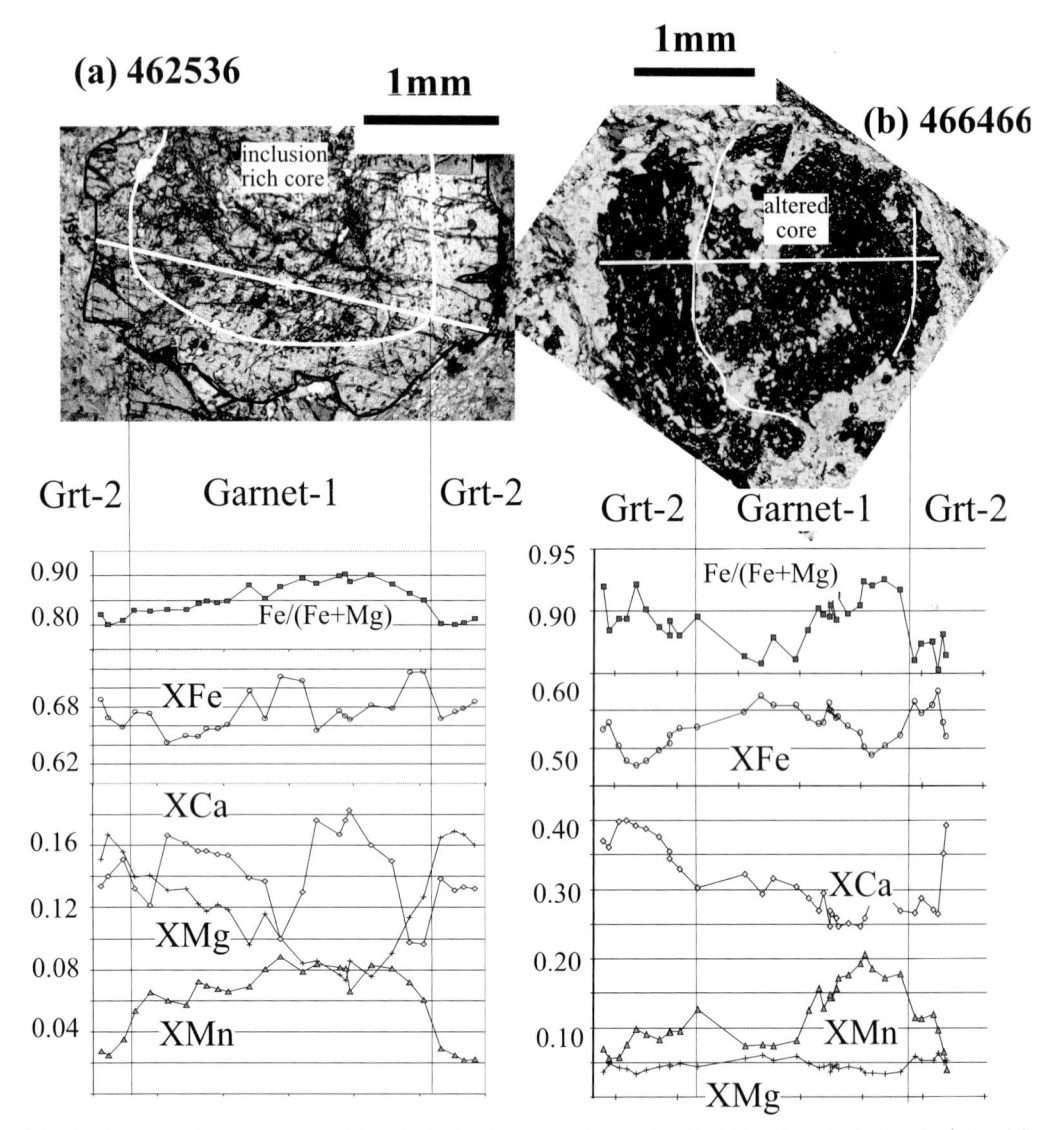

Fig. 7. Garnet zoning profiles from Domain V. (**a**) A subhedral garnet in 462536 with an inclusion-rich, Mn-rich core and a rim with fewer inclusions. (**b**) Deformed and altered garnet in 466466 showing Garnet-1 and Garnet-2 growth (garnet shown in crossed polars).

metamorphism in both early and late Archaean time.

(3) Garnets from Domains II, III, IV and V all contain a newly nucleated, later generation of garnet. In each case it corresponds to the growth of Garnet-2.

(4) Domains II and IV are the only areas where kyanite has been recorded to date. No other aluminosilicate phases are reported from the Isua Greenstone Belt.

Correlation and timing of the metamorphism within the five domains

Although it is not easy to make correlations between garnet growth histories across the domains,

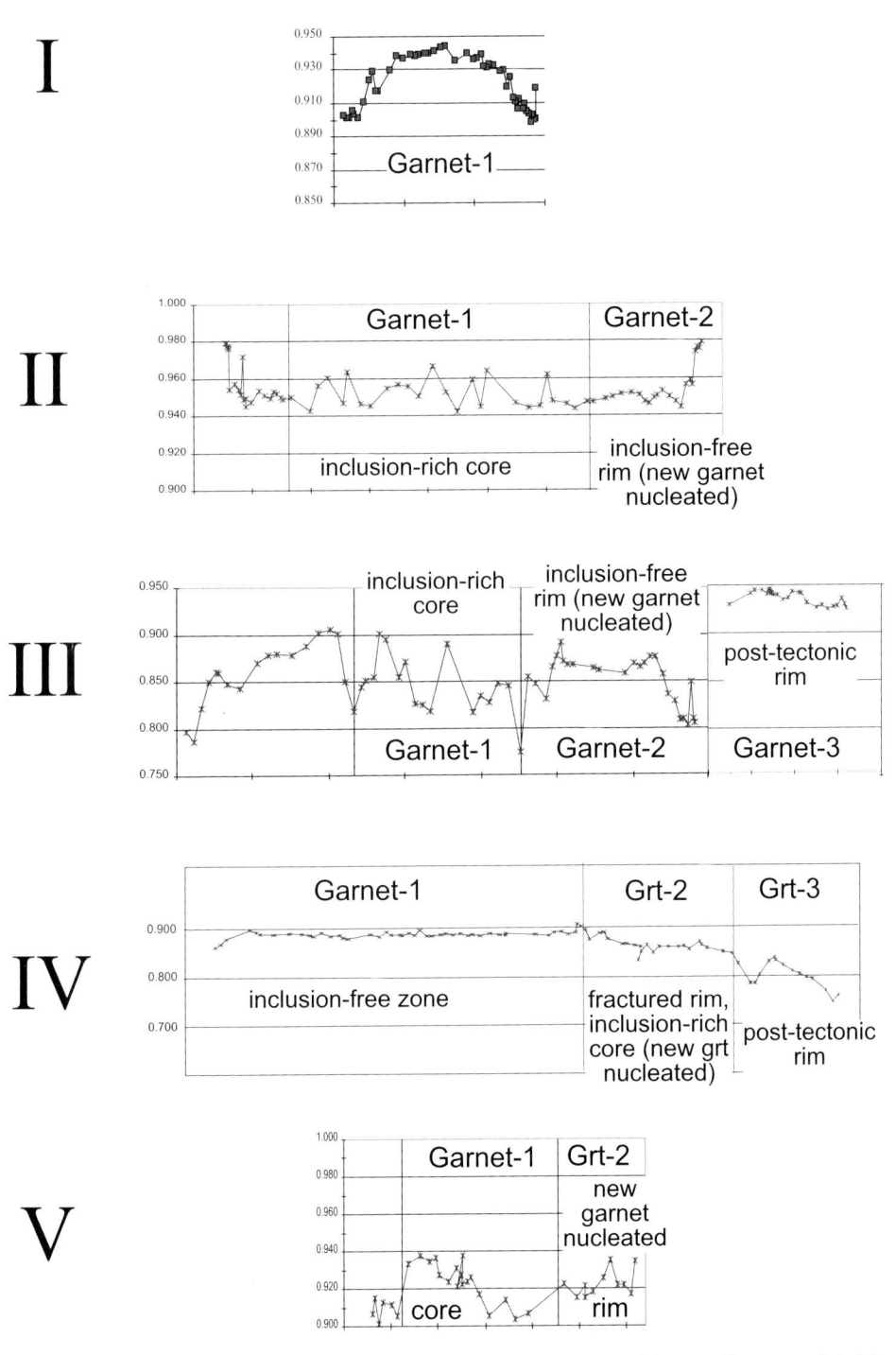

Fig. 8. A summary of garnet zoning profiles for the five structural domains of the Isua Greenstone Belt. The data are presented for fe-number, as an indicator of the thermal history of the grains, and show the generations of garnet in each domain, identified on the basis of their morphological, compositional and chemical-zonation differences.

it is possible that the episode of newly nucleated, late garnet identified in Domains II, III, IV and V is the same event. If this is the case, then Domains II–V preserve a common metamorphic history and record two episodes of garnet growth. These are thought to reflect two separate metamorphic events. Domains III and IV both preserve evidence for a later metamorphic event as indicated by the presence of a third generation of garnet. As both these domains preserve good mineral-isotopic evidence for a late Archaean metamorphic event at c. 2.8 Ga (Grau et al. 1996; Frei et al. 1999) it is possible that Garnet-3 grew during this late Archaean event. It is not clear, however, why this event is preferentially recorded in only these domains. The status of the single metamorphic event in Domain I is unclear, when compared with the relative chronology found in Domains II–V.

The recent isotopic work of Frei et al. (1999) and Frei & Rosing (2001) allows some absolute ages to be applied to the relative chronology established here. In Domain I, Frei et al. (1999) calculated a Pb–Pb, step-leaching age on magnetite from iron formation of 3691 ± 22 Ma, which they interpreted to be the age of amphibolite-facies metamorphism. This establishes an early Archaean age for the metamorphism in Domain I. In Domain V, Frei & Rosing calculated Pb–Pb step-leaching ages on each of the phases garnet, tourmaline and sphalerite. They concluded that these minerals grew at 3.74 Ga during a post-deformational hydrothermal–metasomatic event, which is indistinguishable from the earliest metamorphic overprint. Of particular interest is an age of 3739 ± 21 Ma, measured on a post-tectonic, calcic garnet in a metasomatized garnet amphibolite. The results of this present study show that garnets in Domain V are composite, suggesting that garnet in sample 4660046 analysed by Frei & Rosing (2001) comprises two separate garnet generations. Indeed, only five of the seven leachates are used to define the isochron. If this is the case, the precise timings of Garnet-1 and Garnet-2 growth cannot be determined from these data, although an early Archaean age for the metamorphism in Domain V is not in doubt.

On the basis of these correlations, the metamorphic history of the five domains of the Isua Greenstone Belt may be summarized as follows:

(1) Domains II–V record two early metamorphic events. Evidence from Domain V suggests that the average age of the two events is 3.74 Ga.

(2) Domains III and IV also record a late Archaean metamorphism, at c. 2.8 Ga.

(3) Domain I records a single, early Archaean metamorphic event at 3.69 Ga. At present it is not clear how this correlates with metamorphic events elsewhere in the greenstone belt.

Thermobarometry

A detailed discussion of the P–T evolution of each of the five domains of the Isua Greenstone Belt is beyond the scope of this study. However, the results of garnet–biotite thermometry in each of the five domains are presented in Table 1. Results are presented here using the calibration of Perchuk & Lavrent'eva (1983) as recommended in recent reviews of garnet–biotite Fe–Mg exchange thermometry (Chipera & Perkins 1988; Kleemann & Reinhardt 1994). For each sample the range of calculated results is reported. Average temperatures are reported as the mid-point of the modal class (20°C intervals). The garnet generations identified are the relative generations of garnet reported above.

In Domains II–V it has not been possible to differentiate between Garnet-1 and Garnet-2 biotite temperatures. However, in Domains III and IV the data suggest that temperatures recorded for Garnet-2 may be higher (c. 610°C) than those recorded for Garnet-3 (550–570°C). Higher temperatures for the earlier metamorphism are supported by preliminary hornblende–plagioclase thermometry (Holland & Blundy 1994) for early-grown hornblende in Domain III, which suggests temperatures in the range 636–731°C.

In Domain I early-formed, rapidly grown garnets in manganiferous pelite 462387 are thought to equate to the conditions of growth in the cores of more complex grains. Garnet–biotite temperatures in this sample are c. 470°C. Garnet rims in pelites 462302 and 462303, also from Domain I have garnet–biotite temperatures in the range 530–550°C, consistent with the prograde growth profile preserved in these grains.

Pressure estimates are more difficult to make, and are even more difficult to correlate between domains. Appel et al. (2001) calculated a pressure of c. 3.6–4 kbar for the assemblage quartz–biotite–garnet–muscovite–calcite in Domain I. A minimum pressure estimate may be made for Domain IV using the presence of kyanite. Taking the garnet–biotite temperature estimate of 610°C as a realistic lower limit, a minimum pressure of 6 kbar is calculated. Pressure estimates for both Domain I and Domain IV are for an early Archaean metamorphism, but at present it is not possible to say whether the pressure difference reflects different pressures in different parts of the greenstone belt during the same metamorphic event or two different metamorphic events.

Table 1. *Results of garnet–biotite thermometry (after Perchuk & Lavrent'eva 1983)*

Sample number	Garnet generation	Average temperature (°C)	Temperature range (°C)	Number of grt–bi pairs
Domain I				
462302 & -403	Grt-1	530	461–559	28
462387	Early Grt-1	470	463–533	12
466409	Grt-1	550	514–561	13
462366	?Grt-2	550	497–578	10
Domain II				
462339	Grt-2	530	490–534	20
	Early Grt-2	650	574–654	7
Domain III				
466446	Grt-2	610	582–637	5
466448	Grt-3	550	536–558	5
462350	?Grt-3	550	515–573	11
Domain IV				
466419	?Grt-3	530	480–585	7
466423	?Grt-3	550	517–556	9
466455	Grt-3	570	567–584	3
462556	Grt-3	570	563–590	3
462558	Grt-3	570	569–580	9
466457	Grt-2	610	590-623	12
Dopmain V				
466466	?Grt-2	590	539–590	4
462536	Grt-1	550	510–552	4
462537	?Grt-2	510	496–564	8

All measurements made on garnet–biotite pairs in mutual contact.

Geodynamic considerations

Minimum metamorphic pressures of 6 kbar imply a crustal thickness of at least 20 km during early Archaean time; a point made by Boak & Dymek (1982), who proposed that crustal thickening was due to the intrusion of the Amîtsoq gneisses. A similar approach has been taken by Frei & Rosing (2001), who correlated the 3.74 Ga metasomatism in Domain V with the intrusion of Amîtsoq gneiss tonalite sheets and implied that the associated metamorphic event is associated with an episode of crustal thickening. There is a difficulty with this model, for although older (3.8 Ga) Amîtsoq gneisses are known (Nutman *et al.* 1996), the isotopic evidence for the Amîtsoq gneisses as a whole suggests that they are younger than the greenstone belt and were emplaced at about 3.65 Ga (Kamber & Moorbath 1998). As the current geochronological evidence suggests that the early Archaean metamorphic event took place before 3.65 Ga, an alternative model for thickening the crust has to be found. One possibility, which is consistent with the fault-divided, multi-domain model for the Isua Greenstone belt used in this study, is that crustal thickening was through the stacking of a series of tectonic slices of greenstone belt material.

Implications for the preservation of evidence for early life

One of the principal reasons for the current research emphasis at Isua is that recent carbon isotope studies by Mojzsis *et al.* (1996) and Rosing (1999) appear to confirm the earlier claim by Schidlowski (1988) that there is evidence for early life preserved in the Isua succession. It is important, in the light of these claims, to understand the extent to which the carbon isotope signal might be modified during the metamorphism of the Isua sediments, for this allows a more rigorous evaluation of the carbon isotope data. Two observations may be made from the results presented here. First, the domain in which the carbon isotope signal is likely to be best preserved is that in which the metamorphic effect is least. The results presented here suggest that Domain I is the best place to look. The sample used by Mojzsis *et al.* (1996) was collected from the deformed margin of Domain I (Fig. 1).

Second, carbon isotope evidence for early life has also been reported from Akilia Island, 150 km SW of Isua (Mojzsis et al. 1996). This area is of considerable interest for it is alleged that these rocks are even older than the rocks of the Isua Greenstone Belt and thus preserve a record of the earliest life on Earth. There are, however, two problems with this claim. First, these rocks are metamorphosed to granulite grade (Mojzsis et al. 1996) and so their original carbon isotope signal may be highly disturbed. Second, their age is in dispute (Whitehouse et al. 1999; Nutman et al. 2001). What has been missed in this debate is that recent geochronological work has shown that the oldest part of the Isua Greenstone Belt (Domain V) may be as old as 3.9 Ga. This is not only older than the oldest age proposed for the Akilia Island rocks, but the metamorphic grade at Isua is also lower than that at Akilia. Therefore the possibility of finding the record of the oldest signature of life on Earth lies in Domain V of the Isua Greenstone Belt.

This is a contribution to the Isua Multidisciplinary Research Project (IMRP), supported by the Danish Natural Science Research Council, the Commission for Scientific Research in Greenland and the Minerals Office, Greenland Government. Funding was also provided through a Royal Society European Exchange Programme Grant and a grant from the Leverhulme Trust. The electron microprobe work was carried out at the University of Bristol, through a grant from the EU Geochemical Facility there, and S. Kearns is thanked for his assistance. S. Hanmer and David Waters are thanked for their careful reviews.

Appendix 1

Samples used in this study, listing the mineral assemblage, number of garnet grains analysed per sample and location

Sample number	Rock type	Mineral Assemblage	Garnets	GPS co-ordinates North	West
Domain I					
462302	Amphibolite + pelite	qz–bi–grt–hbl–plag–chlt–czt–cc–ilm	5	065, 10.197	049, 48.825
462303	Amphibolite + pelite	qz–bi–grt–hbl–An$_3$ & An$_{21-27}$–chlt–cc–ilm	3	065, 10.197	049, 48.825
462387	Pelite	qz–bi–grt–chlt–musc–czt–cc–tourm–ilm–allan	4	065, 10.769	049, 48.149
462328	Pelite	qz–grt–staur–An$_{23-30}$–chlt–musc–ilm	1	065, 10.451	049, 49.450
462305	Amphibolite	qz–bi–grt–hbl–grun–chlt–cc–mgt–sulph	1	065, 10.197	049, 48.825
462319	Amphibolite	qz–bi–grt–hbl–An$_{96-97}$–musc–czt–cc–ilm	2	065, 10.360	049, 49.160
466408	Amphibolite	qz–bi–grt–hbl–An$_4$–chlt–ilm	3	065, 10.827	049, 48.130
466409	Amphibolite	qz–bi–grt–hbl–An$_{16-21}$–chlt–ilm–sulph	2	065, 10.827	049, 48.130
462363	Matrix to conglomerate	qz–bi–grt–chlt–cc–mgt	2	065, 10.154	049, 49.327
462366	Pelite	qz–bi–grt–An$_{88-44}$–musc–czt–mgt	5	065, 09.800	049, 49.500
Domain II					
462333	Pelite	qz–bi–grt–chlt–ser–cc–tourm–ilm–ap–zn	2	065, 10.448	049, 49.668
462339	Pelite	qz–bi–grt–grun–chlt–cc–tourm–mgt–ilm	5	065, 10.470	049, 49.820
462340	Pelite	qz–bi–grt–hbl–An$_{23-52}$–chlt–musc–czt–cc–tourm–ilm	3	065, 10.470	049, 49.820
Domain III					
466448	Pelite	qz–bi–grt–hbl–An$_{88-91}$–musc–czt–cc–ilm	2	065, 06.104	049, 58.192
466446	Pelite	qz–bi–grt–An$_{84-92}$–chlt–opq	1	065, 06.080	049, 58.432
466437	Amphibolite	qz–bi–grt–hbl–An$_{26}$ & An$_{74-93}$–chlt–musc–cc–opq–sulph	3	065, 06.065	049, 56.704
462350	Pelite	qz–bi–grt–sphene	1	065, 09.800	049, 48.600
Domain IV					
462555	Amphibolite	qz–grt–hbl–ilm	2	065, 05.740	050, 09.769
462556	Pelite	qz–bi–grt–hbl–trem–plag–chlt–musc	2	065, 05.740	050, 09.769

462558	Pelite	qz–bi–grt–hbl–cc–ilm	2	065, 05.711	050, 09.869
466455	Pelite	qz–bi–grt–An$_{28}$ & An$_{88-93}$–chlt–czt–cc	2	065, 05.629	050, 08.963
466457	Pelite	qz–bi–grt–An$_{58}$–chlt–czt–cc	2	065, 05.629	050, 08.963
466419	Amphibolite	qz–bi–grt–hbl–chlt	2	065, 08.857	050, 10.013
466423	Pelite	qz–bi–grt–An$_{82-93}$–chlt–czt–ilm	3	065, 07.691	050, 10.385
466474	Pelite	qz–bi–grt–staur–ky–chlt–musc	2	065, 07.606	050, 10.951

Domain V

462536	Pelite	qz–bi–grt–hbl–An$_{93-97}$–chlt–musc–czt–tourm	2	065, 05.606	050, 10.649
462537	Pelite	qz–bi–grt–staur–An$_{87-94}$–musc–ilm	1	065, 05.606	050, 10.649
466466	Pelite	qz–bi–grt–hbl–An$_{92-97}$–czt–cc–opq	3	065, 05.268	050, 10.848

Western shear zone

| 462541 | Pelite | qz–bi–grt–plag–musc–cc–opq | 2 | 065, 05.758 | 050, 10.729 |

Qz, quartz; bi, biotite; grt, garnet; hbl, hornblende; trem, tremolite; grun, grunerite; ky, kyanite; staur, staurolite; plag, plagioclase; An$_{58}$, anorthite content of plagioclase; musc, muscovite; ser, sericite; cc, calcite; chlt, chlorite; czt, clinozoisite; ilm, ilmenite; mgt, magnetite; opq, opaque phase; sulph, sulphide mineral; tourm, tourmaline; allan, allanite; ap, apatite; zn, zircon.

References

APPEL, P. W. U., FEDO, C. M., MOORBATH, S. & MYERS, J. S. 1998. Well-preserved volcanic and sedimentary features from a low-strain domain in the ~3.7–3.8 Ga Isua Greenstone Belt, West Greenland. *Terra Nova*, **10**, 57–62.

APPEL, P. W. U., ROLLINSON, H. R. & TOURET, J. L. R. 2001. Remnants of an early Archaean (>3.75 Ga) sea-floor, hydrothermal system in the Isua Greenstone Belt. *Precambrian Research*, **112**, 27–49.

AZOR, A., SIMANCAS, F., EXPOSITO, I., LODEIRO, F. O. & POYATOS, D. J. M. 1997. Deformation of garnets in a low-grade shear zone. *Journal of Structural Geology*, **19**, 1137–1148.

BOAK, J. L. & DYMEK, R. F. 1982. Metamorphism of the ca. 3800 Ma supracrustal rocks of Isua, West Greenland: implications for early Archaean crustal evolution. *Earth and Planetary Science Letters*, **59**, 155–176.

CHIPERA, S. J. & PERKINS, D. 1988. Evaluation of biotite–garnet geothermometers: application to the English River subprovince, Ontario. *Contributions to Mineralogy and Petrology*, **98**, 40–48.

DANIEL, C. G. & SPEAR, F. S. 1998. Three-dimensional patterns of garnet nucleation and growth. *Geology*, **26**, 503–506.

DEN BROK, B. & KRUHL, J. H. 1996. Ductility of garnet as an indicator of extremely high temperature deformation: discussion. *Journal of Structural Geology*, **18**, 1367–1373.

DROOP, G. T. R. 1987. A general equation for estimating Fe^{3+} concentrations in ferromagnesian silicates and oxides from microprobe analyses using stoichiometric criteria. *Mineralogical Magazine*, **51**, 431–435.

DYMEK, R. F., BROTHERS, S. C. & SCHIFFRIES, C. M. 1988. Petrogenesis of ultramafic metamorphic rocks from the 3800 Ma Isua supracrustal belt, West Greenland. *Journal of Petrology* **29**, 1353–1397.

FEDO, C. M., MYERS, J. S. & APPEL, P. W. U. 2001. Depositional setting and palaeogeographic implications of earth's oldest supracrustal rocks, the >3.7 Ga Isua Greenstone Belt, west Greenland. *Sedimentary Geology*, **141/142**, 61–77.

FERRY, J. M. & SPEAR, F. S. 1978. Experimental calibration of the partitioning of Fe and Mg between biotite and garnet. *Contributions to Mineralogy and Petrology*, **66**, 113–117.

FREI, R. & ROSING, M. T. 2001. The least terrestrial leads; implications for the early Archaean crustal evolution and hydrothermal–metasomatic processes in the Isua supracrustal belt (west Greenland). *Chemical Geology*, **181**, 47–66.

FREI, R., BRIDGWATER, D., ROSING, M. & STECHER, O. 1999. Controversial Pb–Pb and Sm–Nd isotope results in the early Archaean Isua (west Greenland) oxide iron formation: preservation of primary signatures versus secondary disturbances. *Geochimica et Cosmochimica Acta*, **63**, 473–488.

FREI, R., ROSING, M. T., WAIGHT, T. E. & ULFBECK, D. G. 2002. Hydrothermal–metasomatic and tectono-metamorphic processes in the Isua supracrustal belt (west Greenland): a multiple-isotopic investigation of their effects on the Earth's oldest oceanic crustal sequence. *Geochimica et Cosmochimica Acta*, **66**, 467–480.

GRAU, G., ROSING, M., BRIDGWATER, D. & GILL, R. C. O. 1996. Resetting of Sm–Nd systematics during metamorphism of >3.7 Ga rocks: implications for isotopic models of early earth differentiation. *Chemical Geology*, **133**, 225–240.

GREGG, W. 1978. The production of tabular grain shapes in metamorphic rocks. *Tectonophysics*, **49**, 19–24.

HOLLAND, T. & BLUNDY, J. 1994. Non-ideal interactions in calcic amphiboles and their bearing on amphibole–plagioclase thermometry. *Contributions to Mineralogy and Petrology*, **116**, 433–447.

JI, S. & MARTIGNOLE, J. 1994. Ductility of garnet as an indicator of extremely high temperature deformation. *Journal of Structural Geology*, **16**, 985–996.

JI, S. & MARTIGNOLE, J. 1996 Ductility of garnet as an indicator of extremely high temperature deformation: reply. *Journal of Structural Geology*, **18**, 1375–1379.

KAMBER, B. S. & MOORBATH, S. 1998. Initial Pb of the Amîtsoq gneiss revisited: implications for the timing of early Archaean crustal evolution in west Greenland. *Chemical Geology*, **150**, 19–41.

KLEEMANN, U. & REINHARDT, J. 1994 Garnet–biotite thermometry revisited: the effect of Al^{VI} and Ti in biotite. *European Journal of Mineralogy*, **6**, 925–941.

KOMIYA, T., MARUYAMA, S., MASUDA, T., NOHADA, S., HAYASHI, M. & OKAMOTO, K. 1999. Plate tectonics at 3.8–3.7 Ga: field evidence from the Isua Accretionary Complex, southern west Greenland. *Journal of Geology*, **107**, 515–554.

MOJZSIS, S. J., ARRHENIUS, G., McKEEGAN, K. D., HARRISON, T. M., NUTMAN, A. P. & FRIEND, C. R. L. 1996. Evidence for life on Earth before 3800 million years ago. *Nature*, **384**, 55–59.

MOORBATH, S. & KAMBER, B. S. 1998. Re-appraisal of the age of the oldest water-lain sediments, West Greenland. *In*: CHELA-FLORES, J. & RAULIN, F. (eds) *Exobiology: Matter, Energy and Information in the Origin and Evolution of Life in the Universe.* Kluwer Academic, Dordrecht, 81–86.

MOORBATH, S., ALLAART, J. H., BRIDGWATER, D. & MCGREGOR, V. R. 1977. Rb–Sr ages of early Archaean supracrustal rocks and Amîtsoq gneisses at Isua. *Nature*, **270**, 43–45.

MOORBATH, S., O'NIONS, R. K. & PANKHURST, R. J. 1973. Early Archaean age for the Isua Iron Formation, West Greenland. *Nature*, **245**, 138–139.

MOORBATH, S., TAYLOR, P. N. & JONES, N. W. 1986. Dating the oldest terrestrial rocks – fact and fiction. *Chemical Geology*, **57**, 63–86.

MOORBATH, S., WHITEHOUSE, M. J. & KAMBER, B. S. 1997. Extreme Nd-isotope heterogeneity in the early Archaean – fact or fiction? Case histories from northern Canada and West Greenland. *Chemical Geology*, **135**, 213–231.

MYERS, J. 2001. Protoliths of the *ca*. 3.7–3.8 Ga Isua Greenstone Belt, West Greenland. *Precambrian Research*, **105**, 129–141.

NUTMAN, A. P. 1986. *The Early Archaean to Proterozoic History of the Isukasia Area, Southern West Greenland*. Grønlands Geologiske Undersøgelse Bulletin, **154**.

NUTMAN, A. P., BENNETT, V. C., FRIEND, C. R. L. & ROSING, M. T. 1997. ~3710 and >/=3790 Ma volcanic sequences in the Isua (Greenland) supracrustal belt: structural and Nd isotopic implications. *Chemical Geology*, **141**, 271–289.

NUTMAN, A. P., McGREGGOR, V. R., BENNETT, V. C., & FRIEND, C. R. L. 2001. Age significance of U–Th–Pb zircon data from early Archaean rocks of west Greenland – a reassessment based on combined ion-microprobe and imaging studies – comment. *Chemical Geology*, **175**, 191–199.

NUTMAN, A. P., McGREGGOR, V. R., FRIEND, C. R. L., BENNETT, V. C. & KINNY, P. D. 1996. The Itsaq Gneiss complex of southern West Greenland: the world's most extensive record of early crustal evolution (3900–3600 Ma). *Precambrian Research*, **78**, 1–39.

PERCHUK, L. L. & LAVRENT'EVA, I. V. 1983. Experimental investigation of exchange equilibria in the system Cordierite–Garnet–Biotite. *In*: SAXENA, S. K. (ed.) *Kinetics and Equilibrium in Mineral Reactions.* Springer, New York 199–240.

ROSE, N. M., ROSING, M. T. & BRIDGWATER, D. 1996. The origin of metacarbonate rocks in the Archaean Isua supracrustal belt, West Greenland. *American Journal of Science*, **296**, 1004–1044.

ROSING, M. T. 1999. 13C-depleted carbon microparticles in >3700-Ma sea-floor sedimentary rocks from west Greenland. *Science*, **283**, 674–676.

ROSING, M. T. & FREI, R. 1999. Late Archaean metasomatism and kyanite formation in the >3700 Ma Isua supracrustals, west Greenland. *Journal of Conference Abstracts*, **4**, 144.

ROSING, M. T., ROSE, N. M., BRIDGWATER, D. & THOMSEN, H. S. 1996. Earliest part of Earth's stratigraphic record: a reappraisal of the >3.7 Ga Isua (Greenland) supracrustal sequence. *Geology*, **24**, 43–46.

SAKAI, C., BANNO, S., TORIUMI, M. & HIGASHINO, T. 1985. Growth history of garnet in pelitic schists of the Sanbagawa metamorphic terrain in central Shikoku. *Lithos*, **18**, 81–95.

SCHIDLOWSKI, M. 1988. A 3800-million-year isotopic record of life from carbon in sedimentary rocks. *Nature*, **333**, 313–318.

SHIMIZU, H., UMEMOTO, N., MASUDA, A. & APPEL, P. W. U. 1990. Sources of iron-formations in the Archaean Isua and Marlene supracrustals, west Greenland: evidence from La–Ce and Sm–Nd isotopic data and REE abundances. *Geochimica et Cosmochimica Acta*, **54**, 1147–1154.

SPEAR, F. S. 1993. *Metamorphic Phase Equilibria and Pressure–Temperature–Time Paths*. Mineralogical Society of America Monograph, **1**.

WHITEHOUSE, M. J., KAMBER, B. S. & MOORBATH, S. 1999. Age significance of U–Th–Pb zircon data from early Archaean rocks of west Greenland – a reassessment based on combined ion-microprobe and imaging studies. *Chemical Geology*, **160**, 201–224.

Index

Note: Page numbers in *italic* type refer to illustrations; those in **bold** type refer to tables.

Abitibi
  mafic-felsic sequences 161
  Pb ratios 109
absolute plate motion, Western Superior Province 35
Acasta gneisses 155, 262
adakites 188
Adirondack Mountains 128
  V_P/V_S ratios *129*
advective thickening 22
aeolian dust, as ocean fertilizer 294
Akilia Island 348
alkali basalts 46, 66
Ameralik dykes 153
Amîtsoq Gneiss 329, 331, 332, 347
ammonia 276
Ancient Gneiss terrane 4
Andean margin 19
Andes, accretionary events 22
anisotropy
  shear-wave splitting 136, 144–147
  surface waves 31–32
  and velocity structure 53
  vertical variation 32
  Western Superior Province 27–44
anorthosite-mangerite-charnockite-granites 125
anorthosites
  Archaean 125
  evolution *131*
  mantle source 125–134
  Proterozoic 125
Aquaspirillum 324
Aquifecales 288
Ar/Ar ratios 214
arc accretion 162
arc magmatism, Zimbabwe Craton 195–199
arc volcanism, CO_2 source 236, 261
archaea
  methanogenesis 280
  symbiotic 293
Archaean, rock record 153–155
Archaean cratons
  basement-cover relationships 156–158
  formation 1–26
  lateral accretion 171
  lateral heterogeneity 164–165
  U/Pb ratios *118*
Archaean crust
  reworking 19
  thickness 19
Archaean provinces *107*
Archaean sea floor, microbial mats *293*
Archaean tectonics, review 151–181
Archaean-Proterozoic boundary 1
Archean ocean, model section *323*
Arcturus Formation 199, 201, 203
argon, atmospheric 214, 269–270, 280
asteroids, carbonaceous 224
asthenosphere, freezing 92
Atlantic, opening 40

atmosphere
  bathtub model 294–295, *298*
  chronological constraints 220
  controls on 294–297, *303*
  formation and evolution 213–229
  hydrogen-rich 220
  influence of mantle and crust 300–301
  inorganic controls 302
  modern 276–280
  organic controls 302–304
  vertical structure 279
  volatile recycling 219–220
atmosphere and ocean
  biology and temperature 299–300
  carbonates in 232
atmospheres, Earth and Venus *279*
atmospheric escape 220, 221
Australian Shield, waveform models 62
Avalon Formation 198
azimuthal anisotropy 30, 32, 53

Bababudan hills 163
bacteria
  denitrifying 276
  planktonic 286
  purple 293, 299
  sulphate reducing 321
bacteriochlorophylls 290
Baltic Shield 117
banded iron formations 158, 331
Barberton greenstone belt 98, 155
  platinum group elements 115
Barberton terrane 4
basalt
  carbonatized 242
  hydrothermal alteration 253
  spilitization 276, 284
basalt-eclogite transition 23
basaltic crust, melting 22–23
Beggiatoa 322
Belingwe Greenstone Belt 85, 106, 158
  age and structure 312
  continental rifting 191–195
  geochemistry 161
  life in 292–294
  map *310*
  sampling 312–313
  sedimentary rocks 190
  stable isotopes 309–328
Belingwean Supergroup 190
Bend Formation 191
Bengal Fan 270
biochemical pathways, microbial mats *289*
biofilms 288
biogeochemical processes, Archaean 285–291
biological control, speed of 295
biosphere
  Belingwe belt 292
  development of 297–299
  Walkerworld model 296
biotic weathering 234
Black Range dykes 153

Black Sea, environment 296, 299, 315, 325
body-wave tomography 58
boninites 5, 171
Bouguer gravity anomalies, cratonic crusts 19
Buhwa greenstone belt 190
Bulawayo greenstone belt 195, 198
buoyancy-driven processes 169
Burwash Formation 170
Bushveld event
  effect on craton keel 3
  emplacement 94
  low velocity anomaly 76
  mafic body 20
  as plume event 302
Bushveld event layered intrusion 77
Bushveld Province 5, 12
  geotherms 18
  seismic velocities 20

C4 metabolism 278
calc-alkaline volcanics, Zimbabwe Craton 191, 198
calcite compensation depth 270
Calvin cycle 278
Canadian Lithoprobe 29
Canadian National Seismograph Network 29
Canadian Shield
  high-velocity root 29
  lithospheric thickness 136
  surface-wave analysis 30–32
  velocity structure 29
Cape Fold Belt 83
carbon, recycling 218, 296, 325, 326
carbon dioxide
  cycling 231–257
  in present atmosphere 276–278
  pressures 270
  production 294
  reservoirs 251, 264
carbon dioxide drawdown 259–274
  capacity 271
  and chemical weathering 260–261
  and crustal recycling 270–272
carbon isotopes
  Belingwe data **317**, 318
  controls 277
  fractionation 320
  Isua 331, 347–348
  Jimmy Member 322
  in shallow-water sediments 320–321
  split 278
carbon-sulphide layers, Belingwe 292
carbonate cycles 232–233
  ancient 239–253
  Archaean 240–247
carbonate fluxes 252, 253
carbonate metamorphism 233
carbonate subduction 254
carbonate weathering 232, 233, 260
carbonates, reactions 232
catastrophic events 282, 302
Ce anomalies, and redox conditions 314
Central Slave Basement Complex 164
Central Slave Cover Group 159
Chemical Index of Alteration 260–261, 271

chemical weathering, and CO_2 drawdown 260–261
Cheshire Formation 191, 193, 195, 312
Chilimanzi Suite 166, 169, 204
Chinamora batholith 204
Chingezi Suite 191, 206
chloroplasts 293, 299
chondrites
  carbonaceous 217, 224, 226
  enstatite 223
chondritic veneer 218, 226
circum-cratonic areas 80
clastic basins 169, 170
climate, and carbon dioxide cycling 231–257
climate buffering 233
CO_2 *see* carbon dioxide
comets, composition 213, **223**
conglomerates
  late-kinematic 169–170
  Slave Craton *167*
continent denudation 270
continental area, and weathering rate 235
continental crust
  formation 91–103, *93*
  intermediate 21
continental flood basalts 106, 116–117, 161
continental freeboard 261
continental growth
  accretionary 4
  progressive 85
continental lithospheric mantle 83–86
continents, carbonates in 233–236
convection, sub-cratonic 39
Coonterunah Group 153
core, formation 215, 221
core complexes 168
core-mantle interactions 113, 114, 115, 152
Coté Township, tonalite 162
craton formation 21–23
  southern Africa 1–26
craton margins
  magmatism at 136
  tectonism 147
craton-mantle boundaries 18
cratonic crust, composition 19
cratonic keels
  loss of 136
  and mantle plumes 135–150, *137*
crust
  Archaean 2
  delamination 115, 117
  dry 301
  first 221
  initial 97, 261
  recycling 261, 269
crust formation, Kimberley region 82
crust mass 266
  and time 267–270
crust-mantle decoupling 207
crust-mantle discontinuity, Archaean 2, 3
crust-mantle evolution, Kaapvaal Craton 82–83
crust-mantle recycling 266, 267, 270
crust-root fusion 21
crustal composition, and seismic velocities 127–128
crustal contamination 116, 161

INDEX

crustal differentiation 86
crustal formation ages, Kaapvaal 6
crustal growth
    models 261–263, *265*
    Zimbabwe Craton *192, 202, 205*, 206–207
crustal melting 207
crustal reworking 172
crustal structure
    Kaapvaal Craton 19–20
    southern Africa seismic experiment 14–17
crustal thickness
    Isua 347
    Limpopo Belt 4, 17
cumulates 84, 93
cyanobacteria
    blooms 297
    chloroplasts as 299
    global spread 294
    planktonic 290, 291
cyanobacterial mats 290, 292

D/H ratios 216, *222*, 223–225, 285
decompression melting 57
degassing
    catastrophic 215
    impact 221
Dharwar Craton 158, 163
diamond mines, Okwa-Magondi Belt 4
diamond-graphite phase boundary 46
diamonds
    abundance 19
    Siberian 70
dimethyl sulphide 280
Dindi greenstone belt 204, 205
dinitrogen *see* nitrogen
diopside 59
dioxygen *see* oxygen
Dismal Ashrock 163
dome structures 165
drip tectonics 172
Duluth igneous complex 171
DUPAL anomaly 106, 116–117
dyke swarms 136
    mafic 158, 163, 164

early Earth, events and processes **154–155**
Earth
    as 'Goldilocks' planet 301
    Hadean temperature 285
    secular evolution 153
    'snowball' 254, 284, 300
    water inventory 281
Earth evolution
    Hadean 282–285
    models *283, 284*
East African plume 140
East African rift systems 141, *142*, 146–147
Eastern Goldfields domain 164, 170
eclogite residues 172
eclogites
    compositional layering 40
    Kaapvaal Craton 5, 18, 94
    production 188
ejecta weathering 251

English River belt 165
enstatite 59, 96
erosion, and recycling 264
erosion law 264, 267, 268
erosional unroofing, Limpopo Belt 21
Ethiopia, plume 144
Eu anomalies, hydrothermal activity 314, 315
eukaryotes
    earliest 288, 299
    as recyclers 294
extensional basins, intraplate 172
extreme ultraviolet radiation 221, 222

Farm Lowrencia kimberlite 73, 83
feeder dykes 158
felsic complexes 164
felsic magmatism, Zimbabwe Craton 190
ferruginization 195
Fig Tree Group 170
final granite bloom events 169
Finland, Pb ratios 110
fires, and methane 279
folding, regional 168
fractional crystallization 93, 188
Fresnel zones 33, 37

gabbro 94
garnet zoning profiles 333–346, *334, 338, 340, 342, 344*
garnets
    elliptical 333, 337
    fe-number plots *336, 345*
    Isua 333–343
    metamorphic history 343–344
geochemical constraints, greenstone belts 161–162
geochronological data, Zimbabwe Craton **184–187**
geothermal gradient, tectosphere 2
geothermal gradients, Kaapvaal 7
glaciation, Archaean 259, 272
Gondwana, flood basalts 116
grain boundaries 62
granite-greenstone terranes 156, 160
granodiorite-granite-monzogranites 106
granodiorites, Vredefort 5
granulite facies, Zimbabwe Craton 190
graphite, Isua 331
Great Dyke, emplacement 205–206, 207
greenhouse gases 243–246, 259, 280–281
Greenland, Pb ratios 109
Greenland Craton, peridotites 66–67, 85
greenstone belts
    erosion 156
    geochronological constraints 162–163
    interpretation of 158
    komatiites 98
    stratigraphic constraints 163–164
    structural geology 158–160
greenstone successions
    Type 1 191–195
    Type 2 195–199
Grenville Province
    map *126*
    meta-anorthosites 125
growth zoning 335

INDEX

Hadean, Earth evolution 282–285
Hadean ages 83–84
Hadean cycle, and impact ejecta 248–253
Hadean temperatures 254
Harare greenstone belt 195, 199, 201
harzburgite 22, 96, 98
Hawaii 153
He fluxes 237
He/Ne ratio, mantle 218
heat flow
    and CO_2 242
    mantle plumes 140, 141
    tectosphere 2
heat production, Archaean 183, 188
heavy rare earth elements, depletion 116
heterocysts 291
Hf data 162
Hf-W chronometer 221
high field strength elements, anomalies 96
high-velocity lid, southern Africa 47, 48, 51, 62
Himalayan margin 19
Huronian Supergroup 259, 272
hydrogen loss, atmospheric 279, 285, 295, 298
hydrothermal carbonatization 237–238
hydrothermal habitats 288, 290, 292, 296
    Belingwe belt 315, 324
hydroxyl 278–279
hypersaline environments 295
hyperthermophiles 286

Iceland 153
ID-TIMS 161, 162
imbricate stacks 207
impact events 115
impact mass *225*
impact rates, lunar basins 248–250
incompatible elements
    Belingwe belt 199
    crustal residue 99
India, Pb ratios 109
interplanetary dust particles 213
inverted ecologies 296
iron content, lithosphere 98–99
iron deposition, Archaean 315, 316
Iron Mask Formation 199, 201, 203
ironstones, deposition 316, 317
island arcs, accretion 28, 40
Isua Greenstone Belt
    map *330*
    metamorphic history 329–350
    samples **348–349**
    tectonic domains 331–333
    thermobarometry 346
    timing of metamorphism 344–346
Isua supracrustals 153
    evidence of life in 290

Jagersfontein, peridotites 67
Jericho peridotites, Re depletion ages *73*
Jimmy Member 292, 311, 312, 316, 318, 322–326
Juan de Fuca ridge, CO_2 237

Kaap Valley pluton 170
Kaapvaal Craton 2, 46
    crust-mantle evolution 82–83
    eclogites 5
    greenstone belts 98
    keel modification 3
    keel structures 20
    kimberlites *84*
    magmatic events 93–94
    mantle root 5, 136
    mineralogy 91
    Pb ratios 110
    peridotites 66–67, 76–77
    Re-Os and PGE case study 80–83
    stabilization 85
    xenoliths 66
Kaapvaal Project 2–3
    location map *3*
Kaapvaal shield 4
Kaapvaal-Zimbabwe collision 17
Kalahari Craton 46
    peridotites 85
    petrology 59–60
Kam Group 159, 161
Kambalda volcanic sequence 163
Kamiskotia Complex 162
Karelian Craton 172
Karoo volcanism 94, 136
keel formation 21
Kensington Formation 198
kerogens, carbon signatures 292
Keweenawan plume 39
Keweenawan rift 136, 171, 172
Kheis overthrust belt 4, 17
Kidd-Munro assemblage 161
Kimberley 7, 76
kimberlite pipes
    Kaapvaal Craton 2, 5, 19
    mantle nodules 46
kimberlites
    Kaapvaal Craton *84*
    Lesotho 80
    production mechanism 148
    rarity 100
    Slave Craton 66
knallgas reaction 288
komatiites
    abundance 92
    Barberton 4
    Kaapvaal Craton 5
    melt depletion 22
    orthopyroxene cumulates 96
    Pb isotope ratios 115
Koodoovale Formation 191
Kostomuksha greenstone belt 172
Kudryavy 236
Kuiper belt 224
Kurile arc 236
Kwekwe complex 190, 198
kyanite, Isua 331, 332, 335, 341

Lake Superior 39
Lalla Rookh basin 170

large ion lithophile patterns, Belingwe belt 199
late bombardment 224–226
lattice preferred orientation, olivine 144, 147
layered ultramafics, Zimbabwe Craton 201
Leeward Antilles arc 28
Lesotho, kimberlites 80
lherzolites, high-T 12
life
   Belingwe study 292–294
   influence of 275–307
   pre-photosynthetic 297
light rare earth element signatures, greenstones 161
lighting, effect on nitrogen 276, 291
Limpopo Belt 4, 12, 46, 190
   crustal thickness 17, 19
   mantle root 21
   North Marginal Zone 201, 207
Limpopo Orogeny 205
LITHOPROBE SNORCLE transect 160
lithosphere, thermal evolution *145*
lithosphere-asthenosphere boundary 147
lithospheric composition, secular variation 99–100
lithospheric keels, development 65–90
lithospheric mantle
   earliest continental 83–86
   evolution 86
   formation 91–103, *93*
   isotopic studies 65
   thickness 85, 116
lithospheric roots, imbricate 29
lithospheric stacking 22
lithospheric stretching 57
lithospheric thickness
   Africa *143*
   Canadian Shield 136
   Kalahari Craton 59
   Tanzania Craton 144
location maps, for tomographic inversions 7
Love waves
   southern Africa 12, 48
   Western Superior 31, *32*
low-velocity zone, absence of 12, 18
low-velocity zones, southern Africa 53
Lower Bulawayan Group 191
lower crust
   below cratons 2
   bulk composition 126
   mafic 132
   tectonothermal evolution 6
Lower Greenstones 190
Lower Shamva Formation 199
lunar basins, impact rates 248–250
lunar regolith 224

M-discontinuity 14, 15, 20
Mackenzie dyke swarm 136
Mafic Formation 199, 203
mafic granulite 94
mafic rocks, velocity signatures 127
magma chamber, stratified 130
magma ocean, Hadean 97, 100, 218, 221
magmatic arcs, Zimbabwe Craton 201

magmatism, Kaapvaal Craton 21
Magondi Belt 4, 12, 20
Makaha belt 202, 204, 205, 206
Maliyami Formation 190, 199, 201
manganese, sea bed 291
Manjeri Formation 191, 201, 309–310
mantle
   chondritic composition 216, 217
   CO_2 exchange 236–239
   conductive layer 2
   degassing 271
   depleted 262
   harzburgitic 98
   heat storage 300
   temperature 99–100
   volatiles in 216–218
mantle and crust, interactions with atmosphere 300–301
mantle flow 148–149
mantle keel, Kaapvaal Craton 6
mantle lithosphere *see* lithospheric mantle
mantle melting
   flux 271
   residue 92
mantle models, global 53–57
mantle nodules, pressure-temperature diagrams *59*
mantle overturn events 152, 155
mantle plumes 93, 96, 113, 117
   and cratonic keels 135–150, *137*
   origin 152
   thermal halo 144
mantle recycling 188
mantle roots 1
   Limpopo Belt 21
mantle structure, tomographic analysis 2
mantle temperatures 22
mantle transition zone 55
mantle velocities, high 12
mantle-atmosphere exchange 216, 221
Maparu Formation 199
Marcy Anorthosite 125, 128
MARCY model 188
Mars 152, 220
   atmosphere 281
Mashaba Ultramafic Suite 201
Mashaba-Chibi dykes 201
Masvingo belt 205
mean degassing duration 216
melt depletion 22
melt depletion histories, peridotites 66–67
melt removal, Kimberley 77
melt residues, proportions **95**, 100–101
melt-rock interactions 80, 85
melting, high-degree 92, 99, 113
metal-silicate partitioning 115
metamorphic conditions, Isua 346, **347**
metamorphic grade, Belingwe 312
metasomatism
   Isua 331, 346
   lithospheric base 98
   peridotites 18, 67–68, 85
   rheological effects 85
meteor impact crater, Vredefort dome 5

meteorites
  ice vaporization 285
  volatiles in 216
methane
  as greenhouse gas 244–246, 254, 259, 280, 295
  in present atmosphere 279–280
  production 295
  in reduced atmosphere 296
methane hydrate 280
methanogenesis, archaea 280
methanogens 290, 291, 292, 297, 320
methanotrophs 280, 295
Michipoten ironstones 321
microbes, endolithic 319
microbial ecosystems 285
microbial evolution 286, 287
microbial mats
  Archaean sea floor 293, 322, 325
  biochemical pathways 289
  early 288
  redox power 291
mid-ocean ridge basalt 106
mid-ocean ridge magmatism, Barberton 4
mid-ocean ridges 57
  life at 296
  and water depth 301
Midlands greenstone belt 195, 198
Minnesota River valley terrane 162
mitochondria 293, 299
mobile belts, Kaapvaal craton 4
modal mineralogy 60, 67, 68
model parameters, CO_2 inventories 240, 241, 245, 247
Moho
  depth to 16
  depth and topography 2
  Kaapvaal 15
  Western Superior 31
Mont d'Or 'granite' 190
Montcalm complex 162
Moodies Group 170
Moon, origin of 115, 221
Morose Suite 165, 169
Mount Darwin greenstone belt 195
Mozaan Group 259, 272
Mtshingwe fault 205
Mungari Formation 199
Murehwa batholith 203
Mushandike granite 190
mylonites, Zimbabwe Craton 202, 203, 204

N/Ar ratios 219
Nain craton 153
Namaqua-Natal Belt 4, 83
  crustal structure 19
  deformation event 6
  velocities 12
Namibian peridotites 78–80, 78
nappe structures 204
Nb depletion, Iceland 161
Nd signatures 161, 262
  crustal models 266
  model erosion 265, 268
neon, atmospheric and mantle 218, 219, 220
Nercmar drillhole 311, 312

Newlands kimberlite 81
Ngezi Group 161, 190, 191, 309
nitrogen
  fixation 276, 291, 295
  isotope systematics 223
  in present atmosphere 276
  recycling 219
nitrogen cycle 288
nitrogen oxides 276
nitrogenase 276, 291, 295
North America, allochthonous terranes 22
North Atlantic Craton see Greenland Craton
North Caribou terrane 164
Northern Slave Craton, melting event 75

obduction 158
ocean chemistry 316
ocean island basalt 106
ocean temperatures 279
oceanic crust
  carbonates in 232
  eclogitic 39
  hydrated 172
oceanic crust thickness, Archaean 29, 39
oceanic lithosphere, subducted slab 39
oceanic pH 239
oceanic plateaux 92
oceans, depth of 301
oil, Archaean 295, 318
Okwa Belt 4, 12, 20
olivine, magnesian 96–97
olivine mg-number 67, 68
Ontario-New York-New England transect 128
Oort cloud 224
Opatica gneiss belt 162
ophiolites
  obducted 158
  stratigraphy 164
organic compounds, delivery of 226
orthopyroxene
  excess 98
  magnesian 96–97
orthopyroxene-olivine ratios 94–96
Os isotope data **72**, 79
  initial crust 97
  and Pb isotope ratios 115
Os isotope evolution 69
outgassing, mantle 231
oxidation, surface processes 295
oxygen
  and carbon dioxide 276
  production 294, 296
oxygen excess 285
oxygen levels
  Archaean 299
  early atmosphere 244, 298–299
ozone, in present atmosphere 278–279

P- and S-wave events, for tomography 37
P- and S-wave velocity ratios 126
P-wave velocities
  anomalous 18, 37
  average crust 14
  Sierra Nevada 130

P-wave velocity models, cross-sections *9–10*
palaeoenvironment, Belingwe belt 324–325
palaeontology, molecular 285, 309
palaeosols
　CO_2 239, 243, 259
　mobile element depletion 260
partial melting 93, 98
　lower crust 126
particle motion analysis, Western Pacific *33*
particle-motion anomalies 30, 32
Passaford Formation 199, 201
Pb isotopes
　data 106
　evolution 108–117
　mantle 113
　methodology 106–108
　provinciality 112–113
　ratios 108, *109*
　survival of 115–117
　variation 105–124, *111*
Pb/Pb isotope ratios *265, 266, 268*
pelagic carbonates 232, 233, 236
peridotites
　Greenland Craton 66–67
　highly depleted 2, 18, 91
　Kaapvaal 66–67
　off-craton 83
　Re–Os studies 66
　redepletion ages *6*
　sample suites 66–68
　Siberian Craton 66–67
　Slave Craton 66–67
　southern Africa *81*
　Tanzanian Craton 66–67
　Vredefort 5
　water content 96
petrology, Kalahari Craton 59–60
phase velocities, TWiST *30*
phasing depth images, southern Africa *15*
phosphorus, as nutrient 294
photic zone 317, 321
photosynthesis 278, 288, 290, 297
　anoxygenic 297, 320
　oxygenic 297, 320
phytoplankton 320
Pickle Lake 29
picoplankton 290, 294, 321, 322, 324
picrites 99
Pilbara Craton 152, 153
　dome growth 168
　folds 169
　palaeomagnetism 171
　satellite image *165*
pillow lavas
　Slave Craton 159
　Zimbabwe Craton 191
Pirella 288
Planctomycetales 288, 290, 299
planetary accretion 218
planetary atmospheres 213
planetary tectonics 152
plants
　and carbon dioxide 246–247, 278
　and weathering 235

plate tectonics
　paradigm 152, 171
　planetary 281–282
　regulation 304
platinum group elements 66
　Barberton Greenstone Belt 115
　data **72**
　fractionation *71, 75*
　Jericho and Kaapvaal peridotites *74*
plum pudding model 22
plume head flow 137–141
plume material
　spreading 140
　thickness *139*
plume tails 141
plume-lithosphere interactions 137–141
polar anisotropy 30, 53
ponding, plume material 138
Pongola basin 4
Popoteke fault 205
Porcupine Group 162, 170
prebiotic molecules 226
Preliminary Reference Earth Model *see* PREM
PREM 51, 55
Premier Mine 76, 81
pressure-temperature diagrams, mantle nodules *59*
prokaryote ecosystem 325–326
Proterozoic crust, thickness 19
protosolar nebula 213, 216, 218, 221
Ps conversions 14, 15, 19

Quetico belt 165

rare earth elements
　Belingwe 314–317, **315**, *316*
　harzburgites 96
rare gases
　depletion 213
　fractionation 216, 221
　models 221–222
ray density maps, velocity sections *11*
Rayleigh waves
　modes *50*
　southern Africa 12, *13*, 48
　Western Superior 31, *32*
Rayleigh–Taylor instabilities 166, *167*
Razi suite 205
Rb–Sr systematics 70
Re depletion ages 69–70
　Jericho peridotites *73*
　Siberian peridotites *70*
Re–Os isotope studies 5, 66
　Kaapvaal Craton **76**, 92
　Okwa-Magondi 12
　and PGE 71–80
　southern Africa 18
　systematics 68–71
receiver function analysis 2
red beds 299
redepletion ages, southern Africa *6*
redox contrasts 288, 290
　boundaries 295
　microbial mats *291*
redux reactions, carbonatization 238–239

reflection seismic profiles, Slave Craton 160–161
refractory mantle 5, 21, 93
  proportions *94*
Rehoboth Subprovince 79
Reliance Formation 191, 193, 312
residue elimination 93
residue segregation, models *97*
resolution problems 46, 55
restites 93, 130
rheological layering 188
Rhodesdale batholith 191, 198
ridge outgassing 236–237, 270
rift systems 138
rifting
  craton margins 86
  Slave Craton 164
  Zimbabwe Craton 201
Rocas Verdes Complex 172
root structures, Kaapvaal 12
rRNA phylogenies 286, 309, 326
rubisco 278, 288, 292, 295, 320, 322
Rupemba 310

S-wave velocities 12, 18
  average lithosphere 51
Sachigo block 28, 40
São Francisco Craton 41
Sauerdale thrust stack *195*, 203
scale factors, cratonic evolution 173
Scotland, Pb ratios 109
sea water, crustal flux 237
sea-floor spreading 242
sea-floor weathering 239, 240, 245
Sebakwe Poort 202, 204
Sebakwe proto-craton 190, 191
sedimentary rocks, Zimbabwe greenstone belts 190, 201–203
sediments
  juvenile 261
  redeposition 261
  reinjection 117
seismic heterogeneity, Western Superior Province 27–44
seismic lithosphere, average thickness 51
seismic refraction
  O-NYNEX 128
  Western Superior 39
seismic studies, Kaapvaal Project 2, 7–17
seismic velocities
  below shields 46
  and crustal composition 127–128
seismograms, southern Africa *52*, *54*, *56*
Semail ophiolite 159
sensitivity tests, velocity models *49*
serpentinites, Isua 331
serpentinization 280, 285, 298
Sesombi Suite 199, 201, 206
Shamva greenstone belt 190
Shamva Mine *198*, 203
Shamvaian Supergroup 199, 201, 202
Shangani batholith 191, 201
Shaw granitoid complex 169
shear modulus, and temperature *61*

shear zones
  Isua 332
  Zimbabwe Craton 195, 198, 199, 202, 203, 204
shear-wave splitting 33
  anisotropy 136, 144–147
  Kaapvaal Craton 53
shear-wave velocity structure
  Kalahari Craton 57
  TWiST 30–31, *31*
Sherwood shear zone 198
SHRIMP 161, 162
Shurugwi greenstone belt 190
sialic basement 163
Siberian Craton, peridotites 66–67
siderite stability field 243–244, 245
siderophile elements, and impacts 224
siderophores 290
Sierra Nevada, P-wave velocities 130
silicate weathering 232, 233–235, 240, 259, 272
simple asthenospheric flow 33, 136
SKS splitting
  southern Africa 12, 41
  Tanzanian Craton *146*, *147*
  Western Superior 32–36, *35*
slab melting 188
slab structures
  imbricate 40
  in mantle 29
slab-like anomaly, Western Superior 37
Slave Craton 40
  basement complex 155
  complex history 152
  conglomerates *167*
  cross-section 159
  detrital zircons *156*
  geology *157*
  peridotites 66, 73–76
  reflection seismic profiles 160–161
  structural-stratigraphic mapping 159
  turbidites 170
  see also Jericho peridotites, Northern Slave Craton
Sleepy Dragon Complex 168, 170
Sm–Nd fractionation 264
Sm–Nd modelling 269
Sm–Nd systematics 70, 79, 92, 262, 263
Sm–Sm systematics 221
Solar System
  condensation 220
  impact history 155
solar wind 218, 223
South-American–Caribbean plate boundary 28
southern Africa
  crustal subdivisions *47*
  density and velocity profiles *48*
  geological outline 4–5
  low-velocity zones 53
  seismograms *52*, *54*, *56*
  upper mantle 45–64
  velocity and density 60–62, *60*
  velocity models 47, 51, 55
southern Africa seismic array, location map *14*
southern Africa seismic experiment 7–17

Southern Cross domain 163, 164
splitting analysis, upper mantle *34*
Sr/Y fractionation *114*
stable isotopes
 analytical methods 313–314
 Belingwe belt 309–328, *319*
 evidence 317–320
 Jimmy Member 322–324
staurolite, Isua 341
Steep Rock Group 163
Stillwater Complex 130
strain regimes 204
strike-slip faults 171
stromatolites
 Belingwe belt 292, 310, 312, 317–318, 324
 Steep Rock 299
structural geology, greenstone belts 158–160
sub-Moho layer, Western Superior 40
subcontinental lithosphere, models 92–99
subduction
 Archaean 20, 84, 188
 fluid transport 96, 97
 Sachigo block 28
subduction zones
 Barberton 4
 imbrication 22
 Kaapvaal Craton 5
sulphate reduction 288, 292
sulphide separates, peridotites 70
sulphides, Jimmy Member 313
sulphur bacteria 325
sulphur cycle 298, 322
sulphur isotopes
 Belingwe belt 292, **317**
 Jimmy Member 322–324
 in shallow-water sediments 321–322
Sun, luminosity changes 231, 234, 236, 272, 282, 300
Superior Craton 39
 dyke swarm 136
 tectonic systems 152
supracrustal rocks, abundance 155
surface temperature
 abiological model 300
 biological control 294
 and CO_2 234, *242*, *243*, *244*, *246*, *248*, *250*
surface water inventory 284
surface-wave analysis
 Canadian Shield 30–32
 southern Africa 12–14

Taba-Mali shear zone 199, 202
Tahawus Complex 128, 130
Tanzania Broadband Seismic Experiment 51
Tanzanian Craton
 lithospheric mantle 82
 lithospheric thickness 144
 margins 136, 141
 peridotites 66–67
tectonic stacking 190, 347
tectonic styles, Archaean 188
tectonism, at cratonic margins 147
tectosphere concept 1–2, 27

tectospheric roots
 Kaapvaal 17–19
 Western Superior 39
Teleseismic Western Superior Transect *see* TWiST
terrane accretion 204
terrane docking 164
Th–U–Pb systematics 268
Th/Nb ratios, volcanics 266
Th/Pb ratios 263
Th/U ratios
 Archaean *112*, 115
 mantle 266
thermal blanketing 22
thrusting 159
Thunder Bay 29
Timiskaming-type deposits 170
Tokwe Segment 190
tomographic analysis
 location maps 7
 mantle structure 2
 South African Craton 57–59
 southern Africa 7–12
 Western Superior Province 36–39, *38*
tonalite–trondhjemite–granodiorites 106, 116, 170
 production 183
 Zimbabwe Craton 190
tonalites, Abitibi 162
tonalitic gneiss, Sachigo 28
Tonga slab 39
trace gases, in present atmosphere 279–280
Trans-Hudson shear zone 29, 35, 40
transport balance modelling *262*, 263–267
 data **263**
trapped melts 84
travel times 58
trench rollback 22
Trojan Mine *197*
turbidites, Slave Craton 170
TWiST 29
two-layer model, Western Superior Province *36*

U/Pb fractionation *114*, 116
U/Pb ratios
 Archaean cratons *118*
 core formation 263
 crustal models 265, 266, 269
 variation 113, 117
Udachnaya kimberlite 70
ultramylonites 203
Umzingwane Formation 198
Umzingwane shear zone 203
underplating
 eclogitic 40
 Grenville 130
 Kaapvaal craton 20, 81, 94
 Proterozoic 132
uniformitarian systems 282
unroofing, erosional 169
Upper Bulawayo Group 201
Upper Greenstones 191
upper mantle
 anisotropy 136
 cooling 22

upper mantle structure
  southern Africa 7–12, 45–64
  vertical resolution 55
  Western Superior 41
Urey cycle 241, 253, 260
V_P/V_S ratios 126, 127
  Adirondack Mountains 129
  continental crust 130
velocity and density, southern Africa 60–62, 60
velocity perturbation model, 3D grid 8
velocity structure
  global models 46
  inversion method 8
Venetia mine 81
vent circulation 237–238
Ventersdorp, volcanism 4, 6, 85, 94
Venus 152
  atmosphere 280–281
vertically coherent deformation 33, 34, 136
Vitim alkali basalt field 73
Vitim peridotites 78–80, 78, 79
volatile elements
  abundance 217
  depletion 217
  major 222–223
  recycling 214
  reservoirs 215–220
Vöring margin 172
Vredefort Structure 5, 21, 82, 183
Vumba greenstone belt 195

Wabigoon greenstone belt 163
Warrawoona Group 153
water
  chondritic origin 223
  as flux 284, 301
  liquid 221
water cycle 281
water-rock reactions 238
Wawa gneiss 162
weathering
  biotic enhancement 234
  and carbonates 232
  depth 302

Wedza suite 204
Western Australia, Pb ratios 110
Western Pacific, convergent zone 28
Western Superior Province 27–44
  absolute plate motion 35
  map 28
  tomographic analysis 36–39, 38
  upper mantle structure 41
wet melting 100
What Cheer Formation 199
Winnipeg River terrane 164
Witwatersrand basin, development 4
Witwatersrand succession 299
Wyoming Craton 31, 41

xenoliths, Kaapvaal Craton 2, 66
xenon isotopic ratios 215, 220, 271

Yarlung–Zangpo ophiolites 117
Yellowknife 160, 170
Yilgarn craton 159, 163, 164
Yule dome 170

Zambezi belt 190, 204
Zeederbergs Formation 191, 193, 195, 312
Zimbabwe Craton 2, 4–5, 46, 188–206
  2.75–2.58 Ga period 191–206
  3.57–2.75 Ga period 190–191
  accretion tectonics 206
  crustal growth 192, 202, 205, 206–207
  geochronological data **184–187**
  greenstone geochemistry 199–201, 200
  late granites 204–205
  maps 189, 194–198
  Moho 15
  Pb ratios 110
  tectonic evolution 183–211
zircons
  detrital 155
  xenocrystic 155